CW01543370

PRICE BOOK

A full and indispensable reference
to materials and labour costs

46th EDITION

Published by:
Glenigan Cost Information Services,
41-47 Seabourne Road, Bournemouth, BH5 2HU

First Edition 1954
46th Edition 2000

© 2000 Glenigan Cost Information Services

Printed in Great Britain by Hillman Printers (Frome) Ltd, Frome, Somerset

ISBN 0 900417 45 5

ISSN 0142 713X

Glenigan Cost Information Services welcomes observations to improve the content and usefulness of the *Griffiths Building Price Book*; comments, suggestions and any queries should be addressed to the General Manager at the address below.

Cost Information Services

**41 - 47 Seabourne Road
Bournemouth BH5 2HU**

**Tel: 01202 432121
Fax: 01202 421807**

Preface

46th Edition 2000

For this edition of the Griffiths Building Price Book all constants, waste factors and material prices used for the calculation of measured rates have been reviewed to reflect more accurately present day market conditions.

The Griffiths Building Price Book is prepared as a guide to pricing typical mid-scale new work projects (i.e. up to £1million) and mid- to small-scale alterations and repairs projects (i.e. up to £50k) with a reasonable balance of work in most trades.

Although SMM7 - the Seventh Edition of The Standard Method of Measurement of Building Works - was published in 1988, SMM6 units and trade sections have been retained to comply with users' current preferences. Suggestions for the layout of future editions, either in the 'Common arrangement of work sections for building works' or in the traditional trade sections, will be welcomed and carefully considered.

Prices for building work generally are based on the CIJC three year agreement promulgated on 23 July 1997 and cover the third year of the agreement, i.e. from 28 June 1999 as set out in detail in the introduction.

Prices for plumbing work are based on the cost of an advanced plumber and third year apprentice at JIBPMES rates payable from 23 August 1999 also set out in the introduction.

Materials costs included are those ruling in the second quarter of 1999. Prices for the major materials used in the calculations are listed under 'Basic prices for materials' after the preamble to each section.

In compiling this price book, it is recognised that prices for materials quoted by suppliers, can vary quite substantially; similarly, the effect of supply and demand will impact on the going rate for labour. For these reasons it is considered that adjustments by simple regional variations can be extremely risky and inappropriate when tendering. In using any price book it is essential that adjustments be made to take account of local market conditions, having careful regard to workload, availability of labour, any changes in wage rates and incidental labour costs and locally prevailing prices for materials.

Glenigan Cost Information Services and the Denley King Partnership acknowledge with thanks all the help received in preparing this edition including consultants and staff, manufacturers, merchants and specialists and also to users of the book for their continued encouragement and suggestions.

Denley King Partnership
Chartered Quantity Surveyors

August 1999

Contents

Index to advertisers

Introduction

Generally

The 2000 edition of the Griffiths Building Price Book is prepared as a guide to pricing the medium to smaller contract (new work to about £1million, alterations and repairs about £50k) with a reasonable balance of work in most trades, in accordance with the rules of measurement and units of billing provided by the Sixth Edition of The Standard Method of Measurement of Building Works. Where departures from SMM6 occur, these have been explained in the preambles to the sections or noted against the specific items concerned.

Unless otherwise specifically stated, figures given in all sections include for the following:
- a.   labour and all costs in connection other than supervision and travel allowances;
- b.   materials and all costs in connection;
- c.   fitting and fixing materials in position;
- d.   plant and all costs in connection other than scaffolding;
- e.   waste of materials;
- f.   square cutting;
- g.   establishment charges, overheads and profit.

It must be emphasised that figures do not include for VAT, nor for the following "Preliminary" items which have been dealt with separately:
- a.   transport of plant and tools;
- b.   scaffolding;
- c.   site administration and security;
- d.   transport for work people;
- e.   protecting the works from inclement weather;
- f.   water for the works;
- g.   lighting and power for the works;
- h.   temporary roads, hardstandings, crossings and similar items;
- j.   temporary accommodation;
- k.   temporary telephones;
- l.   traffic regulations;
- m.   safety, health and welfare of work people;
- n.   travel disbursements arising from the employment of work people;
- p.   maintenance of public and private roads;
- q.   removing rubbish, protective casings and coverings and cleaning the works on completion;
- r.   drying the works or maintenance of specific temperature and humidity levels;
- s.   temporary fencing, hoardings, screens etc.;
- t.   control of noise, pollution and all other statutory obligations;
- u.   insurance of the works;
- v.   local authorities' fees and charges;
- w.   rates on temporary buildings.

Introduction

Presentation of figures

Figures generally are presented in six or seven columns depending on whether there is an element of plant in the prices. Minor elements of plant have been included in the "Materials net" column.

The first two columns give the labour allowance in hours, "C" and "L" for craftsmen and labourers (general operatives) and "P" and "A" for plumbers and apprentices. The "Labour net" column gives figures for the net cost of labour which includes allowances for all costs incidental to the employment of labour. The "Plant net" column gives the net cost of plant including drivers and operators where applicable. The "Materials net" column gives the net cost of materials including an allowance for waste except where specifically stated. The "Price net" column gives figures for the net cost of labour, plant and materials, thus allowing an estimate to be worked out at net cost to which may be added a lump sum or a percentage to cover establishment charges, overheads and profit. The final column gives figures from the "Price net" column increased by 15% in the "New work" section, and by 25% in the "Alterations and repairs" section, these are for establishment charges, overheads and profit.

Figures for specialist work are presented in two columns: the first gives the specialist's guide prices and the second gives the specialist's prices increased by 15% in the "New work" section, by 25% in the "Alterations and repairs" section, for the builder's attendance, establishment charges, overheads and profit.

Labour costs

The all-in rates used in this edition are based upon the three year agreement promulgated by the CIJC on 23 July 1997. The calculations below cover the third year of the agreement with effect on and from Monday 28 June 1999. The rates have been calculated as follows:

Craftsmen and General Operatives

			Craftsman £			General Operative £
Flat time (paid hours)	1883.8 hours	6.05	11,396.99	1883.8 hour	4.55	8,571.29
Non-productive overtime	68 hours	6.05	411.40	68 hour	4.55	309.40
Public holidays	71 hours	6.05	429.55	71 hour	4.55	323.05
Sick pay	5 days	12.10	60.50	5 day	12.10	60.50
			12,298.44			9,264.24
NIC Employers' contribution	12.2%		1,500.41	12.2		1,130.24
CITB levy	0.25%		30.75	0.25		23.16
Holidays with pay	47 weeks	21.30	1,001.10	47 week	21.30	1001.10
			14,830.70			11,418.74
Severance pay and other statutory costs	2%		296.61	2		228.37
			15,127.31			11,647.11
Employers' liability and third party insurance	2%		302.55	2		232.94
Total cost of 1846 Productive hours per annum			£ 15,429.86			£ 11,880.05
Total cost per hour			£ 8.36			£ 6.44

Introduction

Labour costs *(continued)*

Plumbers and Apprentices

The all-in rates for plumbers and apprentices are based upon the Joint Industry Board for Plumbing Mechanical Engineering Services (JIBPMES) payable from 23 August 1999. The rates shown are for advanced plumbers and third year apprentices.

			Advanced Plumber £			Third Year Apprentice £
Flat time (paid hours)	1883.8 hour	7.20	13,563.36	1883.8 hours	3.87	7,290.31
Non-productive overtime	68 hour	7.20	489.60	68 hours	3.87	263.16
Public holidays	71 hour	7.20	511.20	71 hours	3.87	274.77
		£	14,564.16		£	7,828.24
NIC Employers' contribution	12.2		1,776.83	12.2%		955.05
CITB levy	0.25		36.41	0.25%		19.57
JIB Stamp benefit scheme	52 week	28.55	1,484.60	52 weeks	9.80	509.60
JIB Pension scheme	6.50		946.67	6.50%		508.84
			18,808.67			9,821.30
Severance pay and other statutory costs	2		376.17	2%		196.43
			19,184.84			10,017.73
Employers' liability and third party insurance	2		383.70	2%		200.35
Total cost of 1846 Productive hours per annum		£	19,568.54		£	10,218.08
Total cost per hour		£	10.60		£	5.54

Bonus payments

When there is a shortage of craftsmen and general operatives, Employers sometimes make bonus payments to attract and retain staff. These payments are not related to performance and are usually expressed as weekly sums. The following all-in hourly rates reflect the effect of including the stated weekly bonus amounts into the preceding calculations.

Craftsman		General Operative		Advanced Plumber		Third Year Apprentice	
Weekly bonus £	All-in rate £	Weekly bonus £	All-in rate £	Weekly bonus £	All-in rate £	Weekly bonus £	All-in rate £
15.00	8.81	5.00	6.59	15.00	11.10	10.00	5.87
20.00	8.96	10.00	6.74	20.00	11.26	15.00	6.04
25.00	9.11	15.00	6.89	25.00	11.43	20.00	6.20
30.00	9.26	20.00	7.04	30.00	11.59	-	-

Introduction

Labour costs (*continued*)

Rates used in this edition

It has been assumed that a £20 weekly bonus is paid to craftsmen and advanced plumbers and £10 to general operatives and third year apprentices. The all-in hourly rates included in this edition are:

	£
Craftsman	8.96
General operative	6.74
Advanced plumber	11.26
Third Year Apprentice plumber	5.87

Materials costs

Prices have been calculated on the costs of materials ruling in the second quarter of 1999. These costs are given PC (prime cost) with the relevant items, if suitable, or are listed under "Basic prices for materials" after the preambles to sections.

Allowance for waste has been included in the measured rates except where specifically stated.

It cannot be too strongly emphasised that prices actually to be paid for materials in the particular district should be checked against the basic prices and adjustments made as necessary.

Value Added Tax

Prices do not include Value Added Tax. The main provisions current in June 1999 relating to building work are set out in the following H.M. Customs and Excise publications:

Leaflet 700/18/91	: "Relief from VAT on bad debts";
Leaflet 701/16/92	: "Sewerage services and water";
Leaflet 701/23/90	: "Protective boots and helmets";
Notice 708 1/3/95	: "Buildings and Construction"
Notice 742 November 1994	: "Land and Property"(Update 1, Update 2 Supplement)
Notice 742C June 1992	: "Land and Property; Law"
Notice 701/1/95	: "Charities"
Notice 701/7/94	: "VAT relief to people with disabilities"
Notice 701/19/95	: "Fuel and Power"
Information Sheet 10/95	: "Changes to the liability of buildings"

British Standards Institution numbers

BS numbers have been used where still in existence.
BS EN numbers have only been used where no BS reference exists.

Preliminaries

The following items represent typical expenses met on contracts and which cannot properly be included in the unit rates. It should be noted that these concern items of significance, others minor items more likely to go unpriced, or which rarely occur, have been ignored.

Contractor's liability

Generally this can be understood as employers' liability, third party and fire etc. insurances. The first two are allowed for in the all-in hourly costs used in calculating prices; fire etc. insurance, which should be for the full value of the works plus allowances for inflation and for professional fees, can usually be effected from about 0.25% with a minimum charge of approximately £500.00.

Local Authorities' fees and charges

These should be covered by provisional sums when tenders are based on bills of quantities. Where bills of quantities, or provisional sums are not provided, rates on temporary buildings and costs for licences for hoardings etc., should be ascertained from the Local Authority and could be included under this heading or with the items concerned.

Obligations and restrictions imposed by the Employer (the Building Owner)

These include:
- a. access to and possession or use of the site;
- b. limitations of working space;
- c. limitations of working hours;
- d. the use or disposal of any materials found on site;
- e. hoardings, fences, screens, temporary roofs, temporary name boards and advertising rights;
- f. the maintenance of existing live drainage, water, gas, and other mains or power services on or over the site;
- g. the execution or completion of the work in any specific order or in sections or phases;
- h. maintenance of specific temperature and humidity levels;
- j. temporary accommodation and facilities for the use of the Employer including heating, lighting, furnishing and attendance;
- k. the installation of telephones for the use of the Employer and the cost of calls;
- l. any other obligation or restriction.

Costs of such obligations and restrictions can only be assessed in the light of the particular circumstances.

Works by nominated sub-contractors

Where nominated sub-contractors are provided for in tender documents appropriate allowance for the cost of general attendance, and for any other specific attendance items, should be made.

It should be noted that the term "general attendance" is deemed to include the use of the contractor's temporary roads, pavings and paths, standing scaffolding, standing power-operated hoisting plant, the provision of temporary lighting and water supplies, clearing away rubbish, provision of space for the sub-contractor's own offices and for the storage of plant and materials and the use of messrooms, sanitary accommodation and welfare facilities.

Preliminaries

Works by Statutory Authorities, Government Agencies

These should be covered by provisional sums when tenders are based on bills of quantities. Where bills of quantities, or provisional sums, are not provided, costs should be ascertained from the Public Bodies concerned.

Plant, tools and vehicles

Costs for plant and tools are included in the unit prices, but allowances should be made under this heading to cover the cost of transport of plant to and from the site and loading and unloading.

Scaffolding

Pricing scaffolding is made difficult by the fact that the period it may be required to stand is uncertain. Furthermore, scaffolding cannot properly be priced without some degree of measurement, however prices have been included at the end of this section from a specialist sub-contractor.

Site administration and security

Where a general foreman or site agent is fully employed in supervision and organisation, similarly site clerks, timekeepers etc., the appropriate allowances would be their salaries and all other expenses connected with their employment, multiplied by the appropriate periods, plus any addition required for establishment charges, overheads and profit.

Where a working foreman, or perhaps a visiting foreman, would be adequate for smaller contracts, it would be a matter of allowing a proportion of the salary, expenses etc. as above.

Where necessary, similar allowance should be made for a watchman; alternatively, quotations may be obtained from security companies.

Transport for workpeople

Costs can only be assessed in the light of the particular circumstances, based on numbers of workpeople, distances involved, the contract period and the Working Rule travel and lodging allowances.

Protecting the works from inclement weather

Costs can only be assessed in the light of the particular circumstances, based on the time of year and the nature of the work involved.

Water for the works

Building water is generally charged by the water authorities on a cubic metre basis for metered supplies. Where a special connection is necessary for building purposes, allowance may be made based on guide prices included in the "Plumbing and engineering installations" section, to which should be added the cost of temporary plumbing and storage arrangements assessed to take account of the building and site layout.

Lighting and power for the works

Costs can only be assessed in the light of the particular circumstances, based on the time of year and the nature of the work involved, to which should be added the cost of temporary distribution arrangements assessed to take account of the building and site layout.

Preliminaries

Temporary roads, hardstandings and crossings

These cannot properly be priced without some degree of measurement, therefore items have been included under "Unit rates" on page 17.

Temporary accommodation on site

This includes site offices and huts that may be necessary for the general foreman, site agent, site clerks, timekeeper or watchman, also storage sheds.

Costs can only be assessed in the light of the particular circumstances, based on the number and type of temporary buildings required and the periods involved, to which should be added the cost of transport to and from the site, loading and unloading, erecting, cleaning and dismantling.

Temporary telephones on site

Telephone charges in the second quarter of 1999 were: exchange line installation £99.00, line rental £37.37 and pay phone £43.50 per quarter and local calls 3.36p per minute. For an externally mounted bell, the charges are as follows: telephone bell extension cable £89.00, £10.00 installation charge and £2.50 rental per quarter. For a twelve month contract, an allowance of £535.00 upwards is suggested for each line required plus any charges for external bells.

Traffic regulations

Waiting, unloading and other traffic regulations can have the most serious implications. Costs can only be assessed in the light of the particular circumstances, having special regard to possible substantial costs for double handling materials or for dealing with deliveries which may be made only before or after normal working hours.

Safety, health and welfare of workpeople

This item carries similar implications to the earlier one on temporary accommodation and a similar approach is suggested to provide for adequate messrooms, sanitary accommodation and welfare facilities.

Regarding health and safety, the regulations which came into force on 1 January 1993 introduced many changes to accepted health and safety practice: furthermore the Construction Design and Management Regulations which came into force on 31 March 1995 make further demands on those with health and safety responsibility. Complying with the regulations can be time consuming and expensive although costs may be contained to some extent by membership of a safety group. An allowance of about 0.75% is suggested as a minimum.

Disbursements arising from the employment of workpeople

With the exception of travel and lodging allowances noted under the earlier item "Transport for workpeople", no separate addition is necessary on account of these disbursements for which allowance has been included in the all-in hourly costs used in calculating unit prices.

Maintenance of public and private roads

Costs can only be assessed in the light of the particular circumstances and the nature of the work involved.

Removing rubbish and cleaning the works on completion

Depending on circumstances, an allowance of about 0.25% is suggested as an average.

Drying the works

Differing methods of construction, heating installations and weather conditions influence the cost of this item; an allowance of about 0.10% is suggested as an average.

Preliminaries

Temporary fencing, hoardings, screens, footways and gantries

These cannot properly be priced without some degree of measurement, therefore items have been included under "Unit rates" at the end of this section.

Control of noise, pollution and all other statutory obligations

These can have the most serious implications. Costs can only be assessed in the light of the particular circumstances, having special regard to possible substantial costs for using special tools and equipment or for interrupting and postponing work, for example, to suit court sessions, school examinations, security arrangements etc.

Defects after completion

Some making good, adjusting and touching up is invariably called for after completion and an allowance of about 0.50% is suggested as an average.

New work	Unit	Specialist price net	Price with 15%
Scaffolding		£	£
			VAT not included

The following "Specialist price net" figures are guide prices provided by D & G Scaffolding & Power Cradles Co Ltd.

Due to variation of structural application, requirements vary considerably and therefore a quotation should be obtained in each case.

Scaffolding up to 10 m high

Putlog scaffolding, erected, standing four weeks and dismantled

	Unit	Specialist price net	Price with 15%
hired equipment/materials	m2	4.20	4.83
each additional four weeks	m2	1.65	1.90

Independent scaffolding, erected, standing four weeks and dismantled

	Unit	Specialist price net	Price with 15%
hired equipment/materials	m2	4.60	5.29
each additional four weeks	m2	1.70	1.96

Independent birdcage scaffolding over 3.50 m high, erected, standing four weeks and dismantled

	Unit	Specialist price net	Price with 15%
hired equipment/materials	m3	1.52	1.75
each additional four weeks	m3	0.48	0.55

For incidental scaffolding in connection with demolitions and alterations, see "Demolition " section.

Enquiries about the foregoing specialist prices should be made to D & G Scaffolding & Power Cradles Co Ltd., Access House, 5 Upton Road, Fleetsbridge, Poole, Dorset BH17 7AA, tel (01202) 665586, fax (01202) 661615

Preliminaries

New work	Unit	Hours C	Hours L	Labour net	Plant net	Material net	Price net	Price with 15%
Unit rates				£	£	£	£	£
					VAT not included			

Temporary roads, hardstandings and crossings

Excavate topsoil by machine, supply and lay hardcore temporary paving, blind with ashes, maintain allowing 25% additional material, remove on completion and reinstate topsoil

	Unit	Hours C	Hours L	Labour net	Plant net	Material net	Price net	Price with 15%
150 mm thick	m2	-	0.87	5.86	5.25	3.05	14.16	16.28
225 mm thick	m2	-	1.18	7.95	7.70	4.31	19.96	22.95

Excavate topsoil by hand, supply and lay hardcore temporary paving, blind with ashes, maintain allowing 25% additional material, remove on completion and reinstate topsoil

	Unit	Hours C	Hours L	Labour net	Plant net	Material net	Price net	Price with 15%
150 mm thick	m2	-	1.97	13.27	3.74	3.05	20.06	23.07
225 mm thick	m2	-	2.89	19.48	5.45	4.31	29.24	33.63

Temporary fencing, hoardings, footways and gantries

	Unit	Hours C	Hours L	Labour net	Plant net	Material net	Price net	Price with 15%
1200 mm cleft chestnut pale fencing, erected and subsequently dismantled - based on four uses	m	-	0.35	2.36	-	1.16	3.52	4.05

Extra for softwood framed gates including additional posts and fastenings

	Unit	Hours C	Hours L	Labour net	Plant net	Material net	Price net	Price with 15%
1000 x 1200 mm single gates	each	4.00	0.50	39.21	-	11.21	50.42	57.98
3000 x 1200 mm pairs of gates	pair	7.50	0.50	70.57	-	21.49	92.06	105.87

	Unit	Hours C	Hours L	Labour net	Plant net	Material net	Price net	Price with 15%
2000 mm fencing of heavy gauge steel fabric reinforcement with 50 x 75 mm sawn softwood rails, posts and struts, erected and subsequently dismantled - based on four uses	m2	1.50	0.20	14.79	-	5.52	20.31	23.36

Extra for softwood framed gates including additional posts and fastenings

	Unit	Hours C	Hours L	Labour net	Plant net	Material net	Price net	Price with 15%
1000 x 2000 mm single gates	each	6.50	1.00	64.98	-	16.05	81.03	93.18
3000 x 2000 mm pairs of gates	pair	12.00	1.00	114.26	-	31.91	146.17	168.10

	Unit	Hours C	Hours L	Labour net	Plant net	Material net	Price net	Price with 15%
2000 mm corrugated steel fencing with 50 x 75 mm sawn softwood rails, posts and struts, erected and subsequently dismantled - based on four uses	m2	1.75	0.25	17.37	-	7.95	25.32	29.12

Extra for softwood framed gates including additional posts and fastenings

	Unit	Hours C	Hours L	Labour net	Plant net	Material net	Price net	Price with 15%
1000 x 2000 mm single gates	each	7.50	1.00	73.94	-	16.05	89.99	103.49
3000 x 2000 mm pairs of gates	pair	14.00	1.00	132.18	-	31.91	164.09	188.70

Preliminaries

New work	Unit	Hours C	Hours L	Labour net	Plant net	Material net	Price net	Price with 15%
Unit rates				£	£	£	£	£
					VAT not included			

**Temporary fencing, hoardings,
footways and gantries** (*continued*)

	Unit	Hours C	Hours L	Labour net	Plant net	Material net	Price net	Price with 15%
2000 mm hoarding of 18 mm plywood and 13 x 50 mm wrought softwood battens, with 50 x 75 mm sawn softwood rails, posts and struts, erected and subsequently dismantled, including allowance for maintenance and painting between uses - based on ten uses	m2	2.15	0.25	20.95	-	6.97	27.92	32.11
Extra for softwood framed gates including additional posts and fastenings								
1000 x 2000 mm single gates	each	7.00	1.00	69.46	-	6.57	76.03	87.43
3000 x 2000 mm pairs of gates	pair	13.00	1.00	123.22	-	13.00	136.22	156.65
Sawn softwood in footways and gantries, erected and subsequently dismantled - based on four uses	m3	30.00	-	268.80	-	93.33	362.13	416.45

New work

Figures in this section are intended as a guide to pricing the typical mid-scale contracts with a reasonable balance of work in most trades. However, the contract sum is only one factor to be considered; in most cases the two most significant factors are availability of labour and prices for materials in the locality of the work. It must be emphasised that the locality in which a project occurs should be thoroughly investigated and quotations for materials should be obtained so as to ensure that a tender takes full account of local market conditions.

Throughout this section, in addition to net prices, there are shown prices with 15% added to cover general establishment charges, attendance and profit.

Excavation and earthwork

Preamble

"Labour net" figures include allowances for all costs incidental to the employment of labour.

"Plant net" figures include for all costs of plant including drivers and operators where applicable.

"Materials net" figures include for all costs of materials including an allowance for waste except where specifically stated.

"Price net" figures are the totals of the "Labour net", "Plant net" and "Materials net" figures. Prices are for a builder employing his own labour; according to the amount and nature of the work involved; it may well be possible to secure more advantageous prices from specialist sub-contractors.

Excavation prices are for work in firm soil. For other soils the following adjustments should be made:

    clay - add 25%
    hard gravel - add 50%
    chalk - add 100 to 150%
    rock - add 300 to 400%

Earthwork support is to a large extent a risk item. The poling boards may need to be close together in poor ground, they may be well apart in good ground or they may have to be driven as the depth increases. It may be possible for timbering as a whole to be put in afterwards or at wide intervals or not at all, the cost can vary widely. Figures given are for semi-close boarding with conditions assumed as poor to moderate. Material for earthwork support is assumed as being timber used ten times.

Figures for disposal of excavated material include allowance for 25% increase in bulk after excavation.

Excavation and earthwork

New work

	Unit	Price
Basic prices for materials		**£**
Ashes	m3	9.50
	tonne	7.60
Binding gravel (hoggin)	m3	15.84
	tonne	8.55
Hardcore	m3	12.25
	tonne	8.16
Sand	m3	17.96
	tonne	11.22
Timber for earthwork support	m3	253.37

Prices actually to be paid for materials must be checked against the above basic prices and adjustments must be made as necessary.

Excavation and earthwork

New work	Unit	Hours C	Hours L	Labour net	Plant net	Material net	Price net	Price with 15%
Generally				£	£	£	£	£
					VAT not included			

Site preparation

	Unit	Hours C	Hours L	Labour net	Plant net	Material net	Price net	Price with 15%
Cut down hedge and grub up roots								
privet hedge 1.00 m high	m	-	1.25	8.43	-	-	8.43	9.69
field hedge 2.00 m height including								
small trees	m	-	5.50	37.07	-	-	37.07	42.63
Clear site of bushes, scrub and undergrowth and grub up roots	m2	-	0.12	0.81	-	-	0.81	0.93
Lift and roll turf to be preserved, remove 25 m and stack	m2	-	0.30	2.02	-	-	2.02	2.32
Excavate by machine topsoil to be preserved, average depth								
0.150 m	m2	-	-	-	0.90	-	0.90	1.03
0.225 m	m2	-	-	-	1.33	-	1.33	1.53
Excavate by hand topsoil to be preserved, average depth								
0.150 m	m2	-	0.31	2.09	-	-	2.09	2.40
0.225 m	m2	-	0.50	3.37	-	-	3.37	3.88

For tree felling see "Soft landscaping work" in the External works section

Excavation by machine

	Unit	Hours C	Hours L	Labour net	Plant net	Material net	Price net	Price with 15%
Excavate to reduce levels, maximum depth not exceeding								
0.25 m	m3	-	-	-	1.14	-	1.14	1.31
1.00 m	m3	-	-	-	1.27	-	1.27	1.46
2.00 m	m3	-	-	-	1.27	-	1.27	1.46
4.00 m	m3	-	-	-	1.40	-	1.40	1.61
Excavate basements, maximum depth not exceeding								
1.00 m	m3	-	0.06	0.37	1.40	-	1.77	2.04
2.00 m	m3	-	0.11	0.74	2.79	-	3.53	4.06
4.00 m	m3	-	0.21	1.42	5.34	-	6.76	7.77
Excavate pits to receive bases of stanchions etc, maximum depth not exceeding								
0.25 m	m3	-	0.25	1.69	4.69	-	6.38	7.34
1.00 m	m3	-	0.17	1.15	3.19	-	4.34	4.99
2.00 m	m3	-	0.18	1.21	3.38	-	4.59	5.28
4.00 m	m3	-	0.20	1.35	3.75	-	5.10	5.87

Excavation and earthwork

New work	Unit	Hours C	Hours L	Labour net	Plant net	Material net	Price net	Price with 15%
Generally				£	£	£	£	£
					VAT not included			

Excavation by machine (*continued*)

Excavate pits less than 1.25 x 1.25 m
on plan to receive bases of stanchions
etc, maximum depth not exceeding

0.25 m	m3	-	0.38	2.53	7.04	-	9.57	11.01
1.00 m	m3	-	0.25	1.69	4.69	-	6.38	7.34
2.00 m	m3	-	0.26	1.72	4.79	-	6.51	7.49

Excavate trenches not exceeding
0.30 m in width to receive foundations,
average depth

0.25 m	m	-	0.03	0.17	0.47	-	0.64	0.74
0.50 m	m	-	0.05	0.34	0.94	-	1.28	1.47
0.75 m	m	-	0.07	0.44	1.22	-	1.66	1.91
1.00 m	m	-	0.08	0.54	1.50	-	2.04	2.35

Excavate trenches over 0.30 m in width
to receive foundations, average depth

0.25 m	m3	-	0.28	1.89	5.26	-	7.15	8.22
1.00 m	m3	-	0.25	1.65	4.60	-	6.25	7.19
2.00 m	m3	-	0.26	1.75	4.88	-	6.63	7.62
4.00 m	m3	-	0.28	1.92	5.35	-	7.27	8.36

Excavate trenches to receive service
pipes, cables etc, grade bottom, fill in
and compact and remove surplus –
earthwork support not included –
average depth not exceeding

0.25 m	m	-	0.08	0.54	0.64	-	1.18	1.36
0.50 m	m	-	0.12	0.81	1.23	-	2.04	2.35
0.75 m	m	-	0.18	1.21	1.84	-	3.05	3.51
1.00 m	m	-	0.24	1.62	2.48	-	4.10	4.71

Excavate and fill basement working
space, maximum depth not exceeding

1.00 m	m3	-	0.34	2.29	6.56	-	8.85	10.18
2.00 m	m3	-	0.39	2.63	7.83	-	10.46	12.03
4.00 m	m3	-	0.48	3.24	12.66	-	15.90	18.29

Excavate and fill pit working space,
maximum depth not exceeding

0.25 m	m3	-	0.52	3.50	5.83	-	9.33	10.73
1.00 m	m3	-	0.46	3.10	9.10	-	12.20	14.03
2.00 m	m3	-	0.46	3.10	9.10	-	12.20	14.03
4.00 m	m3	-	0.48	3.24	9.47	-	12.71	14.62

Excavate and fill trench working space,
maximum depth not exceeding

0.25 m	m3	-	0.53	3.57	10.41	-	13.98	16.08
1.00 m	m3	-	0.46	3.10	9.10	-	12.20	14.03
2.00 m	m3	-	0.46	3.10	9.10	-	12.20	14.03
4.00 m	m3	-	0.48	3.24	9.47	-	12.71	14.62

Excavation and earthwork

New work	Unit	Hours C	Hours L	Labour net	Plant net	Material net	Price net	Price with 15%
Generally				£	£	£	£	£
					VAT not included			
Excavation by hand								
Excavate to reduce levels, maximum depth not exceeding								
0.25 m	m3	-	2.40	16.18	-	-	16.18	18.61
1.00 m	m3	-	2.55	17.19	-	-	17.19	19.77
Excavate basements, maximum depth not exceeding								
0.25 m	m3	-	2.40	16.18	-	-	16.18	18.61
1.00 m	m3	-	2.55	17.19	-	-	17.19	19.77
2.00 m	m3	-	3.40	22.92	-	-	22.92	26.36
Excavate pits to receive bases of stanchions etc, maximum depth not exceeding								
0.25 m	m3	-	3.40	22.92	-	-	22.92	26.36
1.00 m	m3	-	3.55	23.93	-	-	23.93	27.52
2.00 m	m3	-	4.05	27.30	-	-	27.30	31.40
Excavate pits less than 1.25 x 1.25 m on plan to receive bases of stanchions etc, maximum depth not exceeding								
0.25 m	m3	-	4.42	29.79	-	-	29.79	34.26
1.00 m	m3	-	4.62	31.14	-	-	31.14	35.81
2.00 m	m3	-	5.27	35.52	-	-	35.52	40.85
Excavate trenches not exceeding 0.30 m in width to receive foundations, average depth								
0.25 m	m	-	0.25	1.69	-	-	1.69	1.94
0.50 m	m	-	0.53	3.57	-	-	3.57	4.11
0.75 m	m	-	0.79	5.32	-	-	5.32	6.12
1.00 m	m	-	1.05	7.08	-	-	7.08	8.14
Excavate trenches over 0.30 m in width to receive foundations, maximum depth not exceeding								
0.25 m	m3	-	3.00	20.22	-	-	20.22	23.25
1.00 m	m3	-	3.30	22.24	-	-	22.24	25.58
2.00 m	m3	-	4.15	27.97	-	-	27.97	32.17
4.00m	m3	-	6.62	44.62	-	-	44.62	51.31
Excavate trenches to receive service pipes, cables etc, grade bottom, fill in and compact and remove surplus – earthwork support not included – average depth not exceeding								
0.25 m	m	-	0.55	3.71	0.08	-	3.79	4.36
0.50 m	m	-	1.10	7.41	0.10	-	7.51	8.64
0.75 m	m	-	1.60	10.78	0.15	-	10.93	12.57
1.00 m	m	-	2.15	14.49	0.23	-	14.72	16.93

Excavation and earthwork

New work	Unit	Hours C	Hours L	Labour net	Plant net	Material net	Price net	Price with 15%
Generally				£	£	£	£	£
					VAT not included			
Excavation by hand (continued)								
Excavate and fill basement working space, maximum depth not exceeding								
0.25 m	m3	-	4.25	28.65	0.46	-	29.11	33.48
1.00 m	m3	-	4.75	32.02	0.46	-	32.48	37.35
2.00 m	m3	-	6.30	42.46	0.46	-	42.92	49.36
Excavate and fill pit working space, maximum depth not exceeding								
0.25 m	m3	-	5.25	35.38	0.46	-	35.84	41.22
1.00 m	m3	-	5.85	39.43	0.46	-	39.89	45.87
2.00 m	m3	-	7.40	49.88	0.46	-	50.34	57.89
Excavate and fill trench working space, maximum depth not exceeding								
0.25 m	m3	-	4.60	31.00	0.46	-	31.46	36.18
1.00 m	m3	-	5.15	34.71	0.46	-	35.17	40.45
2.00 m	m3	-	6.70	45.16	0.46	-	45.62	52.46
Breaking up								
Breaking up by machine - excluding reinstatement								
Break up surface concrete, average thickness								
100 mm	m2	-	0.22	1.48	0.89	-	2.37	2.73
150 mm	m2	-	0.28	1.89	1.14	-	3.03	3.48
200 mm	m2	-	0.42	2.83	1.71	-	4.54	5.22
Break up reinforced surface concrete, average thickness								
100 mm	m2	-	0.31	2.09	1.26	-	3.35	3.85
150 mm	m2	-	0.39	2.63	1.59	-	4.22	4.85
200 mm	m2	-	0.59	3.98	2.40	-	6.38	7.34
Break up tarmacadam paving, average thickness								
100 mm	m2	-	0.10	0.67	0.41	-	1.08	1.24
150 mm	m2	-	0.16	1.08	0.65	-	1.73	1.99
200 mm	m2	-	0.22	1.48	0.89	-	2.37	2.73
Break up obstructions in excavations								
concrete	m3	-	5.32	35.86	21.64	-	57.50	66.13
reinforced concrete	m3	-	7.44	50.15	30.26	-	80.41	92.47
brick, block or stonework in lime mortar	m3	-	0.48	3.24	1.95	-	5.19	5.97
brick, block or stonework in cement mortar	m3	-	0.64	4.31	2.60	-	6.91	7.95

Excavation and earthwork

New work	Unit	Hours C	Hours L	Labour net	Plant net	Material net	Price net	Price with 15%
Generally				£	£	£	£	£
					VAT not included			

Breaking up (*continued*)

Breaking up by machine - excluding reinstatement (*continued*)

Break up surface concrete for service trenches 300 mm wide, average thickness

100 mm	m	-	0.07	0.47	0.28	-	0.75	0.86
150 mm	m	-	0.08	0.54	0.33	-	0.87	1.00
200 mm	m	-	0.13	0.88	0.53	-	1.41	1.62

Break up reinforced surface concrete for service trenches 300 mm wide, average thickness

100 mm	m	-	0.09	0.61	0.37	-	0.98	1.13
150 mm	m	-	0.12	0.81	1.10	-	1.91	2.20
200 mm	m	-	0.18	1.21	0.73	-	1.94	2.23

Break up tarmacadam paving for service trenches 300 mm wide, average thickness

100 mm	m	-	0.03	0.20	0.12	-	0.32	0.37
150 mm	m	-	0.05	0.34	0.20	-	0.54	0.62
200 mm	m	-	0.07	0.47	0.28	-	0.75	0.86

Break up hardcore

hardcore	m3	-	0.33	2.22	7.54	-	9.76	11.22

Breaking up by machine - including reinstatement

Breaking up, hardcore and surface concrete for service trenches 300 mm wide and reinstate concrete paving and 150 mm hardcore bed, average concrete thickness

100mm	m	-	5.95	40.10	8.33	70.90	119.33	137.23
150mm	m	-	5.56	37.48	8.38	70.90	116.76	134.27
200mm	m	-	4.71	31.75	8.58	70.90	111.23	127.91

Breaking up, hardcore and reinforced concrete for service trenches 300 mm wide and reinstate concrete paving and 150 mm hardcore bed, average concrete thickness

100mm	m	-	5.97	40.24	8.42	71.18	119.84	137.82
150mm	m	-	5.60	37.75	9.15	71.18	118.08	135.79
200mm	m	-	4.76	32.08	8.78	71.18	112.04	128.85

Excavation and earthwork

New work	Unit	Hours C	Hours L	Labour net	Plant net	Material net	Price net	Price with 15%
Generally				£	£	£	£	£
					VAT not included			

Breaking up (*continued*)

Breaking up by machine - including reinstatement (*continued*)

Breaking up, hardcore, concrete bed
and tarmacadam paving for service
trenches 300 mm wide and reinstate
concrete paving and 150 mm hardcore
bed average tarmacadam thickness

	Unit	Hours C	Hours L	Labour net	Plant net	Material net	Price net	Price with 15%
100mm	m	-	0.52	3.50	1.35	20.69	25.54	29.37
150mm	m	-	0.52	3.52	1.36	29.77	34.65	39.85
200mm	m	-	0.56	3.80	1.56	38.70	44.06	50.67

Earthwork support

Earthwork support to sides of trenches
not exceeding 2.00 m between opposing
faces, in firm soil maximum depth not
exceeding

	Unit	Hours C	Hours L	Labour net	Plant net	Material net	Price net	Price with 15%
1.00 m	m2	-	0.11	0.74	-	0.87	1.61	1.85
2.00 m	m2	-	0.13	0.88	-	0.87	1.75	2.01
4.00 m	m2	-	0.19	1.28	-	0.87	2.15	2.47

Earthwork support to sides of pits not
exceeding 2.00 m between opposing
faces, in firm soil, maximum depth not
exceeding

	Unit	Hours C	Hours L	Labour net	Plant net	Material net	Price net	Price with 15%
1.00 m	m2	-	0.11	0.74	-	0.87	1.61	1.85
2.00 m	m2	-	0.13	0.88	-	0.87	1.75	2.01
4.00 m	m2	-	0.19	1.28	-	0.87	2.15	2.47

Earthwork support to sides of trenches
not exceeding 2.00 m between opposing
faces, in moderately firm soil, maximum
depth not exceeding,

	Unit	Hours C	Hours L	Labour net	Plant net	Material net	Price net	Price with 15%
1.00 m	m2	-	0.44	2.97	-	2.32	5.29	6.08
2.00 m	m2	-	0.57	3.84	-	2.32	6.16	7.08
4.00 m	m2	-	0.70	4.72	-	2.32	7.04	8.10

Earthwork support to sides of pits not
exceeding 2.00 m between opposing
faces in moderately firm soil, maximum
depth not exceeding,

	Unit	Hours C	Hours L	Labour net	Plant net	Material net	Price net	Price with 15%
1.00 m	m2	-	0.44	2.97	-	2.32	5.29	6.08
2.00 m	m2	-	0.57	3.84	-	2.32	6.16	7.08
4.00 m	m2	-	0.70	4.72	-	2.32	7.04	8.10

Excavation and earthwork

New work	Unit	Hours C	Hours L	Labour net	Plant net	Material net	Price net	Price with 15%
Generally				£	£	£	£	£
					VAT not included			

Earthwork support (*continued*)

Earthwork support to sides of trenches not exceeding 2.00 m between opposing faces in loose soil, maximum depth not exceeding,

	Unit	Hours C	Hours L	Labour net	Plant net	Material net	Price net	Price with 15%
1.00 m	m2	-	1.00	6.74	-	3.19	9.93	11.42
2.00 m	m2	-	1.50	10.11	-	3.19	13.30	15.30
4.00 m	m2	-	2.00	13.48	-	3.19	16.67	19.17

Earthwork support to sides of pits not exceeding 2.00 m between opposing faces in loose soil, maximum depth not exceeding,

	Unit	Hours C	Hours L	Labour net	Plant net	Material net	Price net	Price with 15%
1.00 m	m2	-	1.00	6.74	-	3.19	9.93	11.42
2.00 m	m2	-	1.50	10.11	-	3.19	13.30	15.30
4.00 m	m2	-	2.00	13.48	-	3.19	16.67	19.17

Disposal of excavated material

	Unit	Hours C	Hours L	Labour net	Plant net	Material net	Price net	Price with 15%
Excavated material moved by machine and deposited on site in spoil heaps average 100 m from excavation	m3	-	-	-	2.03	-	2.03	2.33
Add or deduct for every 25 m difference in distance	m3	-	-	-	0.51	-	0.51	0.59
Excavated material moved by hand and deposited on site in spoil heaps average 50 m from excavation	m3	-	1.50	10.11	-	-	10.11	11.63
Add or deduct for every 10 m difference in distance	m3	-	0.30	2.02	-	-	2.02	2.32
Excavated material moved by machine and spread on site average 100 m from excavation	m3	-	0.25	1.69	1.52	-	3.21	3.69
Add or deduct for every 25 m difference in distance	m3	-	-	-	0.51	-	0.51	0.59
Excavated material moved by hand and spread on site average 50 m from excavation	m3	-	2.50	16.85	-	-	16.85	19.38
Add or deduct for every 10 m difference in distance	m3	-	0.30	2.02	-	-	2.02	2.32

Excavation and earthwork

New work	Unit	Hours C	Hours L	Labour net	Plant net	Material net	Price net	Price with 15%
Generally				£	£	£	£	£
					VAT not included			

Disposal of excavated material
(*continued*)

Excavated material removed from site
to tip average 15 km from site, £50.00
per load tipping charge

loaded by machine direct from								
excavations	m3	-	-	-	13.85	-	13.85	15.93
loaded by machine from spoil heaps	m3	-	-	-	14.76	-	14.76	16.97
loaded by hand from spoil heaps	m3	-	1.65	11.12	18.38	-	29.50	33.92

Add or deduct for every 1 km difference in distance	m3	-	-	-	0.10	-	0.10	0.12

Filling

With excavated material

Excavated material filling to excavations deposited and compacted in 225 mm layers	m3	-	1.25	8.43	1.59	-	10.02	11.52

Excavated material filling to make up
levels deposited and compacted in
225 mm layers

over 250 mm thick	m3	-	1.60	10.78	2.54	-	13.32	15.32
average 100 mm thick	m2	-	0.16	1.08	0.20	-	1.28	1.47
average 150 mm thick	m2	-	0.24	1.62	0.31	-	1.93	2.22
average 225 mm thick	m2	-	0.36	2.43	0.46	-	2.89	3.32

63 mm layer or binding gravel (hogging) rolled and consolidated	m2	-	0.25	1.69	2.61	17.42	21.72	24.98

With hardcore

Hardcore filling to make up levels by
hand, deposited and compacted in
150 mm layers

over 250 mm thick	m3	-	2.00	13.48	2.54	13.47	29.49	33.91
average 75 mm thick	m2	-	0.25	1.69	0.32	1.01	3.02	3.47
average 100 mm thick	m2	-	0.30	2.02	0.38	1.35	3.75	4.31
average 150 mm thick	m2	-	0.40	2.70	0.51	2.02	5.23	6.01
average 225 mm thick	m2	-	0.55	3.71	0.70	3.03	7.44	8.56

Excavation and earthwork

New work	Unit	Hours C	Hours L	Labour net	Plant net	Material net	Price net	Price with 15%
Generally				£	£	£	£	£
					VAT not included			
Surface treatments								
Level and compact surface of filling or bottom of excavation	m2	-	0.05	0.34	0.13	-	0.47	0.54
Grade and compact surface of filling or bottom of excavation								
to falls	m2	-	0.10	0.67	0.25	-	0.92	1.06
to cross-falls	m2	-	0.20	1.35	0.51	-	1.86	2.14
50 mm blinding on hardcore								
sand	m2	-	0.07	0.47	-	0.99	1.46	1.68
ashes	m2	-	0.08	0.54	-	0.52	1.06	1.22

Concrete work

Preamble

"Labour net" figures include allowances for all costs incidental to the employment of labour.

"Plant net" figures include for all costs of plant including drivers and operators where applicable.

"Materials net" figures include for all costs of materials including an allowance for waste except where specifically stated.

"Price net" figures are the totals of the "Labour net", "Plant net", where applicable, and "Materials net" figures . Prices are for a builder employing his own labour; according to the amount and nature of the work involved, it may well be possible to secure more advantageous prices from specialist sub-contractors.

Figures for site mixed concrete are based on the use of a hired 7/5 mixer.

Figures for formwork are based on the assumptions that timber is used and that each use of material requires the full labour content; if the work is repetitive, permitting the re-use of made up sections, some reduction of the figures could be made.

Prices do not include any allowance for scaffolding, ladders or other plant necessary to reach the work. The "Preliminaries" section includes prices for scaffolding which must be considered and allowance included to suit the particular circumstances of a tender.

Specialist prices

"Price with 15%" figures are all-in guide prices and include 15% for the builder's overheads, profit, unloading materials and general attendance (to include free use of standing scaffolding and hoists, temporary lighting and water and clearing away rubbish).

The amount of attendance required varies between the various trades and also with the circumstances of specific jobs; the percentage addition must always be considered and adjusted as necessary to suit the terms and conditions of the quotation being used.

Quantities and delivery distances are usually the most significant of the many factors which influence prices and it must be emphasised that quotations should always be obtained when preparing a tender.

Piling

Specialist prices for piling have been included at the end of this section.

Concrete work

New work

Basic prices for materials		£
Aggregates		
40 mm	m3	19.75
	tonne	13.16
20 mm	m3	19.91
	tonne	13.27
10 mm	m3	20.07
	tonne	13.38
Ready mixed concrete		
1:3:6 - 40 mm aggregate	m3	53.00
1:2:4 - 20 mm aggregate	m3	56.18
1:1½:3 - 10 mm aggregate	m3	60.42
Bituminous emulsion waterproofing liquid	5 litre	11.16
Cement		
Portland	tonne	89.04
rapid hardening	tonne	150.33
sulphate resisting	tonne	92.66
Formwork		
softwood	m3	244.80
25 mm sawn boarding	m2	7.65
25 mm wrought boarding	m2	13.77
18 mm plywood	m2	10.20
3.2 mm tempered hardboard	m2	1.50
Liquid surface hardener	5 litre	8.95
Polythene building sheet		
medium	m2	0.14
heavy	m2	0.42
Retarder liquid	25 litre	83.00
Sand	m3	17.96
	tonne	11.22
Self-adhesive damp-proof membrane	m2	6.50
Sub-soil building paper	m2	0.62
Waterproofer		
liquid	5 litre	16.79
powder	5 kg	12.42

Prices actually to be paid for materials must be checked against the above basic prices and adjustments made as necessary.

Concrete work

New work	Unit	Hours C	Hours L	Labour net	Plant net	Material net	Price net	Price with 15%
In-situ concrete				£	£	£	£	£
					VAT not included			

Ready mixed

Note: Allowances must be added for waiting time and part loads

Concrete 1:3:6 - 40 mm aggregate

	Unit	Hours C	Hours L	Labour net	Plant net	Material net	Price net	Price with 15%
Foundations in trenches; thickness								
150 - 300 mm	m3	-	1.90	12.81	-	55.65	68.46	78.73
over 300 mm	m3	-	1.70	11.46	-	55.65	67.11	77.18
Isolated foundation bases to columns and piers; thickness								
150 - 300 mm	m3	-	2.15	14.49	-	55.65	70.14	80.66
over 300 mm	m3	-	2.00	13.48	-	55.65	69.13	79.50
Beds; thickness								
not exceeding 100 mm	m3	-	2.38	16.04	-	55.65	71.69	82.44
100 - 150 mm	m3	-	2.38	16.04	-	55.65	71.69	82.44
150 - 300 mm	m3	-	1.66	11.19	-	55.65	66.84	76.87

Concrete 1:2:4 - 20 mm aggregate

	Unit	Hours C	Hours L	Labour net	Plant net	Material net	Price net	Price with 15%
Foundations in trenches; thickness								
150 - 300 mm	m3	-	1.90	12.81	-	58.99	71.80	82.57
over 300 mm	m3	-	1.70	11.46	-	58.99	70.45	81.02
Isolated foundation bases to columns and piers; thickness								
150 - 300 mm	m3	-	2.15	14.49	-	58.99	73.48	84.50
over 300 mm	m3	-	2.00	13.48	-	58.99	72.47	83.34
Foundations and haunching to kerbs; sectional area								
not exceeding 0.03 m2	m3	-	4.70	31.68	-	58.99	90.67	104.27
0.03 - 0.10 m2	m3	-	4.30	28.98	-	58.99	87.97	101.17
Ground beams; sectional area 0.10 - 0.25 m2	m3	-	2.10	14.15	-	58.99	73.14	84.11
Pile caps	m3	-	2.00	13.48	-	58.99	72.47	83.34
Beds; thickness								
not exceeding 100 mm	m3	-	2.38	16.04	-	58.99	75.03	86.28
100 - 150 mm	m3	-	2.38	16.04	-	58.99	75.03	86.28
150 - 300 mm	m3	-	1.66	11.19	-	58.99	70.18	80.71
Suspended slabs and attached beams; thickness								
100 - 150 mm	m3	-	4.52	30.46	-	58.99	89.45	102.87
150 - 300 mm	m3	-	3.75	25.28	-	58.99	84.27	96.91
Upstands and kerbs; sectional area not exceeding 0.03 m2	m3	-	4.50	30.33	-	58.99	89.32	102.72

Concrete work

New work	Unit	Hours C	Hours L	Labour net	Plant net	Material net	Price net	Price with 15%
In-situ concrete				£	£	£	£	£
					VAT not included			
Ready mixed (continued)								
Concrete 1:2:4 - 20 mm aggregate *(continued)*								
Walls thickness								
100 - 150 mm	m3	-	4.45	29.99	-	58.99	88.98	102.33
150 - 300 mm	m3	-	4.30	28.98	-	58.99	87.97	101.17
Isolated beams and casings to isolated steel beams; sectional area								
not exceeding 0.03 m2	m3	-	4.82	32.49	-	58.99	91.48	105.20
0.03 - 0.10 m2	m3	-	4.50	30.33	-	58.99	89.32	102.72
Isolated columns and isolated casings to steel column; sectional area								
not exceeding 0.03 m2	m3	-	6.60	44.48	-	58.99	103.47	118.99
0.03 - 0.10 m2	m3	-	5.96	40.17	-	58.99	99.16	114.03
Steps; staircases and landings	m3	-	7.25	48.87	-	58.99	107.86	124.04
Filling to hollow walls; thickness not exceeding 100 mm	m3	-	4.35	29.32	-	58.99	88.31	101.56
Concrete 1:1½:3 - 10 mm aggregate								
Ground beams; sectional area								
0.10 - 0.25 m2	m3	-	2.10	14.15	-	56.76	70.91	81.55
Pile caps	m3	-	2.00	13.48	-	56.76	70.24	80.78
Suspended slabs and attached beams; thickness								
100 - 150 mm	m3	-	4.52	30.46	-	56.76	87.22	100.30
150 - 300 mm	m3	-	3.75	25.28	-	56.76	82.04	94.35
Upstands and kerbs; sectional area not exceeding 0.03 m2	m3	-	4.50	30.33	-	56.76	87.09	100.15
Walls; thickness								
100 - 150 mm	m3	-	4.45	29.99	-	56.76	86.75	99.76
150 - 300 mm	m3	-	4.30	28.98	-	56.76	85.74	98.60
Isolated beams and casings to isolated steel beams; sectional area								
not exceeding 0.03 m2	m3	-	4.82	32.49	-	56.76	89.25	102.64
0.03 - 0.10 m2	m3	-	4.50	30.33	-	56.76	87.09	100.15
Isolated columns and isolated casings to steel columns; sectional area								
not exceeding 0.03 m2	m3	-	6.60	44.48	-	56.76	101.24	116.43
0.03 - 0.10 m2	m3	-	5.96	40.17	-	56.76	96.93	111.47
Steps; staircases and landings	m3	-	7.25	48.87	-	56.76	105.63	121.47

Concrete work

New work	Unit	Hours C	Hours L	Labour net	Plant net	Material net	Price net	Price with 15%
In-situ concrete				£	£	£	£	£
					VAT not included			
Site mixed								
Concrete 1:3:6 - 40 mm aggregate								
Foundations in trenches; thickness								
150 - 300 mm	m3	-	3.15	21.23	-	68.88	90.11	103.63
over 300 mm	m3	-	2.95	19.88	-	68.88	88.76	102.07
Isolated foundation bases to columns and piers thickness								
150 - 300 mm	m3	-	3.40	22.92	-	68.88	91.80	105.57
over 300 mm	m3	-	3.25	21.91	-	68.88	90.79	104.41
Beds; thickness								
not exceeding 100 mm	m3	-	5.15	34.71	-	68.88	103.59	119.13
100 - 150 mm	m3	-	4.75	32.02	-	68.88	100.90	116.04
150 - 300 mm	m3	-	3.85	25.95	-	68.88	94.83	109.05
Concrete 1:2:4 - 20 mm aggregate								
Foundations in trenches; thickness								
150 - 300 mm	m3	-	3.70	24.94	-	73.19	98.13	112.85
over 300 mm	m3	-	3.50	23.59	-	73.19	96.78	111.30
Isolated foundation bases to columns and piers; thickness								
150 - 300 mm	m3	-	3.95	26.62	-	73.19	99.81	114.78
over 300 mm	m3	-	3.80	25.61	-	73.19	98.80	113.62
Foundations and haunching to kerbs; sectional area								
not exceeding 0.03 m2	m3	-	6.50	43.81	-	73.19	117.00	134.55
0.03 - 0.10 m2	m3	-	6.10	41.11	-	73.19	114.30	131.44
Ground beams; sectional area								
0.10 - 0.25 m2	m3	-	3.90	26.29	-	73.19	99.48	114.40
Pile caps	m3	-	3.80	25.61	-	73.19	98.80	113.62
Beds; thickness								
not exceeding 100 mm	m3	-	5.70	38.42	-	73.19	111.61	128.35
100 - 150 mm	m3	-	5.30	35.72	-	73.19	108.91	125.25
150 - 300 mm	m3	-	4.40	29.66	-	73.19	102.85	118.28
Suspended slabs and attached beams; thickness								
100 - 150 mm	m3	-	10.20	68.75	-	73.19	141.94	163.23
150 - 300 mm	m3	-	9.30	62.68	-	73.19	135.87	156.25
Upstands and kerbs; sectional area not exceeding 0.03 m2	m3	-	11.40	76.84	-	73.19	150.03	172.53
Walls; thickness								
100 - 150 mm	m3	-	10.70	72.12	-	73.19	145.31	167.11
150 - 300 mm	m3	-	10.40	70.10	-	73.19	143.29	164.78

Concrete work

New work	Unit	Hours C	Hours L	Labour net	Plant net	Material net	Price net	Price with 15%
In-situ concrete				£	£	£	£	£
					VAT not included			

Concrete 1:2:4 - 20 mm aggregate
(*continued*)

New work	Unit	Hours C	Hours L	Labour net	Plant net	Material net	Price net	Price with 15%
Isolated beams and casings to isolated steel beams; sectional area								
not exceeding 0.03 m2	m3	-	12.00	80.88	-	73.19	154.07	177.18
0.03 - 0.10 m2	m3	-	10.80	72.79	-	73.19	145.98	167.88
Isolated columns and isolated casings to steel columns; sectional area								
not exceeding 0.03 m2	m3	-	12.60	84.92	-	73.19	158.11	181.83
0.03 - 0.10 m2	m3	-	11.20	75.49	-	73.19	148.68	170.98
Steps; staircases and landings	m3	-	10.50	70.77	-	73.19	143.96	165.55
Filling to hollow walls; thickness not exceeding 100 mm	m3	-	6.15	41.45	-	73.19	114.64	131.84
Concrete 1:1½:3 - 10 mm aggregate								
Ground beams; sectional area								
0.10 - 0.25 m2	m3	-	4.25	28.65	-	78.55	107.20	123.28
Pile caps	m3	-	4.15	27.97	-	78.55	106.52	122.50
Suspend slabs and attached beams; thickness								
100 - 150 mm	m3	-	10.55	71.11	-	78.55	149.66	172.11
150 - 300 mm	m3	-	9.65	65.04	-	78.55	143.59	165.13
Upstands and kerbs; sectional area not exceeding 0.03 m2	m3	-	11.75	79.20	-	78.55	157.75	181.41
Walls; thickness								
100 - 150 mm	m3	-	11.05	74.48	-	78.55	153.03	175.98
150 - 300 mm	m3	-	10.75	72.45	-	78.55	151.00	173.65
Isolated beams and casings to isolated steel beams; sectional area								
not exceeding 0.03 m2	m3	-	12.35	83.24	-	78.55	161.79	186.06
0.03 - 0.10 m2	m3	-	11.15	75.15	-	78.55	153.70	176.76
Isolated columns and isolated casings to steel columns; sectional area								
not exceeding 0.03 m2	m3	-	12.95	87.28	-	78.55	165.83	190.70
0.03 - 0.10 m2	m3	-	11.55	77.85	-	78.55	156.40	179.86
Steps; staircases and landings	m3	-	10.85	73.13	-	78.55	151.68	174.43
No-fines concrete 1:8 - 10 mm aggregate								
Walls; thickness 150 - 300 mm	m3	-	9.65	65.04	-	85.34	150.38	172.94

Concrete work

New work	Unit	Hours C	Hours L	Labour net	Plant net	Material net	Price net	Price with 15%
In-situ concrete				£	£	£	£	£
					VAT not included			
Joints								
Impregnated fibreboard expansion joints in concrete including formwork (four uses)								
12 mm in 100 mm slab	m	0.25	-	2.24	-	2.15	4.39	5.05
12 mm in 125 mm slab	m	0.25	-	2.24	-	2.32	4.56	5.24
12 mm in 150 mm slab	m	0.25	-	2.24	-	2.46	4.70	5.41
Prime and seal top of expansion joint with bituminous compound	m	-	0.18	1.21	-	0.01	1.22	1.40
Labours on concrete								
Extra over beds etc. for								
levelling surface	m2	-	0.03	0.20	-	-	0.20	0.23
tamping surface as paving	m2	-	0.03	0.20	-	-	0.20	0.23
trowelling with a steel float	m2	-	0.33	2.22	-	-	2.22	2.55
power floating, level	m2	-	0.17	1.15	0.26	-	1.41	1.62
grading to falls	m2	-	0.04	0.27	-	-	0.27	0.31
grading to cross-falls	m2	-	0.06	0.40	-	-	0.40	0.46
laying in bays including temporary fillets and jointing and pointing	m2	0.07	0.07	1.10	-	0.58	1.68	1.93
Mortices								
Cut mortices 100 mm deep in concrete for rag bolts and grout in	m2	-	0.25	1.69	1.02	0.01	2.72	3.13

Concrete work

New work	Unit	Hours C	Hours L	Labour net	Plant net	Material net	Price net	Price with 15%
In-situ concrete				£	£	£	£	£
					VAT not included			
Sundries								
Grout under steel stanchion bases including grouting in four anchor bolts; base size								
300 x 300 mm	each	0.50	0.50	7.85	-	0.59	8.44	9.71
450 x 300 mm	each	0.60	0.65	9.76	-	0.79	10.55	12.13
450 x 450 mm	each	0.75	0.85	12.45	-	0.99	13.44	15.46
Fix only anchor bolts including template and								
temporary boxing	each	0.80	-	7.17	-	2.96	10.13	11.65
expanded metal boxing	each	0.38	-	3.40	-	2.56	5.96	6.85
Form 150 x 150 mm holes for pipes through 150 mm concrete floor and make good	each	0.60	0.33	7.60	-	0.59	8.19	9.42
Abrasive grain applied at 0.75 kg per m2 and trowelled in to surface	m2	0.10	0.10	1.57	-	1.59	3.16	3.63
Three coats of liquid surface hardener to concrete floors	m2	-	0.10	0.67	-	1.41	2.08	2.39
Liquid bituminous membrane to concrete surfaces and blinded with sand								
two coats	m2	-	0.40	2.70	-	2.73	5.43	6.24
three coats	m2	-	0.60	4.04	-	3.92	7.96	9.15
Building paper underlay lapped 150 mm at joints	m2	-	0.03	0.20	-	0.68	0.88	1.01
Polythene sheet underlay lapped 150 mm at joints								
medium weight	m2	-	0.03	0.20	-	0.15	0.35	0.40
heavy weight	m2	-	0.04	0.27	-	0.46	0.73	0.84
Self-adhesive damp-proof membrane on concrete lapped 50 mm at joints	m2	0.15	0.15	2.35	-	7.15	9.50	10.93

Concrete work

New work	Unit	Hours C	Hours L	Labour net	Plant net	Material net	Price net	Price with 15%
Reinforcement				£	£	£	£	£
					VAT not included			

Mild steel

Plain round mild steel bar reinforcement
to BS 4449 delivered cut, bent and
bundled and fixed including tying wire,
distance blocks and ordinary spacers

6 mm	tonne	60.00	2.00	551.08	-	499.38	1050.46	1208.03
6 mm	m	0.01	-	0.12	-	0.15	0.27	0.31
8 mm	tonne	50.00	2.00	461.48	-	457.87	919.35	1057.25
8 mm	m	0.02	-	0.18	-	0.24	0.42	0.48
10 mm	tonne	42.00	2.00	389.80	-	422.51	812.31	934.16
10 mm	m	0.03	-	0.22	-	0.32	0.54	0.62
12 mm	tonne	36.00	2.00	336.04	-	395.33	731.37	841.08
12 mm	m	0.03	-	0.29	-	0.38	0.67	0.77
16 mm	tonne	30.00	2.00	282.28	-	366.13	648.41	745.67
16 mm	m	0.05	-	0.42	-	0.57	0.99	1.14
20 mm	tonne	25.00	2.00	237.48	-	357.93	595.41	684.72
20 mm	m	0.06	-	0.54	-	0.89	1.43	1.64
25 mm	tonne	20.00	2.00	192.68	-	357.93	550.61	633.20
25 mm	m	0.08	-	0.69	-	1.30	1.99	2.29

High yield

Deformed high yield steel bar
reinforcement to BS 4449 delivered cut,
bent and bundled and fixed including
tying wire, distance blocks and ordinary
spacers

8 mm	tonne	50.00	2.00	461.48	-	432.25	893.73	1027.79
8 mm	m	0.02	-	0.18	-	0.15	0.33	0.38
10 mm	tonne	42.00	2.00	389.80	-	403.03	792.83	911.75
10 mm	m	0.03	-	0.22	-	0.32	0.54	0.62
12 mm	tonne	36.00	2.00	336.04	-	386.63	722.67	831.07
12 mm	m	0.03	-	0.29	-	0.38	0.67	0.77
16 mm	tonne	30.00	2.00	282.28	-	362.03	644.31	740.96
16 mm	m	0.05	-	0.42	-	0.57	0.99	1.14
20 mm	tonne	25.00	2.00	237.48	-	362.03	599.51	689.44
20 mm	m	0.06	-	0.54	-	0.79	1.33	1.53
25 mm	tonne	20.00	2.00	192.68	-	362.03	554.71	637.92
25 mm	m	0.08	-	0.69	-	1.09	1.78	2.05

Links, stirrups and binders

Mild steel bar links, stirrups, binders and
special spacers delivered cut, bent and
bundled and fixed including tying wire

6 mm	tonne	80.00	2.00	730.28	-	676.30	1406.58	1617.57
6 mm	m	0.02	-	0.16	-	0.12	0.28	0.32
8 mm	tonne	70.00	2.00	640.68	-	614.28	1254.96	1443.20
8 mm	m	0.03	-	0.25	-	0.20	0.45	0.52

Concrete work

New work	Unit	Hours C	Hours L	Labour net	Plant net	Material net	Price net	Price with 15%
Reinforcement				£	£	£	£	£
					VAT not included			

Fabric

Fabric reinforcement to BS 4483 in slabs including tying wire and distance blocks and allowance for 200 mm laps

	Unit	Hours C	Hours L	Labour net	Plant net	Material net	Price net	Price with 15%
A98 - 1.54 kg/m2	m2	0.06	-	0.54	-	0.91	1.45	1.67
A142 - 2.22 kg/m2	m2	0.07	-	0.63	-	1.06	1.69	1.94
A193 - 3.02 kg/m2	m2	0.08	-	0.72	-	1.34	2.06	2.37
A252 - 3.95 kg/m2	m2	0.11	-	0.99	-	1.82	2.81	3.23
A393 - 6.16 kg/m2	m2	0.14	-	1.25	-	2.68	3.93	4.52
B283 - 3.73 kg/m2	m2	0.10	-	0.90	-	1.56	2.46	2.83
B385 - 4.53 kg/m2	m2	0.12	-	1.08	-	2.11	3.19	3.67
B503 - 5.93 kg/m2	m2	0.14	-	1.25	-	2.76	4.01	4.61
B785 - 8.14 kg/m2	m2	0.16	-	1.43	-	3.64	5.07	5.83
C283 - 2.61 kg/m2	m2	0.08	-	0.72	-	1.56	2.28	2.62
C385 - 3.41 kg/m2	m2	0.09	-	0.81	-	1.64	2.45	2.82
C503 - 4.34 kg/m2	m2	0.11	-	0.99	-	2.05	3.04	3.50

Fabric reinforcement to BS 4483 in casings to steel columns and beams including bending, tying wire and distance blocks and allowance for 200 mm laps

	Unit	Hours C	Hours L	Labour net	Plant net	Material net	Price net	Price with 15%
D49 - 0.77 kg/m2	m2	0.15	-	1.34	-	1.53	2.87	3.30
D98 - 1.54 kg/m2	m2	0.20	-	1.79	-	2.20	3.99	4.59

Self Centering

Self-centering combined reinforcement and formwork including temporary supports and allowance for laps but excluding bar reinforcement

	Unit	Hours C	Hours L	Labour net	Plant net	Material net	Price net	Price with 15%
21 mm rib x 0.575 mm	m2	0.85	-	7.62	1.00	14.35	22.97	26.42
21 mm rib x 0.750 mm	m2	0.90	-	8.06	1.00	17.43	26.49	30.46

Concrete work

New work	Unit	Hours C	Hours L	Labour net	Plant net	Material net	Price net	Price with 15%
Generally				£	£	£	£	£
					VAT not included			

Formwork

Figures are based on the use of sawn softwood unless otherwise stated

Formwork to edges and faces of foundations; ground beams and beds (four uses)								
over 1.00 m high	m2	1.63	-	14.60	-	8.26	22.86	26.29
not exceeding 250 mm high	m	0.44	-	3.94	-	2.22	6.16	7.08
250 - 500 mm high	m	0.84	-	7.57	-	4.12	11.69	13.44
500 mm - 1.00 m high	m	1.45	-	12.99	-	8.26	21.25	24.44
Formwork to horizontal soffits of slabs (six uses)	m2	2.65	0.15	24.75	1.00	12.24	37.99	43.69
Formwork to sloping soffits of staircases over 15 deg from horizontal (six uses)	m2	3.30	0.18	30.78	1.00	12.24	44.02	50.62
Formwork to attached beams (four uses)								
400 mm girth	m	1.50	-	13.44	-	4.72	18.16	20.88
600 mm girth	m	1.75	-	15.68	-	5.31	20.99	24.14
Formwork to soffits of projecting eaves and edges of slabs (six uses)								
250 mm girth	m	0.90	0.05	8.40	-	1.36	9.76	11.22
325 mm girth	m	1.10	0.06	10.26	-	1.76	12.02	13.82
Formwork not exceeding 250 mm high to both sides of kickers to walls (four uses)	m	1.50	-	13.44	-	7.83	21.27	24.46
Formwork to edges of slabs (six uses)								
not exceeding 250 mm deep	m	0.90	0.05	8.40	-	1.36	9.76	11.22
250 - 500 mm deep	m	1.65	0.10	15.45	-	2.71	18.16	20.88
Formwork to risers of staircases not exceeding 250 mm deep (six uses)	m	0.50	0.03	4.68	-	1.36	6.04	6.95
Formwork to edges of staircase flights (three uses); maximum width								
200 mm	m	0.85	-	7.62	-	1.67	9.29	10.68
300 mm	m	1.25	-	11.20	-	2.35	13.55	15.58
Formwork to wall faces (six uses)								
vertical	m2	2.90	0.16	27.06	-	7.80	34.86	40.09
battering	m2	3.20	0.20	30.02	-	9.28	39.30	45.20
Formwork to pilasters or attached columns (four uses)								
400 mm girth	m	1.50	-	13.44	-	2.55	15.99	18.39
500 mm girth	m	1.60	-	14.34	-	3.53	17.87	20.55
600 mm girth	m	1.70	-	15.23	-	4.12	19.35	22.25

Concrete work

New work	Unit	Hours C	Hours L	Labour net	Plant net	Material net	Price net	Price with 15%
Generally				£	£	£	£	£
					VAT not included			

Formwork (*continued*)

Formwork to isolated beams (four uses)								
400 mm girth	m	1.50	-	13.44	-	3.53	16.97	19.52
500 mm girth	m	1.65	-	14.78	-	4.42	19.20	22.08
600 mm girth	m	1.75	-	15.68	-	5.31	20.99	24.14
700 mm girth	m	1.85	-	16.58	-	6.21	22.79	26.21
800 mm girth	m	2.00	-	17.92	-	6.25	24.17	27.80
Formwork to isolated columns (four uses)								
600 mm girth	m	1.68	-	15.05	-	5.31	20.36	23.41
750 mm girth	m	1.85	-	16.58	-	6.65	23.23	26.71
900 mm girth	m	2.00	-	17.92	-	7.97	25.89	29.77
1050 mm girth	m	2.15	-	19.26	-	9.30	28.56	32.84
1200 mm girth	m	2.35	-	21.06	-	10.62	31.68	36.43
Cutting formwork								
raking	m	0.15	-	1.34	-	0.59	1.93	2.22
curved	m	0.40	-	3.58	-	0.59	4.17	4.80
Extra over formwork for								
wrought face timber (four uses)	m2	0.45	-	4.03	-	0.83	4.86	5.59
hardboard lining (two uses)	m2	0.40	-	3.58	-	0.89	4.47	5.14
Retarder liquid applied to formwork and brushing surface of concrete								
as key for plastering	m2	-	0.30	2.02	-	3.65	5.67	6.52
to provide exposed aggregate finish	m2	-	1.10	7.41	-	4.57	11.98	13.78

Precast concrete

300 x 75 mm weathered and throated copings bedded in gauged mortar and pointed								
straight	m	0.25	0.25	3.93	-	12.39	16.32	18.77
Extra for								
fair ends	each	0.10	0.10	1.57	-	2.54	4.11	4.73
angles	each	0.10	0.10	1.57	-	9.18	10.75	12.36
intersections	each	0.10	0.10	1.57	-	12.24	13.81	15.88
356 x 75 mm weathered and throated copings bedded in gauged mortar and pointed								
straight	m	0.35	0.35	5.50	-	15.28	20.78	23.90
Extra for								
fair ends	each	0.10	0.10	1.57	-	3.04	4.61	5.30
angles	each	0.10	0.10	1.57	-	11.82	13.39	15.40
intersections	each	0.10	0.10	1.57	-	15.76	17.33	19.93

Concrete work

VAT not included

New work	Unit	Hours C	Hours L	Labour net	Plant net	Material net	Price net	Price with 15%
Generally				£	£	£	£	£
Precast concrete (*continued*)								
450 x 100 mm weathered and throated copings bedded in gauged mortar and pointed								
straight	m	0.45	0.45	7.06	-	20.25	27.31	31.41
Extra for								
fair ends	each	0.10	0.10	1.57	-	3.91	5.48	6.30
angles	each	0.10	0.10	1.57	-	12.02	13.59	15.63
intersections	each	0.10	0.10	1.57	-	16.02	17.59	20.23
Weathered and throated pier caps bedded in gauged mortar								
406 x 406 x 75 mm	each	0.40	0.40	6.28	-	9.95	16.23	18.66
450 x 450 x 75 mm	each	0.40	0.40	6.28	-	12.56	18.84	21.67
530 x 530 x 75 mm	each	0.55	0.55	8.64	-	16.39	25.03	28.78
Weathered; throated and grooved sills with stooled ends bedded in gauged mortar and pointed								
150 x 67 x 900 mm	each	0.20	0.20	3.14	-	13.41	16.55	19.03
150 x 67 x 1200 mm	each	0.27	0.27	4.24	-	17.75	21.99	25.29
150 x 67 x 1500 mm	each	0.33	0.33	5.18	-	22.12	27.30	31.40
203 x 67 x 900 mm	each	0.23	0.23	3.61	-	16.23	19.84	22.82
203 x 67 x 1200 mm	each	0.32	0.32	5.03	-	21.28	26.31	30.26
203 x 67 x 1500 mm	each	0.38	0.38	5.96	-	26.71	32.67	37.57
Padstones bedded in gauged mortar								
215 x 225 x 150 mm	each	0.35	0.35	5.50	-	4.96	10.46	12.03
328 x 225 x 150 mm	each	0.45	0.45	7.06	-	6.36	13.42	15.43
Reinforced lintels bedded in gauged mortar								
75 x 150 x 900 mm	each	0.23	0.23	3.61	-	6.90	10.51	12.09
100 x 150 x 900 mm	each	0.23	0.23	3.61	-	8.23	11.84	13.62
100 x 150 x 1200 mm	each	0.28	0.28	4.40	-	10.22	14.62	16.81
215 x 150 x 900 mm	each	0.26	0.26	4.08	-	13.28	17.36	19.96
215 x 150 x 1200 mm	each	0.35	0.35	5.50	-	17.45	22.95	26.39
100 x 225 x 900 mm	each	0.27	0.27	4.24	-	10.70	14.94	17.18
100 x 225 x 1200 mm	each	0.35	0.35	5.50	-	13.60	19.10	21.97
100 x 225 x 1500 mm	each	0.44	0.44	6.91	-	17.94	24.85	28.58
215 x 225 x 900 mm	each	0.50	0.50	7.85	-	18.33	26.18	30.11
215 x 225 x 1200 mm	each	0.66	0.66	10.36	-	24.90	35.26	40.55
215 x 225 x 1500 mm	each	0.83	0.83	13.03	-	31.14	44.17	50.80
Reinforced boot lintels bedded in gauged mortar								
250 x 225 x 900 mm	each	0.60	0.60	9.42	-	19.22	28.64	32.94
250 x 225 x 1200 mm	each	0.80	0.80	12.56	-	26.08	38.64	44.44
275 x 225 x 900 mm	each	0.60	0.60	9.42	-	20.10	29.52	33.95
275 x 225 x 1200 mm	each	0.80	0.80	12.56	-	27.35	39.91	45.90

Concrete work

New work

	Unit	Hours C	Hours L	Labour net	Plant net	Material net	Price net	Price with 15%
Generally				£	£	£	£	£
					VAT not included			

Precast concrete (*continued*)

Prestressed lintels bedded in gauged
mortar; including strutting

	Unit	Hours C	Hours L	Labour net	Plant net	Material net	Price net	Price with 15%
100 x 66 x 900 mm	each	0.23	0.23	3.61	-	3.96	7.57	8.71
100 x 66 x 1200 mm	each	0.28	0.28	4.40	-	5.16	9.56	10.99
150 x 66 x 900 mm	each	0.23	0.23	3.61	-	5.07	8.68	9.98
150 x 66 x 1200 mm	each	0.28	0.28	4.40	-	6.71	11.11	12.78
215 x 66 x 900 mm	each	0.27	0.27	4.24	-	8.00	12.24	14.08
215 x 66 x 1200 mm	each	0.35	0.35	5.50	-	10.52	16.02	18.42
250 x 66 x 900 mm	each	0.27	0.27	4.24	-	9.39	13.63	15.67
250 x 66 x 1200 mm	each	0.36	0.36	5.66	-	12.35	18.01	20.71
105 x 145 x 1200 mm	each	0.30	0.30	4.71	-	10.15	14.86	17.09
105 x 145 x 1500 mm	each	0.38	0.38	5.96	-	12.68	18.64	21.44
105 x 220 x 1500 mm	each	0.44	0.44	6.91	-	18.43	25.34	29.14
105 x 220 x 1800 mm	each	0.53	0.53	8.32	-	22.13	30.45	35.02

For steel lintels, see "Metalwork"

For precast concrete kerbs, channels
and edgings, see "External works".

**Chasing, cutting and drilling
brickwork, blockwork and concrete**

Cut chase 25 mm wide in

	Unit	Hours C	Hours L	Labour net	Plant net	Material net	Price net	Price with 15%
brickwork	m	-	0.40	2.70	-	0.25	2.95	3.39
blockwork	m	-	0.35	2.36	-	0.25	2.61	3.00
concrete	m	-	0.50	3.37	-	0.25	3.62	4.16

Cut chase 50 mm wide in

	Unit	Hours C	Hours L	Labour net	Plant net	Material net	Price net	Price with 15%
brickwork	m	-	0.50	3.37	-	0.35	3.72	4.28
blockwork	m	-	0.45	3.03	-	0.35	3.38	3.89
concrete	m	-	0.65	4.38	-	0.35	4.73	5.44

Form recess for lighting switch in
exposed position in

	Unit	Hours C	Hours L	Labour net	Plant net	Material net	Price net	Price with 15%
brickwork	each	-	0.30	2.02	-	0.15	2.17	2.50
blockwork	each	-	0.25	1.69	-	0.15	1.84	2.12
concrete	each	-	0.40	2.70	-	0.15	2.85	3.28

Form recess for socket outlet point in
exposed position in

	Unit	Hours C	Hours L	Labour net	Plant net	Material net	Price net	Price with 15%
brickwork	each	-	0.30	2.02	-	0.15	2.17	2.50
blockwork	each	-	0.25	1.69	-	0.15	1.84	2.12
concrete	each	-	0.40	2.70	-	0.15	2.85	3.28

Form recess for lighting switch in
concealed position in

	Unit	Hours C	Hours L	Labour net	Plant net	Material net	Price net	Price with 15%
brickwork	each	-	0.40	2.70	-	0.15	2.85	3.28
blockwork	each	-	0.35	2.36	-	0.15	2.51	2.89
concrete	each	-	0.50	3.37	-	0.15	3.52	4.05

Concrete work

Generally				£	£	£	£	£
					VAT not included			

Chasing, cutting and drilling brickwork, blockwork and concrete (*continued*)

Form recess for socket outlet point in concealed position in

brickwork	each	-	0.50	3.37	-	0.15	3.52	4.05
blockwork	each	-	0.45	3.03	-	0.15	3.18	3.66
concrete	each	-	0.60	4.04	-	0.15	4.19	4.82

Cut hole for 25 mm diameter pipe through

112 mm brickwork	each	-	0.30	2.02	-	0.15	2.17	2.50
215 mm brickwork	each	-	0.50	3.37	-	0.15	3.52	4.05
100 mm blockwork	each	-	0.25	1.69	-	0.15	1.84	2.12
150 mm blockwork	each	-	0.45	3.03	-	0.15	3.18	3.66
100 mm concrete	each	-	1.00	6.74	-	0.50	7.24	8.33
125 mm concrete	each	-	1.25	8.43	-	0.50	8.93	10.27
150 mm concrete	each	-	1.50	10.11	-	0.50	10.61	12.20

Cut hole for 50 mm diameter pipe through

112 mm brickwork	each	-	0.40	2.70	-	0.20	2.90	3.34
215 mm brickwork	each	-	0.60	4.04	-	0.20	4.24	4.88
100 mm blockwork	each	-	0.35	2.36	-	0.20	2.56	2.94
150 mm blockwork	each	-	0.55	3.71	-	0.20	3.91	4.50
100 mm concrete	each	-	1.25	8.43	-	0.75	9.18	10.56
125 mm concrete	each	-	1.50	10.11	-	0.75	10.86	12.49
150 mm concrete	each	-	1.75	11.80	-	0.75	12.55	14.43

Cut hole for 100 mm diameter pipe through

112 mm brickwork	each	-	0.50	3.37	-	0.25	3.62	4.16
215 mm brickwork	each	-	0.70	4.72	-	0.25	4.97	5.72
100 mm blockwork	each	-	0.45	3.03	-	0.25	3.28	3.77
150 mm blockwork	each	-	0.65	4.38	-	0.25	4.63	5.32
100 mm concrete	each	-	1.50	10.11	-	1.00	11.11	12.78
125 mm concrete	each	-	1.75	11.80	-	1.00	12.80	14.72
150 mm concrete	each	-	2.00	13.48	-	1.00	14.48	16.65

Concrete work

	Unit	Specialist price net	Price with 15%

"Techspan" floors

£   £

VAT not included

Precast prestressed concrete beam and block suspended ground floors

The following "Specialist price net" figures are guide prices provided by Ashton Concrete Floors Ltd for quantities of about 500 metres of supply only beams, or a minimum area of 180 square metres/visit of supply and fix floors, within 60 kilometres of their works.

Prices do not include for cash discount.

See the preamble notes for builder's profit and attendance.

For supply only items, components are to be unloaded by the builder.

For supply and fix items, unloading and hoisting have been allowed for in the guide prices in accordance with the Precast Flooring Federation's conditions.

Supply only "Techspan" ground floors

150 mm deep beams only, for use with 440 x 215 x 100 mm concrete blocks, 1350 kg/m3 maximum density, average 8 per m2 supplied by the builder, for ground floors superimposed load 3.00 kN/m2 including finishes, maximum span

	Unit	Specialist price net	Price with 15%
4.00 m - beams at 520 mm centres	m2	7.77	8.94
4.50 m - beams at 290 and 520 mm alternate centres	m2	9.62	11.06
5.00 m - beams at 290 mm centres	m2	12.54	14.42

150 mm deep beams and concrete blocks for ground floors superimposed load 3.00 kN/m2 including finishes, maximum span

	Unit	Specialist price net	Price with 15%
4.00 m	m2	13.15	15.12
4.50 m	m2	14.36	16.51
5.00 m	m2	16.82	19.34

Supply and fix "Techspan" ground floors

150 mm deep beams and concrete blocks grouted in cement and sand for ground floors superimposed load 3.00 kN/m2 including finishes, maximum span

	Unit	Specialist price net	Price with 15%
4.00 m	m2	17.25	19.84
4.50 m	m2	18.24	20.98
5.00 m	m2	20.30	23.34

200mm deep beams and concrete blocks for ground or upper floors superimposed load 3.00 kN/m2 including finishes, maximum span

	Unit	Specialist price net	Price with 15%
6.00 m - beams at 550mm centres	m2	20.49	23.56
6.50 m - beams at 615mm and 390mm alternate centres	m2	24.06	27.67
7.60 m - beams at 390mm centres	m2	29.84	34.32

Concrete work

New work	Unit	Specialist price net	Price with 15%

"Techspan" floors

£   £
VAT not included

Precast prestressed concrete beam and pot suspended floors

For minimum quantities of 60 square metres of supply only, or a minimum of 180 square metres/visit of supply and fix floors, within 60 kilometres of their works.

Supply only "Techspan" suspended floors

Beams, pots and special edge units for suspended floors superimposed load 3.00 kN/m2 including finishes
150mm deep flooring, maximum span

	Unit	Specialist price net	Price with 15%
3.75 m	m2	17.09	19.65
4.50 m	m2	18.36	21.11

Supply and fix "Techspan" suspended floors

Beams, pots and special edge units grouted in cement and sand for suspended floors superimposed load 3.00 kN/m2 including finishes
150 mm deep flooring, maximum span

	Unit	Specialist price net	Price with 15%
3.75 m	m2	19.50	22.43
4.50 m	m2	20.98	24.13

Enquiries about the foregoing specialist prices should be made to Ashton Concrete Floors Ltd, Units A-C, The Old Brickyard, Ashton Keynes, Swindon, Wiltshire, SN6 6QR, tel (01285) 862344 fax (01285) 862655.

LUXCRETE LTD
The UK's Leading
Glass Block Supplier and Installer

Glass blocks by Luxcrete Ltd are one of the most exciting building materials in use today. Outstanding visual appeal, excellent heat and sound insulation, privacy whilst transmitting a high percentage of the available light, negligible solar heat gain and minimal maintenance are just some of their many virtues.

They offer greater compressive strength than other glass blocks, enabling larger individual panels to be constructed, thus reducing the number of sub-frames required.

Luxcrete provide wide-ranging technical back-up services, with CAD and full construction facilities including on-site installation and pre-casting of panels at their own extensive factory in North London where other Luxcrete products such as pavement lights, smoke outlet panels, roof lights, cell windows and bullet-proof panels are constructed.

 **Luxcrete Ltd.**
Premier House, Disraeli Road, Park Royal, London NW10 7BT
Tel: 0181 965 7292   Fax: 0181-961 6337

Concrete work

	Unit	Specialist price net	Price with 15%

Reinforced concrete pavement lights

		£	£
		VAT not included	

The following "Specialist price net" figures are guide prices provided by Luxcrete Ltd for quantities of 50 m2 supplied and fixed in central London, in addition supply only prices are provided for precast glass block window panels; additional charges must be added for sites outside London, for non-standard contract terms and for special performance or insurance requirements.

Prices for pavement lights are for work at ground floor level, those for roof lights are for work at first or second floor level, all with good access.

Prices for glass block window panels are for straight panels at ground floor level with good access, in sheltered positions with perimeter restraint provided by rebates at jambs.

Prices include 2½% main contractor's cash discount.

See the preamble notes for builder's profit and attendance.

Pavement lights with 100 x 100 mm lenses

	Unit	Specialist price net	Price with 15%
Ref. P150/100 for 20 kN/m2 loading	m2	425.00	488.75
Ref. P150/115 for 60 kN concentrated loading	m2	437.00	502.55

Smoke outlet panels including identification plates and brass "Terrabond" demarcation strips

	Unit	Specialist price net	Price with 15%
Ref. S150/100 for 20 kN/m2 loading	m2	401.00	461.15
Ref. S165/165 for 60 kN concentrated loading	m2	425.00	488.75

Smoke outlet panels for surface finishes by others

	Unit	Specialist price net	Price with 15%
Ref. SG150/100 for 20 kN/m2 loading	m2	389.00	447.35
Ref. SG150/115 for 60 kN concentrated loading	m2	396.00	455.40
Ref. SG test panels (required by Fire Brigade for testing surface finishes)	each	207.00	238.05
identification plates for Ref. SG panels	each	20.63	23.72
brass "Terrabond" demarcation strip for Ref. SG panels	m	7.64	8.79

760 x 610 mm clear opening escape hatch with counter-balance mechanism - including escape ladder at 60 deg for 2440 mm vertical height

	Unit	Specialist price net	Price with 15%
	each	4,422.00	5,085.30

Concrete work

New work	Unit	Specialist price net	Price with 15%
Reinforced concrete pavement lights		£	£
		VAT not included	
Rooflights for 2.5 kN/m2 loading			
Ref. R200/75 with 165 x 165 mm lenses	m2	513.00	589.95
Ref. R254/125 with 198 x 198 mm lenses	m2	496.00	570.40
Ref. RC254/125 with 200 mm diameter lenses	m2	508.00	584.20
Ref. R254/B191 with 190 x 190 mm hollow glass blocks	m2	513.00	589.95
Home Office cell rooflights double glazed with 127 x 127 mm lenses	m2	637.00	732.55
In-situ hollow glass block window panels (based on 8 panels each 2500 x 2500 mm)			
Type L115 with 115 x 115 x 80 mm Flemish "Luxblocks"	m2	496.00	570.40
Type L190 with 190 x 190 x 80 mm Flemish, Cross Reeded or Clear "Luxblocks"	m2	230.00	264.50
Type L240 with 240 x 240 x 80 mm Flemish, Cross Reeded or Clear "Luxblocks"	m2	218.00	250.70
Type L2415 with 240 x 115 x 80 mm Flemish or Clear "Luxblocks"	m2	348.00	400.20
Type L300 with 300 x 300 x 100 mm Flemish, Cross Reeded or Clear "Luxblocks"	m2	236.00	271.40
Supply only precast hollow glass block window panels (based on 20 panels each 1580 x 1580 mm)		Supply only	Supply only
Type L115 with 115 x 115 x 80 mm Flemish "Luxblocks"	m2	365.49	420.31
Type L190 with 190 x 190 x 80 mm Flemish, Cross Reeded or Clear "Luxblocks"	m2	188.49	216.76
Type L240 with 240 x 240 x 80 mm Flemish, Cross Reeded or clear "Luxblocks"	m2	159.65	183.60
Type L2415 with 240 x 115 x 80 mm Flemish or Clear "Luxblocks"	m2	218.36	251.11
Type L300 with 300 x 300 x 100 mm Flemish, Cross Reeded or Clear "Luxblocks"	m2	207.03	238.08

Enquiries about the foregoing specialist prices, and about technical matters concerning the Building Control Officer or the Fire Brigade, should be made to Luxcrete Ltd, Disraeli Road, Park Royal, London, NW10 7BT, tel (0181) 965 7292, fax (0181) 961 6337.

Concrete work

	Unit	Specialist price net	Price with 15%

New work

Prefabricated buildings

		£	£
			VAT not included

Precast concrete framed buildings

The following "Specialist price net" figures are guide prices provided by Atcost Buildings Ltd, for work of about £20,000 value within about 350 kilometres of a specialist sub-contractor's works.

Prices are per square metre of floor area, for frames delivered and erected upon foundations constructed by others.

Prices do not include for cash discount.

See preamble notes for builder's profit and attendance.

Precast reinforced concrete portal frames, purlins and gutters, 3 m high to eaves, 14 deg roof pitch, delivered and erected on prepared foundations, span:

	Unit	Specialist price net	Price with 15%
9.00 m	m2	55.65	64.00
12.00 m	m2	49.10	56.46
15.00 m	m2	46.85	53.88
18.00 m	m2	46.30	53.25
21.00 m	m2	43.60	50.14

Add to the foregoing for each additional 300 mm in height, span:

	Unit	Specialist price net	Price with 15%
9.00 m	m2	1.10	1.26
12.00 m	m2	1.00	1.15
15.00 m	m2	0.95	1.10
18.00 m	m2	0.90	1.04
21.00 m	m2	0.75	0.86

Twin and multi-spans including valley gutters priced pro-rata to the foregoing less 10 - 15%.

New Technology non-asbestos roof cladding complete with close fitting ridge capping, barge boards and finials.

	Unit	Specialist price net	Price with 15%
single skin	m2	15.50	17.83
double insulated cladding	m2	34.90	40.14

0.7mm thick E.P I000/32 Profile PVC-U coated galvanised steel sheeting with fibre glass insulation and an internal lining panel of polyester coated steel.

	Unit	Specialist price net	Price with 15%
	m2	28.00	32.20

Enquiries about the foregoing specialist prices should be made to Atcost Buildings Ltd, Spa House, Wadhurst Business Park, Faircrouch Lane, Wadhurst, East Sussex, TN5 6PT, tel (01892) 526288/784488, fax (01892) 784464, e-mail: atcostd&b@dfairc.demon.co.uk.

Concrete work

New work

Prefabricated buildings

£   £
VAT not included

Enterprise budget buildings

The following "Specialist price net" figures are guide prices provided by Leofric Building Systems Ltd for supply and construction of buildings on an in-situ base by others on the mainland of England, Wales and Scotland, excluding the area to the North and West of the Caledonian Canal

Budget buildings are of precast concrete wall panels with aggregate face, steel frames supporting steel purlins and plastic coated profiled galvanised steel sheet roofing, or the same insulated with quilt and white polyester profiled steel lining concealing purlins and frames. Vinyl gutters and downpipes as standard

Height to eaves is 2388 mm externally and minimum clear height internally is 2210 mm.

Lengths start at 5093 mm and are constructed of standard larger bays of 3658 mm (unlimited quantity) with optional smaller bays of 2438 mm.

Prices per m2 are calculated on the internal floor area.

Prices include for cash discount.

See the preamble notes for builder's profit and attendance.

	Unit	Specialist price net	Price with 15%
EB0800 Budget Buildings 2438 mm internal width 2591 mm externally			
5093 mm long externally (2 small bays)	each	2,370.38	2,725.93
	m2	179.77	206.74
7531 mm long externally (2 large bays)	each	2,751.84	3,164.62
	m2	141.17	162.35
11252 mm long externally (3 large bays)	each	3,899.54	4,484.47
	m2	137.81	158.48
3658 mm long additional bays	each	1,147.70	1,319.86
2438 mm long additional bays	each	956.97	1,100.52
EB0800 Budget Buildings 2286 mm internal width 2591 mm externally, with insulated roof and walls			
5093 mm long externally (2 small bays)	each	3,769.45	4,334.86
	m2	285.61	328.45
7531 mm long externally (2 large bays)	each	4,518.05	5,195.75
	m2	231.53	266.25
1252 mm long externally (3 large bays)	each	6,217.00	7,149.55
	m2	213.23	245.22
3658 mm long additional bays	each	1,698.95	1,953.80
2438 mm long additional bays	each	1,326.31	1,525.25

Concrete work

New work

Prefabricated buildings

	£	£
	VAT not included	

Enterprise budget buildings *(continued)*

EB1008 Budget Buildings 3251 mm internal width 3404 mm externally

	Unit	Specialist price net	Price with 15%
5093 mm long externally (2 small bays)	each	2,618.44	3,011.20
	m2	151.04	173.70
7531 mm long externally (2 large bays)	each	3,007.62	3,458.76
	m2	117.34	134.94
11252 mm long externally (3 large bays)	each	4,260.06	4,899.07
	m2	111.21	127.89
3658 mm long additional bays	each	1,251.34	1.439.04
2438 mm long additional bays	each	1,056.20	1,214.62

EB1008 Budget Buildings 3099 mm internal width 3404 mm externally, with insulated roof and walls

	Unit	Specialist price net	Price with 15%
5093 mm long externally (2 small bays)	each	4,250.14	4,887.66
	m2	245.18	281.95
7531 mm long externally (2 large bays)	each	5,073.71	5,834.76
	m2	197.89	227.58
11252 mm long externally (3 large bays)	each	6,974.42	8,020.58
	m2	182.08	209.39
3658 mm long additional bays	each	1,900.71	2,185.82
2438 mm long additional bays	each	1,492.79	1,716.70

EB1304 Budget Buildings 4064 mm internal width 4216 mm externally

	Unit	Specialist price net	Price with 15%
5093 mm long externally (2 small bays)	each	2,866.50	3,296.48
	m2	133.49	153.51
7531 mm long externally (2 large bays)	each	3,263.40	3,752.91
	m2	102.78	118.20
11252 mm long externally (3 large bays)	each	4,616.17	5,308.59
	m2	97.29	111.89
3658 mm long additional bays	each	1,351.67	1,554.41
2438 mm long additional bays	each	1,152.11	1,324.93

EB1304 Budget Buildings 3912 mm internal width 4216 mm externally, with insulated roof and walls

	Unit	Specialist price net	Price with 15%
5093 mm long externally (2 small bays)	each	4,733.03	5,442.99
	m2	220.44	253.50
7531 mm long externally (2 large bays)	each	5,630.47	6,475.04
	m2	177.32	203.92
11252 mm long externally (3 large bays)	each	7,732.94	8,892.88
	m2	162.78	187.20
3658 mm long additional bays	each	2,103.57	2,419.11
2438 mm long additional bays	each	1,649.34	1,896.74

Concrete work

New work	Unit	Specialist price net	Price with 15%

Prefabricated buildings

£   £

VAT not included

Enterprise budget buildings *(continued)*

EB1600 Budget Buildings 4877 mm internal width 5029 mm externally

	Unit	Specialist price net	Price with 15%
5093 mm long externally (2 small bays)	each	3,449.72	3,967.18
	m2	134.68	154.89
7531 mm long externally (2 large bays)	each	3,911.67	4,498.42
	m2	103.29	118.78
11252 mm long externally (3 large bays)	each	5,344.92	6,146.66
	m2	94.46	108.63
3658 mm long additional bays	each	1,433.25	1,648.24
2438 mm long additional bays	each	1,202.83	1,383.25

EB1600 Budget Buildings 4724 mm internal width 5029 mm externally, with insulated roof and walls

	Unit	Specialist price net	Price with 15%
5093 mm long externally (2 small bays)	each	5,551.09	6,383.75
	m2	216.09	248.50
7531 mm long externally (2 large bays)	each	6,605.08	7,595.84
	m2	173.79	199.85
11252 mm long externally (3 small bays)	each	8,867.41	10,197.52
	m2	156.71	180.22
3658 mm long additional bays	each	2,285.48	2,628.30
2438 mm long additional bays	each	1,770.62	2,036.21

EB1808 Budget Buildings 5690 mm internal width 5842 mm externally

	Unit	Specialist price net	Price with 15%
5093 mm long externally (2 small bays)	each	3,728.66	4,287.95
	m2	125.33	144.13
7531 mm long externally (2 large bays)	each	4,228.09	4,862.30
	m2	96.10	110.51
11252 mm long externally (3 large bays)	each	5,782.61	6,650.00
	m2	87.98	101.18
3658 mm long additional bays	each	1,555.63	1,788.97
2438 mm long additional bays	each	1,304.26	1,499.90

EB1808 Budget Buildings 5537 mm internal width 5842 mm externally, with insulated roof and walls

	Unit	Specialist price net	Price with 15%
5093 mm long externally (2 small bays)	each	6,065.96	6,975.85
	m2	203.88	234.46
7531 mm long externally (2 large bays)	each	7,199.33	8,279.22
	m2	163.63	188.18
11252 mm long externally (3 large bays)	each	9,708.62	11,164.91
	m2	147.69	169.85
3658 mm long additional bays	each	2,509.29	2,885.68
2438 mm long additional bays	each	1,940.40	2,231.46

Concrete work

New work	Unit	Specialist price net	Price with 15%

Prefabricated buildings

£ £

VAT not included

Enterprise budget buildings *(continued)*

EB2104 Budget Buildings 6502 mm internal width 6655 mm externally

	Unit	Specialist price net	Price with 15%
5093 mm long externally (2 small bays)	each	4,007.59	4,608.73
	m2	118.24	135.98
7531 mm long externally (2 large bays)	each	4,545.61	5,227.45
	m2	90.70	104.30
11252 mm long externally (3 large bays)	each	6,219.20	7,152.08
	m2	82.47	94.84
3658 mm long additional bays	each	1,674.70	1,925.90
2438 mm long additional bays	each	1,403.48	1,614.00

EB2104 Budget Buildings 6350 mm internal width 6655 mm externally, with insulated roof and walls

	Unit	Specialist price net	Price with 15%
5093 mm long externally (2 small bays)	each	6,579.72	7,566.68
	m2	194.11	223.23
7531 mm long externally (2 large bays)	each	7,821.14	8,994.31
	m2	156.06	179.47
11252 mm long externally (3 large bays)	each	10,546.52	12,128.49
	m2	140.84	161.96
3658 mm long additional bays	each	2,725.38	3,134.19
2438 mm long additional bays	each	2,105.78	2,421.64

Accessories

	Unit	Specialist price net	Price with 15%
698 x 1918 mm "Galvasteel" single doors	each	186.32	214.27
1105 x 1918 mm "Galvasteel" single doors	each	220.50	253.58
1410 x 1918 mm "Galvasteel" double doors	each	294.37	338.52
2222 x 1918 mm "Galvasteel" double doors	each	395.80	455.17
concrete vented panels (with two 152 x 229 mm vents)	each	29.77	34.23
single skin rooflights	each	59.54	68.47
double skin rooflights	each	128.99	148.34
aluminium window with fanlight (unlined walls only)	each	84.89	97.63
white PVC-U frame and double glazing (lined walls only)	each	130.10	149.61

An extended range of large up and over double canopy doors is available.

Concrete work

New work	Unit	Specialist price net	Price with 15%
Prefabricated buildings		**£**	**£**
		VAT not included	

Mayfair garages

Single model "Classic 800" garages 2438 mm internal width 2616 mm externally, of precast wall panels with aggregate face, single skin fibre cement sheet mono-pitch roof, "Galvasteel" up and over door, aluminium window with opening fanlight and vinyl coated galvanised steel fascias

	Unit	Specialist price net	Price with 15%
5029 mm long externally	each	1,765.10	2,029.87
5435 mm long externally	each	1,832.36	2,107.21
5842 mm long externally	each	1,926.07	2,214.98

Twin model "Classic 800" garages 2438 mm internal width each bay, 5131 mm overall externally, of precast wall panels with aggregate face, dividing partition wall, single skin fibre cement sheet mono-pitch roof, "Galvasteel" up and over door, aluminium window with opening fanlight and vinyl coated galvanised steel fascias

	Unit	Specialist price net	Price with 15%
5029 mm long externally	each	3,026.36	3,480.32
5435 mm long externally	each	3,137.72	3,608.37
5842 mm long externally	each	3,250.17	3,737.70

Windsor 904 garages 2845 mm internal width, 3023 mm externally, of "Brick Effect" front posts, precast wall panels with aggregate face dark red Marley "Ludlow Major" tiled 25 degrees pitched roof, "Regency" galvanised steel up and over door, aluminium window with opening fanlight, softwood front and rear gable end trusses, Marley dry verge system and vinyl gutters and downpipes

	Unit	Specialist price net	Price with 15%
5029 mm long externally	each	3,139.92	3,610.91
5842 mm long externally	each	3,329.55	3,828.98
6655 mm long externally	each	3,521.39	4,049.59
7467 mm long externally	each	3,809.14	4,380.51

Windsor 1600 garages 4877 mm internal width, 5055 mm externally, of "Brick Effect" front posts, precast wall panels with aggregate face, dark red Marley "Ludlow Major" tiled 25 degrees pitched roof, full width "Regency" galvanised steel up and over door, aluminium window with opening fanlight, softwood front and rear gable end trusses, Marley dry verge system and vinyl gutters and downpipes

	Unit	Specialist price net	Price with 15%
5105 mm long externally	each	4,708.78	5,415.09
5969 mm long externally	each	5,034.02	5,789.12
6756 mm long externally	each	5,357.05	6,106.60
7569 mm long externally	each	5,777.10	6,643.67

Concrete work

New work	Unit	Specialist price net	Price with 15%

Prefabricated buildings

£   £

VAT not included

Mayfair garages (*continued*)

Heritage garages (based on a block of four garages), of "Brick Effect" front posts, precast wall panels with aggregate face, Marley tiled pitched roof, "Galvasteel" up and over door and vinyl gutters and downpipes

	Unit	Specialist price net	Price with 15%
2438 mm internal width and 5093 mm long externally	each	2,021.99	2,325.28
2438 mm internal width and 5969 mm long externally	each	2,315.25	2,662.54
2845 mm internal width and 5093 mm long externally	each	2,304.23	2,649.86
2845 mm internal width and 5969 mm long externally	each	2,598.59	2,988.38

Extra for 152 mm step between garages

	Unit	Specialist price net	Price with 15%
5093 mm long externally	each	171.99	197.79
5969 mm long externally	each	171.99	197.79

Accessories

Extra for

	Unit	Specialist price net	Price with 15%
813 x 1829 mm cedar personal doors	each	192.94	221.88
813 x 1829 mm galvanised steel personal doors	each	144.43	166.09
1219 x 1829 mm galvanised steel personal doors	each	176.40	202.86
813 x 698 mm aluminium window with opening fanlight	each	86.00	98.89

Enquiries about the foregoing specialist prices should be made to Leofric Building Systems Ltd, Oxford Road, Ryton-on-Dunsmore, Coventry, CV8 3ED, tel (01203) 301301, fax (01203) 301148.

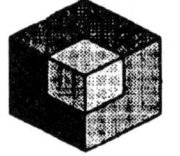

Concrete work

	Unit	Specialist price net	Price with 15%

Concrete piling

£ £

VAT not included

The following "Specialist price net" figures are guide prices provided by Westpile Ltd, for work within 100 kilometres of central London.

Prices are based on work on open sites with unlimited headroom and average soft sub-strata conditions.

Maximum loads quoted are subject to sub-strata conditions.

Prices do not include for cash discount.

See the preamble notes for builder's profit and attendance.

Guide prices for about 500 m of piling

West's CFA Bored Piling*

	Unit	Specialist price net	Price with 15%
diameter 300 mm and maximum load 400 kN	500 m	12,000.00	13,800.00
extra for length in excess of 500 m	m	16.00	18.40
diameter 400 mm and maximum load 700 kN	500 m	13,500.00	15,525.00
extra for length in excess of 500 m	m	19.00	21.85
diameter 450 mm and maximum load 1000 kN	500 m	16,000.00	18,400.00
extra for length in excess of 500 m	m	24.00	27.60
diameter 600 mm and maximum load 1500 kN	500 m	22,000.00	25,300.00
extra for length in excess of 500 m	m	35.00	40.25

West's driven "Hardrive" piles

	Unit	Specialist price net	Price with 15%
270 x 270 mm and maximum load 1000 kN	500m	14,000.00	16,100.00
extra for length in excess of 500 m	m	20.00	23.00

*Bored spoil removal is extra to the rates quoted.

Concrete work

New work	Unit	Specialist price net	Price with 15%
Concrete piling		£	£
			VAT not included

Guide prices for about 250 m piling

West's CFA Bored Piling*

diameter 300 mm and maximum load 400 kN	250 m	10,000.00	11,500.00
extra for length in excess of 250 m	m	22.00	25.30
diameter 400 mm and maximum load 700 kN	250 m	11,500.00	13,225.00
extra for length in excess of 250 m	m	26.00	29.90
diameter 450 mm and maximum load 1000 kN	250 m	13,000.00	14,950.00
extra for length in excess of 250 m	m	32.00	36.80
diameter 600 mm and maximum load 1500 kN	250 m	16,000.00	18,400.00
extra for length in excess of 250 m	m	45.00	51.75

West's driven "Hardrive" piles

270 x 270 mm and maximum load 1000 kN	250 m	12,200.00	14,030.00
extra for length in excess of 250 m	m	30.00	34.50

*Bored spoil removal is extra to the rates quoted.

Enquiries about the foregoing specialist prices should be made to Westpile Ltd, Dolphin Bridge House, Rockingham Road, Uxbridge, UB8 2UB, tel (01895) 258266, fax (01895) 271805.

Brickwork and blockwork

Preamble

"Labour net" figures include allowances for all costs incidental to the employment of labour. Except in the case of the heavier blocks, the labour for brickwork and blockwork has generally been based on a team of three bricklayers to two labourers which is considered a suitable ratio for work not too high above ground.

"Materials net" figures include for all costs of materials including an allowance for waste except where specifically stated.

"Price net" figures are the totals of the "Labour net" and "Materials net" figures. Prices are for a builder employing his own labour; according to the amount and nature of the work involved, it may well be possible to secure more advantageous prices from specialist sub-contractors.

Prices do not include any allowance for scaffolding, ladders or other plant necessary to reach the work. The "Preliminaries" section includes prices for scaffolding which must be considered and allowance included to suit the particular circumstances of a tender.

Specialist prices

"Price with 15%" figures are all-in guide prices and include 15% for the builder's overheads, profit, unloading materials and general attendance (to include free use of standing scaffolding and hoists, temporary lighting and water and clearing away rubbish).

The amount of attendance required varies between the various trades and also with the circumstances of specific jobs; the percentage addition must always be considered and adjusted as necessary to suit the terms and conditions of the quotation being used.

Quantities and delivery distances are usually the most significant of the many factors which influence prices and it must be emphasised that quotations should always be obtained when preparing a tender.

Composite walls

Although not in accordance with the Standard Method of Measurement, figures for a range of composite walls has been included at the end of this section.

Brickwork and blockwork

	Unit	Price
Basic prices for materials		**£**
Bricks		
class B engineering	1,000	360.00
commons	1,000	170.00
refractory	1,000	920.00
facings	1,000	325.00
Blocks		
75mm dense aggregate blocks	m2	5.29
100mm dense aggregate blocks	m2	5.87
140mm dense aggregate blocks	m2	11.52
215mm dense aggregate blocks	m2	13.61
75mm lightweight aggregate loadbearing blocks	m2	7.04
100mm lightweight aggregate loadbearing blocks	m2	8.44
150mm lightweight aggregate loadbearing blocks	m2	14.09
215mm lightweight aggregate loadbearing blocks	m2	20.20
Bituminous emulsion waterproofing liquid	5 litre	10.94
Cavity insulation retaining clips	1,000	55.61
Clay flue linings		
185 mm square x 300 mm - straight	each	5.51
185 mm square - curved	each	14.60
Clay tiles - 265 x 165 mm		
creasing	100	42.78
roofing	100	43.69
Damp-proof courses		
fibre base		
112.5 mm wide	8 m roll	4.88
150 mm wide	8 m roll	6.50
225 mm wide	8 m roll	9.76
fibre base lead lined		
112.5 mm wide	8 m roll	17.34
225 mm wide	8 m roll	34.63
hessian base		
112.5 mm wide	8 m roll	6.65
150 mm wide	8 m roll	8.86
225 mm wide	8 m roll	13.30
hessian base lead lined		
112.5 mm wide	8 m roll	18.25
225 mm wide	8 m roll	36.50
pitch polymer		
112.5 mm wide	20 m roll	18.48
150 mm wide	20 m roll	24.63
225 mm wide	20 m roll	36.95
polyethylene		
112.5 mm wide	30 m roll	4.78
225 mm wide	30 m roll	9.56
slates		
350 x 112.5 mm	100	56.07
350 x 225 mm	100	119.99

Prices actually to be paid for materials must be checked against the above basic prices and adjustments made as necessary.

Brickwork and blockwork

	Unit	Price
Basic prices for materials		£
Fireclay	tonne	255.00
Glass blocks		
190 x 190 x 80 mm	each	3.92
240 x 240 x 80 mm	each	6.75
Hydrated lime	tonne	139.84
Mesh reinforcement		
64 mm wide	25 m coil	9.32
178 mm wide	25 m coil	25.16
Portland cement	tonne	89.04
Typex Twin Wall metal flue pipes	m	26.32
Sand	m3	17.96
	tonne	11.22
Wall-ties - 200 mm long		
galvanised butterfly	1,000	101.40
stainless steel butterfly	1,000	121.68
3 mm galvanised vertical-twist	400	75.36
6 mm stainless steel-pressed	250	62.34

Prices actually to be paid for materials must be checked against the above basic prices and adjustments made as necessary.

Brickwork and blockwork

New work	Unit	Hours C	Hours L	Labour net	Material net	Price net	Price with 15%
Brickwork				£	£	£	£
					VAT not included		
Class B engineering bricks in cement mortar 1:3							
Walls							
102.5 mm	m2	1.10	0.55	13.57	27.08	40.65	46.75
215 mm	m2	2.20	1.10	27.12	53.78	80.90	93.03
327.5 mm	m2	2.95	1.49	36.47	79.85	116.32	133.77
Skins of hollow walls							
102.5 mm	m2	1.10	0.55	13.57	27.08	40.65	46.75
215 mm	m2	2.20	1.10	27.12	53.78	80.90	93.03
327.5 mm	m2	2.95	1.49	36.47	79.85	116.32	133.77
Common bricks in gauged mortar 1:1:6							
Walls							
102.5 mm	m2	1.00	0.50	12.33	14.28	26.61	30.60
215 mm	m2	2.00	1.00	24.66	28.39	53.05	61.01
327.5 mm	m2	2.70	1.35	33.29	41.83	75.12	86.39
Walls - curved to 3.00 m radius							
102.5 mm	m2	1.25	0.63	15.45	14.28	29.73	34.19
215 mm	m2	2.50	1.25	30.83	28.39	59.22	68.10
327.5 mm	m2	3.38	1.69	41.67	41.83	83.50	96.03
Honeycomb walls							
102.5 mm	m2	0.90	0.45	11.09	9.53	20.62	23.71
Filling existing openings							
102.5 mm	m2	1.10	0.55	13.57	14.28	27.85	32.03
215 mm	m2	2.20	1.10	27.12	28.39	55.51	63.84
327.5 mm	m2	3.00	1.50	36.99	41.83	78.82	90.64
Skins of hollow walls							
102.5 mm	m2	1.20	0.60	14.79	14.28	29.07	33.43
215 mm	m2	2.20	1.10	27.12	28.39	55.51	63.84
327.5 mm	m2	3.24	1.62	39.95	41.83	81.78	94.05
Projections of footings and chimney-breasts							
102.5 mm	m2	1.45	1.12	20.54	14.28	34.82	40.04
215 mm	m2	2.90	2.25	41.15	28.39	69.54	79.97
Isolated piers and chimney-stacks							
215 mm	m2	2.60	1.30	32.06	28.39	60.45	69.52
327.5 mm	m2	3.51	1.76	43.31	41.83	85.14	97.91
Backing to masonry including cutting and bonding to masonry							
102.5 mm	m2	1.45	0.73	17.91	16.49	34.40	39.56
215 mm	m2	2.65	1.33	32.70	30.60	63.30	72.80
327.5 mm	m2	3.90	1.95	48.08	44.04	92.12	105.94

Brickwork and blockwork

New work	Unit	Hours C	Hours L	Labour net	Material net	Price net	Price with 15%
Brickwork				£	£	£	£
					VAT not included		

Common bricks in gauged mortar 1:1:6
(*continued*)

Projections of attached piers; plinths; bands;
oversailing courses and the like

215 x 102.5 mm	m	0.40	0.20	4.93	3.09	8.02	9.22
215 x 215 mm	m	0.80	0.40	9.87	6.16	16.03	18.43
327.5 x 102.5 mm	m	0.60	0.30	7.40	4.67	12.07	13.88
327.5 x 215 mm	m	1.10	0.55	13.57	9.35	22.92	26.36

Thickening existing walls including cutting and
bonding new to existing and extra material for
bonding

102.5 mm	m2	2.50	1.25	30.83	16.49	47.32	54.42
215 mm	m2	3.65	1.33	41.63	30.60	72.23	83.06

Projections on existing walls of attached piers;
chimney-breasts and the like including cutting and
bonding new to existing and extra material for
bonding

215 x 102.5 mm	m	0.67	0.50	9.37	3.09	12.46	14.33
215 x 215 mm	m	1.20	0.90	16.82	6.16	22.98	26.43
327.5 x 102.5 mm	m	0.95	0.75	13.56	4.67	18.23	20.96
327.5 x 215 mm	m	1.75	1.33	24.64	9.35	33.99	39.09

Form 50 - 100 mm cavities in hollow walls with
200 mm wall-ties at 5 per m2 using

galvanised butterfly ties	m2	0.12	0.06	1.48	0.56	2.04	2.35
stainless steel butterfly ties	m2	0.12	0.06	1.48	0.67	2.15	2.47
3 mm galvanised vertical-twist ties	m2	0.12	0.06	1.48	1.03	2.51	2.89
0.6 mm stainless steel pressed ties	m2	0.12	0.06	1.48	1.37	2.85	3.28
0.6 mm stainless steel pressed ties with insulation retaining clips	m2	0.15	0.07	1.81	1.66	3.47	3.99

Close 50 - 100 mm cavities at ends of hollow
walls and at jambs or sills of openings with

102.5 mm brickwork and additional ties	m	0.40	0.20	4.93	1.63	6.56	7.54
102.5 mm brickwork; 150 mm asbestos base bitumen damp-proof course bedded in gauged mortar 1:1:6 and additional ties	m	0.50	0.25	6.17	2.48	8.65	9.95

Extra for rough arches 215 mm thick

in one half brick ring	m	0.60	0.30	7.40	3.10	10.50	12.07
in two half brick rings	m	0.90	0.45	11.09	6.19	17.28	19.87

Block bond ends of new walls to existing including
cutting pockets in existing work and extra material
for bonding

102.5 mm	m	0.42	0.21	5.18	0.89	6.07	6.98
215 mm	m	0.80	0.40	9.87	1.77	11.64	13.39

Brickwork and blockwork

New work	Unit	Hours C	Hours L	Labour net	Material net	Price net	Price with 15%
Brickwork				£	£	£	£
					VAT not included		
Refractory bricks in fireclay mortar							
102.5 mm linings to flues flush pointed one side and bonded to surrounding brickwork	m2	2.05	1.02	25.28	128.71	153.99	177.09
215 mm boiler flues flush pointed one side and fair faced and flush pointed the other side	m2	3.33	1.67	41.06	256.45	297.51	342.14
Brick facework							
Common bricks in gauged mortar 1:1:6							
Extra over common brickwork for fair face and flush pointing as the work proceeds							
stretcher bond	m2	0.45	0.23	5.55	0.08	5.63	6.47
Flemish bond	m2	0.50	0.25	6.17	0.08	6.25	7.19
margins	m	0.08	0.04	0.99	0.08	1.07	1.23
flat arch 215 mm on face and 50 mm on exposed soffit	m	0.40	0.25	5.27	0.08	5.35	6.15
Extra over common brickwork for fair face; raking out joints and flush pointing in cement mortar 1:3							
stretcher bond	m2	0.30	0.15	3.70	0.10	3.80	4.37
Flemish bond	m2	0.35	0.17	4.32	0.10	4.42	5.08
margins	m	0.09	0.04	1.11	0.10	1.21	1.39
flat arch 215 mm on face and 50 mm on exposed soffit	m	0.40	0.20	4.93	0.10	5.03	5.78
102.5 mm walls in stretcher bond fair faced and flush pointed both sides as the work proceeds	m2	1.60	0.80	19.73	14.28	34.01	39.11
102.5 mm walls in stretcher bond fair faced and flush pointed both sides in cement mortar 1:3 including raking out joints	m2	2.45	1.23	30.21	14.48	44.69	51.39
215 mm walls in Flemish bond fair faced and flush pointed both sides in cement mortar 1:3 including raking out joints	m2	3.80	1.90	46.86	28.59	75.45	86.77
Fair squint angles in purpose made squint bricks	m	0.25	0.12	3.07	3.61	6.68	7.68
Fair rounded angles in purpose made bullnose bricks	m	0.25	0.13	3.08	3.61	6.69	7.69
Extra over 215 mm wall for projecting double course tile creasings 20 mm set-forward and with cement fillets both sides	m	0.65	0.33	8.01	7.47	15.48	17.80
215 x 102.5 mm brick on edge copings fair faced and flush pointed as the work proceeds	m	0.80	0.40	9.87	3.18	13.05	15.01

New work	Unit	Hours C	Hours L	Labour net	Material net	Price net	Price with 15%
Brickwork				£	£	£	£
					VAT not included		

Brick facework (*continued*)

Facing bricks in gauged mortar 1:1:6

Extra over common brickwork for facework and flush pointing as the work proceeds

	Unit	Hours C	Hours L	Labour net	Material net	Price net	Price with 15%
Flemish bond	m2	0.66	0.33	8.13	13.19	21.32	24.52
margins	m	0.08	0.04	0.99	0.08	1.07	1.23
flat arch 215 mm on face and 50 mm on exposed soffit	m	0.46	0.23	5.67	1.74	7.41	8.52
segmental arch 215 mm on face and 50 mm on exposed soffit	m	1.50	0.75	18.49	1.74	20.23	23.26
semicircular arch 215 mm on face and 50 mm on exposed soffit	m	1.50	0.75	18.49	1.74	20.23	23.26

Extra over common brickwork for facework; raking out joints and flush pointing in cement mortar 1:3

	Unit	Hours C	Hours L	Labour net	Material net	Price net	Price with 15%
Flemish bond	m2	0.80	0.40	9.87	0.10	9.97	11.47
margins	m	0.13	0.07	1.60	0.10	1.70	1.96
flat arch 215 mm on face and 50 mm on exposed soffit	m	0.50	0.25	6.17	0.10	6.27	7.21
segmental arch 215 mm on face and on exposed soffit	m	1.60	8.00	68.26	0.10	68.36	78.61
semicircular arch 215 mm on face and on exposed soffit	m	1.60	0.80	19.73	0.10	19.83	22.80

102.5 mm walls in stretcher bond faced and flush pointed as the work proceeds

	Unit	Hours C	Hours L	Labour net	Material net	Price net	Price with 15%
one side	m2	1.50	0.75	18.49	24.16	42.65	49.05
both sides	m2	1.80	0.90	22.20	24.16	46.36	53.31

102.5 mm walls in stretcher bond faced and flush pointed in cement mortar 1:3 including raking out joints

	Unit	Hours C	Hours L	Labour net	Material net	Price net	Price with 15%
one side	m2	1.80	0.90	22.20	24.26	46.46	53.43
both sides	m2	2.40	1.20	29.59	24.36	53.95	62.04

	Unit	Hours C	Hours L	Labour net	Material net	Price net	Price with 15%
215 mm walls in Flemish bond faced and flush pointed both sides as the work proceeds	m2	3.20	1.60	39.45	48.16	87.61	100.75
215 mm walls in Flemish bond faced and flush pointed both sides in cement mortar 1:3 including raking out joints	m2	3.92	1.96	48.33	48.19	96.52	111.00
102.5 mm skins of hollow walls in stretcher bond faced and flush pointed one side as the work proceeds	m2	1.80	0.90	22.20	24.16	46.36	53.31
102.5 mm skins of hollow walls in stretcher bond faced and flush pointed one side in cement mortar 1:3 including raking out joints	m2	2.16	1.08	26.63	24.26	50.89	58.52

Brickwork and blockwork

New work	Unit	Hours C	Hours L	Labour net	Material net	Price net	Price with 15%
Brickwork				£	£	£	£
					VAT not included		
Brick facework *(continued)*							
Facing bricks in gauged mortar 1:1:6 *(continued)*							
Fair squint angles in purpose made squint bricks	m	0.25	0.13	3.08	6.75	9.83	11.30
Fair rounded angles in purpose made bullnose bricks	m	0.25	0.13	3.08	6.01	9.09	10.45
Extra for facework to							
65 mm recessed bands 25 mm set-back	m	0.25	0.13	3.08	-	3.08	3.54
65 mm projecting bands 25 mm set-forward	m	0.22	0.11	2.71	-	2.71	3.12
Extra over 215 mm wall for projecting double course tile creasings 20 mm set-forward and with cement fillets both sides	m	0.65	0.33	8.01	7.47	15.48	17.80
215 x 102.5 mm brick on edge copings faced and flush pointed in cement mortar 1:3 including raking out joints	m	0.80	0.40	9.87	5.44	15.31	17.61
215 x 102.5 mm brick on edge sills faced and flush pointed in cement mortar 1:3 including raking out joints	m	0.75	0.38	9.25	5.54	14.79	17.01
Engineering bricks in cement mortar 1:3							
Brick on edge steps fair faced and flush pointed as the work proceeds							
215 x 102.5 mm	m	0.83	0.41	10.24	5.96	16.20	18.63
327.5 x 102.5 mm	m	1.40	0.70	17.26	8.95	26.21	30.14
Extra for rounded edges formed with							
bullnosed bricks	m	0.25	0.13	3.08	3.71	6.79	7.81
angles	each	0.10	0.05	1.24	2.21	3.45	3.97

Brickwork and blockwork

New work	Unit	Hours C	Hours L	Labour net	Material net	Price net	Price with 15%
Blockwork				£	£	£	£
					VAT not included		

Precast concrete dense aggregate blocks to BS 6073 in gauged mortar 1:1:6

	Unit	Hours C	Hours L	Labour net	Material net	Price net	Price with 15%
Walls and partitions							
75 mm solid	m2	0.65	0.49	9.12	5.97	15.09	17.35
100 mm solid	m2	0.80	0.60	11.21	6.67	17.88	20.56
140 mm solid	m2	1.00	0.75	14.01	12.77	26.78	30.80
215 mm hollow	m2	1.20	0.90	16.82	15.47	32.29	37.13
Honeycomb walls							
100 mm solid	m2	0.70	0.53	9.81	4.41	14.22	16.35
Filling existing openings							
75 mm solid	m2	0.84	0.63	11.78	5.97	17.75	20.41
100 mm solid	m2	1.04	0.78	14.58	6.67	21.25	24.44
140 mm solid	m2	1.30	0.97	18.22	12.77	30.99	35.64
215 mm hollow	m2	1.30	0.97	18.22	15.30	33.52	38.55
Skins of hollow walls							
75 mm solid	m2	0.78	0.58	10.93	5.97	16.90	19.43
100 mm solid	m2	0.96	0.72	13.45	6.67	20.12	23.14
140 mm solid	m2	1.20	0.90	16.82	12.77	29.59	34.03
215 mm hollow	m2	1.22	0.92	17.10	15.47	32.57	37.46
Walls and partitions fair faced and flush pointed one side as the work proceeds							
75 mm solid	m2	0.83	0.62	11.64	5.97	17.61	20.25
100 mm solid	m2	0.98	0.73	13.73	6.67	20.40	23.46
140 mm solid	m2	1.18	0.89	16.53	12.77	29.30	33.70
Walls and partitions fair faced and flush pointed both sides as the work proceeds							
75 mm solid	m2	1.01	0.76	14.16	5.97	20.13	23.15
100 mm solid	m2	1.16	0.87	16.25	6.67	22.92	26.36
140 mm solid	m2	1.36	1.02	19.06	12.77	31.83	36.60
Extra for filling every fourth void of 215 mm hollow blocks with concrete 1:2:4 and reinforcing with one 6 mm bar	m2	0.08	0.29	2.67	1.01	3.68	4.23
Form 50 - 100 mm cavities in hollow walls with 200 mm wall-ties at 5 per m2 using							
galvanised butterfly ties	m2	0.12	0.06	1.48	0.56	2.04	2.35
stainless steel butterfly ties	m2	0.12	0.06	1.48	0.67	2.15	2.47
3 mm galvanised vertical-twist ties	m2	0.12	0.06	1.48	1.03	2.51	2.89
0.6 mm stainless steel pressed ties	m2	0.12	0.06	1.48	1.37	2.85	3.28
0.6 mm stainless steel pressed ties with insulation retaining clips	m2	0.12	0.08	1.62	1.66	3.28	3.77

Brickwork and blockwork

New work	Unit	Hours C	Hours L	Labour net	Material net	Price net	Price with 15%
Blockwork				£	£	£	£
					VAT not included		

Precast concrete dense aggregate blocks to BS 6073 in gauged mortar 1:1:6 *(continued)*

	Unit	Hours C	Hours L	Labour net	Material net	Price net	Price with 15%
Close 50 - 100 mm cavities at ends of hollow walls and at jambs or sills of openings with							
100 mm blockwork and additional ties	m	0.10	0.07	1.41	0.91	2.32	2.67
100 mm blockwork; 150 mm fibre base bitumen damp-proof course bedded in gauged mortar 1:1:6 and additional ties	m	0.20	0.15	2.80	1.76	4.56	5.24
Close 50 - 100 mm cavities at tops of hollow walls with single course of blocks laid flat in gauged mortar 1:1:6							
100 mm blocks	m	0.20	0.15	2.80	0.57	3.37	3.88
140 mm blocks	m	0.18	0.14	2.52	1.04	3.56	4.09
Bond ends of new walls to other types of construction including forming pockets in new construction and extra material for bonding							
75 mm	m	0.25	0.19	3.52	0.37	3.89	4.47
100 mm	m	0.30	0.23	4.21	0.41	4.62	5.31
140 mm	m	0.35	0.26	4.91	0.72	5.63	6.47
215 mm	m	0.40	0.30	5.60	0.98	6.58	7.57
Bond ends of new walls to other types of construction including cutting pockets in existing construction and extra material for bonding							
75 mm	m	0.40	0.30	5.60	0.37	5.97	6.87
100 mm	m	0.45	0.34	6.31	0.41	6.72	7.73
140 mm	m	0.55	0.41	7.71	0.72	8.43	9.69
215 mm	m	0.60	0.45	8.41	0.98	9.39	10.80

Precast concrete clinker aggregate blocks to BS 6073 in gauged mortar 1:1:6

	Unit	Hours C	Hours L	Labour net	Material net	Price net	Price with 15%
Walls and partitions							
60 mm	m2	0.45	0.34	6.31	5.22	11.53	13.26
75 mm	m2	0.50	0.38	7.01	5.22	12.23	14.06
100 mm	m2	0.55	0.41	7.71	6.20	13.91	16.00
140 mm	m2	0.70	0.53	9.81	7.57	17.38	19.99
215 mm	m2	0.95	0.71	13.32	13.91	27.23	31.31
Skins of hollow walls							
60 mm	m2	0.54	0.41	7.60	5.22	12.82	14.74
75 mm	m2	0.60	0.45	8.41	5.22	13.63	15.67
100 mm	m2	0.66	0.50	9.25	6.20	15.45	17.77
140 mm	m2	0.84	0.63	11.78	7.57	19.35	22.25
215 mm	m2	1.14	0.85	15.97	13.91	29.88	34.36

Brickwork and blockwork

New work	Unit	Hours C	Hours L	Labour net	Material net	Price net	Price with 15%
Blockwork				£	£	£	£
					VAT not included		

Precast concrete clinker aggregate blocks to BS 6073 in gauged mortar 1:1:6 (*continued*)

Form 50 - 100 mm cavities in hollow walls with 200 mm wall-ties at 5 per m2 using

	Unit	Hours C	Hours L	Labour net	Material net	Price net	Price with 15%
galvanised butterfly ties	m2	0.12	0.06	1.48	0.56	2.04	2.35
stainless steel butterfly ties	m2	0.12	0.06	1.48	0.67	2.15	2.47
3 mm galvanised vertical-twist ties	m2	0.12	0.06	1.48	1.03	2.51	2.89
0.6 mm stainless steel pressed ties	m2	0.12	0.06	1.48	1.37	2.85	3.28
0.6 mm stainless steel pressed ties with insulation retaining clips	m2	0.12	0.08	1.62	1.66	3.28	3.77

Close 50 - 100 mm cavities at ends of hollow walls and at jambs or sills of openings with

	Unit	Hours C	Hours L	Labour net	Material net	Price net	Price with 15%
100 mm blockwork and additional ties	m	0.10	0.05	1.24	0.87	2.11	2.43
100 mm blockwork; 150 mm fibre base bitumen damp-proof course bedded in gauged mortar 1:1:6 and additional ties	m	0.20	0.10	2.46	1.72	4.18	4.81

Close 50 - 100 mm cavities at tops of hollow walls with single course of blocks laid flat in gauged mortar 1:1:6

	Unit	Hours C	Hours L	Labour net	Material net	Price net	Price with 15%
100 mm blocks	m	0.20	0.10	2.46	0.53	2.99	3.44
140 mm blocks	m	0.23	0.12	2.87	0.63	3.50	4.03

Bond ends of new walls to other types of construction including forming pockets in new construction and extra material for bonding

	Unit	Hours C	Hours L	Labour net	Material net	Price net	Price with 15%
60 mm	m	0.20	0.10	2.46	0.33	2.79	3.21
75 mm	m	0.25	0.13	3.08	0.33	3.41	3.92
100 mm	m	0.30	0.15	3.70	0.38	4.08	4.69
140 mm	m	0.33	0.17	4.07	0.45	4.52	5.20
215 mm	m	0.40	0.20	4.93	0.76	5.69	6.54

Bond ends of new walls to other types of construction including cutting pockets in existing construction and extra material for bonding

	Unit	Hours C	Hours L	Labour net	Material net	Price net	Price with 15%
60 mm	m	0.28	0.21	3.93	0.33	4.26	4.90
75 mm	m	0.30	0.23	4.24	0.33	4.57	5.26
100 mm	m	0.35	0.26	4.89	0.38	5.27	6.06
140 mm	m	0.40	0.30	5.60	0.45	6.05	6.96
215 mm	m	0.45	0.34	6.31	0.76	7.07	8.13

Precast concrete lightweight aggregate loadbearing blocks to BS 6073 in gauged mortar 1:1:6

Walls and partitions

	Unit	Hours C	Hours L	Labour net	Material net	Price net	Price with 15%
75 mm	m2	0.45	0.34	6.32	7.81	14.13	16.25
100 mm	m2	0.50	0.38	7.04	9.36	16.40	18.86
150 mm	m2	0.65	0.49	9.12	15.56	24.68	28.38
215 mm	m2	0.90	0.68	12.64	22.22	34.86	40.09

Brickwork and blockwork

New work	Unit	Hours C	Hours L	Labour net	Material net	Price net	Price with 15%
Blockwork				£	£	£	£
					VAT not included		

Precast concrete lightweight aggregate loadbearing blocks to BS 6073 in gauged mortar 1:1:6 (continued)

	Unit	Hours C	Hours L	Labour net	Material net	Price net	Price with 15%
Honeycomb walls							
100 mm	m2	0.60	0.45	8.41	6.19	14.60	16.79
Skins of hollow walls							
75 mm	m2	0.54	0.41	7.60	7.81	15.41	17.72
100 mm	m2	0.60	0.45	8.41	9.36	17.77	20.44
150 mm	m2	0.78	0.59	10.97	15.56	26.53	30.51
215 mm	m2	1.08	0.81	15.14	22.22	37.36	42.96
Form 50 - 100 mm cavities in hollow walls with 200 mm wall-ties at 5 per m2 using							
galvanised butterfly ties	m2	0.12	0.06	1.48	0.56	2.04	2.35
stainless steel butterfly ties	m2	0.12	0.06	1.48	0.67	2.15	2.47
3 mm galvanised vertical-twist ties	m2	0.12	0.06	1.48	1.03	2.51	2.89
0.6 mm stainless steel pressed ties	m2	0.12	0.06	1.48	1.37	2.85	3.28
0.6 mm stainless steel pressed ties with insulation retaining clips	m2	0.12	0.08	1.62	1.66	3.28	3.77
Close 50 - 100 mm cavities at ends of hollow walls and at jambs or sills of openings with							
100 mm blockwork and additional ties	m	0.08	0.06	1.12	1.12	2.24	2.58
100 mm blockwork; 150 mm fibre base bitumen damp-proof course bedded in gauged mortar 1:1:6 and additional ties	m	0.18	0.14	2.55	1.97	4.52	5.20
Close 50 - 100 mm cavities at tops of hollow walls with single course of blocks laid flat in gauged mortar 1:1:6							
100 mm blocks	m	0.15	0.11	2.08	0.78	2.86	3.29
150 mm blocks	m	0.20	0.15	2.80	1.25	4.05	4.66
Bond ends of new walls to other types of construction including forming pockets in new construction and extra material for bonding							
75 mm	m	0.25	0.19	3.52	0.47	3.99	4.59
100 mm	m	0.30	0.23	4.24	0.55	4.79	5.51
150 mm	m	0.33	0.25	4.65	0.86	5.51	6.34
215 mm	m	0.40	0.30	5.60	1.20	6.80	7.82
Bond ends of new walls to other types of construction including cutting pockets in existing construction and extra material for bonding							
75 mm	m	0.40	0.30	5.60	0.47	6.07	6.98
100 mm	m	0.45	0.34	6.32	0.55	6.87	7.90
150 mm	m	0.55	0.41	7.69	0.86	8.55	9.83
215 mm	m	0.65	0.49	9.12	1.20	10.32	11.87

Brickwork and blockwork

New work	Unit	Hours C	Hours L	Labour net	Material net	Price net	Price with 15%
Blockwork				£	£	£	£
					VAT not included		
290 x 290 mm perforated precast concrete screen walling blocks in gauged mortar 1:1:6							
90 mm walls flush pointed both sides as the work proceeds	m2	1.05	0.79	14.73	22.33	37.06	42.62
Tie ends of 90 mm walls to brickwork with galvanised butterfly ties at 300 mm centres vertically	m	0.11	0.08	1.53	0.34	1.87	2.15
Glass blocks							
80 mm partitions in gauged mortar 1:1:6 flush pointed both sides as the work proceeds							
in 190 x 190 mm blocks	m2	2.40	1.80	33.63	113.12	146.75	168.76
in 240 x 240 mm blocks	m2	1.80	1.35	25.23	153.78	179.01	205.86

For specialist prices for glass block window panels, see "Concrete work".

Brickwork and blockwork

New work	Unit	Hours C	Hours L	Labour net	Material net	Price net	Price with 15%
Damp proof courses				£	£	£	£
					VAT not included		
Fibre based bitumen							
Horizontal damp-proof courses bedded in cement mortar 1:3							
over 225 mm wide	m2	0.25	0.13	3.08	5.76	8.84	10.17
112.5 mm wide	m	0.03	0.01	0.34	0.64	0.98	1.13
225 mm wide	m	0.06	0.03	0.69	1.28	1.97	2.27
150 mm vertical damp-proof courses bedded in gauged mortar 1:1:6	m	0.07	0.03	0.84	0.85	1.69	1.94
Lead lined fibre base bitumen							
Horizontal damp-proof courses bedded in cement mortar 1:3							
over 225 mm wide	m2	0.33	0.17	4.07	20.45	24.52	28.20
112.5 mm wide	m	0.04	0.02	0.49	2.28	2.77	3.19
225 mm wide	m	0.07	0.04	0.91	4.54	5.45	6.27
Hessian base bitumen							
Horizontal damp-proof courses bedded in cement mortar 1:3							
over 225 mm wide	m	0.25	0.13	3.08	7.86	10.94	12.58
112.5 mm wide	m	0.03	0.01	0.34	0.87	1.21	1.39
225 mm wide	m	0.06	0.03	0.69	1.75	2.44	2.81
150 mm vertical damp-proof courses bedded in gauged mortar 1:1:6	m	0.07	0.03	0.83	1.16	1.99	2.29
Lead lined hessian base bitumen							
Horizontal damp-proof courses bedded in cement mortar 1:3							
over 225 mm wide	m2	0.33	0.17	4.07	21.56	25.63	29.47
112.5 mm wide	m	0.04	0.02	0.49	2.40	2.89	3.32
225 mm wide	m	0.07	0.04	0.91	4.79	5.70	6.55
Pitch polymer							
Horizontal damp-proof courses bedded in cement mortar 1:3							
over 225 mm wide	m2	0.25	0.13	3.08	8.73	11.81	13.58
112.5 mm wide	m	0.03	0.01	0.34	0.97	1.31	1.51
225 mm wide	m	0.06	0.03	0.69	1.94	2.63	3.02
150 mm vertical damp-proof courses bedded in gauged mortar 1:1:6	m	0.07	0.03	0.83	1.29	2.12	2.44

Brickwork and blockwork

New work	Unit	Hours C	Hours L	Labour net	Material net	Price net	Price with 15%
Damp proof courses				£	£	£	£
					VAT not included		
Polythene							
Horizontal damp-proof courses bedded in cement mortar 1:3							
over 225 mm wide	m2	0.25	0.13	3.08	1.51	4.59	5.28
112.5 mm wide	m	0.03	0.01	0.34	0.17	0.51	0.59
225 mm wide	m	0.06	0.03	0.69	0.34	1.03	1.18
Slate							
Double course slate horizontal damp-proof courses bedded in cement mortar 1:3							
over 225 mm wide	m2	1.10	0.55	13.57	41.73	55.30	63.59
112.5 mm wide	m	0.15	0.07	1.85	2.81	4.66	5.36
225 mm wide	m	0.25	0.13	3.08	9.45	12.53	14.41

Brickwork and blockwork

New work	Unit	Hours C	Hours L	Labour net	Material net	Price net	Price with 15%
Sundries				£	£	£	£
					VAT not included		
Cavities							
Form 50 mm cavities in hollow walls with wall-ties at 5 per m2 using							
galvanised butterfly ties	m2	0.12	0.06	1.48	0.56	2.04	2.35
stainless steel butterfly ties	m2	0.12	0.06	1.48	0.67	2.15	2.47
3 mm galvanised vertical-twist ties	m2	0.12	0.06	1.48	1.03	2.51	2.89
0.6 mm stainless steel pressed ties	m2	0.12	0.06	1.48	1.37	2.85	3.28
with insulation retaining clips	m2	0.12	0.08	1.62	1.66	3.28	3.77
Seal 50 - 100 mm cavities at eaves and sills in hollow walls with single course of slates in gauged mortar 1:1:6	m	0.15	0.07	1.85	2.79	4.64	5.34
Reinforcement							
Mesh reinforcement in walls							
64 mm	m	0.03	0.01	0.37	0.41	0.78	0.90
178 mm	m	0.03	0.01	0.37	1.11	1.48	1.70
Joints							
Rake out joints of brickwork to form key for plastering etc.	m2	0.20	0.10	2.46	-	2.46	2.83
Chases							
Horizontal rough chases							
25 x 25 mm	m	0.20	0.10	2.46	0.50	2.96	3.40
50 x 50 mm	m	0.30	0.15	3.70	0.60	4.30	4.95
75 x 75 mm	m	0.40	0.20	4.93	0.70	5.63	6.47
100 x 100 mm	m	0.50	0.25	6.17	0.80	6.97	8.02
Vertical rough chases							
25 x 25 mm	m	0.30	0.15	3.70	0.50	4.20	4.83
50 x 50 mm	m	0.45	0.23	5.55	0.60	6.15	7.07
75 x 75 mm	m	0.60	0.30	7.40	0.70	8.10	9.31
100 x 100 mm	m	0.75	0.38	9.25	0.80	10.05	11.56
Waterproofing							
Three coats of bituminous emulsion water-proofing liquid on brick or block walls including blinding final coat with sand	m2	-	0.60	4.04	3.62	7.66	8.81
Angle fillets							
Mortar angle-fillets							
50 mm	m	0.25	0.13	3.08	0.30	3.38	3.89
75 mm	m	0.45	0.23	5.55	0.60	6.15	7.07

Brickwork and blockwork

New work	Unit	Hours C	Hours L	Labour net	Material net	Price net	Price with 15%
Sundries				£	£	£	£
					VAT not included		
Weatherings							
150 mm sills or weatherings to projections of two courses of plain clay roofing tiles set weathering and breaking joint and bedded; jointed and pointed in cement mortar 1:3							
straight lengths	m	0.55	0.28	6.78	5.04	11.82	13.59
cut and fitted ends	each	0.10	0.05	1.24	0.69	1.93	2.22
Bedding							
Bed plates 100 mm wide in mortar	m	0.05	0.03	0.62	0.10	0.72	0.83
Bed wood frames and sills in mortar and point one side (includes allowance for unloading and hoisting)	m	0.09	0.04	1.11	0.60	1.71	1.97
Wedge and pin							
Wedge and pin up to underside of existing construction with slates in cement mortar 1:3							
102.5 mm walls	m	0.85	0.42	10.48	2.53	13.01	14.96
215 mm walls	m	1.40	0.70	17.26	4.82	22.08	25.39
327.5 mm walls	m	1.95	0.97	24.04	6.13	30.17	34.70
Rake out							
Rake out joints for turned-in edges of flashings and point							
horizontal	m	0.25	0.13	3.08	0.10	3.18	3.66
stepped	m	0.35	0.17	4.32	0.20	4.52	5.20
Build in							
Build in metal windows including building in lugs at jambs; plugging and screwing frames at head and sill; filling backs of frames with cement mortar 1:3 and pointing one side; window area approximately							
0.50 m2	each	0.67	0.34	8.26	1.03	9.29	10.68
1.00 m2	each	1.10	0.55	13.57	1.51	15.08	17.34
1.50 m2	each	1.45	0.72	17.88	1.98	19.86	22.84
2.00 m2	each	1.65	0.82	20.34	2.45	22.79	26.21
2.50 m2	each	2.15	1.08	26.51	2.93	29.44	33.86
3.00 m2	each	2.33	1.17	28.73	3.42	32.15	36.97
Build in 800 x 2100 mm metal door frames including building in lugs and filling backs of frames with cement mortar 1:3	each	0.85	0.42	10.48	0.80	11.28	12.97
Cut and pin							
Cut and pin ends of steel sections and make good							
small (not exceeding 250 mm deep)	each	0.60	0.30	7.40	0.10	7.50	8.63
large (250 - 500 mm deep)	each	0.90	0.45	11.09	0.20	11.29	12.98
extra large (over 500 mm deep)	each	1.20	0.60	14.79	0.20	14.99	17.24

Brickwork and blockwork

New work	Unit	Hours C	Hours L	Labour net	Material net	Price net	Price with 15%
Sundries				£	£	£	£
					VAT not included		
Holes							
Holes for small pipes (not exceeding 55 mm) through walls and make good							
75 mm	each	0.17	0.09	2.09	0.10	2.19	2.52
102.5 mm	each	0.30	0.15	3.70	0.20	3.90	4.49
215 mm	each	0.45	0.23	5.55	0.30	5.85	6.73
Holes for large pipes (55 - 110 mm) through walls and make good							
75 mm	each	0.27	0.14	3.33	0.10	3.43	3.94
102.5 mm	each	0.37	0.19	4.57	0.20	4.77	5.49
215 mm	each	0.50	0.25	6.17	0.30	6.47	7.44
Holes for extra large pipes (over 110 mm) through walls and make good							
75 mm	each	0.38	0.19	4.68	0.20	4.88	5.61
102.5 mm	each	0.45	0.23	5.55	0.30	5.85	6.73
215 mm	each	0.75	0.38	9.25	0.40	9.65	11.10
Holes for small pipes (not exceeding 55 mm) through walls and make good facings							
102.5 mm	each	0.40	0.20	4.93	0.10	5.03	5.78
215 mm	each	0.55	0.28	6.78	0.10	6.88	7.91
Holes for large pipes (55 - 110 mm) through walls and make good facings							
102.5 mm	each	0.55	0.28	6.78	0.20	6.98	8.03
215 mm	each	0.80	0.40	9.87	0.20	10.07	11.58
Holes for extra large pipes (over 110 mm) through walls and make good facings							
102.5 mm	each	0.67	0.34	8.26	0.20	8.46	9.73
215 mm	each	1.00	0.50	12.33	0.30	12.63	14.52
Mortices							
Mortices for bolts and run with mortar	each	0.35	0.17	4.32	0.10	4.42	5.08
Openings							
225 x 75 mm openings through walls with slate lintels and make good							
102.5 mm	each	0.12	0.06	1.48	0.78	2.26	2.60
215 mm	each	0.20	0.10	2.46	1.51	3.97	4.57
225 x 150 mm openings through walls with slate lintels and make good							
102.5 mm	each	0.15	0.07	1.85	0.78	2.63	3.02
215 mm	each	0.25	0.13	3.08	1.51	4.59	5.28
225 x 225 mm openings through walls with slate lintels and make good							
102.5 mm	each	0.18	0.09	2.22	0.78	3.00	3.45
215 mm	each	0.30	0.15	3.70	1.51	5.21	5.99

Brickwork and blockwork

New work	Unit	Hours C	Hours L	Labour net	Material net	Price net	Price with 15%
Sundries				£	£	£	£
					VAT not included		

Openings (*continued*)

New work	Unit	Hours C	Hours L	Labour net	Material net	Price net	Price with 15%
250 x 250 mm openings through walls with slate lintels and make good							
102.5 mm	each	0.30	0.15	3.70	0.78	4.48	5.15
215 mm	each	0.50	0.25	6.17	1.51	7.68	8.83
225 x 75 mm openings through walls with slate lintels and make good facings							
102.5 mm	each	0.18	0.09	2.22	0.78	3.00	3.45
215 mm	each	0.27	0.14	3.33	1.51	4.84	5.57
225 x 150 mm openings through walls with slate lintels and make good facings							
102.5 mm	each	0.23	0.12	2.84	0.78	3.62	4.16
215 mm	each	0.33	0.17	4.07	1.51	5.58	6.42
225 x 225 mm openings through walls with slate lintels and make good facings							
102.5 mm	each	0.27	0.14	3.33	0.78	4.11	4.73
215 mm	each	0.40	0.20	4.93	1.51	6.44	7.41
250 x 250 mm openings through walls with slate lintels and make good facings							
102.5 mm	each	0.42	0.21	5.18	0.78	5.96	6.85
215 mm	each	0.63	0.32	7.76	1.51	9.27	10.66

Air bricks and soot doors

New work	Unit	Hours C	Hours L	Labour net	Material net	Price net	Price with 15%
Provide and build in air-bricks - nominal size							
229 x 76 mm galvanised	each	0.08	0.04	0.99	2.54	3.53	4.06
229 x 152 mm galvanised	each	0.08	0.04	0.99	4.67	5.66	6.51
229 x 229 mm galvanised	each	0.08	0.04	0.99	6.84	7.83	9.00
229 x 76 mm terra-cotta	each	0.08	0.04	0.99	1.60	2.59	2.98
229 x 152 mm terra-cotta	each	0.08	0.04	0.99	2.22	3.21	3.69
229 x 229 mm terra-cotta	each	0.08	0.04	0.99	6.10	7.09	8.15
Provide and build in terra-cotta air brick extension cavity liners 300 mm long - nominal size							
225 x 75 mm horizontal	each	0.10	0.05	1.24	2.97	4.21	4.84
225 x 150 mm horizontal	each	0.10	0.05	1.24	3.22	4.46	5.13
225 x 225 mm horizontal	each	0.10	0.05	1.24	8.59	9.83	11.30
225 x 75 mm inclined	each	0.10	0.05	1.24	19.94	21.18	24.36
225 x 150 mm inclined	each	0.10	0.05	1.24	23.79	25.03	28.78
225 x 225 mm inclined	each	0.10	0.05	1.24	46.81	48.05	55.26
Provide and build in 250 x 250 mm (nominal) cast iron soot-doors with double covers	each	0.25	0.13	3.08	18.45	21.53	24.76

Brickwork and blockwork

New work	Unit	Hours C	Hours L	Labour net	Material net	Price net	Price with 15%
Sundries				£	£	£	£
					VAT not included		
Clay linings and pots							
185 x 185 mm clay flue linings to BS 1181 with rebated joints							
straight lengths	m	0.40	0.20	4.93	19.77	24.70	28.41
extra for bends	each	0.11	0.06	1.36	15.33	16.69	19.19
Clay chimney-pots set and flaunched in mortar							
300 mm	each	0.50	0.25	6.17	18.69	24.86	28.59
450 mm	each	0.50	0.25	6.17	22.93	29.10	33.47
600 mm	each	0.55	0.28	6.78	33.71	40.49	46.56
Fix only							
Fix only in fireplace openings							
continuous burning fires	each	0.75	0.38	9.25	16.07	25.32	29.12
back boilers	each	1.50	0.75	18.49	16.07	34.56	39.74
Fix only average fireplace surrounds and hearths including assembling and jointing and setting in mortar	each	4.00	2.00	49.32	16.07	65.39	75.20
Gas flue blocks							
Typex HP gas flue blocks bedded and jointed in "Fluejoint" mortar							
Gas fire recess set comprising three 405 x 147 x 222 mm recess blocks HP1	set	0.30	0.15	3.70	27.75	31.45	36.17
385 x 140 x 222 mm cover block HP2	each	0.12	0.06	1.48	10.57	12.05	13.86
280 x 140 x 222 mm standard block HP3	each	0.11	0.06	1.36	9.01	10.37	11.93
380 x 140 x 222 mm closer block HP4	each	0.12	0.06	1.48	8.92	10.40	11.96
400 x 140 x 222 mm side offset block HP5	each	0.12	0.06	1.48	9.61	11.09	12.75
280 x 210 x 222 mm back offset block HP6	each	0.17	0.09	2.09	8.90	10.99	12.64
280 x 181 x 222 mm vertical exit block HP7	each	0.15	0.07	1.85	11.96	13.81	15.88
280 x 240 x 230 mm angled entry/exit block HP8	each	0.15	0.07	1.85	11.96	13.81	15.88
280 x 140 x 222 mm double rebate block HP9	each	0.11	0.06	1.36	10.50	11.86	13.64
280 x 262 x 222 mm corbel block HP10	each	0.15	0.07	1.85	11.65	13.50	15.53
Metal flue pipes							
Typex Twin Wall metal flue pipes							
125 mm pipes in roof space	m	0.25	0.13	3.08	23.35	26.43	30.39
Extra for adjustable bends	each	0.33	0.17	4.07	12.14	16.21	18.64
Joints of 125 mm pipe to terminal block	each	0.17	0.09	2.09	0.10	2.19	2.52
Type ridge adapters bolted to ridge terminal and jointed to 125 mm pipe	each	0.25	0.13	3.08	12.94	16.02	18.42
Gas ridge terminal bedded and pointed in cement mortar 1:3	each	0.13	0.07	1.60	63.10	64.70	74.41

Brickwork and blockwork

New work	Unit	Hours C	Hours L	Labour net	Material net	Price net	Price with 15%
Sundries				£	£	£	£
					VAT not included		
Centering							
Centering for flat brick arches not exceeding 2.00 m span; 102.5 mm wide							
first use	m	0.30	0.30	4.71	18.05	22.76	26.17
subsequent uses	m	0.20	0.20	3.14	0.02	3.16	3.63
Centering for segmental brick arches 1.00 m span; 102.5 mm wide and 25 mm rise							
first use	each	0.78	0.78	12.25	5.06	17.31	19.91
subsequent uses	each	0.50	0.50	7.85	0.33	8.18	9.41
Centering for segmental brick arches 2.00 m span; 102.5 mm wide and 25 mm rise							
first use	each	1.10	1.10	17.27	6.28	23.55	27.08
subsequent uses	each	0.66	0.66	10.36	0.66	11.02	12.67
Centering for semicircular brick arches 1.00 m span and 215 mm wide							
first use	each	3.00	3.00	47.10	8.01	55.11	63.38
subsequent uses	each	1.80	1.80	28.26	0.34	28.60	32.89
Centering for semicircular brick arches 2.00 m span and 215 mm wide							
first use	each	4.50	4.50	70.65	13.70	84.35	97.00
subsequent uses	each	2.70	2.70	42.39	0.69	43.08	49.54
Cavity insulation							
Expanded polystyrene cavity batts in cavities of hollow walls - retaining ties included with forming cavities							
25 mm	m2	0.25	0.13	3.08	1.62	4.70	5.41
40 mm	m2	0.25	0.13	3.08	2.60	5.68	6.53
50 mm	m2	0.25	0.13	3.08	3.38	6.46	7.43
Glass fibre cavity batts filling cavities of hollow walls							
50 mm	m2	0.25	0.13	3.08	3.21	6.29	7.23
75 mm	m2	0.25	0.13	3.08	4.27	7.35	8.45
100 mm	m2	0.25	0.13	3.08	5.68	8.76	10.07

Brickwork and blockwork

New work	Unit	Specialist price net	Price with 15%
UF foam cavity wall insulation		£	£
		VAT not included	

The following "Specialist price net" figures are guide prices provided by The Cavity Foam Bureau.

Prices do not include for cash discount.

	Unit	Specialist price net	Price with 15%
UF foam insulation to hollow walls in areas 50 - 100 m2, cavity widths			
up to 65 mm	m2	3.20	3.68
65 - 75 mm	m2	3.45	3.97
UF foam insulation to hollow walls in areas 100 - 500 m2, cavity widths			
up to 65 mm	m2	3.00	3.45
65 - 75 mm	m2	3.25	3.74
UF foam insulation to hollow walls in areas over 500 m2, cavity widths			
up to 65 mm	m2	2.70	3.10
65 - 75 mm	m2	2.95	3.39

Enquiries about the foregoing specialist prices and technical information about UF foam should be made to The Cavity Foam Bureau, PO Box 79, Oldbury, Warley, West Midlands, B69 4PW, tel (0121) 544 4949.

Brickwork and blockwork

New work	Unit	Hours C	Hours L	Labour net	Material net	Price net	Price with 15%
Composite walls				£	£	£	£
					VAT not included		

Facing brick external skin

Hollow walls in gauged mortar 1:1:6 of 102.5 mm facing brick external skin pointed as the work proceeds; 50 - 100 mm cavity with galvanised butterfly wall-ties and internal skin of

	Unit	Hours C	Hours L	Labour net	Material net	Price net	Price with 15%
100 mm dense aggregate blocks	m2	2.88	1.68	37.14	31.40	68.54	78.82
140 mm dense aggregate blocks	m2	2.92	1.71	37.70	37.50	75.20	86.48
100 mm clinker blocks	m2	2.58	1.46	32.93	30.93	63.86	73.44
140 mm clinker blocks	m2	2.76	1.59	35.45	32.30	67.75	77.91
100 mm lightweight aggregate blocks	m2	2.52	1.41	32.09	34.09	66.18	76.11
150 mm lightweight aggregate blocks	m2	2.70	1.55	34.64	40.28	74.92	86.16

Dense aggregate block external skin

Hollow walls in gauged mortar 1:1:6 of 100 mm dense aggregate block external skin; 50 - 100 mm cavity with galvanised butterfly wall-ties and internal skin of

	Unit	Hours C	Hours L	Labour net	Material net	Price net	Price with 15%
100 mm dense aggregate blocks	m2	1.96	1.44	27.27	13.91	41.18	47.36
140 mm dense aggregate blocks	m2	2.28	1.68	31.75	20.01	51.76	59.52
100 mm clinker aggregate blocks	m2	1.74	1.27	24.18	13.44	37.62	43.26
140 mm clinker aggregate blocks	m2	1.92	1.41	26.70	14.81	41.51	47.74
100 mm lightweight aggregate blocks	m2	1.68	1.23	23.34	16.60	39.94	45.93
140 mm lightweight aggregate blocks	m2	1.86	1.37	25.90	22.79	48.69	55.99

Lightweight aggregate block external skin

Hollow walls in gauged mortar 1:1:6 of 100 mm lightweight aggregate block external skin; 50 - 100 mm cavity with galvanised butterfly wall-ties and internal skin of

	Unit	Hours C	Hours L	Labour net	Material net	Price net	Price with 15%
100 mm dense aggregate blocks	m2	1.68	1.23	23.34	16.68	40.02	46.02
140 mm dense aggregate blocks	m2	1.92	1.41	26.70	22.70	49.40	56.81
100 mm clinker aggregate blocks	m2	1.38	1.00	19.13	16.13	35.26	40.55
140 mm clinker aggregate blocks	m2	1.56	1.14	21.66	17.50	39.16	45.03
100 mm lightweight aggregate blocks	m2	1.32	0.96	18.30	19.30	37.60	43.24
150 mm lightweight aggregate blocks	m2	1.50	1.10	20.85	25.48	46.33	53.28

Insulation

Extra for expanded polystyrene cavity batts in cavities and 0.6 mm stainless steel pressed ties with insulation retaining clips (ties & clips measured in form cavity)

	Unit	Hours C	Hours L	Labour net	Material net	Price net	Price with 15%
25 mm batts	m2	0.25	0.13	3.08	1.62	4.70	5.41
40 mm batts	m2	0.25	0.13	3.08	2.60	5.68	6.53
50 mm batts	m2	0.25	0.13	3.08	3.38	6.46	7.43

Extra for glass fibre cavity batts filling cavities

	Unit	Hours C	Hours L	Labour net	Material net	Price net	Price with 15%
50 mm batts	m2	0.20	0.10	2.46	3.21	5.67	6.52
75 mm batts	m2	0.20	0.10	2.46	4.27	6.73	7.74
100 mm batts	m2	0.20	0.10	2.46	5.68	8.14	9.36

Brickwork and blockwork

New work	Unit	Specialist price net	Price with 15%
Radon protection barriers		£	£
		VAT not included	
		Supply only	Supply only

The following "Specialist price net" figures are guide prices supplied by Cavity Trays Ltd of Yeovil

Prices do not include for cash discount.

Supply and deliver the following radon protection barriers, membranes and associated protection measures for new-build applications:

Type N Sitesealer gas grade oversite membrane, for damp-proofing and protection against rising gases.

Sitesealer membrane	m2	5.20	5.98

Petheleyne radon protection barriers for use in cavity walls to arrest rising gases within cavity.   Barriers link with oversite membrane to create an integrity shield across walls and floor.

Barriers	m	7.42	8.53
Barrier angles	each	13.25	15.24
Service entry points	each	18.00	20.70

Discharge cavibricks for use in cavity wall under barrier level, to permit rising gas to discharge into the open air.

Cavibricks, high performance airbricks	each	1.17	1.35

Type W combined caviweeps/cavivents, for use in perpendicular joints to discharge rainwater arrested above and on cavity barrier.

Type W Caviweeps/Cavivents	each	0.41	0.47

Petheleyne gas reception sump.   Incorporated within the granular fill under the ground floor, the sump passively exhausts gas via a standard 110m stack pipe which discharges at roof level.

Petheleyne Radon Reception Sump	each	45.58	52.42

Enquiries about the foregoing specialist prices, free design and advisory service, should be made to Cavity Trays Ltd, Administration Centre, Lufton Trading Estate, Yeovil, Somerset BA22 8HU, tel (01935) 474769, fax (01935) 428223

Underpinning

Preamble

"Labour net" figures include allowances for all costs incidental to the employment of labour.

"Plant net" figures include for all costs of plant including drivers and operators where applicable.

"Materials net" figures include for all costs of materials including an allowance for waste except where specifically stated.

"Price net" figures are the totals of the "Labour net", "Plant net" and "Materials net" figures. Prices are for a builder employing his own labour; according to the amount and nature of the work involved, it may well be possible to secure more advantageous prices from specialist sub-contractors.

Prices do not include any allowance for scaffolding, ladders or other plant necessary to reach the work. The "Preliminaries" section includes prices for scaffolding which must be considered and allowance included to suit the particular circumstances of a tender.

Underpinning

New work

Basic prices for materials		£
Aggregate		
40 mm	m3	19.75
	tonne	13.16
20 mm	m3	19.91
	tonne	13.27
Damp-proof courses		
fibre base - 225 mm wide	8 m roll	9.76
pitch polymer - 225 mm wide	20 m roll	36.95
slates - 350 x 225 mm	100	119.99
Portland cement	tonne	89.04
Sand	m3	17.96
	tonne	11.22

Prices actually to be paid for materials must be checked against the above basic prices and adjustments made as necessary.

Underpinning

New work	Unit	Hours C	Hours L	Labour net	Plant net	Material net	Price net	Price with 15%
Work in all trades				£	£	£	£	£
					VAT not included			

For temporary supports to work to be underpinned, see "Demolition" in the Alterations and repairs" section

Preliminary trenches

Excavate preliminary trenches down to the level of the base of the existing foundation, maximum depth not exceeding

1.00 m	m3	-	5.20	35.05	-	-	35.05	40.31
2.00 m	m3	-	7.00	47.18	-	-	47.18	54.26

Excavate below foundations

Excavate below the level of the base of the existing foundation, maximum depth not exceeding

1.00 m	m3	-	5.90	39.77	-	-	39.77	45.74
2.00 m	m3	-	7.80	52.57	-	-	52.57	60.46

Working space

Excavate and fill working space, maximum depth not exceeding

1.00 m	m3	-	6.90	46.51	-	-	46.51	53.49
2.00 m	m3	-	8.60	57.96	-	-	57.96	66.65

Projecting foundations

Cut away projecting foundations
two courses of footings and 600 x

225 mm concrete	m	-	2.60	17.52	-	-	17.52	20.15

three courses of footings and 825 x

300 mm concrete	m	-	4.90	33.03	-	-	33.03	37.98

Prepare underside

Prepare the underside of the existing work to receive the pinning up of the new

300 mm wide	m	-	0.30	2.02	-	-	2.02	2.32
450 mm wide	m	-	0.45	3.03	-	-	3.03	3.48
600 mm	m	-	0.65	4.38	-	-	4.38	5.04
900 mm	m	-	0.85	5.73	-	-	5.73	6.59
1200 mm	m	-	1.15	7.75	-	-	7.75	8.91

Underpinning

New work	Unit	Hours C	Hours L	Labour net	Plant net	Material net	Price net	Price with 15%
Work in all trades				£	£	£	£	£
					VAT not included			

Earthwork support

Earthwork support to sides of
preliminary trenches not exceeding
2.00 m between opposing faces, in stiff
soil, maximum depth not exceeding

	Unit	Hours C	Hours L	Labour net	Plant net	Material net	Price net	Price with 15%
1.00 m	m2	-	0.15	1.01	-	0.87	1.88	2.16
2.00 m	m2	-	0.17	1.15	-	0.87	2.02	2.32

Earthwork support to sides of
excavation below the level of the base
of existing foundation not exceeding
2.00 m between opposing faces, in stiff
soil, maximum depth not exceeding

1.00 m	m2	-	0.16	1.08	-	0.87	1.95	2.24
2.00 m	m2	-	0.19	1.28	-	0.87	2.15	2.47

Earthwork support to sides of
preliminary trenches, not exceeding
2.00 m between opposing faces, in
moderately firm soil, maximum depth
not exceeding,

1.00 m	m2	-	0.59	3.98	-	2.32	6.30	7.25
2.00 m	m2	-	0.76	5.12	-	4.63	9.75	11.21

Earthwork support to sides of
excavation below the level of the base
of existing foundation, not exceeding
2.00m between opposing faces, in
moderately firm soil, maximum depth
not exceeding,

1.00 m	m2	-	0.65	4.38	-	2.32	6.70	7.71
2.00 m	m2	-	0.84	5.66	-	2.32	7.98	9.18

Earthwork support to sides of
preliminary trenches not exceeding 2.00
m between opposing faces, in soft soil,
maximum depth not exceeding,

1.00 m	m2	-	1.17	7.89	-	3.19	11.08	12.74
2.00 m	m2	-	1.37	9.23	-	3.19	12.42	14.28

Earthwork support to sides of
excavation below the level of existing
foundation, not exceeding 2.00 m
between opposing faces, in soft soil,
maximum depth not exceeding,

1.00 m	m2	-	1.29	8.69	-	3.19	11.88	13.66
2.00 m	m2	-	1.51	10.18	-	3.19	13.37	15.38

Underpinning

New work	Unit	Hours C	Hours L	Labour net	Plant net	Material net	Price net	Price with 15%
Work in all trades				£	£	£	£	£
					VAT not included			

Earthwork support (*continued*)

Earthwork support left in excavation below level of base ,

	Unit	Hours C	Hours L	Labour net	Plant net	Material net	Price net	Price with 15%
1.00 m deep - stiff soil	m2	-	0.11	0.74	-	5.79	6.53	7.51
2.00 m deep - stiff soil	m2	-	0.13	0.88	-	11.59	12.47	14.34
1.00 m deep - moderately stiff soil	m2	-	0.43	2.90	-	13.90	16.80	19.32
2.00 m deep - moderately stiff soil	m2	-	0.56	3.77	-	13.90	17.67	20.32
1.00 m deep - soft soil	m2	-	0.86	5.80	-	19.12	24.92	28.66
2.00 m deep - soft soil	m2	-	1.01	6.81	-	19.12	25.93	29.82

Disposal

	Unit	Hours C	Hours L	Labour net	Plant net	Material net	Price net	Price with 15%
Excavated material removed from site to tip average 15 km from site - loaded by hand	m3	-	1.76	11.86	18.53	-	30.39	34.95
Add or deduct for every 1 km difference in distance	m3	-	-	-	0.20	-	0.20	0.23

Filling

	Unit	Hours C	Hours L	Labour net	Plant net	Material net	Price net	Price with 15%
Excavated material filling to excavations deposited and compacted in 225 mm layers	m3	-	1.63	10.99	3.48	-	14.47	16.64

Trench bottoms

	Unit	Hours C	Hours L	Labour net	Plant net	Material net	Price net	Price with 15%
Level and compact bottom of excavation	m2	-	0.25	1.69	0.64	-	2.33	2.68

Concrete work

Foundations in trenches over 200 mm thick

	Unit	Hours C	Hours L	Labour net	Plant net	Material net	Price net	Price with 15%
1:3:6 concrete	m3	-	5.25	35.38	-	68.88	104.26	119.90
1:2:4 concrete	m3	-	5.25	35.38	-	73.19	108.57	124.86

Brickwork and blockwork

Walls in common bricks in cement mortar 1:3

	Unit	Hours C	Hours L	Labour net	Plant net	Material net	Price net	Price with 15%
215 mm	m2	4.00	2.00	49.32	-	29.70	79.02	90.87
327.5 mm	m2	5.40	2.70	66.58	-	43.63	110.21	126.74

Underpinning

New work	Unit	Hours C	Hours L	Labour net	Plant net	Material net	Price net	Price with 15%
Work in all trades				£	£	£	£	£
						VAT not included		
Damp proof courses								
Fibre base bitumen horizontal damp-proof courses bedded in cement mortar 1:3								
over 225 mm wide	m2	0.25	0.13	3.08	-	5.76	8.84	10.17
225 mm wide	m	0.06	0.03	0.69	-	1.28	1.97	2.27
Pitch polymer horizontal damp-proof courses bedded in cement mortar 1:3								
over 225 mm wide	m2	0.25	0.13	3.08	-	8.73	11.81	13.58
225 mm wide	m	0.06	0.03	0.69	-	1.94	2.63	3.02
Double course slate horizontal damp-proof courses bedded in cement mortar 1:3								
over 225 mm wide	m2	1.65	0.82	20.34	-	41.73	62.07	71.38
225 mm wide	m	0.38	0.19	4.68	-	9.45	14.13	16.25
Wedge and pin								
Wedge and pin up to underside of existing construction with slates in cement mortar 1:3								
215 mm walls	m	0.48	0.24	5.92	-	4.36	10.28	11.82
327.5 mm walls	m	0.74	0.37	9.12	-	6.39	15.51	17.84

Rubble walling

Preamble

"Labour net" figures include allowances for all costs incidental to the employment of labour. The labour for rubble walling has been based generally on a team of three masons to two labourers as for brickwork.

"Materials net" figures include for all costs of materials including an allowance for waste except where specifically stated.

"Price net" figures are the totals of the "Labour net" and "Materials net" figures. Prices are for a builder employing his own labour; according to the amount and nature of the work involved, it may well be possible to secure more advantageous prices from specialist sub-contractors.

Prices do not include any allowance for scaffolding, ladders or other plant necessary to reach the work. The "Preliminaries" section includes prices for scaffolding which must be considered and allowance included to suit the particular circumstances of a tender.

Rubble walling

New work

Basic prices for materials		£
Galvanised butterfly wall-ties	1000	101.40
Hydrated lime	tonne	139.84
Portland cement	tonne	89.04
Sand	m3	17.96
	tonne	11.22
Walling stone		
150 mm on bed	tonne	94.50
225 mm on bed	tonne	80.50
random bed width	tonne	30.50

Prices actually to be paid for materials must be checked against the above basic prices and adjustments made as necessary.

Rubble walling

New work							
Stone rubble work				£	£	£	£
					VAT not included		

Dry walling

300 mm random rubble walls laid dry (stones 75-100mm average)

	Unit	Hours C	Hours L	Labour net	Material net	Price net	Price with 15%
300 mm random rubble walls laid dry (stones 75-100mm average)	m2	2.28	2.28	35.80	21.75	57.55	66.18

Walling laid in mortar

	Unit	Hours C	Hours L	Labour net	Material net	Price net	Price with 15%
300 mm random rubble walls (stones 75-100mm average) in gauged mortar 1:1:6 flush pointed both sides	m2	2.40	2.40	37.68	31.49	69.17	79.55
Random rubble walls in gauged mortar 1:1:6 coursed average 450 mm high and flush pointed one side							
150 mm	m2	2.10	2.10	32.97	38.86	71.83	82.60
225 mm	m2	2.25	2.25	35.33	50.69	86.02	98.92
450 mm	m2	2.45	2.45	38.46	46.61	85.07	97.83
Random rubble walls in gauged mortar 1:1:6 coursed average 450 mm high, built against backing of other material, secured to backing with galvanised butterfly ties (5 per m2) and flush pointed one side							
150 mm	m2	3.36	3.36	52.76	39.42	92.18	106.01
225 mm	m2	3.37	3.37	52.91	51.08	103.99	119.59
Extra for dressed face to walls	m2	1.80	-	16.13	-	16.13	18.55
Level uncoursed rubble work for damp-proof courses etc							
225 mm	m	0.25	0.17	3.39	0.25	3.64	4.19
300 mm	m	0.33	0.22	4.44	0.34	4.78	5.50
Fair return on rubble work							
150 mm	m	1.00	-	8.96	-	8.96	10.30
225 mm	m	1.25	-	11.20	-	11.20	12.88
300 mm	m	1.50	-	13.44	-	13.44	15.46
300 x 150 mm rough stone coping in gauged mortar 1:1:6 flush pointed all round	m	0.54	0.36	7.27	9.51	16.78	19.30

Build in

	Unit	Hours C	Hours L	Labour net	Material net	Price net	Price with 15%
Build in ends of steel sections							
small (not exceeding 250 mm deep)	each	0.50	0.33	6.70	-	6.70	7.71
large (250 - 500 mm deep)	each	0.75	0.50	10.09	-	10.09	11.60
extra large (over 500 mm deep)	each	1.00	0.67	13.48	-	13.48	15.50

Rubble walling

New work	Unit	Hours C	Hours L	Labour net	Material net	Price net	Price with 15%
Stone rubble work				£	£	£	£
					VAT not included		
Holes							
Holes for small pipes (not exceeding 55 mm) through rubble work and make good							
150 mm	each	0.75	0.50	10.09	-	10.09	11.60
225 mm	each	1.00	0.67	13.48	-	13.48	15.50
300 mm	each	1.25	0.83	16.79	-	16.79	19.31
Holes for large pipes (55 - 110 mm) through rubble work and make good							
150 mm	each	1.25	0.83	16.79	0.17	16.96	19.50
225 mm	each	1.75	1.17	23.57	0.17	23.74	27.30
300 mm	each	2.25	1.50	30.27	0.17	30.44	35.01
Holes for extra large pipes (over 110 mm) through rubble work and make good							
150 mm	each	2.25	1.50	30.27	0.25	30.52	35.10
225 mm	each	3.25	2.17	43.75	0.25	44.00	50.60
300 mm	each	4.25	2.83	57.15	0.25	57.40	66.01
Mortices							
Mortices for bolts and run with mortar	each	0.50	0.33	6.70	0.25	6.95	7.99

Masonry

Preamble

"Labour net" figures include allowances for all costs incidental to the employment of labour.

"Materials net" figures include for all costs of materials including an allowance for waste except where specifically stated.

"Price net" figures are totals of the "Labour net" and "Materials net" figures. Prices are for a builder employing his own labour; according to the amount and nature of the work involved, it may well be possible to secure more advantageous prices from specialist sub-contractors.

Prices do not include any allowance for scaffolding, ladders or other plant necessary to reach the work. The "Preliminaries" section includes prices for scaffolding which must be considered and allowance included to suit the particular circumstances of a tender.

Specialist prices

"Price with 15%" figures are all-in guide prices and include 15% for the builder's overheads, profit, unloading materials and general attendance (to include free use of standing scaffolding and hoists, temporary lighting and water and clearing away rubbish).

The amount of attendance required varies between the various trades and also with the circumstances of specific jobs; the percentage addition must always be considered and adjusted as necessary to suit the terms and conditions of the quotation being used.

Quantities and delivery distances are usually the most significant of the many factors which influence prices and it must be emphasised that quotations should always be obtained when preparing a tender.

	Unit	Price

New work

Basic prices for materials £

Hydrated lime	tonne	139.84
Portland cement	tonne	89.04
Sand	m3	17.96
	tonne	11.22

Prices actually to be paid for materials must be checked against the above basic prices and adjustments made as necessary.

Masonry

New work

Natural stonework

| | | £ | £ |
| | | VAT not included | |

The following "Specialist price net" figures are guide prices provided by Hanson Bath & Portland Stone, for 15 tonne lorry loads delivered to site within about 80 kilometres of a specialist sub-contractor's depot. Prices do not include for cash discount.

	Unit	Specialist price net	Price with 15%
		Supply only	Supply only
Supply only Portland Whitbed stone			
Plain ashlar			
50 mm	m2	154.68	177.88
63 mm	m2	177.02	203.58
75 mm	m2	198.58	228.37
100 mm	m2	223.24	256.73
Weathered and twice throated copings			
225 x 50 mm	m	40.68	46.78
300 x 50 mm	m	50.54	58.13
375 x 50 mm	m	64.37	74.03
300 x 75 mm	m	61.94	71.23
375 x 75 mm	m	76.74	88.25
375 x 100 mm	m	90.96	104.60
Extra for angles on copings			
225 x 50 mm	each	65.33	75.13
300 x 50 mm	each	104.53	120.21
375 x 50 mm	each	130.65	150.24
300 x 75 mm	each	104.53	120.21
375 x 75 mm	each	130.65	150.24
375 x 100 mm	each	130.65	150.24
Band-courses moulded 100 mm girth			
100 x 200 mm	m	85.61	98.45
125 x 225 mm	m	104.41	120.07
150 x 250 mm	m	138.69	159.49
150 x 300 mm	m	144.19	165.82
175 x 300 mm	m	156.30	179.74
225 x 300 mm	m	173.27	199.26
Band-courses moulded 200 mm girth			
100 x 200 mm	m	100.13	115.15
175 x 225 mm	m	118.96	136.81
150 x 250 mm	m	153.23	176.21
150 x 300 mm	m	159.20	183.08
175 x 300 mm	m	170.80	205.34
225 x 300 mm	m	187.77	215.94
Splayed surrounds			
175 x 75 mm	m	55.60	63.95
200 x 100 mm	m	65.01	74.76
225 x 125 mm	m	78.92	90.76

Masonry

New work	Unit	Specialist price net	Price with 15%
Natural stonework		£	£
		VAT not included	
Supply only Portland Whitbed stone (continued)		Supply only	Supply only
Sunk splayed surrounds			
200 x 75 mm	m	59.50	68.42
250 x 75 mm	m	67.77	77.94
300 x 75 mm	m	75.45	86.77
375 x 75 mm	m	90.96	104.60
200 x 125 mm	m	82.11	94.43
250 x 125 mm	m	96.69	106.36
300 x 125 mm	m	114.62	131.81
375 x 125 mm	m	132.88	152.81
Surrounds moulded 100 mm girth			
200 x 100 mm	m	84.82	97.54
225 x 125 mm	m	104.41	120.07
250 x 150 mm	m	127.69	146.84
300 x 150 mm	m	147.93	170.12
300 x 175 mm	m	156.30	179.74
300 x 225 mm	m	175.47	201.79
Surrounds moulded 200 mm girth			
200 x 100 mm	m	100.10	115.11
225 x 125 mm	m	118.96	136.81
250 x 150 mm	m	153.16	176.14
300 x 150 mm	m	159.20	183.08
300 x 175 mm	m	170.80	196.42
300 x 225 mm	m	187.77	215.94
Extra over surrounds for			
grooves	m	9.23	10.61
throats	m	9.23	10.61
rebates	m	11.09	12.75
stoolings	each	52.24	60.07
Mortices for dowels	each	0.67	0.77
Sinkings for cramps	each	1.44	1.66
Holes for small pipes (not exceeding 55 mm) through ashlar			
50 mm	each	2.41	2.77
63 mm	each	2.75	3.16
75 mm	each	3.01	3.47
100 mm	each	3.51	4.04
Holes for large pipes (55 - 110 mm) through ashlar			
50 mm	each	2.74	3.15
63 mm	each	3.01	3.47
75 mm	each	3.67	4.23
100 mm	each	4.45	5.12

Masonry

	Unit	Specialist price net	Price with 15%

Natural stonework

		£	£
		VAT not included	

Supply only Portland Whitbed stone (continued)

		Supply only	Supply only
Holes for extra large pipes (over 110 mm) through ashlar			
50 mm	each	3.82	4.39
63 mm	each	4.45	5.12
75 mm	each	5.06	5.82
100 mm	each	6.65	7.65
50 mm paving in 900 x 600 mm slabs	m2	133.06	153.01
350 x 50 mm treads in 900 mm lengths	m	46.53	53.51
38 x 100 mm risers in 900 mm lengths	m	13.82	15.89

		Fix only	Fix only

The following "Specialist price net" figures are guide prices provided by Hanson Bath & Portland Stone. Prices for fixing stonework vary greatly dependent on the quantity to be fixed and therefore it is essential that quotations are obtained on each occasion

Prices for fixing allow for bedding in mason's mortar, flush pointing as the work proceeds, slurrying with weak lime mortar and cleaning down on completion but do not include for wall ties, anchors, dowels, cramps, mortices or cutting and pinning.

Prices do not include for cash discount.

See the preamble notes for builder's profit and attendance.

The builder would be required to provide a mechanical hoist and all necessary scaffolding for the use of the masons.

Fix only Portland Whitbed stone

Plain ashlar			
50 mm	m2	96.47	110.94
63 mm	m2	104.40	120.06
75 mm	m2	111.75	128.51
100 mm	m2	128.03	147.23
Weathered and twice throated copings			
225 x 50 mm	m	29.00	33.35
300 x 50 mm	m	33.32	38.32
375 x 50 mm	m	37.67	43.33
300 x 75 mm	m	40.34	46.39
375 x 75 mm	m	44.73	51.43
375 x 100 mm	m	49.98	57.48

Masonry

VAT not included

New work	Unit	Specialist price net	Price with 15%
Natural stonework		£	£
Fix only Portland Whitbed stone *(continued)*		Fix only	Fix only
Band-courses moulded 100 mm girth			
100 x 200 mm	m	42.10	48.41
125 x 225 mm	m	50.89	58.52
150 x 250 mm	m	60.51	69.59
150 x 300 mm	m	67.54	77.67
175 x 300 mm	m	71.07	81.73
225 x 300 mm	m	78.03	89.74
Band-courses moulded 200 mm girth			
100 x 200 mm	m	43.20	49.68
125 x 225 mm	m	50.89	58.52
150 x 250 mm	m	60.51	69.59
150 x 300 mm	m	67.54	77.67
175 x 300 mm	m	71.07	81.73
225 x 300 mm	m	78.03	89.74
Splayed surrounds			
175 x 75 mm	m	36.00	41.40
200 x 100 mm	m	40.10	48.41
225 x 125 mm	m	50.89	58.52
Sunk splayed surrounds			
200 x 75 mm	m	37.63	43.27
250 x 75 mm	m	41.23	47.41
300 x 75 mm	m	44.73	51.43
375 x 75 mm	m	49.13	56.49
200 x 125 mm	m	47.16	54.23
250 x 125 mm	m	53.50	61.53
300 x 125 mm	m	58.77	67.59
375 x 125 mm	m	66.65	76.65
Surrounds moulded 100 mm girth			
200 x 100 mm	m	42.10	48.41
225 x 125 mm	m	50.89	58.52
250 x 150 mm	m	60.51	69.59
300 x 150 mm	m	67.61	77.75
300 x 175 mm	m	71.07	81.73
300 x 225 mm	m	78.03	89.74
Surrounds moulded 200 mm girth			
200 x 100 mm	m	42.10	48.41
225 x 125 mm	m	50.89	58.52
250 x 150 mm	m	60.51	69.59
300 x 150 mm	m	67.54	77.67
300 x 175 mm	m	71.07	81.73
300 x 225 mm	m	78.03	89.74

Masonry

New work	Unit	Specialist price net	Price with 15%
Natural stonework		£	£
		VAT not included	
Fix only Portland Whitbed stone *(continued)*		Fix only	Fix only
50 mm paving in 900 x 600 mm slabs	m2	82.63	95.03
350 x 50 mm treads in 900 mm lengths	m	28.91	33.24
38 x 100 mm risers in 900 mm lengths	m	11.63	13.37

Enquiries about the foregoing specialist prices should be made to Hanson Bath & Portland Stone, Bumpers Lane, Wakeham, Portland, Dorset, DT5 1HY, tel (01305) 820207, fax (01305) 860275.

Masonry

New work	Unit	Hours C	Hours L	Labour net	Material net	Price net	Price with 15%
Cast stonework				£	£	£	£
					VAT not included		
Bradstone cast stone in gauged mortar 1:1:6 flush pointed on exposed faces							
100 mm walls in walling blocks pointed one side							
tooled finish	m2	1.85	1.45	26.35	21.43	47.78	54.95
squared and pitched	m2	2.25	1.75	31.96	18.84	50.80	58.42
100 mm square reveals in walling blocks							
tooled finish	m	0.08	0.05	1.06	-	1.06	1.22
squared and pitched	m	0.08	0.05	1.06	-	1.06	1.22
Extra for quoin blocks with tooled finish	m	-	-	-	2.08	2.08	2.39
100 mm walls in masonry blocks (to simulate random rubble) pointed one side	m2	2.25	1.75	31.96	18.84	50.80	58.42
Lintels							
102 x 152 mm	m	0.25	0.35	4.60	16.65	21.25	24.44
102 x 229 mm	m	0.25	0.50	5.61	21.05	26.66	30.66
Weathered, throated and grooved sills							
197 x 140 mm	m	0.40	0.75	8.63	25.11	33.74	38.80
Weathered and throated copings							
191 x 76 mm	m	0.25	0.60	6.28	9.75	16.03	18.43
305 x 76 mm	m	0.30	0.70	7.41	16.98	24.39	28.05
Pier caps							
381 x 381 mm	each	0.25	0.25	3.93	10.91	14.84	17.07
533 x 533 mm	each	0.35	0.35	5.50	21.02	26.52	30.50
Chimney caps							
533 x 533 mm with one opening	each	0.40	0.40	6.28	23.94	30.22	34.75
Window surrounds comprising 146 x 143 mm rebated and splayed sill and head, 105 x 102 mm label mould with mitred and returned ends and rebated and splayed jamb blocks to bond with walling blocks							
for 508 x 629 mm window	each	1.75	2.25	30.85	94.50	125.35	144.15
for four 508 x 1219 mm windows and with three rebated and splayed mullions	set	4.50	6.00	80.76	273.46	354.22	407.35
Door surrounds for 939 x 2032 mm door frame comprising 181 x 102 mm head, 162 x 105 mm label mould with mitred and returned ends and moulded jamb blocks to bond with walling blocks	each	3.00	4.00	53.84	228.81	282.65	325.05

Masonry

New work	Unit	Specialist price net	Price with 15%

Cast stonework

		£	£
		VAT not included	

The following "Specialist price net" figures are guide prices provided by Mexboro Concrete Ltd for 15 tonne lorry loads delivered to site within about 60 kilometres of their depot.

Prices do not include for cash discount.

Supply only the following cast stone:

	Unit	Specialist price net	Price with 15%
		Supply only	Supply only
plain cladding			
50 mm	m2	40.00	46.00
75 mm	m2	42.00	48.30
plain face ashlar 100 mm	m2	44.00	50.06
plain string 100 x 300 mm	m	48.00	55.20
plain keystone lintels 100 x 215 mm	m	32.00	36.80
plain sills, jambs or heads			
150 x 75 mm	m	25.00	28.75
200 x 75 mm	m	30.00	34.50
280 x 100 mm	m	40.00	46.00
copings			
300 x 75 mm	m	16.00	18.40
325 x 100 mm	m	20.00	23.00
400 x 125 mm	m	30.00	34.50
boot lintels 325 x 150 mm overall comprising 225 x 150 mm reinforced concrete lintel with 100 x 75 mm cast stone projecting toe	m	50.00	57.50
pier caps 525 x 525 x 100 mm weathered and throated all round			
one only	each	46.00	52.90
eight or more	each	40.00	46.00
chimney caps 940 x 600 x 100 mm weathered and throated all round and holed for two 225 x 225 mm flues			
one only	each	82.00	94.30
eight or more	each	70.00	80.50
tuscan columns 260 mm diameter tapering to 220 x 2300 mm high overall	each	270.00	310.50

Enquiries about the foregoing specialist prices should be made to Mexboro Concrete Ltd, Yalberton Industrial Estate, Alders Way, Paignton, Devon, TQ4 7QQ, tel (01803) 558025, fax (01803) 524717.

Masonry

New work	Unit	Specialist price net	Price with 15%

Cast stonework

£   £

VAT not included

The following "Specialist price net" figures are guide prices for work on sites within about 60 kilometres of a specialist depot.

Fix only   Fix only

Prices allow for all bedding and pointing materials and for cleaning down on completion but do not include for cramps or dowels.

Prices do not include for cash discount.

See the preamble notes for builder's profit and attendance.

The builder would be required to provide a mechanical hoist and all necessary scaffolding for the use of the masons.

Fix only the following cast stone:

	Unit	Specialist price net	Price with 15%
plain cladding			
50 mm	m2	81.90	94.18
75 mm	m2	110.67	127.27
plain face ashlar 100 mm	m2	81.90	94.18
plain string 100 x 300 mm	m	21.00	24.15
plain sills, jambs or heads			
150 x 75 mm	m	9.45	10.87
200 x 75 mm	m	11.55	13.28
280 x 100 mm	m	22.05	25.36
copings			
300 x 75 mm	m	15.75	18.11
325 x 100 mm	m	22.05	25.36
400 x 125 mm	m	34.65	39.85
boot lintels 325 x 150 mm overall	m	33.60	38.64
pier caps 525 x 525 x 100 mm	each	18.90	21.73
chimney caps 940 x 600 x 100 mm	each	38.85	44.68
tuscan columns 260 mm diameter tapering to 220 x 2300 mm high overall	each	441.00	507.15

Asphalt work

Preamble

Asphalt work is the province of specialist sub-contractors, consequently prices in this section are based on those of specialists.

Prices do not include any allowance for scaffolding, ladders or other plant necessary to reach the work. The "Preliminaries" section includes prices for scaffolding which must be considered and allowance included to suit the particular circumstances of a tender.

Specialist prices

"Price with 15%" figures are all-in guide prices and include 15% for the builder's overheads, profit, unloading materials and general attendance (to include free use of standing scaffolding and hoists, temporary lighting and water and clearing away rubbish).

The amount of attendance required varies between the various trades and also with the circumstances of specific jobs; the percentage addition must always be considered and adjusted as necessary to suit the terms and conditions of the quotation being used.

Quantities and delivery distances are usually the most significant of the many factors which influence prices and it must be emphasised that quotations should always be obtained when preparing a tender.

Asphalt work

New work	Unit	Specialist price net	Price with 15%

Generally

			£	£

VAT not included

The following "Specialist price net" figures are guide prices provided by Asphaltic Contracts for quantities of about 250 square metres within 25 kilometres of a branch depot. For smaller quantities, see "Alterations and repairs".

Prices do not include for cash discount.

See the preamble notes for builder's profit and attendance.

Damp-proofing and tanking

Mastic asphalt (limestone aggregate) to BS 6925 Type T1097

	Unit	Specialist price net	Price with 15%
13 mm one coat horizontal coverings on concrete			
over 300 mm wide	m2	9.22	10.60
not exceeding 150 mm wide	m	3.05	3.51
150 - 300 mm wide	m	4.35	5.00
20 mm two coat horizontal coverings on concrete			
over 300 mm wide	m2	11.39	13.10
not exceeding 150 mm wide	m	3.05	3.51
150 - 300 mm wide	m	4.35	5.00
30 mm three coat horizontal coverings on concrete			
over 300 mm wide	m2	17.22	19.80
not exceeding 150 mm wide	m	4.69	5.39
150 - 300 mm wide	m	7.63	8.77
13 mm two coat vertical coverings on brickwork			
over 300 mm wide	m2	30.08	34.59
not exceeding 150 mm wide	m	5.88	6.76
150 - 300 mm wide	m	9.41	10.82
20 mm three coat vertical coverings on brickwork			
over 300 mm wide	m2	37.38	42.99
not exceeding 150 mm wide	m	7.49	8.61
150 - 300 mm wide	m	12.32	14.17
Internal angle fillets	m	2.87	3.30
Turning nibs into grooves	m	1.95	2.24
Working into outlets	each	20.50	23.58
Collars and internal angle fillets around			
small pipes (not exceeding 55 mm)	each	14.35	16.50
large pipes (55 - 110 mm)	each	16.40	18.86

Asphalt work

New work

Generally

		£	£
		VAT not included	

Flooring

Mastic asphalt (limestone aggregate) to BS 6925 Type F1076

	Unit	Specialist price net	Price with 15%
15 mm one coat light duty flooring and isolating membrane			
over 300 mm wide	m2	11.29	12.98
not exceeding 150 mm wide	m	3.26	3.75
150 - 300 mm wide	m	5.51	6.34
20 mm one coat medium duty flooring and isolating membrane			
over 300 mm wide	m2	12.77	14.69
not exceeding 150 mm wide	m	3.71	4.27
150 - 300 mm wide	m	6.17	7.10
30 mm one coat heavy duty flooring and isolating membrane			
over 300 mm wide	m2	18.23	20.96
not exceeding 150 mm wide	m	5.36	6.16
150 - 300 mm wide	m	7.75	8.91
Working against metal frames	m	2.87	3.30
Extra for working flooring into recessed covers not exceeding 1.00 m2	each	15.38	17.69
13 x 150 mm two coat skirtings with fair edge, coved angle fillet and nib turned into groove - including angles	m	4.57	5.26

Coloured mastic asphalt (limestone aggregate) to BS 6925 Type F1451

	Unit	Specialist price net	Price with 15%
15 mm one coat light duty brown flooring and isolating membrane			
over 300 mm wide	m2	14.10	16.22
not exceeding 150 mm wide	m	4.43	5.09
150 - 300 mm wide	m	7.59	8.73
Working against metal frames	m	1.95	2.24
Extra for working flooring into recessed covers not exceeding 1.00 m2	each	15.38	17.69
13 x 150 mm two coat brown skirtings with fair edge, coved angle fillet and nib turned into groove - including angles	m	6.43	7.39

Asphalt work

New work

Generally

£   £

VAT not included

Roofing

Mastic asphalt (limestone aggregate) to BS 6925 Type R988

	Unit	Specialist price net	Price with 15%
20 mm two coat flat coverings and isolating membrane			
over 300 mm wide	m2	12.24	14.08
not exceeding 150 mm wide	m	4.50	5.18
150 - 300 mm wide	m	7.10	8.17
25 mm two coat flat coverings and isolating membrane			
over 300 mm wide	m2	15.53	17.86
not exceeding 150 mm wide	m	4.58	5.27
150 - 300 mm wide	m	7.63	8.77
Turning nibs into grooves	m	1.85	2.13
Working to metal flashings	m	1.95	2.24
Working into outlets	each	18.45	21.22
Two coat skirtings with fair edge, internal angle fillet and nib turned into groove - including angles			
13 x 150 mm	m	7.00	8.05
13 x 250 mm	m	8.20	9.43
Two coat aprons with undercut drip edge and rounded arris - including angles			
13 x 75 mm	m	7.00	8.05
13 x 100 mm	m	7.60	8.74
13 x 300 mm two coat linings to gutter with two rounded arrises and two internal angle fillets - including angles and intersections	m	18.50	21.28
ends	each	3.69	4.24
outlets	each	18.45	21.22
Collars and internal angle fillets around			
small pipes (not exceeding 55 mm)	each	14.35	16.50
large pipes (55 - 110 mm)	each	16.40	18.86

Accessories

	Unit	Specialist price net	Price with 15%
50 x 65 mm aluminium edge trims including butt straps and working asphalt to trim	m	11.53	13.26
Extra for right angle corner pieces			
Internal	each	10.46	12.03
external	each	10.46	12.03
75 x 65 mm aluminium edge trims including butt straps and working asphalt to trim	m	12.05	13.86

Asphalt work

New work	Unit	Specialist price net	Price with 15%
Generally		£	£
		VAT not included	

Roofing *(continued)*

Accessories *(continued)*

	Unit	Specialist price net	Price with 15%
Extra for right angle corner pieces			
internal	each	10.56	12.14
external	each	10.56	12.14
Pressure release breather ventilators including asphalt collars	each	30.75	35.36

Enquiries about the foregoing specialist prices should be made to Asphaltic Contracts Ltd., Meesons Wharf, 1-15 High Street, Stratford, London E15 2QQ, tel (0181) 519 9555, fax (0181) 519 9666.

If you're waiting for your old contacts to bear fruit you may be waiting for a long time. Most of your competitors are already taking a more scientific approach to generating a regular supply of fresh sales leads. They've picked Glenigan. The leads we supply are ripe and ready for your team to convert into target-busting sales.

By the time you've heard it here you've missed the pick of the bunch

But don't just take our word for it.

Call freephone 0800 373771

and we'll give you a taste of what you've been missing.

GLENIGAN
g
GROUP

GLENIGAN LEADS, BUSINESS FOLLOWS

41-47 Seabourne Road, Bournemouth, Dorset BH5 2HU Tel: 01202 432121 Fax: 01202 423411 e mail: info@glenigan.emap.co.uk

Roofing

Preamble

"Labour net" figures include allowances for all costs incidental to the employment of labour. "Labour net" figures for sheet lead work are based on the labour costs of an advanced plumber working with an apprentice in the third year of training.

"Materials net" figures include for all costs of materials including an allowance for waste except where specifically stated.

"Price net" figures are the totals of the "Labour net" and "Materials net" figures.

Prices do not include any allowance for scaffolding, ladders or other plant necessary to reach the work. The "Preliminaries" section includes prices for scaffolding which must be considered and allowance included to suit the particular circumstances of a tender.

Specialist prices

"Price with 15%" figures are all-in guide prices and include 15% for the builder's overheads, profit, unloading materials and general attendance (to include free use of standing scaffolding and hoists, temporary lighting and water and clearing away rubbish).

The amount of attendance required varies between the various trades and also with the circumstances of specific jobs; the percentage addition must always be considered and adjusted as necessary to suit the terms and conditions of the quotation being used.

Quantities and delivery distances are usually the most significant of the many factors which influence prices and it must be emphasised that quotations should always be obtained when preparing a tender.

Roofing

New work

Basic prices for materials £

Slate and tiles
510 x 255 mm best Welsh slates	1,000	3,211.00
random Westmorland green slates	tonne	1,350.00
500 x 250 mm asbestos-free blue/black slates	1,000	1,054.00
random reconstructed Cotswold stone slates	m2	20.81
430 x 380 mm concrete interlocking slating	1,000	799.00
265 x 165 mm hand-made sand-faced clay plain tiles	1,000	888.48
265 x 165 mm best quality Staffordshire machine-made sand-faced clay plain tiles	1,000	528.98
265 x 165 mm concrete plain tiles	1,000	285.00
265 x 165 mm hand-made sand-faced clay ornamental tiles	1,000	656.00
265 x 165 mm machine-made black sand-faced clay ornamental tiles	1,000	630.00
265 x 165 mm concrete ornamental tiles	1,000	635.18
340 x 280 mm machine-made clay pantiles	1,000	862.16
420 x 332 mm concrete double pantiles	1,000	620.00
420 x 330 mm concrete double Roman tiles	1,000	580.00
impregnated sawn softwood shingles	m2	32.51

Corrugated and troughed sheeting
standard 75 mm natural grey reinforced-cement sheets	m2	7.35
standard 150 mm natural grey reinforced-cement sheets	m2	7.74
0.7 mm mill finish profiled aluminium coloured sheets	m2	12.05
standard 75 mm corrugated galvanised steel sheets	m2	13.08
standard 75 mm corrugated vinyl translucent sheets	m2	9.65
standard 150 mm corrugated vinyl translucent sheets	m2	14.47
fire resisting standard 75 mm corrugated glass fibre reinforced plastics translucent sheets	m2	12.06
fire resisting standard 150 mm corrugated glass fibre reinforced plastics translucent sheets	m2	17.37

Roof decking
0.7 x 35 mm galvanised steel troughed decking	m2	10.09
0.7 x 63 mm galvanised steel troughed decking	m2	10.40
50 mm wood wool slabs size 1800 x 600 mm	per slab	7.35
75 mm wood wool slabs size 2700 x 600 mm	per slab	11.73

Flexible sheet finishings
glass fibre felt type 3B 18 kg	m2	1.00
"Asbex" glass fibre mineral surfaced felt type 3E 28 kg	m2	1.47
high performance polyester based roofing (Ruberglas 120 - GP) 28 kg	m2	4.08
high performance polyester based mineral surfaced roofing (Ruberfort) 32 kg	m2	5.76

Sheet metal roofing
milled sheet lead code 4 300 mm wide	m	5.06
milled sheet lead code 5	m2	24.92
sheet zinc 0.65 mm	m2	10.23
sheet zinc 0.80 mm	m2	12.08
sheet aluminium 0.60 mm	m2	4.50
sheet aluminium 0.80 mm	m2	5.72
sheet copper .55 mm	m2	17.25
sheet copper .70 mm	m2	22.08

Prices actually to be paid for materials must be checked against the above basic prices and adjustments made as necessary.

Roofing

New work

Basic prices for materials

£

	Unit	Price
Impregnated softwood battens		
19 x 38 mm	100 m	19.00
19 x 50 mm	100 m	35.15
Reinforced bituminous felt		
BS 747 type 1F	15 m2	26.00
BS 747 type 1F with aluminium face	15 m2	27.00
Brown sheathing felt to BS 747	m2	2.60
Bitumen emulsion	25 litre	35.00

Prices actually to be paid for materials must be checked against the above basic prices and adjustments made as necessary.

Roofing

New work	Unit	Hours C	Hours L	Labour net	Material net	Price net	Price with 15%
Slate and tile roofing				£	£	£	£
					VAT not included		
510 x 225 mm best Welsh slating							
Coverings to 75 mm lap fixed with alloy nails including 50 x 19 mm impregnated softwood battens							
roof coverings	m2	0.75	0.37	9.21	65.44	74.65	85.85
mansard and vertical coverings	m2	0.98	0.49	12.08	69.56	81.64	93.89
Extra for 50 x 25 mm impregnated softwood battens	m2	-	-	-	0.18	0.18	0.21
Labours							
square cutting to large openings	m	0.20	0.10	2.46	12.22	14.68	16.88
raking cutting	m	0.30	0.15	3.70	15.27	18.97	21.82
close mitred hips, valleys or angles	m	0.60	0.30	7.40	12.22	19.62	22.56
holes for pipes not exceeding 110 mm diameter	each	0.50	0.25	6.17	-	6.17	7.10
forming small openings not exceeding 0.50 m2	each	0.40	0.20	4.93	24.43	29.36	33.76
Extra for							
double course at eaves	m	0.30	0.15	3.70	19.32	23.02	26.47
verge and undercloak bedded and pointed	m	0.45	0.23	5.58	13.10	18.68	21.48
blue Staffordshire plain angle ridge or hip tiles bedded and pointed	m	0.18	0.09	2.22	16.09	18.31	21.06
Ancillaries							
galvanised hip irons screwed on	each	0.10	0.05	1.24	1.56	2.80	3.22
fixing metal slates	each	0.30	-	3.38	-	3.38	3.89
fixing soakers	each	0.20	-	2.25	-	2.25	2.59
Random Westmorland green slating							
Coverings in random lengths 450 - 250 mm laid to diminishing courses to 75 mm lap fixed with alloy nails including 50 x 25 mm impregnated softwood battens nailed to timber							
roof coverings	m2	1.12	0.56	13.81	97.18	110.99	127.64
mansard and vertical coverings	m2	1.45	0.73	17.91	86.02	103.93	119.52
Extra for 50 x 38 mm impregnated softwood battens	m	-	-	-	0.75	0.75	0.86
Labours							
square cutting to large openings	m	0.20	0.10	2.46	14.16	16.62	19.11
raking cutting	m	0.35	0.17	4.29	18.37	22.66	26.06
close mitred hips, valleys or angles	m	3.75	1.80	45.73	28.32	74.05	85.16
holes for pipes not exceeding 110 mm diameter	each	0.80	0.40	9.87	-	9.87	11.35
forming small openings not exceeding 0.50 m2	each	0.40	0.20	4.93	28.32	33.25	38.24
Extra for							
double course at eaves	m	0.30	0.15	3.70	29.26	32.96	37.90
verge and undercloak bedded and pointed	m	0.45	0.23	5.58	15.04	20.62	23.71

Roofing

New work	Unit	Hours C	Hours L	Labour net	Material net	Price net	Price with 15%

Slate and tile roofing

				£	£	£	£
					VAT not included		

Random Westmorland green slating
(*continued*)

Ancillaries

fixing metal slates	each	0.30	-	3.38	-	3.38	3.89
fixing soakers	each	0.20	-	2.25	-	2.25	2.59

500 x 250 mm asbestos-free blue/black slating

Coverings to 90 mm lap fixed with copper nails
and disc rivets including 38 x 25 mm impregnated
softwood battens nailed to timber

roof coverings	m2	0.60	0.30	7.40	24.21	31.61	36.35
mansard and vertical coverings	m2	0.90	0.45	11.09	25.22	36.31	41.76

Extra for 50 x 25 mm impregnated softwood
battens

	m	-	-	-	0.50	0.50	0.57

Labours

square cuttings to large openings	m	0.20	0.10	2.46	5.73	8.19	9.42
raking cutting	m	0.25	0.12	3.05	5.73	8.78	10.10
close mitred hips, valleys or angles	m	2.20	1.10	27.12	9.17	36.29	41.73
holes for pipes not exceeding 110 mm diameter	each	0.50	0.25	6.17	-	6.17	7.10
forming small openings not exceeding 0.50 m2	each	0.40	0.20	4.93	11.46	16.39	18.85

Extra for

double course at eaves	m	0.30	0.15	3.70	10.17	13.87	15.95
verge and undercloak bedded and pointed	m	0.35	0.17	4.29	6.73	11.02	12.67
socketed ridge or hip bedded and pointed	m	0.20	0.10	2.46	11.66	14.12	16.24
blue Staffordshire plain angle ridge or hip tiles bedded and pointed	m	0.18	0.09	2.22	16.09	18.31	21.06

Ancillaries

galvanised hip irons screwed on	each	0.10	0.05	1.24	1.56	2.80	3.22
fixing metal slates	each	0.30	-	3.38	-	3.38	3.89
fixing metal soakers	each	0.20	-	2.25	-	2.25	2.59

Random reconstructed Cotswold stone slating

Coverings in diminishing courses to 80 mm lap
fixed with alloy nails including 38 x 19 mm
impregnated softwood battens nailed to timber

roof coverings	m2	1.20	0.60	14.79	24.15	38.94	44.78
mansard and vertical coverings	m2	1.56	0.78	19.24	24.40	43.64	50.19

Extra for 50 x 25 mm impregnated softwood
battens

	m	-	-	-	1.41	1.41	1.62

Roofing

New work	Unit	Hours C	Hours L	Labour net	Material net	Price net	Price with 15%
Slate and tile roofing				£	£	£	£
					VAT not included		

Random reconstructed Cotswold stone slating
(*continued*)

	Unit	Hours C	Hours L	Labour net	Material net	Price net	Price with 15%
Labours							
square cuttings to large openings	m	0.40	0.20	4.93	5.64	10.57	12.16
raking cutting	m	0.50	0.25	6.17	11.38	17.55	20.18
holes for pipes not exceeding 110 mm diameter	each	0.50	0.25	6.17	-	6.17	7.10
forming small openings not exceeding 0.50 m2	each	0.80	0.40	9.87	11.29	21.16	24.33
Extra for							
double course at eaves	m	0.51	0.25	6.26	7.80	14.06	16.17
verge and undercloak bedded and pointed	m	0.63	0.32	7.80	7.80	15.60	17.94
angle ridge units bedded and pointed	m	0.64	0.37	8.22	14.30	22.52	25.90
angle hip units bedded and pointed	m	0.65	0.33	8.04	14.06	22.10	25.41
Ancillaries							
galvanised hip irons screwed on	each	0.10	0.05	1.24	1.56	2.80	3.22
fixing metal slates	each	0.30	-	3.38	-	3.38	3.89
fixing soakers	each	0.20	-	2.25	-	2.25	2.59

430 x 380 mm concrete interlocking slating

	Unit	Hours C	Hours L	Labour net	Material net	Price net	Price with 15%
Roof coverings to 75 mm lap all slates clipped including 38 x 19 mm impregnated softwood battens nailed to timber	m2	0.34	0.17	4.20	7.85	12.05	13.86
Extra for 38 x 25 mm impregnated softwood battens	m2	-	-	-	0.41	0.41	0.47
Labours							
square cutting to large openings	m	0.20	0.10	2.46	1.71	4.17	4.80
raking cutting	m	0.25	0.12	3.05	8.69	11.74	13.50
holes for pipes not exceeding 110 mm diameter	each	0.50	0.25	6.17	-	6.17	7.10
forming small openings not exceeding 0.50 m2	each	0.40	0.20	4.93	4.28	9.21	10.59
Extra for							
purpose made fixing clips at eaves	m	0.06	0.03	0.74	0.51	1.25	1.44
verge, verge clipping and non asbestos undercloak bedded and pointed	m	0.30	0.15	3.70	9.10	12.80	14.72
valley trough units	m	0.60	0.30	7.40	26.30	33.70	38.76
angle ridge or hip units bedded and pointed	m	0.18	0.09	2.22	16.09	18.31	21.06
ventilating/gas flue ridge units bedded and pointed - adaptors not included	each	0.33	0.16	4.04	68.23	72.27	83.11
abutment flashing units	m	0.20	1.00	8.53	15.15	23.68	27.23
Ancillaries							
galvanised hip irons screwed on	each	0.10	0.05	1.24	1.56	2.80	3.22
fixing metal slates	each	0.30	-	3.38	-	3.38	3.89

Roofing

New work	Unit	Hours C	Hours L	Labour net	Material net	Price net	Price with 15%

Slate and tile roofing

				£	£	£	£
					VAT not included		

265 x 165 mm hand-made sand-faced clay plain tiling

Roof coverings to 65 mm lap alloy nailed every fourth course including 25 x 19 mm impregnated softwood battens nailed to timber	m2	1.00	0.50	12.33	67.05	79.38	91.29
Extra for 32 x 25 mm impregnated softwood battens	m2	-	-	-	1.15	1.15	1.32
Mansard and vertical coverings to 38 mm lap alloy nailed every course including 38 x 19 mm impregnated softwood battens nailed to timber	m2	1.25	0.62	15.38	59.54	74.92	86.16
Extra for 38 x 25 mm impregnated softwood	m2	-	-	-	0.68	0.68	0.78

Labours							
square cutting to large openings	m	0.20	0.10	2.46	10.70	13.16	15.13
raking cutting	m	0.25	0.12	3.05	10.55	13.60	15.64
holes for pipes not exceeding 110 mm diameter	each	0.70	0.35	8.63	-	8.63	9.92
forming small openings not exceeding 0.50 m2	each	0.40	0.20	4.93	21.40	26.33	30.28

Extra for							
double course at eaves	m	0.25	0.12	3.05	10.70	13.75	15.81
verge and undercloak bedded and pointed	m	0.42	0.21	5.18	11.70	16.88	19.41
valley tiles	m	0.62	0.31	7.65	52.58	60.23	69.26
half round ridge tiles bedded and pointed	m	0.50	0.25	6.17	20.26	26.43	30.39
bonnet hip tiles bedded and pointed	m	0.76	0.38	9.37	53.73	63.10	72.56
vertical angle tiles	m	0.45	0.23	5.58	39.66	45.24	52.03

Ancillaries							
fixing metal slates	each	0.30	-	3.38	-	3.38	3.89
fixing soakers	each	0.20	-	2.25	-	2.25	2.59

265 x 165 best quality Staffordshire machine-made sand-faced clay plain tiling

Roof coverings to 65 mm lap alloy nailed every fourth course including 25 x 19 mm impregnated softwood battens nailed to timber	m2	1.00	0.50	12.33	31.30	43.63	50.17
Extra for 32 x 25 mm impregnated softwood battens	m2	-	-	-	1.15	1.15	1.32
Mansard and vertical coverings to 38 mm lap nailed every course including 38 x 19 mm impregnated softwood battens nailed to timber	m2	1.25	0.62	15.38	27.48	42.86	49.29
Extra for 38 x 25 mm impregnated softwood battens	m2	-	-	-	0.98	0.98	1.13

Roofing

New work	Unit	Hours C	Hours L	Labour net	Material net	Price net	Price with 15%
Slate and tile roofing				£	£	£	£
					VAT not included		

265 x 165 best quality Staffordshire machine-made sand-faced clay plain tiling (*continued*)

	Unit	Hours C	Hours L	Labour net	Material net	Price net	Price with 15%
Labours							
square cutting to large openings	m	0.20	0.10	2.46	4.80	7.26	8.35
raking cutting	m	0.25	0.12	3.05	4.80	7.85	9.03
holes for pipes not exceeding 110 mm diameter	each	0.70	0.35	8.63	-	8.63	9.92
forming small openings not exceeding 0.50 m2	each	0.40	0.20	4.93	9.60	14.53	16.71
Extra for							
double course at eaves	m	0.25	0.12	3.05	5.80	8.85	10.18
verge and undercloak bedded and pointed	m	0.42	0.21	5.18	5.80	10.98	12.63
valley tiles	m	0.65	0.33	8.04	51.15	59.19	68.07
half round ridge tiles bedded and pointed	m	0.50	0.25	6.17	19.79	25.96	29.85
bonnet hip tiles bedded and pointed	m	0.76	0.38	9.37	52.30	61.67	70.92
vertical angle tiles	m	0.45	0.23	5.58	38.58	44.16	50.78
Ancillaries							
fixing metal slates	each	0.30	-	3.38	-	3.38	3.89
fixing soakers	each	0.20	-	2.25	-	2.25	2.59

265 x 165 mm concrete plain tiling

	Unit	Hours C	Hours L	Labour net	Material net	Price net	Price with 15%
Roof coverings to 65 mm lap alloy nailed every fourth course including 25 x 19 mm impregnated softwood battens nailed to timber	m2	1.00	0.50	12.33	22.03	34.36	39.51
Extra for 32 x 25 mm impregnated softwood battens	m2	-	-	-	1.78	1.78	2.05
Mansard and vertical coverings to 38 mm lap alloy nailed every course including 38 x 19 mm impregnated softwood battens nailed to timber	m2	1.25	0.62	15.38	19.80	35.18	40.46
Extra for 38 x 25 mm impregnated softwood battens	m2	-	-	-	1.78	1.78	2.05
Labours							
square cutting to large openings	m	0.20	0.10	2.46	3.27	5.73	6.59
raking cutting	m	0.25	0.12	3.05	2.97	6.02	6.92
holes for pipes not exceeding 110 mm diameter	each	0.70	0.35	8.63	-	8.63	9.92
forming small openings not exceeding 0.50 m2	each	0.40	0.20	4.93	6.54	11.47	13.19
Extra for							
double course at eaves	m	0.25	0.12	3.05	3.97	7.02	8.07
verge and undercloak bedded and pointed	m	0.43	0.21	5.27	3.97	9.24	10.63
valley tiles	m	0.65	0.33	8.04	23.88	31.92	36.71
half round ridge tiles bedded and pointed	m	0.50	0.25	6.17	12.53	18.70	21.50
bonnet hip tiles bedded and pointed	m	0.76	0.38	9.37	49.63	59.00	67.85
vertical angle tiles	m	0.45	0.23	5.58	18.13	23.71	27.27

Roofing

New work	Unit	Hours C	Hours L	Labour net	Material net	Price net	Price with 15%
Slate and tile roofing				£	£	£	£
					VAT not included		

265 x 165 mm concrete plain tiling (*continued*)

	Unit	Hours C	Hours L	Labour net	Material net	Price net	Price with 15%
Ancillaries							
fixing metal slates	each	0.30	-	3.38	-	3.38	3.89
fixing soakers	each	0.20	-	2.25	-	2.25	2.59

265 x 165 mm hand-made sand-faced clay ornamental tiling

	Unit	Hours C	Hours L	Labour net	Material net	Price net	Price with 15%
Vertical coverings to 38 mm lap alloy nailed every course including 38 x 19 mm impregnated softwood battens nailed to timber	m2	1.33	0.66	16.37	40.63	57.00	65.55
Extra for 38 x 25 mm impregnated softwood battens	m2	-	-	-	1.10	1.10	1.26
Labours							
square cutting to large openings	m	0.20	0.10	2.46	7.19	9.65	11.10
raking cutting	m	0.25	0.12	3.05	7.19	10.24	11.78
holes for pipes not exceeding 110 mm diameter	each	0.70	0.35	8.63	-	8.63	9.92
forming small openings not exceeding 0.50 m2	each	0.40	0.20	4.93	14.38	19.31	22.21
Extra for							
double course at eaves	m	0.25	0.12	3.05	7.19	10.24	11.78
vertical angle tiles	m	0.43	0.23	5.40	38.58	43.98	50.58
Ancillaries							
fixing soakers	each	0.20	-	2.25	-	2.25	2.59

265 x 165 mm machine-made black sand-faced clay ornamental tiling

	Unit	Hours C	Hours L	Labour net	Material net	Price net	Price with 15%
Vertical coverings to 38 mm lap alloy nailed every course including 38 x 19 mm impregnated softwood battens nailed to timber	m2	1.33	0.66	16.37	39.22	55.59	63.93
Extra for 38 x 25 mm impregnated battens	m2	-	-	-	1.10	1.10	1.26
Labours							
square cutting to large openings	m	0.20	0.10	2.46	6.92	9.38	10.79
raking cutting	m	0.25	0.12	3.05	6.92	9.97	11.47
holes for pipes not exceeding 110 mm diameter	each	0.70	0.35	8.63	-	8.63	9.92
forming small openings not exceeding 0.50 m2	each	0.40	0.20	4.93	13.84	18.77	21.59
Extra for							
double course at eaves	m	0.25	0.13	3.12	6.92	10.04	11.55
vertical angle tiles	m	0.45	0.23	5.58	38.58	44.16	50.78
Ancillaries							
fixing soakers	each	0.20	-	2.25	-	2.25	2.59

Roofing

New work	Unit	Hours C	Hours L	Labour net	Material net	Price net	Price with 15%
Slate and tile roofing				£	£	£	£
					VAT not included		
265 x 165 concrete ornamental tiling							
Vertical coverings to 38 mm lap alloy nailed every course including 38 x 19 mm impregnated softwood battens nailed to timber	m2	1.33	0.66	16.37	39.48	55.85	64.23
Extra for 38 x 25 mm impregnated battens	m2	-	-	-	1.10	1.10	1.26
Labours							
square cutting to large openings	m	0.20	0.10	2.46	6.97	9.43	10.84
raking cutting	m	0.25	0.12	3.05	6.97	10.02	11.52
holes for pipes not exceeding 110 mm diameter	each	0.70	0.35	8.63	-	8.63	9.92
forming small openings not exceeding 0.50 m2	each	0.40	0.20	4.93	13.94	18.87	21.70
Extra for							
double course at eaves	m	0.25	0.12	3.05	6.97	10.02	11.52
vertical angle tiles	m	0.45	0.23	5.58	18.13	23.71	27.27
Ancillaries							
fixing soakers	each	0.20	-	2.25	-	2.25	2.59
340 x 280 mm machine-made clay pantiling							
Roof coverings to 75 mm lap alloy nailed every alternate course including 38 x 19 mm impregnated softwood battens nailed to timber	m2	0.51	0.26	6.32	18.22	24.54	28.22
Extra for 38 x 25 mm impregnated softwood battens	m2	-	-	-	0.34	0.34	0.39
Labours							
square cutting to large openings	m	0.16	0.08	1.97	1.87	3.84	4.42
raking cutting	m	0.30	0.15	3.70	2.71	6.41	7.37
holes for pipes not exceeding 110 mm diameter	each	0.50	0.25	6.17	-	6.17	7.10
forming small openings not exceeding 0.50 m2	each	0.20	0.10	2.46	3.74	6.20	7.13
Extra for							
undercourse at eaves bedded and pointed	m	0.35	0.17	4.29	4.68	8.97	10.32
verge and undercloak bedded and pointed	m	0.30	0.15	3.70	4.68	8.38	9.64
half round ridge or hip tiles with dentil slips bedded and pointed both sides	m	0.55	0.27	6.75	20.58	27.33	31.43
Ancillaries							
galvanised hip irons screwed on	each	0.10	0.05	1.24	1.56	2.80	3.22
fixing metal slates	each	0.30	-	3.38	-	3.38	3.89

Roofing

New work	Unit	Hours C	Hours L	Labour net	Material net	Price net	Price with 15%
Slate and tile roofing				£	£	£	£
					VAT not included		
420 x 332 mm concrete double pantiling							
Roof coverings to 75 mm lap clipped every alternate course including 38 x 19 mm impregnated softwood battens nailed to timber	m2	0.25	0.12	3.05	7.21	10.26	11.80
Extra for 38 x 25 mm impregnated battens	m2	-	-	-	0.34	0.34	0.39
Labours							
square cutting to large openings	m	0.10	0.05	1.24	1.30	2.54	2.92
raking cutting	m	0.30	0.15	3.70	1.30	5.00	5.75
holes for pipes not exceeding 110 mm diameter	each	0.50	0.25	6.17	-	6.17	7.10
forming small openings not exceeding 0.50 m2	each	0.20	0.10	2.46	2.60	5.06	5.82
Extra for							
undercourse at eaves bedded and pointed	m	0.30	0.15	3.70	4.32	8.02	9.22
verge tiles and clips and undercloak bedded and pointed	m	0.30	0.15	3.70	4.32	8.02	9.22
half round ridge or hip tiles with dentil slips bedded and pointed both sides	m	0.55	0.27	6.75	20.58	27.33	31.43
Ancillaries							
galvanised metal hip iron screwed on	each	0.10	0.05	1.24	1.56	2.80	3.22
fixing metal slates	each	0.30	-	3.38	-	3.38	3.89
420 x 330 mm concrete Double Roman tiling							
Roof coverings to 75 mm lap alloy nailed every alternate course including 38 x 19 mm impregnated softwood battens nailed to timber	m2	0.25	0.12	3.05	6.66	9.71	11.17
Extra for 38 x 25 mm impregnated battens	m2	-	-	-	0.36	0.36	0.41
Labours							
square cutting to large openings	m	0.10	0.05	1.24	1.22	2.46	2.83
raking cutting	m	0.30	0.15	3.70	1.22	4.92	5.66
holes for pipes not exceeding 110 mm diameter	each	0.50	0.25	6.17	-	6.17	7.10
forming small openings not exceeding 0.50 m2	each	0.20	0.11	2.53	3.65	6.18	7.11
Extra for							
verge tiles and undercloak bedded and pointed	m	0.30	0.15	3.70	4.04	7.74	8.90
half round ridge or hip tiles with dentil slips bedded and pointed both sides	m	0.55	0.27	6.75	20.58	27.33	31.43
Ancillaries							
galvanised hip irons screwed on	each	0.10	0.05	1.24	1.56	2.80	3.22
fixing metal slates	each	0.30	-	3.38	-	3.38	3.89

Roofing

New work	Unit	Hours C	Hours L	Labour net	Material net	Price net	Price with 15%
Slate and tile roofing				£	£	£	£
					VAT not included		
Impregnated sawn cedar shingles							
Coverings to 125 mm lap fixed with silicone bronze nails including 38 x 19 mm impregnated softwood battens nailed to timber							
roof coverings	m2	1.25	0.63	15.45	23.65	39.10	44.97
mansard and vertical coverings	m2	1.63	0.81	20.06	23.95	44.01	50.61
Extra for 38 x 25 mm impregnated battens	m2	-	-	-	0.50	0.50	0.57
Labours							
square cutting to large openings	m	0.35	0.17	4.29	8.76	13.05	15.01
raking cutting	m	0.46	0.23	5.67	8.76	14.43	16.59
holes for pipes not exceeding 110 mm diameter	each	0.20	0.10	2.46	-	2.46	2.83
forming small openings not exceeding 0.50 m2	each	0.70	0.34	8.56	17.52	26.08	29.99
Extra for							
double course at eaves	m	0.25	0.13	3.08	8.76	11.84	13.62
laced shingles to ridges or hips	m	2.00	1.00	24.66	22.54	47.20	54.28
Ancillaries							
fixing metal slates	each	0.30	-	3.38	-	3.38	3.89
fixing soakers	each	0.20	-	2.25	-	2.25	2.59
Counter-battening							
Impregnated softwood counter-battening at 600 mm centres nailed to timber							
50 x 19 mm	m2	0.10	0.05	1.24	0.71	1.95	2.24
50 x 25 mm	m2	0.10	0.05	1.24	0.78	2.02	2.32
Impregnated softwood counter-battening at 450 mm centres nailed to timber							
50 x 19 mm	m2	0.17	0.08	2.06	0.98	3.04	3.50
50 x 25 mm	m2	0.17	0.08	2.06	1.08	3.14	3.61
Reinforced bituminous felt to BS 747 Type 1F							
Underlay lapped 150 mm horizontally and 300mm vertically and fixed with galvanised felt nails	m2	0.04	0.02	0.49	2.91	3.40	3.91
Labours							
raking cutting	m	0.04	0.02	0.49	0.03	0.52	0.60
holes for pipes	each	0.04	0.02	0.49	-	0.49	0.56

Roofing

New work	Unit	Hours C	Hours L	Labour net	Material net	Price net	Price with 15%
Slate and tile roofing				£	£	£	£
					VAT not included		
Reinforced bituminous felt to BS 747 Type 1F with aluminium foil face							
Underlay lapped 150 mm horizontally and 300 mm vertically and fixed with galvanised felt nails	m2	0.06	0.03	0.74	3.02	3.76	4.32
Labours							
raking cutting	m	0.05	0.03	0.65	0.03	0.68	0.78
holes for pipes	each	0.05	0.03	0.65	-	0.65	0.75
Brown sheathing felt to BS 747							
Underlay close butted at joints and fixed with galvanised felt nails	m2	0.06	0.03	0.74	3.00	3.74	4.30
Roof ventilators							
Glidevale rafter ventilators, fixed with nails							
type RV601 to suit 600 mm rafter centres	m	0.10	-	0.90	2.96	3.86	4.44
type RV451 to suit 450 mm rafter centres	m	0.12	-	1.08	3.65	4.73	5.44
type RV401 to suit 400 mm rafter centres	m	0.12	-	1.08	4.07	5.15	5.92
Glidevale soffit ventilators, fixed with nails							
type SV603	m	0.15	-	1.34	5.02	6.36	7.31
type SV606	m	0.15	-	1.34	4.41	5.75	6.61

New work	Unit	Hours C	Hours L	Labour net	Material net	Price net	Price with 15%
Corrugated and troughed sheeting				£	£	£	£
					VAT not included		

Standard 75 mm natural grey corrugated reinforced-cement sheeting

Coverings with 1.5 corrugation side laps and 150 mm end laps fixed with drive screws to timber purlins or rails

	Unit	Hours C	Hours L	Labour net	Material net	Price net	Price with 15%
roof coverings	m2	0.25	0.12	3.05	8.79	11.84	13.62
vertical coverings	m2	0.33	0.17	4.11	8.79	12.90	14.84

Coverings with 1.5 corrugation side laps and 150 mm end laps fixed with hook bolts to steel purlins or rails

	Unit	Hours C	Hours L	Labour net	Material net	Price net	Price with 15%
roof coverings	m2	0.30	0.15	3.70	9.70	13.40	15.41
vertical coverings	m2	0.39	0.19	4.77	9.70	14.47	16.64

Ancillaries

	Unit	Hours C	Hours L	Labour net	Material net	Price net	Price with 15%
square cutting to large openings	m	0.15	0.07	1.81	0.79	2.60	2.99
raking cutting	m	0.20	0.10	2.46	3.96	6.42	7.38
holes for pipes not exceeding 150 mm diameter	each	0.50	0.25	6.17	-	6.17	7.10
eaves filler pieces	m	0.14	0.07	1.72	5.53	7.25	8.34
eaves closure pieces	m	0.14	0.07	1.72	5.53	7.25	8.34
two-piece close fitting ridge cappings	m	0.25	0.17	3.39	11.58	14.97	17.22
one piece finials	each	0.40	0.20	4.93	9.19	14.12	16.24

300 mm barge boards

	Unit	Hours C	Hours L	Labour net	Material net	Price net	Price with 15%
straight	m	0.10	0.05	1.24	8.44	9.68	11.13
mitred angles	each	0.40	0.20	4.93	14.28	19.21	22.09

Extra for

	Unit	Hours C	Hours L	Labour net	Material net	Price net	Price with 15%
1225 x 762 mm translucent sheets Grade 1 EXT SAA	each	0.30	0.15	3.70	12.30	16.00	18.40
soaker flanges for pipes not exceeding 150 mm diameter	each	0.70	0.35	8.63	22.71	31.34	36.04

Standard 150 mm natural grey corrugated reinforced-cement sheeting

Coverings with three-quarter corrugation side laps and 150 mm end laps fixed with drive screws to timber purlins or rails

	Unit	Hours C	Hours L	Labour net	Material net	Price net	Price with 15%
roof coverings	m2	0.25	0.12	3.05	8.98	12.03	13.83
vertical coverings	m2	0.33	0.17	4.11	8.98	13.09	15.05

Coverings with three-quarter corrugation side laps and 150 mm end laps fixed with hook bolts to steel purlins or rails

	Unit	Hours C	Hours L	Labour net	Material net	Price net	Price with 15%
roof coverings	m2	0.30	0.15	3.70	9.89	13.59	15.63
vertical coverings	m2	0.39	0.19	4.77	9.89	14.66	16.86

Roofing

New work	Unit	Hours C	Hours L	Labour net	Material net	Price net	Price with 15%
Corrugated and troughed sheeting				£	£	£	£
					VAT not included		
Standard 150 mm natural grey corrugated reinforced-cement sheeting (*continued*)							
Ancillaries							
square cutting to large openings	m	0.15	0.07	1.81	0.81	2.62	3.01
raking cutting	m	0.20	0.10	2.46	4.06	6.52	7.50
holes for pipes not exceeding 150 mm diameter	each	0.50	0.35	6.84	-	6.84	7.87
eaves filler piece	m	0.14	0.07	1.72	8.40	10.12	11.64
eaves closure piece	m	0.14	0.07	1.72	9.04	10.76	12.37
underglazing flashing pieces	m	0.12	0.06	1.48	7.96	9.44	10.86
two-piece close fitting ridge cappings	m	0.16	0.08	1.97	15.92	17.89	20.57
one-piece finials	each	0.40	0.20	4.93	9.19	14.12	16.24
300 mm barge boards							
straight	m	0.10	0.05	1.24	8.44	9.68	11.13
mitred angle	each	0.40	0.20	4.93	14.28	19.21	22.09
Extra for							
1525 x 1016 mm translucent sheets Grade 1 EXT SAA	each	0.47	0.23	5.76	16.16	21.92	25.21
soaker flanges for pipes not exceeding 150 mm diameter	each	0.70	0.35	8.63	22.71	31.34	36.04
Standard 75 mm corrugated galvanised steel sheeting							
24G coverings with 1.5 corrugation side laps and 150 mm end laps fixed with drive screws to timber purlins or rails							
roof coverings	m2	0.16	0.16	2.51	15.69	18.20	20.93
vertical coverings	m2	0.20	0.20	3.14	15.69	18.83	21.65
24G coverings with 1.5 corrugation side laps and 150 mm end laps fixed with hook bolts to steel purlins or rails							
roof coverings	m2	0.24	0.24	3.77	15.69	19.46	22.38
vertical coverings	m2	0.30	0.30	4.71	15.69	20.40	23.46
Ancillaries							
square cutting to large openings	m	0.33	0.16	4.04	1.44	5.48	6.30
raking cutting	m	0.36	0.18	4.44	7.19	11.63	13.37
holes for pipes not exceeding 110 mm diameter	each	0.65	0.33	8.04	-	8.04	9.25
375 mm x 26G ridge cappings	m	0.35	0.17	4.29	10.38	14.67	16.87
Extra for 1225 x 762 mm translucent sheets Grade 1EXT SAA	each	0.30	0.15	3.70	12.30	16.00	18.40

Roofing

New work	Unit	Hours C	Hours L	Labour net	Material net	Price net	Price with 15%
Corrugated and troughed sheeting				£	£	£	£
					VAT not included		
Mill finish profiled aluminium roofing							
0.7 mm roof coverings fixed with top seal stainless steel screws to steel purlins to minimum 10 degree pitch, side laps stitched at 450 mm centres							
roof coverings	m2	2.00	1.00	28.39	15.09	43.48	50.00
Ancillaries							
square cutting to large openings	m	0.25	0.25	4.28	1.30	5.58	6.42
raking cutting	m	0.33	0.33	5.66	6.48	12.14	13.96
holes for pipes not exceeding 110 mm diameter	each	0.50	0.25	7.10	-	7.10	8.16
0.9 x 150 mm girth eaves flashing including foam filler pieces	m	0.65	0.65	10.20	4.29	14.49	16.66
0.9 x 400 mm girth capping three times bent and stitched to sheeting at verges	m	1.20	1.20	20.55	6.14	26.69	30.69
0.9 x 450 mm girth plain angular ridge cappings including foam filler pieces	m	0.90	0.90	15.41	11.58	26.99	31.04
finials	each	1.20	1.20	20.55	61.21	81.76	94.02
Standard 75 mm corrugated vinyl translucent sheeting to BS 4203							
Coverings with 1.5 corrugation side laps and 150 mm end laps fixed with drive screws to timber purlins or rails							
roof coverings	m2	0.25	0.12	3.05	10.99	14.04	16.15
vertical coverings	m2	0.33	0.17	4.11	10.99	15.10	17.36
Coverings with 1.5 corrugated side laps and 150 mm end laps fixed with hook bolts to steel purlins or rails							
roof coverings	m2	0.30	0.15	3.70	11.90	15.60	17.94
vertical coverings	m2	0.39	0.19	4.77	11.90	16.67	19.17
Ancillaries							
square cutting to large openings	m	0.20	0.10	2.46	1.01	3.47	3.99
raking cutting	m	0.25	0.13	3.12	5.07	8.19	9.42
holes for pipes not exceeding 110 mm diameter	each	0.50	0.25	6.17	-	6.17	7.10
foam filler pieces at eaves	m	0.05	0.02	0.58	1.08	1.66	1.91
Standard 150 mm corrugated vinyl translucent sheeting to BS 4203							
Coverings with three-quarter corrugation side laps and 150 mm end laps fixed with drive screws to timber purlins or rails							
roof coverings	m2	0.25	0.12	3.05	16.06	19.11	21.98
vertical coverings	m2	0.30	0.15	3.70	16.06	19.76	22.72

Roofing

New work	Unit	Hours C	Hours L	Labour net	Material net	Price net	Price with 15%

Corrugated and troughed sheeting

£ £ £ £

VAT not included

Standard 150 mm corrugated vinyl translucent sheeting to BS 4203 (continued)

Coverings with three-quarter corrugation side laps and 150 mm end laps fixed with hook bolts to steel purlins or rails

	Unit	Hours C	Hours L	Labour net	Material net	Price net	Price with 15%
roof coverings	m2	0.30	0.15	3.70	16.97	20.67	23.77
vertical coverings	m2	0.39	0.19	4.77	16.97	21.74	25.00

Ancillaries

square cutting to large openings	m	0.20	0.10	2.46	1.52	3.98	4.58
raking cutting	m	0.25	0.13	3.12	7.60	10.72	12.33
holes for pipes not exceeding 110 mm diameter	each	0.50	0.25	6.17	-	6.17	7.10
foam filler pieces at eaves	m	0.05	0.02	0.58	1.08	1.66	1.91

Fire resisting standard 75 mm corrugated glass fibre reinforced plastics translucent sheeting to BS

Coverings with 1.5 corrugation side laps and 150 mm end laps fixed with drive screws to timber purlins or rails

roof coverings	m2	0.25	0.12	3.05	13.52	16.57	19.06
vertical coverings	m2	0.33	0.17	4.11	13.52	17.63	20.27

Coverings with 1.5 corrugation side laps and 150 mm end laps fixed with hook bolts to steel purlins or rails

roof coverings	m2	0.30	0.15	3.70	14.43	18.13	20.85
vertical coverings	m2	0.39	0.19	4.77	14.43	19.20	22.08

Ancillaries

square cutting to large openings	m	0.20	0.10	2.46	1.27	3.73	4.29
raking cutting	m	0.25	0.13	3.12	6.33	9.45	10.87
holes for pipes not exceeding 110 mm diameter	each	0.50	0.25	6.17	-	6.17	7.10
foam filler pieces at eaves	m	0.05	0.02	0.58	1.08	1.66	1.91

Fire resisting standard 150 mm corrugated glass fibre reinforced plastics translucent sheeting to BS

Coverings with three-quarter corrugation side laps and 150 mm end laps fixed with drive screws to timber purlins or rails

roof coverings	m2	0.25	0.12	3.05	19.09	22.14	25.46
vertical coverings	m2	0.33	0.17	4.11	19.09	23.20	26.68

Coverings with three-quarter corrugation side laps and 150 mm end laps fixed with hook bolts to steel purlins or rails

roof coverings	m2	0.30	0.15	3.70	20.00	23.70	27.25
vertical coverings	m2	0.39	0.19	4.77	20.00	24.77	28.49

Roofing

New work	Unit	Hours C	Hours L	Labour net	Material net	Price net	Price with 15%

Corrugated and troughed sheeting

£ £ £ £

VAT not included

Fire resisting standard 150 mm corrugated glass fibre reinforced plastics translucent sheeting to BS *(continued)*

	Unit	Hours C	Hours L	Labour net	Material net	Price net	Price with 15%
Ancillaries							
square cutting to large openings	m	0.20	0.10	2.46	1.82	4.28	4.92
raking cutting	m	0.25	0.13	3.12	9.12	12.24	14.08
holes for pipes not exceeding 110 mm diameter	each	0.50	0.25	6.17	-	6.17	7.10
foam filler pieces at eaves	m	0.05	0.02	0.58	1.08	1.66	1.91
Roof Decking							
0.7 x 35 mm galvanised steel troughed decking							
decking bolted to steelwork	m2	0.24	0.24	3.77	12.87	16.64	19.14
square cutting to large openings	m	0.15	0.07	1.81	1.11	2.92	3.36
holes for pipes	each	0.70	0.35	8.63	-	8.63	9.92
0.7 x 63 mm galvanised steel troughed decking							
decking bolted to steelwork	m2	0.24	0.24	3.77	13.21	16.98	19.53
square cutting to large openings	m	0.15	0.07	1.81	1.14	2.95	3.39
holes for pipes	each	0.70	0.35	8.63	-	8.63	9.92
50 mm wood wool slabs							
decking in 1800 x 600 mm standard slabs							
nailed to timber joists	m2	0.15	0.07	1.81	7.55	9.36	10.76
square cutting to large openings	m	0.20	0.10	2.46	0.75	3.21	3.69
raking cutting	m	0.24	0.12	2.96	3.75	6.71	7.72
holes for pipes	each	0.70	0.35	8.63	-	8.63	9.92
75 mm woodwool slabs							
decking in 2700 x 600 mm standard slabs							
nailed to timber joists	m2	0.20	0.10	2.46	8.53	10.99	12.64
square cutting to large openings	m	0.20	0.10	2.46	0.80	3.26	3.75
raking cutting	m	0.24	0.12	2.96	3.98	6.94	7.98
holes for pipes	each	0.70	0.35	8.63	-	8.63	9.92

Roofing

New work	Unit	Hours C	Hours L	Labour net	Material net	Price net	Price with 15%
Flexible sheet finishings				£	£	£	£
					VAT not included		
Bitumen-felt roofing							
Two layer glass fibre felt coverings and stone chippings surfacing							
Flat coverings over 300 mm wide	m2	0.28	0.14	3.45	5.35	8.80	10.12
Working outlets	m2	0.65	0.33	8.04	2.10	10.14	11.66
Three layer glass fibre felt coverings and stone chippings surfacing							
Flat coverings over 300 mm wide	m2	0.32	0.16	3.95	7.53	11.48	13.20
Working into outlets	each	0.70	0.35	8.63	2.43	11.06	12.72
Two layer "Asbex" glass fibre felt mineral-surfaced coverings							
Sloping coverings over 300 mm wide	m2	0.35	0.17	4.29	5.05	9.34	10.74
Working into outlets	each	0.65	0.37	8.31	2.07	10.38	11.94
Three layer "Asbex" glass fibre felt mineral-surfaced coverings							
Sloping coverings over 300 mm wide	m2	0.47	0.23	5.76	7.85	13.61	15.65
Working into outlets	each	0.70	0.35	8.63	2.91	11.54	13.27
Mineral-surfaced "Asbex" glass fibre felt finishes							
Aprons with fair drip edge at eaves or verges							
75 mm	m	0.18	0.09	2.22	0.92	3.14	3.61
150 mm	m	0.22	0.11	2.71	1.33	4.04	4.65
Skirtings dressed over angle fillet							
150 mm girth	m	0.12	0.06	1.48	1.23	2.71	3.12
300 mm girth	m	0.24	0.12	2.96	2.45	5.41	6.22
200 mm three coat linings to gutters dressed over two angle fillets							
straight	m	0.30	0.15	3.70	1.78	5.48	6.30
ends	each	0.40	0.20	4.93	1.71	6.64	7.64
outlets	each	0.50	0.25	6.17	2.54	8.71	10.02
300 mm three coat linings to gutters dressed over two angle fillets							
straight	m	0.45	0.22	5.51	2.66	8.17	9.40
ends	each	0.52	0.26	6.41	2.73	9.14	10.51
outlets	each	0.60	0.30	7.40	3.55	10.95	12.59
Collars around							
small pipes (not exceeding 55 mm)	each	1.25	0.63	15.45	1.43	16.88	19.41
large pipes (55 - 110 mm)	each	1.60	0.80	19.73	1.99	21.72	24.98

Roofing

New work	Unit	Hours C	Hours L	Labour net	Material net	Price net	Price with 15%

Flexible sheet finishings

				£	£	£	£
				VAT not included			

Underlays

	Unit	Hours C	Hours L	Labour net	Material net	Price net	Price with 15%
19 mm fibre insulation board bedded in hot bitumen	m2	0.10	0.05	1.24	4.06	5.30	6.09
Holes for pipes	each	0.30	0.15	3.70	-	3.70	4.25
25 mm resin-bonded glass fibre slabs bedded in hot bitumen	m2	0.12	0.06	1.48	4.90	6.38	7.34
Holes for pipes	each	0.31	0.15	3.79	-	3.79	4.36
Felt vapour barrier bedded in hot bitumen	m2	0.07	0.04	0.90	1.78	2.68	3.08

Accessories

	Unit	Hours C	Hours L	Labour net	Material net	Price net	Price with 15%
40 x 65 mm aluminium edge trims including butt straps and working feltwork to trim	m	0.30	0.15	3.70	5.22	8.92	10.26
Extra for right angle corner pieces							
internal	each	0.35	0.17	4.29	4.50	8.79	10.11
external angle	each	0.35	0.17	4.29	5.04	9.33	10.73
75 x 65 mm aluminium edge trims including butt straps and working feltwork to trim	m	0.33	0.16	4.04	6.10	10.14	11.66
Extra for right angle corner pieces							
internal angle	each	0.35	0.17	4.29	4.88	9.17	10.55
external angle	each	0.35	0.17	4.29	4.88	9.17	10.55

High performance polyester based roofing

	Unit	Hours C	Hours L	Labour net	Material net	Price net	Price with 15%
Two layer coverings and stone chippings surfacing							
flat coverings over 300 mm wide	m2	0.28	0.14	3.45	12.43	15.88	18.26
working into outlets	each	0.65	0.37	8.31	3.13	11.44	13.16
Three layer flat coverings and stone chippings over 300 mm wide							
with Type 2B (BS 747) base layer	m2	0.32	0.16	3.95	14.46	18.41	21.17
with Type 3B (BS 747) base layer	m2	0.32	0.16	3.95	15.61	19.56	22.49
working into outlets	each	0.70	0.35	8.63	4.38	13.01	14.96
Two layer mineral surfaced coverings							
flat coverings over 300 mm wide	m2	0.28	0.14	3.45	14.58	18.03	20.73
working into outlets	each	0.65	0.37	8.31	4.94	13.25	15.24
Three layer mineral surfaced coverings over 300 mm wide							
with Type 2B (BS 747) base layer	m2	0.32	0.16	3.95	15.90	19.85	22.83
with Type 3B (BS 747) base layer	m2	0.32	0.16	3.95	16.05	20.00	23.00
working into outlets	each	0.70	0.35	8.63	5.54	14.17	16.30

Roofing

New work	Unit	Hours C	Hours L	Labour net	Material net	Price net	Price with 15%
Flexible sheet finishings				£	£	£	£
					VAT not included		
Mineral surfaced finishes							
Aprons with fair drip edge at eaves or verges							
75 mm wide	m	0.18	0.09	2.22	2.41	4.63	5.32
150 mm	m	0.22	0.11	2.71	3.55	6.26	7.20
Skirtings dressed over angle fillet							
150 mm	m	0.12	0.06	1.48	4.44	5.92	6.81
300 mm girth	m	0.24	0.12	2.96	6.90	9.86	11.34
200 mm three coat linings to gutters dressed over two angle fillets							
straight	m	0.30	0.15	3.70	3.62	7.32	8.42
ends	each	0.35	0.17	4.29	3.70	7.99	9.19
outlets	each	0.50	0.25	6.17	5.51	11.68	13.43
300 mm three coat linings to gutters dressed over two angle fillets							
straight	m	0.45	0.22	5.51	5.46	10.97	12.62
ends	each	0.52	0.25	6.35	5.53	11.88	13.66
outlets	each	0.60	0.30	7.40	7.29	14.69	16.89
Collars around							
small pipes (not exceeding 55 mm)	each	1.25	0.63	15.45	5.94	21.39	24.60
large pipes (55 - 110 mm)	each	1.60	0.80	19.73	2.79	22.52	25.90

Roofing

New work

Roof Ventilation

£   £

VAT not included

The following "Specialist price net" are guide prices provided by Cavity Trays Ltd.

Prices do not include for cash discount.

Supply and deliver the following:

	Unit	Specialist price net	Price with 15%
Type CSV Circular Soffit Ventilator; white, brown or black, size 79/79mm x 15mm (required 70mm hole), BS polypropylene, 2100mm² free airflow	each	0.30	0.35
Type CV Corbel Ventilator; black PVCU, 42mm high x 28mm deep	2400mm length	6.95	7.99
Type ECF Eaves Comb Filler; 1000mm x 55mm; polypropylene black	m	0.52	0.60
Type EROV Eaves Roll-Out Ventilator; black, standard six metre rolls	m	1.00	1.15
Type OEVWF Open Eaves Ventilator with Flyscreen; black PVCU, for use with 400,450 and 600mm truss centres	each	1.95	2.24
Type OFV Over Facia Ventilator; CFV10 - 1000mm x 400mm wide and 20mm high, black polypropylene	m	1.50	1.72
Type PV Panel Ventilator; for use with 400,450 and 600mm truss centres	each	0.75	0.86
Type REV Refurbishment Eaves Ventilator; two sizes available, REV600 for 600mm rafter spacing and REV400 for 400 and 450mm rafter spacing	each	1.01	1.16
Type RAV-FL Roof Abutment Ventilators; 1.2m x approx. 200mm o/a x 50mm o/a x 3mm thickness, integral ventilation grille equivalent of 25mm continuous gap, grey grade A1 compound PVCU	1200mm length	10.59	12.18
Roof Ventilators for tile roofs	each	21.15	24.32
Roof Ventilators for slate roofs	each	16.91	19.45
Strip Soffit Ventilators; 500mm wide x 29mm high	2400mm length	3.80	4.37
Strip Soffit Ventilators for pitches below 15°; 80mm x approx 29mm high	2400mm length	6.31	7.26
5mm Reduced Upstand Strip Soffit Ventilators; can be fitted into facia groove	2400mm length	3.80	4.37
Flat Strip Soffit Ventilators	2400mm length	5.95	6.84
Reversible Angled Strip Soffit Ventilators	2400mm length	6.31	7.26

Enquiries about the foregoing specialist prices should be made to Cavity Trays Ltd, Administration Centre, Lufton Trading Estate, Yeovil, Somerset BA22 8HU, tel (01935) 474769 fax (01935) 428223)

Roofing

New work	Unit	Hours C	Hours L	Labour net	Material net	Price net	Price with 15%
Sheet metal roofing and flashings				£	£	£	£
					VAT not included		
Milled sheet lead							
2.24 mm (BS Code 5) flat roof and gutter coverings	m2	3.00	3.00	51.39	36.12	87.51	100.64
2.24 mm (BS Code 5) sloping roof coverings							
over 50 deg	m2	3.60	3.60	61.67	29.25	90.92	104.56
bossed ends to rolls	each	0.50	0.50	8.57	0.29	8.86	10.19
bossed intersections to rolls	each	1.00	1.00	17.13	0.51	17.64	20.29
copper nailing at 50 mm centres	m	0.25	0.25	4.28	0.42	4.70	5.41
soldered dots with brass screws	each	0.50	0.50	8.57	0.74	9.31	10.71
1.80 mm (BS Code 4) flashings lapped 100 mm and lead wedged							
150 mm girth	m	0.66	0.66	11.30	3.45	14.75	16.96
210 mm girth	m	0.85	0.85	14.56	4.68	19.24	22.13
1.80 mm (BS Code 4) stepped flashings lapped 100 mm and lead wedged							
180 mm girth	m	1.00	1.00	17.13	4.06	21.19	24.37
240 mm girth	m	1.15	1.14	19.64	5.38	25.02	28.77
1.80 mm (BS Code 4) sloping gutters lapped 100 mm lead wedged and dressed over tilting fillet							
240 mm girth	m	1.25	1.25	21.41	5.66	27.07	31.13
300 mm girth	m	1.40	1.40	23.98	6.89	30.87	35.50
1.80 mm (Code 4) soakers (for fixing by slater or tiler)							
175 x 165 mm	each	1.50	1.50	25.69	0.66	26.35	30.30
300 x 165 mm	each	1.50	1.50	25.69	1.10	26.79	30.81
1.80 mm (BS Code 4) slates (for fixing by slater or tiler)							
600 x 450 mm	each	2.00	2.00	34.26	5.95	40.21	46.24
600 x 600 mm	each	2.00	2.00	34.26	7.93	42.19	48.52
Pipe flashings							
Aluminium slates (for fixing by asphalter or roofer to flat roofs) with synthetic rubber cone and elastomeric seal around							
75 mm pipes	each	0.25	0.25	4.28	15.02	19.30	22.20
100 mm pipes	each	0.25	0.25	4.28	20.07	24.35	28.00
Aluminium slates (for fixing by slater or tiler to sloping roofs) with synthetic rubber cone and elastomeric seal around							
75 mm pipes	each	0.25	0.25	4.28	15.02	19.30	22.20
100 mm pipes	each	0.25	0.25	4.28	20.07	24.35	28.00

Roofing

New work	Unit	Hours C	Hours L	Labour net	Material net	Price net	Price with 15%
Sheet metal roofing and flashings				£	£	£	£
				VAT not included			

Sheet zinc

	Unit	Hours C	Hours L	Labour net	Material net	Price net	Price with 15%
Zinc sheeting 12 gauge (0.65 mm thick) in							
flat roofing	m2	3.00	1.00	33.62	12.10	45.72	52.58
sloping roofing 10 - 50 degrees	m2	3.25	1.25	37.55	12.10	49.65	57.10
sloping roofing over 50 degrees	m2	3.75	1.25	42.03	12.10	54.13	62.25
Flashings							
100 mm girth	m	0.45	0.10	4.70	1.31	6.01	6.91
150 mm girth	m	0.50	0.10	5.15	1.61	6.76	7.77
200 mm girth	m	0.60	0.15	6.39	2.15	8.54	9.82
Wedging with zinc wedges	m	0.20	0.05	2.13	0.26	2.39	2.75
Zinc sheeting 14 gauge (0.8 mm thick) in							
flat roofing	m2	3.00	1.00	33.62	14.28	47.90	55.09
sloping roofing 10 - 50 degrees	m2	3.25	1.25	37.55	14.28	51.83	59.60
sloping roofing over 50 degrees	m2	3.75	1.25	42.03	14.28	56.31	64.76
Flashings							
100 mm girth	m	0.45	0.10	4.70	3.32	8.02	9.22
150 mm girth	m	0.50	0.10	5.15	1.89	7.04	8.10
200 mm girth	m	0.60	0.15	6.39	2.54	8.93	10.27
Wedging with zinc wedges	m	0.20	0.05	2.13	0.32	2.45	2.82

Sheet aluminium

	Unit	Hours C	Hours L	Labour net	Material net	Price net	Price with 15%
Aluminium sheeting commercial grade (0.6 mm thick) in							
flat roofing	m2	3.20	1.00	35.41	12.08	47.49	54.61
sloping roofing 10 - 50 degrees	m2	3.50	1.25	39.79	12.08	51.87	59.65
sloping roofing over 50 degrees	m2	4.00	1.25	44.27	12.08	56.35	64.80
Flashings							
100 mm girth	m	0.45	0.10	4.70	1.37	6.07	6.98
150 mm girth	m	0.50	0.10	5.15	1.69	6.84	7.87
200 mm girth	m	0.60	0.15	6.39	2.25	8.64	9.94
Wedging with aluminium wedges	m	0.25	0.05	2.58	0.32	2.90	3.34
Aluminium sheeting commercial grade (0.8 mm thick) in							
flat roofing	m2	3.20	1.00	35.41	16.37	51.78	59.55
sloping roofing 10 - 50 degrees	m2	3.50	1.25	39.79	16.37	56.16	64.58
sloping roofing over 50 degrees	m2	4.00	1.25	44.27	16.37	60.64	69.74
Flashings							
100 mm girth	m	0.45	0.10	4.70	1.89	6.59	7.58
150 mm girth	m	0.50	0.10	5.15	2.29	7.44	8.56
200 mm girth	m	0.60	0.15	6.39	3.04	9.43	10.84
Wedging with aluminium wedges	m	0.25	0.05	2.58	0.37	2.95	3.39

Roofing

New work	Unit	Hours C	Hours L	Labour net	Material net	Price net	Price with 15%
Sheet metal roofing and flashings				£	£	£	£
					VAT not included		
Sheet copper							
Copper sheeting 0.55 mm thick in							
flat roofing	m2	3.20	1.00	35.41	28.59	64.00	73.60
sloping roofing 10 - 50 degrees	m2	3.50	1.25	39.79	28.59	68.38	78.64
sloping roofing over 50 degrees	m2	4.00	1.25	44.27	28.59	72.86	83.79
Flashings							
100 mm girth	m	0.45	0.10	4.70	3.15	7.85	9.03
150 mm girth	m	0.50	0.10	5.15	3.99	9.14	10.51
200 mm girth	m	0.60	0.15	6.39	5.32	11.71	13.47
Wedging with copper wedges	m	0.25	0.05	2.58	0.42	3.00	3.45
Copper sheeting 0.7 mm thick in							
flat roofing	m2	3.20	1.00	35.41	36.40	71.81	82.58
sloping roofing 10 - 50 degrees	m2	3.50	1.25	39.79	36.40	76.19	87.62
sloping roofing over 50 degrees	m2	4.00	1.25	44.27	36.40	80.67	92.77
Flashings							
100 mm girth	m	0.45	0.10	4.70	4.20	8.90	10.23
150 mm girth	m	0.50	0.10	5.15	5.08	10.23	11.76
200 mm girth	m	0.60	0.15	6.39	6.77	13.16	15.13
Wedging with copper wedges	m	0.25	0.05	2.58	0.42	3.00	3.45

Roofing

	Unit	Specialist price net	Price with 15%
New work			

Thatching

£   £
VAT not included

The following "Specialist price net" figures are guide prices for average size jobs provided by the National Council of Master Thatchers Associations. Quantities and locations may affect prices, therefore it is advisable to submit details and request a firm quotation for any proposed work.

Prices do not include for cash discount.

See the preamble notes for builder's profit and attendance.

Thatching to a thickness of about 300 mm, fixed with iron hooks and finished with a block cut, patterned and saddled ridge

	Unit	Specialist price net	Price with 15%
with best quality water reed	m2	92.64	106.54
with best quality combed wheat reed	m2	84.32	96.97
with best quality long straw	m2	77.27	88.86
Wiring over thatching with 1200 x 19 mm x 20G galvanised wire netting	m2	6.04	6.95

Enquiries about the foregoing specialist prices should be made to the National Council of Master Thatchers Associations, tel 07000 781909.

THE NATIONAL COUNCIL OF MASTER THATCHERS ASSOCIATIONS

The NCMTA can provide information on all aspects of thatching
- *Information provided on represented Master Thatcher Associations*
- *Location of individual thatchers*
- *Details on a comprehensive range of specialist insurance companies*
- *Roof construction details provided*
- *Free information pack*
- *Assistance provided on local planning, conservation and grant policies*
- *Advice on Fire Preventative measures*
- *Seminars and informative talks arranged for architects/ surveyors/conservation officers and other interested parties.*

The NCMTA represents the county-based master thatchers association and can offer advice on all aspects on the thatching industry.

NATIONAL COUNCIL OF MASTER THATCHERS ASSOCIATIONS

Tel: 07000 781909

Woodwork

Preamble

"Labour net" figures refer to site labour only and include allowances for all costs incidental to the employment of labour. Except for unloading and helping with heavy lifting, no labourer assistance has been allowed in compiling this section, carpenters and joiners usually being able to work together and assist each other more economically.

The cost of workshop labour, manufacturing the joinery etc., has been included in figures entered under the "Materials net" heading.

"Materials net" figures include for all costs of materials including workshop labour and an allowance for waste except where specifically stated.

"Price net" figures are the totals of the "Labour net" and "Materials net" figures. Prices are for a builder employing his own labour; according to the amount and nature of the work involved, it may well be possible to secure more advantageous prices from specialist sub-contractors.

Stated sizes of woodwork sections are basic (nominal) sizes before planing and prices include for fixing with nails except where another method of fixing has been described.

Prices for boarding and flooring include allowance for reduced coverage resulting from machined or tongued and grooved edges to boards.

For curved work, add from 100% to the stated figures according to the types of curves required.

Although not in accordance with the Standard Method of Measurement, figures per square or linear metre have been included for certain composite items, such as casements and frames, staircases etc., from which figures for any particular enumerated items can be calculated.

Prices do not include any allowance for scaffolding, ladders or other plant necessary to reach the work. The "Preliminaries" section includes prices for scaffolding which must be considered and allowance included to suit the particular circumstances of a tender.

Specialist prices

"Price with 15%" figures are all-in guide prices and include 15% for the builder's overheads, profit, unloading materials and general attendance (to include free use of standing scaffolding and hoists, temporary lighting and water and clearing away rubbish).

The amount of attendance required varies between the various trades and also with the circumstances of specific jobs; the percentage addition must always be considered and adjusted as necessary to suit the terms and conditions of the quotation being used.

Quantities and delivery distances are usually the most significant of the many factors which influence prices and it must be emphasised that quotations should always be obtained when preparing a tender.

Woodwork

New work

Basic prices for materials		£
Carcassing softwood		
Smaller scantlings	m3	244.80
Larger scantlings	m3	244.80
Extra for stress grading		
SC3 grade	m3	10.20
SC4 grade	m3	20.40
Joinery softwood		
Smaller scantlings	m3	494.00
Larger scantlings	m3	428.40
Boarding		
Softwood sawn boarding		
19 mm	m2	5.87
25 mm	m2	7.65
19 mm softwood plain edged and tongued and grooved flooring		
in 150 mm widths	m2	12.75
in 100 mm widths	m2	12.75
25 mm softwood plain edged and tongued and grooved flooring		
in 150 mm widths	m2	13.77
in 100 mm widths	m2	13.77
Softwood tongued and grooved and V jointed boarding		
19 mm	m2	13.72
25 mm	m2	15.45
19 mm sawn softwood feather edged boarding	m2	5.41
Softwood matchboarding or shiplap boarding		
19 mm	m2	13.12
25 mm	m2	14.34
Western red cedar matchboarding or shiplap boarding		
19 mm	m2	34.88
25 mm	m2	46.67
19 mm waney edge larch boarding	m2	23.28

Prices actually to be paid for materials must be checked against the above basic prices and adjustments made as necessary.

Woodwork

New work

Basic prices for materials
£

Wrought softwood battens and mouldings

Battens

	Unit	Price
13 x 38 mm	m	0.42
13 x 50 mm	m	0.58
13 x 75 mm	m	0.76
16 x 50 mm	m	0.55
19 x 38 mm	m	0.56
19 x 50 mm	m	0.72
19 x 75 mm	m	1.14
25 x 38 mm	m	0.59
25 x 50 mm	m	0.65
25 x 75 mm	m	1.31
25 x 100 mm	m	1.80
25 x 125 mm	m	2.28
25 x 175 mm	m	3.24
25 x 225 mm	m	4.55
32 x 75 mm	m	1.53
32 x 100 mm	m	1.86
32 x 150 mm	m	2.80
38 x 50 mm	m	1.09
50 x 50 mm	m	1.09
50 x 75 mm	m	1.59
50 x 100 mm	m	2.11

Stock pattern skirtings

	Unit	Price
19 x 75 mm	m	0.83
19 x 100 mm	m	1.71
25 x 150 mm	m	3.28

Stock pattern architraves

	Unit	Price
19 x 50 mm	m	1.06
25 x 25 mm	m	0.64
25 x 75 mm	m	1.86

Quadrants

	Unit	Price
19 mm	m	0.48
25 mm	m	0.65

	Unit	Price
16 x 50 mm stock pattern moulded stop	m	0.86

Prices actually to be paid for materials must be checked against the above basic prices and adjustments made as necessary.

Woodwork

New work

Basic prices for materials £

Hardwood - minimum 1 m3 lots based on 50 mm thickness

Kiln dried

	Unit	Price
Afrormosia	m3	1,142.40
Ash, American	m3	990.61
Iroko	m3	791.83
Shorea Species	m3	760.05
Brazilian Mahogany	m3	1,219.95
Oak, American Red	m3	1,479.00
Oak, American White	m3	1,299.96
Sapele	m3	882.00
Utile	m3	969.00

Naturally seasoned

	Unit	Price
Keruing	m3	413.53
Oak, English	m3	1,122.00

Smaller lots subject to increase according to quantity and specification.

Blockboard, plywood and sundries

	Unit	Price
Blockboard		
18 mm	m2	19.11
25 mm	m2	24.89
Blockboard veneered both sides		
18 mm	m2	41.96
25 mm	m2	58.80
Plywood WBP grade		
4 mm	m2	4.98
6 mm	m2	7.57
12 mm	m2	12.51
Plywood veneered both sides		
6 mm	m2	27.73
12 mm	m2	33.67
Plywood in 1500 x 600 mm tongued and grooved panels		
12 mm	m2	13.70
15 mm	m2	17.12
18 mm	m2	18.44
25 mm chipboard standard grade	m2	6.22

Prices actually to be paid for materials must be checked against the above basic prices and adjustments made as necessary.

Woodwork

New work

Basic prices for materials £

Blockboard, plywood and sundries *(Continued)*

	Unit	Price
Chipboard flooring grade Type C4		
18 mm	m2	4.22
22 mm	m2	4.62
Chipboard tongued and grooved all edges Type C4		
18 mm	m2	4.25
22 mm	m2	4.84
18 mm melamine faced chipboard	m2	6.36
Medium density fibreboard		
6 mm	m2	2.76
12 mm	m2	4.64
18 mm	m2	6.10
25 mm	m2	7.38
3.2 mm hardboard	m2	1.31
3.2 mm decorated hardboard	m2	1.99
Plastic laminate sheet		
covering	m2	18.02
balancing	m2	2.83
13 mm sanded finish fibre insulation boards	m2	1.66
58 mm strawboard	m2	9.44

Insulating materials

	Unit	Price
Expanded polystyrene ISD grade		
25 mm	m2	1.41
40 mm	m2	2.26
13 mm sound deadening quilt	m2	1.66
Thermal insulating quilt		
80 mm	m2	2.91
100 mm	m2	3.46
150 mm	m2	5.38
Granular loose fill	110 litre bag	7.57
Building paper		
standard grade	m2	1.07
reflective grade	m2	1.61

Prices actually to be paid for materials must be checked against the above basic prices and adjustments made as necessary.

Woodwork

New work	Unit	Specialist price net	Price with 15%
Preservation		£	£
		VAT not included	

C.C.A. and "Vac-Vac" treatments

The following "Specialist price net" figures are guide prices only

C.C.A. and "Vac-Vac" are general purpose wood preservatives against insect and fungal attack.

C.C.A. vacuum/pressure impregnation treatment to			
interior building timbers to a dry salt retention of 4 kg per m3	m3	32.02	36.82
general exterior building timbers to a dry salt retention of 5.3 kg per m3	m3	34.65	39.85
special timbers to a dry salt retention of 8 kg per m3	m3	52.50	60.37
Vacsol Aqua double vacuum and pressure impregnation treatment to interior building timbers	m3	48.30	55.54

"Tanalith" and "Vac-Vac" treatments

The following "Specialist price net" figures are guide prices provided by Hickson Timber Products Ltd.

Tanalith treatment is for sawn carcassing and fencing and "Vac-Vac" treatment for window frames and structural components such as trussed rafters.

Tanalith vacuum/pressure impregnation treatment to			
interior building timbers with Lifetime Guarantee	m3	32.20	37.03
exterior building timbers and cladding above DPC with Lifetime Guarantee	m3	34.65	39.85
fencing, gates and posts (20 years desired service life)	m3	34.65	39.85
fencing, gates and posts (40 years desired service life)	m3	44.00	50.60
TANAtone colour processing vacuum/pressure impregnation treatment to			
fencing, gates and posts (20 years desired service life)	m3	40.20	46.23
fencing, gates and posts (40 years desired service life)	m3	54.60	62.79
Vac-Vac double vacuum impregnation treatment to exterior joinery cladding and building components	m3	39.80	45.77

Woodwork

New work	Unit	Specialist price net	Price with 15%

Preservation

£ £
VAT not included

Flame retardant treatments

DRICON vacuum/pressure impregnation treatment inclusive of kiln drying to interior and weather protected timbers.

Solid timbers			
BS 476 Part 7 Class 1 SSF	m3	240.00	276.00
BS 476 Parts 6 & 7 Class 0 in accordance with The Building Regulations	m3	480.00	552.00
Plywoods			
BS 476 Part 7 Class 1 SSF	m3	272.00	312.80
BS 476 Parts 6 & 7 Class 0 in accordance with The Building Regulations	m3	545.00	626.75
Non-Com exterior vacuum/pressure impregnation treatment to			
interior and exterior building timber to BS 476, Class 1 SSF	m3	390.00	448.50
WBP plywood to BS 476, Class 1 SSF	m3	430.00	494.50
Western Red Cedar Shingles to AA rating, BS 476, Part 3	bundle	16.80	19.32

Enquiries about the foregoing specialist prices should be made to Hickson Timber Products Ltd, A1 Business Park, Knottingley, West Yorkshire, WF11 0BU, tel (01977) 671771, fax (01977) 671701/2/3.

Woodwork

New work	Unit	Hours C	Hours L	Labour net	Material net	Price net	Price with 15%
Carcassing				£	£	£	£
					VAT not included		
Impregnated sawn softwood - SC3 Grade							
Floors							
38 x 75 mm	m	0.07	-	0.63	0.97	1.60	1.84
38 x 100 mm	m	0.07	-	0.63	1.27	1.90	2.19
38 x 125 mm	m	0.08	-	0.72	1.55	2.27	2.61
38 x 150 mm	m	0.09	-	0.77	1.82	2.59	2.98
38 x 175 mm	m	0.10	-	0.90	2.14	3.04	3.50
38 x 200 mm	m	0.11	-	0.99	2.45	3.44	3.96
38 x 225 mm	m	0.13	-	1.16	2.80	3.96	4.55
50 x 75 mm	m	0.08	-	0.72	0.98	1.70	1.96
50 x 100 mm	m	0.09	-	0.76	1.30	2.06	2.37
50 x 125 mm	m	0.09	-	0.81	1.63	2.44	2.81
50 x 150 mm	m	0.11	-	0.99	1.95	2.94	3.38
50 x 175 mm	m	0.13	-	1.16	2.29	3.45	3.97
50 x 200 mm	m	0.15	-	1.34	2.62	3.96	4.55
50 x 225 mm	m	0.15	-	1.34	2.94	4.28	4.92
75 x 150 mm	m	0.15	-	1.34	3.12	4.46	5.13
75 x 175 mm	m	0.17	-	1.52	3.65	5.17	5.95
75 x 200 mm	m	0.20	-	1.79	4.17	5.96	6.85
75 x 225 mm	m	0.22	-	1.97	4.68	6.65	7.65
Partitions							
38 x 75 mm	m	0.13	-	1.16	0.97	2.13	2.45
38 x 100 mm	m	0.17	-	1.52	1.27	2.79	3.21
50 x 75 mm	m	0.17	-	1.52	0.98	2.50	2.88
50 x 100 mm	m	0.23	-	2.06	1.30	3.36	3.86
75 x 75 mm	m	0.22	-	1.97	1.62	3.59	4.13
75 x 100 mm	m	0.29	-	2.60	2.07	4.67	5.37
100 x 100 mm	m	0.39	-	3.49	2.97	6.46	7.43
Flat roofs							
38 x 75 mm	m	0.07	-	0.63	0.97	1.60	1.84
38 x 100 mm	m	0.10	-	0.90	1.27	2.17	2.50
38 x 125 mm	m	0.12	-	1.08	1.55	2.63	3.02
38 x 150 mm	m	0.14	-	1.25	1.82	3.07	3.53
38 x 175 mm	m	0.16	-	1.43	2.14	3.57	4.11
38 x 200 mm	m	0.18	-	1.61	2.45	4.06	4.67
50 x 75 mm	m	0.10	-	0.90	0.98	1.88	2.16
50 x 100 mm	m	0.13	-	1.16	1.30	2.46	2.83
50 x 125 mm	m	0.15	-	1.34	1.63	2.97	3.42
50 x 150 mm	m	0.18	-	1.61	1.95	3.56	4.09
50 x 175 mm	m	0.21	-	1.88	2.29	4.17	4.80
50 x 200 mm	m	0.24	-	2.15	2.62	4.77	5.49
50 x 225 mm	m	0.25	-	2.24	2.94	5.18	5.96
75 x 150 mm	m	0.25	-	2.24	3.12	5.36	6.16
75 x 175 mm	m	0.29	-	2.60	3.65	6.25	7.19
75 x 200 mm	m	0.33	-	2.96	4.17	7.13	8.20
75 x 225 mm	m	0.37	-	3.32	4.68	8.00	9.20

Woodwork

New work	Unit	Hours C	Hours L	Labour net	Material net	Price net	Price with 15%
Carcassing				£	£	£	£
					VAT not included		

Impregnated sawn softwood – SC3 Grade
(*continued*)

Pitched roofs including ceiling joists

	Unit	Hours C	Hours L	Labour net	Material net	Price net	Price with 15%
25 x 150 mm	m	0.14	-	1.25	1.20	2.45	2.82
32 x 175 mm	m	0.19	-	1.70	2.21	3.91	4.50
38 x 75 mm	m	0.10	-	0.90	0.98	1.88	2.16
38 x 100 mm	m	0.14	-	1.25	1.30	2.55	2.93
38 x 125 mm	m	0.17	-	1.52	1.58	3.10	3.57
38 x 150 mm	m	0.19	-	1.70	1.86	3.56	4.09
38 x 175 mm	m	0.23	-	2.06	2.18	4.24	4.88
38 x 200 mm	m	0.26	-	2.33	2.49	4.82	5.54
38 x 225 mm	m	0.29	-	2.60	2.85	5.45	6.27
50 x 75 mm	m	0.14	-	1.25	0.99	2.24	2.58
50 x 100 mm	m	0.18	-	1.61	1.32	2.93	3.37
50 x 125 mm	m	0.21	-	1.88	1.66	3.54	4.07
75 x 75 mm	m	0.19	-	1.70	1.64	3.34	3.84
75 x 100 mm	m	0.26	-	2.33	2.10	4.43	5.09
75 x 125 mm	m	0.32	-	2.87	2.62	5.49	6.31
75 x 150 mm	m	0.34	-	3.05	3.14	6.19	7.12
100 x 100 mm	m	0.34	-	3.05	2.99	6.04	6.95
100 x 150 mm	m	0.45	-	4.03	3.85	7.88	9.06
100 x 175 mm	m	0.53	-	4.75	4.38	9.13	10.50

Kerbs, bearers etc.

	Unit	Hours C	Hours L	Labour net	Material net	Price net	Price with 15%
38 x 75 mm	m	0.04	-	0.36	0.95	1.31	1.51
38 x 100 mm	m	0.05	-	0.45	1.26	1.71	1.97
50 x 100 mm	m	0.07	-	0.63	1.29	1.92	2.21
75 x 100 mm	m	0.09	-	0.81	2.06	2.87	3.30
50 x 100 mm - bolted	m	0.13	-	1.16	1.48	2.64	3.04

Noggings between joists

	Unit	Hours C	Hours L	Labour net	Material net	Price net	Price with 15%
38 x 50 mm	m	0.17	-	1.52	0.66	2.18	2.51
50 x 50 mm	m	0.18	-	1.61	0.67	2.28	2.62
50 x 75 mm	m	0.20	-	1.79	1.02	2.81	3.23

Herringbone strutting between (size 38 x 38 mm)
to joists, (measured over joists)

	Unit	Hours C	Hours L	Labour net	Material net	Price net	Price with 15%
100 mm deep	m	0.30	-	2.69	0.82	3.51	4.04
150 mm deep	m	0.30	-	2.69	0.80	3.49	4.01
175 mm deep	m	0.33	-	2.96	0.86	3.82	4.39
200 mm deep	m	0.36	-	3.23	0.90	4.13	4.75
225 mm deep	m	0.39	-	3.49	0.93	4.42	5.08

Herringbone strutting (size 50 x 50 mm) to joists,
(measured over joists)

	Unit	Hours C	Hours L	Labour net	Material net	Price net	Price with 15%
100 mm deep	m	0.30	-	2.69	1.10	3.79	4.36
150 mm deep	m	0.30	-	2.69	1.10	3.79	4.36
175 mm deep	m	0.33	-	2.96	1.14	4.10	4.72
200 mm deep	m	0.36	-	3.23	1.19	4.42	5.08
225 mm deep	m	0.39	-	3.49	1.25	4.74	5.45

Woodwork

New work	Unit	Hours C	Hours L	Labour net	Material net	Price net	Price with 15%

Carcassing

£ £ £ £

VAT not included

Impregnated sawn softwood – SC3 Grade
(*continued*)

Solid bridging between joists (measured over the joists)

	Unit	Hours C	Hours L	Labour net	Material net	Price net	Price with 15%
38 x 100 mm	m	0.20	-	1.79	1.29	3.08	3.54
50 x 100 mm	m	0.20	-	1.79	1.36	3.15	3.62
50 x 150 mm	m	0.20	-	1.79	1.61	3.40	3.91
50 x 175 mm	m	0.22	-	1.97	1.87	3.84	4.42
50 x 200 mm	m	0.24	-	2.15	2.13	4.28	4.92
50 x 225 mm	m	0.26	-	2.33	2.39	4.72	5.43
50 x 75 x 450 mm sprockets	each	0.15	-	1.34	0.46	1.80	2.07
Notch and fit ends of members to metal	each	0.25	-	2.24	-	2.24	2.58

Trim members around openings with four skew nailed butt joints

	Unit	Hours C	Hours L	Labour net	Material net	Price net	Price with 15%
50 x 75 mm	each	0.67	-	6.00	0.24	6.24	7.18
50 x 100 mm	each	0.75	-	6.72	0.24	6.96	8.00

Trim members around openings with six skew nailed butt joints

	Unit	Hours C	Hours L	Labour net	Material net	Price net	Price with 15%
50 x 75 mm	each	1.00	-	8.96	0.35	9.31	10.71
50 x 100 mm	each	1.15	-	10.30	0.35	10.65	12.25

Trim members around openings with four framed joints

	Unit	Hours C	Hours L	Labour net	Material net	Price net	Price with 15%
50 x 150 mm	each	4.00	-	35.84	0.24	36.08	41.49
50 x 175 mm	each	4.10	-	36.74	0.24	36.98	42.53
50 x 200 mm	each	4.20	-	37.63	0.36	37.99	43.69
50 x 225 mm	each	4.20	-	37.63	0.36	37.99	43.69

Trim members around openings with five framed joints

	Unit	Hours C	Hours L	Labour net	Material net	Price net	Price with 15%
50 x 150 mm	each	4.50	-	40.32	0.30	40.62	46.71
50 x 175 mm	each	4.63	-	41.48	0.30	41.78	48.05
50 x 200 mm	each	4.75	-	42.56	0.45	43.01	49.46
50 x 225 mm	each	4.88	-	43.72	0.45	44.17	50.80

Trim members around openings with six framed joints

	Unit	Hours C	Hours L	Labour net	Material net	Price net	Price with 15%
50 x 150 mm	each	5.00	-	44.80	0.35	45.15	51.92
50 x 175 mm	each	5.15	-	46.14	0.35	46.49	53.46
50 x 200 mm	each	5.30	-	47.49	0.54	48.03	55.23
50 x 225 mm	each	5.45	-	48.83	0.54	49.37	56.78

Trim members around openings with seven framed joints

	Unit	Hours C	Hours L	Labour net	Material net	Price net	Price with 15%
50 x 150 mm	each	5.50	-	49.28	0.42	49.70	57.16
50 x 175 mm	each	5.68	-	50.89	0.42	51.31	59.01
50 x 200 mm	each	5.85	-	52.42	0.62	53.04	61.00
50 x 225 mm	each	6.03	-	54.03	0.62	54.65	62.85

Woodwork

New work	Unit	Hours C	Hours L	Labour net	Material net	Price net	Price with 15%
Carcassing				£	£	£	£
					VAT not included		

Impregnated sawn softwood – SC3 Grade
(*continued*)

Trim members around openings with eight framed joints

	Unit	Hours C	Hours L	Labour net	Material net	Price net	Price with 15%
50 x 150 mm	each	6.00	-	53.76	0.47	54.23	62.36
50 x 175 mm	each	6.20	-	55.55	0.47	56.02	64.42
50 x 200 mm	each	6.40	-	57.34	0.72	58.06	66.77
50 x 225 mm	each	6.60	-	59.14	0.72	59.86	68.84

Trim members around openings with four steel joist hangers

	Unit	Hours C	Hours L	Labour net	Material net	Price net	Price with 15%
50 x 150 mm	each	2.50	-	22.40	4.60	27.00	31.05
50 x 175 mm	each	2.60	-	23.30	4.90	28.20	32.43
50 x 200 mm	each	2.70	-	24.19	5.59	29.78	34.25
50 x 225 mm	each	2.80	-	25.09	5.94	31.03	35.68

Trim members around openings with five steel joist hangers

	Unit	Hours C	Hours L	Labour net	Material net	Price net	Price with 15%
50 x 150 mm	each	3.13	-	28.04	1.51	29.55	33.98
50 x 175 mm	each	3.25	-	29.12	6.13	35.25	40.54
50 x 200 mm	each	3.38	-	30.28	6.99	37.27	42.86
50 x 225 mm	each	3.50	-	31.36	7.43	38.79	44.61

Trim members around openings with six steel joist hangers

	Unit	Hours C	Hours L	Labour net	Material net	Price net	Price with 15%
50 x 150 mm	each	3.75	-	33.60	6.90	40.50	46.58
50 x 175 mm	each	3.90	-	34.94	7.35	42.29	48.63
50 x 200 mm	each	4.05	-	36.29	8.38	44.67	51.37
50 x 225 mm	each	4.20	-	37.63	8.91	46.54	53.52

Trim members around openings with seven steel joist hangers

	Unit	Hours C	Hours L	Labour net	Material net	Price net	Price with 15%
50 x 150 mm	each	4.38	-	39.24	1.69	40.93	47.07
50 x 175 mm	each	4.55	-	40.77	8.58	49.35	56.75
50 x 200 mm	each	4.73	-	42.38	9.78	52.16	59.98
50 x 225 mm	each	4.90	-	43.90	10.40	54.30	62.45

Trim members around openings with eight steel joist hangers

	Unit	Hours C	Hours L	Labour net	Material net	Price net	Price with 15%
50 x 150 mm	each	5.00	-	44.80	9.20	54.00	62.10
50 x 175 mm	each	5.20	-	46.59	9.81	56.40	64.86
50 x 200 mm	each	5.40	-	48.38	11.18	59.56	68.49
50 x 225 mm	each	5.60	-	50.18	11.89	62.07	71.38

Woodwork

New work	Unit	Hours C	Hours L	Labour net	Material net	Price net	Price with 15%

Carcassing

£ £ £ £

Gang-nailed trussed rafters, stress-graded, "Protim" treated, 450 mm overhang both sides

Fink 22½ degree pitch, length

5.00 m	each	1.00	1.00	15.70	21.16	36.86	42.39
7.00 m	each	1.10	1.10	17.27	27.31	44.58	51.27
8.00 m	each	1.20	1.20	18.84	32.28	51.12	58.79

Fink 35 degree pitch, length

7.00 m	each	1.10	1.10	17.27	28.33	45.60	52.44
8.00 m	each	1.20	1.20	18.84	34.10	52.94	60.88
10.00 m	each	1.40	1.40	21.98	47.15	69.13	79.50

Fink 45 degree pitch, length

7.00 m	each	1.10	1.10	17.27	48.28	65.55	75.38
8.00 m	each	1.20	1.20	18.84	55.21	74.05	85.16
9.00 m	each	1.40	1.40	21.98	62.30	84.28	96.92

Glued laminated timber beams

Glued laminated timber beam, planed finish, lifted and placed in position, size

270 x 65 mm	m	0.25	0.25	3.93	14.08	18.01	20.71
315 x 65 mm	m	0.30	0.30	4.71	16.42	21.13	24.30
360 x 90 mm	m	0.40	0.40	6.28	23.66	29.94	34.43
405 x 90 mm	m	0.45	0.45	7.06	26.62	33.68	38.73
450 x 115 mm	m	0.50	0.50	7.85	29.58	37.43	43.04
495 x 115 mm	m	0.70	0.70	10.99	50.61	61.60	70.84

Gang-nailed trussed rafters – composite items

Unload and hoist 32 trusses two storeys high, for roof approximately 15.00 m long and with a span of 8.00 m. Set out spacings on wall plate at 450mm centres and place in position, plumb up and secure temporarily with light battens. Supply and fix two lengths of 75 x 38 mm treated sawn softwood structural grade to top side of ceiling joists to maintain spacings and to act as permanent binders (light battens used to secure apex of trusses being removed when tiling or slating battens fixed). Nail to plates, line out and cut feet of rafters as required to take eaves fascia.

Labour and material for the whole	Item	34.50	10.00	376.52	956.69	1333.21	1533.20
Labour and material per truss	each	1.08	0.31	11.77	29.89	41.66	47.91
Labour and material per square metre on plan	m2	0.29	0.08	3.14	7.97	11.11	12.78

Woodwork

New work	Unit	Hours C	Hours L	Labour net	Material net	Price net	Price with 15%
First fixings				£	£	£	£
					VAT not included		
Floors							
25mm softwood plain edged board flooring							
in 150 mm widths	m2	0.55	-	4.93	10.55	15.48	17.80
in 125 mm widths	m2	0.61	-	5.47	9.88	15.35	17.65
25mm softwood tongued and grooved board flooring							
in 150 mm widths	m2	0.66	-	5.91	9.76	15.67	18.02
in 100 mm widths	m2	0.77	-	6.90	8.01	14.91	17.15
Plywood flooring in 1500 x 600mm panels tongued and grooved all edges							
12 mm	m2	0.30	-	2.69	13.01	15.70	18.06
15 mm	m2	0.33	-	2.96	16.29	19.25	22.14
18 mm	m2	0.35	-	3.14	17.57	20.71	23.82
Chipboard Type C4 flooring							
18 mm	m2	0.25	-	2.24	5.78	8.02	9.22
22 mm	m2	0.30	-	2.69	7.03	9.72	11.18
Chipboard Type C4 flooring tongued and grooved all edges							
18 mm	m2	0.33	-	2.96	5.78	8.74	10.05
22 mm	m2	0.40	-	3.58	6.93	10.51	12.09
External walls							
19mm impregnated softwood feather edged boarding							
over 1.00m2	m2	0.60	-	5.38	4.43	9.81	11.28
not exceeding 1.00m2	each	0.75	-	6.72	4.43	11.15	12.82
raking cutting	m	0.07	-	0.63	0.44	1.07	1.23
19mm impregnated softwood matchboarding or shiplap boarding							
over 1.00m2	m2	0.65	-	5.82	10.78	16.60	19.09
not exceeding 1.00m2	each	0.80	-	7.17	10.78	17.95	20.64
raking cutting	m	0.07	-	0.63	1.07	1.70	1.96
25mm impregnated softwood matchboarding or shiplap boarding							
over 1.00m2	m2	0.70	-	6.27	12.26	18.53	21.31
not exceeding 1.00m2	each	0.88	-	7.88	12.26	20.14	23.16
raking cutting	m	0.08	-	0.72	1.22	1.94	2.23
19mm western red cedar matchboarding or shiplap boarding, fixed with galvanised nails							
over 1.00m2	m2	0.70	-	6.27	26.21	32.48	37.35
not exceeding 1.00m2	each	0.88	-	7.88	26.21	34.09	39.20
raking cutting	m	0.08	-	0.72	2.61	3.33	3.83

Woodwork

New work	Unit	Hours C	Hours L	Labour net	Material net	Price net	Price with 15%
First fixings				£	£	£	£
					VAT not included		

External walls (*continued*)

25mm western red cedar matchboarding or
shiplap boarding, fixed with galvanised nails

over 1.00m2	m2	0.75	-	6.72	33.90	40.62	46.71
not exceeding 1.00m2	each	0.95	-	8.51	33.90	42.41	48.77
raking cutting	m	0.08	-	0.72	3.38	4.10	4.71

19mm waney edge elm boarding lapped average
50mm, fixed with galvanised nails

over 1.00m2	m2	0.80	-	7.17	16.96	24.13	27.75
not exceeding 1.00m2	each	1.00	-	8.96	16.96	25.92	29.81
raking cutting	m	0.09	-	0.81	1.69	2.50	2.88

Cellular PVCU cladding of 150 mm (cover width)
shiplap planks , fixed with galvanised nails

over 1.00m2	m2	0.70	-	6.27	23.28	29.55	33.98
not exceeding 1.00m2	each	1.05	-	9.41	23.28	32.69	37.59
raking cutting	m	0.07	-	0.63	0.35	0.98	1.13

Cellular PVCU cladding of 100 mm (cover width)
shiplap planks , fixed with galvanised nails

over 1.00m2	m2	1.00	-	8.96	25.24	34.20	39.33
not exceeding 1.00m2	each	1.50	-	13.44	25.24	38.68	44.48
raking cutting	m	0.10	-	0.90	0.25	1.15	1.32

PVCU cladding of 150 mm (cover width) shiplap
planks , fixed with clips and aluminium nails

over 1.00m2	m2	0.60	-	5.38	14.00	19.38	22.29
not exceeding 1.00m2	each	0.90	-	8.06	14.00	22.06	25.37
raking cutting	m	0.08	-	0.72	0.21	0.93	1.07

PVCU cladding of 100 mm (cover width) shiplap
planks , fixed with clips and aluminium nails

over 1.00m2	m2	0.90	-	8.06	14.82	22.88	26.31
not exceeding 1.00m2	each	1.35	-	12.10	14.82	26.92	30.96
raking cutting	m	0.10	-	0.90	0.15	1.05	1.21

Ancillaries for PVCU cladding

starter strip fixed with aluminium nails	m	0.07	-	0.63	0.65	1.28	1.47
top channel fixed with aluminium nails	m	0.10	-	0.90	0.55	1.45	1.67
side channel fixed with aluminium nails	m	0.12	-	1.08	1.12	2.20	2.53
top cover plank snap fixed to top channel	m	0.07	-	0.63	1.70	2.33	2.68

Roofs

19 mm impregnated sawn softwood boarding

flat	m2	0.50	-	4.48	7.02	11.50	13.22
sloping	m2	0.75	-	6.72	7.02	13.74	15.80
raking cutting	m	0.07	-	0.63	0.10	0.73	0.84

25 mm impregnated sawn softwood boarding

flat	m2	0.55	-	4.93	8.08	13.01	14.96
sloping	m2	0.61	-	5.47	8.08	13.55	15.58
raking cutting	m	0.07	-	0.63	0.12	0.75	0.86

Woodwork

New work	Unit	Hours C	Hours L	Labour net	Material net	Price net	Price with 15%
First fixings				£	£	£	£
					VAT not included		
Roofs (*continued*)							
19 mm impregnated softwood tongued and grooved boarding							
flat	m2	0.60	-	5.38	5.94	11.32	13.02
sloping	m2	0.90	-	8.06	5.94	14.00	16.10
raking cutting	m	0.07	-	0.63	0.09	0.72	0.83
25 mm impregnated softwood tongued and grooved boarding							
flat	m2	0.66	-	5.91	6.66	12.57	14.46
sloping	m2	0.99	-	8.87	6.66	15.53	17.86
raking cutting	m	0.08	-	0.72	0.10	0.82	0.94
12 mm WBP grade plywood boarding							
flat	m2	0.70	-	6.27	8.50	14.77	16.99
sloping	m2	0.87	-	7.80	8.50	16.30	18.75
raking cutting	m	0.08	-	0.72	0.84	1.56	1.79
18 mm chipboard boarding tongued and grooved all edges							
flat	m2	0.25	-	2.24	4.48	6.72	7.73
sloping	m2	0.30	-	2.69	4.48	7.17	8.25
raking cutting	m	0.07	-	0.63	0.44	1.07	1.23
22 mm chipboard boarding tongued and grooved all edges							
flat	m2	0.30	-	2.69	5.46	8.15	9.37
sloping	m2	0.38	-	3.40	5.46	8.86	10.19
raking cutting	m	0.08	-	0.72	0.54	1.26	1.45
Labour chamfered, rebated or rounded edge to boarding	m	0.11	-	0.99	-	0.99	1.14
Impregnated sawn softwood gutter boarding etc							
25 mm gutter and lay boarding	m2	1.55	-	13.89	8.02	21.91	25.20
Splayed chimney gutter or valley boards							
25 x 150 mm	m	0.25	-	2.24	1.21	3.45	3.97
25 x 225 mm	m	0.38	-	3.40	1.86	5.26	6.05
25 mm boxed cesspools 225 x 225 x 100 mm	each	2.50	-	22.40	1.96	24.36	28.01

Woodwork

New work	Unit	Hours C	Hours L	Labour net	Material net	Price net	Price with 15%
First fixings				£	£	£	£
					VAT not included		
Eaves and verge boarding							
19 mm impregnated softwood matchboarded soffits							
over 300 mm wide	m2	1.40	-	12.54	6.80	19.34	22.24
225 mm wide	m	0.39	-	3.49	2.04	5.53	6.36
19 mm Western red cedar matchboarded soffits							
over 300 mm wide	m2	1.20	-	10.75	26.23	36.98	42.53
225 mm wide	m	0.28	-	2.51	5.91	8.42	9.68
6 mm sanded finish asbestos-free insulation board soffits							
over 300 mm wide	m2	0.80	-	7.17	10.75	17.92	20.61
225 mm wide	m	0.22	-	1.97	2.48	4.45	5.12
Impregnated wrought softwood eaves and verge boarding							
25 x 150 mm square							
fascias	m	0.25	-	2.24	1.58	3.82	4.39
returned ends	each	0.05	-	0.45	-	0.45	0.52
mitres	each	0.20	-	1.79	-	1.79	2.06
32 x 175 mm moulded							
fascias	m	0.30	-	2.69	2.34	5.03	5.78
returned ends	each	0.25	-	2.24	-	2.24	2.58
mitres	each	0.33	-	2.96	-	2.96	3.40
25 x 150 mm square							
barge boards	m	0.25	-	2.24	1.58	3.82	4.39
returned ends with sprocket piece	each	0.38	-	3.40	0.49	3.89	4.47
mitres	each	0.25	-	2.24	-	2.24	2.58
32 x 175 mm moulded							
barge boards	m	0.29	-	2.60	2.34	4.94	5.68
returned ends with sprocket piece	each	1.00	-	8.96	1.05	10.01	11.51
mitres	each	0.33	-	2.96	-	2.96	3.40
32 x 225 mm moulded							
barge boards	m	0.45	-	4.03	2.93	6.96	8.00
returned ends with sprocket piece	each	1.06	-	9.50	1.39	10.89	12.52
mitres	each	0.33	-	2.96	-	2.96	3.40
25 x 63 mm bed moulding including ends and mitres	m	0.13	-	1.16	3.94	5.10	5.87

Woodwork

New work	Unit	Hours C	Hours L	Labour net	Material net	Price net	Price with 15%
First fixings				£	£	£	£
					VAT not included		
Impregnated sawn softwood firrings, bearers etc							
38 mm firrings, average depth							
38 mm	m	0.06	-	0.54	0.55	1.09	1.25
50 mm	m	0.06	-	0.54	0.90	1.44	1.66
63 mm	m	0.07	-	0.63	1.05	1.68	1.85
50 mm firrings, average depth							
38 mm	m	0.06	-	0.54	0.82	1.36	1.56
50 mm	m	0.07	-	0.63	1.34	1.97	2.27
63 mm	m	0.08	-	0.72	1.57	2.29	2.63
Bearers							
25 x 50 mm	m	0.12	-	1.08	0.43	1.51	1.74
38 x 50 mm	m	0.12	-	1.08	0.63	1.71	1.97
50 x 50 mm	m	0.13	-	1.16	1.86	3.02	3.47
38 x 63 mm	m	0.13	-	1.16	0.69	1.85	2.13
50 x 63 mm	m	0.13	-	1.16	0.96	2.12	2.44
50 x 75 mm	m	0.15	-	1.34	1.07	2.41	2.77
Bearers not exceeding 300 mm long							
25 x 50 mm	each	0.05	-	0.45	0.13	0.58	0.67
38 x 50 mm	each	0.05	-	0.45	0.19	0.64	0.74
50 x 50 mm	each	0.06	-	0.54	0.56	1.10	1.26
38 x 63 mm	each	0.06	-	0.54	0.21	0.75	0.86
50 x 63 mm	each	0.06	-	0.54	0.29	0.83	0.95
50 x 75 mm	each	0.07	-	0.63	0.32	0.95	1.09
Angle fillets							
38 x 38 mm	m	0.13	-	1.16	0.26	1.42	1.63
50 x 50 mm	m	0.14	-	1.25	0.38	1.63	1.87
75 x 75 mm	m	0.15	-	1.34	0.79	2.13	2.45
Tilting fillets							
25 x 50 mm	m	0.07	-	0.63	0.61	1.24	1.43
38 x 75 mm	m	0.10	-	0.90	0.89	1.79	2.06
50 x 50 mm rolls for metal roofing	m	0.15	-	1.34	0.93	2.27	2.61
Impregnated sawn softwood grounds and battens							
Open spaced battening at 400 mm centres one way							
13 x 38 mm	m2	0.25	-	2.24	0.46	2.70	3.11
25 x 50 mm	m2	0.28	-	2.51	0.99	3.50	4.03
Open spaced battening at 400 mm centres both ways							
13 x 38 mm	m2	0.62	-	5.56	0.94	6.50	7.47
25 x 50 mm	m2	0.70	-	6.27	2.00	8.27	9.51

Woodwork

New work

First fixings

£ £ £ £

VAT not included

Impregnated sawn softwood grounds and battens (*continued*)

	Unit	Hours C	Hours L	Labour net	Material net	Price net	Price with 15%
Individual grounds and battens							
13 x 38 mm	m	0.10	-	0.90	0.19	1.09	1.25
13 x 75 mm	m	0.10	-	0.90	0.33	1.23	1.41
25 x 38 mm	m	0.10	-	0.90	0.35	1.25	1.44
25 x 50 mm	m	0.11	-	0.99	0.40	1.39	1.60
Impregnated floor fillets fixed to clips in concrete							
50 x 50 mm	m	0.07	-	0.63	0.65	1.28	1.47
50 x 75 mm	m	0.09	-	0.81	0.97	1.78	2.05
Fixing blocks wedged into steelwork							
50 x 50 x 150 mm long	each	0.10	-	0.90	0.10	1.00	1.15
50 x 75 x 300 mm long	each	0.11	-	0.99	0.29	1.28	1.47
Impregnated sawn softwood framework							
Cradling to steel beams with 50 x 75 mm wedged blocks and 25 x 50 mm members at 400 mm centres	m2	2.15	-	19.26	3.04	22.30	25.65
50 x 50 mm bracketing to false ceilings etc	m	0.20	-	1.79	0.68	2.47	2.84
38 x 50 mm bath panel framing							
to front only	set	1.24	-	11.11	3.48	14.59	16.78
to front and one end	set	1.94	-	17.38	4.82	22.20	25.53

Woodwork

New work	Unit	Hours C	Hours L	Labour net	Material net	Price net	Price with 15%
Second fixings				£	£	£	£
					VAT not included		
Softwood							
Stock pattern skirtings including ends and mitres							
19 x 75 mm	m	0.10	-	0.90	0.82	1.72	1.98
19 x 100 mm	m	0.11	-	0.99	1.24	2.23	2.56
25 x 150 mm							
stock pattern skirtings	m	0.12	-	1.08	2.35	3.43	3.94
returned ends	each	0.10	-	0.90	-	0.90	1.03
mitres	each	0.15	-	1.34	-	1.34	1.54
25 x 175 mm							
skirtings moulded to detail	m	0.14	-	1.25	2.72	3.97	4.57
returned ends	each	0.10	-	0.90	-	0.90	1.03
mitres	each	0.17	-	1.52	-	1.52	1.75
25 x 50 mm dado rails moulded to detail including ends and mitres	m	0.10	-	0.90	1.35	2.25	2.59
25 x 75 mm							
dado rails moulded to detail	m	0.11	-	0.99	1.59	2.58	2.97
returned ends	each	0.10	-	0.90	-	0.90	1.03
mitres	each	0.15	-	1.34	-	1.34	1.54
Stock pattern architraves including ends and mitres							
19 x 50 mm	m	0.13	-	1.16	0.77	1.93	2.22
25 x 75 mm	m	0.15	-	1.34	1.34	2.68	3.08
Architraves moulded to detail including ends and mitres							
19 x 50 mm	m	0.13	-	1.16	0.77	1.93	2.22
25 x 75 mm	m	0.15	-	1.34	1.36	2.70	3.11
Quadrant cover fillets including ends and mitres							
19 mm	m	0.07	-	0.63	0.33	0.96	1.10
25 mm	m	0.07	-	0.63	0.46	1.09	1.25
13 x 50 mm square edged cover fillets including ends and mitres	m	0.07	-	0.63	0.53	1.16	1.33
13 x 50 mm twice moulded cover fillets including ends and mitres	m	0.07	-	0.63	1.25	1.88	2.16
19 x 75 mm chamfered cover fillets including ends and mitres	m	0.07	-	0.63	1.95	2.58	2.97
Stops including ends and mitres							
13 x 50 mm	m	0.13	-	1.16	0.57	1.73	1.99
16 x 50 mm	m	0.13	-	1.16	0.62	1.78	2.05
25 x 50 mm	m	0.13	-	1.16	1.45	2.61	3.00

Woodwork

New work

Second fixings

				£	£	£	£
					VAT not included		

Softwood (*continued*)

Glazing beads including mitres

	Unit	Hours C	Hours L	Labour net	Material net	Price net	Price with 15%
13 x 19 mm	m	0.10	-	0.90	0.21	1.11	1.28
13 x 25 mm	m	0.10	-	0.90	0.22	1.12	1.29
13 x 32 mm	m	0.11	-	0.99	0.25	1.24	1.43
19 x 32 mm	m	0.12	-	1.08	0.32	1.40	1.61
19 x 38 mm	m	0.12	-	1.08	0.38	1.46	1.68

Glazing beads fixed with brass screws and cups including mitres

13 x 19 mm	m	0.25	-	2.24	0.41	2.65	3.05
13 x 25 mm	m	0.25	-	2.24	0.42	2.66	3.06
13 x 32 mm	m	0.26	-	2.33	0.45	2.78	3.20

Slat shelves of 25 x 50 mm slats 25 mm apart	m2	1.35	-	12.10	8.80	20.90	24.04

19 mm shelves

150 mm wide	m	0.15	-	1.34	1.19	2.53	2.91
200 mm wide	m	0.16	-	1.43	2.80	4.23	4.86

19 mm crosstongued shelves

over 300 mm wide	m2	2.25	-	20.16	14.78	34.94	40.18
250 mm wide	m	0.17	-	1.52	2.65	4.17	4.80

25 mm shelves

150 mm wide	m	0.18	-	1.61	1.31	2.92	3.36
200m wide	m	0.19	-	1.70	3.09	4.79	5.51

25 mm crosstongued shelves

over 300 mm wide	m2	2.35	-	21.06	16.23	37.29	42.88
250 mm wide	m	0.20	-	1.79	3.95	5.74	6.60

Crosstongued worktops button blocked to framing

25 mm	m2	3.85	-	34.50	16.77	51.27	58.96
32 mm	m2	4.00	-	35.84	28.19	64.03	73.63
38 mm	m2	4.20	-	37.63	21.91	59.54	68.47

Bearers

19 x 38 mm	m	0.15	-	1.34	0.47	1.81	2.08
25 x 50 mm	m	0.17	-	1.52	0.59	2.11	2.43
38 x 50 mm	m	0.17	-	1.52	0.90	2.42	2.78
50 x 50 mm	m	0.18	-	1.61	0.94	2.55	2.93
50 x 75 mm	m	0.20	-	1.79	1.34	3.13	3.60

Bearers not exceeding 300 mm long

19 x 38 mm	each	0.07	-	0.63	0.14	0.77	0.89
25 x 50 mm	each	0.08	-	0.72	0.19	0.91	1.05
38 x 50 mm	each	0.08	-	0.72	0.28	1.00	1.15
50 x 50 mm	each	0.08	-	0.72	0.32	1.04	1.20
50 x 75 mm	each	0.09	-	0.81	0.44	1.25	1.44

Legs and bearers framed on site

50 x 50 mm	m	0.67	-	6.00	0.94	6.94	7.98
50 x 75 mm	m	0.83	-	7.44	1.34	8.78	10.10

Woodwork

New work	Unit	Hours C	Hours L	Labour net	Material net	Price net	Price with 15%
Second fixings				£	£	£	£
					VAT not included		
Softwood (*continued*)							
32 x 50 mm window nosings tongued on including ends	m	0.28	-	2.51	1.21	3.72	4.28
25 x 150 mm							
rounded window boards tongued on	m	0.30	-	2.69	2.55	5.24	6.03
returned ends	each	0.12	-	1.08	-	1.08	1.24
25 x 200 mm							
rounded window boards tongued on	m	0.33	-	2.96	3.20	6.16	7.08
returned ends	each	0.17	-	1.52	-	1.52	1.75
50 x 75 mm							
moulded handrails screwed on	m	0.20	-	1.79	2.08	3.87	4.45
returned ends	each	0.60	-	5.38	-	5.38	6.19
Hardwood							
Skirtings moulded to detail including ends and mitres							
19 x 75 mm	m	0.14	-	1.25	4.44	5.69	6.54
19 x 100 mm	m	0.15	-	1.34	4.86	6.20	7.13
25 x 150 mm							
skirtings moulded to detail	m	0.16	-	1.43	9.05	10.48	12.05
returned ends	each	0.25	-	2.24	-	2.24	2.58
mitres	each	0.25	-	2.24	-	2.24	2.58
25 x 200 mm							
skirtings moulded to detail	m	0.19	-	1.70	12.08	13.78	15.85
returned ends	each	0.30	-	2.69	-	2.69	3.09
mitres	each	0.40	-	3.58	-	3.58	4.12
25 x 63 mm dado rails moulded to detail including ends and mitres	m	0.14	-	1.25	3.80	5.05	5.81
25 x 75 mm							
dado rails moulded to detail	m	0.15	-	1.34	4.54	5.88	6.76
returned ends	each	0.30	-	2.69	-	2.69	3.09
mitres	each	0.34	-	3.05	-	3.05	3.51
25 x 75 mm architraves moulded to detail including ends and mitres	m	0.18	-	1.61	4.54	6.15	7.07
32 x100 mm							
architraves moulded to detail	m	0.20	-	1.79	5.69	7.48	8.60
returned ends	each	0.30	-	2.69	-	2.69	3.09
mitres	each	0.34	-	3.05	-	3.05	3.51
Plinth blocks							
32 x 88 mm x 175 mm high	each	0.25	-	2.24	1.91	4.15	4.77
38 x 113 mm x 225 mm high	each	0.35	-	3.14	4.44	7.58	8.72

Woodwork

New work	Unit	Hours C	Hours L	Labour net	Material net	Price net	Price with 15%
Second fixings				£	£	£	£
					VAT not included		
Hardwood (*continued*)							
Quadrant cover fillets including ends and mitres							
19 mm	m	0.10	-	0.90	0.59	1.49	1.71
25 mm	m	0.11	-	0.99	1.34	2.33	2.68
13 x 50 mm square edged cover fillets including ends and mitres	m	0.10	-	0.90	0.82	1.72	1.98
13 x 50 twice moulded cover fillets including ends and mitres	m	0.10	-	0.90	1.68	2.58	2.97
19 x 75 mm chamfered cover fillets including ends and mitres	m	0.10	-	0.90	2.77	3.67	4.22
25 x 32 mm scotia moulds including ends	m	0.18	-	1.61	2.39	4.00	4.60
Stops including ends and mitres							
13 x 50 mm	m	0.17	-	1.52	0.95	2.47	2.84
16 x 50 mm	m	0.18	-	1.61	1.02	2.63	3.02
25 x 50 mm	m	0.20	-	1.79	2.13	3.92	4.51
Glazing beads including mitres							
13 x 19 mm	m	0.13	-	1.16	0.83	1.99	2.29
13 x 25 mm	m	0.13	-	1.16	1.04	2.20	2.53
13 x 32 mm	m	0.14	-	1.25	1.27	2.52	2.90
19 x 32 mm	m	0.16	-	1.43	1.49	2.92	3.36
19 x 38 mm	m	0.16	-	1.43	1.56	2.99	3.44
Glazing beads fixed with brass screws and cups including mitres							
13 x 19 mm	m	0.32	-	2.87	1.03	3.90	4.49
13 x 25 mm	m	0.32	-	2.87	1.24	4.11	4.73
13 x 32 mm	m	0.34	-	3.05	1.47	4.52	5.20
19 mm shelves							
150 mm wide	m	0.21	-	1.88	4.95	6.83	7.85
200 mm wide	m	0.22	-	1.97	6.60	8.57	9.86
19 mm crosstongued shelves							
over 300 mm wide	m2	2.95	-	26.43	34.35	60.78	69.90
250 mm wide	m	0.23	-	2.06	8.52	10.58	12.17
25 mm shelves							
150 mm wide	m	0.25	-	2.24	6.34	8.58	9.87
200m wide	m	0.26	-	2.33	8.44	10.77	12.39
25 mm crosstongued shelves							
over 300 mm wide	m2	3.05	-	27.33	43.56	70.89	81.52
250 mm wide	m	0.27	-	2.42	10.82	13.24	15.23

Woodwork

New work	Unit	Hours C	Hours L	Labour net	Material net	Price net	Price with 15%
Second fixings				£	£	£	£
					VAT not included		
Hardwood (continued)							
Crosstongued worktops button blocked to framing							
25 mm	m2	6.00	-	53.76	45.26	99.02	113.87
32 mm	m2	6.00	-	53.76	61.25	115.01	132.26
38 mm	m2	6.00	-	53.76	73.05	126.81	145.83
Bearers							
19 x 38 mm	m	0.20	-	1.79	1.27	3.06	3.52
25 x 50 mm	m	0.23	-	2.06	2.14	4.20	4.83
Bearers not exceeding 300 mm long							
19 x 38 mm	each	0.09	-	0.81	0.39	1.20	1.38
25 x 50 mm	each	0.10	-	0.90	0.65	1.55	1.78
25 x 150 mm							
rounded window boards tongued on	m	0.39	-	3.49	3.75	7.24	8.33
returned ends	each	0.25	-	2.24	-	2.24	2.58
25 x 200 mm							
rounded window boards tongued on	m	0.49	-	4.39	4.90	9.29	10.68
returned ends	each	0.30	-	2.69	-	2.69	3.09
50 x 75 mm							
moulded handrails screwed on	m	0.29	-	2.60	5.54	8.14	9.36
returned ends	each	0.90	-	8.06	-	8.06	9.27
Sundry worktops							
25 mm blockboard worktops screwed on	m2	0.90	-	8.06	21.46	29.52	33.95
25 mm chipboard worktops screwed on	m2	0.70	-	6.27	8.13	14.40	16.56
Extra for lippings to 25 mm worktop							
softwood	m	0.20	-	1.79	0.18	1.97	2.27
hardwood	m	0.24	-	2.15	0.20	2.35	2.70
Laminated plastic sheet fixed to worktop with adhesive							
balancers	m2	0.90	-	8.06	20.15	28.21	32.44
coverings	m2	0.90	-	8.06	20.15	28.21	32.44
edgings 25 mm	m	0.20	-	1.79	0.49	2.28	2.62
edgings 32 mm	m	0.20	-	1.79	0.65	2.44	2.81
edgings 38 mm	m	0.20	-	1.79	0.77	2.56	2.94

Woodwork

New work	Unit	Hours C	Hours L	Labour net	Material net	Price net	Price with 15%
Second fixings				£	£	£	£
					VAT not included		

Sheet linings and casings

7 mm pre-finished veneered plywood, grooved to resemble plank panelling, pinned to studding or battens to walls							
over 300 mm wide	m2	0.75	-	6.72	5.66	12.38	14.24
not exceeding 100 mm wide	m	0.19	-	1.70	0.57	2.27	2.61
100 - 200 mm wide	m	0.25	-	2.24	1.14	3.38	3.89
200 - 300 mm wide	m	0.30	-	2.69	1.71	4.40	5.06
3.2 mm hardboard pinned to studding or battens to walls							
over 300 mm wide	m2	0.27	-	2.42	1.70	4.12	4.74
not exceeding 100 mm wide	m	0.07	-	0.63	0.18	0.81	0.93
100 - 200 mm wide	m	0.09	-	0.81	0.35	1.16	1.33
200 - 300 mm wide	m	0.11	-	0.99	0.52	1.51	1.74
3.2 mm hardboard bath panel (finished white), pinned to studding or battens to walls							
over 300 mm wide	m2	0.50	-	4.48	2.18	6.66	7.66
not exceeding 100 mm wide	m	0.13	-	1.16	0.22	1.38	1.59
100 - 200 mm wide	m	0.17	-	1.52	0.45	1.97	2.27
200 - 300 mm wide	m	0.20	-	1.79	0.66	2.45	2.82
3.2 mm hardboard bath panels with melamine surface fixed with chromium plated screws to framing							
to front only	each	0.30	-	2.69	9.60	12.29	14.13
to front and one end including stainless steel angle strip	set	0.45	-	4.03	16.51	20.54	23.62
Moulded acrylic bath panels fixed with chromium plated screws to framing							
to front only	each	0.30	-	2.69	24.30	26.99	31.04
to front and one end	set	0.45	-	4.03	38.05	42.08	48.39
18 mm melamine faced chipboard pinned to studding or battens to walls							
over 300 mm wide	m2	0.22	-	1.97	7.21	9.18	10.56
not exceeding 100 mm wide	m	0.15	-	1.34	0.73	2.07	2.38
100 - 200 mm wide	m	0.19	-	1.70	1.46	3.16	3.63
200 - 300 mm wide	m	0.23	-	2.06	2.17	4.23	4.86
13 mm sanded finish fibre insulation board pinned to studding or battens to walls							
over 300 mm wide	m2	0.27	-	2.42	11.01	13.43	15.44
not exceeding 100 mm wide	m	0.07	-	0.63	1.11	1.74	2.00
100 - 200 mm wide	m	0.09	-	0.81	2.20	3.01	3.46
200 - 300 mm wide	m	0.11	-	0.99	3.30	4.29	4.93

Woodwork

New work	Unit	Hours C	Hours L	Labour net	Material net	Price net	Price with 15%
Second fixings				£	£	£	£
					VAT not included		

Sheet linings and casing (*continued*)

6 mm sanded finish asbestos free insulation board pinned to studding or battens to walls

over 300 mm wide	m2	0.46	-	4.12	10.70	14.82	17.04
not exceeding 100 mm wide	m	0.12	-	1.08	1.08	2.16	2.48
100 - 200 mm wide	m	0.15	-	1.34	2.14	3.48	4.00
200 - 300 mm wide	m	0.18	-	1.61	3.21	4.82	5.54

9 mm sanded finish asbestos free insulation board pinned to studding or battens to walls

over 300 mm wide	m2	0.50	-	4.48	20.94	25.42	29.23
not exceeding 100 mm wide	m	0.13	-	1.16	2.10	3.26	3.75
100 - 200 mm wide	m	0.15	-	1.34	4.19	5.53	6.36
200 - 300 mm wide	m	0.20	-	1.79	6.28	8.07	9.28

25 mm sterlingboard nailed to studding or battens to walls

over 300 mm wide	m2	0.42	-	3.76	10.90	14.66	16.86
not exceeding 100 mm wide	m	0.11	-	0.99	1.09	2.08	2.39
100 - 200 mm wide	m	0.14	-	1.25	2.18	3.43	3.94
200 - 300 mm wide	m	0.17	-	1.52	3.27	4.79	5.51

Extra for fixing sheet linings with adhesive

to battens	m2	0.10	-	0.90	0.25	1.15	1.32
to flat wall surfaces	m2	0.18	-	1.61	1.23	2.84	3.27

50 mm paper faced cotton tape pasted over joints	m	0.05	-	0.45	0.05	0.50	0.57
19 mm plastic cover strips fixed with adhesive	m	0.04	-	0.36	0.54	0.90	1.03
19 x 19 mm white plastic angle cover strips fixed with adhesive	m	0.07	-	0.63	0.74	1.37	1.58

Black or white plastic joint holder for 3 mm sheets

to intermediate joints	m	0.33	-	2.96	0.84	3.80	4.37
to internal angles	m	0.33	-	2.96	0.84	3.80	4.37
to external angles	m	0.33	-	2.96	0.84	3.80	4.37

Angle pipe casings 225 mm girth with 25 mm softwood side, 19 x 25 mm backings and front fixed with brass screws and cups

with 19 mm softwood beaded front	m	1.10	-	9.86	7.10	16.96	19.50
with 6 mm plywood front	m	1.10	-	9.86	6.74	16.60	19.09
with 3.2 mm hardboard front	m	1.10	-	9.86	6.14	16.00	18.40

Boxed pipe casings 300 mm girth with 25 mm softwood sides, 19 x 25 mm backings and front fixed with brass screws and cups

with 19 mm softwood beaded front	m	1.17	-	10.48	8.73	19.21	22.09
with 6 mm plywood front	m	1.17	-	10.48	7.23	17.71	20.37
with 3.2 mm hardboard front	m	1.17	-	10.48	6.26	16.74	19.25

Woodwork

New work	Unit	Hours C	Hours L	Labour net	Material net	Price net	Price with 15%
Purpose made composite items				£	£	£	£
					VAT not included		
Hardwood panelling							
19 mm panelling with 63 mm square stiles and rails and 6 mm veneered plywood panels	m2	0.81	-	7.26	13.38	20.64	23.74
25 mm panelling with 63 mm moulded stiles and rails and 6 mm veneered plywood panels	m2	0.81	-	7.26	14.79	22.05	25.36
32 x 63 mm							
moulded cappings	m	0.25	-	2.24	4.76	7.00	8.05
mitres	each	0.25	-	2.24	-	2.24	2.58
32 x 75 mm							
moulded cappings	m	0.25	-	2.24	5.53	7.77	8.94
mitres	each	0.33	-	2.96	-	2.96	3.40
Softwood doors							
762 x 1981 mm ledged and braced doors of 25 mm matchboarding and 25 mm ledges and braces	each	0.85	-	7.62	69.03	76.65	88.15
Framed, ledged and braced doors of 44 mm thick, 25 mm matchboarding and 25 mm ledges and braces							
762 x 1981 mm	each	1.10	-	9.86	84.92	94.78	109.00
826 x 2040 mm	each	1.20	-	10.75	86.65	97.40	112.01
762 x 1981 x 44 mm casement doors open for glass and divided into eight panes	each	1.10	-	9.86	110.14	120.00	138.00
838 x 2057 x 50 mm one panel doors moulded one side, bolection moulded the other side and with 12 mm plywood panel	each	1.25	-	11.20	167.92	179.12	205.99
762 x 1981 x 50 mm three panel doors, the lower panels bead butt and square and the upper panel with diminished stiles, open for glass and divided into six panes	each	1.15	-	10.30	127.06	137.36	157.96
Four panel doors square both sides with 6 mm plywood panels							
762 x 1981 x 38 mm	each	1.10	-	9.86	226.54	236.40	271.86
762 x 1981 x 50 mm	each	1.33	-	11.92	226.54	238.46	274.23
Four panel doors moulded on solid both sides and with 6 mm plywood panels							
762 x 1981 x 38 mm	each	1.10	-	9.86	113.66	123.52	142.05
762 x 1981 x 50 mm	each	1.33	-	11.92	119.03	130.95	150.59
762 x 1981 x 50 mm four panel solid doors bead butt both sides	each	1.40	-	12.54	208.62	221.16	254.33

Woodwork

New work	Unit	Hours C	Hours L	Labour net	Material net	Price net	Price with 15%
Purpose made composite items				£	£	£	£
					VAT not included		

Softwood doors (*continued*)

	Unit	Hours C	Hours L	Labour net	Material net	Price net	Price with 15%
762 x 1981 x 50 mm three panel doors, the lower panels solid bead butt both sides and the upper panel open for glass and divided into four panes	each	1.40	-	12.54	143.18	155.72	179.08
Extra for							
rebated meeting stiles (both sides measured)	m	-	-	-	0.99	0.99	1.14
rebated and beaded meeting stiles (both sides measured)	m	-	-	-	2.28	2.28	2.62
rounded heels or stiles	m	-	-	-	1.99	1.99	2.29
32 x 100 mm twice splayed weatherboards not exceeding 900 mm long screwed on including notching frames	each	0.30	-	2.69	1.48	4.17	4.80
50 x 75 mm moulded and throated weatherboards not exceeding 900 mm long screwed on including notching frames	each	0.30	-	2.69	1.52	4.21	4.84
Hardwood doors							
50 mm casement doors open for glass in one pane							
762 x 1981 mm	each	2.60	-	23.30	123.16	146.48	168.43
838 x 2057 mm	each	2.75	-	24.64	123.16	147.80	169.97
762 x 1981 x 50 mm casement doors moulded one side, open for glass and divided into eight panes	each	2.50	-	22.40	181.83	204.23	234.86
838 x 2057 x 50 mm one panel doors moulded one side, bolection moulded the other side and with 12 mm plywood panel	each	3.00	-	26.88	318.47	345.35	397.15
762 x 1981 x 50 mm four panel doors moulded both sides with 6 mm veneered plywood panels	each	2.50	-	22.40	254.02	276.42	317.88
762 x 1981 x 50 mm four panel solid doors bead butt both sides	each	2.50	-	22.40	345.47	367.87	423.05
762 x 1981 x 50 mm three panel doors, the lower panels solid bead butt both sides and the upper panel open for glass and divided into four panes	each	2.50	-	22.40	303.49	325.89	374.77
Extra for							
rebated meeting stiles (both sides measured)	m	-	-	-	0.50	0.50	0.58
rebated and beaded meeting stiles (both sides measured)	m	-	-	-	5.94	5.94	6.83
rounded heels or stiles	m	-	-	-	1.99	1.99	2.29
50 x 75 mm moulded and grooved weatherboards not exceeding 900 mm long screwed on and pelleted including notching frames	each	0.30	-	2.69	8.75	11.44	13.16

Woodwork

New work	Unit	Hours C	Hours L	Labour net	Material net	Price net	Price with 15%
Purpose made composite items				£	£	£	£
					VAT not included		
Flush doors							
762 x 1981 x 38 mm flush doors of 32 mm softwood skeleton framed core covered on both sides with 3.2 mm hardboard							
unlipped	each	1.00	-	8.96	21.89	30.85	35.48
lipped on long edges	each	1.00	-	8.96	21.89	30.85	35.48
762 x 1981 x 40 mm flush doors of 32 mm softwood skeleton framed core covered on both sides with 4 mm plywood							
unlipped	each	1.00	-	8.96	23.83	32.79	37.71
lipped on long edges	each	1.00	-	8.96	23.83	32.79	37.71
762 x 1981 x 40 mm flush doors of 32 mm softwood skeleton framed core covered on both sides with 4 mm hardwood veneered plywood and hardwood lipped on all edges	each	1.00	-	8.96	23.83	32.79	37.71
Softwood door frames and linings							
Frames							
50 x 75 mm	m	0.22	-	1.97	1.29	3.26	3.75
50 x 100 mm	m	0.22	-	1.97	1.66	3.63	4.17
50 x 150 mm	m	0.22	-	1.97	3.33	5.30	6.09
Rebated frames							
50 x 75 mm	m	0.24	-	2.15	1.56	3.71	4.27
50 x 100 mm	m	0.24	-	2.15	1.93	4.08	4.69
50 x 125 mm	m	0.24	-	2.15	2.98	5.13	5.90
50 x 150 mm	m	0.25	-	2.24	3.60	5.84	6.72
75 x 75 mm	m	0.24	-	2.15	2.72	4.87	5.60
75 x 100 mm	m	0.24	-	2.15	3.48	5.63	6.47
75 x 150 mm	m	0.25	-	2.24	5.19	7.43	8.54
Rebated and moulded frames							
50 x 100 mm	m	0.24	-	2.15	2.20	4.35	5.00
50 x 150 mm	m	0.25	-	2.24	3.87	6.11	7.03
75 x 100 mm	m	0.24	-	2.15	3.75	5.90	6.79
75 x 150 mm	m	0.25	-	2.24	5.46	7.70	8.86
Linings tongued at angles							
25 x 75 mm	m	0.20	-	1.79	1.05	2.84	3.27
25 x 100 mm	m	0.22	-	1.97	1.45	3.42	3.93
25 x 125 mm	m	0.22	-	1.97	1.84	3.81	4.38
25 x 150 mm	m	0.23	-	2.06	1.32	3.38	3.89
32 x 100 mm	m	0.22	-	1.97	1.51	3.48	4.00
32 x 125 mm	m	0.22	-	1.97	2.27	4.24	4.88
32 x 150 mm	m	0.23	-	2.06	2.27	4.33	4.98

Woodwork

New work	Unit	Hours C	Hours L	Labour net	Material net	Price net	Price with 15%
Purpose made composite items				£	£	£	£
					VAT not included		
Softwood door frames and linings (*continued*)							
Rebated linings tongued at angles							
38 x 100 mm	m	0.22	-	1.97	1.78	3.75	4.31
38 x 125 mm	m	0.22	-	1.97	2.54	4.51	5.19
38 x 150 mm	m	0.24	-	2.15	2.54	4.69	5.39
Extra for hollow groove in frames for heel of swing door	m	-	-	-	0.27	0.27	0.31
Hardwood door frames and linings							
Rebated and moulded frames							
63 x 75 mm	m	0.36	-	3.23	12.01	15.24	17.53
75 x 100 mm	m	0.41	-	3.67	15.47	19.14	22.01
75 x 100 mm twice moulded frames	m	0.42	-	3.76	15.85	19.61	22.55
Rebated linings tongued at angles							
38 x 125 mm	m	0.39	-	3.49	10.98	14.47	16.64
38 x 150 mm	m	0.39	-	3.49	13.17	16.66	19.16
Extra for hollow groove in frames for heel of swing door	m	-	-	-	0.38	0.38	0.44
Sunk weathered and grooved thresholds							
75 x 100 mm	m	0.28	-	2.51	15.47	17.98	20.68
75 x 175 mm	m	0.28	-	2.51	22.84	25.35	29.15
Softwood casements							
Moulded casements in one pane 0.50 - 1.00m2							
38 mm	m2	-	-	-	38.95	38.95	44.79
50 mm	m2	-	-	-	41.00	41.00	47.15
Moulded casements divided into panes 0.10 - 0.50m2							
38 mm	m2	-	-	-	59.45	59.45	68.37
50 mm	m2	-	-	-	61.50	61.50	70.72
Moulded casements divided into panes not exceeding 0.10m2							
38 mm	m2	-	-	-	88.15	88.15	101.37
50 mm	m2	-	-	-	92.25	92.25	106.09
Fitting and fixing casements	each	0.33	-	2.96	0.07	3.03	3.48
Fitting and hanging casements on 63 mm light steel butts	each	0.67	-	6.00	0.63	6.63	7.62
Fitting and hanging casements on stove enamelled pivots	each	1.15	-	10.30	7.91	18.21	20.94

Woodwork

New work	Unit	Hours C	Hours L	Labour net	Material net	Price net	Price with 15%
Purpose made composite items				£	£	£	£
					VAT not included		
Hardwood casements							
Moulded casements in one pane 0.50 - 1.00m2							
38 mm	m2	-	-	-	75.51	75.51	86.84
50 mm	m2	-	-	-	81.31	81.31	93.51
Moulded casements divided into panes 0.10 - 0.50m2							
38 mm	m2	-	-	-	110.36	110.36	126.91
50 mm	m2	-	-	-	116.16	116.16	133.58
Moulded casements divided into panes not exceeding 0.10m2							
38 mm	m2	-	-	-	162.63	162.63	187.02
50 mm	m2	-	-	-	174.25	174.25	200.39
Fitting and fixing casements	each	0.50	-	4.48	0.07	4.55	5.23
Fitting and hanging casements on 63 mm brass butts	each	1.13	-	10.12	5.10	15.22	17.50
Fitting and hanging casements on brass pivots	each	1.75	-	15.68	25.09	40.77	46.89
Softwood frames							
Rebated casement frames							
50 x 63 mm	m	0.22	-	1.97	1.76	3.73	4.29
63 x 75 mm	m	0.22	-	1.97	2.32	4.29	4.93
75 x 100 mm	m	0.22	-	1.97	3.48	5.45	6.27
Rebated and moulded casement frames							
50 x 75 mm	m	0.22	-	1.97	1.83	3.80	4.37
63 x 75 mm	m	0.22	-	1.97	2.59	4.56	5.24
75 x 100 mm	m	0.22	-	1.97	3.75	5.72	6.58
Twice rebated mullions or transoms							
50 x 63 mm	m	0.22	-	1.97	2.03	4.00	4.60
63 x 75 mm	m	0.22	-	1.97	2.59	4.56	5.24
75 x 100 mm	m	0.22	-	1.97	3.75	5.72	6.58
Twice rebated and twice moulded mullions or transoms							
50 x 63 mm	m	0.22	-	1.97	2.57	4.54	5.22
63 x 75 mm	m	0.22	-	1.97	3.13	5.10	5.87
75 x 100 mm	m	0.22	-	1.97	4.29	6.26	7.20
13 x 25 mm cut and mitred beads to pivot hung casement and frame	m	0.25	-	2.24	0.22	2.46	2.83

Woodwork

New work	Unit	Hours C	Hours L	Labour net	Material net	Price net	Price with 15%
Purpose made composite items				£	£	£	£
					VAT not included		
Hardwood frames							
Rebated and grooved cills							
63 x 63 mm	m	0.28	-	2.51	15.57	18.08	20.79
63 x 75 mm	m	0.28	-	2.51	16.68	19.19	22.07
Rebated, sunk weathered, throated, check throated and grooved cills							
63 x 150 mm	m	0.28	-	2.51	21.16	23.67	27.22
75 x 150 mm	m	0.28	-	2.51	23.97	26.48	30.45
75 x 175 mm	m	0.28	-	2.51	27.60	30.11	34.63
Rebated, moulded, sunk weathered, throated, check throated and grooved cills							
63 x 150 mm	m	0.28	-	2.51	21.54	24.05	27.66
75 x 150 mm	m	0.28	-	2.51	24.35	26.86	30.89
75 x 175 mm	m	0.28	-	2.51	27.98	30.49	35.06
Mitres to cill including handrail bolt	each	0.35	-	3.14	1.68	4.82	5.54
Softwood window surrounds							
19 x 125 mm window linings tongued at angles and tongued on	m	0.35	-	3.14	1.19	4.33	4.98
Pelmets of 19 x 100 mm top, 13 x 150 mm front							
straight	m	0.60	-	5.38	2.36	7.74	8.90
returned ends	each	0.20	-	1.79	0.12	1.91	2.20
Hardwood window surrounds							
19 x 125 mm window linings tongued at angles and tongued on	m	0.47	-	4.21	5.49	9.70	11.15
38 x 75 mm rebated and moulded shop window framing	m	0.70	-	6.27	7.32	13.59	15.63
50 x 75 mm twice rebated and twice moulded mullion or transom in shop window framing	m	0.80	-	7.17	12.76	19.93	22.92
50 x 100 mm rebated, weathered and moulded cill in shop window framing	m	0.90	-	8.06	16.14	24.20	27.83
Softwood screens and borrowed lights							
Square framing							
25 x 100 mm	m	0.14	-	1.25	1.44	2.69	3.09
50 x 50 mm	m	0.14	-	1.25	0.89	2.14	2.46
50 x 75 mm	m	0.14	-	1.25	1.29	2.54	2.92
50 x 100 mm	m	0.15	-	1.34	1.66	3.00	3.45

Woodwork

New work	Unit	Hours C	Hours L	Labour net	Material net	Price net	Price with 15%
Purpose made composite items				£	£	£	£
					VAT not included		
Softwood screens and borrowed lights (*continued*)							
Rebated and moulded framing							
50 x 75 mm	m	0.14	-	1.25	1.83	3.08	3.54
50 x 100 mm	m	0.15	-	1.34	2.20	3.54	4.07
75 x 100 mm	m	0.16	-	1.43	3.75	5.18	5.96
Twice rebated and twice moulded framing							
50 x 75 mm	m	0.14	-	1.25	2.37	3.62	4.16
50 x 100 mm	m	0.15	-	1.34	2.74	4.08	4.69
75 x 100 mm	m	0.16	-	1.43	4.29	5.72	6.58
Square glazing bars							
25 x 50 mm	m	0.10	-	0.90	0.57	1.47	1.69
25 x 75 mm	m	0.10	-	0.90	1.05	1.95	2.24
Twice rebated and twice moulded glazing bars							
32 x 75 mm	m	0.10	-	0.90	2.32	3.22	3.70
32 x 100 mm	m	0.10	-	0.90	2.58	3.48	4.00
38 x 75 mm	m	0.10	-	0.90	2.35	3.25	3.74
38 x 100 mm	m	0.10	-	0.90	2.74	3.64	4.19
16 x 50 mm stock pattern moulded stop	m	0.13	-	1.16	0.63	1.79	2.06
Twice rebated feature rail							
50 x 100 mm	m	0.15	-	1.34	2.20	3.54	4.07
50 x 150 mm	m	0.16	-	1.43	3.87	5.30	6.09
Softwood closed tread staircases							
Straight flight staircases, 13 up, total going 2700 mm							
2600 mm rise, 855 mm wide	each	9.00	9.00	141.30	336.02	477.32	548.92
2600 mm rise, 864 mm wide	each	9.00	9.00	141.30	341.02	482.32	554.67
2600 mm rise, 914 mm wide	each	9.00	9.00	141.30	343.02	484.32	556.97
Straight half flight staircases, 7 up, total going 1350 mm							
1421 mm rise, 864 mm wide	each	4.00	4.00	62.80	220.68	283.48	326.00
Landings of 25 mm tongued and grooved flooring with 32 x 100 mm rounded nosing, including framed bearers	m2	1.58	0.60	18.20	11.11	29.31	33.71
Solid balustrade fillings of 25 x 50 mm uprights and 19 x 38 mm fillets covered on both sides with 3 mm hardboard							
square	m2	1.85	-	16.58	6.45	23.03	26.48
raking	m2	2.44	-	21.86	6.45	28.31	32.56
19 x 200 mm beaded apron linings	m	0.30	-	2.69	1.82	4.51	5.19
25 x 50 mm chamfered floor fillets	m	0.10	-	0.90	0.84	1.74	2.00

Woodwork

New work	Unit	Hours C	Hours L	Labour net	Material net	Price net	Price with 15%

Purpose made composite items

				£	£	£	£
				VAT not included			

Softwood closed tread staircases (*continued*)

	Unit	Hours C	Hours L	Labour net	Material net	Price net	Price with 15%
32 x 32 mm							
balusters	m	0.05	-	0.45	3.30	3.75	4.31
fitted ends	each	0.02	-	0.18	-	0.18	0.21
fitted ends on rake	each	0.03	-	0.27	-	0.27	0.31
32 x 50 mm moulded cappings	m	0.17	-	1.52	1.39	2.91	3.35
32 x 100 mm rebated nosings to flooring including scotia moulding under	m	0.30	-	2.69	3.05	5.74	6.60
32 x 225 mm							
wall strings	m	0.60	-	5.38	4.20	9.58	11.02
fitted ends	each	0.30	-	2.69	-	2.69	3.09
extra for ramps	each	0.35	-	3.14	-	3.14	3.61
38 x 225 mm							
outer strings	m	0.60	-	5.38	4.83	10.21	11.74
ends framed to newel	each	0.40	-	3.58	-	3.58	4.12
50 x 100 mm half newels	m	0.63	-	5.64	24.69	30.33	34.88
100 x 100 mm							
newels	m	0.90	-	8.06	11.29	19.35	22.25
shaped or rounded ends	each	0.75	-	6.72	-	6.72	7.73
63 x 150 x 150 mm moulded newel caps	each	0.50	-	4.48	12.27	16.75	19.26
63 x 75 mm							
moulded handrails	each	0.20	-	1.79	2.92	4.71	5.42
ends framed to newel	each	0.33	-	2.96	-	2.96	3.40
ends framed on rake to newel	each	0.50	-	4.48	-	4.48	5.15

Softwood open tread staircases

	Unit	Hours C	Hours L	Labour net	Material net	Price net	Price with 15%
Straight flight staircases, 13 up, total going 2700 mm							
2600 mm rise, 864 mm wide	each	9.00	9.00	141.30	316.02	457.32	525.92
2600 mm rise, 914 mm wide	each	9.00	9.00	141.30	318.02	459.32	528.22
Strings							
50 x 225 mm	m	0.25	-	2.24	5.46	7.70	8.86
fitted ends	each	0.30	-	2.69	-	2.69	3.09
50 x 75 mm newels bolted to string	m	0.55	-	4.93	1.29	6.22	7.15
M10 x 110 mm steel carriage bolts and washer including boring softwood	each	0.25	-	2.24	0.47	2.71	3.12

Woodwork

New work	Unit	Hours C	Hours L	Labour net	Material net	Price net	Price with 15%
Purpose made composite items				£	£	£	£
					VAT not included		
Softwood open tread staircases (*continued*)							
Handrails							
50 x 75 mm, with rounded top	m	0.25	-	2.24	2.05	4.29	4.93
returned ends	each	0.30	-	2.69	-	2.69	3.09
ends housed to newel	each	0.35	-	3.14	-	3.14	3.61
Rails							
32 x 150 mm, screwed on	m	0.25	-	2.24	2.30	4.54	5.22
returned ends on rake	each	0.35	-	3.14	-	3.14	3.61
Hardwood handrails to staircases							
Handrails							
63 x 75 mm moulded	m	0.28	-	2.51	11.73	14.24	16.38
63 x 75 mm, moulded, grooved for core rail	m	0.35	-	3.14	12.51	15.65	18.00
ends framed to newel	each	0.50	-	4.48	-	4.48	5.15
ends framed on rake to newel	each	0.75	-	6.72	-	6.72	7.73
heading joints including handrail screw	each	1.88	-	16.84	-	16.84	19.37
mitres including handrail screw	each	1.75	-	15.68	-	15.68	18.03
Softwood built-in cupboards							
Cupboard fronts, outer framing 32 x 50 mm moulded one side, 100 x 25 mm softwood plinth, door framing 25 x 19 mm with 4 mm plywood panels; doors hung on pairs 63 mm pressed steel butts							
full height	m2	0.85	-	7.62	11.99	19.61	22.55
dwarf	m2	1.10	-	9.86	14.36	24.22	27.85
Cupboard ends moulded one side, with 4 mm plywood panels							
full height	m2	1.34	-	12.01	9.83	21.84	25.12
dwarf	m2	2.00	-	17.92	11.52	29.44	33.86
Flush cupboard fronts of 32 mm outer frame and 25 mm door framing, covered one side with 3.2 mm hardboard; doors hung on pairs 63 mm pressed steel butts							
full height	m2	0.85	-	7.62	7.35	14.97	17.22
dwarf	m2	1.11	-	9.95	9.45	19.40	22.31
Flush cupboard ends of 25 mm framing covered on one side with 3.2 mm hardboard							
full height	m2	1.34	-	12.01	4.07	16.08	18.49
dwarf	m2	2.00	-	17.92	5.39	23.31	26.81
Flush cupboard fronts with blockboard doors lipped on long edges and hung on pairs 63 mm pressed steel butts							
full height with 32 mm outer frame and 25 mm doors	m2	0.85	-	7.62	25.84	33.46	38.48
dwarf with 25 mm outer frame and 18 mm doors	m2	1.11	-	9.95	21.65	31.60	36.34

Woodwork

New work	Unit	Hours C	Hours L	Labour net	Material net	Price net	Price with 15%

Purpose made composite items

£ £ £ £

VAT not included

Softwood built-in cupboards (*continued*)

	Unit	Hours C	Hours L	Labour net	Material net	Price net	Price with 15%
Flush cupboard ends of 18 mm blockboard							
full height	m2	1.34	-	12.01	18.59	30.60	35.19
dwarf	m2	2.00	-	17.92	18.58	36.50	41.98
Crosstongued shelves or divisions							
19 mm	m2	2.80	-	25.09	16.23	41.32	47.52
25 mm	m2	3.00	-	26.88	16.23	43.11	49.58
Crosstongued tops to dwarf cupboards, fixed with buttons							
25 mm	m2	2.16	-	19.35	16.30	35.65	41.00
32 mm	m2	2.30	-	20.61	18.03	38.64	44.44
Extra for moulded edges to tops	m	-	-	-	0.54	0.54	0.62
Blockboard tops to dwarf cupboards							
18 mm	m2	2.16	-	19.35	16.29	35.64	40.99
25 mm	m2	2.30	-	20.61	21.39	42.00	48.30
Extra for softwood lippings to tops							
18 mm	m	0.20	-	1.79	0.14	1.93	2.22
25 mm	m	0.20	-	1.79	0.18	1.97	2.27

Hardwood built-in cupboards

	Unit	Hours C	Hours L	Labour net	Material net	Price net	Price with 15%
Cupboard fronts, outer framing 32 x 50 mm moulded one side, 100 x 25 mm softwood plinth, door framing 25 x 19 mm, with 6 mm veneered plywood panels; doors hung on pairs 63 mm brass butts							
full height	m2	1.13	-	10.12	25.34	35.46	40.78
dwarf	m2	1.40	-	12.54	33.40	45.94	52.83
Cupboard ends moulded one side, with 6 mm veneered plywood panels							
full height	m2	1.67	-	14.96	16.37	31.33	36.03
dwarf	m2	2.50	-	22.40	12.72	35.12	40.39
Flush cupboard fronts with veneered blockboard doors lipped on all edges and doors hung on pairs 63 mm brass butts							
full height with 32 mm outer frame and 25 mm doors	m2	1.13	-	10.12	148.97	159.09	182.95
dwarf with 25 mm outer frame and 18 mm doors	m2	1.40	-	12.55	35.03	47.58	54.72

Woodwork

New work	Unit	Hours C	Hours L	Labour net	Material net	Price net	Price with 15%
Purpose made composite items				£	£	£	£
					VAT not included		
Hardwood built-in cupboards (*continued*)							
Flush cupboard ends of 18 mm veneered blockboard							
full height	m2	1.67	-	14.96	25.61	40.57	46.66
dwarf	m2	2.50	-	22.40	30.38	52.78	60.70
Crosstongued tops to dwarf cupboards, fixed with buttons							
25 mm	m2	2.25	-	20.16	49.66	69.82	80.29
32 mm	m2	2.25	-	20.16	64.19	84.35	97.00
Extra for moulded edges to tops	m	-	-	-	0.75	0.75	0.86

Woodwork

New work	Unit	Specialist price net	Price with 15%

"Astraseal" PVC-U windows

£   £

VAT not included

The following "Specialist price net" figures are guide prices provided by Graham-Holmes Plastics Ltd for work over £700 in value within 30 miles of an "Astraseal" manufacturing and installation base.

Prices do not include for cash discount.

See the preamble notes for builder's profit and attendance.

Inward opening "tilt and turn" system

Self-finished windows complete with sealed double glazed units and PVC-U external sills fitted in builder's prepared openings, including all necessary fixings, and pointed around externally with silicone rubber mastic

	Unit	Specialist price net	Price with 15%
Windows in one fixed light, size			
600 x 600 mm	each	79.00	90.85
600 x 900 mm	each	86.00	98.90
600 x 1200 mm	each	93.00	106.95
900 x 900 mm	each	95.00	109.25
900 x 1200 mm	each	106.00	121.90
900 x 1500 mm	each	113.00	129.95
Windows in one "tilt and turn" light, size			
600 x 600 mm	each	140.00	161.00
600 x 900 mm	each	140.00	161.00
600 x 1200 mm	each	158.00	181.70
900 x 900 mm	each	154.00	177.10
900 x 1200 mm	each	174.00	200.10
900 x 1500 mm	each	189.00	217.35
1200 x 900 mm	each	175.00	201.25
1200 x 1200 mm	each	197.00	226.55
1200 x 1500 mm	each	214.00	246.10
Windows in two lights, one fixed beside one "tilt and turn", size			
1500 x 900 mm	each	187.00	215.05
1500 x 1200 mm	each	214.00	246.10
1800 x 900 mm	each	203.00	233.45
1800 x 1200 mm	each	232.00	266.80
1800 x 1500 mm	each	254.00	292.10
2400 x 900 mm	each	232.00	266.80
2400 x 1200 mm	each	264.00	303.60
2400 x 1500 mm	each	290.00	333.50
Windows in three lights, one fixed between two "tilt and turn", size			
1800 x 900 mm	each	262.00	301.30
1800 x 1200 mm	each	302.00	347.30
2400 x 900 mm	each	288.00	331.20
2400 x 1200 mm	each	332.00	381.80
2400 x 1500 mm	each	367.00	422.05
2700 x 900 mm	each	307.00	353.05
2700 x 1200 mm	each	354.00	407.10
2700 x 1500 mm	each	390.00	448.50

Woodwork

New work	Unit	Specialist price net	Price with 15%

"Astraseal" PVC-U windows

£   £

VAT not included

Inward opening "tilt and turn" system *(continued)*

Self-finished windows complete with sealed double glazed units and PVC-U external sills fitted in builder's prepared openings, including all necessary fixings, and pointed around externally with silicone rubber mastic *(continued)*

	Unit	Specialist price net	Price with 15%
Extra for			
Georgian effect double glazed units (glazing bars sealed between panes)	m2	38.60	44.39
leaded light effect double glazed units	m2	40.25	46.29

Outward opening casement system

Self-finished windows complete with sealed double glazed units and PVC-U external sills fitted in builder's prepared opening, including all necessary fixings, and pointed around externally with silicone rubber mastic

	Unit	Specialist price net	Price with 15%
Windows in one top hung light, size			
600 x 600 mm	each	101.00	116.15
600 x 900 mm	each	111.00	127.65
900 x 600 mm	each	110.00	126.50
900 x 900 mm	each	123.00	141.45
1000 x 1000 mm (maximum)	each	135.00	155.25
Windows in two lights, one fixed beside one side hung, size			
1200 x 600 mm	each	126.00	144.90
1200 x 900 mm	each	139.00	159.85
1200 x 1050 mm	each	148.00	170.20
Windows in two lights, one fixed beside one top hung, size			
1200 x 600 mm	each	126.00	144.90
1200 x 900 mm	each	142.00	163.30
1800 x 600 mm	each	143.00	164.45
1800 x 900 mm	each	165.00	189.75
Windows in two lights, one top hung over one fixed, size			
600 x 600 mm	each	113.00	129.95
600 x 900 mm	each	116.00	133.40
600 x 1200 mm	each	124.00	142.60
Windows in three lights, one fixed between two side hung, size			
1800 x 600 mm	each	173.00	198.95
1800 x 900 mm	each	193.00	221.95
1800 x 1050 mm	each	205.00	235.75
1800 x 1200 mm	each	217.00	249.55
Windows in three lights, one fixed between two top hung, size			
2400 x 600 mm	each	189.00	217.35
2400 x 900 mm	each	222.00	255.30
2700 x 600 mm	each	199.00	228.85
2700 x 900 mm	each	233.00	267.95

Woodwork

New work	Unit	Specialist price net	Price with 15%

"Astraseal" PVC-U windows

£ £

VAT not included

Outward opening casement system *(continued)*

Self-finished windows complete with sealed double glazed units and PVC-U external sills fitted in builder's prepared opening, including all necessary fixings, and pointed around externally with silicone rubber mastic *(continued)*

	Unit	net	with 15%
Windows in three lights, one fixed beside one top hung over one side hung, size			
1200 x 900 mm	each	169.00	194.35
1200 x 1050 mm	each	176.00	202.40
1200 x 1200 mm	each	182.00	209.30
1200 x 1500 mm	each	197.00	226.55
Windows in three lights, one side hung beside one top hung over one fixed, size			
1200 x 900 mm	each	170.00	195.50
1200 x 1050 mm	each	179.00	205.85
1200 x 1200 mm	each	186.00	213.90
Windows in four lights, one top hung over one fixed between two side hung, size			
1800 x 900 mm	each	224.00	257.60
1800 x 1050 mm	each	237.00	272.55
1800 x 1200 mm	each	248.00	285.20
Windows in five lights, one top hung over one side hung on both sides of one fixed, size			
1800 x 900 mm	each	252.00	289.80
1800 x 1050 mm	each	261.00	300.15
1800 x 1200 mm	each	271.00	311.65
1800 x 1200 mm	each	294.00	338.10
Extra for			
Georgian effect double glazed units (glazing bars sealed between panes)	m2	38.60	44.39
leaded light effect double glazed units	m2	40.25	46.29

Enquiries about the foregoing specialist prices should be made to Graham-Holmes Plastics Ltd., Astraseal House, Paterson Road, Finedon Road Industrial Estate, Wellingborough, Northants, NN8 4EX, tel (01933) 227233, fax (01933) 228951.

Woodwork

New work	Unit	Hours C	Hours L	Labour net	Material net	Price net	Price with 15%
Standard composite items				£	£	£	£
				VAT not included			

Redwood overhead garage doors

Vertically boarded garage doors with galvanised
steel gear screwed to timber frame

2135 x 1980 mm	each	4.00	-	35.84	424.62	460.46	529.53
2135 x 2135 mm	each	4.00	-	35.84	444.00	479.84	551.82
4270 x 1980 mm	each	6.00	-	53.76	1028.77	1082.53	1244.91
4270 x 2135 mm	each	6.00	-	53.76	1028.77	1082.53	1244.91

Vertically boarded garage doors with galvanised
steel gear screwed and integral frame plugged
and screwed

2135 x 1980 mm	each	4.00	-	35.84	477.66	513.50	590.52
2135 x 2135 mm	each	4.00	-	35.84	511.32	547.16	629.23

Cedar overhead garage doors

Vertically boarded garage doors with galvanised
steel gear screwed to timber frame

2135 x 1980 mm	each	4.00	-	35.84	464.40	500.24	575.28
2135 x 2135 mm	each	4.00	-	35.84	490.92	526.76	605.77
4270 x 1980 mm	each	6.00	-	53.76	1098.13	1151.89	1324.67
4270 x 2135 mm	each	6.00	-	53.76	1098.13	1151.89	1324.67

Horizontally boarded garage doors with
galvanised steel gear screwed to timber frame

2135 x 1980 mm	each	4.00	-	35.84	464.40	500.24	575.28
2135 x 2135 mm	each	4.00	-	35.84	490.92	526.76	605.77

Diagonally boarded garage doors with galvanised
steel gear screwed to timber frame

2135 x 1980 mm	each	4.00	-	35.84	589.85	625.69	719.54
2135 x 2135 mm	each	4.00	-	35.84	607.19	643.03	739.48

Vertically boarded garage doors with galvanised
steel gear screwed and integral frame plugged
and screwed

2135 x 1980 mm	each	4.00	-	35.84	511.32	547.16	629.23
2135 x 2135 mm	each	4.00	-	35.84	543.95	579.79	666.76

Horizontally boarded garage doors with
galvanised steel gear screwed and integral frame
plugged and screwed

2135 x 1980 mm	each	4.00	-	35.84	511.32	547.16	629.23
2135 x 2135 mm	each	4.00	-	35.84	543.95	579.79	666.76

Diagonally boarded garage doors with galvanised
steel gear screwed and integral frame plugged
and screwed

2135 x 1980 mm	each	4.00	-	35.84	642.89	678.73	780.54
2135 x 2135 mm	each	4.00	-	35.84	660.23	696.07	800.48

Woodwork

New work	Unit	Hours C	Hours L	Labour net	Material net	Price net	Price with 15%
Standard composite items				£	£	£	£
					VAT not included		

Roof windows

Velux roof windows in treated Nordic red pine;
exterior aluminium cladding; factory double
glazed sealed unit with 3 mm clear float glass
panes; screwed to softwood

550 x 780 mm GGL 102	each	0.80	-	7.17	145.69	152.86	175.79
550 x 980 mm GGL 104	each	0.80	-	7.17	162.10	169.27	194.66
780 x 980 mm GGL 304	each	0.90	-	8.06	185.07	193.13	222.10
940 x 1600 mm GGL 410	each	1.00	-	8.96	273.66	282.62	325.01
1340 x 980 mm GGL 804	each	1.00	-	8.96	257.25	266.21	306.14
550 x 980 mm GHL 104	each	0.80	-	7.17	210.32	217.49	250.11
780 x 980 mm GHL 304	each	0.90	-	8.06	236.34	244.40	281.06
1340 x 980 mm GHL 804	each	1.00	-	8.96	302.93	311.89	358.67

Velux roof windows in treated Nordic red pine;
exterior aluminium cladding; factory double
glazed sealed unit with 4 mm toughened outer
pane and 3 mm clear float glass inner pane;
screwed to softwood

550 x 780 mm GGL 102	each	0.80	-	7.17	190.54	197.71	227.37
550 x 980 mm GGL 104	each	0.80	-	7.17	210.22	217.39	250.00
780 x 980 mm GGL 304	each	0.90	-	8.06	209.72	217.78	250.45
940 x 1600 mm GGL 410	each	1.00	-	8.96	329.81	338.77	389.59
1340 x 980 mm GGL 804	each	1.00	-	8.96	309.90	318.86	366.69
550 x 980 mm GHL 104	each	0.80	-	7.17	252.15	259.32	298.22
780 x 980 mm GHL 304	each	0.90	-	8.06	286.59	294.65	338.85
1340 x 980 mm GHL 804	each	1.00	-	8.96	377.61	386.57	444.56

Velux flashings for roof windows

EDZ 102	each	0.80	-	7.17	31.72	38.89	44.72
EDZ 104	each	0.85	-	7.62	33.90	41.52	47.75
EDZ 304	each	0.90	-	8.06	37.19	45.25	52.04
EDZ 410	each	1.00	-	8.96	47.04	56.00	64.40
EDZ 804	each	1.00	-	8.96	48.13	57.09	65.65
EDH 102	each	0.80	-	7.17	38.29	45.46	52.28
EDH 104	each	0.85	-	7.62	39.38	47.00	54.05
EDH 304	each	0.90	-	8.06	44.84	52.90	60.84
EDH 410	each	1.00	-	8.96	49.79	58.75	67.56
EDH 804	each	1.00	-	8.96	54.77	63.73	73.29
EDL 102	each	0.80	-	7.17	27.35	34.52	39.70
EDL 104	each	0.85	-	7.62	29.53	37.15	42.72
EDL 304	each	0.90	-	8.06	32.81	40.87	47.00
EDL 410	each	1.00	-	8.96	41.56	50.52	58.10
EDL 804	each	1.00	-	8.96	39.38	48.34	55.59

Woodwork

New work	Unit	Hours C	Hours L	Labour net	Material net	Price net	Price with 15%
Standard composite items				£	£	£	£
					VAT not included		

Roof windows (*continued*)

Velux roller blinds to suit roof windows

	Unit	Hours C	Hours L	Labour net	Material net	Price net	Price with 15%
GGL 102	each	0.25	-	2.24	35.05	37.29	42.88
GGL 104	each	0.25	-	2.24	35.05	37.29	42.88
GGL 304	each	0.30	-	2.69	47.92	50.61	58.20
GGL 410	each	0.36	-	3.23	54.94	58.17	66.90
GGL 804	each	0.36	-	3.23	72.49	75.72	87.08
GHL 104	each	0.25	-	2.24	35.05	37.29	42.88
GHL 304	each	0.30	-	2.69	47.92	50.61	58.20
GHL 804	each	0.36	-	3.23	72.49	75.72	87.08

Velux portable rods for roof windows

	Unit	Hours C	Hours L	Labour net	Material net	Price net	Price with 15%
800 mm long ZCZ 080	each	0.15	-	1.34	14.04	15.38	17.69
1000 - 1800 mm long telescopic ZCT 200	each	0.06	-	0.54	26.89	27.43	31.54
1000 mm long extension ZCT 100	each	0.06	-	0.54	5.98	6.52	7.50

Fixing fittings

Fixing floor units including allowance for scribing to floor and/or walls and plugging and screwing units

	Unit	Hours C	Hours L	Labour net	Material net	Price net	Price with 15%
600 x 500 x 900 mm high	each	0.75	-	6.72	0.37	7.09	8.15
1200 x 500 x 900 mm high	each	1.25	-	11.20	0.37	11.57	13.31
1500 x 500 x 900 mm high	each	1.50	-	13.44	0.37	13.81	15.88

Fixing wall units including allowance for scribing to walls and plugging and screwing units

	Unit	Hours C	Hours L	Labour net	Material net	Price net	Price with 15%
600 x 300 x 900 mm high	each	1.00	-	8.96	0.37	9.33	10.73
1200 x 300 x 900 mm high	each	1.75	-	15.68	0.37	16.05	18.46

Woodwork

New work	Unit	Hours C	Hours L	Labour net	Material net	Price net	Price with 15%
Sundries				£	£	£	£
					VAT not included		
Plugging etc							
Plugging brickwork or concrete							
for open spaced members at 400 mm centres							
one way	m2	0.83	-	7.44	0.17	7.61	8.75
for open spaced members at 400 mm centres							
both ways	m2	1.90	-	17.02	0.34	17.36	19.96
for individual members	m	0.33	-	2.96	0.09	3.05	3.51
Extra over fixing with nails for fixing with							
steel screws	m	0.15	-	1.34	0.07	1.41	1.62
steel screws including sinking heads and							
pelleting	m	0.55	-	4.93	0.91	5.84	6.72
brass screws and cups	m	0.30	-	2.69	0.88	3.57	4.11
Rawlnuts to plasterboard	each	0.10	-	0.90	0.39	1.29	1.48
Extra over fixing with nails for fixing with hardened masonry pins to brickwork or concrete							
open spaced members at 400 mm centres							
one way	m2	0.25	-	2.24	0.02	2.26	2.60
open spaced members at 400 mm centres							
both ways	m2	0.60	-	5.38	0.04	5.42	6.23
individual members	m	0.10	-	0.90	0.01	0.91	1.05
Holes in timber							
Holes for bolts through timber							
25 mm	each	0.08	-	0.72	-	0.72	0.83
50 mm	each	0.12	-	1.08	-	1.08	1.24
75 mm	each	0.18	-	1.61	-	1.61	1.85
100 mm	each	0.24	-	2.15	-	2.15	2.47
Holes for pipes not exceeding 110 mm diameter through timber							
25 mm	each	0.33	-	2.96	-	2.96	3.40
50 mm	each	0.50	-	4.48	-	4.48	5.15
75 mm	each	0.75	-	6.72	-	6.72	7.73
100 mm	each	1.00	-	8.96	-	8.96	10.30
Holes for ducting etc not exceeding 0.025m2 through timber							
25 mm	each	0.44	-	3.94	-	3.94	4.53
50 mm	each	0.66	-	5.91	-	5.91	6.80
75 mm	each	1.00	-	8.96	-	8.96	10.30
100 mm	each	1.33	-	11.92	-	11.92	13.71

Woodwork

New work	Unit	Hours C	Hours L	Labour net	Material net	Price net	Price with 15%
Sundries				£	£	£	£
					VAT not included		
Insulating materials							
13 mm sound deadening quilts cut and laid under flooring between joists at 400 mm centres (measured overall)	m2	0.10	-	0.90	2.04	2.94	3.38
Thermal insulating quilts cut and laid in roof between joists at 400 mm centres (measured overall)							
80 mm	m2	0.12	-	1.08	2.98	4.06	4.67
100 mm	m2	0.12	-	1.08	3.54	4.62	5.31
150 mm	m2	0.12	-	1.08	5.51	6.59	7.58
Expanded polystyrene ISD grade laid with close butted joints taped with waterproof tape on concrete							
25 mm	m2	0.12	-	1.08	1.50	2.58	2.97
40 mm	m2	0.12	-	1.08	2.40	3.48	4.00
Granular loose fill spread 50 mm thick between joists at 400 mm centres (measured overall)	m2	0.16	-	1.43	7.76	9.19	10.57
Building paper lapped 150 mm and fixed with galvanised clout nails							
standard grade	m2	0.20	-	1.79	1.27	3.06	3.52
reflective grade	m2	0.20	-	1.79	1.91	3.70	4.25
Metalwork							
Galvanised steel dowels							
10 x 50 mm	each	0.08	-	0.72	0.20	0.92	1.06
12 x 50 mm	each	0.08	-	0.72	0.31	1.03	1.18
10 x 50 mm non-ferrous dowels	each	0.08	-	0.72	0.20	0.92	1.06
Steel carriage bolts with nut and two washers							
M10 x 75 mm	each	0.10	-	0.90	0.32	1.22	1.40
M12 x 75 mm	each	0.10	-	0.90	0.52	1.42	1.63
M12 x 100 mm	each	0.12	-	1.08	0.64	1.72	1.98
M12 x 150 mm	each	0.15	-	1.34	0.89	2.23	2.56
Galvanised steel water bars including grooves in timber							
3 x 25 mm	each	0.25	-	2.24	2.06	4.30	4.95
5 x 30 mm	each	0.26	-	2.33	3.02	5.35	6.15
6 x 40 mm	each	0.27	-	2.42	4.71	7.13	8.20
25 x 3.2 x 250 mm girth galvanised steel cramps, the other end built in	each	0.10	-	0.90	0.31	1.21	1.39

Woodwork

Sundries

				£	£	£	£
					VAT not included		

Metalwork (*continued*)

30 x 2.5 mm galvanised steel holding down
straps, twice bent, one end drilled and screwed to
timber, the other end built in

	Unit	Hours C	Hours L	Labour net	Material net	Price net	Price with 15%
400 mm girth	each	0.30	-	2.69	0.74	3.43	3.94
600 mm girth	each	0.33	-	2.96	1.12	4.08	4.69
900 mm girth	each	0.40	-	3.58	1.65	5.23	6.01

Galvanised steel joist hangers - built in

	Unit	Hours C	Hours L	Labour net	Material net	Price net	Price with 15%
38 x 100 mm	each	0.10	-	0.90	2.13	3.03	3.48
38 x 125 mm	each	0.10	-	0.90	2.15	3.05	3.51
38 x 150 mm	each	0.12	-	1.08	2.23	3.31	3.81
38 x 175 mm	each	0.12	-	1.08	2.24	3.32	3.82
38 x 200 mm	each	0.13	-	1.16	2.37	3.53	4.06
38 x 225 mm	each	0.13	-	1.16	2.72	3.88	4.46
50 x 100 mm	each	0.10	-	0.90	2.13	3.03	3.48
50 x 125 mm	each	0.12	-	1.08	2.15	3.23	3.71
50 x 150 mm	each	0.12	-	1.08	2.23	3.31	3.81
50 x 175 mm	each	0.12	-	1.08	2.24	3.32	3.82
50 x 200 mm	each	0.13	-	1.16	2.48	3.64	4.19
50 x 225 mm	each	0.13	-	1.16	2.76	3.92	4.51
75 x 150 mm	each	0.13	-	1.16	3.05	4.21	4.84
75 x 175 mm	each	0.13	-	1.16	3.11	4.27	4.91
75 x 200 mm	each	0.14	-	1.25	3.17	4.42	5.08
75 x 225 mm	each	0.15	-	1.34	3.26	4.60	5.29

Galvanised steel truss clips, fixed with nails to suit

	Unit	Hours C	Hours L	Labour net	Material net	Price net	Price with 15%
38 mm thick members	each	0.20	-	1.79	0.32	2.11	2.43
50 mm thick members	each	0.20	-	1.79	0.34	2.13	2.45

Galvanised steel square toothed plate timber
connectors to BS 1579 Table 4

	Unit	Hours C	Hours L	Labour net	Material net	Price net	Price with 15%
38 mm diameter, single sided	each	0.03	-	0.27	0.16	0.43	0.49
38 mm diameter, double sided	each	0.03	-	0.27	0.25	0.52	0.60
50 mm diameter, single sided	each	0.03	-	0.27	0.21	0.48	0.55
50 mm diameter, double sided	each	0.03	-	0.27	0.26	0.53	0.61
63 mm diameter, single sided	each	0.03	-	0.27	0.30	0.57	0.66
63 mm diameter, double sided	each	0.03	-	0.27	0.35	0.62	0.71
75 mm diameter, single sided	each	0.03	-	0.27	0.32	0.59	0.68
75 mm diameter, double sided	each	0.03	-	0.27	0.40	0.67	0.77

Galvanised steel herringbone joist struts, fixed
with nails, to suit joist centres of

	Unit	Hours C	Hours L	Labour net	Material net	Price net	Price with 15%
400 mm	each	0.27	-	2.42	0.44	2.86	3.29
450 mm	each	0.26	-	2.33	0.48	2.81	3.23
600 mm	each	0.22	-	1.97	0.53	2.50	2.88

Galvanised steel pelmet brackets

	Unit	Hours C	Hours L	Labour net	Material net	Price net	Price with 15%
75 x 100 mm, screwed to timber	each	0.15	-	1.34	0.96	2.30	2.65
Plugged and screwed to brickwork or concrete	each	0.25	-	2.24	1.09	3.33	3.83

Enamelled shelf brackets

	Unit	Hours C	Hours L	Labour net	Material net	Price net	Price with 15%
150 x 200 mm grey, screwed to timber	each	0.15	-	1.34	0.38	1.72	1.98
plugged and screwed to brickwork or concrete	each	0.25	-	2.24	0.55	2.79	3.21

Woodwork

New work	Unit	Hours C	Hours L	Labour net	Material net	Price net	Price with 15%
Ironmongery				£	£	£	£
					VAT not included		
Butts etc, including hanging doors							
Light steel butts and labour hanging doors							
50 mm to small cupboard doors	pair	0.50	-	4.48	0.56	5.04	5.80
75 mm to 38 mm softwood doors1	pair	1.10	-	9.86	0.66	10.52	12.10
100 mm to 50 mm softwood doors	pair	1.33	-	11.92	1.32	13.24	15.23
Light steel sheradised butts and labour hanging doors							
50 mm to small cupboard doors	pair	0.50	-	4.48	0.72	5.20	5.98
75 mm to 38 mm softwood doors	pair	1.10	-	9.86	1.56	11.42	13.13
100 mm to 50 mm softwood doors	pair	1.33	-	11.92	2.92	14.84	17.07
Cast iron butts and labour hanging doors							
75 mm to 38 mm softwood doors	pair	1.10	-	9.86	2.00	11.86	13.64
100 mm to 50 mm softwood doors	pair	1.33	-	11.92	2.77	14.69	16.89
100 mm steel rising butts and labour hanging 50 mm softwood doors	pair	1.60	-	14.34	5.14	19.48	22.40
Steel washered brass butts and labour hanging doors							
75 mm to 38 mm hardwood doors	pair	2.00	-	17.92	3.31	21.23	24.41
100 mm to 50 mm hardwood doors	pair	2.50	-	22.40	6.96	29.36	33.76
450 mm japanned tee hinges and labour hanging softwood doors	pair	0.85	-	7.62	4.04	11.66	13.41
450 mm light reversible hinges and labour hanging softwood doors	pair	1.10	-	9.86	9.58	19.44	22.36
600 mm heavy cast iron reversible hinges with cast or malleable cups and bolts and labour hanging garage doors	pair	1.50	-	13.44	16.31	29.75	34.21
125 mm double action regulating spring hinges and labour hanging 50 mm softwood doors	pair	3.00	-	26.88	38.14	65.02	74.77
Check action floor springs with all fittings and loose box and labour hanging 50 mm hardwood door including setting box in floor							
single action	each	5.50	-	49.28	211.66	260.94	300.08
double action	each	6.00	-	53.76	227.88	281.64	323.89
Overhead sliding door tracks with hangers and steel pelmet etc and labour hanging 38 mm or 50 mm single softwood door (side fixing to door head)	set	3.25	-	29.12	20.72	49.84	57.32
Overhead garage door gear and fixing to softwood frame and labour hanging 50 mm softwood door	set	3.00	-	26.88	109.23	136.11	156.53

Woodwork

New work	Unit	Hours C	Hours L	Labour net	Material net	Price net	Price with 15%
Ironmongery				£	£	£	£
					VAT not included		
Fixing only to softwood							
Barrel or tower bolts							
small	each	0.25	-	2.24	-	2.24	2.58
medium	each	0.33	-	2.96	-	2.96	3.40
large	each	0.42	-	3.76	-	3.76	4.32
Necked bolts							
small	each	0.33	-	2.96	-	2.96	3.40
medium	each	0.42	-	3.76	-	3.76	4.32
Flush bolts							
small	each	0.75	-	6.72	-	6.72	7.73
medium	each	1.00	-	8.96	-	8.96	10.30
large	each	1.33	-	11.92	-	11.92	13.71
Espagnolette bolt sets	each	1.25	-	11.20	-	11.20	12.88
Panic bolt sets	each	1.50	-	13.44	-	13.44	15.46
Ball catches							
small	each	0.33	-	2.96	-	2.96	3.40
medium	each	0.50	-	4.48	-	4.48	5.15
Cupboard catches	each	0.17	-	1.52	-	1.52	1.75
Roller bolt catches	each	0.83	-	7.44	-	7.44	8.56
Fanlight catches	each	0.17	-	1.52	-	1.52	1.75
Casement fasteners							
with hook plate	each	0.17	-	1.52	-	1.52	1.75
with mortice plate	each	0.50	-	4.48	-	4.48	5.15
Casement stays	each	0.25	-	2.24	-	2.24	2.58
Sash fasteners	each	0.33	-	2.96	-	2.96	3.40
Quadrant stays	each	0.33	-	2.96	-	2.96	3.40
Thumb latches	each	0.60	-	5.38	-	5.38	6.19
Rim night latches	each	1.00	-	8.96	-	8.96	10.30
Cupboard and drawer locks							
surface pattern	each	0.50	-	4.48	-	4.48	5.15
mortice pattern	each	1.00	-	8.96	-	8.96	10.30
Rim locks and furniture	each	0.75	-	6.72	-	6.72	7.73
Mortice locks and furniture							
shallow pattern	each	1.50	-	13.44	-	13.44	15.46
deep pattern	each	2.00	-	17.92	-	17.92	20.61

Woodwork

New work	Unit	Hours C	Hours L	Labour net	Material net	Price net	Price with 15%
Ironmongery				£	£	£	£
					VAT not included		
Fixing only to softwood (*continued*)							
Locking bars	each	0.50	-	4.48	-	4.48	5.15
Bow handles							
medium	each	0.13	-	1.16	-	1.16	1.33
large	each	0.33	-	2.96	-	2.96	3.40
Drawer pulls	each	0.10	-	0.90	-	0.90	1.03
Flush pulls	each	0.60	-	5.38	-	5.38	6.19
Finger plates	each	0.13	-	1.16	-	1.16	1.33
Helical door springs	each	0.25	-	2.24	-	2.24	2.58
Overhead door closers							
surface fixing	each	0.75	-	6.72	-	6.72	7.73
mortice fixing	each	1.75	-	15.68	-	15.68	18.03
Letter plates in 50 mm doors	each	1.50	-	13.44	-	13.44	15.46
Coat hooks	each	0.10	-	0.90	-	0.90	1.03
Flap table brackets	each	0.60	-	5.38	-	5.38	6.19
Fixing only to hardwood							
Barrel or tower bolts							
small	each	0.38	-	3.40	-	3.40	3.91
medium	each	0.50	-	4.48	-	4.48	5.15
large	each	0.63	-	5.64	-	5.64	6.49
Necked bolts							
small	each	0.50	-	4.48	-	4.48	5.15
medium	each	0.63	-	5.64	-	5.64	6.49
Flush bolts							
small	each	1.13	-	10.12	-	10.12	11.64
medium	each	1.50	-	13.44	-	13.44	15.46
large	each	2.00	-	17.92	-	17.92	20.61
Espagnolette bolt sets	each	1.88	-	16.84	-	16.84	19.37
Panic bolt sets	each	2.25	-	20.16	-	20.16	23.18
Ball catches							
small	each	0.50	-	4.48	-	4.48	5.15
medium	each	0.75	-	6.72	-	6.72	7.73
Cupboard catches	each	0.26	-	2.33	-	2.33	2.68
Roller bolt catches	each	1.25	-	11.20	-	11.20	12.88

Woodwork

New work	Unit	Hours C	Hours L	Labour net	Material net	Price net	Price with 15%
Ironmongery				£	£	£	£
					VAT not included		
Fixing only to hardwood (*continued*)							
Fanlight catches	each	0.26	-	2.33	-	2.33	2.68
Casement fasteners							
with hook plate	each	0.26	-	2.33	-	2.33	2.68
with mortice plate	each	0.75	-	6.72	-	6.72	7.73
Casement stays	each	0.38	-	3.40	-	3.40	3.91
Sash fasteners	each	0.50	-	4.48	-	4.48	5.15
Quadrant stays	each	0.50	-	4.48	-	4.48	5.15
Thumb latches	each	0.90	-	8.06	-	8.06	9.27
Rim night latches	each	1.50	-	13.44	-	13.44	15.46
Cupboard and drawer locks							
surface pattern	each	0.75	-	6.72	-	6.72	7.73
mortice pattern	each	1.50	-	13.44	-	13.44	15.46
Rim locks and furniture	each	1.13	-	10.12	-	10.12	11.64
Mortice locks and furniture							
shallow pattern	each	2.25	-	20.16	-	20.16	23.18
deep pattern	each	3.00	-	26.88	-	26.88	30.91
Locking bars	each	0.75	-	6.72	-	6.72	7.73
Bow handles							
medium	each	0.20	-	1.79	-	1.79	2.06
large	each	0.50	-	4.48	-	4.48	5.15
Drawer pulls	each	0.15	-	1.34	-	1.34	1.54
Flush pulls	each	0.90	-	8.06	-	8.06	9.27
Finger plates	each	0.20	-	1.79	-	1.79	2.06
Helical door springs	each	0.38	-	3.40	-	3.40	3.91
Overhead door closers							
surface fixing	each	1.13	-	10.12	-	10.12	11.64
mortice fixing	each	2.63	-	23.56	-	23.56	27.09
Letter plates in 50 mm doors	each	2.25	-	20.16	-	20.16	23.18
Coat hooks	each	0.15	-	1.34	-	1.34	1.54
Flap table brackets	each	0.90	-	8.06	-	8.06	9.27

Woodwork

New work	Unit	Hours C	Hours L	Labour net	Material net	Price net	Price with 15%
Ironmongery				£	£	£	£
					VAT not included		
Sundries							
Fibre sliding cupboard door tracks including groove	m	0.10	-	0.90	3.24	4.14	4.76
Nylon sliders including fitting doors	pair	0.75	-	6.72	0.43	7.15	8.22
Aluminium bookcase							
strips including groove	m	0.28	-	2.51	2.43	4.94	5.68
set of four adjustable studs	set	0.10	-	0.90	0.52	1.42	1.63
Standard curtain tracks and runners							
anodised aluminium	m	0.67	-	6.00	3.73	9.73	11.19
plastics	m	0.67	-	6.00	4.41	10.41	11.97
Rubber door stops							
screwed to timber	each	0.13	-	1.16	0.36	1.52	1.75
plugged and screwed to concrete	each	0.25	-	2.24	0.40	2.64	3.04

DENLEY KING
G R O U P

PROVIDING SPECIALIST SERVICES TO THE CONSTRUCTION INDUSTRY

PROJECT MANAGEMENT • PROJECT COORDINATION • LOTTERY/PFI SUBMISSIONS • QUANTITY SURVEYING • BUILDING SURVEYING • PLANNING SUPERVISION • PARTY WALL SURVEYORS • EXPERT WITNESS ADJUDICATION • FACILITIES MANAGEMENT

LONDON

17 Red Lion Square London WC1R 4QH
Tel 0171 4044913 Fax 0171 4044913 Email london@denleyking.co.uk

POOLE

Tower House 45 Commercial Road Poole BH14 0JA
Tel 01202 715300 Fax 01202 715411 Email poole@denleyking.co.uk

WINCHESTER

12 Parchment Street Winchester SO23 8AZ
Tel 01962 840345 Fax 01962 841801 Email winchester@denleyking.co.uk

WEYMOUTH

Catherine House 40a St Thomas Street Weymouth DT4 8EH
Tel 01305 760175 Fax 01305 760662 Email weymouth@denleyking.co.uk

Structural steelwork

Preamble

"Labour net" figures include allowances for all costs incidental to the employment of labour.

"Materials net" figures include for all costs of materials including off-site fabrication and an allowance for waste except where specifically stated.

"Price net" figures are the totals of the "Labour net" and "Materials net" figures. Prices are for a builder employing his own labour; according to the amount and nature of the work involved, it may well be possible to secure more advantageous prices from specialist sub-contractors.

The cost of steel and steelwork is governed by such factors as size of sections and quantities required, and in the case of fabricated work, by the nature of the job. It will be seen, therefore, that quotations are necessary at all times when definite requirements are known, but as a guide, some figures are given for simple items such as single joists to span openings in brickwork or joists bolted together or to stanchions. The prices in this case, are for delivery to site by the structural engineers and for unloading, hoisting and fixing by builder's own labour.

Prices do not include any allowance for scaffolding, ladders or other plant necessary to reach the work. The "Preliminaries" section includes prices for scaffolding which must be considered and allowance included to suit the particular circumstances of a tender.

Specialist prices

"Price with 15%" figures are all-in guide prices and include 15% for the builder's overheads, profit, unloading materials and general attendance (to include free use of standing scaffolding and hoists, temporary lighting and water and clearing away rubbish).

The amount of attendance required varies between the various trades and also with the circumstances of specific jobs; the percentage addition must always be considered and adjusted as necessary to suit the terms and conditions of the quotation being used.

Quantities and delivery distances are usually the most significant of the many factors which influence prices and it must be emphasised that quotations should always be obtained when preparing a tender.

Structural steelwork

New work	Unit	Hours C	Hours L	Labour net	Material net	Price net	Price with 15%
Fabricated steelwork				£	£	£	£
					VAT not included		

Columns, joists and beams

Universal column stanchions with cap and base
plates bolted at ground level

	Unit	Hours C	Hours L	Labour net	Material net	Price net	Price with 15%
152 x 152 mm	tonne	15.00	27.50	319.75	1066.45	1386.20	1594.13

Universal beams fixed at not exceeding 6.00 m
above ground level, using simple tackle

	Unit	Hours C	Hours L	Labour net	Material net	Price net	Price with 15%
305 x 127 mm	tonne	12.50	27.50	297.35	1043.85	1341.20	1542.38
305 x 102 mm	tonne	13.50	29.50	319.79	1147.17	1466.96	1687.00
254 x 146 mm	tonne	12.75	28.00	302.96	1125.64	1428.60	1642.89
254 x 102 mm	tonne	13.75	30.25	327.09	1061.07	1388.16	1596.38
203 x 133 mm	tonne	13.75	30.25	327.09	-	327.09	376.15

Rolled joist beams fixed at not exceeding 6.00 m
above ground level, using simple tackle

	Unit	Hours C	Hours L	Labour net	Material net	Price net	Price with 15%
203 x 102 mm	tonne	13.75	30.25	327.09	1087.97	1415.06	1627.32
178 x 102 mm	tonne	14.00	31.00	334.38	1055.69	1390.07	1598.58
152 x 89 mm	tonne	14.25	31.50	339.99	1233.27	1573.26	1809.25
127 x 76 mm	tonne	14.50	31.75	343.92	-	343.92	395.51

Universal beams drilled and bolted to steelwork at
not exceeding 6.00 m above ground level, using
simple tackle

	Unit	Hours C	Hours L	Labour net	Material net	Price net	Price with 15%
305 x 127 mm	tonne	15.00	30.00	336.60	1043.85	1380.45	1587.52
305 x 102 mm	tonne	16.13	32.25	361.89	1211.74	1573.63	1809.67
254 x 146 mm	tonne	15.38	30.75	345.06	1190.22	1535.28	1765.57
254 x 102 mm	tonne	16.50	33.00	370.26	1233.27	1603.53	1844.06
203 x 133 mm	tonne	16.50	33.00	370.26	-	370.26	425.80

Rolled joist beams drilled and bolted to steelwork
at not exceeding 6.00 m above ground level,
using simple tackle

	Unit	Hours C	Hours L	Labour net	Material net	Price net	Price with 15%
203 x 102 mm	tonne	16.50	33.00	370.26	1157.93	1528.19	1757.42
178 x 102 mm	tonne	16.88	33.75	378.71	1125.64	1504.35	1730.00
152 x 89 mm	tonne	17.13	34.25	384.33	1330.13	1714.46	1971.63
127 x 76 mm	tonne	17.25	34.50	387.09	-	387.09	445.15

Holes

10 mm holes made on site through steelwork

	Unit	Hours C	Hours L	Labour net	Material net	Price net	Price with 15%
6 mm	each	0.18	0.18	2.82	-	2.82	3.24
10 mm	each	0.20	0.20	3.14	-	3.14	3.61
15 mm	each	0.22	0.22	3.45	-	3.45	3.97
20 mm	each	0.24	0.24	3.77	-	3.77	4.34

16 mm holes made on site through steelwork

	Unit	Hours C	Hours L	Labour net	Material net	Price net	Price with 15%
6 mm	each	0.23	0.23	3.61	-	3.61	4.15
10 mm	each	0.25	0.25	3.93	-	3.93	4.52
15 mm	each	0.27	0.27	4.24	-	4.24	4.88
20 mm	each	0.29	0.29	4.55	-	4.55	5.23

New work	Unit	Hours C	Hours L	Labour net	Material net	Price net	Price with 15%
Fabricated steelwork				£	£	£	£
					VAT not included		
Prefabricated units							
Metsec joists							
Metsec open web steel joists including end fixings, delivered and fixed							
Type B22, 220 mm deep	m	0.05	0.10	1.12	24.38	25.50	29.32
Type B30, 300 mm deep	m	0.06	0.12	1.35	27.49	28.84	33.17
Type B35, 350 mm deep	m	0.07	0.13	1.51	29.42	30.93	35.57
Type D35, 350 mm deep	m	0.07	0.15	1.64	45.24	46.88	53.91

Structural steelwork

	Unit	Specialist price net	Price with 15%

Prefabricated units

£ £
VAT not included

The following "Specialist price net" figures are ex works guide prices provided by Snashall Steel Fabrications Co Ltd for quantities of not less than four identical units.

Allowance should be added for delivery charges.

Prices do not include for cash discount.

Roof trusses designed for 3.6 m maximum spacing including angle cleats for purlins and cleaning and painting one coat of primer at works		Ex works	Ex works
Segmental trusses for sheeted roof with rise of 1/5 span, span:			
6 m	each	332.00	381.80
9 m	each	555.00	638.25
12 m	each	901.00	1,036.15
15 m	each	1,064.00	1,223.60
18 m	each	1,510.00	1,736.50
Trusses for slated roof with 30 deg pitch, span:			
6 m	each	387.00	445.05
9 m	each	586.00	673.90
12 m	each	910.00	1,046.50
15 m	each	1,229.00	1,413.35
18 m	each	1,675.00	1,926.25
Trusses for sheeted roof with pitch of ¼ span, span:			
6 m	each	330.00	379.50
9 m	each	512.00	588.80
12 m	each	790.00	908.50
15 m	each	1,080.00	1,242.00
18 m	each	1,562.00	1,796.30

Portal frames designed to BS 449, with 15 deg roof pitch and 6.0 m maximum spacing

Allowance must be added for steel ties to portals at eaves level

	Unit		
6 m span portal frames, height to eaves:			
2.40 m	each	499.00	573.85
3.60 m	each	694.00	798.10
4.80 m	each	869.00	999.35
9 m span portal frames, height to eaves:			
2.40 m	each	738.00	848.70
3.60 m	each	930.00	1,069.50
4.80 m	each	1,158.00	1,331.70
12 m span portal frames, height to eaves:			
2.40 m	each	1,083.00	1,245.45
3.60 m	each	1,383.00	1,590.45
4.80 m	each	1,554.00	1,787.10

Structural steelwork

New work	Unit	Specialist price net	Price with 15%
Prefabricated units		£	£
		VAT not included	
Portal frames designed to BS 449, with 15 deg roof pitch and 6.0 m maximum spacing *(continued)*		Ex works	Ex works
15 m span portal frames, height to eaves:			
2.40 m	each	1,360.00	1,564.00
3.60 m	each	1,643.00	1,889.45
4.80 m	each	1,918.00	2,205.70
18 m span portal frames, height to eaves:			
2.40 m	each	1,895.00	2,179.25
3.60 m	each	2,139.00	2,459.85
4.80 m	each	2,325.00	2,673.75

Enquiries about the foregoing specialist prices should be made to Snashall Steel Fabrications Co Ltd, Pulham Business Park, Pulham, Dorchester, Dorset, DT2 7DX, tel (01300) 345588, fax, (01300) 345533.

Structural steelwork

New work	Unit	Specialist price net	Price with 15%

Prefabricated units

Metsec lattice joists

The following " Specialist price net" figures are guide prices provided by Metsec
Building Products Ltd for quantities of about 100 metres.

Due to variation of structural application, requirements for associated fittings vary
considerably and therefore a quotation should be obtained in each case. In
addition a much wider range of joists and trusses is available upon application
covering the various ranges up to the spans shown

	Unit	Specialist price net	Price with 15%
Metsec open web steel joists - short span range up to 6 m long			
Type B22	m	22.00	25.30
Type B30	m	24.20	27.83
Type D35	m	40.72	46.83
Metsec open web steel joists - intermediate span range up to 17 m long			
Type B50	m	40.53	46.61
Type D50	m	54.30	62.44
Type G50	m	63.45	72.97
Metsec open web steel joists - long span range up to 38 m long			
Type J100	m	80.88	93.01
Type L100	m	92.09	105.90
Type L200	m	134.71	154.92

The foregoing rates include the "Metsec" standard finish of one coat Zinc
Phosphate red oxide travel primer.

Timber inserts can be factory fitted to either chord - p.o.a.

Enquiries about foregoing specialist prices should be made to Metsec Building
Products Ltd, Lattice Joists Division, Broadwell Road, Oldbury, Warley, West
Midlands, B69 4HE, tel (0121) 601 6000, fax (0121) 601 6109.

Metalwork

Preamble

"Labour net" figures include allowances for all costs incidental to the employment of labour.

"Materials net" figures include for all costs of materials including an allowance for waste except where specifically stated.

"Price net" figures are the totals of the "Labour net" and "Materials net" figures. Prices are for a builder employing his own labour; according to the amount and nature of the work involved, it may well be possible to secure more advantageous prices from specialist sub-contractors.

Prices do not include any allowance for scaffolding, ladders or other plant necessary to reach the work. The "Preliminaries" section includes prices for scaffolding which must be considered and allowance included to suit the particular circumstances of a tender.

Specialist prices

"Price with 15%" figures are all-in guide prices and include 15% for the builder's overheads, profit, unloading materials and general attendance (to include free use of standing scaffolding and hoists, temporary lighting and water and clearing away rubbish).

The amount of attendance required varies between the various trades and also with the circumstances of specific jobs; the percentage addition must always be considered and adjusted as necessary to suit the terms and conditions of the quotation being used.

Quantities and delivery distances are usually the most significant of the many factors which influence prices and it must be emphasised that quotations should always be obtained when preparing a tender.

Metalwork

	Unit	Specialist price net	Price with 15%
New work			

Composite items

		£	£
		VAT not included	

Windows, doors and sidelights

The following "Specialist price net" figures are guide prices provided by Crittall Windows Ltd.		Supply only	Supply only

Prices include for delivery carriage paid to sites in the United Kingdom, except for destinations quoted below.

> For delivery to the Highland and Grampian Regions of Scotland and to the former County of Argyll in the Strathclyde Region 6½% should be added to the prices.

> For delivery to islands off the Scottish mainland, i.e. Orkneys, Shetland, Inner and Outer Hebrides and all other islands including those in the Firth of Clyde, 10% should be added to the prices - prices will then be Free on Quay Port of Entry.

> For delivery to Northern Ireland, Isle of Man and the Channel Islands 6½% should be added to the prices - prices will then be Free on Quay Port of Entry.

A cash discount of 2½% will be allowed for payment made up to 30 days from the end of the month in which goods are dispatched.

Non-standard size windows and doors are priced by adding 30% to the price of the nearest larger standard size by length and height, up to the largest standard size shown for the elevational design concerned, e.g. "as NCO13F but 1000 x 1000 mm" is priced as standard NCO13F plus 30%.

Steel windows

Standard galvanised steel window types including weatherstripping opening lights

	Unit	Specialist price net	Price with 15%
508 x 292 mm NG5	each	14.37	16.53
508 x 292 mm NG1	each	48.21	55.44
628 x 292 mm ZNG5	each	16.16	18.58
628 x 292 mm ZNG1	each	52.22	60.05
997 x 292 mm NG13	each	22.09	25.40
997 x 292 mm NG13G	each	66.17	76.10
1237 x 292 mm ZNG13	each	24.87	28.60
1237 x 292 mm ZNG13G	each	74.18	85.31
1486 x 292 mm NG14	each	26.07	29.98
1486 x 292 mm NG4	each	113.17	130.15
1846 x 292 mm ZNG14	each	32.44	37.31
1846 x 292 mm ZNG4	each	125.77	144.64
508 x 457 mm NH5	each	15.71	18.07
508 x 457 mm NH1	each	56.00	64.40
628 x 457 mm ZNH5	each	17.62	20.26
628 x 457 mm ZNH1	each	60.37	69.43
997 x 457 mm NH13	each	24.33	27.98
997 x 457 mm NH13H	each	70.03	80.53

Metalwork

	Unit	Specialist price net	Price with 15%
Composite items		£	£
		VAT not included	

Steel windows *(continued)*

Standard galvanised steel window types including weatherstripping opening lights *(continued)*		Supply only	Supply only
1237 x 457 mm ZNH13	each	26.77	30.79
1237 x 457 mm ZNH13H	each	77.46	89.08
1486 x 457 mm NH14	each	28.94	33.28
1486 x 457 mm NH4	each	128.24	147.48
1846 x 457 mm ZNH14	each	34.11	39.23
1846 x 457 mm ZNH4	each	137.45	158.07
279 x 628 mm NE6F	each	49.94	57.43
508 x 628 mm NE5	each	16.92	19.46
508 x 628 mm NES1	each	61.57	70.81
628 x 628 mm ZNE5	each	20.82	23.94
628 x 628 mm ZNES1	each	66.60	76.59
997 x 628 mm NE13	each	26.01	29.91
1237 x 628 mm ZNE13	each	29.11	33.48
1486 x 628 mm NE14	each	31.13	35.80
1846 x 628 mm ZNE14	each	36.29	41.73
279 x 923 mm NC6F	each	49.94	57.43
508 x 923 mm NC5	each	19.70	22.66
508 x 923 mm NC1	each	66.16	76.08
508 x 923 mm NC5F	each	58.54	67.32
628 x 923 mm ZNC5	each	22.67	26.07
628 x 923 mm ZNC1	each	70.82	81.44
628 x 923 mm ZNC5F	each	63.78	73.35
997 x 923 mm NC13	each	30.45	35.02
997 x 923 mm NC13R	each	189.98	218.48
997 x 923 mm NC13F	each	85.53	98.36
1237 x 923 mm ZNC13	each	32.37	37.23
1237 x 923 mm ZNC13R	each	208.99	240.34
1237 x 923 mm ZNC13F	each	92.28	106.12
1237 x 923 mm ZNC2F	each	130.24	149.78
1237 x 923 mm ZNC2V	each	156.25	179.69
1486 x 923 mm NC14	each	34.99	40.24
1486 x 923 mm NC14R	each	214.19	246.32
1486 x 923 mm NC10F	each	163.48	188.00
1846 x 923 mm ZNC14	each	39.86	45.84
1846 x 923 mm ZNC10F	each	175.59	201.93
508 x 1067 mm NCO5	each	21.97	25.27
508 x 1067 mm NC01	each	70.13	80.65
508 x 1067 mm NC05F	each	59.93	68.92
628 x 1067 mm ZNC05	each	25.72	29.58
628 x 1067 mm ZNC01	each	82.49	94.86
628 x 1067 mm ZNC05F	each	66.84	76.87
997 x 1067 mm NC013	each	33.23	38.21
997 x 1067 mm NC013R	each	201.39	231.60
997 x 1067 mm NC013F	each	88.43	101.69

Metalwork

New work	Unit	Specialist price net	Price with 15%

Composite items

			£	£
			VAT not included	

Steel windows *(continued)*

Standard galvanised steel window types including weatherstripping opening lights *(continued)*

	Unit	Supply only	Supply only
1237 x 1067 mm ZNC013	each	36.17	41.60
1237 x 1067 mm ZNC013R	each	216.74	249.25
1237 x 1067 mm ZNC013F	each	95.14	109.41
1237 x 1067 mm ZNC02F	each	146.10	168.01
1237 x 1067 mm ZNC02V	each	161.29	185.48
1486 x 1067 mm NC014	each	37.96	43.65
1486 x 1067 mm NC014R	each	223.79	257.36
1486 x 1067 mm NC010F	each	171.98	197.78
1846 x 1067 mm ZNC010F	each	180.40	207.46
508 x 1218 mm ND5	each	22.49	25.86
508 x 1218 mm ND1	each	74.65	85.85
508 x 1218 mm ND5F	each	61.41	70.62
628 x 1218 mm ZND5	each	26.69	30.69
628 x 1218 mm ZND1	each	87.51	100.64
628 x 1218 mm ZND5F	each	67.77	77.94
997 x 1218 mm ND13	each	35.31	40.61
997 x 1218 mm ND13R	each	209.77	241.24
997 x 1218 mm ND13F	each	91.29	104.98
1237 x 1218 mm ZND13	each	38.73	44.54
1237 x 1218 mm ZND13R	each	227.47	261.59
1237 x 1218 mm ZND13F	each	96.98	111.53
1237 x 1218 mm ZND2F	each	152.98	175.93
1237 x 1218 mm ZND2V	each	169.34	194.74
1486 x 1218 mm ND14	each	39.60	45.54
1486 x 1218 mm ND14RS	each	238.90	274.74
1486 x 1218 mm ND10F	each	180.84	207.97
1846 x 1218 mm ZND10F	each	191.35	220.05
508 x 1513 mm NDV5	each	33.04	38.00
508 x 1513 mm NDV5F	each	66.47	76.44
628 x 1513 mm ZNDV5	each	33.23	38.21
628 x 1513 mm ZNDV5F	each	74.86	86.09
628 x 1513 mm ZNDV1S	each	104.86	120.59
997 x 1513 mm NDV13	each	40.55	46.63
997 x 1513 mm NDV13RS	each	232.27	267.11
997 x 1513 mm NDV13F	each	98.52	113.30
1237 x 1513 mm ZNDV13	each	43.45	49.97
1237 x 1513 mm ZNDV13RS	each	245.70	282.55
1237 x 1513 mm ZNDV2V	each	186.96	215.00

Metalwork

New work	Unit	Specialist price net	Price with 15%
Composite items		£	£
		VAT not included	
Steel doors		Supply only	Supply only
Standard galvanised steel door types including weatherstripping			
761 x 2056 mm NA15	each	506.49	582.46
997 x 2056 mm NA2	each	762.36	876.71
1143 x 2056 mm NA25	each	776.78	893.30
1237 x 2056 mm ZNA25	each	815.42	937.73
Standard galvanised steel sidelight types			
279 x 2056 mm NA6	each	66.30	76.24
508 x 2056 mm NA5	each	80.58	92.67
628 x 2056 mm ZNA5	each	83.74	96.30
997 x 2056 mm NA13F	each	153.30	176.29

Enquiries about the foregoing specialist prices should be made to Crittall
Windows Ltd, Springwood Drive, Braintree, Essex, CM7 2YN, tel (01376) 324106,
fax (01376) 349662, E-mail hq@crittall-windows.co.uk

For the cost of building in metal windows, see "Brickwork and blockwork".

Metalwork

New work	Unit	Hours C	Hours L	Labour net	Material net	Price net	Price with 15%
Composite items				£	£	£	£
					VAT not included		

Overhead garage doors

Primed galvanised steel garage doors and gear
screwed to timber frame

2135 x 1980 mm	each	4.00	-	35.84	257.04	292.88	336.81
2135 x 2135 mm	each	4.00	-	35.84	280.04	315.88	363.26
4270 x 1980 mm	each	6.00	-	53.76	839.74	893.50	1027.53
4270 x 2135 mm	each	6.00	-	53.76	839.74	893.50	1027.53

Primed galvanised steel garage doors, gear and
integral frame plugged and screwed to wall

2135 x 1980 mm	each	4.00	-	35.84	314.04	349.88	402.36
2135 x 2135 mm	each	4.00	-	35.84	335.04	370.88	426.51

Self-finish GRP garage doors and primed
galvanised steel gear screwed to timber frame

2135 x 1980 mm	each	4.00	-	35.84	782.04	817.88	940.56
2135 x 2135 mm	each	4.00	-	35.84	782.04	817.88	940.56
4270 x 1980 mm	each	6.00	-	53.76	1236.74	1290.50	1484.08
4270 x 2135 mm	each	6.00	-	53.76	1236.74	1290.50	1484.08

Self-finish GRP garage doors and primed
galvanised steel gear and integral frame plugged
and screwed to wall

2135 x 1980 mm	each	4.00	-	35.84	838.04	873.88	1004.96
2135 x 2135 mm	each	4.00	-	35.84	838.04	873.88	1004.96

Railings

900 mm high tubular railings of 25 mm galvanised medium steel tube , with two horizontal rails and standards at 1350 mm centres, incl. patent railing fittings, joints and fanged ends to standards - mortices not included	m	1.25	1.25	19.63	9.99	29.62	34.06

Galvanised steel balustrades

50 x 8 mm flat rail	m	0.40	-	3.58	17.57	21.15	24.32
fanged ends	each	0.60	-	5.38	4.39	9.77	11.24
mitred and welded angles	each	0.60	-	5.38	4.39	9.77	11.24
level bends	each	0.60	-	5.38	4.39	9.77	11.24
ramps	each	0.60	-	5.38	4.39	9.77	11.24
wreaths	each	1.20	-	10.75	26.35	37.10	42.66
38 x 12 mm convex rails	m	0.40	-	3.58	17.57	21.15	24.32
fanged ends	each	0.60	-	5.38	4.39	9.77	11.24
mitred and welded angles	each	0.60	-	5.38	4.39	9.77	11.24
level bends	each	0.60	-	5.38	4.39	9.77	11.24
ramps	each	0.60	-	5.38	4.39	9.77	11.24
wreaths	each	1.20	-	10.75	26.35	37.10	42.66

Metalwork

New work	Unit	Hours C	Hours L	Labour net	Material net	Price net	Price with 15%
Composite items				£	£	£	£
					VAT not included		
Galvanised steel balustrades (*continued*)							
16 mm diameter balusters	m	0.20	-	1.79	10.45	12.24	14.08
fanged ends	each	0.30	-	2.69	3.13	5.82	6.69
welded ends	each	0.30	-	2.69	1.57	4.26	4.90
20 x 20 mm balusters	m	0.20	-	1.79	15.77	17.56	20.19
fanged ends	each	0.30	-	2.69	4.73	7.42	8.53
welded ends	each	0.30	-	2.69	2.37	5.06	5.82
Galvanised steel ladders							
10 x 65 mm flat strips	m	0.60	-	5.38	32.73	38.11	43.83
fanged ends	each	0.60	-	5.38	6.55	11.93	13.72
welded ends	each	0.60	-	5.38	6.55	11.93	13.72
20 mm diameter rungs	m	0.60	-	5.38	9.33	14.71	16.92
welded ends	each	0.60	-	5.38	3.73	9.11	10.48
50 x 6 mm safety hoops	each	0.60	-	5.38	29.93	35.31	40.61
Galvanised steel staircases							
Fire escapes comprising 15 treads, 3.0 m high, 180 x 10 mm flat strips, 6 mm chequer plate treads, 25 x 25 mm standards, 40 x 8 mm flat bottom rail, 38 x 12 mm convex handrail and 12 x 12 mm balusters at 100 mm centres, and all cleats and angles	each	40.00	20.00	493.20	4264.00	4757.20	5470.78

Metalwork

New work	Unit	Hours C	Hours L	Labour net	Material net	Price net	Price with 15%
Plate, bars, etc				£	£	£	£
					VAT not included		
Mild steel							
Flat arch-bars							
25 x 6 mm	m	0.20	0.10	2.46	1.22	3.68	4.23
30 x 6 mm	m	0.25	0.13	3.08	1.38	4.46	5.13
Angle bearers							
40 x 40 x 6 mm	m	0.25	0.13	3.08	3.03	6.11	7.03
50 x 50 x 6 mm	m	0.28	0.14	3.45	3.81	7.26	8.35
75 x 50 x 8 mm	m	0.30	0.15	3.70	6.66	10.36	11.91
190 x 65 x 10 mm plate corbels drilled and countersunk for wood screws	each	0.30	-	2.69	1.26	3.95	4.54
Galvanised steel lintels							
Lintels 25 mm high to support 75 mm partitions							
900 mm	each	0.07	0.07	1.10	3.24	4.34	4.99
1050 mm	each	0.08	0.08	1.26	3.73	4.99	5.74
1200 mm	each	0.09	0.09	1.42	4.21	5.63	6.47
Lintels 25 mm high to support 100 mm partitions							
900 mm	each	0.07	0.07	1.10	3.95	5.05	5.81
1050 mm	each	0.08	0.08	1.26	4.70	5.96	6.85
1200 mm	each	0.09	0.09	1.42	5.19	6.61	7.60
Lintels, standard duty box profile to support 100 mm partitions							
1050 mm	each	0.15	0.15	2.37	12.85	15.22	17.51
1200 mm	each	0.17	0.17	2.68	14.69	17.37	19.98
1500 mm	each	0.19	0.19	2.99	19.74	22.73	26.14
1800 mm	each	0.20	0.20	3.15	30.14	33.29	38.29
2400 mm	each	0.24	0.24	3.77	41.35	45.12	51.89
3000 mm	each	0.30	0.30	4.72	76.46	81.18	93.36
3600 mm	each	0.36	0.36	5.66	92.35	98.01	112.72
4200 mm	each	0.42	0.42	6.60	126.42	133.02	152.98
Lintels, standard duty angle profile to support 102.5 mm external solid wall							
900 mm	each	0.13	0.13	2.05	10.73	12.78	14.70
1200 mm	each	0.17	0.17	2.68	14.28	16.96	19.51
1500 mm	each	0.19	0.19	2.99	17.86	20.85	23.98
1800 mm	each	0.20	0.20	3.15	24.11	27.26	31.35
2100 mm	each	0.22	0.22	3.47	28.11	31.58	36.32
2400 mm	each	0.24	0.24	3.77	39.32	43.09	49.56
2700 mm	each	0.27	0.27	4.24	60.47	64.71	74.42

Metalwork

New work	Unit	Hours C	Hours L	Labour net	Material net	Price net	Price with 15%

Plate, bars, etc

| | | | | | £ | £ | £ | £ |

VAT not included

Galvanised steel lintels *(continued)*

Combined lintels and cavity trays standard duty to
support hollow walls of 100 mm inner and
102.5 mm outer skins and 50 mm cavity

900 mm	each	0.14	0.14	2.21	18.60	20.81	23.94
1200 mm	each	0.18	0.18	2.84	24.15	26.99	31.04
1500 mm	each	0.22	0.22	3.47	31.23	34.70	39.91
1800 mm	each	0.25	0.25	3.93	38.84	42.77	49.19
2100 mm	each	0.28	0.28	4.40	44.62	49.02	56.38
2400 mm	each	0.31	0.31	4.87	54.66	59.53	68.46
2700 mm	each	0.34	0.34	5.35	62.04	67.39	77.50

Combined lintels and cavity trays standard duty to
support hollow walls of 100 mm inner and
102.5 mm outer skins and 50 mm cavity

3000 mm	each	0.39	0.39	6.13	85.24	91.37	105.08
3300 mm	each	0.42	0.42	6.60	95.73	102.33	117.68
3600 mm	each	0.45	0.45	7.08	105.05	112.13	128.95
3900 mm	each	0.47	0.47	7.39	151.48	158.87	182.70
4200 mm	each	0.49	0.49	7.70	160.80	168.50	193.78
4575 mm	each	0.51	0.51	8.01	173.01	181.02	208.18
4800 mm	each	0.53	0.53	8.33	181.84	190.17	218.70

Combined lintels and cavity trays standard duty to
support hollow walls of 140 mm inner and
102.5 mm outer skins and 50 mm cavity

900 mm	each	0.15	0.15	2.37	18.06	20.43	23.50
1200 mm	each	0.20	0.20	3.15	24.11	27.26	31.35
1500 mm	each	0.24	0.24	3.77	31.03	34.80	40.02
1800 mm	each	0.28	0.28	4.40	38.53	42.93	49.37
2100 mm	each	0.31	0.31	4.87	44.51	49.38	56.79
2400 mm	each	0.34	0.34	5.35	54.23	59.58	68.52
2700 mm	each	0.37	0.37	5.82	62.04	67.86	78.03

Combined lintels and cavity trays standard duty to
support hollow walls of 100 mm inner and
102.5 mm outer skins and 50 mm cavity

3000 mm	each	0.43	0.43	6.76	85.47	92.23	106.07
3300 mm	each	0.46	0.46	7.23	96.16	103.39	118.90
3600 mm	each	0.49	0.49	7.70	104.82	112.52	129.40
3900 mm	each	0.51	0.51	8.01	142.21	150.22	172.75
4200 mm	each	0.53	0.53	8.33	149.63	157.96	181.66
4575 mm	each	0.55	0.55	8.64	168.73	177.37	203.98
4800 mm	each	0.57	0.57	8.96	180.13	189.09	217.45

Metalwork

	Unit	Hours C	Hours L	Labour net	Material net	Price net	Price with 15%
Plate, bars, etc				£	£	£	£
					VAT not included		
Aluminium							
Mat-frames of 25 x 25 x 3 mm angle with lugs welded on							
450 x 750 mm	each	2.45	-	21.95	21.00	42.95	49.39
600 x 900 mm	each	2.45	-	21.95	26.30	48.25	55.49
Bolts etc							
Steel bolts with nut and washer							
M10 x 50 mm	each	0.10	-	0.90	0.31	1.21	1.39
M10 x 70 mm	each	0.10	-	0.90	0.36	1.26	1.45
M10 x 100 mm	each	0.12	-	1.08	0.48	1.56	1.79
M12 x 70 mm	each	0.10	-	0.90	0.53	1.43	1.64
M12 x 100 mm	each	0.12	-	1.08	0.65	1.73	1.99
M12 x 120 mm	each	0.12	-	1.08	0.74	1.82	2.09
Steel "Bolt Projecting Rawlbolts"							
M10 30P	each	0.10	-	0.90	1.34	2.24	2.58
M12 30P	each	0.10	-	0.90	1.56	2.46	2.83
M16 35P	each	0.12	-	1.08	2.62	3.70	4.25
Steel, "Rawlbolt", loose bolt type							
M10 25L	nr	0.10	-	0.90	1.60	2.50	2.88
M12 10L	nr	0.10	-	0.90	2.57	3.47	3.99
M12 60L	nr	0.10	-	0.90	6.25	7.15	8.22
Resin Bonded Fixings							
Chemical anchors							
Kemfix chemical anchors with standard studs, drilling masonry							
capsule type bolt 12 x 160	nr	0.20	-	1.79	2.69	4.48	5.15

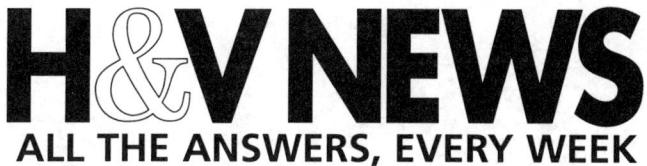

H&V NEWS

ALL THE ANSWERS, EVERY WEEK

The only source of up-to-date, high quality information for the hvac industry.

The business journal for specifiers and buyers, to find out all you need to know about your industry.
Our readers are kept updated in a rapidly changing marketplace.

★ **Exclusive news coverage**

★ **In-depth news features**

★ **Live project details and tender information**

★ **Regular bulletins on latest products**

★ **Market research on developments and financial news**

You could be entitled to receive H&V News free of charge if you meet our terms of control.

call

0181 277 5208

to receive a registration card.

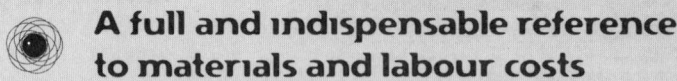
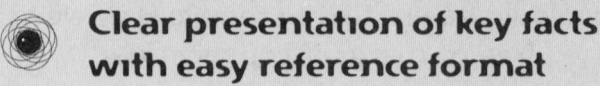

Plumbing and engineering installations

Preamble

"Labour net" figures for plumbing and engineering work are based on the labour costs of an advanced plumber working with an apprentice in the third year of training and include allowances for all costs incidental to the employment of labour.

"Plant net" figures include for all costs of plant including drivers and operators where applicable.

"Materials net" figures include for all costs of materials including an allowance for waste except where specifically stated.

"Price net" figures are the totals of the "Labour net", "Plant net", where applicable, and "Materials net" figures. Prices are for a builder employing his own labour; according to the amount and nature of the work involved, it may well be possible to secure more advantageous prices from specialist sub-contractors.

Although not in accordance with the Standard Method of Measurement, allowance has been included in pipework prices for cutting and pinning holderbats.

Prices do not include any allowance for scaffolding, ladders or other plant necessary to reach the work. The "Preliminaries" section includes prices for scaffolding which must be considered and allowance included to suit the particular circumstances of a tender.

Specialist prices

"Price with 15%" figures are all-in guide prices and include 15% for the builder's overheads, profit, unloading materials and general attendance (to include free use of standing scaffolding and hoists, temporary lighting and water and clearing away rubbish).

The amount of attendance required varies between the various trades and also with the circumstances of specific jobs; the percentage addition must always be considered and adjusted as necessary to suit the terms and conditions of the quotation being used.

Quantities and delivery distances are usually the most significant of the many factors which influence prices and it must be emphasised that quotations should always be obtained when preparing a tender.

Plumbing and engineering installations

	Unit	Price

New work

Basic prices for materials

£

Gutters

Aluminium - 1830 mm lengths

	Unit	Price
100 mm half round	each	14.00
125 mm half round	each	17.49
115 mm ogee	each	21.36
125 mm ogee	each	24.35

Asbestos-free reinforced cement - 1800 mm lengths

125 mm half round	each	13.52
150 mm half round	each	16.25
127 x 152 mm box	each	39.18
457 x 127 x 152 mm gutter	each	43.91

Cast iron - 1830 mm lengths

100 mm half round	each	12.34
125 mm half round	each	15.04
150 mm half round	each	25.71
115 mm ogee	each	15.14
125 mm ogee	each	15.88

Pressed steel - in purpose-made lengths

375 mm girth x 12G	m	10.27
525 mm girth x 3.2 mm	m	17.85

Galvanised pressed steel - in purpose-made lengths

375 mm girth x 12G	m	9.98
525 mm girth x 3.2 mm	m	19.40

PVC-U

76 mm half round - 2000 mm lengths	each	3.09
112 mm half round - 4000 mm lengths	each	7.76
150 mm half round - 4000 mm lengths	each	21.88
112 mm square section - 4000 mm lengths	each	7.84

Rainwater pipes

Aluminium - 1830 mm lengths

63 mm	each	19.38
75 mm	each	22.55

Asbestos-free reinforced cement - 1800 mm lengths

75 mm	each	15.89
100 mm	each	21.10

Cast iron - 1830 mm lengths

50 mm	each	24.26
65 mm	each	24.26
75 mm	each	24.26
100 mm	each	32.56

Prices actually to be paid for materials must be checked against the above basic prices and adjustments made as necessary.

Plumbing and engineering installations

New work

Basic prices for materials		£

Rainwater pipes *(continued)*

PVC-U

50 mm - 2000 mm lengths	each	4.04
68 mm - 2000 mm lengths	each	4.22
110 mm - 3000 mm lengths	each	12.58
65 mm square section - 2500 mm lengths	each	5.29

Overflow, soil, waste and vent pipes

Ensign cast iron - 3000 mm lengths

50 mm pipes	each	26.83
50 mm couplings	each	3.79
70 mm pipes	each	31.05
70 mm couplings	each	4.17
100 mm pipes	each	37.30
100 mm couplings	each	5.45

Copper to BS 2871 Table X

35 mm	m	6.28
42 mm	m	7.67
54 mm	m	9.89

Thinwall copper to BS 2871 Table Z

35 mm	m	5.55
42 mm	m	7.10
54 mm	m	9.60

Polypropylene - 3000 mm lengths

32 mm	each	1.99
38 mm	each	2.45
50 mm	each	3.35

PVC-U 3000 mm lengths

19 mm	each	1.28
32 mm	each	2.98
38 mm	each	3.67
50 mm	each	5.39
100 mm	each	11.33

Service pipes

Copper to BS 2871 Table X

15 mm	m	1.01
22 mm	m	2.03
28 mm	m	2.78
35 mm	m	6.28
42 mm	m	7.67
54 mm	m	9.89

Prices actually to be paid for materials must be checked against the above basic prices and adjustments made as necessary.

Plumbing and engineering installations

	Unit	Price
New work		

Basic prices for materials		£

Service pipes *(continued)*

	Unit	Price
Thinwall copper to BS 2871 Table Z		
15 mm	m	1.06
22 mm	m	1.98
28 mm	m	2.55
35 mm	m	5.55
42 mm	m	7.10
54 mm	m	9.60
Underground copper to BS 2871 Table Y		
15 mm	m	2.38
22 mm	m	4.19
28 mm	m	5.57
Medium density polyethylene to BS 6730 - black		
20 mm	m	0.96
25 mm	m	1.22
32 mm	m	1.93
50 mm	m	4.43
63 mm	m	6.90
Medium density polyethylene to BS 6572 - blue		
20 mm	m	0.62
25 mm	m	0.80
32 mm	m	1.33
Unplasticised PVC to BS 3505, Class E		
3/8 "	m	0.90
½"	m	1.09
¾"	m	1.46
1"	m	1.88
1¼"	m	2.76
1½"	m	3.48
Hepworth Hep20 Polybutylene Standard		
15 mm - 50 m coils	each	61.48
15 mm - 3000 mm cut lengths	each	3.74
15 mm - 6000 mm cut lengths	each	7.43
22 mm - 50 m coils	each	116.45
22 mm - 3000 mm cut lengths	each	7.07
22 mm - 6000 mm cut lengths	each	14.12
Black medium steel to BS 1387		
15 mm	m	2.79
20 mm	m	3.28
25 mm	m	4.69
32 mm	m	5.79
40 mm	m	6.73
50 mm	m	9.46

Prices actually to be paid for materials must be checked against the above basic prices and adjustments made as necessary.

Plumbing and engineering installations

New work

Basic prices for materials £

Service pipes *(continued)*

Black heavy steel to BS 1387

15 mm	m	3.27
20 mm	m	3.87
25 mm	m	5.62
32 mm	m	6.97
40 mm	m	8.13
50 mm	m	11.28

Galvanised medium steel to BS 1387

15 mm	m	4.25
20 mm	m	4.80
25 mm	m	6.68
32 mm	m	8.28
40 mm	m	9.62
50 mm	m	13.48

Galvanised heavy steel to BS 1387

15 mm	m	4.95
20 mm	m	5.62
25 mm	m	7.99
32 mm	m	9.92
40 mm	m	11.57
50 mm	m	16.02

Prices actually to be paid for materials must be checked against the above basic prices and adjustments made as necessary.

Plumbing and engineering installations

New work	Unit	Hours C	Hours L	Labour net	Material net	Price net	Price with 15%
Gutterwork				£	£	£	£
					VAT not included		
Aluminium to BS 2997							
100 mm half round gutters jointed and fixed with brackets screwed to timber	m	0.30	0.30	5.14	11.98	17.12	19.69
Extra for							
angles	each	0.25	0.25	4.28	9.37	13.65	15.70
stopped ends	each	0.17	0.17	2.91	6.10	9.01	10.36
outlets	each	0.25	0.25	4.28	10.33	14.61	16.80
125 mm half round gutters jointed and fixed with brackets screwed to timber	m	0.35	0.35	5.99	16.04	22.03	25.33
Extra for							
angles	each	0.25	0.25	4.28	12.15	16.43	18.89
stopped ends	each	0.17	0.17	2.91	8.14	11.05	12.71
outlets	each	0.25	0.25	4.28	12.38	16.66	19.16
115 mm ogee gutters jointed and fixed with mushroom head screws to timber	m	0.35	0.35	5.99	18.18	24.17	27.80
Extra for							
angles	each	0.25	0.25	4.28	12.19	16.47	18.94
stopped ends	each	0.17	0.17	2.91	7.04	9.95	11.44
outlets	each	0.25	0.25	4.28	12.39	16.67	19.17
125 mm ogee gutters jointed and fixed with mushroom head screws to timber	m	0.38	0.38	6.51	20.50	27.01	31.06
Extra for							
angles	each	0.33	0.33	5.66	14.47	20.13	23.15
stopped ends	each	0.20	0.20	3.42	7.89	11.31	13.01
outlets	each	0.33	0.33	5.66	14.38	20.04	23.05
Asbestos-free reinforced cement							
125 mm half round gutters jointed with butyl pads and fixed with galvanised brackets screwed to timber	m	0.30	0.30	5.14	10.03	15.17	17.45
Extra for							
angles	each	0.25	0.25	4.28	8.34	12.62	14.51
stopped ends	each	0.17	0.17	2.91	2.71	5.62	6.46
outlets	each	0.25	0.25	4.28	6.84	11.12	12.79
150 mm half round gutters jointed with butyl pads and fixed with galvanised brackets screwed to timber	m	0.35	0.35	5.99	11.74	17.73	20.39
Extra for							
angles	each	0.33	0.33	5.66	9.59	15.25	17.54
stopped ends	each	0.20	0.20	3.42	3.13	6.55	7.53
outlets	each	0.33	0.33	5.66	9.68	15.34	17.64

Plumbing and engineering installations

New work	Unit	Hours C	Hours L	Labour net	Material net	Price net	Price with 15%
Gutterwork				£	£	£	£
					VAT not included		
Asbestos-free reinforced cement (*continued*)							
127 x 152 mm box gutters jointed with butyl pads – bearers not included	m	0.35	0.35	5.99	25.01	31.00	35.65
Extra for							
stopped ends	each	0.33	0.33	5.66	7.70	13.36	15.36
outlets	each	0.50	0.50	8.57	20.82	29.39	33.80
457 x 127 x 152 mm valley gutters jointed with butyl pads - bearers not included	m	0.60	0.60	10.28	31.16	41.44	47.66
Extra for							
stopped ends	each	0.50	0.50	8.57	7.70	16.27	18.71
outlets	each	0.75	0.75	12.85	20.90	33.75	38.81
Cast iron to BS 460							
100 mm half round gutters jointed and fixed with galvanised brackets screwed to timber	m	0.30	0.30	5.14	12.01	17.15	19.72
Extra for							
angles	each	0.25	0.25	4.28	11.49	15.77	18.14
stopped ends	each	0.17	0.17	2.91	6.93	9.84	11.32
outlets	each	0.25	0.25	4.28	9.64	13.92	16.01
125 mm half round gutters jointed and fixed with galvanised brackets screwed to timber	m	0.35	0.35	5.99	14.57	20.56	23.64
Extra for							
angles	each	0.25	0.25	4.28	14.82	19.10	21.97
stopped ends	each	0.17	0.17	2.91	9.47	12.38	14.24
outlets	each	0.25	0.25	4.28	13.90	18.18	20.91
150 mm half round gutters jointed and fixed with galvanised brackets screwed to timber	m	0.40	0.40	6.85	21.64	28.49	32.76
Extra for							
angles	each	0.33	0.33	5.66	22.50	28.16	32.38
stopped ends	each	0.20	0.20	3.42	11.27	14.69	16.89
outlets	each	0.33	0.33	5.66	21.47	27.13	31.20
115 mm ogee gutters jointed and fixed with mushroom head screws to timber	m	0.35	0.35	5.99	13.82	19.81	22.78
Extra for							
angles	each	0.25	0.25	4.28	12.39	16.67	19.17
stopped ends	each	0.17	0.17	2.91	7.40	10.31	11.86
outlets	each	0.25	0.25	4.28	10.08	14.36	16.51

Plumbing and engineering installations

New work	Unit	Hours C	Hours L	Labour net	Material net	Price net	Price with 15%
Gutterwork				£	£	£	£
					VAT not included		
Cast iron to BS460 (*continued*)							
125 mm ogee gutters jointed and fixed with mushroom head screws to timber	m	0.40	0.40	6.85	15.16	22.01	25.31
Extra for							
angles	each	0.33	0.33	5.66	14.26	19.92	22.91
stopped ends	each	0.20	0.20	3.42	9.57	12.99	14.94
outlets	each	0.33	0.33	5.66	12.26	17.92	20.61
Pressed steel box gutters in purpose-made lengths							
375 mm girth x 12G gutters with bolted socketed joints - bearers not included	m	0.35	0.35	5.99	10.78	16.77	19.29
Extra for							
stopped ends welded in	each	-	-	-	7.10	7.10	8.16
75 mm outlets welded in	each	-	-	-	9.47	9.47	10.89
525 mm girth x 3.2 mm gutters with bolted socketed joints - bearers not included	m	0.45	0.45	7.71	18.74	26.45	30.42
Extra for							
stopped ends welded in	each	-	-	-	9.47	9.47	10.89
100 mm outlets welded in	each	-	-	-	12.63	12.63	14.52
Galvanised pressed steel box gutters in purpose- made lengths							
375 mm girth x 12G gutters with bolted socketed joints - bearers not included	m	0.35	0.35	5.99	10.48	16.47	18.94
Extra for							
stopped ends welded in	each	-	-	-	7.10	7.10	8.16
75 mm outlets welded in	each	-	-	-	9.47	9.47	10.89
525 mm girth x 3.2 mm gutters with bolted socketed joints - bearers not included	m	0.45	0.45	7.71	39.11	46.82	53.84
Extra for							
stopped ends welded in	each	-	-	-	9.47	9.47	10.89
100 mm outlets welded in	each	-	-	-	12.63	12.63	14.52
PVCU to BS 4576							
76 mm half round gutters jointed and fixed with support and union brackets screwed to timber	m	0.25	0.25	4.28	3.20	7.48	8.60
Extra for							
angles	each	0.25	0.25	4.28	2.93	7.21	8.29
stopped ends	each	0.17	0.17	2.91	1.61	4.52	5.20
outlets	each	0.25	0.25	4.28	2.55	6.83	7.85

Plumbing and engineering installations

New work	Unit	Hours C	Hours L	Labour net	Material net	Price net	Price with 15%
Gutterwork				£	£	£	£
					VAT not included		
PVCU to BS 4576 (*continued*)							
112 mm half round gutters jointed and fixed with support and union brackets screwed to timber	m	0.27	0.27	4.62	3.81	8.43	9.69
Extra for							
angles	each	0.25	0.25	4.28	2.88	7.16	8.23
stopped ends	each	0.17	0.17	2.91	1.61	4.52	5.20
outlets	each	0.25	0.25	4.28	2.55	6.83	7.85
150 mm half round gutters jointed with floating unions and fixed with support brackets screwed to timber	m	0.30	0.30	5.14	7.51	12.65	14.55
Extra for							
angles	each	0.25	0.25	4.28	10.83	15.11	17.38
stopped ends	each	0.17	0.17	2.91	4.03	6.94	7.98
outlets	each	0.25	0.25	4.28	8.31	12.59	14.48
112 mm square section gutters jointed and fixed with support and union brackets screwed to timber	m	0.30	0.30	5.14	3.83	8.97	10.32
Extra for							
angles	each	0.25	0.25	4.28	3.45	7.73	8.89
stopped ends	each	0.17	0.17	2.91	1.52	4.43	5.09
outlets	each	0.25	0.25	4.28	3.52	7.80	8.97

Plumbing and engineering installations

New work	Unit	Hours C	Hours L	Labour net	Material net	Price net	Price with 15%
Rainwater pipework				£	£	£	£
					VAT not included		
Aluminium to BS 2997							
63 mm pipes with ears plugged to walls	m	0.28	0.28	4.79	11.73	16.52	19.00
Extra for							
bends	each	0.25	0.25	4.28	9.56	13.84	15.92
150 mm offsets	each	0.25	0.25	4.28	28.94	33.22	38.20
300 mm offsets	each	0.30	0.30	5.14	41.37	46.51	53.49
branches	each	0.33	0.33	5.66	12.53	18.19	20.92
shoes	each	0.25	0.25	4.28	8.34	12.62	14.51
63 mm flat back hopper heads plugged to walls	each	0.50	0.50	8.57	17.19	25.76	29.62
75 mm pipes with ears plugged to walls	m	0.30	0.30	5.14	13.55	18.69	21.49
Extra for							
bends	each	0.25	0.25	4.28	9.56	13.84	15.92
150 mm offsets	each	0.25	0.25	4.28	31.72	36.00	41.40
300 mm offsets	each	0.30	0.30	5.14	47.39	52.53	60.41
branches	each	0.33	0.33	5.66	15.55	21.21	24.39
shoes	each	0.25	0.25	4.28	11.59	15.87	18.25
75 mm flat back hopper heads plugged to walls	each	0.54	0.54	9.25	17.94	27.19	31.27
Asbestos-free reinforced cement							
75 mm pipes fixed with galvanised clips plugged to walls	m	0.25	0.25	4.28	11.56	15.84	18.22
Extra for							
bends	each	0.17	0.17	2.91	8.55	11.46	13.18
150 mm offsets	each	0.25	0.25	4.28	12.76	17.04	19.60
300 mm offsets	each	0.25	0.25	4.28	16.47	20.75	23.86
branches	each	0.30	0.30	5.14	12.76	17.90	20.59
shoes	each	0.20	0.20	3.42	7.39	10.81	12.43
75 mm rectangular hopper heads	each	0.60	0.60	10.28	22.23	32.51	37.39
100 mm pipes fixed with galvanised clips plugged to walls	m	0.30	0.30	5.14	14.74	19.88	22.86
Extra for							
bends	each	0.20	0.20	3.42	12.76	16.18	18.61
150 mm offsets	each	0.25	0.25	4.28	14.61	18.89	21.72
300 mm offsets	each	0.25	0.25	4.28	20.09	24.37	28.03
450 mm offsets	each	0.30	0.30	5.14	28.37	33.51	38.54
branches	each	0.35	0.35	5.99	17.16	23.15	26.62
shoes	each	0.30	0.30	5.14	9.68	14.82	17.04
100 mm rectangular hopper heads	each	0.67	0.67	11.47	27.62	39.09	44.95

Plumbing and engineering installations

New work	Unit	Hours C	Hours L	Labour net	Material net	Price net	Price with 15%
Rainwater pipework				£	£	£	£
					VAT not included		
Cast iron to BS 460							
50 mm pipes with ears plugged to walls	m	0.25	0.25	4.28	13.94	18.22	20.95
65 mm pipes with ears plugged to walls	m	0.43	0.43	7.36	13.97	21.33	24.53
Extra for							
bends	each	0.13	0.13	2.22	9.11	11.33	13.03
150 mm offsets	each	0.13	0.13	2.22	13.83	16.05	18.46
300 mm offsets	each	0.17	0.17	2.91	18.78	21.69	24.94
branches	each	0.17	0.17	2.91	17.58	20.49	23.56
shoes	each	0.54	0.54	9.25	14.75	24.00	27.60
65 mm flat back hopper heads plugged to walls	each	0.54	0.54	9.25	12.77	22.02	25.32
75 mm pipes with ears plugged to walls	m	0.45	0.45	7.71	14.04	21.75	25.01
Extra for							
bends	each	0.17	0.17	2.91	9.18	12.09	13.90
150 mm offsets	each	0.17	0.17	2.91	13.90	16.81	19.33
300 mm offsets	each	0.20	0.20	3.42	18.85	22.27	25.61
branches	each	0.20	0.20	3.42	17.71	21.13	24.30
shoes	each	0.60	0.60	10.28	14.82	25.10	28.86
75 mm flat back hopper heads plugged to walls	each	0.60	0.60	10.28	12.84	23.12	26.59
100 mm pipes with ears plugged to walls	m	0.50	0.50	8.57	18.58	27.15	31.22
Extra for							
bends	each	0.20	0.20	3.42	15.58	19.00	21.85
150 mm offsets	each	0.20	0.20	3.42	26.02	29.44	33.86
300 mm offsets	each	0.25	0.25	4.28	31.45	35.73	41.09
branches	each	0.25	0.25	4.28	23.12	27.40	31.51
shoes	each	0.67	0.67	11.47	19.25	30.72	35.33
100 mm flat back hopper heads plugged to walls	each	0.67	0.67	11.47	30.13	41.60	47.84
PVCU to BS 4576							
50 mm pipes fixed with brackets plugged to walls	m	0.25	0.25	4.28	4.16	8.44	9.71
Extra for							
bends	each	0.15	0.15	2.57	1.76	4.33	4.98
offset bends	each	0.15	0.15	2.57	1.76	4.33	4.98
shoes	each	0.30	0.30	5.14	1.72	6.86	7.89
68 mm pipes fixed with brackets plugged to walls	m	0.25	0.25	4.28	4.77	9.05	10.41
Extra for							
bends	each	0.15	0.15	2.57	5.11	7.68	8.83
offset bends	each	0.15	0.15	2.57	2.85	5.42	6.23
branches	each	0.20	0.20	3.42	9.08	12.50	14.38
shoes	each	0.30	0.30	5.14	3.00	8.14	9.36

Plumbing and engineering installations

New work	Unit	Hours C	Hours L	Labour net	Material net	Price net	Price with 15%
Rainwater pipework				£	£	£	£
					VAT not included		
PVCU to BS 4576 (continued)							
68 mm flat back hopper heads plugged to walls	each	0.50	0.50	8.57	12.77	21.34	24.54
110 mm pipes fixed with brackets plugged to walls	m	0.30	0.30	5.14	6.71	11.85	13.63
Extra for							
bends	each	0.25	0.25	4.28	6.81	11.09	12.75
top offset bends	each	0.25	0.25	4.28	7.04	11.32	13.02
bottom offset bends	each	0.25	0.25	4.28	5.52	9.80	11.27
branches	each	0.20	0.20	3.42	10.82	14.24	16.38
shoes	each	0.30	0.30	5.14	9.04	14.18	16.31
65 mm square section pipes fixed with brackets plugged to walls	m	0.25	0.25	4.28	4.38	8.66	9.96
Extra for							
bends	each	0.15	0.15	2.57	1.48	4.05	4.66
offset bends	each	0.15	0.15	2.57	1.48	4.05	4.66
44 mm offsets	each	0.15	0.15	2.57	3.61	6.18	7.11
branches	each	0.20	0.20	3.42	4.76	8.18	9.41
shoes	each	0.30	0.30	5.14	2.00	7.14	8.21
Flat back hopper heads for 65 mm square section pipes plugged to walls	each	0.50	0.50	8.57	8.38	16.95	19.49

Plumbing and engineering installations

New work	Unit	Hours C	Hours L	Labour net	Material net	Price net	Price with 15%
Overflow, soil, waste and vent pipework				£	£	£	£
					VAT not included		
Ensign cast iron system							
50 mm pipes jointed with flexible couplings and fixed with brackets bolted to walls	m	0.17	0.17	2.83	15.50	18.33	21.08
Extra for							
access pipes with round door	each	0.14	0.14	2.48	17.91	20.39	23.45
bends	each	0.14	0.14	2.48	6.72	9.20	10.58
single equal branches	each	0.18	0.18	3.09	10.80	13.89	15.97
P traps with access door	each	0.18	0.18	3.09	25.42	28.51	32.79
plain blank ends	each	0.08	0.08	1.37	2.71	4.08	4.69
70 mm pipes jointed with flexible couplings and fixed with brackets bolted to walls	m	0.19	0.19	3.17	17.18	20.35	23.40
Extra for							
access pipes with round door	each	0.16	0.16	2.74	18.96	21.70	24.95
bends	each	0.16	0.16	2.74	7.57	10.31	11.86
single equal branches	each	0.20	0.20	3.42	11.40	14.82	17.04
70 x 50 mm taper pipes	each	0.16	0.16	2.74	10.35	13.09	15.05
P traps with access door	each	0.20	0.20	3.42	25.42	28.84	33.17
plain blank ends	each	0.10	0.10	1.72	2.86	4.58	5.27
expansion plugs	each	0.10	0.10	1.72	13.07	14.79	17.01
100 mm pipes jointed with flexible couplings and fixed with brackets bolted to walls	m	0.20	0.20	3.42	20.02	23.44	26.96
Extra for							
access pipes with round door	each	0.17	0.17	3.00	20.85	23.85	27.43
bends	each	0.17	0.17	3.00	8.97	11.97	13.77
75 mm offsets	each	0.17	0.17	3.00	13.80	16.80	19.32
single equal branches	each	0.21	0.21	3.59	16.04	19.63	22.57
single unequal branches	each	0.21	0.21	3.59	16.40	19.99	22.99
single equal branches with access door	each	0.21	0.21	3.59	27.61	31.20	35.88
double equal branches	each	0.25	0.25	4.28	20.88	25.16	28.93
100 x 50 taper pipes	each	0.17	0.17	3.00	12.18	15.18	17.46
100 x 70 taper pipes	each	0.17	0.17	3.00	12.18	15.18	17.46
plain "P" traps	each	0.25	0.25	4.28	16.65	20.93	24.07
P traps with access door	each	0.25	0.25	4.28	27.52	31.80	36.57
roof connectors for asphalt	each	0.17	0.17	3.00	25.27	28.27	32.51
expansion plugs	each	0.09	0.09	1.54	14.11	15.65	18.00
Extra for 100 mm blank ends							
plain	each	0.09	0.09	1.54	3.34	4.88	5.61
tapped for 50 mm connection	each	0.09	0.09	1.54	6.88	8.42	9.68
with rubber plug for push-fit connection	each	0.09	0.09	1.54	6.18	7.72	8.88
Extra for 100 mm boss pipes tapped for 50 mm connection							
pipes with one boss	each	0.17	0.17	3.00	18.53	21.53	24.76
pipes with two bosses	each	0.17	0.17	3.00	23.94	26.94	30.98

Plumbing and engineering installations

New work	Unit	Hours C	Hours L	Labour net	Material net	Price net	Price with 15%
Overflow, soil, waste and vent pipework				£	£	£	£
					VAT not included		

Copper to BS 2871 Table X

	Unit	Hours C	Hours L	Labour net	Material net	Price net	Price with 15%
Pipes jointed with compression or capillary fittings (measured separately) and fixed with two-piece copper spacing clips plugged to walls							
35 mm	m	0.31	0.31	5.31	6.88	12.19	14.02
42 mm	m	0.33	0.33	5.66	8.36	14.02	16.12
54 mm	m	0.35	0.35	5.99	10.76	16.75	19.26
Pipes jointed with compression or capillary fittings (measure separately) and fixed with pressed brass brackets plugged to walls							
35 mm	m	0.31	0.31	5.31	7.08	12.39	14.25
42 mm	m	0.33	0.33	5.66	8.62	14.28	16.42
54 mm	m	0.35	0.35	5.99	11.05	17.04	19.60
Extra for made bends							
35 mm	each	0.28	0.28	4.79	-	4.79	5.51
42 mm	each	0.30	0.30	5.14	-	5.14	5.91
54 mm	each	0.38	0.38	6.51	-	6.51	7.49

Thinwall copper to BS 2871 Table Z

	Unit	Hours C	Hours L	Labour net	Material net	Price net	Price with 15%
Pipes jointed with compression or capillary fittings (measure separately) and fixed with two-piece copper spacing clips plugged to walls							
35 mm	m	0.31	0.31	5.31	6.12	11.43	13.14
42 mm	m	0.33	0.33	5.66	7.76	13.42	15.43
54 mm	m	0.35	0.35	5.99	10.46	16.45	18.92
Pipes jointed with compression or capillary fittings (measure separately) and fixed with pressed brass brackets plugged to walls							
35 mm	m	0.31	0.31	5.31	6.32	11.63	13.37
42 mm	m	0.33	0.33	5.66	8.02	13.68	15.73
54 mm	m	0.35	0.35	5.99	10.75	16.74	19.25

Note: Thinwall copper is not suitable for made bends

Capillary fittings for copper

	Unit	Hours C	Hours L	Labour net	Material net	Price net	Price with 15%
Straight couplings							
35 mm	each	0.27	0.27	4.62	1.80	6.42	7.38
42 mm	each	0.33	0.33	5.66	2.67	8.33	9.58
54 mm	each	0.37	0.37	6.34	5.84	12.18	14.01
Bends							
35 mm	each	0.27	0.27	4.62	3.71	8.33	9.58
42 mm	each	0.33	0.33	5.66	6.28	11.94	13.73
54 mm	each	0.37	0.37	6.34	13.87	20.21	23.24

Plumbing and engineering installations

New work	Unit	Hours C	Hours L	Labour net	Material net	Price net	Price with 15%
Overflow, soil, waste and vent pipework				£	£	£	£
					VAT not included		
Capillary fittings for copper (*continued*)							
Tees							
35 mm	each	0.35	0.35	5.99	6.43	12.42	14.28
42 mm	each	0.38	0.38	6.51	9.49	16.00	18.40
54 mm	each	0.45	0.45	7.71	18.48	26.19	30.12
Adjustable two-piece tubular "P" traps with 38 mm seal and screwed and capillary joints							
35 mm	each	0.30	0.30	5.14	8.40	13.54	15.57
42 mm	each	0.35	0.35	5.99	12.43	18.42	21.18
Adjustable two-piece tubular "P" traps with 76 mm seal and screwed and capillary joints							
35 mm	each	0.30	0.30	5.14	9.04	14.18	16.31
42 mm	each	0.35	0.35	5.99	12.88	18.87	21.70
Adjustable two-piece tubular "S" traps with 38 mm seal and screwed and capillary joints							
35 mm	each	0.30	0.30	5.14	9.04	14.18	16.31
42 mm	each	0.35	0.35	5.99	12.94	18.93	21.77
Adjustable two-piece tubular "S" traps with 76 mm seal and screwed and capillary joints							
35 mm	each	0.30	0.30	5.14	8.98	14.12	16.24
42 mm	each	0.35	0.35	5.99	13.39	19.38	22.29
Compression fittings for copper							
Straight couplings							
35 mm	each	0.25	0.25	4.28	6.78	11.06	12.72
42 mm	each	0.27	0.27	4.62	8.69	13.31	15.31
54 mm	each	0.31	0.31	5.31	13.00	18.31	21.06
Bends							
35 mm	each	0.25	0.25	4.28	8.78	13.06	15.02
42 mm	each	0.27	0.27	4.62	12.37	16.99	19.54
54 mm	each	0.31	0.31	5.31	20.33	25.64	29.49
Tees							
35 mm	each	0.29	0.29	4.97	11.61	16.58	19.07
42 mm	each	0.34	0.34	5.83	19.32	25.15	28.92
54 mm	each	0.39	0.39	6.68	30.30	36.98	42.53
Adjustable two-piece tubular "P" traps with 38 mm seal and screwed and compression joints							
35 mm	each	0.30	0.30	5.14	15.92	21.06	24.22
42 mm	each	0.35	0.35	5.99	18.39	24.38	28.04
Adjustable two-piece tubular "S" traps with 38 mm seal and screwed and compression joints							
35 mm	each	0.30	0.30	5.14	16.87	22.01	25.31
42 mm	each	0.35	0.35	5.99	19.92	25.91	29.80

Plumbing and engineering installations

New work	Unit	Hours C	Hours L	Labour net	Material net	Price net	Price with 15%
Overflow, soil, waste and vent pipework				£	£	£	£
					VAT not included		
Compression fittings for copper (*continued*)							
Adjustable two-piece tubular "P" traps with 76 mm seal and screwed and compression joints							
35 mm	each	0.30	0.30	5.14	17.06	22.20	25.53
42 mm	each	0.35	0.35	5.99	19.91	25.90	29.79
Adjustable two-piece tubular "S" traps with 76 mm seal and screwed and compression joints							
35 mm	each	0.30	0.30	5.14	18.01	23.15	26.62
42 mm	each	0.35	0.35	5.99	21.42	27.41	31.52
Polypropylene							
32 mm pipes jointed with seal ring fittings and fixed with clips plugged to walls	m	0.20	0.20	3.42	1.16	4.58	5.27
Extra for							
bends	each	0.13	0.13	2.22	0.60	2.82	3.24
branches	each	0.15	0.15	2.57	0.60	3.17	3.65
38 mm pipes jointed with seal ring fittings and fixed with clips plugged to walls	m	0.23	0.23	3.94	1.33	5.27	6.06
Extra for							
bends	each	0.13	0.13	2.22	0.60	2.82	3.24
branches	each	0.20	0.20	3.42	0.60	4.02	4.62
50 mm pipes jointed with seal ring fittings and fixed with clips plugged to walls	m	0.26	0.26	4.46	1.78	6.24	7.18
Extra for							
bends	each	0.17	0.17	2.91	1.06	3.97	4.57
branches	each	0.25	0.25	4.28	1.06	5.34	6.14
Tubular "P" traps with 75 mm seal and screwed and seal ring joints							
32 mm	each	0.25	0.25	4.28	1.60	5.88	6.76
38 mm	each	0.33	0.33	5.66	1.84	7.50	8.63
Tubular "S" traps with 75 mm seal and screwed and seal ring joints							
32 mm	each	0.25	0.25	4.28	2.02	6.30	7.25
38 mm	each	0.33	0.33	5.66	2.38	8.04	9.25
Bottle "P" traps with 75 mm seal and screwed and seal ring joints							
32 mm	each	0.25	0.25	4.28	1.78	6.06	6.97
38 mm	each	0.33	0.33	5.66	2.13	7.79	8.96
Bottle "S" traps with 75 mm seal and screwed and seal ring joints							
32 mm	each	0.25	0.25	4.28	2.14	6.42	7.38
38 mm	each	0.33	0.33	5.66	2.61	8.27	9.51

Plumbing and engineering installations

New work	Unit	Hours C	Hours L	Labour net	Material net	Price net	Price with 15%
Overflow, soil, waste and vent pipework				£	£	£	£
					VAT not included		
PVCU							
19 mm pipes jointed with solvent welded fittings and fixed with clips plugged to walls	m	0.20	0.20	3.42	0.75	4.17	4.80
Extra for							
bends	each	0.18	0.18	3.09	0.47	3.56	4.09
branches	each	0.21	0.21	3.59	0.47	4.06	4.67
straight tank connectors	each	0.21	0.21	3.59	0.57	4.16	4.78
bent tank connectors	each	0.21	0.21	3.59	0.65	4.24	4.88
32 mm pipes jointed with solvent welded fittings and fixed with clips plugged to walls	m	0.23	0.23	3.94	1.37	5.31	6.11
Extra for							
bends	each	0.21	0.21	3.59	0.80	4.39	5.05
branches	each	0.23	0.23	3.94	1.23	5.17	5.95
38 mm pipes jointed with solvent welded fittings and fixed with clips plugged to walls	m	0.26	0.26	4.46	1.62	6.08	6.99
Extra for							
bends	each	0.25	0.25	4.28	0.91	5.19	5.97
branches	each	0.29	0.29	4.97	1.49	6.46	7.43
50 mm pipes jointed with solvent welded fittings and fixed with clips plugged to walls	m	0.29	0.29	4.97	2.24	7.21	8.29
Extra for							
bends	each	0.27	0.27	4.62	1.31	5.93	6.82
branches	each	0.31	0.31	5.31	2.50	7.81	8.98
100 mm pipes jointed with solvent welded fittings and fixed with brackets plugged to walls	m	0.34	0.34	5.83	3.93	9.76	11.22
Extra for							
seal ring expansion joint adaptors	each	0.20	0.20	3.42	0.91	4.33	4.98
access doors	each	0.50	0.50	8.57	5.39	13.96	16.05
bends	each	0.32	0.32	5.48	6.11	11.59	13.33
bends with access door	each	0.32	0.32	5.48	13.72	19.20	22.08
branches	each	0.37	0.37	6.34	8.10	14.44	16.61
branches with access door	each	0.37	0.37	6.34	15.67	22.01	25.31
WC connectors	each	0.32	0.32	5.48	4.37	9.85	11.33
WC connecting bends	each	0.33	0.33	5.66	6.51	12.17	14.00
WC connecting branches	each	0.36	0.36	6.16	27.40	33.56	38.59
32 mm boss connector	each	0.37	0.37	6.34	3.49	9.83	11.30
38 mm boss connector	each	0.37	0.37	6.34	3.65	9.99	11.49
50 mm boss connector	each	0.37	0.37	6.34	4.34	10.68	12.28
100 mm vent cowls solvent welded to pipe	each	0.20	0.20	3.42	1.27	4.69	5.39
100 mm weathering aprons solvent welded to pipe	each	0.25	0.25	4.28	1.47	5.75	6.61

Plumbing and engineering installations

New work	Unit	Hours C	Hours L	Labour net	Material net	Price net	Price with 15%
Overflow, soil, waste and vent pipework				£	£	£	£
					VAT not included		
PVCU (continued)							
Aluminium weathering slates (for fixing by asphalter or roofer to flat and sloping roofs) with synthetic rubber cone and elastomeric seal round pipe, slate size							
406 x 406 mm	each	0.25	0.25	4.28	13.39	17.67	20.32
457 x 457 mm	each	0.25	0.25	4.28	13.72	18.00	20.70
610 x 610 mm	each	0.25	0.25	4.28	18.53	22.81	26.23
100 mm flat or domed roof outlets and solvent welded joint to pipe	each	0.33	0.33	5.66	16.33	21.99	25.29

Plumbing and engineering installations

New work	Unit	Hours C	Hours L	Labour net	Material net	Price net	Price with 15%
Service pipework				£	£	£	£
					VAT not included		

Copper to BS 2871 Table X

Pipes jointed with compression or capillary fittings (measured separately) and fixed with two-piece copper spacing clips plugged to walls

15 mm	m	0.20	0.20	3.42	1.28	4.70	5.41
22 mm	m	0.21	0.21	3.59	2.36	5.95	6.84
28 mm	m	0.23	0.23	3.94	3.17	7.11	8.18
35 mm	m	0.28	0.28	4.79	6.88	11.67	13.42
42 mm	m	0.30	0.30	5.14	8.36	13.50	15.53
54 mm	m	0.32	0.32	5.48	10.76	16.24	18.68

Pipes jointed with compression or capillary fittings and fixed with pressed brass brackets plugged to walls

15 mm	m	0.20	0.20	3.42	1.41	4.83	5.55
22 mm	m	0.21	0.21	3.59	2.49	6.08	6.99
28 mm	m	0.23	0.23	3.94	3.33	7.27	8.36
35 mm	m	0.28	0.28	4.79	7.08	11.87	13.65
42 mm	m	0.30	0.30	5.14	8.62	13.76	15.82
54 mm	m	0.32	0.32	5.48	11.05	16.53	19.01

Extra for made bends

15 mm	each	0.11	0.11	1.89	-	1.89	2.17
22 mm	each	0.16	0.16	2.74	-	2.74	3.15
28 mm	each	0.21	0.21	3.59	-	3.59	4.13
35 mm	each	0.26	0.26	4.46	-	4.46	5.13
42 mm	each	0.29	0.29	4.97	-	4.97	5.72
54 mm	each	0.36	0.36	6.16	-	6.16	7.08

Thinwall copper to BS 2871 Table Z

Pipes jointed with compression or capillary fittings (measured separately) and fixed with two-piece copper spacing clips plugged to walls

15 mm	m	0.20	0.20	3.42	1.33	4.75	5.46
22 mm	m	0.21	0.21	3.59	2.30	5.89	6.77
28 mm	m	0.25	0.25	4.28	2.93	7.21	8.29
35 mm	m	0.28	0.28	4.79	6.12	10.91	12.55
42 mm	m	0.30	0.30	5.14	7.76	12.90	14.84
54 mm	m	0.32	0.32	5.48	10.46	15.94	18.33

Pipes jointed with compression or capillary fittings (measured separately) and fixed with pressed brass brackets plugged to walls

15 mm	m	0.20	0.20	3.42	1.46	4.88	5.61
22 mm	m	0.21	0.21	3.59	2.43	6.02	6.92
28 mm	m	0.23	0.23	3.94	3.09	7.03	8.08
35 mm	m	0.28	0.28	4.79	6.32	11.11	12.78
42 mm	m	0.30	0.30	5.14	8.02	13.16	15.13
54 mm	m	0.32	0.32	5.48	10.75	16.23	18.66

Note: Thinwall copper is not suitable for made bends

Plumbing and engineering installations

New work	Unit	Hours C	Hours L	Labour net	Material net	Price net	Price with 15%
Service pipework				£	£	£	£
					VAT not included		

Capillary fittings for copper

	Unit	Hours C	Hours L	Labour net	Material net	Price net	Price with 15%
Straight couplings							
15 mm	each	0.19	0.19	3.26	0.13	3.39	3.90
22 mm	each	0.23	0.23	3.94	0.30	4.24	4.88
28 mm	each	0.25	0.25	4.28	0.71	4.99	5.74
35 mm	each	0.29	0.29	4.97	1.80	6.77	7.79
42 mm	each	0.35	0.35	5.99	2.67	8.66	9.96
54 mm	each	0.38	0.38	6.51	5.84	12.35	14.20
Elbows							
15 mm	each	0.20	0.20	3.42	0.25	3.67	4.22
22 mm	each	0.23	0.23	3.94	0.57	4.51	5.19
28 mm	each	0.26	0.26	4.46	1.18	5.64	6.49
35 mm	each	0.29	0.29	4.97	3.71	8.68	9.98
42 mm	each	0.35	0.35	5.99	6.28	12.27	14.11
54 mm	each	0.39	0.39	6.68	13.87	20.55	23.63
Slow bends							
15 mm	each	0.22	0.22	3.77	1.06	4.83	5.55
22 mm	each	0.25	0.25	4.28	2.06	6.34	7.29
28 mm	each	0.27	0.27	4.62	3.44	8.06	9.27
Tees							
15 mm	each	0.25	0.25	4.28	0.44	4.72	5.43
22 mm	each	0.28	0.28	4.79	1.00	5.79	6.66
28 mm	each	0.33	0.33	5.66	2.44	8.10	9.31
35 mm	each	0.38	0.38	6.51	6.43	12.94	14.88
42 mm	each	0.40	0.40	6.85	9.49	16.34	18.79
54 mm	each	0.47	0.47	8.05	18.48	26.53	30.51
Straight tank connectors							
15 mm x ½"	each	0.29	0.29	4.97	2.60	7.57	8.71
22 mm x ¾"	each	0.31	0.31	5.31	3.98	9.29	10.68
28 mm x 1"	each	0.32	0.32	5.48	5.31	10.79	12.41
35 mm x 1¼"	each	0.37	0.37	6.34	6.86	13.20	15.18
42 mm x 1½"	each	0.43	0.43	7.36	8.98	16.34	18.79
54 mm x 2"	each	0.47	0.47	8.05	13.72	21.77	25.04
Straight tap connectors							
15 mm x ½"	each	0.23	0.23	3.94	0.99	4.93	5.67
22 mm x ¾"	each	0.25	0.25	4.28	1.34	5.62	6.46
Bent tap connectors							
15 mm x ½"	each	0.23	0.23	3.94	1.26	5.20	5.98
22 mm x ¾"	each	0.25	0.25	4.28	1.34	5.62	6.46
Gunmetal stopcocks							
15 mm	each	0.20	0.20	3.42	6.24	9.66	11.11
22 mm	each	0.23	0.23	3.94	10.82	14.76	16.97
28 mm	each	0.26	0.26	4.46	18.03	22.49	25.86
35 mm	each	0.30	0.30	5.14	28.28	33.42	38.43
42 mm	each	0.35	0.35	5.99	37.57	43.56	50.09
54 mm	each	0.40	0.40	6.85	56.14	62.99	72.44

Plumbing and engineering installations

New work	Unit	Hours C	Hours L	Labour net	Material net	Price net	Price with 15%
Service pipework				£	£	£	£
					VAT not included		

Capillary fittings for copper (*continued*)

Gunmetal double union stopcocks
15 mm	each	0.20	0.20	3.42	13.13	16.55	19.03
22 mm	each	0.23	0.23	3.94	16.18	20.12	23.14
28 mm	each	0.26	0.26	4.46	29.89	34.35	39.50
35 mm	each	0.30	0.30	5.14	49.86	55.00	63.25
42 mm	each	0.35	0.35	5.99	68.39	74.38	85.54
54 mm	each	0.40	0.40	6.85	107.61	114.46	131.63

Gunmetal combined stopcocks and draincocks
15 mm	each	0.20	0.20	3.42	13.84	17.26	19.85
22 mm	each	0.23	0.23	3.94	17.04	20.98	24.13

Gunmetal gatevalves
15 mm	each	0.20	0.20	3.42	7.81	11.23	12.91
22 mm	each	0.23	0.23	3.94	9.04	12.98	14.93
28 mm	each	0.26	0.26	4.46	12.58	17.04	19.60
35 mm	each	0.30	0.30	5.14	28.03	33.17	38.15
42 mm	each	0.35	0.35	5.99	35.09	41.08	47.24
54 mm	each	0.40	0.40	6.85	41.74	48.59	55.88

Compression fittings for copper

Straight couplings
15 mm	each	0.17	0.17	2.91	1.10	4.01	4.61
22 mm	each	0.18	0.18	3.09	1.71	4.80	5.52
28 mm	each	0.23	0.23	3.94	3.27	7.21	8.29
35 mm	each	0.27	0.27	4.62	6.78	11.40	13.11
42 mm	each	0.29	0.29	4.97	8.69	13.66	15.71
54 mm	each	0.33	0.33	5.66	13.00	18.66	21.46

Elbows
15 mm	each	0.17	0.17	2.91	1.28	4.19	4.82
22 mm	each	0.17	0.17	2.91	2.08	4.99	5.74
28 mm	each	0.23	0.23	3.94	4.23	8.17	9.40
35 mm	each	0.27	0.27	4.62	8.78	13.40	15.41
42 mm	each	0.29	0.29	4.97	12.37	17.34	19.94
54 mm	each	0.33	0.33	5.66	20.33	25.99	29.89

Slow bends
15 mm	each	0.20	0.20	3.42	4.23	7.65	8.80
22 mm	each	0.23	0.23	3.94	6.94	10.88	12.51
28 mm	each	0.26	0.26	4.46	9.34	13.80	15.87

Tees
15 mm	each	0.21	0.21	3.59	1.77	5.36	6.16
22 mm	each	0.21	0.21	3.59	2.99	6.58	7.57
28 mm	each	0.27	0.27	4.62	6.34	10.96	12.60
35 mm	each	0.31	0.31	5.31	11.61	16.92	19.46
42 mm	each	0.36	0.36	6.16	19.32	25.48	29.30
54 mm	each	0.41	0.41	7.03	30.30	37.33	42.93

Plumbing and engineering installations

New work	Unit	Hours C	Hours L	Labour net	Material net	Price net	Price with 15%
Service pipework				£	£	£	£
					VAT not included		

Compression fittings for copper (*continued*)

	Unit	Hours C	Hours L	Labour net	Material net	Price net	Price with 15%
Straight tank connectors							
15 mm	each	0.27	0.27	4.62	2.58	7.20	8.28
22 mm	each	0.29	0.29	4.97	2.90	7.87	9.05
28 mm	each	0.31	0.31	5.31	4.95	10.26	11.80
35 mm	each	0.35	0.35	5.99	8.66	14.65	16.85
42 mm	each	0.41	0.41	7.03	14.07	21.10	24.27
54 mm	each	0.45	0.45	7.71	18.54	26.25	30.19
Straight tap connectors							
15 mm x ½"	each	0.21	0.21	3.59	1.96	5.55	6.38
22 mm x ¾"	each	0.23	0.23	3.94	3.53	7.47	8.59
Bent tap connectors							
15 mm x ½"	each	0.21	0.21	3.59	2.35	5.94	6.83
22 mm x ¾"	each	0.23	0.23	3.94	4.07	8.01	9.21
Gunmetal stopcocks							
15 mm	each	0.20	0.20	3.42	7.74	11.16	12.83
22 mm	each	0.23	0.23	3.94	12.72	16.66	19.16
28 mm	each	0.26	0.26	4.46	21.03	25.49	29.31
35 mm	each	0.30	0.30	5.14	38.69	43.83	50.40
42 mm	each	0.35	0.35	5.99	55.36	61.35	70.55
54 mm	each	0.40	0.40	6.85	75.40	82.25	94.59
Gunmetal combined stopcocks and draincocks							
15 mm	each	0.20	0.20	3.42	14.38	17.80	20.47
22 mm	each	0.23	0.23	3.94	21.06	25.00	28.75
Gunmetal gatevalves							
15 mm	each	0.20	0.20	3.42	9.38	12.80	14.72
22 mm	each	0.23	0.23	3.94	11.10	15.04	17.30
28 mm	each	0.26	0.26	4.46	15.06	19.52	22.45
35 mm	each	0.30	0.30	5.14	21.93	27.07	31.13
42 mm	each	0.35	0.35	5.99	30.31	36.30	41.74
54 mm	each	0.40	0.40	6.85	50.87	57.72	66.38

Underground copper to BS 2871 Table Y

	Unit	Hours C	Hours L	Labour net	Material net	Price net	Price with 15%
Pipes jointed with gunmetal compression fittings and laid in trench							
15 mm	m	0.10	0.10	1.72	2.50	4.22	4.85
22 mm	m	0.12	0.12	2.05	4.39	6.44	7.41
28 mm	m	0.15	0.15	2.57	5.84	8.41	9.67

Plumbing and engineering installations

New work	Unit	Hours C	Hours L	Labour net	Material net	Price net	Price with 15%
Service pipework				£	£	£	£
					VAT not included		

Gunmetal compression fittings for "underground" copper

Straight couplings

15 mm	each	0.17	0.17	2.91	4.29	7.20	8.28
22 mm	each	0.17	0.17	2.91	5.67	8.58	9.87
28 mm	each	0.23	0.23	3.94	10.05	13.99	16.09

Elbows

15 mm	each	0.17	0.17	2.91	4.48	7.39	8.50
22 mm	each	0.17	0.17	2.91	6.04	8.95	10.29
28 mm	each	0.23	0.23	3.94	11.01	14.95	17.19

Tees

15 mm	each	0.21	0.21	3.59	6.57	10.16	11.68
22 mm	each	0.21	0.21	3.59	8.94	12.53	14.41
28 mm	each	0.27	0.27	4.62	16.51	21.13	24.30

Stopcocks

15 mm	each	0.25	0.25	4.28	12.26	16.54	19.02
22 mm	each	0.33	0.33	5.66	16.68	22.34	25.69
28 mm	each	0.40	0.40	6.85	23.94	30.79	35.41

Black medium steel to BS 1387

Pipes jointed with screwed sockets and fixed with galvanised clips plugged to walls

15 mm	m	0.30	0.30	5.14	3.53	8.67	9.97
20 mm	m	0.37	0.37	6.34	4.07	10.41	11.97
25 mm	m	0.45	0.45	7.71	5.59	13.30	15.30
32 mm	m	0.55	0.55	9.42	6.83	16.25	18.69
40 mm	m	0.65	0.65	11.14	7.93	19.07	21.93
50 mm	m	0.75	0.75	12.85	10.94	23.79	27.36

Pipes jointed with screwed sockets and fixed with malleable schoolboard brackets plugged to walls

15 mm	m	0.30	0.30	5.14	3.76	8.90	10.23
20 mm	m	0.37	0.37	6.34	4.33	10.67	12.27
25 mm	m	0.45	0.45	7.71	5.91	13.62	15.66
32 mm	m	0.55	0.55	9.42	7.30	16.72	19.23
40 mm	m	0.65	0.65	11.14	8.57	19.71	22.67
50 mm	m	0.75	0.75	12.85	11.84	24.69	28.39

Black heavy steel to BS 1387

Pipes jointed with screwed sockets and fixed with galvanised clips plugged to walls

15 mm	m	0.30	0.30	5.14	4.03	9.17	10.55
20 mm	m	0.37	0.37	6.34	4.70	11.04	12.70
25 mm	m	0.45	0.45	7.71	6.57	14.28	16.42
32 mm	m	0.55	0.55	9.42	8.07	17.49	20.11
40 mm	m	0.65	0.65	11.14	9.40	20.54	23.62
50 mm	m	0.75	0.75	12.85	12.85	25.70	29.56

Plumbing and engineering installations

New work	Unit	Hours C	Hours L	Labour net	Material net	Price net	Price with 15%
Service pipework				£	£	£	£
					VAT not included		

Black heavy steel to BS 1387 (*continued*)

Pipes jointed with screwed sockets and fixed with
malleable schoolboard brackets plugged to walls

	Unit	Hours C	Hours L	Labour net	Material net	Price net	Price with 15%
15 mm	m	0.30	0.30	5.14	4.26	9.40	10.81
20 mm	m	0.37	0.37	6.34	4.96	11.30	12.99
25 mm	m	0.45	0.45	7.71	6.89	14.60	16.79
32 mm	m	0.55	0.55	9.42	8.54	17.96	20.65
40 mm	m	0.65	0.65	11.14	10.04	21.18	24.36
50 mm	m	0.75	0.75	12.85	13.75	26.60	30.59

Black malleable iron beaded fittings for steel

Conical unions

15 mm	each	0.40	0.40	6.85	2.04	8.89	10.22
20 mm	each	0.50	0.50	8.57	2.32	10.89	12.52
25 mm	each	0.63	0.63	10.79	2.71	13.50	15.53
32 mm	each	0.80	0.80	13.71	3.98	17.69	20.34
40 mm	each	1.00	1.00	17.13	4.63	21.76	25.02
50 mm	each	1.25	1.25	21.41	7.66	29.07	33.43

Elbows

15 mm	each	0.33	0.33	5.66	0.53	6.19	7.12
20 mm	each	0.40	0.40	6.85	0.74	7.59	8.73
25 mm	each	0.50	0.50	8.57	1.13	9.70	11.15
32 mm	each	0.63	0.63	10.79	1.91	12.70	14.61
40 mm	each	0.80	0.80	13.71	3.15	16.86	19.39
50 mm	each	1.00	1.00	17.13	3.73	20.86	23.99

Bends

15 mm	each	0.33	0.33	5.66	1.13	6.79	7.81
20 mm	each	0.40	0.40	6.85	1.75	8.60	9.89
25 mm	each	0.50	0.50	8.57	2.44	11.01	12.66
32 mm	each	0.63	0.63	10.79	4.29	15.08	17.34
40 mm	each	0.80	0.80	13.71	5.22	18.93	21.77
50 mm	each	1.00	1.00	17.13	7.30	24.43	28.09

Tees

15 mm	each	0.40	0.40	6.85	0.74	7.59	8.73
20 mm	each	0.50	0.50	8.57	1.07	9.64	11.09
25 mm	each	0.63	0.63	10.79	1.55	12.34	14.19
32 mm	each	0.80	0.80	13.71	2.55	16.26	18.70
40 mm	each	1.00	1.00	17.13	3.51	20.64	23.74
50 mm	each	1.25	1.25	21.41	5.04	26.45	30.42

Longscrew connectors with socket and backnut

15 mm	each	0.17	0.17	2.91	1.65	4.56	5.24
20 mm	each	0.20	0.20	3.42	1.96	5.38	6.19
25 mm	each	0.25	0.25	4.28	2.73	7.01	8.06
32 mm	each	0.33	0.33	5.66	3.53	9.19	10.57
40 mm	each	0.40	0.40	6.85	4.26	11.11	12.78
50 mm	each	0.50	0.50	8.57	6.32	14.89	17.12

Plumbing and engineering installations

New work	Unit	Hours C	Hours L	Labour net	Material net	Price net	Price with 15%
Service pipework				£	£	£	£
					VAT not included		

Galvanised medium steel to BS 1387

Pipes jointed with screwed sockets and fixed with
galvanised clips plugged to walls

15 mm	m	0.30	0.30	5.14	5.13	10.27	11.81
20 mm	m	0.37	0.37	6.34	5.73	12.07	13.88
25 mm	m	0.45	0.45	7.71	7.76	15.47	17.79
32 mm	m	0.55	0.55	9.42	9.55	18.97	21.82
40 mm	m	0.65	0.65	11.14	11.10	22.24	25.58
50 mm	m	0.75	0.75	12.85	15.35	28.20	32.43

Pipes jointed with screwed sockets and fixed with
galvanised malleable schoolboard brackets
plugged to walls

15 mm	m	0.30	0.30	5.14	5.30	10.44	12.01
20 mm	m	0.37	0.37	6.34	5.93	12.27	14.11
25 mm	m	0.45	0.45	7.71	8.01	15.72	18.08
32 mm	m	0.55	0.55	9.42	9.92	19.34	22.24
40 mm	m	0.65	0.65	11.14	11.61	22.75	26.16
50 mm	m	0.75	0.75	12.85	16.07	28.92	33.26

Galvanised heavy steel to BS 1387

Pipes jointed with sockets and laid in trench

15 mm	m	0.13	0.13	2.22	5.20	7.42	8.53
20 mm	m	0.16	0.16	2.74	5.90	8.64	9.94
25 mm	m	0.20	0.20	3.42	8.39	11.81	13.58

Pipes jointed with screwed sockets and fixed with
galvanised clips plugged to walls

15 mm	m	0.30	0.30	5.14	5.86	11.00	12.65
20 mm	m	0.37	0.37	6.34	6.59	12.93	14.87
25 mm	m	0.45	0.45	7.71	9.13	16.84	19.37
32 mm	m	0.55	0.55	9.42	11.26	20.68	23.78
40 mm	m	0.65	0.65	11.14	13.14	24.28	27.92
50 mm	m	0.75	0.75	12.85	18.01	30.86	35.49

Pipes jointed with screwed sockets and fixed with
galvanised malleable schoolboard brackets
plugged to walls

15 mm	m	0.30	0.30	5.14	6.03	11.17	12.85
20 mm	m	0.37	0.37	6.34	6.79	13.13	15.10
25 mm	m	0.45	0.45	7.71	9.38	17.09	19.65
32 mm	m	0.55	0.55	9.42	11.63	21.05	24.21
40 mm	m	0.65	0.65	11.14	13.65	24.79	28.51
50 mm	m	0.75	0.75	12.85	18.73	31.58	36.32

Plumbing and engineering installations

New work	Unit	Hours C	Hours L	Labour net	Material net	Price net	Price with 15%
Service pipework				£	£	£	£
					VAT not included		

Galvanised malleable iron beaded fittings for steel

Conical unions

15 mm	each	0.40	0.40	6.85	2.85	9.70	11.15
20 mm	each	0.50	0.50	8.57	3.26	11.83	13.60
25 mm	each	0.63	0.63	10.79	3.79	14.58	16.77
32 mm	each	0.80	0.80	13.71	5.53	19.24	22.13
40 mm	each	1.00	1.00	17.13	6.53	23.66	27.21
50 mm	each	1.25	1.25	21.41	10.74	32.15	36.97

Elbows

15 mm	each	0.33	0.33	5.66	0.72	6.38	7.34
20 mm	each	0.40	0.40	6.85	1.00	7.85	9.03
25 mm	each	0.50	0.50	8.57	1.51	10.08	11.59
32 mm	each	0.63	0.63	10.79	2.57	13.36	15.36
40 mm	each	0.80	0.80	13.71	4.33	18.04	20.75
50 mm	each	1.00	1.00	17.13	4.98	22.11	25.43

Bends

15 mm	each	0.33	0.33	5.66	1.55	7.21	8.29
20 mm	each	0.40	0.40	6.85	2.36	9.21	10.59
25 mm	each	0.50	0.50	8.57	3.33	11.90	13.69
32 mm	each	0.63	0.63	10.79	5.88	16.67	19.17
40 mm	each	0.80	0.80	13.71	7.18	20.89	24.02
50 mm	each	1.00	1.00	17.13	9.79	26.92	30.96

Tees

15 mm	each	0.40	0.40	6.85	1.00	7.85	9.03
20 mm	each	0.50	0.50	8.57	1.46	10.03	11.53
25 mm	each	0.63	0.63	10.79	2.08	12.87	14.80
32 mm	each	0.80	0.80	13.71	3.44	17.15	19.72
40 mm	each	1.00	1.00	17.13	4.86	21.99	25.29
50 mm	each	1.25	1.25	21.41	6.76	28.17	32.40

Longscrew connectors with socket and backnut

15 mm	each	0.17	0.17	2.91	2.23	5.14	5.91
20 mm	each	0.20	0.20	3.42	2.65	6.07	6.98
25 mm	each	0.25	0.25	4.28	3.68	7.96	9.15
32 mm	each	0.33	0.33	5.66	4.76	10.42	11.98
40 mm	each	0.40	0.40	6.85	5.75	12.60	14.49
50 mm	each	0.50	0.50	8.57	8.55	17.12	19.69

Brasswork for steel

Brass stopcocks

15 mm	each	0.33	0.33	5.66	5.09	10.75	12.36
20 mm	each	0.40	0.40	6.85	7.59	14.44	16.61
25 mm	each	0.50	0.50	8.57	12.15	20.72	23.83

Plumbing and engineering installations

New work	Unit	Hours C	Hours L	Labour net	Material net	Price net	Price with 15%
Service pipework				£	£	£	£
					VAT not included		

Brasswork for steel (*continued*)

Brass gatevalves

15 mm	each	0.33	0.33	5.66	3.81	9.47	10.89
20 mm	each	0.40	0.40	6.85	4.61	11.46	13.18
25 mm	each	0.50	0.50	8.57	6.68	15.25	17.54
32 mm	each	0.63	0.63	10.79	9.81	20.60	23.69
40 mm	each	0.80	0.80	13.71	13.77	27.48	31.60
50 mm	each	1.00	1.00	17.13	19.05	36.18	41.61

Gunmetal gatevalves

15 mm	each	0.33	0.33	5.66	12.13	17.79	20.46
20 mm	each	0.40	0.40	6.85	16.01	22.86	26.29
25 mm	each	0.50	0.50	8.57	20.89	29.46	33.88
32 mm	each	0.63	0.63	10.79	30.81	41.60	47.84
40 mm	each	0.80	0.80	13.71	39.51	53.22	61.20
50 mm	each	1.00	1.00	17.13	57.71	74.84	86.07

Medium density polyethylene to BS 6730 – black

Pipes jointed with compression fittings and fixed with galvanised clips plugged to walls

20 mm	m	0.21	0.21	3.59	1.21	4.80	5.52
25 mm	m	0.23	0.23	3.94	1.50	5.44	6.26
32 mm	m	0.28	0.28	4.79	2.24	7.03	8.08
50 mm	m	0.32	0.32	5.48	4.89	10.37	11.93
63 mm	m	0.36	0.36	6.16	7.49	13.65	15.70

Brass and gunmetal compression fittings for black polyethylene to BS 6730

Straight couplings

20 mm	each	0.23	0.23	3.94	2.29	6.23	7.16
25 mm	each	0.25	0.25	4.28	3.28	7.56	8.69
32 mm	each	0.29	0.29	4.97	5.65	10.62	12.21
50 mm	each	0.39	0.39	6.68	13.65	20.33	23.38
63 mm	each	0.44	0.44	7.53	18.70	26.23	30.16

Elbows

20 mm	each	0.23	0.23	3.94	2.81	6.75	7.76
25 mm	each	0.25	0.25	4.28	4.13	8.41	9.67
32 mm	each	0.29	0.29	4.97	7.24	12.21	14.04
50 mm	each	0.39	0.39	6.68	16.87	23.55	27.08
63 mm	each	0.44	0.44	7.53	20.29	27.82	31.99

Tees

20 mm	each	0.29	0.29	4.97	3.84	8.81	10.13
25 mm	each	0.33	0.33	5.66	6.36	12.02	13.82
32 mm	each	0.37	0.37	6.34	9.14	15.48	17.80
50 mm	each	0.47	0.47	8.05	22.45	30.50	35.08
63 mm	each	0.55	0.55	9.42	32.99	42.41	48.77

Plumbing and engineering installations

New work	Unit	Hours C	Hours L	Labour net	Material net	Price net	Price with 15%
Service pipework				£	£	£	£
					VAT not included		

Brass and gunmetal compression fittings for black polyethylene to BS 6730 *(continued)*

	Unit	Hours C	Hours L	Labour net	Material net	Price net	Price with 15%
Straight swivel connectors							
20 mm x ½"	each	0.23	0.23	3.94	3.23	7.17	8.25
25 mm x ¾"	each	0.25	0.25	4.28	5.56	9.84	11.32
Bent swivel connectors							
20 mm x ½"	each	0.25	0.25	4.28	3.86	8.14	9.36
25 mm x ¾"	each	0.29	0.29	4.97	6.21	11.18	12.86
Gunmetal stopcocks							
20 mm	each	0.23	0.23	3.94	9.73	13.67	15.72
25 mm	each	0.26	0.26	4.46	15.50	19.96	22.95
32 mm	each	0.30	0.30	5.14	20.74	25.88	29.76
50 mm	each	0.40	0.40	6.85	51.91	58.76	67.57
63 mm	each	0.50	0.50	8.57	72.89	81.46	93.68
20 x 15 mm gunmetal combined stopcocks and draincocks	each	0.20	0.20	3.42	13.43	16.85	19.38

Medium density polyethylene to BS 6572 – blue

	Unit	Hours C	Hours L	Labour net	Material net	Price net	Price with 15%
Pipes jointed with compression fittings and laid in trench							
20 mm	m	0.05	0.05	0.85	0.65	1.50	1.73
25 mm	m	0.06	0.06	1.03	0.84	1.87	2.15
32 mm	m	0.07	0.07	1.20	1.40	2.60	2.99
Pipes jointed with compression fittings and fixed with galvanised clips plugged to walls							
20 mm	m	0.35	0.35	5.99	0.86	6.85	7.88
25 mm	m	0.35	0.35	5.99	1.06	7.05	8.11
32 mm	m	0.40	0.40	6.85	1.62	8.47	9.74

Brass and gunmetal compression fittings for blue polyethylene to BS 6572

	Unit	Hours C	Hours L	Labour net	Material net	Price net	Price with 15%
Straight couplings							
20 mm	each	0.23	0.23	3.94	2.29	6.23	7.16
25 mm	each	0.25	0.25	4.28	3.28	7.56	8.69
32 mm	each	0.29	0.29	4.97	5.65	10.62	12.21
Elbows							
20 mm	each	0.23	0.23	3.94	2.81	6.75	7.76
25 mm	each	0.25	0.25	4.28	4.13	8.41	9.67
32 mm	each	0.29	0.29	4.97	7.24	12.21	14.04
Tees							
20 mm	each	0.29	0.29	4.97	3.84	8.81	10.13
25 mm	each	0.33	0.33	5.66	6.36	12.02	13.82
32 mm	each	0.37	0.37	6.34	9.14	15.48	17.80

Plumbing and engineering installations

New work	Unit	Hours C	Hours L	Labour net	Material net	Price net	Price with 15%
Service pipework				£	£	£	£
					VAT not included		

Brass and gunmetal compression fittings for blue polyethylene to BS 6572 (*continued*)

New work	Unit	Hours C	Hours L	Labour net	Material net	Price net	Price with 15%
Straight swivel connectors							
20 mm x ½"	each	0.23	0.23	3.94	3.23	7.17	8.25
25 mm x ¾"	each	0.25	0.25	4.28	5.56	9.84	11.32
Bent swivel connectors							
20 mm x ½"	each	0.25	0.25	4.28	3.86	8.14	9.36
25 mm x ¾"	each	0.29	0.29	4.97	6.21	11.18	12.86
Gunmetal stopcocks							
20 mm	each	0.23	0.23	3.94	9.73	13.67	15.72
25 mm	each	0.26	0.26	4.46	15.50	19.96	22.95
32 mm	each	0.30	0.30	5.14	20.74	25.88	29.76
Gunmetal combined stopcocks and draincocks							
20 x 15 mm	each	0.23	0.23	3.94	18.33	22.27	25.61
25 x 15 mm	each	0.26	0.26	4.46	19.21	23.67	27.22
25 x 22 mm	each	0.26	0.26	4.46	22.10	26.56	30.54

PVCU to BS 3505 Class E

New work	Unit	Hours C	Hours L	Labour net	Material net	Price net	Price with 15%
Pipes jointed with solvent welded fittings and fixed with polypropylene clips plugged to walls							
⅜"	m	0.20	0.20	3.42	1.19	4.61	5.30
½"	m	0.20	0.20	3.42	1.40	4.82	5.54
¾"	m	0.21	0.21	3.59	1.80	5.39	6.20
1"	m	0.23	0.23	3.94	2.27	6.21	7.14
1¼"	m	0.28	0.28	4.79	3.23	8.02	9.22
1½"	m	0.28	0.28	4.79	4.04	8.83	10.15

Solvent welded fittings for PVCU

New work	Unit	Hours C	Hours L	Labour net	Material net	Price net	Price with 15%
Straight couplings							
⅜"	each	0.18	0.18	3.09	0.37	3.46	3.98
½"	each	0.18	0.18	3.09	0.41	3.50	4.03
¾"	each	0.20	0.20	3.42	0.45	3.87	4.45
1"	each	0.21	0.21	3.59	0.53	4.12	4.74
1¼"	each	0.21	0.21	3.59	0.86	4.45	5.12
1½"	each	0.23	0.23	3.94	1.06	5.00	5.75
Elbows							
⅜"	each	0.18	0.18	3.09	0.49	3.58	4.12
½"	each	0.18	0.18	3.09	0.56	3.65	4.20
¾"	each	0.18	0.18	3.09	0.66	3.75	4.31
1"	each	0.18	0.18	3.09	0.82	3.91	4.50
1¼"	each	0.18	0.18	3.09	1.51	4.60	5.29
1½"	each	0.20	0.20	3.42	1.92	5.34	6.14

Plumbing and engineering installations

New work	Unit	Hours C	Hours L	Labour net	Material net	Price net	Price with 15%
Service pipework				£	£	£	£
					VAT not included		

Solvent welded fittings for PVCU (*continued*)

	Unit	Hours C	Hours L	Labour net	Material net	Price net	Price with 15%
Tees							
⅜"	each	0.20	0.20	3.42	0.53	3.95	4.54
½"	each	0.20	0.20	3.42	0.63	4.05	4.66
¾"	each	0.20	0.20	3.42	0.80	4.22	4.85
1"	each	0.23	0.23	3.94	1.16	5.10	5.87
1¼"	each	0.23	0.23	3.94	2.46	6.40	7.36
1½"	each	0.25	0.25	4.28	2.46	6.74	7.75
Straight tank connectors							
½"	each	0.18	0.18	3.09	1.61	4.70	5.41
¾"	each	0.28	0.28	4.79	1.80	6.59	7.58
Straight tap connectors							
⅜"x ½"	each	0.18	0.18	3.09	1.51	4.60	5.29
½" x ½"	each	0.18	0.18	3.09	1.72	4.81	5.53
½" x ¾"	each	0.20	0.20	3.42	1.81	5.23	6.01
Bent tap connectors							
⅜"x ½"	each	0.18	0.18	3.09	1.61	4.70	5.41
½" x ½"	each	0.18	0.18	3.09	1.81	4.90	5.63
½" x ¾"	each	0.20	0.20	3.42	1.92	5.34	6.14
Hepworth "Hep2 O" polybutylene flexible plumbing system with Slimline fittings							
15 mm standard pipes jointed with straight connectors and fixed with clips plugged to walls	m	0.20	0.20	3.42	1.75	5.17	5.95
Extra for							
elbows	each	0.11	0.11	1.89	1.44	3.33	3.83
equal tees	each	0.14	0.14	2.40	2.02	4.42	5.08
straight tap connectors	each	0.11	0.11	1.89	1.51	3.40	3.91
22 mm standard pipes jointed with straight connectors and fixed with clips plugged to walls	m	0.22	0.22	3.77	2.94	6.71	7.72
Extra for							
elbows	each	0.12	0.12	2.05	2.09	4.14	4.76
equal tees	each	0.15	0.15	2.57	2.58	5.15	5.92

Plumbing and engineering installations

New work	Unit	Specialist price net	Price with 15%
Connections to public mains		£	£
			VAT not included

Connections to public mains, making good public highways and any other work which may only be carried out by a Statutory Water Service Company should, when tenders are based on bills of quantities, be covered by provisional sums.

Where bills of quantities, or provisional sums, are not provided, the following "Specialist price net" figures indicate the order of charges made by water authorities. Confirmation of local charges should be obtained when preparing a tender.

Prices do not include for cash discount.

	Unit	Specialist price net	Price with 15%
Connections only - up to 28 mm (1") pipes			
Single connections including installation of water meters	each	170.00	195.50
Additional connections in the same trench up to a maximum of 5 connections including installation of water meters	each	142.00	163.30
Additional charges for pipes	m	1.00	1.15
Connections and trenchwork - up to 28 mm (1") pipes			
Excavation to main, connections and installation of water meters including 3 metres of service pipe			
single connections	each	365.00	419.75
additional connections in the same trench up to a maximum of 5 connections	each	142.00	163.30
Additional charges for			
each metre, or part, of trench and service pipe over 3 metres	m	47.00	54.05
each metre, or part, of permanent reinstatement of carriageways and footpaths (excluding any highway authority inspection charge)	m	50.00	57.50
Infrastructure charges			
Domestic connections			
water	each	226.00	259.90
sewerage	each	226.00	259.90

Plumbing and engineering installations

New work	Unit	Hours C	Hours L	Labour net	Material net	Price net	Price with 15%
Ancillaries				£	£	£	£
					VAT not included		
Prices include for screwed joints to fittings, adaptors, etc							
Brass bib valves							
13 mm	each	0.10	0.10	1.72	6.22	7.94	9.13
19 mm	each	0.12	0.12	2.05	9.01	11.06	12.72
Chromium plated bib valves							
13 mm	each	0.10	0.10	1.72	8.99	10.71	12.32
19 mm	each	0.12	0.12	2.05	13.67	15.72	18.08
Chromium plated basin or bath pillar valves							
13 mm	each	0.15	0.15	2.57	9.15	11.72	13.48
19 mm	each	0.17	0.17	2.91	12.24	15.15	17.42
Plastics basin or bath pillar valves							
13 mm	each	0.15	0.15	2.57	5.70	8.27	9.51
19 mm	each	0.17	0.17	2.91	7.61	10.52	12.10
13 mm chromium plated high necked sink pillar valves	each	0.15	0.15	2.57	13.21	15.78	18.15
13 mm plastics high necked sink pillar valves	each	0.15	0.15	2.57	7.24	9.81	11.28
Brass drain cocks							
13 mm	each	0.10	0.10	1.72	2.16	3.88	4.46
19 mm	each	0.12	0.12	2.05	3.44	5.49	6.31
Brass high pressure Portsmouth pattern ball valves with copper float							
13 mm	each	0.17	0.17	2.91	5.56	8.47	9.74
19 mm	each	0.20	0.20	3.42	14.80	18.22	20.95
25 mm	each	0.25	0.25	4.28	32.27	36.55	42.03
Brass low pressure Portsmouth pattern ball valves with plastic float							
13 mm	each	0.17	0.17	2.91	5.43	8.34	9.59
19 mm	each	0.20	0.20	3.42	12.10	15.52	17.85
25 mm	each	0.25	0.25	4.28	29.20	33.48	38.50
13 mm plastics diaphragm pattern ball valves with plastic float							
high pressure	each	0.17	0.17	2.91	4.65	7.56	8.69
low pressure	each	0.17	0.17	2.91	4.30	7.21	8.29
Chromium plated straight union radiator valves							
13 mm	each	0.20	0.20	3.42	12.37	15.79	18.16
19 mm	each	0.25	0.25	4.28	13.73	18.01	20.71
Chromium plated angle union radiator valves							
13 mm	each	0.20	0.20	3.42	8.41	11.83	13.60
19 mm	each	0.25	0.25	4.28	9.52	13.80	15.87

Plumbing and engineering installations

New work	Unit	Hours C	Hours L	Labour net	Material net	Price net	Price with 15%
Ancillaries				£	£	£	£
					VAT not included		
Chromium plated straight union radiator valves with compression joint for copper							
13 mm	each	0.15	0.15	2.57	7.87	10.44	12.01
19 mm	each	0.17	0.17	2.91	10.25	13.16	15.13
Chromium plated angle union radiator valves with compression joint for copper							
13 mm	each	0.15	0.15	2.57	6.76	9.33	10.73
19 mm	each	0.17	0.17	2.91	8.88	11.79	13.56
Brass spring safety valves							
13 mm	each	0.10	0.10	1.72	2.88	4.60	5.29
19 mm	each	0.12	0.12	2.05	3.17	5.22	6.00
Brass main gas cocks							
13 mm	each	0.20	0.20	3.42	5.68	9.10	10.47
19 mm	each	0.22	0.22	3.77	9.22	12.99	14.94

Plumbing and engineering installations

New work	Unit	Hours C	Hours L	Labour net	Material net	Price net	Price with 15%
Equipment				£	£	£	£
					VAT not included		
Galvanised steel cisterns							
Galvanised steel cisterns to BS 417 - connections not included							
18 litre (4 gallon) BS type SCM45	each	0.50	0.50	8.57	32.57	41.14	47.31
114 litre (25 gallon) BS type SCM180	each	0.80	0.80	13.71	52.09	65.80	75.67
191 litre (42 gallon) BS type SCM270	each	0.85	0.85	14.56	78.57	93.13	107.10
327 litre (72 gallon) BS type SCM450/1	each	1.30	1.30	22.27	100.89	123.16	141.63
709 litre (156 gallon) BS type SCM910	each	1.60	1.60	27.41	192.85	220.26	253.30
Lids for cisterns capacities							
18 litre	each	0.17	0.17	2.91	6.86	9.77	11.24
114 litre	each	0.17	0.17	2.91	8.40	11.31	13.01
191 litre	each	0.17	0.17	2.91	11.03	13.94	16.03
327 litre	each	0.25	0.25	4.28	14.40	18.68	21.48
709 litre	each	0.25	0.25	4.28	33.38	37.66	43.31
Drillings							
13 - 25 mm	each	0.12	0.12	2.05	-	2.05	2.36
30 - 54 mm	each	0.20	0.20	3.42	-	3.42	3.93
Plastics cisterns							
Plastics cisterns to BS 4213 - connections not included							
18 litre (4 gallon) BS type PC4	each	0.60	0.60	10.28	6.71	16.99	19.54
68 litre (15 gallon) BS type PC15	each	0.70	0.70	11.99	31.89	43.88	50.46
114 litre (25 gallon) BS type PC25	each	0.80	0.80	13.71	36.97	50.68	58.28
227 litre (50 gallon) BS type PC50	each	1.00	1.00	17.13	80.53	97.66	112.31
Lids for cisterns capacities							
18 litre	each	0.17	0.17	2.91	4.50	7.41	8.52
68 litre	each	0.17	0.17	2.91	10.55	13.46	15.48
114 litre	each	0.17	0.17	2.91	13.19	16.10	18.52
227 litre	each	0.17	0.17	2.91	23.11	26.02	29.92
Drillings							
13 - 25 mm	each	0.12	0.12	2.05	-	2.05	2.36
30 - 54 mm	each	0.20	0.20	3.42	-	3.42	3.93
Galvanised steel tanks							
Galvanised steel tanks to BS 417 Grade A with bolted hand hole cover - connections not included							
95 litre (21 gallon) BS type T25/1	each	0.65	0.65	11.14	106.05	117.19	134.77
114 litre (25 gallon) BS type T30/1	each	0.80	0.80	13.71	113.62	127.33	146.43
123 litre (27 gallon) BS type T30/2	each	0.80	0.80	13.71	115.03	128.74	148.05
155 litre (34 gallon) BS type T40	each	0.90	0.90	15.41	139.31	154.72	177.93
Drillings							
13 - 25 mm	each	0.12	0.12	2.05	-	2.05	2.36
30 - 54 mm	each	0.20	0.20	3.42	-	3.42	3.93

Plumbing and engineering installations

New work	Unit	Hours C	Hours L	Labour net	Material net	Price net	Price with 15%
Equipment				£	£	£	£
					VAT not included		

Galvanised steel direct cylinders

Galvanised steel direct cylinders to BS 417 Grade
A with 5 bosses, immersion heater boss and
bolted hand hole cover - connections not included

100 litre (22 gallon) BS type Y25	each	0.50	0.50	8.57	176.12	184.69	212.39
123 litre (27 gallon) BS type Y31	each	0.55	0.55	9.42	185.02	194.44	223.61
136 litre (30 gallon) BS type Y33	each	0.65	0.65	11.14	189.03	200.17	230.20
159 litre (35 gallon) BS type Y39	each	0.80	0.80	13.71	197.94	211.65	243.40

Galvanised steel indirect cylinders

Galvanised steel indirect cylinders to BS 1565
Class B with 5 bosses and immersion heater boss
– connections not included

109 litre (24 gallon) BS size BSG1M	each	0.50	0.50	8.57	263.09	271.66	312.41
136 litre (30 gallon) BS size BSG2M	each	0.65	0.65	11.14	280.25	291.39	335.10
159 litre (35 gallon) BS size BSG3M	each	0.80	0.80	13.71	298.14	311.85	358.63

Copper direct cylinders

Integral foam lagged copper direct cylinders to BS
699 Grade 3 with 4 bosses and immersion heater
boss - connections not included

116 litre (25 gallon) BS type 3	each	0.55	0.55	9.42	77.04	86.46	99.43
120 litre (26 gallon) BS type 7	each	0.55	0.55	9.42	73.98	83.40	95.91
144 litre (32 gallon) BS type 8	each	0.65	0.65	11.14	80.83	91.97	105.77
166 litre (37 gallon) BS type 9	each	0.85	0.85	14.56	96.15	110.71	127.32

Copper indirect cylinders

Integral foam lagged copper indirect cylinders to
BS 1566 Grade 3 with 4 bosses and immersion
heater boss - connections not included

104 litre (23 gallon) BS type 3	each	0.55	0.55	9.42	88.55	97.97	112.67
108 litre (24 gallon) BS type 7	each	0.55	0.55	9.42	87.79	97.21	111.79
130 litre (29 gallon) BS type 8	each	0.65	0.65	11.14	100.18	111.32	128.02
152 litre (33 gallon) BS type 9	each	0.70	0.70	11.99	130.55	142.54	163.92

Copper direct combination tanks

Integral foam lagged direct pattern copper
combination tanks - connections not included

115 litre (25 gallon) hot and 25 litre (5 gallon) cold	each	0.77	0.77	13.19	105.70	118.89	136.72
115 litre (25 gallon) hot and 45 litre (10 gallon) cold	each	0.77	0.77	13.19	111.70	124.89	143.62
115 litre (25 gallon) hot and 115 litre (25 gallon) cold	each	0.89	0.89	15.24	175.87	191.11	219.78

Plumbing and engineering installations

New work	Unit	Hours C	Hours L	Labour net	Material net	Price net	Price with 15%
Equipment				£	£	£	£
					VAT not included		

Copper indirect combination tanks

Integral foam lagged indirect pattern copper
combination tanks - connections not included

	Unit	Hours C	Hours L	Labour net	Material net	Price net	Price with 15%
115 litre (25 gallon) hot and 25 litre (5 gallon) cold	each	0.77	0.77	13.19	154.79	167.98	193.18
115 litre (25 gallon) hot and 45 litre (10 gallon) cold	each	0.77	0.77	13.19	160.98	174.17	200.30
115 litre (25 gallon) hot and 115 litre (25 gallon) cold	each	0.89	0.89	15.24	225.52	240.76	276.87

Steel oil storage tanks

12G primed steel oil storage tanks with fill and
vent assemblies, cap and chain, sight gauge and
sludge cock, tanks nominal capacity

	Unit	Hours C	Hours L	Labour net	Material net	Price net	Price with 15%
1362 litre (300 gallon)	each	2.00	2.00	34.26	187.79	222.05	255.36
1816 litre (400 gallon)	each	2.25	2.25	38.55	211.13	249.68	287.13
2725 litre (600 gallon)	each	3.00	3.00	51.39	255.22	306.61	352.60

**Fixing only equipment – connections to
service and waste pipes not included**

Prices include for unloading, storing, hoisting and
fixing in position

	Unit	Hours C	Hours L	Labour net	Material net	Price net	Price with 15%
Vitreous china lavatory basins with waste fitting, plug, chain and stay and pair of brackets plugged to wall	each	1.75	1.75	29.97	1.21	31.18	35.86
Vitreous china lavatory basins with waste fitting, plug, chain and stay, bracket plugged to wall and pedestal plugged to floor	each	2.00	2.00	34.26	1.61	35.87	41.25
Acrylic baths with cradle and brackets, waste, and overflow fittings, plug and chain	each	2.55	2.55	43.68	2.02	45.70	52.56
Vitreous enamelled pressed steel baths with cradle feet, waste and overflow fittings, plug and chain	each	2.65	2.65	45.40	2.02	47.42	54.53
Fireclay sinks with waste fitting, plug, chain and stay and pair of brackets cut and pinned to wall	each	2.50	2.50	42.83	1.21	44.04	50.65
Stainless steel sink units with waste and overflow fittings, plug and chain - fitted to sink units	each	2.00	2.00	34.26	1.21	35.47	40.79
WC suites with pan plugged to floor, bolted seat and cover, high or low level flushing cistern and brackets plugged to wall and flush pipe connected to cistern and to pan	each	2.50	2.50	42.83	1.61	44.44	51.11

Plumbing and engineering installations

New work	Unit	Hours C	Hours L	Labour net	Material net	Price net	Price with 15%
Equipment				£	£	£	£
					VAT not included		

Fixing only equipment – connections to service and waste pipes not included
(*continued*)

	Unit	Hours C	Hours L	Labour net	Material net	Price net	Price with 15%
Domestic central heating boilers plugged to floor with flue pipe connected to boiler							
40,000 Btu/h - 11.7 kWh	each	4.25	4.25	72.80	1.21	74.01	85.11
80,000 Btu/h - 23.4 kWh	each	4.50	4.50	77.08	1.21	78.29	90.03
Domestic central heating circulating pump kits connected to pipework - electrical work not included	each	1.00	1.00	17.13	-	17.13	19.70
Single panel radiators with brackets and stays plugged to wall							
946 x 685 mm	each	0.60	0.60	10.28	0.81	11.09	12.75
1759 x 685 mm	each	0.70	0.70	11.99	0.81	12.80	14.72

Supplying and fixing equipment - connections to service and waste pipes not included

Prices include for unloading, storing, hoisting and fixing in position.

	Unit	Hours C	Hours L	Labour net	Material net	Price net	Price with 15%
560 x 405 mm vitreous china lavatory basins with 32 mm chromium plated waste, plug, chain and stay and pair of towel rail brackets plugged to wall							
white basins	each	1.75	1.75	29.97	36.65	66.62	76.61
coloured basins	each	1.75	1.75	29.97	46.65	76.62	88.11
560 x 405 mm vitreous china lavatory basins with 32 mm chromium plated waste, plug, chain and stay, bracket plugged to wall and pedestal plugged to floor							
white basin and pedestal sets	each	2.00	2.00	34.26	47.28	81.54	93.77
coloured basin and pedestal sets	each	2.00	2.00	34.26	62.28	96.54	111.02
1700 mm acrylic baths with cradle and brackets, 38 mm chromium plated waste and overflow fittings, plug and chain							
white baths	each	2.55	2.55	43.68	115.07	158.75	182.56
coloured baths	each	2.55	2.55	43.68	120.07	163.75	188.31
1700 mm vitreous enamelled pressed steel baths with cradle feet, 38 mm chromium plated waste and overflow fittings, plug and chain							
white baths	each	2.65	2.65	45.40	96.07	141.47	162.69
coloured baths	each	2.65	2.65	45.40	107.07	152.47	175.34
610 x 535 x 255 mm shelf pattern white fireclay sinks with 38 mm chromium plated waste fitting, plug, chain and stay and pair of cantilever brackets cut and pinned to wall	each	2.50	2.50	42.83	175.03	217.86	250.54

Plumbing and engineering installations

New work	Unit	Hours C	Hours L	Labour net	Material net	Price net	Price with 15%
Equipment				£	£	£	£
					VAT not included		

Supplying and fixing equipment - connections to service and waste pipes not included (*continued*)

Stainless steel sinks with 38 mm chromium plated waste and overflow fittings, plug and chain - fitted to sink units

1000 x 500 mm single drainer tops	each	2.00	2.00	34.26	80.26	114.52	131.70
1500 x 500 mm double drainer tops	each	2.00	2.00	34.26	106.26	140.52	161.60

High level WC suites with white vitreous china pan plugged to floor, BS pattern black seat and cover with plastics fittings, 7.5 litre standard finish black plastics flushing cistern with ball valve and all fittings, flush pipe connector

cistern brackets plugged to wall and flush pipe connected to cistern and pan	each	3.00	3.00	51.39	124.90	176.29	202.73

Low level WC suites with white vitreous china pan plugged to floor BS pattern seat and cover with plastics fittings, 7.5 litre streamlined finish plastics flushing cistern with BS ball valve and all fittings, flush bend and connector

cistern brackets plugged to wall and flush bend connected to cistern and pan	each	2.50	2.50	42.83	128.90	171.73	197.49

Low level WC suites with coloured vitreous china pan plugged to floor BS pattern seat and cover with plastics fittings, 7.5 litre streamlined finish plastics flushing cistern with BS ball valve and all fittings, flush bend and connector

cistern brackets plugged to wall and flush bend connected to cistern and pan	each	2.50	2.50	42.83	152.90	195.73	225.09

Wheelchair WC facility vitreous china lavatory basin with CP thermostatic spray mixer, CP chain waste, VC close coupled WC suite & flushing cistern,1 white nylon coated aluminium hinged support rail with toilet roll holder, 4 white nylon coated grab rails (600)

	each	6.08	6.08	104.15	690.83	794.98	914.23

Floor standing gas fired domestic central heating boilers with stove enamel finish casing plugged to floor with flue pipe connected to boiler

40,000 Btu/h - 11.7 kWh	each	4.25	4.25	72.80	552.21	625.01	718.76
80,000 Btu/h - 23.4 kWh	each	4.50	4.50	77.08	834.21	911.29	1047.98
40,000 Btu – Conventional Flue	each	4.25	4.25	72.80	433.21	506.01	581.91
80,000 Btu – Conventional Flue	each	4.50	4.50	77.08	625.21	702.29	807.63

Wall mounted gas fired CH boiler, plugged and screwed to wall

40,000 Btu - Balanced Flue	each	4.25	4.25	72.80	451.21	524.01	602.61
60,000 Btu - Balanced Flue	each	4.50	4.50	77.08	622.21	699.29	804.18
40,000 Btu - Conventional Flue	each	4.25	4.25	72.80	438.21	511.01	587.66
80,000 Btu - Conventional Flue	each	4.50	4.50	77.08	637.21	714.29	821.43

Plumbing and engineering installations

New work	Unit	Hours C	Hours L	Labour net	Material net	Price net	Price with 15%
Equipment				£	£	£	£
					VAT not included		

Supplying and fixing equipment - connections to service and waste pipes not included
(*continued*)

Gas fired combined room heater, surround and boiler unit with conventional flue

	Unit	Hours C	Hours L	Labour net	Material net	Price net	Price with 15%
40,000 Btu	each	4.25	4.25	72.80	650.00	722.80	831.22

Oil fired floor standing boiler, plugged and screwed to floor

40,000 Btu - Balanced Flue	each	4.25	4.25	72.80	876.21	949.01	1091.36
40,000 Btu - Conventional Flue	each	4.25	4.25	72.80	801.21	874.01	1005.11

Solid fuel boilers

solid fuel floor standing boiler	each	4.25	4.25	72.80	926.21	999.01	1148.86
solid fuel free standing room heater with high output boiler	each	4.25	4.25	72.80	431.21	504.01	579.61
solid fuel open fire with high output boiler	each	4.25	4.25	72.80	259.21	332.01	381.81

Domestic central heating circulating pump kits connected to pipework - electrical work not included

	each	1.00	1.00	17.13	74.20	91.33	105.03

Single panel white stove primed steel radiators with plug, air vent and brackets and stays plugged to wall

946 x 685 mm	each	0.60	0.60	10.28	30.81	41.09	47.25
1759 x 685 mm	each	0.70	0.70	11.99	52.81	64.80	74.52

Supplying and fixing equipment including all connections - pipework not included

Prices include taps, traps and valves and allowances for unloading, storing, hoisting and fixing in position.

560 x 405 mm vitreous china lavatory basins with pair of 13 mm chromium plated pillar valves, 32 mm chromium plated waste, plastics trap, plug, chain and stay and pair of towel rail brackets plugged to wall

white basins	each	3.20	3.20	54.83	65.80	120.63	138.73
coloured basins	each	3.20	3.20	54.83	77.69	132.52	152.40

560 x 405 mm vitreous china lavatory basins with pair of 13 mm chromium plated pillar valves, 32 mm chromium plated waste, plastics trap, plug, chain and stay, bracket plugged to wall and pedestal plugged to floor

white basin and pedestal sets	each	2.95	2.95	50.54	85.81	136.35	156.81
coloured basin and pedestal sets	each	2.95	2.95	50.54	105.28	155.82	179.20

Plumbing and engineering installations

New work	Unit	Hours C	Hours L	Labour net	Material net	Price net	Price with 15%
Equipment				£	£	£	£
					VAT not included		

Supplying and fixing equipment including all connections - pipework not included
(continued)

1700 mm acrylic baths with cradle and brackets, pair of 19 mm chromium plated pillar valves, 38 mm chromium plated waste and overflow fittings, plastics trap and overflow connection, plug and chain

	Unit	Hours C	Hours L	Labour net	Material net	Price net	Price with 15%
white baths	each	4.63	4.63	79.32	165.42	244.74	280.88
coloured baths	each	4.63	4.63	79.32	170.83	250.15	287.68

1700 mm vitreous enamelled pressed steel baths with cradle feet, pair of 19 mm chromium plated pillar valves, 38 mm chromium plated waste and overflow fittings, plastics trap and overflow connection, plug and chain

	Unit	Hours C	Hours L	Labour net	Material net	Price net	Price with 15%
white baths	each	4.63	4.63	79.32	158.93	238.25	273.99
coloured baths	each	4.63	4.63	79.32	172.99	252.31	290.16

610 x 535 x 255 mm shelf pattern white fireclay sinks with pair of 13 mm chromium plated high necked pillar valves, 38 mm chromium plated waste fitting, plastics trap, plug, chain and stay and pair of cantilever brackets cut and pinned to wall

	Unit	Hours C	Hours L	Labour net	Material net	Price net	Price with 15%
wall	each	3.53	3.53	60.48	232.90	293.38	337.39

Stainless steel sinks with pair of 13 mm chromium plated high necked pillar valves, 38 mm chromium plated waste and overflow fittings, plastics trap and overflow connection, plug and chain - fitted to sink units

	Unit	Hours C	Hours L	Labour net	Material net	Price net	Price with 15%
1000 x 500 mm single drainer tops	each	3.03	3.03	51.91	121.08	172.99	198.94
1500 x 500 mm double drainer tops	each	3.03	3.03	51.91	149.20	201.11	231.28

High level WC suites with white vitreous china pan plugged to floor, BS pattern black seat and cover with plastics fittings, 7.5 litre standard finish black plastics flushing cistern with BS ball valve and all fittings, flush pipe and connector, cistern

	Unit	Hours C	Hours L	Labour net	Material net	Price net	Price with 15%
	each	4.03	4.03	69.04	120.02	189.06	217.42

Low level WC suites with vitreous china pan plugged to floor, BS pattern seat and cover with plastics flushing cistern with BS ball valve and all fittings, flush bend and connector, cistern brackets plugged to wall and flush bend connected to cistern and

	Unit	Hours C	Hours L	Labour net	Material net	Price net	Price with 15%
white suites	each	4.03	4.03	69.04	122.18	191.22	219.91
coloured suites	each	4.03	4.03	69.04	152.46	221.50	254.73

Plumbing and engineering installations

New work	Unit	Hours C	Hours L	Labour net	Material net	Price net	Price with 15%
Equipment				£	£	£	£
					VAT not included		

Supplying and fixing equipment including all connections - pipework not included
(continued)

Floor standing gas-fired domestic central heating boilers with stove enamel finish casing plugged to floor with flue pipe connected to boiler

	Unit	Hours C	Hours L	Labour net	Material net	Price net	Price with 15%
40,000 BTU/h - 11.7 Kw/h	each	5.75	5.75	98.51	433.23	531.74	611.51
80,000 BTU/h - 23.4 Kw/h	each	6.15	6.15	105.35	633.17	738.52	849.30

Domestic central heating circulating pump kits connected to pipework - electrical work not included

	Unit	Hours C	Hours L	Labour net	Material net	Price net	Price with 15%
	each	1.00	1.00	17.13	86.52	103.65	119.20

Single panel white stove primed steel radiators with plug, air vent, pair of 13 mm chromium plated angle valves, and brackets and stays plugged to wall

	Unit	Hours C	Hours L	Labour net	Material net	Price net	Price with 15%
946 x 685 mm (3144 BTU/h - 0.921 Kw/h)	each	2.30	2.30	39.41	53.44	92.85	106.78
1759 x 685 mm (5712 BTU/h - 1.674 Kw/h)	each	2.55	2.55	43.69	82.64	126.33	145.28

Plumbing and engineering installations

New work	Unit	Hours C	Hours L	Labour net	Material net	Price net	Price with 15%
Insulation				£	£	£	£
					VAT not included		
Jute felt mesh pipe lagging 100 mm wide wrapped with minimum overlap and secured with galvanised wire around pipes, external diameters							
15 mm	m	0.10	0.10	1.72	0.15	1.87	2.15
22 mm	m	0.11	0.11	1.89	0.20	2.09	2.40
28 mm	m	0.12	0.12	2.05	0.24	2.29	2.63
35 mm	m	0.13	0.13	2.22	0.28	2.50	2.88
42 mm	m	0.14	0.14	2.40	0.33	2.73	3.14
48 mm	m	0.16	0.16	2.74	0.36	3.10	3.56
54 mm	m	0.18	0.18	3.09	0.40	3.49	4.01
60 mm	m	0.20	0.20	3.42	0.46	3.88	4.46
13 mm flexible foam sectional pipe lagging split side and joints sealed with adhesive tape around pipes, external diameters							
15 mm	m	0.07	0.07	1.20	1.17	2.37	2.73
22 mm	m	0.07	0.07	1.20	1.38	2.58	2.97
28 mm	m	0.08	0.08	1.37	1.57	2.94	3.38
35 mm	m	0.08	0.08	1.37	1.77	3.14	3.61
42 mm	m	0.09	0.09	1.54	1.17	2.71	3.12
48 mm	m	0.09	0.09	1.54	1.38	2.92	3.36
54 mm	m	0.10	0.10	1.72	1.57	3.29	3.78
60 mm	m	0.10	0.10	1.72	1.77	3.49	4.01
Anti-corrosive pipe wrapping 100 mm wide roll wrapped with minimum overlap around pipes, external diameters							
15 mm	m	0.10	0.10	1.72	0.71	2.43	2.79
22 mm	m	0.11	0.11	1.89	0.95	2.84	3.27
28 mm	m	0.12	0.12	2.05	1.14	3.19	3.67
35 mm	m	0.13	0.13	2.22	1.32	3.54	4.07
42 mm	m	0.14	0.14	2.40	1.56	3.96	4.55
48 mm	m	0.16	0.16	2.74	1.70	4.44	5.11
54 mm	m	0.18	0.18	3.09	1.89	4.98	5.73
60 mm	m	0.20	0.20	3.42	2.18	5.60	6.44
50 mm expanded polystyrene lagging sets to sides and top of galvanised steel cisterns complete with Byelaw 30 kit, seal lid - lid not included							
457 x 305 x 305 mm deep	each	0.50	0.50	8.57	2.56	11.13	12.80
686 x 508 x 508 mm deep	each	0.75	0.75	12.85	4.20	17.05	19.61
762 x 584 x 610 mm deep	each	0.75	0.75	12.85	6.66	19.51	22.44
1219 x 610 x 610 mm deep	each	1.00	1.00	17.13	10.25	27.38	31.49
50 mm expanded polystyrene lagging sets to bottom, sides and top of galvanised steel cisterns complete with Byelaw 30 kit, seal lid - lid not included							
457 x 305 x 305 mm deep	each	0.65	0.65	11.14	3.08	14.22	16.35
686 x 508 x 508 mm deep	each	0.90	0.90	15.41	5.04	20.45	23.52
762 x 584 x 610 mm deep	each	0.90	0.90	15.41	8.00	23.41	26.92
1219 x 610 x 610 mm deep	each	1.25	1.25	21.41	12.30	33.71	38.77

Plumbing and engineering installations

New work	Unit	Hours C	Hours L	Labour net	Material net	Price net	Price with 15%
Insulation				£	£	£	£
					VAT not included		
Byelaw 30 kit complete with insulation jacket and lid to sides and top of plastic cisterns							
458 x 305 x 305 mm deep	each	0.33	0.33	5.66	2.46	8.12	9.34
736 x 585 x 533 mm deep	each	0.50	0.50	8.57	6.25	14.82	17.04
1016 x 635 x 533 mm deep	each	0.50	0.50	8.57	9.53	18.10	20.82
1118 x 585 x 585 mm deep	each	0.67	0.67	11.47	11.39	22.86	26.29
80 mm glass fibre filled PVCU insulating jackets to sides and top of cylinders							
400 x 1050 mm	each	0.50	0.50	8.57	7.40	15.97	18.37
450 x 750 mm	each	0.50	0.50	8.57	5.88	14.45	16.62
450 x 900 mm	each	0.50	0.50	8.57	6.34	14.91	17.15
450 x 1050 mm	each	0.50	0.50	8.57	7.40	15.97	18.37
450 x 1200 mm	each	0.50	0.50	8.57	8.40	16.97	19.52

Plumbing and engineering installations

New work	Unit	Hours C	Hours L	Labour net	Plant net	Material net	Price net	Price with 15%
Builders work				£	£	£	£	£
					VAT not included			

Pipe trenches

Excavate by machine trenches to
receive service pipes, grade bottom, fill
in and compact and remove surplus -
earthwork support not included, average
depth not exceeding

0.25 m	m	-	0.06	0.40	0.64	-	1.04	1.20
0.50 m	m	-	0.12	0.81	1.23	-	2.04	2.35
0.75 m	m	-	0.18	1.21	1.84	-	3.05	3.51
1.00 m	m	-	0.24	1.62	2.48	-	4.10	4.71

Excavate by hand trenches to receive
service pipes, grade bottom, fill in and
compact and remove surplus –
earthwork support not included, average
depth not exceeding

0.25 m	m	-	0.55	3.71	0.08	-	3.79	4.36
0.50 m	m	-	1.10	7.41	0.10	-	7.51	8.64
0.75 m	m	-	1.60	10.78	0.15	-	10.93	12.57
1.00 m	m	-	2.15	14.49	0.23	-	14.72	16.93

Stopcock pits

Stopcock pit 1000 mm deep with 100 mm concrete base, 150 mm clayware pipe shaft and cast iron hinged cover flaunched in mortar	each	-	3.00	20.22	-	17.29	37.51	43.14
Stopcock pit 1000 mm deep with 100 mm concrete base, half brick sides from 215 x 215 mm to 150 x 150 mm in clear and cast iron hinged cover flaunched in mortar	each	-	4.50	30.33	-	45.36	75.69	87.04

Holes

Form 150 x 150 mm holes for pipes through 150 mm concrete floor and make good	each	0.60	0.33	7.60	-	5.91	13.51	15.54

Cut holes for pipes through 150 mm
concrete floor and make good

small (not exceeding 55 mm)	each	-	1.00	6.74	-	1.00	7.74	8.90
large (55 - 110 mm)	each	-	1.30	8.76	-	3.01	11.77	13.54
extra large (over 110 mm)	each	-	1.80	12.13	-	5.02	17.15	19.72

Holes for small pipes (not exceeding
55 mm) through brick or block walls and
make good

75 mm	each	0.17	0.10	2.19	-	-	2.19	2.52
102.5 mm	each	0.30	0.18	3.90	-	-	3.90	4.49
215 mm	each	0.45	0.28	5.92	-	-	5.92	6.81

Plumbing and engineering installations

New work	Unit	Hours C	Hours L	Labour net	Plant net	Material net	Price net	Price with 15%
Builders work				£	£	£	£	£
					VAT not included			

Holes (*continued*)

Holes for large pipes (55 - 110 mm) through brick or block walls and make good

75 mm	each	0.27	0.18	3.63	-	0.10	3.73	4.29
102.5 mm	each	0.37	0.25	5.01	-	0.20	5.21	5.99
215 mm	each	0.50	0.40	7.18	-	0.30	7.48	8.60

Holes for extra large pipes (over 110 mm) through brick or block walls and make good

75 mm	each	0.38	0.25	5.09	-	0.20	5.29	6.08
102.5 mm	each	0.45	0.30	6.05	-	0.30	6.35	7.30
215 mm	each	0.75	0.50	10.09	-	0.40	10.49	12.06

Holes for small pipes (not exceeding 55 mm) through brick walls and make good facings

102.5 mm	each	0.40	0.25	5.27	-	0.10	5.37	6.18
215 mm	each	0.55	0.35	7.29	-	0.20	7.49	8.61

Holes for large pipes (55 - 110 mm) through brick walls and make good facings

102.5 mm	each	0.55	0.35	7.29	-	0.10	7.39	8.50
215 mm	each	0.80	0.50	10.54	-	0.20	10.74	12.35

Holes for extra large pipes (over 110 mm) through brick walls and make good facings

102.5 mm	each	0.67	0.45	9.03	-	0.20	9.23	10.61
215 mm	each	1.00	0.67	13.48	-	0.30	13.78	15.85

Chases

Horizontal rough chases in brickwork

25 x 25 mm	m	0.20	0.15	2.80	-	-	2.80	3.22
50 x 50 mm	m	0.30	0.20	4.04	-	-	4.04	4.65
75 x 75 mm	m	0.40	0.30	5.60	-	-	5.60	6.44
100 x 100 mm	m	0.50	0.35	6.84	-	-	6.84	7.87

Vertical rough chases in brickwork

25 x 25 mm	m	0.30	0.20	4.04	-	-	4.04	4.65
50 x 50 mm	m	0.45	0.30	6.05	-	-	6.05	6.96
75 x 75 mm	m	0.60	0.40	8.08	-	-	8.08	9.29
100 x 100 mm	m	0.75	0.50	10.09	-	-	10.09	11.60

For prices for chasing, cutting and drilling concrete and brickwork, see "Concrete work"
For specialist prices for holes and collars for pipes, see "Asphalt work"

Plumbing and engineering installations

New work	Unit	Hours C	Hours L	Labour net	Plant net	Material net	Price net	Price with 15%
Builders work				£	£	£	£	£
					VAT not included			

Lead slates (for roofing)

	Unit	Hours C	Hours L	Labour net	Plant net	Material net	Price net	Price with 15%
1.80 mm (BS Code 4) sheet lead slates (for fixing by slater or tiler) with 150 mm collar for large pipe, slate size								
600 x 450 mm	each	2.00	2.00	34.26	-	-	34.26	39.40
600 x 600 mm	each	2.00	2.00	34.26	-	-	34.26	39.40

Aluminium slates (for roofing)

	Unit	Hours C	Hours L	Labour net	Plant net	Material net	Price net	Price with 15%
Aluminium slates (for fixing by asphalter or roofer to flat roofs) with synthetic rubber cone and elastomeric seal around								
75 mm pipes	each	0.25	0.25	4.28	-	12.20	16.48	18.95
100 mm pipes	each	0.25	0.25	4.28	-	12.20	16.48	18.95
Aluminium slates (for fixing by slater or tiler to sloping roofs) with synthetic rubber cone and elastomeric seal around								
75 mm pipes	each	0.25	0.25	4.28	-	-	4.28	4.92

Floor joists

	Unit	Hours C	Hours L	Labour net	Plant net	Material net	Price net	Price with 15%
Notching floor joists for pipe not exceeding 25 mm diameter	m	0.15	-	1.34	-	-	1.34	1.54

Bath panels

	Unit	Hours C	Hours L	Labour net	Plant net	Material net	Price net	Price with 15%
38 x 50 mm softwood bath panel framing								
to front only	set	1.24	-	11.11	-	3.48	14.59	16.78
to front and one end	set	1.94	-	17.38	-	4.82	22.20	25.53
3.2 mm hardboard bath panels with melamine surface fixed with chromium plated screws to framing								
to front only	each	0.30	-	2.69	-	9.60	12.29	14.13
to front and one end including stainless steel angle strip	set	0.45	-	4.03	-	14.43	18.46	21.23
Moulded acrylic bath panels fixed with chromium plated screws to framing								
to front only	each	0.30	-	2.69	-	24.30	26.99	31.04
to front and one end	set	0.45	-	4.03	-	38.05	42.08	48.39

Plasterwork

	Unit	Hours C	Hours L	Labour net	Plant net	Material net	Price net	Price with 15%
Make good plastering around pipes								
not exceeding 0.30 m girth	each	0.17	0.10	2.19	-	0.37	2.56	2.94
0.30 - 1.00 m girth	each	0.25	0.15	3.25	-	0.75	4.00	4.60

Plumbing and engineering installations

New work	Unit	Hours C	Hours L	Labour net	Plant net	Material net	Price net	Price with 15%
Builders work				£	£	£	£	£
					VAT not included			
Painting								
Prime and paint one undercoat and one coat of gloss enamel on metal pipes – internally								
over 300 mm girth	m2	0.66	-	5.91	-	1.34	7.25	8.34
not exceeding 150 mm girth	m	0.13	-	1.16	-	0.22	1.38	1.59
150 - 300 mm girth	m	0.22	-	1.97	-	0.43	2.40	2.76
Prime and paint two undercoats and one coat of gloss enamel on metal pipes – internally								
over 300 mm girth	m2	0.88	-	7.88	-	1.37	9.25	10.64
not exceeding 150 mm girth	m	0.17	-	1.52	-	0.63	2.15	2.47
150 - 300 mm girth	m	0.33	-	2.96	-	0.44	3.40	3.91
Two coats of aluminium heat resisting paint on radiators and pipes - internally								
over 300 mm girth	m2	0.60	-	5.38	-	1.07	6.45	7.42
not exceeding 150 mm girth	m	0.15	-	1.34	-	0.16	1.50	1.73
150 - 300 mm girth	m	0.24	-	2.15	-	0.32	2.47	2.84
Prime and paint one undercoat and one coat of gloss enamel on metal gutters and pipes - externally								
over 300 mm girth	m2	1.05	-	9.41	-	1.42	10.83	12.45
not exceeding 150 mm girth	m	0.26	-	2.33	-	0.22	2.55	2.93
150 - 300 mm girth	m	0.42	-	3.76	-	0.44	4.20	4.83
Prime and paint two undercoats and one coat of gloss enamel on metal gutters and pipes - externally								
over 300 mm girth	m2	1.35	-	12.10	-	1.85	13.95	16.04
not exceeding 150 mm girth	m	0.34	-	3.05	-	0.29	3.34	3.84
150 - 300 mm girth	m	0.54	-	4.84	-	0.57	5.41	6.22

Electrical installations

Preamble

Electrical work is now almost exclusively the province of specialist sub-contractors, consequently, with the exception of builder's work, prices in this section are based on those of specialists.

"Labour net" figures for builder's work include allowances for all costs incidental to the employment of labour.

"Materials net" figures for builder's work include for all costs of materials including an allowance for waste except where specifically stated.

"Price net" figures for builder's work are the totals of the "Labour net" and "Materials net" figures. Prices are for a builder employing his own labour; according to the amount and nature of the work involved, it may well be possible to secure more advantageous prices from specialist sub-contractors.

Specialist prices

"Price with 15%" figures are all-in guide prices and include 15% for the builder's overheads, profit, unloading materials and general attendance (to include free use of standing scaffolding and hoists, temporary lighting and water and clearing away rubbish).

The amount of attendance required varies between the various trades and also with the circumstances of specific jobs; the percentage addition must always be considered and adjusted as necessary to suit the terms and conditions of the quotation being used.

Quantities and delivery distances are usually the most significant of the many factors which influence prices and it must be emphasised that quotations should always be obtained when preparing a tender.

Electrical installations

	Unit	Specialist price net	Price with 15%

Electrical work

£   £

VAT not included

The following "Specialist price net" figures are guide prices provided by Buckman & Hayward Ltd for an installation minimum of about 20 points within 55 kilometres of the specialist sub-contractor's works.

Prices include for fluorescent tubes, but not Electricity Board charges or cash discount.

See the preamble notes for builder's profit and attendance.

Lighting points

PVC-U insulated and sheathed cables in domestic houses or flats, installed in floor cavities and roof voids, flush in walls

	Unit	Specialist price net	Price with 15%
single lighting points - excluding switches - see below	each	15.51	17.84
one gang - one way switch	each	16.42	18.88
two gang - one way switch	each	22.98	26.43
one gang - two way switch	each	18.49	21.26
one gang - intermediate switch	each	22.20	25.53

PVC-U insulated and sheathed cables in commercial offices, shops, schools, etc., installed above false ceilings on cable trays where necessary, flush in walls

single lighting points - excluding switches - see below	each	24.26	27.90
one gang - one way switch	each	26.62	30.61
two gang - one way switch	each	36.44	41.91
four gang - one way switch	each	57.85	66.53
eight gang - one way switch	each	133.09	153.05
one gang - two way switch	each	29.88	34.36
one gang - intermediate switch	each	31.80	36.57

PVC-U insulated, single core cables in commercial offices, shops, schools, etc., drawn into heavy gauge high impact plastic conduit and conduit fittings, installed above false ceilings, flush in walls

single lighting points - excluding switches - see below	each	31.80	36.57
one gang - one way switch	each	29.99	34.49
two gang - one way switch	each	35.19	40.47
four gang - one way switch	each	74.96	86.20
eight gang - one way switch	each	172.42	198.28
one gang - two way switch	each	31.55	36.28
one gang - intermediate switch	each	42.26	48.60

PVC-U insulated, single core cables in commercial offices, shops, schools, etc., drawn into heavy gauge black enamel screwed conduit and conduit fittings, installed above false ceilings, flush in walls

single lighting points - excluding switches - see below	each	37.49	43.11
one gang - one way switch	each	41.95	48.24
two gang - one way switch	each	45.58	52.42
four gang - one way switch	each	91.41	105.12
eight gang - one way switch	each	210.24	241.78
one gang - two way switch	each	41.27	47.46
one gang - intermediate switch	each	49.96	57.45

Electrical installations

New work	Unit	Specialist price net	Price with 15%

Electrical work

£ £

VAT not included

Lighting points *(continued)*

PVC-U insulated, single core cables in commercial offices, shops, schools, etc., drawn into heavy gauge galvanised screwed conduit and conduit fittings, installed above false ceilings, flush in walls

	Unit	net	15%
single lighting points - excluding switches - see below	each	40.97	47.11
one gang - one way switch	each	42.92	49.36
two gang - one way switch	each	47.67	54.82
four gang - one way switch	each	93.90	107.98
eight gang - one way switch	each	215.97	248.36
one gang - two way switch	each	45.62	52.46
one gang - intermediate switch	each	51.34	59.04

Mineral insulated copper sheathed cables in commercial offices, shops, schools, etc., installed on the surface

	Unit	net	15%
single lighting points - excluding switches - see below	each	43.76	50.32
one gang - one way switch	each	61.33	70.53
two gang - one way switch	each	68.34	78.59
one gang - two way switch	each	64.30	73.94
one gang - intermediate switch	each	84.16	96.78

Lighting fittings

N.B. Prices for lighting fittings include tubes, fixing and connections, prices are based on Fitzgerald fittings

Fluorescent drum fitting, ceiling or wall mounted

	Unit	net	15%
16 watt	each	26.90	30.93
28 watt	each	33.90	38.98
38 watt	each	35.82	41.19

Fluorescent fittings single, complete with tube, ceiling mounted

		Unit	net	15%
600 mm (Lightpack Range)	18 watt	each	15.79	18.16
900 mm (Lightpack Range)	30 watt	each	16.45	18.92
1200 mm (Lightpack Range)	36 watt	each	19.71	22.67
1500 mm (Lightpack Range)	58 watt	each	26.04	29.95
1800 mm (Lightpack Range)	70 watt	each	29.28	33.67
2400 mm (Lightpack Range)	100 watt	each	39.48	45.40

Fluorescent fittings twin, complete with tubes, ceiling mounted

		Unit	net	15%
600 mm (Lightpack Range)	2 x 18 watt	each	21.61	24.85
1200 mm (Lightpack Range)	2 x 36 watt	each	29.94	34.43
1500 mm (Lightpack Range)	2 x 58 watt	each	36.50	41.97
1800 mm (Lightpack Range)	2 x 70 watt	each	42.23	48.56
2400 mm (Lightpack Range)	2 x 100 watt	each	56.78	65.30

Electrical installations

New work

Electrical work

£   £

VAT not included

Lighting fittings *(continued)*

Fluorescent fittings single, prismatic controller, complete with tubes, ceiling mounted

600 mm (Lightpack Range)	18 watt	each	23.44	26.96
900 mm (Lightpack Range)	30 watt	each	25.78	29.65
1200 mm (Lightpack Range)	36 watt	each	27.63	31.78
1500 mm (Lightpack Range)	58 watt	each	33.48	38.50
1800 mm (Lightpack Range)	70 watt	each	41.13	47.30
2400 mm (Lightpack Range)	100 watt	each	54.44	62.61

Fluorescent fittings twin, prismatic controller, complete with tubes, ceiling mounted

600 mm (Lightpack Range)	2 x 18 watt	each	32.06	36.87
1200 mm (Lightpack Range)	2 x 36 watt	each	45.04	51.80
1500 mm (Lightpack Range)	2 x 58 watt	each	52.79	60.70
1800 mm (Lightpack Range)	2 x 70 watt	each	60.64	69.74
2400 mm (Lightpack Range)	2 x 100 watt	each	82.62	95.02

300/500 watt enclosed tungsten halogen security floodlights controlled by a separate 180 deg PIR detector

Domestic installation with PVC-U/PVC-U cables	each	115.40	132.71
Commercial installation with plastic conduit	each	176.73	203.24
Commercial installation with MCIS	each	202.43	232.79

Small power 13 amp general purpose switched socket outlets and spur units, wired on a ring circuit

PVC-U insulated and sheathed cables in domestic houses or flats, installed in floor cavities and roof voids, flush in the walls

13 amp single switched socket outlets	each	37.54	43.17
13 amp twin switched socket outlets	each	41.14	47.31
13 amp switched spur unit complete with flex outlet	each	40.65	46.75

PVC-U insulated and sheathed cables in commercial offices, shops, schools, etc., installed above false ceilings on cable trays where necessary, flush in walls

13 amp single switched socket outlets	each	48.44	55.70
13 amp twin switched socket outlets	each	52.05	59.86
13 amp switched spur unit complete with flex outlet	each	51.30	59.00

PVC-U insulated, single core cables in commercial offices, shops, schools, etc., drawn into heavy gauge high impact plastic conduit and conduit fittings, installed above false ceilings, flush in walls

13 amp single switched socket outlets	each	77.90	89.59
13 amp twin switched socket outlets	each	84.74	97.45
13 amp switched spur unit complete with flex outlet	each	84.76	97.47

PVC-U insulated, single core cables in commercial offices, shops, schools, etc., drawn into heavy gauge black enamel screwed conduit and conduit fittings, installed above false ceilings, flush in walls

13 amp single switched socket outlets	each	96.66	111.16
13 amp twin switched socket outlets	each	100.02	115.03
13 amp switched spur unit complete with flex outlet	each	100.30	115.35

Electrical installations

New work	Unit	Specialist price net	Price with 15%
Electrical work		£	£
			VAT not included

Small power 13 amp general purpose switched socket outlets and spur units, wired on a ring circuit *(continued)*

PVC-U insulated, single core cables in commercial offices, shops, schools, etc., drawn into heavy gauge galvanised screwed conduit and conduit fittings, installed above false ceilings, flush in walls

13 amp single switched socket outlets	each	98.20	112.93
13 amp twin switched socket outlets	each	102.14	117.47
13 amp switched spur unit complete with flex outlet	each	102.16	117.49

Mineral insulated copper sheathed cables in commercial offices, shops, schools, etc., installed on the surface

13 amp single switched socket outlets	each	97.11	111.68
13 amp twin switched socket outlets	each	100.61	115.70
13 amp switched spur unit complete with flex outlet	each	100.63	115.73

Cooker points

PVC-U insulated and sheathed cables in domestic houses or flats, installed in floor cavities and roof voids, flush in walls

30 amp cooker points with panel and outlet plate	each	87.33	100.43
45 amp cooker points with panel and outlet plate	each	92.01	105.81

PVC-U insulated, single core cables in commercial offices, shops, schools, etc., drawn into heavy gauge galvanised screwed conduit and conduit fittings, installed above false ceilings, flush in walls

30 amp cooker points with panel and outlet plate	each	142.08	163.40
45 amp cooker points with panel and outlet plate	each	168.26	193.50

Mineral insulated copper sheathed cables in commercial offices, shops, schools, etc., installed on the surface

30 amp cooker points with panel and outlet plate	each	168.38	193.64
45 amp cooker points with panel and outlet plate	each	233.12	268.09

Immersion heater points - complete with local control switch and heat resisting flexible cable, excluding immersion heater and thermostat

PVC-U insulated and sheathed cables in domestic houses or flats, installed in floor cavities and roof voids, flush in walls

immersion heater points with control switch (excluding heater)	each	42.21	48.54

PVC-U insulated and sheathed cables in commercial offices, shops, schools, etc., installed above false ceilings on cable trays where necessary, flush in walls

immersion heater points with control switch (excluding heater)	each	60.22	69.25

PVC-U insulated, single core cables in commercial offices, shops, schools, etc., drawn into heavy gauge high impact plastic conduit and conduit fittings, installed above false ceilings, flush in walls

immersion heater points with control switch (excluding heater)	each	85.71	98.56

Electrical installations

New work	Unit	Specialist price net	Price with 15%
Electrical work		£	£
			VAT not included

Immersion heater points - complete with local control switch and heat resisting flexible cable, excluding immersion heater and thermostat *(continued)*

	Unit	Specialist price net	Price with 15%
PVC-U insulated, single core cables in commercial offices, shops, schools, etc., drawn into heavy gauge black enamel screwed conduit and conduit fittings, installed above false ceilings, flush in walls			
immersion heater points with control switch (excluding heater)	each	110.70	127.31
Mineral insulated copper sheathed cables in commercial offices, shops, schools, etc., installed on the surface			
immersion heater points with control switch (excluding heater)	each	135.59	155.92
Supply and connect 3 kW immersion heaters complete with thermostat			
single heater	each	50.05	57.56
dual heater	each	74.91	86.15

Infra-red heater points only, excluding heater and earth bonding

	Unit	Specialist price net	Price with 15%
PVC-U insulated and sheathed cables in domestic houses or flats, installed in floor cavities and roof voids, flush in walls			
infra-red heater points with control switch (excluding heater and earth bonding)	each	51.89	59.68
PVC-U insulated and sheathed cables in commercial offices, shops, schools, etc., installed above false ceilings on cable trays where necessary, flush in walls			
infra-red heater points with control switch (excluding heater and earth bonding)	each	62.53	71.91
PVC-U insulated, single core cables in commercial offices, shops, schools, etc., drawn into heavy gauge high impact plastic conduit and conduit fittings, installed above false ceilings, flush in walls			
infra-red heater points with control switch (excluding heater and earth bonding)	each	98.60	113.40
PVC-U insulated, single core cables in commercial offices, shops, schools, etc., drawn into heavy gauge black enamel screwed conduit and conduit fittings, installed above false ceilings, flush in walls			
infra-red heater points with control switch (excluding heater and earth bonding)	each	113.12	130.08
Mineral insulated copper sheathed cables in commercial offices, shops, schools, etc., installed on the surface			
infra-red heater points with control switch (excluding heater and earth bonding)	each	141.48	162.71
Supply, fix and connect infra-red heater with flex			
750 W	each	40.24	46.28
1000 W	each	42.12	48.44

Electrical installations

New work	Unit	Specialist price net	Price with 15%
Electrical work		£	£
		VAT not included	

Shaver points

PVC-U insulated and sheathed cables in domestic houses or flats, installed in floor cavities and roof voids, flush in walls

dual type shaver socket outlets	each	51.91	59.69

PVC-U insulated and sheathed cables in commercial offices, shops, schools, etc., installed above false ceilings on cable trays where necessary, flush in walls

dual type shaver socket outlets	each	56.86	65.39

PVC-U insulated, single core cables in commercial offices, shops, schools, etc., drawn into heavy gauge high impact plastic conduit and conduit fittings, installed above false ceilings, flush in walls

dual type shaver socket outlets	each	73.29	84.28

PVC-U insulated, single core cables in commercial offices, shops, schools, etc., drawn into heavy gauge black enamel screwed conduit and conduit fittings, installed above false ceilings, flush in walls

dual type shaver socket outlets	each	101.67	116.92

Bell installations

PVC-U insulated and sheathed cables in domestic houses or flats, installed in floor cavities and roof voids, flush in walls

bells controlled by front door push including transformer	each	86.26	99.20
bell and buzzer sets controlled by front door push and back door push including transformer	each	103.83	119.41
Hi-lo chimes controlled by front door push and back door push including transformer	each	94.62	108.81

PVC-U insulated and sheathed cables in commercial offices, shops, schools, etc., installed above false ceilings on cable trays where necessary, flush in walls

bells controlled by front door push including transformer	each	148.24	170.48
bell and buzzer sets controlled by front door push and back door push including transformer	each	183.84	211.41
Hi-lo chimes controlled by front door push and back door push including transformer	each	164.49	189.17

TV outlets

Low loss coaxial cable downleads in houses or flats, protected by PVC-U conduit in walls, terminating with a single TV outlet, and allowing 3 metres of cable in roof void for connection to aerial by others

	each	38.56	44.35

Storage heater points

PVC-U insulated and sheathed cables in domestic houses or flats, installed in floor cavities and roof voids, flush in walls protected by steel or PVC-U channels

3kW storage heater points controlled by switch adjacent to heater position	each	55.17	63.44

Electrical installations

New work	Unit	Specialist price net	Price with 15%
Electrical work		£	£
		VAT not included	

Consumer units

Insulated consumer units complete with miniature circuit breakers, 100 A switch control and meter tails

4 way	each	88.87	102.19
6 way	each	115.73	133.09
8 way	each	138.75	159.56

Metalclad consumer units complete with miniature circuit breakers, 100 A switch control and meter tails

4 way	each	91.40	105.11
6 way	each	117.53	135.16
8 way	each	140.29	161.34

Insulated consumer units complete with miniature circuit breakers, RCD controls and meter tails

4 way	each	143.00	164.44
6 way	each	174.10	200.21
8 way	each	200.56	230.65

Metalclad consumer units complete with miniature circuit breakers, RCD controls and meter tails

4 way	each	146.27	168.21
6 way	each	174.83	201.06
8 way	each	200.56	230.65

Equipotential Bonding

Domestic main bonding of incoming gas and water services to supply company's protective multiple earthing terminal	lot	50.34	57.89
Domestic local supplementary bond to or between metalic pipework etc.	each	20.10	23.11
Domestic earth electrode, comprising earth rod, cable clamp, inspection pit, protective conduit and earthing conductor	lot	88.05	101.25

Enquiries about the foregoing specialist prices should be made in writing to Buckman & Hayward Ltd, 145a Ashford Road, Eastbourne, East Sussex, BN21 3UA, tel (01323) 642815, fax (01323) 410225.

Electrical installations

New work	Unit	Hours C	Hours L	Labour net	Material net	Price net	Price with 15%
Builder's work				£	£	£	£
					VAT not included		
Concealed conduits and cables							
Cutting away for and making good after the electrician to points with concealed conduits or cables							
Lighting points and associated switch points	each	0.50	2.00	17.96	0.50	18.46	21.23
Socket outlet points	each	0.25	1.25	10.67	0.20	10.87	12.50
Fitting outlet points	each	0.33	2.33	18.66	0.20	18.86	21.69
Equipment points and control gear points	each	0.25	1.75	14.04	0.30	14.34	16.49
Exposed conduits and cables							
Cutting away for and making good after the electrician to points with exposed conduits or cables							
Lighting points and associated switch points	each	0.50	0.50	7.85	0.10	7.95	9.14
Socket outlet points	each	0.25	0.25	3.93	0.10	4.03	4.63
Fitting outlet points	each	0.33	0.33	5.18	0.10	5.28	6.07
Equipment points and control gear points	each	0.25	0.50	5.61	0.10	5.71	6.57

Electrical installations

Lightning conductors

		£	£
		VAT not included	

The following "Specialist price net" figures are guide prices provided by Omega Furse Contracting Ltd. References in brackets are to component catalogue numbers.

Prices do not include for cash discount.

See the preamble notes for builder's profit and attendance.

Standard range lightning conductors to BS 6651

Light duty air terminals

	Unit	Specialist price net	Price with 15%
Air rods complete with flat base plugged to walls			
15 diameter x 510 mm solid aluminium (RA105)	each	25.19	28.97
15 diameter x 510 mm solid copper (RA305)	each	31.24	35.93
Taper-pointed air rods complete with rod to tape coupling and rod brackets plugged to walls			
15 diameter x 1000 mm solid aluminium (RA025)	each	37.04	42.60
15 diameter x 1000 mm solid copper (RA225)	each	44.00	50.60
15 diameter x 2000 mm solid copper (RA240)	each	49.26	56.65
20 diameter x 1000 mm solid copper (RA255)	each	48.22	55.45
20 diameter x 2000 mm solid copper (RA270)	each	52.90	60.84

Roof and down conductors

	Unit	Specialist price net	Price with 15%
High conductivity aluminium tape complete with direct contact clips plugged to walls at 1000 mm centres			
20 x 3 mm horizontal (TA020)	m	7.94	9.13
20 x 3 mm vertical (TA020)	m	8.84	10.17
25 x 3 mm horizontal (TA030)	m	8.46	9.73
25 x 3 mm vertical (TA030)	m	8.98	10.33
High conductivity copper tape complete with direct contact clips plugged to walls at 1000 mm centres			
20 x 3 mm horizontal (TC020)	m	11.13	12.80
20 x 3 mm vertical (TC020)	m	11.70	13.46
25 x 3 mm horizontal (TC030)	m	11.79	13.56
25 x 3 mm vertical (TC030)	m	12.05	13.86

Earthing

	Unit	Specialist price net	Price with 15%
Electrodes driven into ground			
16 diameter x 2400 mm "Copper-bond" electrodes complete with coupling and driving stud (RB205)	each	57.46	66.08
16 diameter x 2400 mm solid copper electrodes complete with coupling and driving stud (RC010)	each	68.16	78.38
900 x 900 x 1.5 mm solid copper earthplates (PE015) (excluding excavations)	each	150.42	172.98

Electrical installations

New work	Unit	Specialist price net	Price with 15%

Lightning conductors

		£	£
		VAT not included	

Standard range lightning conductors to BS 6651 *(continued)*

Test points

Gunmetal test point (CN105)	each	21.43	26.64
Bi-metallic test point (CN910)	each	26.82	30.84

International range lightning conductors to BS 6651

Air terminals

Taper pointed air rods complete with base plugged to walls vertically or horizontally

10 diameter x 300 mm solid aluminium (RA075)	each	17.82	20.49
10 diameter x 500 mm solid aluminium (RA080)	each	18.73	21.54
10 diameter x 1000 mm solid aluminium (RA2085)	each	19.63	22.58
10 diameter x 300 mm solid copper (RA398)	each	26.77	30.79
10 diameter x 500 mm solid copper (RA400)	each	27.82	31.99
10 diameter x 1000 mm solid copper (RA402)	each	28.93	33.27

Roof and down conductors

High conductivity solid aluminium conductor complete with direct contact clips plugged to walls at 500 mm centres

8 mm diameter horizontal (CD080)	m	8.98	10.33
8 mm diameter vertical (CD080)	m	9.50	10.93

High conductivity solid copper conductor complete with direct contact clips plugged to walls at 500 mm centres

8 mm diameter horizontal (CD035)	m	12.13	13.95
8 mm diameter vertical (CD035)	m	12.76	14.67

Earthing

Electrodes driven into ground

16 diameter x 2400 mm "Copper-bond" electrodes complete with coupling and driving stud (RB205)	each	57.46	66.08
16 diameter x 2400 mm solid copper electrodes complete with coupling and driving stud (RC010)	each	68.16	78.38
900 x 900 x 1.5 mm solid copper earthplates (PE015) (excluding excavations)	each	150.42	172.98

Test points

Gunmetal test point (CN305)	each	17.85	20.53
Bi-metallic test point (CN915)	each	26.82	30.84

Enquiries about the foregoing specialist prices should be made to Omega Furse Contracting Ltd, Private Road No 7, Colwick Industrial Estate, Nottingham, NG4 2JW, tel (0115) 9402001, fax (0115) 9403001.

Floor, wall and ceiling finishings

Preamble

"Labour net" figures include allowances for all costs incidental to the employment of labour.

"Materials net" figures include for all costs of materials including an allowance for waste except where specifically stated.

"Price net" figures are the totals of the "Labour net" and "Materials net" figures. Prices are for a builder employing his own labour; according to the amount and nature of the work involved, it may well be possible to secure more advantageous prices from specialist sub-contractors.

Prices do not include any allowance for scaffolding, ladders or other plant necessary to reach the work. The "Preliminaries" section includes prices for scaffolding which must be considered and allowance included to suit the particular circumstances of a tender.

Specialist prices

"Price with 15%" figures are all-in guide prices and include 15% for the builder's overheads, profit, unloading materials and general attendance (to include free use of standing scaffolding and hoists, temporary lighting and water and clearing away rubbish).

The amount of attendance required varies between the various trades and also with the circumstances of specific jobs; the percentage addition must always be considered and adjusted as necessary to suit the terms and conditions of the quotation being used.

Quantities and delivery distances are usually the most significant of the many factors which influence prices and it must be emphasised that quotations should always be obtained when preparing a tender.

Floor, wall and ceiling finishings

New work

Basic prices for materials		£
Abrasive grain	kg	2.25
Bonding agent	2½ litres	13.67
Cellular core plasterboard dry partitions		
57 mm	m2	5.56
63 mm	m2	5.66
Cement		
Portland	tonne	89.04
white	tonne	281.96
265 x 165 mm clay roofing tiles	100	49.59
Expanded metal lathing		
0.725 mm coated	m2	5.08
0.500 mm galvanised	m2	4.36
0.725 mm galvanised	m2	5.08
Glazed ceramic tiles		
108 x 108 x 4 mm (Group A)	100	15.46
108 x 108 x 4 mm (Group C)	100	19.33
152 x 152 x 5.5 mm (Group A)	100	44.18
152 x 152 x 5.5 mm (Group B)	100	44.42
152 x 152 x 5.5 mm (Group C)	100	54.85
Granite chippings	tonne	29.00
Hydrated lime	tonne	139.84
90 mm jute scrim	100 m roll	3.69
Liquid colouring agent	pack	9.12
Liquid surface hardener	litre	1.79
Liquid waterproofer and hardener	litre	1.83
Plasterboard baseboard		
9.5 mm	m2	1.59
9.5 mm insulating grade	m2	2.18
Plasterboard lath		
9.5 mm	m2	1.74
12.5 mm	m2	2.07
19 mm plasterboard plank	m2	3.15

Prices actually to be paid for materials must be checked against the above basic prices and
adjustments made as necessary.

Floor, wall and ceiling finishings

New work

Basic prices for materials		£
Plasterboard wallboard		
9.5 mm	m2	1.57
12.5 mm	m2	1.82
9.5 mm insulating grade	m2	2.14
12.5 mm insulating grade	m2	2.39
Premixed lightweight ("Carlite") plasters		
browning grade	tonne	175.12
bonding grade	tonne	172.64
finish grade	tonne	136.00
Quarry tiles - Autumn blend		
150 x 150 x 12.5 mm	10	4.69
150 x 150 x 12.5 mm round edge	10	5.27
Quarry tiles - red		
150 x 150 x 12.5 mm	10	4.70
150 x 150 x 19 mm	10	4.41
229 x 229 x 29 mm	10	21.15
150 x 150 x 12.5 mm round edge	10	4.66
150 x 150 x 19 mm round edge	10	5.13
200 x 200 x 19 mm round edge	10	11.25
229 x 229 x 29 mm round edge	10	21.08
Retarded hemihydrate ("Thistle") plasters		
board finish grade	tonne	128.52
hardwall grade	tonne	175.12
multi-finish grade	tonne	128.52
Sand	m3	17.96
	tonne	11.22
Tyrolean finish	tonne	330.00
Vermiculite grade DSF	110 litre bag	6.59

Prices actually to be paid for materials must be checked against the above basic prices and
adjustments made as necessary.

Floor, wall and ceiling finishings

New work	Unit	Hours C	Hours L	Labour net	Material net	Price net	Price with 15%
Internal in-situ finishings				£	£	£	£
					VAT not included		
Retarded hemihydrate gypsum ("Thistle") plasters to BS 1191, Class B							
5 mm one coat plastering on plasterboard walls including scrimming joints							
over 300 mm wide	m2	0.38	0.19	4.68	1.02	5.70	6.55
not exceeding 300 mm wide	m2	0.57	0.28	7.03	1.02	8.05	9.26
5 mm one coat plastering on plasterboard ceilings including scrimming joints							
over 300 mm wide	m2	0.40	0.20	4.93	1.02	5.95	6.84
not exceeding 300 mm wide	m2	0.60	0.30	7.40	1.02	8.42	9.68
13 mm two coat plastering on brick or block walls							
over 300 mm wide	m2	0.55	0.28	6.78	2.15	8.93	10.27
not exceeding 300 mm wide	m2	0.82	0.42	10.21	2.15	12.36	14.21
16 mm three coat plastering on brick or block walls							
over 300 mm wide	m2	0.75	0.38	9.25	2.63	11.88	13.66
not exceeding 300 mm wide	m2	1.13	0.56	13.87	2.63	16.50	18.98
13 mm three coat plastering on metal lathing to ceilings							
over 300 mm wide	m2	0.75	0.38	9.25	3.59	12.84	14.77
not exceeding 300 mm wide	m2	1.13	0.56	13.87	3.59	17.46	20.08
Retarded hemihydrate gypsum ("Thistle") plasters to BS 1191, Class B and cement and sand 1:3 undercoats							
13 mm two coat plastering on brick or block walls							
over 300 mm wide	m2	0.60	0.30	7.40	1.75	9.15	10.52
not exceeding 300 mm wide	m2	0.90	0.45	11.09	1.75	12.84	14.77
Premixed lightweight ("Carlite") plaster to BS 1191							
13 mm two coat plastering on brick or block walls							
over 300 mm wide	m2	0.55	0.28	6.78	1.91	8.69	9.99
not exceeding 300 mm wide	m2	0.82	0.41	10.17	1.91	12.08	13.89
13 mm two coat plastering on wood wool slab walls including reinforcing joints with metal scrim							
over 300 mm wide	m2	0.60	0.35	7.74	2.78	10.52	12.10
not exceeding 300 mm wide	m2	0.90	0.45	11.09	2.77	13.86	15.94
10 mm two coat plastering on concrete ceilings							
over 300 mm wide	m2	0.55	0.28	6.78	1.94	8.72	10.03
not exceeding 300 mm wide	m2	0.82	0.42	10.21	1.81	12.02	13.82

Floor, wall and ceiling finishings

New work	Unit	Hours C	Hours L	Labour net	Material net	Price net	Price with 15%
Internal in-situ finishings				£	£	£	£
					VAT not included		
Premixed lightweight ("Carlite") plaster to BS 1191 (*continued*)							
10 mm two coat plastering on plasterboard ceilings including scrimming joints							
over 300 mm wide	m2	0.65	0.33	8.01	1.86	9.87	11.35
not exceeding 300 mm wide	m2	0.97	0.49	12.03	1.86	13.89	15.97
13 mm three coat plastering on metal lathing to ceilings							
over 300 mm wide	m2	0.75	0.38	9.25	3.60	12.85	14.78
not exceeding 300 mm wide	m2	1.13	0.56	13.87	3.60	17.47	20.09
Plastering sundries							
Treatment to edges							
Arrises	m	0.11	0.06	1.36	-	1.36	1.56
Rounded external angles not exceeding 10 mm radius	m	0.15	0.07	1.85	-	1.85	2.13
Make good plastering around pipes							
not exceeding 0.30 m girth	each	0.17	0.09	2.09	0.37	2.46	2.83
0.30 - 1.00 m girth	each	0.25	0.13	3.08	0.75	3.83	4.40
Galvanised steel angle beads fixed with plaster dabs	m	0.08	-	0.72	0.64	1.36	1.56
100 mm girth plaster core coved cornices fixed with adhesive							
straight	m	0.15	0.07	1.85	1.42	3.27	3.76
angles	each	0.24	-	2.15	0.38	2.53	2.91
127 mm girth plaster core coved cornices fixed with adhesive							
straight	m	0.15	0.07	1.85	1.62	3.47	3.99
angles	each	0.24	-	2.15	0.45	2.60	2.99
Plaster air vents set in plastering							
225 x 150 mm	each	0.14	-	1.25	0.65	1.90	2.19
225 x 225 mm	each	0.14	-	1.25	1.07	2.32	2.67
Flyproof plaster air vents set in plastering							
225 x 150 mm	each	0.14	-	1.25	1.44	2.69	3.09
225 x 225 mm	each	0.14	-	1.25	1.75	3.00	3.45
Coated metal lathing to BS 1369							
0.725 mm lathing stapled to softwood to ceilings							
over 300 mm wide	m2	0.35	-	3.14	5.46	8.60	9.89
not exceeding 300 mm wide	m2	0.42	-	3.81	5.46	9.27	10.66

Floor, wall and ceiling finishings

New work	Unit	Hours C	Hours L	Labour net	Material net	Price net	Price with 15%
Internal in-situ finishings				£	£	£	£
					VAT not included		
Galvanised metal lathing to BS 1369							
0.500 mm lathing stapled to softwood to ceilings							
over 300 mm wide	m2	0.35	-	3.14	4.70	7.84	9.02
not exceeding 300 mm wide	m2	0.42	-	3.81	4.70	8.51	9.79
0.725 mm lathing stapled to softwood to ceilings							
over 300 mm wide	m2	0.35	-	3.14	5.46	8.60	9.89
not exceeding 300 mm wide	m2	0.42	-	3.81	5.46	9.27	10.66
Gypsum plasterboard to BS 1230							
Lathing fixed with galvanised nails to softwood to ceilings over 300 mm wide							
9.5 mm	m2	0.16	0.08	1.97	2.60	4.57	5.26
12.5 mm	m2	0.18	0.09	2.22	2.95	5.17	5.95
Baseboard fixed with galvanised nails to softwood to ceilings over 300 mm wide							
9.5 mm	m2	0.16	0.08	1.97	2.45	4.42	5.08
9.5 mm insulating grade	m2	0.16	0.08	1.97	3.07	5.04	5.80
Cement and sand 1:3							
Steel trowelled pavings, level and to falls, on concrete over 300 mm wide including thoroughly cleaning concrete base							
25 mm	m2	0.39	0.20	4.80	2.51	7.31	8.41
32 mm	m2	0.42	0.21	5.18	3.21	8.39	9.65
38 mm	m2	0.44	0.22	5.42	3.82	9.24	10.63
50 mm	m2	0.49	0.25	6.04	5.02	11.06	12.72
65 mm	m2	0.52	0.26	6.41	6.53	12.94	14.88
75 mm	m2	0.55	0.28	6.78	7.53	14.31	16.46
Steel trowelled pavings, level and to falls, on concrete not exceeding 300 mm wide including thoroughly cleaning concrete base							
25 mm	m2	0.60	0.30	7.40	2.51	9.91	11.40
32 mm	m2	0.63	0.32	7.76	3.21	10.97	12.62
38 mm	m2	0.66	0.33	8.13	3.82	11.95	13.74
50 mm	m2	0.74	0.37	9.12	5.02	14.14	16.26
65 mm	m2	0.78	0.39	9.62	6.53	16.15	18.57
75 mm	m2	0.83	0.41	10.24	7.53	17.77	20.44
19 x 150 mm steel trowelled skirtings with fair edge on brick or block walls							
straight	m	0.27	0.14	3.33	0.10	3.43	3.94
angles	each	0.09	0.04	1.11	-	1.11	1.28
19 x 225 mm steel trowelled skirtings with fair edge on brick or block walls							
straight	m	0.30	0.15	3.70	0.20	3.90	4.49
angles	each	0.10	0.05	1.24	-	1.24	1.43

Floor, wall and ceiling finishings

New work	Unit	Hours C	Hours L	Labour net	Material net	Price net	Price with 15%
Internal in-situ finishings				£	£	£	£
					VAT not included		

Granolithic 1:2½

Steel trowelled pavings, level and to falls, on concrete over 300 mm wide including thoroughly cleaning concrete base

	Unit	Hours C	Hours L	Labour net	Material net	Price net	Price with 15%
25 mm	m2	0.45	0.23	5.55	3.50	9.05	10.41
32 mm	m2	0.50	0.25	6.17	4.48	10.65	12.25
38 mm	m2	0.55	0.28	6.78	5.32	12.10	13.91

Steel trowelled pavings, level and to falls, on concrete not exceeding 300 mm wide including thoroughly cleaning concrete base

	Unit	Hours C	Hours L	Labour net	Material net	Price net	Price with 15%
25 mm	m2	0.68	0.34	8.38	3.50	11.88	13.66
32 mm	m2	0.75	0.38	9.25	4.48	13.73	15.79
38 mm	m2	0.83	0.41	10.24	5.32	15.56	17.89

Steel trowelled treads with internal angle on concrete

	Unit	Hours C	Hours L	Labour net	Material net	Price net	Price with 15%
225 x 25 mm	m	0.33	0.17	4.07	0.84	4.91	5.65
300 x 25 mm	m	0.40	0.20	4.93	1.12	6.05	6.96

19 x 150 mm steel trowelled risers with external angle on concrete

	Unit	Hours C	Hours L	Labour net	Material net	Price net	Price with 15%
	m	0.30	0.15	3.70	0.42	4.12	4.74

19 x 150 mm steel trowelled skirtings on brick or block walls

	Unit	Hours C	Hours L	Labour net	Material net	Price net	Price with 15%
with fair edge	m	0.30	0.15	3.70	0.42	4.12	4.74
angles on skirting with fair edge	each	0.12	0.06	1.48	-	1.48	1.70
with rounded edge and coved junction with paving	m	0.55	0.28	6.78	0.42	7.20	8.28
angles on skirting with rounded edge	each	0.20	0.10	2.46	-	2.46	2.83

Ancillary work

	Unit	Hours C	Hours L	Labour net	Material net	Price net	Price with 15%
Extra for liquid waterproofer and hardener - per 25 mm thickness	m2	0.02	0.01	0.25	0.08	0.33	0.38
Extra for liquid colouring agent at the rate of 0.25 packs per m2 per 25 mm thickness in pavings - per 25 mm thickness	m2	0.04	0.02	0.49	0.02	0.51	0.59
Three coats of liquid surface hardener on pavings	m2	0.05	0.03	0.62	1.14	1.76	2.02
Abrasive grain sprinkled on pavings at the rate of 0.80 kg per m2 and trowelled in	m2	0.10	0.05	1.24	1.58	2.82	3.24
Solution 1:2 of bonding agent on concrete before laying pavings or screeds	m2	-	0.10	0.67	1.03	1.70	1.96

For sprayed fire protection to structural steelwork, see "Sprayed finishes" in "Painting and decorating".

Floor, wall and ceiling finishings

New work	Unit	Specialist price net	Price with 15%
Internal in-situ finishings		**£**	**£**
			VAT not included

Terrazzo work

The following "Specialist price net" figures are guide prices provided by Alpha M & T Ltd, for quantities over 100 square metres in London or the Home Counties.

Prices include for 2½% cash discount.

See the preamble notes for builder's profit and attendance.

In addition to general attendance the builder will be required to provide cement, sand, water, electric power, storage, welfare and to unload and hoist materials and equipment.

In-situ terrazzo on builder's screeded beds or backings

	Unit	Specialist price net	Price with 15%
16 mm pavings divided into panels (dividing strips priced separately)	m2	73.29	84.28
9.5 mm wall linings or dados	m2	119.80	137.77
16 x 450 mm (girth) finish to treads and risers with slightly rounded external angle and coved internal angle	m	89.44	102.86
Extra for "Ferodo OT2" F non-slip nosings to treads	m	13.88	15.96
9.5 x 225 mm finish to edges of landings with slightly rounded external angle	m	36.21	41.64
9.5 x 150 mm skirtings with rounded or square top edge and cove	m	26.07	29.98
internal or external angles	each	2.81	3.23
stop ends	each	2.81	3.23
fair ends	each	2.81	3.23
9.5 x 225 mm wall strings with rounded or square top edge and cove	m	33.82	38.89
9.5 x 300 mm (average) finish to spandril ends of steps with fair edge and arris	m	40.16	46.18
Arrises	m	3.95	4.54
Rounded edges	m	3.38	3.89
Make good pavings around			
mat frames etc	m	4.37	5.03
pipes not exceeding 0.30 m girth	each	4.59	5.28
6 x 25 mm black or coloured plastics dividing strips	m	4.50	5.18

Terrazzo tiling

	Unit	Specialist price net	Price with 15%
300 x 300 x 25 mm hydraulically pressed tiles bedded in cement and sand to 50 mm overall thickness with in-situ terrazzo margins			
pavings	m2	51.17	58.85
pavings to landings	m2	54.96	63.20
Raking cutting 25 mm tiles	m	7.04	8.10

Floor, wall and ceiling finishings

New work	Unit	Specialist price net	Price with 15%
Internal in-situ finishings		£	£
		VAT not included	
Precast terrazzo steps			
300 x 38 mm reinforced terrazzo faced treads and 150 x 38 mm similar risers with square polished front edge, bedded 12.5 mm thick in cement and sand	m	114.16	131.28
Extra for polished fair edges to exposed ends of 300 x 38 mm treads and 150 x 38 mm risers	each	5.49	6.31
Precast terrazzo partitions			
38 mm reinforced WC partitions or slabs terrazzo faced both sides and fixed in chases	m2	197.31	226.91
75 x 63 mm terrazzo faced, reinforced, rebated wall or end posts intermediate posts grooved for 38 mm partition	m	73.01	83.96
150 x 63 mm terrazzo faced, reinforced, twice rebated intermediate posts grooved for 38 mm partition	m	97.25	111.84
75 x 63 mm terrazzo faced, reinforced plain lintels or heads	m	73.01	83.96
Fair polished ends	each	3.53	4.06
Extra for incorporating tapped brass insets to receive			
hinges	pair	18.32	21.07
bolt keeps	each	10.29	11.83
Polished marble and granite etc fixed with adhesive to builder's beds or backings			
305 x 305 x 10 mm white Carrara pure marble tile pavings	m2	101.74	117.00
305 x 150 x 10 mm white Carrara pure marble skirtings with polished top edge	m	32.53	37.41
600 x 300 x 10 mm grey Sardinian pure granite tile pavings	m2	155.03	178.28
600 x 150 x 10 mm grey Sardinian pure granite skirtings with polished top edge	m	45.68	52.53
300 x 300 x 9 mm "Agglosimplex" resin bonded marble conglomerate tiling to walls and floors - Category "A" colours	m2	80.29	92.33
Polished edge at external angles	m	6.36	7.31
300 x 150 x 9 mm "Agglosimplex" resin bonded marble conglomerate skirtings with polished top edge	m	27.14	31.21

Floor, wall and ceiling finishings

New work	Unit	Specialist price net	Price with 15%
Internal in-situ finishings		£	£
		VAT not included	
Polished marble and granite etc fixed with adhesive to builder's beds or backings *(continued)*			
300 x 300 x 9 mm "Marghestone" resin bonded granite conglomerate tiling to walls and floors - Category "A" colours	m2	94.12	108.24
Polished edge at external angles	m	6.36	7.31
300 x 150 x 9 mm "Marghestone" resin bonded granite conglomerate skirtings with polished top edge	m	27.96	32.15

Enquiries about the foregoing specialist prices should be made to Alpha M & T Ltd, Unit 2, Munro Drive, off Cline Road, London, N11 2LZ, tel (0181) 368 2230, fax (0181) 368 2301.

Floor, wall and ceiling finishings

New work	Unit	Specialist price net	Price with 15%

Internal in-situ finishings

£   £

VAT not included

Lathing

The following "Specialist price net" figures are guide prices provided by Hatmet Limited, for quantities of about 160 square metres in rooms not less than about 16 square metres within 50 kilometres of London or Birmingham.

Prices do not include for cash discount.

See the preamble notes for builder's profit and attendance.

	Unit	Specialist price net	Price with 15%
Metal lathing wired to suitable channel grid, hung by means of 25 x 3 mm mild steel strap hangers fixed to underside of concrete floor, total depth of suspension 300 mm	m2	23.89	27.47
Extra for each additional 150 mm of suspension	m2	0.17	0.20
Vertical return in metal lathing 150 mm deep	m	8.07	9.28
Extra labour in trimming aperture in ceiling	m	3.95	4.54
Riblath ribbed metal lathing, nailed to underside of timber joists up to 600 mm centres	m2	13.33	15.33
Metal lathing wired to galvanised steel channels at 350 mm centres complete with alignment channels, to receive insulating plaster, as fire protection to structural steel columns	m2	31.54	36.27
Metal lathing wired to 25 x 3 mm galvanised steel flat bars at 350 mm centres fixed to underside of concrete slab to receive insulating plaster, as fire protection to structural steel beams	m2	41.55	47.78

Enquiries about the foregoing specialist prices should be made to Hatmet Ltd, Interiors House, Lynton Road, Crouch End, London, N8 8SL, tel (0181) 348 9262, fax (0181) 341 9878.

Floor, wall and ceiling finishings

New work	Unit	Hours C	Hours L	Labour net	Material net	Price net	Price with 15%
External in-situ finishings				£	£	£	£
					VAT not included		
16 mm cement and sand 1:3 two coat wood floated finish on brick or block walls							
over 300 mm wide	m2	0.65	0.20	7.17	1.61	8.78	10.10
not exceeding 300 mm wide	m2	0.97	0.49	12.03	1.61	13.64	15.69
16 mm cement and sand 1:3 two coat wood floated finish on brick or block walls lined to imitate plain blocks of about 0.20 m2							
over 300 mm wide	m2	0.85	0.42	10.48	1.61	12.09	13.90
not exceeding 300 mm wide	m2	1.27	0.64	15.72	1.61	17.33	19.93
Extra for waterproofer in backing coat	m2	0.01	0.01	0.12	0.27	0.39	0.45
19 mm two coat finish on brick or block walls of 13 mm cement, lime and sand 1:1:5 rendering with waterproofer and 6 mm white cement, lime and silver sand 1:1:5 lightly scraped finish							
over 300 mm wide	m2	0.70	0.35	8.63	1.65	10.28	11.82
not exceeding 300 mm wide	m2	1.05	0.53	12.95	1.65	14.60	16.79
22 mm three coat finish on brick or block walls of 10 mm cement and sand 1:3 rendering with waterpoofer, 6 mm cement, lime and sand 1:1:5 floating and 6 mm white cement, lime and silver sand 1:1:5 lightly scraped finish							
over 300 mm wide	m2	0.90	0.45	11.09	1.86	12.95	14.89
not exceeding 300 mm	m2	1.35	0.68	16.65	1.86	18.51	21.29
Extra for coloured cement in 6 mm finish coat	m2	-	-	-	0.03	0.03	0.03
15 mm two coat finish on brick or block walls of 10 mm cement, lime and sand 1:1:6 rendering and Tyrolean finish							
over 300 mm wide	m2	0.60	0.30	7.40	2.76	10.16	11.68
not exceeding 300 mm wide	m2	0.90	0.45	11.09	2.76	13.85	15.93
25 mm three coat finish on brick or block walls of 10 mm cement and sand 1:3 rendering with waterproofer, 10 mm cement, lime and sand 1:1:6 floating and Tyrolean finish							
over 300 mm wide	m2	0.85	0.42	10.48	3.76	14.24	16.38
not exceeding 300 mm wide	m2	1.27	0.64	15.72	3.76	19.48	22.40
20 mm three coat finish on brick or block walls of 8 mm cement and sand 1:3 rendering and floating coats and pebbledash finish							
over 300 mm wide	m2	0.70	0.35	8.63	2.37	11.00	12.65
not exceeding 300 mm wide	m2	1.05	0.53	12.95	2.37	15.32	17.62
Arrises on external finishings	m	0.12	0.06	1.48	-	1.48	1.70

Floor, wall and ceiling finishings

New work	Unit	Hours C	Hours L	Labour net	Material net	Price net	Price with 15%

Beds and backings

£ £ £ £

VAT not included

Cement and sand 1:3

Screeded beds, level and to falls, on concrete over 300 mm wide including thoroughly cleaning concrete base

19 mm	m2	0.25	0.13	3.08	1.91	4.99	5.74
25 mm	m2	0.28	0.14	3.45	2.51	5.96	6.85
32 mm	m2	0.31	0.16	3.82	3.21	7.03	8.08
38 mm	m2	0.33	0.17	4.07	3.82	7.89	9.07
50 mm	m2	0.38	0.19	4.68	5.02	9.70	11.15
65 mm	m2	0.41	0.20	5.05	6.53	11.58	13.32
75 mm	m2	0.44	0.22	5.42	7.53	12.95	14.89

Screeded beds, level and to falls, on concrete not exceeding 300 mm wide including thoroughly cleaning concrete base

19 mm	m2	0.38	0.19	4.63	1.91	6.54	7.52
25 mm	m2	0.42	0.21	5.18	2.51	7.69	8.84
32 mm	m2	0.47	0.23	5.79	3.21	9.00	10.35
38 mm	m2	0.50	0.25	6.17	3.82	9.99	11.49
50 mm	m2	0.57	0.28	7.03	5.02	12.05	13.86
65 mm	m2	0.62	0.31	7.65	6.53	14.18	16.31
75 mm	m2	0.66	0.33	8.13	7.53	15.66	18.01

Floated beds, level and to falls, on concrete over 300 mm wide including thoroughly cleaning concrete base

19 mm	m2	0.30	0.15	3.70	1.91	5.61	6.45
25 mm	m2	0.35	0.17	4.32	2.51	6.83	7.85
32 mm	m2	0.40	0.20	4.93	3.21	8.14	9.36
38 mm	m2	0.41	0.20	5.05	3.82	8.87	10.20
50 mm	m2	0.47	0.23	5.79	5.02	10.81	12.43
65 mm	m2	0.48	0.24	5.92	6.53	12.45	14.32
75 mm	m2	0.52	0.26	6.41	7.53	13.94	16.03

Floated beds, level and to falls, on concrete not exceeding 300 mm wide including thoroughly cleaning concrete base

19 mm	m2	0.45	0.23	5.55	1.91	7.46	8.58
25 mm	m2	0.53	0.27	6.54	2.51	9.05	10.41
32 mm	m2	0.60	0.30	7.40	3.21	10.61	12.20
38 mm	m2	0.62	0.31	7.65	3.82	11.47	13.19
50 mm	m2	0.71	0.35	8.75	5.02	13.77	15.84
65 mm	m2	0.72	0.36	8.88	6.53	15.41	17.72
75 mm	m2	0.78	0.39	9.62	7.53	17.15	19.72

Floor, wall and ceiling finishings

New work	Unit	Hours C	Hours L	Labour net	Material net	Price net	Price with 15%
Beds and backings				£	£	£	£
					VAT not included		

Cement and sand 1:3 (*continued*)

Trowelled beds, level and to falls, on concrete
over 300 mm wide including thoroughly cleaning
concrete base

19 mm	m2	0.35	0.17	4.32	1.91	6.23	7.16
25 mm	m2	0.39	0.20	4.80	2.51	7.31	8.41
32 mm	m2	0.42	0.21	5.18	3.21	8.39	9.65
38 mm	m2	0.44	0.22	5.42	3.82	9.24	10.63
50 mm	m2	0.49	0.25	6.04	5.02	11.06	12.72
65 mm	m2	0.52	0.26	6.41	6.53	12.94	14.88
75 mm	m2	0.55	0.28	6.78	7.53	14.31	16.46

Trowelled beds, level and to falls, on concrete not
exceeding 300 mm wide including thoroughly
cleaning concrete base

19 mm	m2	0.53	0.27	6.54	1.91	8.45	9.72
25 mm	m2	0.59	0.29	7.28	2.51	9.79	11.26
32 mm	m2	0.63	0.32	7.76	3.21	10.97	12.62
38 mm	m2	0.66	0.33	8.13	3.82	11.95	13.74
50 mm	m2	0.74	0.37	9.12	5.02	14.14	16.26
65 mm	m2	0.78	0.39	9.62	6.53	16.15	18.57
75 mm	m2	0.83	0.41	10.24	7.53	17.77	20.44

13 mm screeded backings on brick or block walls

over 300 mm wide	m2	0.30	0.15	3.70	1.31	5.01	5.76
not exceeding 300 mm wide	m2	0.45	0.23	5.55	1.31	6.86	7.89

13 mm floated backings on brick or block walls

over 300 mm wide	m2	0.35	0.17	4.32	1.31	5.63	6.47
not exceeding 300 mm wide	m2	0.53	0.26	6.48	1.31	7.79	8.96

13 mm trowelled backings on brick or block walls

over 300 mm wide	m2	0.40	0.20	4.93	1.31	6.24	7.18
not exceeding 300 mm wide	m2	0.60	0.30	7.40	13.06	20.46	23.53

Insulating beds on concrete

Vermiculite concrete 1:8 bed and 13 mm cement
and sand 1:3 screeded topping to falls and
crossfalls over 300 mm wide total average
thickness

63 mm	m2	0.35	0.85	8.87	5.81	14.68	16.88
88 mm	m2	0.35	1.15	10.89	8.06	18.95	21.79
113 mm	m2	0.35	1.47	13.05	19.05	32.10	36.91

Floating beds on insulation

Screeded cement and sand 1:3 beds, level and to
falls, around reinforcement over 300 mm wide

65 mm	m2	0.55	0.28	6.78	6.53	13.31	15.31
75 mm	m2	0.65	0.33	8.01	7.53	15.54	17.87

Floor, wall and ceiling finishings

New work	Unit	Hours C	Hours L	Labour net	Material net	Price net	Price with 15%
Beds and backings				£	£	£	£
					VAT not included		
Floating beds on insulation *(continued)*							
Floated cement and sand 1:3 beds, level and to falls, around reinforcement over 300 mm wide							
65 mm	m2	0.60	0.30	7.40	6.53	13.93	16.02
75 mm	m2	0.70	0.35	8.63	7.53	16.16	18.58
Trowelled cement and sand 1:3 beds, level and to falls, around reinforcement over 300 mm wide							
65 mm	m2	0.65	0.33	8.01	6.53	14.54	16.72
75 mm	m2	0.75	0.38	9.25	7.53	16.78	19.30
Screed reinforcement							
Galvanised wire netting reinforcement including tying wire in beds over 300 mm wide							
25 mm x 20G	m2	0.05	0.03	0.62	2.16	2.78	3.20
38 mm x 19G	m2	0.05	0.03	0.62	2.00	2.62	3.01
50 mm x 19G	m2	0.05	0.03	0.62	1.54	2.16	2.48

Floor, wall and ceiling finishings

New work	Unit	Hours C	Hours L	Labour net	Material net	Price net	Price with 15%
Tile, slab and block finishings				£	£	£	£
					VAT not included		
150 mm sills of two courses of plain clay roofing tiles set weathering and breaking joint and bedded, jointed and pointed in cement mortar 1:3							
straight	m	0.55	0.28	6.78	5.04	11.82	13.59
cut and fitted ends	each	0.10	0.05	1.24	0.69	1.93	2.22
150 x 150 x 12.5 mm clay floor quarries bedded, jointed and pointed in cement mortar 1:3							
Red tile paving							
over 300 mm wide	m2	0.82	0.41	10.11	12.82	22.93	26.37
not exceeding 300 mm wide	m2	1.23	0.61	15.17	12.82	27.99	32.19
Autumn blend tile paving							
over 300 mm wide	m2	0.82	0.41	10.11	19.54	29.65	34.10
not exceeding 300 mm wide	m2	1.23	0.61	15.17	19.54	34.71	39.92
Extra for rounded edge tiles							
red	m	-	-	-	1.46	1.46	1.68
autumn blend	m	-	-	-	1.94	1.94	2.23
Make good or labour finishing including jointing and pointing around ducting etc							
not exceeding 0.30 m girth	each	0.15	0.07	1.85	-	1.85	2.13
0.30 - 1.00 m girth	each	0.22	0.11	2.71	-	2.71	3.12
1.00 - 2.00 m girth	each	0.45	0.23	5.55	-	5.55	6.38
over 2.00 m girth	m	0.33	0.17	4.07	-	4.07	4.68
150 x 150 x 19 mm quarry tiles bedded, jointed and pointed in cement mortar 1:3							
125 mm rounded edge sills							
straight	m	0.27	0.14	3.33	3.93	7.26	8.35
cut and fitted ends	each	0.08	0.04	0.99	1.04	2.03	2.33
150 mm rounded edge sills							
straight	m	0.25	0.13	3.08	3.93	7.01	8.06
cut and fitted ends	each	0.08	0.04	0.99	1.04	2.03	2.33
175 mm rounded edge sills							
straight	m	0.35	0.17	4.32	4.13	8.45	9.72
cut and fitted ends	each	0.09	0.04	1.11	1.04	2.15	2.47
225 mm rounded edge sills							
straight	m	0.43	0.22	5.30	4.13	9.43	10.84
cut and fitted ends	each	0.10	0.05	1.24	1.04	2.28	2.62

Floor, wall and ceiling finishings

New work	Unit	Hours C	Hours L	Labour net	Material net	Price net	Price with 15%
Tile, slab and block finishings				£	£	£	£
					VAT not included		
200 x 200 x 19 mm quarry tiles bedded, jointed and pointed in cement mortar 1:3							
200 mm rounded edge sills							
straight	m	0.33	0.17	4.07	6.04	10.11	11.63
cut and fitted ends	each	0.12	0.06	1.48	2.29	3.77	4.34
229 x 229 x 29 mm quarry tiles bedded, jointed and pointed in cement mortar 1:3							
125 mm rounded edge sills							
straight	m	0.27	0.14	3.33	6.13	9.46	10.88
cut and fitted ends	each	0.12	0.06	1.48	2.92	4.40	5.06
175 mm rounded edge sills							
straight	m	0.35	0.17	4.32	6.13	10.45	12.02
cut and fitted ends	each	0.13	0.07	1.60	2.92	4.52	5.20
229 mm rounded edge sills							
straight	m	0.35	0.17	4.32	6.13	10.45	12.02
cut and fitted ends	each	0.15	0.07	1.85	2.92	4.77	5.49
Glazed ceramic tiles fixed with adhesive to plastered backings and grouted							
108 x 108 x 4 mm wall tiling (Group A)							
over 300 mm wide	m2	0.90	0.45	11.09	19.73	30.82	35.44
not exceeding 300 mm wide	m2	1.35	0.63	16.31	19.73	36.04	41.45
108 x 108 x 4 mm wall tiling (Group C)							
over 300 mm wide	m2	0.90	0.45	11.09	23.59	34.68	39.88
not exceeding 300 mm wide	m2	1.35	0.63	16.31	23.59	39.90	45.88
152 x 152 x 5.5 mm wall tiling (Group A)							
over 300 mm wide	m2	0.85	0.42	10.48	21.63	32.11	36.93
not exceeding 300 mm wide	m2	1.27	0.63	15.69	21.67	37.36	42.96
152 x 152 x 5.5 mm wall tiling (Group B)							
over 300 mm wide	m2	0.85	0.42	10.48	21.76	32.24	37.08
not exceeding 300 mm wide	m2	1.27	0.63	15.69	21.76	37.45	43.07
152 x 152 x 5.5 mm wall tiling (Group C)							
over 300 mm wide	m2	0.85	0.42	10.48	26.44	36.92	42.46
not exceeding 300 mm wide	m2	1.27	0.64	15.72	26.44	42.16	48.48
Extra for rounded edge tiles	m	-	-	-	0.69	0.69	0.79
Extra for external angle bead							
Group A	m	0.15	0.07	1.85	2.85	4.70	5.41
Groups B or C	m	0.15	0.07	1.85	4.17	6.02	6.92

Floor, wall and ceiling finishings

New work	Unit	Hours C	Hours L	Labour net	Material net	Price net	Price with 15%
Tile, slab and block finishings				£	£	£	£
					VAT not included		
Glazed ceramic tiles fixed with adhesive to plastered backings and grouted (*continued*)							
Make good or labour finishing including jointing and pointing around ducting etc							
not exceeding 0.30 m girth	each	0.10	0.05	1.24	-	1.24	1.43
0.30 - 1.00 m girth	each	0.15	0.07	1.85	-	1.85	2.13
1.00 - 2.00 m girth	each	0.30	0.15	3.70	-	3.70	4.25
over 2.00 m girth	m	0.22	0.11	2.71	-	2.71	3.12

Floor, wall and ceiling finishings

New work	Unit	Specialist price net	Price with 15%
Tile, slab and block finishings		**£**	**£**
		VAT not included	

The following "Specialist price net" figures are guide prices provided by East Devon Flooring Ltd for quantities of about 50 square metres where floor finishes are installed on sub floors not requiring preparatory work.

Prices do not include for cash discount.

See the preamble notes for builder's profit and attendance.

	Unit	Specialist price net	Price with 15%
Marleyflex Tiles			
2.00 mm series 2	m2	8.11	9.33
2.00 mm series 4	m2	9.15	10.52
2.00 mm series 8	m2	9.18	10.56
2.50 mm series 2	m2	8.75	10.06
2.50 mm series 4	m2	9.46	10.88
2.50 mm series 8	m2	10.51	12.09
2.00 mm Europa	m2	9.38	10.79
2.50 mm Europa	m2	10.77	12.39
2.00 mm Vylon Plus	m2	9.26	10.65
2.00 mm Traditional Vylon	m2	9.26	10.65
2.50 mm Travertine	m2	12.99	14.94
Fully Flexible Sheet and Tiles			
2.00 mm Marleyflor Plus Tiles	m2	10.82	12.44
2.00 mm Marleyflor Plus Sheet (welded)	m2	11.97	13.77
2.50 mm Marleyflor Plus Tiles	m2	12.71	14.62
2.50 mm Marleyflor Plus Sheet (welded)	m2	13.93	16.02
2.00 mm HD Vinyl Tiles	m2	13.77	15.84
2.00 mm HD Vinyl Sheet (welded)	m2	15.04	17.30
2.50 mm HD Vinyl Tiles	m2	15.07	17.33
2.50 mm HD Vinyl Sheet (welded)	m2	16.41	18.87
Non-Directional Heavy Duty Floor Coverings			
2.00 mm Eclipse Tiles	m2	14.11	16.23
2.00 mm Eclipse Sheet (welded)	m2	15.37	17.68
2.00 mm Elite Xtra Tiles	m2	16.82	19.34
2.00 mm Elite Extra Sheet (welded)	m2	18.23	20.96
Slip-resistant Floor Coverings			
2.00 mm Safetread Universal Sheet (welded)	m2	17.89	20.57
2.50 mm Safetread Universal Sheet (welded)	m2	20.40	23.46
2.00 mm Safetread Dimension (welded)	m2	19.20	22.08
2.00 mm Safetread Aqua (welded)	m2	21.63	24.87
Anti-static Floor Coverings			
2.00 mm HD Hitech Tiles	m2	17.81	20.48
2.00 mm HD Hitech Sheet (welded)	m2	19.25	22.14
2.00 mm Conductive Vinyl Tiles	m2	29.07	33.43

Floor, wall and ceiling finishings

New work	Unit	Specialist price net	Price with 15%
Tile, slab and block finishings		£	£
		VAT not included	

	Unit	Specialist price net	Price with 15%
Acoustic Backed Floor Coverings			
Vynatred Felt Backed Vinyl Sheet	m2	13.00	14.95
3.00 mm HD Acoustic Vinyl Sheet (welded)	m2	15.91	18.30
Contract Carpet Sheet and Tiles			
Marleytex Tiles	m2	8.99	10.34
Marleytex Sheet	m2	8.34	9.59
Marleytex Cord Tiles	m2	9.83	11.30
Marleytex Cord Sheet	m2	9.11	10.48
Marleytex Broadcord Tiles	m2	10.62	12.21
Marleytex Broadcord Sheet	m2	9.85	11.33
Marleytex Velour Tiles	m2	17.20	19.78
Marley Complement Carpet	m2	19.17	22.04
Marley Consort Carpet	m2	20.75	23.86
Accessories			
Minicove	m	1.90	2.18
100 mm x 2.00 mm Sit-on Skirting	m	2.43	2.79
100 mm x 2.00 mm Set-in Skirting	m	3.50	4.02
100 mm x 3.00 mm Set-in Skirting	m	3.72	4.28
100 mm x 2.00 mm Plain Skirting	m	2.26	2.60
2.00 mm "Accoflex" universal tiles	m2	10.16	11.68
2.00 mm "Accoflex" housing tiles	m2	8.72	10.03
2.00 mm "Armstrong Rhinotex" tiles	m2	10.58	12.17
2.00 mm "Polyflor XL" vinyl tiles	m2	10.78	12.40
4.75 mm Presealed Cork Tiles	m2	21.93	25.22
20 mm (nominal) tongued and grooved block flooring bedded in mastic, sanded and 3 coats seal applied			
American Oak	m2	50.52	58.10
Iroko	m2	45.37	52.18
Merbau	m2	41.51	47.74
Teak	m2	56.64	65.14
450 x 450 x 7.5 mm Iroko, Oak, Jatoba and Mahogony flooring bedded in mastic, sanded and 3 coats seal applied	m2	29.55	33.98
Burmatex needlepunch carpet			
Broadloom 2200 grade	m2	7.92	9.11
Tiles 2200 grade	m2	10.00	11.50
Broadloom 3300 grade	m2	9.75	11.21
Broadloom 4400 grade	m2	9.24	10.63
Broadloom 4200 grade	m2	8.59	9.88

Floor, wall and ceiling finishings

New work	Unit	Specialist price net	Price with 15%
Tile, slab and block finishings		£	£
		VAT not included	
Burmatex contract carpet tiles			
Velour	m2	16.01	18.41
Toreador	m2	18.22	20.95
Tivoli 21	m2	16.98	19.53
Academy	m2	13.96	16.05

Enquiries about the foregoing specialist prices should be made to East Devon Flooring Ltd, Units 3 and 5, Kingfisher Court, Venny Bridge, Exeter, EX4 8JN, tel (01392) 462033, fax (01392) 462032.

Floor, wall and ceiling finishings

New work	Unit	Hours C	Hours L	Labour net	Material net	Price net	Price with 15%
Sheet finishings				£	£	£	£
					VAT not included		

Gypsum plasterboard to BS 1230

Square edge wallboard fixed to softwood to walls with galvanised nails including covering joints with 50 mm paper backed cotton scrim and filling nail holes with joint filler

New work	Unit	Hours C	Hours L	Labour net	Material net	Price net	Price with 15%
9.5 mm over 300 mm wide	m2	0.26	0.13	3.21	2.56	5.77	6.64
9.5 mm not exceeding 300 mm wide	m2	0.39	0.20	4.80	2.56	7.36	8.46
12.5 mm over 300 mm wide	m2	0.30	0.15	3.70	2.82	6.52	7.50
12.5 mm not exceeding 300 mm wide	m2	0.45	0.23	5.55	2.82	8.37	9.63

Square edge insulating grade wallboard fixed to softwood to walls with galvanised nails including covering joints with 50 mm paper backed cotton scrim and filling nail holes with joint filler

New work	Unit	Hours C	Hours L	Labour net	Material net	Price net	Price with 15%
9.5 mm over 300 mm wide	m2	0.26	0.13	3.21	3.15	6.36	7.31
9.5 mm not exceeding 300 mm wide	m2	0.45	0.23	5.55	3.15	8.70	10.01
12.5 mm over 300 mm wide	m2	0.30	0.15	3.70	3.42	7.12	8.19
12.5 mm not exceeding 300 mm wide	m2	0.45	0.23	5.55	3.42	8.97	10.32

Tapered edge wallboard fixed to softwood to walls with galvanised nails including flush jointing with joint filler, joint tape and joint finish and filling nail holes with joint filler

New work	Unit	Hours C	Hours L	Labour net	Material net	Price net	Price with 15%
9.5 mm over 300 mm wide	m2	0.31	0.16	3.82	2.80	6.62	7.61
9.5 mm not exceeding 300 mm wide	m2	0.47	0.23	5.79	2.80	8.59	9.88
12.5 mm over 300 mm wide	m2	0.36	0.18	4.44	3.42	7.86	9.04
12.5 mm not exceeding 300 mm wide	m2	0.54	0.27	6.66	3.42	10.08	11.59

New work	Unit	Hours C	Hours L	Labour net	Material net	Price net	Price with 15%
54 mm reinforced paper corner tape bedded in joint filler and flushed over with joint filler and joint finish	m	0.10	0.05	1.24	0.16	1.40	1.61

Plaster core coved cornices fixed with adhesive

100 mm girth cornices

	Unit	Hours C	Hours L	Labour net	Material net	Price net	Price with 15%
straight	m	0.15	0.07	1.85	1.42	3.27	3.76
angles	each	0.25	-	2.24	0.38	2.62	3.01

127 mm girth cornices

	Unit	Hours C	Hours L	Labour net	Material net	Price net	Price with 15%
straight	m	0.15	0.07	1.85	1.62	3.47	3.99
angles	each	0.25	-	2.24	0.45	2.69	3.09

Floor, wall and ceiling finishings

New work	Unit	Hours C	Hours L	Labour net	Material net	Price net	Price with 15%
Dry linings and partitions				£	£	£	£
					VAT not included		

Dry linings

Tapered edge wallboard fixed "Gyproc Dri-wall System" method to brick or block walls including flush jointing with joint filler, joint tape and joint finish and filling nail holes with joint filler

9.5 mm over 300 mm wide	m2	0.35	0.40	5.84	2.80	8.64	9.94
9.5 mm not exceeding 300 mm wide	m2	0.70	0.80	11.66	3.01	14.67	16.87
12.5 mm over 300 mm wide	m2	0.35	0.40	5.84	3.42	9.26	10.65
12.5 mm not exceeding 300 mm wide	m2	0.70	0.80	11.66	3.63	15.29	17.58

Cellular partitions

Cellular core plasterboard dry partitions of tapered edge panels with 37 x 37 mm softwood battens at vertical joints including flush jointing with joint filler, joint tape and joint finish and filling nail holes with joint filler

57 mm	m2	0.60	0.30	7.40	8.62	16.02	18.42
63 mm	m2	0.70	0.35	8.63	8.62	17.25	19.84

Labours

Fair ends to partitions including 37 x 37 mm softwood batten in core	m	0.20	0.10	2.46	0.86	3.32	3.82
Angles of partitions including 19 x 37 mm and 37 x 37 mm vertical battens and cutting panel to form rebate	m	0.30	0.15	3.70	1.48	5.18	5.96
Intersections of partitions including 19 x 37 mm softwood vertical batten fixed to 37 x 37 x 100 mm plugs at 450 mm centres in core	m	0.18	0.09	2.22	0.66	2.88	3.31
Extra for square edge panels to partitions and filling joints with neat board finish plaster, reinforcing with 90 mm jute scrim and setting with 5 mm board finish plaster - both sides	m2	0.55	0.28	6.78	1.25	8.03	9.23

Softwood

19 x 37 mm wall or ceiling battens	m	0.10	0.05	1.24	0.20	1.44	1.66
37 x 37 mm fixing battens at openings	m	0.12	0.06	1.48	0.44	1.92	2.21
19 x 57 mm sole plates with 300 x 19 x 37 mm blockings at panel joints	m	0.25	0.13	3.08	0.36	3.44	3.96
19 x 63 mm sole plates with 300 x 19 x 37 mm blockings at panel joints	m	0.25	0.13	3.08	0.51	3.59	4.13
100 x 37 x 37 mm plugs in core	each	0.08	-	0.72	0.04	0.76	0.87
Extra for plugging battens etc to brickwork or concrete	m	0.33	-	2.96	0.61	3.57	4.11

Floor, wall and ceiling finishings

New work	Unit	Hours C	Hours L	Labour net	Material net	Price net	Price with 15%
Dry linings and partitions				£	£	£	£
					VAT not included		

Laminated partitions

	Unit	Hours C	Hours L	Labour net	Material net	Price net	Price with 15%
50 mm laminated partitions of two layers of 12.5 mm tapered edge wallboard and one layer of 19 mm square edge plank bonded with adhesive including flush jointing with joint filler, joint tape and joint finish and filling nail holes with joint filler	m2	0.90	0.45	11.09	14.84	25.93	29.82
65 mm laminated partitions of two layers of 19 mm tapered edge plank and one layer of 19 mm square edge plank bonded with adhesive including flush jointing with joint filler, joint tape and joint finish and filling nail holes with joint filler	m2	1.00	0.50	12.33	18.03	30.36	34.91
Fair ends to partitions	m	0.19	0.10	2.34	2.98	5.32	6.12
Angles of partitions including 25 x 38 mm softwood batten in core and 54 mm reinforced paper corner tape bedded in joint filler and flushed over with joint filler and joint finish	m	0.25	0.13	3.08	3.27	6.35	7.30
Intersections of partitions	m	0.19	0.10	2.34	2.96	5.30	6.09
Softwood 25 x 38 mm plates or wall battens	m	0.10	0.05	1.24	0.32	1.56	1.79
25 x 38 mm fixing battens in core including cutting middle layer around							
over 300 mm long	m	0.10	0.05	1.24	0.29	1.53	1.76
not exceeding 300 mm long	each	0.08	0.04	0.99	0.09	1.08	1.24
Extra for plugging plates or battens to brickwork or concrete	m2	0.33	-	2.96	0.24	3.20	3.68

Floor, wall and ceiling finishings

New work	Unit	Specialist price net	Price with 15%
Metal stud partitions		£	£
			VAT not included

The following "Specialist price net" figures are guide prices for work carried out by specialist Contractors.

Prices do not include for cash discount.

See the preamble notes for builder's profit and attendance.

Metal stud partitions

75mm Gyproc metal stud partition 48mm studs at 600 centres. 12.5mm taper edge wallboard each side; fixing with pozidriv head screws to steel frame; butt joints filled with joint filler tape and paint finish, spot filling, surface finished with one coat Gyproc drywall top coat.

	Unit	Specialist price net	Price with 15%
height 2.10 - 2.40m	m	58.27	67.01
height 2.40 - 2.70m	m	67.09	77.15
height 2.70 - 3.00m	m	74.69	85.89
height 3.00 - 3.30m	m	85.01	97.76

Floor, wall and ceiling finishings

New work	Unit	Specialist price net	Price with 15%
		£	£

Suspended ceilings, linings and support work

VAT not included

The following "Specialist price net" figures are guide prices provided by ITA Ceilings Ltd.

Prices do not include for cash discount.

See the preamble notes for builder's profit and attendance.

Suspended ceilings and metal support work to finish not exceeding 300 mm below concrete soffits

	Unit	Specialist price net	Price with 15%
White PVC faced plasterboard panels with aluminium foil backing complete with exposed white support grid and hangers etc.			
1200 x 600 x 7.5 mm	m2	10.30	11.85
600 x 600 x 7.5 mm	m2	12.10	13.92
Predecorated "Supalux" panels complete with exposed white support grid and hangers etc.			
1200 x 600 x 6 mm	m2	16.48	18.95
600 x 600 x 6 mm	m2	20.03	23.04
Undecorated "Supalux" panels complete with concealed support grid and hangers etc.			
1200 x 600 x 9 mm	m2	29.87	34.35
600 x 600 x 6 mm	m2	32.96	37.90
Predecorated mineral fibre panels complete with exposed white support grid and hangers etc. 25 mm table grid			
1200 x 600 x 15 mm	m2	8.76	10.07
600 x 600 x 15 mm	m2	10.30	11.85
Extra for recessed edge profile			
600 x 600 x 15 mm thick	m2	4.12	4.74
1200 x 600 x 17 mm thick	m2	5.77	6.63
600 x 600 x 19 mm thick	m2	24.05	27.66
Predecorated mineral fibre panels complete with exposed white support grid and hangers etc. 15 mm table grid			
600 x 600 x 15 mm	m2	16.94	19.49
600 x 600 x 17 mm	m2	26.21	30.15
Moisture resistant ceiling tiles, in white exposed lay-in grid			
1200 x 600 x 10 mm	m2	16.07	18.48
600 x 600 x 10 mm	m2	17.67	20.33
Predecorated mineral panels complete with concealed support grid and hangers etc.			
300 x 300 x 15 mm BKR	m2	17.30	19.90
600 x 600 x 15 mm BKR	m2	16.92	19.46
300 x 300 x 19 mm SE	m2	29.61	34.05

Floor, wall and ceiling finishings

New work	Unit	Specialist price net	Price with 15%

Suspended ceilings, linings and support work

		£	£
			VAT not included

Suspended ceilings and metal support work to finish not exceeding 300 mm below concrete soffits *(continued)*

Predecorated metal tray acoustic panels with 25 mm mineral wool infill complete with concealed support grid and hangers etc.

	Unit	Specialist price net	Price with 15%
600 x 600 mm plain	m2	30.39	34.94
600 x 600 mm perforated	m2	32.03	36.84

Gyproc M/F system, with tapered edge Gyproc wallboard, joints filled with joint filler and taped to receive direct decoration, fixed in accordance with the manufacturers instructions.

900 x 1800 x 12.5mm	m2	19.00	21.85

Linings and metal support work

9.5 mm PVC faced plasterboard panels in 600 mm widths with aluminium foil backing complete with exposed galvanised support grid and brackets etc. fixed to

wall sheeting rails	m2	17.25	19.84
underside of roof purlins	m2	16.33	18.77

Sundries

Fibreglass overlay - 100 mm	m2	4.69	5.39
19 x 25 mm white angle trim at edges plugged to walls	m	1.39	1.60
19 x 9 x 9 x 19 mm shadowline angle trim at edges plugged to walls	m	1.96	2.25
Extra for ceiling contractor to provide mobile scaffolding	m2	0.98	1.13

Enquiries about the foregoing specialist prices should be made to ITA Ceilings Ltd, Unit 4, 107 Summerway, Exeter, EX4 8DP, tel (01392) 468781, fax (01392) 465476.

Floor, wall and ceiling finishings

New work	Unit	Specialist price net	Price with 15%
		£	£

Fibrous plaster

VAT not included

The following "Specialist price net" figures are guide prices provided by Hodkin & Jones (Sheffield) Ltd. Prices are given for supply only and for supply and fix.

Allowance should be added to ex works prices for delivery charges.

Prices do not include for cash discount.

See the preamble notes for builder's profit and attendance.		Ex works	Ex works
Supply only the following:			
Plain moulded cornices			
150 mm girth	m	6.17	7.10
225 mm girth	m	6.85	7.88
Enriched moulded cornices			
150 mm girth	m	6.85	7.88
225 mm girth	m	8.15	9.37
450 mm girth	m	22.23	25.56
2000 mm fluted pilasters with parallel shaft and plain cap	each	72.23	83.06
2133 mm plain columns with entasised shaft and plain cap	each	205.56	236.39
2133 mm fluted columns with entasised shaft and enriched cap	each	238.82	274.64
Enriched corbels			
100- 200 mm overall width	each	21.67	24.92
275 mm overall width	each	25.00	28.75
Adam style ceiling centres with diminishing flutes with husk ornament and scalloped edge			
600 mm diameter	each	26.12	30.04
1200 mm diameter	each	95.00	109.25
1800 x 1125 mm elliptical	each	95.00	109.25
Supply and fix the following:			
Plain moulded cornices			
150 mm girth	m	14.70	16.91
225 mm girth	m	19.40	22.31
Extra for mitres on plain cornices			
150 mm girth	each	3.50	4.03
225 mm girth	each	5.60	6.44
Enriched moulded cornices			
150 mm girth	m	15.50	17.83
225 mm girth	m	20.70	23.81
450 mm girth	m	39.40	45.31

Floor, wall and ceiling finishings

New work	Unit	Specialist price net	Price net 15%
Fibrous plaster		£	£
		VAT not included	

Supply and fix the following (continued):

	Unit	Specialist price net	Price net 15%
Extra for mitres on enriched cornices			
150mm girth	each	4.12	4.74
225mm girth	each	7.78	8.95
450mm girth	each	11.12	12.79

Enquiries about the foregoing specialist prices should be made to Hodkin & Jones (Sheffield) Ltd, Callywhite Lane, Dronfield, Sheffield, S18 6XP, tel (01246) 290890, fax (01246) 290292

Glazing

Preamble

"Labour net" figures include allowances for all costs incidental to the employment of labour.

"Materials net" figures include for all costs of materials including an allowance for waste except where specifically stated.

"Price net" figures are the totals of the "Labour net" and "Material net" figures. Prices are for a builder employing his own labour; according to the amount and nature of the work involved, it may well be possible to secure more advantageous prices from specialist sub-contractors.

Although not in accordance with the Standard Method of Measurement, figures per square metre have been included for sealed double glazing units, from which figures for any particular enumerated items can be calculated.

Prices do not include any allowance for scaffolding, ladders or other plant necessary to reach the work. The "Preliminaries" section includes prices for scaffolding which must be considered and allowance included to suit the particular circumstances of a tender.

Specialist prices

"Price with 15%" figures are all-in guide prices and include 15% for the builder's overheads, profit, unloading materials and general attendance (to include free use of standing scaffolding and hoists, temporary lighting and water and clearing away rubbish).

The amount of attendance required varies between the various trades and also with the circumstances of specific jobs; the percentage addition must always be considered and adjusted as necessary to suit the terms and conditions of the quotation being used.

Quantities and delivery distances are usually the most significant of the many factors which influence prices and it must be emphasised that quotations should always be obtained when preparing a tender.

Glazing

New work

Basic prices for materials		£

Float glass

3 mm - not exceeding 2400 x 1300 mm	m2	21.91
4 mm - not exceeding 2400 x 1300 mm	m2	21.91
5 mm - not exceeding 3150 x 2050 mm	m2	32.03
6 mm - not exceeding 4550 x 3150 mm	m2	32.03

White patterned glass

4 mm - not exceeding 2100 x 1300 mm	m2	21.76
6 mm polished - not exceeding 2100 x 1300 mm	m2	38.01

Georgian wired glass

7 mm cast - not exceeding 3450 x 1900 mm	m2	33.73
6 mm polished - not exceeding 3250 x 1900 mm	m2	71.61

Georgian wired safety glass

7mm cast	m2	43.89
6mm polished	m2	84.33

Laminated glass

4.4mm	m2	49.24
6.4mm	m2	41.27

Double glazing units incorporating two panes of 4 mm float glass

0.25 - 0.35 m2	m2	40.66
0.35 - 0.50 m2	m2	40.66
0.50 - 0.75 m2	m2	40.66
0.75 - 1.00 m2	m2	40.66
1.00 - 2.00 m2	m2	40.66

Double glazing units incorporating one pane of 4 mm float glass and one pane of 4 mm white patterned glass

0.25 - 0.35 m2	m2	47.43
0.35 - 0.50 m2	m2	47.43
0.50 - 0.75 m2	m2	47.43
0.75 - 1.00 m2	m2	47.43
1.00 - 2.00 m2	m2	47.43

Putty

linseed oil	25 kg	14.08
metal casement	25 kg	14.80
non-setting compound	25 kg	20.12

Prices actually to be paid for materials must be checked against the above basic prices and adjustments made as necessary.

Glazing

New work	Unit	Hours C	Hours L	Labour net	Material net	Price net	Price with 15%
Glass pre-cut to size in openings				£	£	£	£
					VAT not included		
3 mm float glass							
To wood with putty in panes							
not exceeding 0.10 m2	m2	0.80	-	7.17	25.64	32.81	37.73
0.10 - 0.50 m2	m2	0.45	-	4.03	24.93	28.96	33.30
0.50 - 1.00 m2	m2	0.40	-	3.58	24.34	27.92	32.11
To wood with pinned beads in panes							
not exceeding 0.10 m2	m2	0.96	-	8.60	23.75	32.35	37.20
0.10 - 0.50 m2	m2	0.54	-	4.84	23.75	28.59	32.88
0.50 - 1.00 m2	m2	0.48	-	4.30	23.75	28.05	32.26
To wood with screwed beads in panes							
not exceeding 0.10 m2	m2	1.12	-	10.04	23.75	33.79	38.86
0.10 - 0.50 m2	m2	0.63	-	5.64	23.75	29.39	33.80
0.50 - 1.00 m2	m2	0.56	-	5.02	23.75	28.77	33.09
To metal with metal casement putty in panes							
not exceeding 0.10 m2	m2	0.74	-	6.63	25.74	32.37	37.23
0.10 - 0.50 m2	m2	0.41	-	3.67	24.99	28.66	32.96
0.50 - 1.00 m2	m2	0.37	-	3.32	24.37	27.69	31.84
4 mm float glass							
To wood with putty in panes							
not exceeding 0.10 m2	m2	0.80	-	7.17	25.64	32.81	37.73
0.10 - 0.50 m2	m2	0.45	-	4.03	24.93	28.96	33.30
0.50 - 1.00 m2	m2	0.40	-	3.58	24.34	27.92	32.11
over 1.00 m2	m2	0.29	-	2.60	24.05	26.65	30.65
To wood with pinned beads in panes							
not exceeding 0.10 m2	m2	0.96	-	8.60	23.75	32.35	37.20
0.10 - 0.50 m2	m2	0.54	-	4.84	23.75	28.59	32.88
0.50 - 1.00 m2	m2	0.48	-	4.30	23.75	28.05	32.26
over 1.00 m2	m2	0.35	-	3.14	23.75	26.89	30.92
To wood with screwed beads in panes							
not exceeding 0.10 m2	m2	1.12	-	10.04	23.75	33.79	38.86
0.10 - 0.50 m2	m2	0.63	-	5.64	23.75	29.39	33.80
0.50 - 1.00 m2	m2	0.56	-	5.02	23.75	28.77	33.09
over 1.00 m2	m2	0.41	-	3.67	23.75	27.42	31.53
To metal with metal casement putty in panes							
not exceeding 0.10 m2	m2	0.74	-	6.63	25.74	32.37	37.23
0.10 - 0.50 m2	m2	0.41	-	3.67	24.99	28.66	32.96
0.50 - 1.00 m2	m2	0.37	-	3.32	24.37	27.69	31.84
over 1.00 m2	m2	0.27	-	2.42	24.06	26.48	30.45
Float glass to wood or metal with non-setting compound and screwed beads							
5 mm in panes over 1.00 m2	m2	0.53	-	4.75	35.19	39.94	45.93
6 mm in panes over 1.00 m2	m2	0.56	-	5.02	35.32	40.34	46.39

Glazing

New work	Unit	Hours C	Hours L	Labour net	Material net	Price net	Price with 15%
Glass pre-cut to size in openings				£	£	£	£
					VAT not included		
4 mm white patterned glass							
To wood with putty in panes							
not exceeding 0.10 m2	m2	0.91	-	8.15	25.48	33.63	38.67
0.10 - 0.50 m2	m2	0.52	-	4.66	24.77	29.43	33.84
0.50 - 1.00 m2	m2	0.46	-	4.12	24.18	28.30	32.55
To wood with pinned beads in panes							
not exceeding 0.10 m2	m2	1.09	-	9.77	23.59	33.36	38.36
0.10 - 0.50 m2	m2	0.62	-	5.56	23.59	29.15	33.52
0.50 - 1.00 m2	m2	0.55	-	4.93	23.59	28.52	32.80
To wood with screwed beads in panes							
not exceeding 0.10 m2	m2	1.27	-	11.38	23.59	34.97	40.22
0.10 - 0.50 m2	m2	0.73	-	6.54	23.59	30.13	34.65
0.50 - 1.00 m2	m2	0.64	-	5.73	23.59	29.32	33.72
To metal with metal casement putty in panes							
not exceeding 0.10 m2	m2	0.84	-	7.53	25.58	33.11	38.08
0.10 - 0.50 m2	m2	0.48	-	4.30	24.83	29.13	33.50
0.50 - 1.00 m2	m2	0.42	-	3.76	24.21	27.97	32.17
Extra for lining up patterned glass - one way only	m2	-	-	-	11.80	11.80	13.57
6 mm white patterned glass							
To wood with putty in panes							
not exceeding 0.10 m2	m2	1.04	-	9.32	42.95	52.27	60.11
0.10 - 0.50 m2	m2	0.59	-	5.29	42.24	47.53	54.66
0.50 - 1.00 m2	m2	0.52	-	4.66	41.65	46.31	53.26
over 1.00 m2	m2	0.38	-	3.40	41.36	44.76	51.47
To wood with pinned beads in panes							
not exceeding 0.10 m2	m2	1.23	-	11.02	41.06	52.08	59.89
0.10 - 0.50 m2	m2	0.70	-	6.27	41.06	47.33	54.43
0.50 - 1.00 m2	m2	0.62	-	5.56	41.06	46.62	53.61
over 1.00 m2	m2	0.46	-	4.12	41.06	45.18	51.96
To wood with screwed beads in panes							
not exceeding 0.10 m2	m2	1.42	-	12.72	41.06	53.78	61.85
0.10 - 0.50 m2	m2	0.82	-	7.35	41.06	48.41	55.67
0.50 - 1.00 m2	m2	0.72	-	6.45	41.06	47.51	54.64
over 1.00 m2	m2	0.53	-	4.75	41.06	45.81	52.68
To metal with metal casement putty in panes							
not exceeding 0.10 m2	m2	0.96	-	8.60	43.05	51.65	59.40
0.10 - 0.50 m2	m2	0.54	-	4.84	42.30	47.14	54.21
0.50 - 1.00 m2	m2	0.48	-	4.30	41.68	45.98	52.88
over 1.00 m2	m2	0.35	-	3.14	41.37	44.51	51.19
Extra for lining up patterned glass - one way only	m2	-	-	-	20.53	20.53	23.61

Glazing

New work	Unit	Hours C	Hours L	Labour net	Material net	Price net	Price with 15%
Glass pre-cut to size in openings				£	£	£	£
					VAT not included		
7 mm Georgian wired cast glass							
To wood with putty in panes							
not exceeding 0.10 m2	m2	1.00	-	8.96	37.39	46.35	53.30
0.10 - 0.50 m2	m2	0.80	-	7.17	37.15	44.32	50.97
0.50 - 1.00 m2	m2	0.70	-	6.27	36.98	43.25	49.74
over 1.00 m2	m2	0.50	-	4.48	36.92	41.40	47.61
To wood with pinned beads in panes							
not exceeding 0.10 m2	m2	1.20	-	10.75	36.80	47.55	54.68
0.10 - 0.50 m2	m2	0.96	-	8.60	36.80	45.40	52.21
0.50 - 1.00 m2	m2	0.84	-	7.53	36.80	44.33	50.98
over 1.00 m2	m2	0.60	-	5.38	36.80	42.18	48.51
To wood with screwed beads in panes							
not exceeding 0.10 m2	m2	1.40	-	12.54	36.80	49.34	56.74
0.10 - 0.50 m2	m2	1.22	-	10.93	36.80	47.73	54.89
0.50 - 1.00 m2	m2	0.98	-	8.78	36.80	45.58	52.42
over 1.00 m2	m2	0.70	-	6.27	36.80	43.07	49.53
To metal with metal casement putty in panes							
not exceeding 0.10 m2	m2	0.92	-	8.24	37.42	45.66	52.51
0.10 - 0.50 m2	m2	0.73	-	6.54	37.17	43.71	50.27
0.50 - 1.00 m2	m2	0.64	-	5.73	36.99	42.72	49.13
over 1.00 m2	m2	0.46	-	4.12	36.92	41.04	47.20
Extra for lining up wires - one way only	m2	-	-	-	18.40	18.40	21.16
6 mm Georgian wired polished glass							
To wood with putty in panes							
not exceeding 0.10 m2	m2	1.50	-	13.44	77.98	91.42	105.13
0.10 - 0.50 m2	m2	1.20	-	10.75	77.74	88.49	101.76
0.50 - 1.00 m2	m2	1.05	-	9.41	77.57	86.98	100.03
over 1.00 m2	m2	0.75	-	6.72	77.51	84.23	96.86
To wood with pinned beads in panes							
not exceeding 0.10 m2	m2	1.80	-	16.13	77.39	93.52	107.55
0.10 - 0.50 m2	m2	1.44	-	12.90	77.39	90.29	103.83
0.50 - 1.00 m2	m2	1.26	-	11.29	77.39	88.68	101.98
over 1.00 m2	m2	0.90	-	8.06	77.39	85.45	98.27
To wood with screwed beads in panes							
not exceeding 0.10 m2	m2	2.10	-	18.82	77.39	96.21	110.64
0.10 - 0.50 m2	m2	2.00	-	17.92	77.39	95.31	109.61
0.50 - 1.00 m2	m2	1.47	-	13.17	77.39	90.56	104.14
over 1.00 m2	m2	1.05	-	9.41	77.39	86.80	99.82
To metal with metal casement putty in panes							
not exceeding 0.10 m2	m2	1.38	-	12.36	78.01	90.37	103.93
0.10 - 0.50 m2	m2	1.10	-	9.86	77.76	87.62	100.76
0.50 - 1.00 m2	m2	0.97	-	8.69	77.58	86.27	99.21
over 1.00 m2	m2	0.70	-	6.27	77.51	83.78	96.35
Extra for lining up wires - one way only	m2	-	-	-	38.69	38.69	44.49

Glazing

New work	Unit	Hours C	Hours L	Labour net	Material net	Price net	Price with 15%
Glass pre-cut to size in openings				£	£	£	£
				VAT not included			

7 mm Georgian wired cast safety glass

To wood with putty in panes

not exceeding 0.10 m2	m2	1.00	-	8.96	48.87	57.83	66.50
0.10 - 0.50 m2	m2	0.80	-	7.17	48.63	55.80	64.17
0.50 - 1.00 m2	m2	0.70	-	6.27	48.46	54.73	62.94
over 1.00 m2	m2	0.50	-	4.48	48.40	52.88	60.81

To wood with pinned beads in panes

not exceeding 0.10 m2	m2	1.20	-	10.75	48.28	59.03	67.88
0.10 - 0.50 m2	m2	0.96	-	8.60	48.28	56.88	65.41
0.50 - 1.00 m2	m2	0.84	-	7.53	48.28	55.81	64.18
over 1.00 m2	m2	0.60	-	5.38	48.28	53.66	61.71

To wood with screwed beads in panes

not exceeding 0.10 m2	m2	1.40	-	12.54	48.28	60.82	69.94
0.10 - 0.50 m2	m2	1.33	-	11.92	48.28	60.20	69.23
0.50 - 1.00 m2	m2	0.98	-	8.78	48.28	57.06	65.62
over 1.00 m2	m2	0.70	-	6.27	48.28	54.55	62.73

Extra for lining up wires - one way only	m2	0.50	-	4.48	-	4.48	5.15

6 mm Georgian wired polished safety glass

To wood with putty in panes

not exceeding 0.10 m2	m2	1.50	-	13.44	93.35	106.79	122.81
0.10 - 0.50 m2	m2	1.20	-	10.75	93.11	103.86	119.44
0.50 - 1.00 m2	m2	1.05	-	9.41	92.94	102.35	117.70
over 1.00 m2	m2	0.75	-	6.72	92.88	99.60	114.54

To wood with pinned beads in panes

not exceeding 0.10 m2	m2	1.80	-	16.13	92.76	108.89	125.22
0.10 - 0.50 m2	m2	1.44	-	12.90	92.76	105.66	121.51
0.50 - 1.00 m2	m2	1.26	-	11.29	92.76	104.05	119.66
over 1.00 m2	m2	0.90	-	8.06	92.76	100.82	115.94

To wood with screwed beads in panes

not exceeding 0.10 m2	m2	2.10	-	18.82	92.76	111.58	128.32
0.10 - 0.50 m2	m2	2.00	-	17.92	92.76	110.68	127.28
0.50 - 1.00 m2	m2	1.47	-	13.17	92.76	105.93	121.82
over 1.00 m2	m2	1.05	-	9.41	92.76	102.17	117.50

Extra for lining up wires - one way only	m2	0.50	-	4.48	-	4.48	5.15

4.4 mm laminated glass

To wood with pinned beads in panes

not exceeding 0.10 m2	m2	1.56	-	13.98	54.16	68.14	78.36
0.10 - 0.50 m2	m2	1.25	-	11.20	54.16	65.36	75.16
0.50 - 1.00 m2	m2	1.09	-	9.77	54.16	63.93	73.52
over 1.00 m2	m2	0.78	-	6.99	54.16	61.15	70.32

Glazing

New work	Unit	Hours C	Hours L	Labour net	Material net	Price net	Price with 15%
Glass pre-cut to size in openings				£	£	£	£
					VAT not included		
4.4 mm laminated glass (*continued*)							
To wood with screwed beads in panes							
not exceeding 0.10 m2	m2	1.81	-	16.22	54.16	70.38	80.94
0.10 - 0.50 m2	m2	1.45	-	12.99	54.16	67.15	77.22
0.50 - 1.00 m2	m2	1.27	-	11.38	54.16	65.54	75.37
over 1.00 m2	m2	0.90	-	8.06	54.16	62.22	71.55
6.4 mm laminated glass							
To wood with pinned beads in panes							
not exceeding 0.10 m2	m2	1.82	-	16.31	45.40	61.71	70.97
0.10 - 0.50 m2	m2	1.43	-	12.81	45.40	58.21	66.94
0.50 - 1.00 m2	m2	1.21	-	10.84	45.40	56.24	64.68
over 1.00 m2	m2	0.90	-	8.06	45.40	53.46	61.48
To wood with screwed beads in panes							
not exceeding 0.10 m2	m2	2.11	-	18.91	45.40	64.31	73.96
0.10 - 0.50 m2	m2	1.68	-	15.05	45.40	60.45	69.52
0.50 - 1.00 m2	m2	1.41	-	12.63	45.40	58.03	66.73
over 1.00 m2	m2	1.06	-	9.50	45.40	54.90	63.13
Double glazing units incorporating two panes of 4 mm float glass							
To wood or metal with non-setting compound and screwed beads in panes							
0.25 - 0.35 m2	m2	3.50	-	31.36	43.37	74.73	85.94
0.35 - 0.50 m2	m2	3.00	-	26.88	43.37	70.25	80.79
0.50 - 0.75 m2	m2	2.50	-	22.40	43.37	65.77	75.64
0.75 - 1.00 m2	m2	2.00	-	17.92	43.37	61.29	70.48
Double glazing units incorporating one pane of 4 mm float glass and one pane of 4 mm white patterned glass							
To wood or metal with non-setting compound and screwed beads in panes							
0.25 - 0.35 m2	m2	3.50	-	31.36	50.99	82.35	94.70
0.35 - 0.50 m2	m2	3.00	-	26.88	50.99	77.87	89.55
0.50 - 0.75 m2	m2	2.50	-	22.40	50.86	73.26	84.25
0.75 - 1.00 m2	m2	2.00	-	17.92	50.73	68.65	78.95
Bedding edge of glass							
In imitation washleather strip	m	0.10	-	0.90	0.39	1.29	1.48
In self-adhesive plastics sealing strip	m	0.10	-	0.90	0.42	1.32	1.52

Glazing

New work	Unit	Specialist price net	Price with 15%

Patent glazing

£ £
VAT not included

The following "Specialist price net" figures are for quantities between 50 and 100 square metres.

Prices do not include for cash discount.

See the preamble notes for builder's profit and attendance.

Single patent glazing

7 mm wired cast glass to roofs with aluminium bars at 610 mm centres bolted to steel purlins

not exceeding 1000 mm span	m2	178.39	205.15
1000 - 1500 mm span	m2	161.71	185.96
1500 - 2000 mm span	m2	156.15	179.57

7 mm wired cast glass to vertical surfaces with aluminium bars at 610 mm centres bolted to steel rails

not exceeding 1000 mm span	m2	194.78	223.99
1000 - 1500 mm span	m2	178.06	204.77
1500 - 2000 mm span	m2	172.63	198.52

6 mm wired polished glass to roofs with aluminium bars at 610 mm centres bolted to steel purlins

not exceeding 1000 mm span	m2	221.43	254.65
1000 - 1500 mm span	m2	205.28	236.07
1500 - 2000 mm span	m2	198.68	228.48

6 mm wired polished glass to vertical surfaces with aluminium bars at 610 mm centres bolted to steel rails

not exceeding 1000 mm span	m2	238.20	273.94
1000 - 1500 mm span	m2	222.06	255.37
1500 - 2000 mm span	m2	215.47	247.79

Double patent glazing

6 mm float and 6 mm wired polished glass to roofs with aluminium bars at 610 mm centres bolted to steel purlins

not exceeding 1000 mm span	m2	353.75	406.81
1000 - 1500 mm span	m2	351.71	404.47
1500 - 2000 mm span	m2	344.65	396.35

6 mm float and 6 mm wired polished glass to vertical surfaces with aluminium bars at 610 mm centres bolted to steel rails

not exceeding 1000 mm span	m2	366.50	421.47
1000 - 1500 mm span	m2	358.69	412.49
1500 - 2000 mm span	m2	351.25	403.94

6 mm float and 6.4 mm clear laminated safety glass to roofs with aluminium bars at 610 mm centres bolted to steel purlins

not exceeding 1000 mm span	m2	299.72	344.68
1000 - 1500 mm span	m2	297.04	341.60
1500 - 2000 mm span	m2	290.14	333.66

Glazing

New work

	Unit	Specialist price net	Price with 15%

Patent glazing

		£	£

VAT not included

Double patent glazing *(continued)*

6 mm float and 6.4 mm clear laminated safety glass to vertical surfaces with aluminium bars at 610 mm centres bolted to steel rails

	Unit	Specialist price net	Price with 15%
not exceeding 1000 mm span	m2	308.74	355.05
1000 - 1500 mm span	m2	300.83	345.96
1500 - 2000 mm span	m2	296.30	340.74

Glazing

New work	Unit	Hours C	Hours L	Labour net	Material net	Price net	Price with 15%
Domelights				£	£	£	£
					VAT not included		

One piece units screwed to wood kerb

Rough cast glass one-piece domelights with clips

600 x 600mm	each	1.50	1.00	20.18	62.01	82.19	94.52
900 x 900 mm	each	2.00	3.00	38.14	102.81	140.95	162.09
1200 x 900 mm	each	2.50	3.25	44.31	143.23	187.54	215.67
1200 x 1200 mm	each	3.00	3.50	50.47	178.78	229.25	263.64
1800 x 1200 mm	each	3.00	5.00	60.58	254.39	314.97	362.22
600 mm diameter	each	1.50	1.00	20.18	59.51	79.69	91.64
900 mm diameter	each	2.00	2.50	34.77	83.99	118.76	136.57
1200 mm diameter	each	2.50	3.50	45.99	167.25	213.24	245.23
1800 mm diameter	each	3.00	5.00	60.58	442.57	503.15	578.62

Wired cast glass one-piece domelights with clips

600 x 600mm	each	1.50	1.00	20.18	55.31	75.49	86.81
900 x 900 mm	each	2.00	3.00	38.14	102.21	140.35	161.40
1200 x 900 mm	each	2.50	3.25	44.31	146.03	190.34	218.89
1200 x 1200 mm	each	3.00	3.50	50.47	180.68	231.15	265.82
1800 x 1200 mm	each	3.00	5.00	60.58	250.89	311.47	358.19
600 mm diameter	each	1.50	1.00	20.18	52.91	73.09	84.05
900 mm diameter	each	2.00	2.50	34.77	86.29	121.06	139.22
1200 mm diameter	each	2.50	3.50	45.99	169.55	215.54	247.87
1800 mm diameter	each	3.00	5.00	60.58	395.77	456.35	524.80

Single skin polycarbonate one-piece domelights with clips

600 x 600mm	each	1.00	0.50	12.33	72.60	84.93	97.67
900 x 900 mm	each	1.50	0.50	16.81	110.87	127.68	146.83
1200 x 900 mm	each	1.75	1.00	22.42	176.89	199.31	229.21
1200 x 1200 mm	each	2.00	1.25	26.35	204.27	230.62	265.21
1800 x 1200 mm	each	2.50	3.00	42.62	329.66	372.28	428.12
600 mm diameter	each	1.00	0.50	12.33	79.68	92.01	105.81
900 mm diameter	each	1.50	0.50	16.81	150.55	167.36	192.46
1200 mm diameter	each	2.00	1.25	26.35	223.77	250.12	287.64

Double skin polycarbonate one-piece domelights with clips

600 x 600mm	each	1.00	0.50	12.33	143.07	155.40	178.71
900 x 900 mm	each	1.50	0.50	16.81	272.03	288.84	332.17
1200 x 900 mm	each	1.75	1.00	22.42	349.36	371.78	427.55
1200 x 1200 mm	each	2.00	1.25	26.35	404.95	431.30	496.00
1800 x 1200 mm	each	2.50	2.00	35.88	638.68	674.56	775.74
600 mm diameter	each	1.00	0.50	12.33	157.26	169.59	195.03
900 mm diameter	each	1.50	0.50	16.81	298.70	315.51	362.84
1200 mm diameter	each	2.00	1.25	26.35	444.53	470.88	541.51

Painting and decorating

Preamble

"Labour net" figures include allowances for all costs incidental to the employment of labour.

"Plant net" figures include for all costs of plant including drivers and operators where applicable.

"Materials net" figures include for all costs of materials including an allowance for waste except where specifically stated and an extra allowance to cover the cost of wear on brushes.

"Price net" figures are the totals of the "Labour net", "Plant net", where applicable, and "Materials net" figures. Prices are for a builder employing his own labour; according to the amount and nature of the work involved, it may well be possible to secure more advantageous prices from specialist sub-contractors.

Although not specifically mentioned in the various descriptions, allowance has been made for rubbing down, stopping and general preparation to permit high class work.

Where estimates are prepared without bills of quantities, it is essential that painting work be fully measured, extra care should therefore be taken to ensure that full account is taken of narrow widths and of the extra girth of moulded work, edges and returns.

Prices do not include any allowance for scaffolding, ladders or other plant necessary to reach the work. The "Preliminaries" section includes prices for scaffolding which must be considered and allowance included to suit the particular circumstances of a tender.

Specialist prices

"Price with 15%" figures are all-in guide prices and include 15% for the builder's overheads, profit, unloading materials and general attendance (to include free use of standing scaffolding and hoists, temporary lighting and water and clearing away rubbish).

The amount of attendance required varies between the various trades and also with the circumstances of specific jobs; the percentage addition must always be considered and adjusted as necessary to suit the terms and conditions of the quotation being used.

Quantities and delivery distances are usually the most significant of the many factors which influence prices and it must be emphasised that quotations should always be obtained when preparing a tender.

Painting and decorating

New work	Unit	Price	Cover allowed per coat		
			Smooth Surfaces	Textured Surfaces	Rough Surfaces
Basic prices for materials		£	m2	m2	m2
Aluminium heat resisting paint	5 litre	22.46	75	-	-
Anti-condensation paint	5 litre	18.98	20	15	-
Bituminous paint	25 litre	38.86	250	-	-
Cement paint	40 kg	38.40	-	200	100
Creosote	5 litre	3.54	40	-	20
Eggshell paint	5 litre	22.40	65	50	-
Emulsion paint - vinyl matt	5 litre	14.94	60	45	20
Emulsion paint - vinyl silk	5 litre	15.45	60	45	20
Flame retardant					
undercoat	5 litre	33.40	50	-	-
gloss finish	5 litre	33.40	50	-	-
Gloss finish	5 litre	19.25	60	45	-
Linseed oil	5 litre	13.46	80	-	-
Masonry oil paint	5 litre	20.80	-	45	20
Metal primer	5 litre	30.05	60	-	-
Multicolour					
finish	5 litre	35.05	20	15	-
primer	5 litre	27.26	55	40	-
emulsion glaze	5 litre	29.75	75	60	-
Oil varnish stain floor seal	5 litre	19.12	60	-	-
Plaster primer	5 litre	21.53	60	45	20
Polyurethane					
floor seal	5 litre	30.17	65	-	-
varnish	5 litre	24.59	65	-	-
Stone paint					
finish	5 litre	13.73	-	10	5
primer	5 litre	13.11	-	40	20
emulsion	5 litre	17.00	-	45	20
Textured cement paint	5 litre	17.17	-	30	15

Prices actually to be paid for materials must be
checked against the above basic prices and
adjustments made as necessary.

Painting and decorating

New work	Unit	Price	Cover allowed per coat		
			Smooth Surfaces	Textured Surfaces	Rough Surfaces
Basic prices for materials		£	m2	m2	m2
Textured plastic compound ("Artex")					
finish	25 kg	8.39	40	-	-
sealer	5 litre	13.12	60	-	-
Undercoat	5 litre	19.25	60	45	-
Water repellent decorative timber dressing	5 litre	11.65	75	-	-
Wood preservative	5 litre	10.71	30	-	20
Wood primer	5 litre	23.11	50	-	-
Cetol Interior	2.5 litre	30.60	32	-	-
Cetol HLS	5 litre	44.88	70	-	21
Cetol Filter 7	5 litre	43.48	85	-	-
Hammerite metal finish	5 litre	34.09	23	-	-

Prices actually to be paid for materials must be
checked against the above basic prices and
adjustments made as necessary.

Painting and decorating

New work	Unit	Hours C	Hours L	Labour net	Material net	Price net	Price with 15%
Generally				£	£	£	£
					VAT not included		

Before fixing

One coat of wood primer on joinery

Surfaces							
over 300 mm girth	m2	0.15	-	1.34	0.53	1.87	2.15
not exceeding 150 mm girth	m	0.03	-	0.27	0.08	0.35	0.40
150 - 300 mm girth	m	0.05	-	0.45	0.17	0.62	0.71
not exceeding 0.50 m2	each	0.11	-	0.99	0.26	1.25	1.44

One coat of polyurethane sealer on joinery

Surfaces							
over 300 mm girth	m2	0.23	-	2.06	0.53	2.59	2.98
not exceeding 150 mm girth	m	0.04	-	0.40	0.09	0.49	0.56
150 - 300 mm girth	m	0.08	-	0.72	0.17	0.89	1.02
not exceeding 0.50 m2	each	0.17	-	1.52	0.26	1.78	2.05

One coat of creosote on woodwork

Wrought surfaces							
over 300 mm girth	m2	0.10	-	0.90	0.09	0.99	1.14
not exceeding 150 mm girth	m	0.02	-	0.18	0.01	0.19	0.22
150 - 300 mm girth	m	0.03	-	0.30	0.03	0.33	0.38
not exceeding 0.50 m2	each	0.08	-	0.72	0.04	0.76	0.87

Sawn surfaces							
over 300 mm girth	m2	0.13	-	1.16	0.21	1.37	1.58
not exceeding 150 mm girth	m	0.03	-	0.22	0.04	0.26	0.30
150 - 300 mm girth	m	0.04	-	0.39	0.07	0.46	0.53
not exceeding 0.50 m2	each	0.10	-	0.90	0.11	1.01	1.16

One coat of wood preservative on woodwork

Wrought surfaces							
over 300 mm girth	m2	0.10	-	0.90	0.30	1.20	1.38
not exceeding 150 mm girth	m	0.02	-	0.18	0.05	0.23	0.26
150 - 300 mm girth	m	0.03	-	0.30	0.09	0.39	0.45
not exceeding 0.50 m2	each	0.08	-	0.72	0.15	0.87	1.00

Sawn surfaces							
over 300 mm girth	m2	0.13	-	1.16	0.31	1.47	1.69
not exceeding 150 mm girth	m	0.03	-	0.22	0.06	0.28	0.32
150 - 300 mm girth	m	0.04	-	0.39	0.09	0.48	0.55
not exceeding 0.50 m2	each	0.10	-	0.90	0.16	1.06	1.22

One coat of metal primer on metalwork

Surfaces							
over 300 mm girth	m2	0.14	-	1.25	0.80	2.05	2.36
not exceeding 150 mm girth	m	0.03	-	0.25	0.13	0.38	0.44
150 - 300 mm girth	m	0.05	-	0.41	0.25	0.66	0.76
not exceeding 0.50 m2	each	0.11	-	0.99	0.40	1.39	1.60

Painting and decorating

New work	Unit	Hours C	Hours L	Labour net	Material net	Price net	Price with 15%
Generally				£	£	£	£
					VAT not included		

Walls and ceilings internally

Two coats of vinyl matt emulsion paint on walls and ceilings

Surfaces of							
smooth plaster	m2	0.24	-	2.15	0.50	2.65	3.05
rendering, fair face or similar textured surfaces	m2	0.26	-	2.33	0.53	2.86	3.29

Three coats of vinyl matt emulsion paint on walls and ceilings

Surfaces of							
smooth plaster	m2	0.36	-	3.23	0.71	3.94	4.53
rendering, fair face or similar textured surfaces	m2	0.38	-	3.40	0.77	4.17	4.80

Two coats of vinyl silk emulsion paint on walls and ceilings

Surfaces of							
smooth plaster	m2	0.25	-	2.22	0.50	2.72	3.13
rendering, fair face or similar textured surfaces	m2	0.25	-	2.28	0.85	3.13	3.60

Three coats of vinyl silk emulsion paint on walls and ceilings

Surfaces of							
smooth plaster	m2	0.36	-	3.25	0.73	3.98	4.58
rendering, fair face or similar textured surfaces	m2	0.38	-	3.36	0.75	4.11	4.73

Primer and two coats of eggshell paint on walls and ceilings

Surfaces of							
smooth plaster	m2	0.47	-	4.24	1.35	5.59	6.43
rendering, fair face or similar textured surfaces	m2	0.58	-	5.18	1.66	6.84	7.87

Primer, one undercoat and one coat of gloss finish paint on walls and ceilings

Surfaces of							
smooth plaster	m2	0.56	-	4.99	1.14	6.13	7.05
rendering, fair face or similar textured surfaces	m2	0.64	-	5.71	1.36	7.07	8.13

Painting and decorating

New work	Unit	Hours C	Hours L	Labour net	Material net	Price net	Price with 15%
Generally				£	£	£	£
					VAT not included		
Walls and ceilings internally (*continued*)							
Primer, two undercoats and one coat of gloss finish paint on walls and ceilings							
Surfaces of							
smooth plaster	m2	0.73	-	6.58	1.45	8.03	9.23
rendering, fair face or similar textured surfaces	m2	0.85	-	7.61	1.75	9.36	10.76
Primer and two coats of anti-condensation paint finished stippled on walls and ceilings							
Surfaces of							
smooth plaster	m2	0.85	-	7.62	0.79	8.41	9.67
rendering, fair face or similar textured surfaces	m2	1.00	-	8.96	1.06	10.02	11.52
Sundries							
Cutting in edges on flush surface	m	0.10	-	0.90	-	0.90	1.03
25 mm dado lines in one coat	m	0.25	-	2.24	-	2.24	2.58
Textured plastic compound ("Artex") stipple finish on ceilings							
Surfaces of							
smooth plaster	m2	0.22	-	1.97	0.46	2.43	2.79
plasterboard	m2	0.24	-	2.15	0.46	2.61	3.00
fair faced in-situ concrete	m2	0.27	-	2.42	0.56	2.98	3.43
fair faced precast concrete units	m2	0.33	-	2.96	0.60	3.56	4.09
Paperhanging							
Prepare, size, supply and hang wallpaper to walls							
lining paper PC £1.10 per roll	m2	0.30	-	2.69	0.31	3.00	3.45
vinyl paper PC £4.00 per roll	m2	0.30	-	2.69	1.51	4.20	4.83
embossed paper PC £6.00 per roll	m2	0.35	-	3.14	2.75	5.89	6.77
hessian paper PC £8.00 per m2	m2	0.35	-	3.14	3.01	6.15	7.07
Prepare, size, supply and hang wallpaper to ceilings							
lining paper PC £1.10 per roll	m2	0.35	-	3.14	0.31	3.45	3.97
vinyl paper PC £4.00 per roll	m2	0.35	-	3.14	1.51	4.65	5.35
embossed paper PC £6.00 per roll	m2	0.40	-	3.58	2.75	6.33	7.28
hessian paper PC £8.00 per m2	m2	0.40	-	3.58	3.01	6.59	7.58
Trim and hang only paper border	m	0.10	-	0.90	-	0.90	1.03

Painting and decorating

New work	Unit	Hours C	Hours L	Labour net	Material net	Price net	Price with 15%
Generally				£	£	£	£
					VAT not included		

Floors internally

Two coats of polyurethane floor seal

Floors

softwood	m2	0.33	-	2.96	1.07	4.03	4.63
hardwood	m2	0.30	-	2.69	1.07	3.76	4.32

Three coats of polyurethane floor seal

Floors

softwood	m2	0.50	-	4.48	1.60	6.08	6.99
hardwood	m2	0.45	-	4.03	1.60	5.63	6.47

Two coats of oil varnish stain floor seal

Floors

softwood	m2	0.33	-	2.96	0.84	3.80	4.37
hardwood	m2	0.30	-	2.69	0.84	3.53	4.06

Woodwork internally

Preparatory work

Brush fill and rub down to obtain superfine surface for painting	m2	0.40	-	3.58	0.09	3.67	4.22

Knot, prime, stop and paint one undercoat and one coat of gloss finish paint on joinery

General surfaces

over 300 mm girth	m2	0.52	-	4.66	1.14	5.80	6.67
not exceeding 150 mm girth	m	0.10	-	0.90	0.19	1.09	1.25
150 - 300 mm girth	m	0.17	-	1.52	0.36	1.88	2.16
not exceeding 0.50 m2	each	0.20	-	1.79	0.59	2.38	2.74

Glazed doors and screens in panes

small - not exceeding 0.10m2	m2	0.77	-	6.90	0.90	7.80	8.97
medium - 0.10 - 0.50 m2	m2	0.65	-	5.82	0.74	6.56	7.54
large - 0.50 - 1.00 m2	m2	0.57	-	5.11	0.64	5.75	6.61
extra large - over 1.00 m2	m2	0.52	-	4.66	0.45	5.11	5.88

Windows (measured flat overall) in panes

small - not exceeding 0.10 m2	m2	0.78	-	6.99	0.90	7.89	9.07
medium - 0.10 - 0.50 m2	m2	0.65	-	5.82	0.74	6.56	7.54
large - 0.50 - 1.00 m2	m2	0.57	-	5.11	0.64	5.75	6.61
extra large - over 1.00 m2	m2	0.52	-	4.66	0.45	5.11	5.88

Edges of opening casements	m	0.09	-	0.81	0.08	0.89	1.02

Painting and decorating

New work	Unit	Hours C	Hours L	Labour net	Material net	Price net	Price with 15%
Generally				£	£	£	£
					VAT not included		

Woodwork internally (*continued*)

Knot, prime, stop and paint two undercoats and one coat of gloss finish paint on joinery

New work	Unit	Hours C	Hours L	Labour net	Material net	Price net	Price with 15%
General surfaces							
over 300 mm girth	m2	0.66	-	5.91	1.42	7.33	8.43
not exceeding 150 mm girth	m	0.13	-	1.16	0.23	1.39	1.60
150 - 300 mm girth	m	0.22	-	1.97	0.45	2.42	2.78
not exceeding 0.50 m2	each	0.38	-	3.40	0.73	4.13	4.75
Glazed doors and screens in panes							
small - not exceeding 0.10 m2	m2	0.99	-	8.87	1.16	10.03	11.53
medium - 0.10 - 0.50 m2	m2	0.83	-	7.44	0.93	8.37	9.63
large - 0.50 - 1.00 m2	m2	0.73	-	6.54	0.79	7.33	8.43
extra large - over 1.00 m2	m2	0.66	-	5.91	0.56	6.47	7.44
Windows (measured flat overall) in panes							
small - not exceeding 0.10 m2	m2	0.99	-	8.87	1.13	10.00	11.50
medium - 0.10 - 0.50 m2	m2	0.83	-	7.44	0.93	8.37	9.63
large - 0.50 - 1.00 m2	m2	0.73	-	6.54	0.79	7.33	8.43
extra large - over 1.00 m2	m2	0.66	-	5.91	0.56	6.47	7.44
Edges of opening casements	m	0.12	-	1.08	0.10	1.18	1.36

Knot, prime, stop and paint one flame retardant undercoat and one coat of flame retardant gloss finish paint on joinery

New work	Unit	Hours C	Hours L	Labour net	Material net	Price net	Price with 15%
General surfaces							
over 300 mm girth	m2	0.90	-	8.06	1.85	9.91	11.40
not exceeding 150 mm girth	m	0.18	-	1.61	0.30	1.91	2.20
150 - 300 mm girth	m	0.30	-	2.69	0.57	3.26	3.75
not exceeding 0.50 m2	each	0.70	-	6.27	0.94	7.21	8.29
Glazed doors and screens in panes							
small - not exceeding 0.10 m2	m2	2.00	-	17.92	1.85	19.77	22.74
medium - 0.10 - 0.50 m2	m2	1.20	-	10.75	1.85	12.60	14.49
large - 0.50 - 1.00 m2	m	0.80	-	7.17	1.85	9.02	10.37
extra large - over 1.00 m2	m2	0.60	-	5.38	1.85	7.23	8.31

Stain, body in and wax polish hardwood joinery

New work	Unit	Hours C	Hours L	Labour net	Material net	Price net	Price with 15%
General surfaces							
over 300 mm girth	m2	0.90	-	8.06	0.92	8.98	10.33
not exceeding 150 mm girth	m	0.18	-	1.61	0.15	1.76	2.02
150 - 300 mm girth	m	0.30	-	2.69	0.28	2.97	3.42
not exceeding 0.50 m2	each	0.70	-	6.27	0.45	6.72	7.73

Painting and decorating

New work	Unit	Hours C	Hours L	Labour net	Material net	Price net	Price with 15%
Generally				£	£	£	£
					VAT not included		

Woodwork internally (*continued*)

Two coats of polyurethane varnish on joinery

General surfaces
over 300 mm girth	m2	0.39	-	3.49	1.32	4.81	5.53
not exceeding 150 mm girth	m	0.08	-	0.72	0.21	0.93	1.07
150 - 300 mm girth	m	0.13	-	1.16	0.41	1.57	1.81
not exceeding 0.50 m2	each	0.19	-	1.70	0.66	2.36	2.71

Three coats of polyurethane varnish on joinery

General surfaces
over 300 mm girth	m2	0.59	-	5.29	1.98	7.27	8.36
not exceeding 150 mm girth	m	0.12	-	1.08	0.32	1.40	1.61
150 - 300 mm girth	m	0.19	-	1.70	0.63	2.33	2.68
not exceeding 0.50 m2	each	0.29	-	2.60	0.99	3.59	4.13

Three coats of Cetol Interior on joinery

General surfaces
over 300 mm girth	m2	0.58	-	5.20	3.23	8.43	9.69
not exceeding 150 mm girth	m	0.11	-	0.99	0.51	1.50	1.73
150 - 300 mm girth	m	0.19	-	1.70	1.03	2.73	3.14
not exceeding 0.50 m2	each	0.29	-	2.60	3.18	5.78	6.65

Metalwork internally

Prime and paint one undercoat and one coat of gloss finish on metalwork

General surfaces
over 300 mm girth	m2	0.47	-	4.21	1.34	5.55	6.38
not exceeding 150 mm girth	m	0.09	-	0.81	0.22	1.03	1.18
150 - 300 mm girth	m	0.16	-	1.43	0.43	1.86	2.14
not exceeding 0.50 m2	each	0.24	-	2.15	0.68	2.83	3.25

Glazed doors, screens and windows (measured flat overall) in panes
small - not exceeding 0.10 m2	m2	0.71	-	6.36	1.01	7.37	8.48
medium - 0.10 - 0.50 m2	m2	0.58	-	5.20	0.89	6.09	7.00
large - 0.50 - 1.00 m2	m	0.52	-	4.66	0.73	5.39	6.20
extra large - over 1.00 m2	m2	0.47	-	4.21	0.51	4.72	5.43

Edges of opening casements	m	0.08	-	0.72	0.08	0.80	0.92

Structural members and pipes
over 300 mm girth	m2	0.66	-	5.91	1.34	7.25	8.34
not exceeding 150 mm girth	m	0.13	-	1.16	0.22	1.38	1.59
150 - 300 mm girth	m	0.22	-	1.97	0.43	2.40	2.76

Painting and decorating

New work	Unit	Hours C	Hours L	Labour net	Material net	Price net	Price with 15%
Generally				£	£	£	£
					VAT not included		

Metalwork internally (*continued*)

Prime and paint two undercoats and one coat of gloss finish on metalwork

New work	Unit	Hours C	Hours L	Labour net	Material net	Price net	Price with 15%
General surfaces							
over 300 mm girth	m2	0.60	-	5.38	1.62	7.00	8.05
not exceeding 150 mm girth	m	0.12	-	1.08	0.27	1.35	1.55
150 - 300 mm girth	m	0.20	-	1.79	0.52	2.31	2.66
not exceeding 0.50 m2	each	0.30	-	2.69	0.82	3.51	4.04
Glazed doors, screens and windows (measured flat overall) in panes							
small - not exceeding 0.10 m2	m2	0.90	-	8.06	1.23	9.29	10.68
medium - 0.10 - 0.50 m2	m2	0.75	-	6.72	1.08	7.80	8.97
large - 0.50 - 1.00 m2	m2	0.66	-	5.91	0.88	6.79	7.81
extra large - over 1.00 m2	m2	0.60	-	5.38	0.63	6.01	6.91
Edges of opening casements	m	0.13	-	1.16	0.10	1.26	1.45
Structural members and pipes							
over 300 mm girth	m2	0.88	-	7.88	1.37	9.25	10.64
not exceeding 150 mm girth	m	0.17	-	1.52	0.27	1.79	2.06
150 - 300 mm girth	m	0.29	-	2.60	0.52	3.12	3.59

Two coats of aluminium heat resisting paint on metalwork

New work	Unit	Hours C	Hours L	Labour net	Material net	Price net	Price with 15%
Radiators and pipes							
over 300 mm girth	m2	0.60	-	5.38	1.07	6.45	7.42
not exceeding 150 mm girth	m	0.12	-	1.08	0.16	1.24	1.43
150 - 300 mm girth	m	0.20	-	1.79	0.32	2.11	2.43

Walls externally

Two coats of vinyl matt emulsion paint on walls

New work	Unit	Hours C	Hours L	Labour net	Material net	Price net	Price with 15%
Surfaces of							
rendering, fair face or similar textured							
surfaces	m2	0.26	-	2.33	0.53	2.86	3.29
roughcast surfaces	m2	0.39	-	3.49	0.85	4.34	4.99

Primer and two coats of masonry oil paint on walls

New work	Unit	Hours C	Hours L	Labour net	Material net	Price net	Price with 15%
Surfaces of							
rendering, fair face or similar textured							
surfaces	m2	0.47	-	4.21	1.22	5.43	6.24
roughcast surfaces	m2	0.71	-	6.36	2.46	8.82	10.14

Painting and decorating

New work	Unit	Hours C	Hours L	Labour net	Material net	Price net	Price with 15%
Generally				£	£	£	£
					VAT not included		

Walls externally (*continued*)

Two coats of cement paint on walls

Surfaces of							
rendering, fair face or similar textured							
surfaces	m2	0.45	-	4.03	0.65	4.68	5.38
roughcast surfaces	m2	0.68	-	6.09	1.20	7.29	8.38

Two coats of textured cement paint on walls

Surfaces of							
rendering, fair face or similar textured							
surfaces	m2	0.50	-	4.48	1.16	5.64	6.49
roughcast surfaces	m2	0.70	-	6.27	2.38	8.65	9.95

Primer and one coat of stone paint on walls

Surfaces of							
rendering, fair face or similar textured							
surfaces	m2	0.50	-	4.48	0.50	4.98	5.73
roughcast surfaces	m2	0.70	-	6.27	1.21	7.48	8.60

Primer and two coats of stone paint on walls

Surfaces of							
rendering, fair face or similar textured							
surfaces	m2	0.70	-	6.27	0.75	7.02	8.07
roughcast surfaces	m2	0.95	-	8.51	1.96	10.47	12.04

One coat of vinyl matt emulsion paint to stone paint finish on walls

Surfaces of							
rendering, fair face or similar textured							
surfaces	m2	0.20	-	1.79	0.27	2.06	2.37
roughcast surfaces	m2	0.28	-	2.51	0.45	2.96	3.40

Woodwork externally

Knot, prime, stop and paint one undercoat and one coat of gloss finish paint on joinery

General surfaces							
over 300 mm girth	m2	0.59	-	5.29	1.14	6.43	7.39
not exceeding 150 mm girth	m	0.11	-	0.99	0.19	1.18	1.36
150 - 300 mm girth	m	0.19	-	1.70	0.39	2.09	2.40
not exceeding 0.50 m2	each	0.30	-	2.69	0.59	3.28	3.77
Glazed doors and screens in panes							
small - not exceeding 0.10 m2	m2	0.88	-	7.88	0.90	8.78	10.10
medium - 0.10 - 0.50 m2	m2	0.74	-	6.63	0.74	7.37	8.48
large - 0.50 - 1.00 m2	m	0.65	-	5.82	0.64	6.46	7.43
extra large - over 1.00 m2	m2	0.59	-	5.29	0.45	5.74	6.60

Painting and decorating

New work	Unit	Hours C	Hours L	Labour net	Material net	Price net	Price with 15%
Generally				£	£	£	£
					VAT not included		

Woodwork externally (*continued*)

Knot, prime, stop and paint one undercoat and one coat of gloss finish paint on joinery (*continued*)

	Unit	Hours C	Hours L	Labour net	Material net	Price net	Price with 15%
Windows (measured flat overall) in panes							
small - not exceeding 0.10 m2	m2	0.89	-	7.97	0.90	8.87	10.20
medium - 0.10 - 0.50 m2	m2	0.74	-	6.63	0.74	7.37	8.48
large - 0.50 - 1.00 m2	m2	0.65	-	5.82	0.64	6.46	7.43
extra large - over 1.00 m2	m2	0.59	-	5.29	0.45	5.74	6.60
Edges of opening casements	m	0.11	-	0.99	0.08	1.07	1.23

Knot, prime, stop and paint two undercoats and one coat of gloss finish paint on joinery

	Unit	Hours C	Hours L	Labour net	Material net	Price net	Price with 15%
General surfaces							
over 300 mm girth	m2	0.75	-	6.72	1.39	8.11	9.33
not exceeding 150 mm girth	m	0.15	-	1.34	0.22	1.56	1.79
150 - 300 mm girth	m	0.25	-	2.24	0.44	2.68	3.08
not exceeding 0.50 m2	each	0.38	-	3.40	0.71	4.11	4.73
Glazed doors and screens in panes							
small - not exceeding 0.10 m2	m2	1.13	-	10.08	1.13	11.21	12.89
medium - 0.10 - 0.50 m2	m2	0.94	-	8.42	0.90	9.32	10.72
large - 0.50 - 1.00 m2	m2	0.83	-	7.44	0.76	8.20	9.43
extra large - over 1.00 m2	m2	0.75	-	6.72	0.53	7.25	8.34
Windows (measured flat overall) in panes							
small - not exceeding 0.10 m2	m2	1.13	-	10.12	1.10	11.22	12.90
medium - 0.10 - 0.50 m2	m2	0.95	-	8.51	0.90	9.41	10.82
large - 0.50 - 1.00 m2	m2	0.83	-	7.44	0.76	8.20	9.43
extra large - over 1.00 m2	m2	0.75	-	6.72	0.48	7.20	8.28
Edges of opening casements	m	0.14	-	1.25	0.09	1.34	1.54

One coat of creosote on woodwork

	Unit	Hours C	Hours L	Labour net	Material net	Price net	Price with 15%
Wrought surfaces							
over 300 mm girth	m2	0.12	-	1.08	0.12	1.20	1.38
not exceeding 150 mm girth	m	0.03	-	0.27	0.01	0.28	0.32
150 - 300 mm girth	m	0.05	-	0.45	0.04	0.49	0.56
not exceeding 0.50 m2	each	0.09	-	0.81	0.06	0.87	1.00
Sawn surfaces							
over 300 mm girth	m2	0.15	-	1.34	0.13	1.47	1.69
not exceeding 150 mm girth	m	0.04	-	0.36	0.02	0.38	0.44
150 - 300 mm girth	m	0.06	-	0.54	0.04	0.58	0.67
not exceeding 0.50 m2	each	0.11	-	0.99	0.07	1.06	1.22

Painting and decorating

| --- | --- | --- | --- | --- | --- | --- | --- |
New work							
Generally				£	£	£	£
					VAT not included		

Woodwork externally (*continued*)

One coat of wood preservative on woodwork

Wrought surfaces

	Unit	Hours C	Hours L	Labour net	Material net	Price net	Price with 15%
over 300 mm girth	m2	0.12	-	1.08	0.30	1.38	1.59
not exceeding 150 mm girth	m	0.03	-	0.27	0.04	0.31	0.36
150 - 300 mm girth	m	0.05	-	0.45	0.09	0.54	0.62
not exceeding 0.50 m2	each	0.09	-	0.81	0.15	0.96	1.10

Sawn surfaces

over 300 mm girth	m2	0.15	-	1.34	0.31	1.65	1.90
not exceeding 150 mm girth	m	0.04	-	0.36	0.05	0.41	0.47
150 - 300 mm girth	m	0.06	-	0.54	0.09	0.63	0.72
not exceeding 0.50 m2	each	0.11	-	0.99	0.16	1.15	1.32

Two coats of boiled linseed oil rubbed in to hardwood joinery

Surfaces

over 300 mm girth	m2	0.33	-	2.96	1.03	3.99	4.59
not exceeding 150 mm girth	m	0.08	-	0.72	0.15	0.87	1.00
150 - 300 mm girth	m	0.13	-	1.16	0.30	1.46	1.68
not exceeding 0.50 m2	each	0.25	-	2.24	0.51	2.75	3.16

Two coats of water repellent decorative timber dressing on cedar boarding etc

Surfaces

over 300 mm girth	m2	0.45	-	4.03	0.87	4.90	5.63
not exceeding 150 mm girth	m	0.11	-	0.99	0.13	1.12	1.29
150 - 300 mm girth	m	0.18	-	1.61	0.26	1.87	2.15
not exceeding 0.50 m2	each	0.35	-	3.14	0.46	3.60	4.14

Two coats of polyurethane varnish on joinery

General surfaces

over 300 mm girth	m2	0.44	-	3.94	1.32	5.26	6.05
not exceeding 150 mm girth	m	0.09	-	0.81	0.21	1.02	1.17
150 - 300 mm girth	m	0.15	-	1.34	0.41	1.75	2.01
not exceeding 0.50 m2	each	0.22	-	1.97	0.66	2.63	3.02

Three coats of polyurethane varnish on joinery

General surfaces

over 300 mm girth	m2	0.67	-	6.00	1.98	7.98	9.18
not exceeding 150 mm girth	m	0.14	-	1.25	0.32	1.57	1.81
150 - 300 mm girth	m	0.22	-	1.97	0.63	2.60	2.99
not exceeding 0.50 m2	each	0.34	-	3.05	0.99	4.04	4.65

Painting and decorating

| --- | --- | --- | --- | --- | --- | --- | --- |
| **Generally** | | | | £ | £ | £ | £ |
| | | | | | VAT not included | | |

Woodwork externally (*continued*)

One coat of Cetol HLS and two coats of Cetol Filter 7 on Joinery

General surfaces

over 300 mm girth	m2	0.66	-	5.91	1.84	7.75	8.91
not exceeding 150 mm girth	m	0.13	-	1.16	0.28	1.44	1.66
150 - 300 mm girth	m	0.22	-	1.97	0.58	2.55	2.93
not exceeding 0.50 m2	each	0.33	-	2.96	1.79	4.75	5.46

Three coats of Cetol HLS on sawn fencing

General surfaces

over 300 mm girth	m2	0.99	-	8.87	4.72	13.59	15.63
not exceeding 150 mm girth	m	0.19	-	1.70	0.76	2.46	2.83
150 - 300 mm girth	m	0.33	-	2.96	1.51	4.47	5.14

Metalwork externally

Prime and paint one undercoat and one coat of gloss finish on metalwork

General surfaces

over 300 mm girth	m2	0.54	-	4.84	1.34	6.18	7.11
not exceeding 150 mm girth	m	0.11	-	0.99	0.22	1.21	1.39
150 - 300 mm girth	m	0.18	-	1.61	0.43	2.04	2.35
not exceeding 0.50 m2	each	0.27	-	2.42	0.68	3.10	3.56

Glazed doors, screens and windows (measured flat overall) in panes

small - not exceeding 0.10 m2	m2	0.81	-	7.26	1.01	8.27	9.51
medium - 0.10 - 0.50 m2	m2	0.66	-	5.91	0.89	6.80	7.82
large - 0.50 - 1.00 m2	m2	0.59	-	5.29	0.73	6.02	6.92
extra large - over 1.00 m2	m2	0.54	-	4.84	0.51	5.35	6.15

Edges of opening casements	m	0.10	-	0.90	0.08	0.98	1.13

Structural members, gutters and pipes

over 300 mm girth	m2	0.75	-	6.72	1.34	8.06	9.27
not exceeding 150 mm girth	m	0.15	-	1.34	0.22	1.56	1.79
150 - 300 mm girth	m	0.25	-	2.24	0.43	2.67	3.07

Perforated plate landings and treads - each side	m2	1.20	-	10.75	1.42	12.17	14.00
Ornamental railings and gates - each side	m2	1.60	-	14.34	1.42	15.76	18.12

Painting and decorating

Generally

£ £ £ £

VAT not included

Metalwork externally (*continued*)

Prime and paint two undercoats and one coat of gloss finish on metalwork

	Unit	Hours C	Hours L	Labour net	Material net	Price net	Price with 15%
General surfaces							
over 300 mm girth	m2	0.68	-	6.09	1.62	7.71	8.87
not exceeding 150 mm girth	m	0.14	-	1.25	0.27	1.52	1.75
150 - 300 mm girth	m	0.23	-	2.06	0.52	2.58	2.97
not exceeding 0.50 m2	each	0.34	-	3.05	0.82	3.87	4.45
Glazed doors, screens and windows (measured flat overall) in panes							
small - not exceeding 0.10 m2	m2	1.03	-	9.23	1.23	10.46	12.03
medium - 0.10 - 0.50 m2	m2	0.86	-	7.71	1.08	8.79	10.11
large - 0.50 - 1.00 m2	m2	0.75	-	6.72	0.88	7.60	8.74
extra large - over 1.00 m2	m2	0.68	-	6.09	0.63	6.72	7.73
Edges of opening casements	m	0.14	-	1.25	0.10	1.35	1.55
Structural members, gutters and pipes							
over 300 mm girth	m2	1.00	-	8.96	1.15	10.11	11.63
not exceeding 150 mm girth	m	0.19	-	1.70	0.27	1.97	2.27
150 - 300 mm girth	m	0.33	-	2.96	0.52	3.48	4.00
Perforated plate landings and treads - each side	m2	1.50	-	13.44	1.85	15.29	17.58
Ornamental railings and gates - each side	m2	2.00	-	17.92	1.85	19.77	22.74

One coat of bituminous paint on metalwork

	Unit	Hours C	Hours L	Labour net	Material net	Price net	Price with 15%
General surfaces							
over 300 mm girth	m2	0.30	-	2.69	0.43	3.12	3.59
not exceeding 150 mm girth	m	0.06	-	0.54	0.07	0.61	0.70
150 - 300 mm girth	m	0.12	-	1.08	0.13	1.21	1.39
not exceeding 0.50 m2	each	0.25	-	2.24	0.21	2.45	2.82

Two coats of bituminous paint on metalwork

	Unit	Hours C	Hours L	Labour net	Material net	Price net	Price with 15%
General surfaces							
over 300 mm girth	m2	0.50	-	4.48	0.86	5.34	6.14
not exceeding 150 mm girth	m	0.06	-	0.54	0.13	0.67	0.77
150 - 300 mm girth	m	0.10	-	0.90	0.15	1.05	1.21
not exceeding 0.50 m2	each	0.15	-	1.34	0.43	1.77	2.04

One coat of "Hammerite" paint on metalwork

	Unit	Hours C	Hours L	Labour net	Material net	Price net	Price with 15%
General surfaces							
over 300 mm girth	m2	0.25	-	2.24	0.36	2.60	2.99
not exceeding 150 mm girth	m	0.06	-	0.54	0.05	0.59	0.68
150 - 300 mm girth	m	0.10	-	0.90	0.10	1.00	1.15
not exceeding 0.50 m2	each	0.20	-	1.79	0.18	1.97	2.27

Painting and decorating

New work	Unit	Hours C	Hours L	Labour net	Material net	Price net	Price with 15%
French polishing				£	£	£	£
					VAT not included		
Seal and French polish hardwood joinery to an open grain finish							
General surfaces							
over 300 mm girth	m2	1.75	-	15.68	1.50	17.18	19.76
not exceeding 150 mm girth	m	0.39	-	3.49	0.25	3.74	4.30
150 - 300 mm girth	m	0.65	-	5.82	0.47	6.29	7.23
not exceeding 0.50 m2	each	1.30	-	11.65	0.75	12.40	14.26
Glazed doors and screens in panes							
small - not exceeding 0.10 m2	m2	3.85	-	34.50	1.19	35.69	41.04
medium - 0.10 - 0.50 m2	m2	2.30	-	20.61	0.96	21.57	24.81
large - 0.50 - 1.00 m2	m2	1.53	-	13.71	0.80	14.51	16.69
extra large - over 1.00 m2	m2	1.15	-	10.30	0.56	10.86	12.49
Windows (measured flat overall) in panes							
small - not exceeding 0.10 m2	m2	4.24	-	37.99	1.19	39.18	45.06
medium - 0.10 - 0.50 m2	m2	2.69	-	24.10	0.96	25.06	28.82
large - 0.50 - 1.00 m2	m2	1.92	-	17.20	0.80	18.00	20.70
extra large - over 1.00 m2	m2	1.54	-	13.80	0.56	14.36	16.51
Edges of opening casements	m	0.20	-	1.79	0.08	1.87	2.15
Stain, seal and French polish hardwood joinery to an open grain finish							
General surfaces							
over 300 mm girth	m2	1.75	-	15.68	2.15	17.83	20.50
not exceeding 150 mm girth	m	0.39	-	3.49	0.34	3.83	4.40
150 - 300 mm girth	m	0.65	-	5.82	0.67	6.49	7.46
not exceeding 0.50 m2	each	1.30	-	11.65	1.08	12.73	14.64
Glazed doors and screens in panes							
small - not exceeding 0.10 m2	m2	3.85	-	34.50	1.70	36.20	41.63
medium - 0.10 - 0.50 m2	m2	2.30	-	20.61	1.38	21.99	25.29
large - 0.50 - 1.00 m2	m2	1.53	-	13.71	1.14	14.85	17.08
extra large - over 1.00 m2	m2	1.15	-	10.30	0.80	11.10	12.77
Windows (measured flat overall) in panes							
small - not exceeding 0.10 m2	m2	4.24	-	37.99	1.70	39.69	45.64
medium - 0.10 - 0.50 m2	m2	2.69	-	24.10	1.38	25.48	29.30
large - 0.50 - 1.00 m2	m2	1.92	-	17.20	1.14	18.34	21.09
extra large - over 1.00 m2	m2	1.54	-	13.80	0.80	14.60	16.79
Edges of opening casements	m	0.20	-	1.79	0.12	1.91	2.20
Seal, body in and fully French polish hardwood joinery							
General surfaces							
over 300 mm girth	m2	2.55	-	22.85	1.87	24.72	28.43
not exceeding 150 mm girth	m	0.57	-	5.11	0.31	5.42	6.23
150 - 300 mm girth	m	0.95	-	8.51	0.58	9.09	10.45
not exceeding 0.50 m2	each	1.90	-	17.02	0.93	17.95	20.64

Painting and decorating

New work	Unit	Hours C	Hours L	Labour net	Material net	Price net	Price with 15%
French polishing				£	£	£	£
					VAT not included		
Seal, body in and fully French polish hardwood joinery *(continued)*							
Glazed doors and screens in panes							
small - not exceeding 0.10 m2	m2	5.65	-	50.62	1.47	52.09	59.90
medium - 0.10 - 0.50 m2	m2	3.40	-	30.46	1.19	31.65	36.40
large - 0.50 - 1.00 m2	m2	2.27	-	20.34	0.98	21.32	24.52
extra large - over 1.00 m2	m2	1.70	-	15.23	0.69	15.92	18.31
Windows (measured flat overall) in panes							
small - not exceeding 0.10 m2	m2	6.22	-	55.73	1.47	57.20	65.78
medium - 0.10 - 0.50 m2	m2	3.97	-	35.57	1.19	36.76	42.27
large - 0.50 - 1.00 m2	m2	2.84	-	25.45	0.98	26.43	30.39
extra large - over 1.00 m2	m2	2.27	-	20.34	0.69	21.03	24.18
Edges of opening casements	m	0.28	-	2.51	0.10	2.61	3.00
Stain, seal, body in and fully French polish hardwood joinery							
General surfaces							
over 300 mm girth	m2	2.70	-	24.19	2.52	26.71	30.72
not exceeding 150 mm girth	m	0.60	-	5.38	0.42	5.80	6.67
150 - 300 mm girth	m	1.00	-	8.96	0.78	9.74	11.20
not exceeding 0.50 m2	each	2.00	-	17.92	1.26	19.18	22.06
Glazed doors and screens in panes							
small - not exceeding 0.10 m2	m2	6.00	-	53.76	1.98	55.74	64.10
medium - 0.10 - 0.50 m2	m2	3.60	-	32.26	1.61	33.87	38.95
large - 0.50 - 1.00 m2	m2	2.40	-	21.50	1.32	22.82	26.24
extra large - over 1.00 m2	m2	1.80	-	16.13	1.16	17.29	19.88
Windows (measured flat overall) in panes							
small - not exceeding 0.10 m2	m2	6.60	-	59.14	1.98	61.12	70.29
medium - 0.10 - 0.50 m2	m2	4.20	-	37.63	1.61	39.24	45.13
large - 0.50 - 1.00 m2	m2	3.00	-	26.88	1.32	28.20	32.43
extra large - over 1.00 m2	m2	2.40	-	21.50	0.93	22.43	25.79
Edges of opening casements	m	0.30	-	2.69	0.14	2.83	3.25

Painting and decorating

New work	Unit	Hours C	Hours L	Labour net	Material net	Price net	Price with 15%
Sprayed finishes				£	£	£	£
					VAT not included		
Internally							
Two coats of vinyl matt emulsion paint on smooth plaster, surfaces over 300 mm							
walls	m2	0.18	-	1.61	0.61	2.22	2.55
ceilings	m2	0.22	-	1.97	0.61	2.58	2.97
One coat Artex preparation, one coat Artex AX finish, stippled finish, surfaces over 300 mm girth							
brickwork	m2	0.40	-	3.58	2.46	6.04	6.95
blockwork	m2	0.40	-	3.58	2.22	5.80	6.67
smooth plaster	m2	0.35	-	3.14	2.03	5.17	5.95
plasterboards	m2	0.35	-	3.14	2.03	5.17	5.95
cement render	m2	0.36	-	3.23	2.46	5.69	6.54
concrete	m2	0.40	-	3.58	2.22	5.80	6.67
One coat Artex preparation, one coat Artex XL finish, stippled finish, surfaces over 300 mm girth							
brickwork	m2	0.40	-	3.58	3.26	6.84	7.87
blockwork	m2	0.40	-	3.58	3.02	6.60	7.59
smooth plaster	m2	0.35	-	3.14	2.83	5.97	6.87
plasterboards	m2	0.35	-	3.14	2.83	5.97	6.87
cement render	m2	0.36	-	3.23	3.26	6.49	7.46
concrete	m2	0.40	-	3.58	3.02	6.60	7.59
One coat Mosaico XL groundcoat, one coat Mosaico XL glaze finish, surfaces over 300 mm girth							
brickwork	m2	0.28	-	2.51	1.35	3.86	4.44
smooth plaster	m2	0.25	-	2.24	4.93	7.17	8.25
cement render	m2	0.46	-	4.12	1.24	5.36	6.16
concrete	m2	0.28	-	2.51	1.24	3.75	4.31
Two coats Mosaico XL groundcoat, one coat Mosaico XL glaze finish, surfaces over 300 mm girth							
brickwork	m2	0.42	-	3.76	2.00	5.76	6.62
smooth plaster	m2	0.37	-	3.32	1.27	4.59	5.28
cement render	m2	0.69	-	6.18	1.82	8.00	9.20
concrete	m2	0.42	-	3.76	1.82	5.58	6.42
One coat Portabond (brush applied), one coat Portaflek, surfaces over 300 mm girth							
brickwork	m2	0.28	-	2.51	5.68	8.19	9.42
smooth plaster	m2	0.25	-	2.24	5.52	7.76	8.92
cement render	m2	0.46	-	4.12	5.61	9.73	11.19
concrete	m2	0.28	-	2.51	5.61	8.12	9.34
Two coats Portabond (brush applied), one coat Portaflek, surfaces over 300 mm girth							
brickwork	m2	0.40	-	3.58	6.27	9.85	11.33
smooth plaster	m2	0.38	-	3.40	5.94	9.34	10.74
cement render	m2	0.56	-	5.02	6.13	11.15	12.82
concrete	m2	0.40	-	3.58	6.13	9.71	11.17

Painting and decorating

	Unit	Hours C	Hours L	Labour net	Material net	Price net	Price with 15%
Sprayed finishes				£	£	£	£
					VAT not included		

Internally (*continued*)

One coat Portabond (brush applied), one coat
Portatone, surfaces over 300 mm girth

	Unit	Hours C	Hours L	Labour net	Material net	Price net	Price with 15%
brickwork	m2	0.38	-	3.40	2.30	5.70	6.55
smooth plaster	m2	0.35	-	3.14	2.14	5.28	6.07
cement render	m2	0.50	-	4.48	2.23	6.71	7.72
concrete	m2	0.38	-	3.40	2.23	5.63	6.47

Two coats Portabond (brush applied), one coat
Portatone, surfaces over 300 mm girth

	Unit	Hours C	Hours L	Labour net	Material net	Price net	Price with 15%
brickwork	m2	0.40	-	3.58	2.89	6.47	7.44
smooth plaster	m2	0.38	-	3.40	2.56	5.96	6.85
cement render	m2	0.56	-	5.02	2.75	7.77	8.94
concrete	m2	0.40	-	3.58	2.75	6.33	7.28

Externally

One coat Artex Force 8 masonry paint, surfaces
over 300 mm girth

	Unit	Hours C	Hours L	Labour net	Material net	Price net	Price with 15%
brickwork	m2	0.16	-	1.43	0.69	2.12	2.44
cement render	m2	0.16	-	1.43	1.08	2.51	2.89
concrete	m2	0.15	-	1.34	0.69	2.03	2.33

One coat Artex Hyclad finish, surfaces over
300 mm girth

	Unit	Hours C	Hours L	Labour net	Material net	Price net	Price with 15%
brickwork	m2	0.32	-	2.87	1.64	4.51	5.19
cement render	m2	0.32	-	2.87	2.05	4.92	5.66
concrete	m2	0.30	-	2.69	1.64	4.33	4.98

Painting and decorating

New work	Unit	Hours C	Hours L	Labour net	Material net	Price net	Price with 15%
Signwriting				£	£	£	£
					VAT not included		
Writing in one coat of oil paint on painted backgrounds							
Plain letters or numerals							
up to 50 mm high	each	0.14	-	1.25	-	1.25	1.44
per additional 25 mm high	each	0.06	-	0.54	-	0.54	0.62
Plain letters or numerals with shading							
up to 50 mm high	each	0.20	-	1.79	0.01	1.80	2.07
per additional 25 mm high	each	0.08	-	0.72	0.01	0.73	0.84
Plain letters or numerals with outlining							
up to 50 mm high	each	0.28	-	2.51	0.01	2.52	2.90
per additional 25 mm high	each	0.12	-	1.08	0.01	1.09	1.25
Ornamental letters or numerals							
up to 50 mm high	each	0.21	-	1.88	-	1.88	2.16
per additional 25 mm high	each	0.09	-	0.81	-	0.81	0.93
Ornamental letters or numerals with shading							
up to 50 mm high	each	0.29	-	2.60	0.01	2.61	3.00
per additional 25 mm high	each	0.13	-	1.16	0.01	1.17	1.35
Ornamental letters or numerals with outlining							
up to 50 mm high	each	0.42	-	3.76	0.01	3.77	4.34
per additional 25 mm high	each	0.18	-	1.61	0.01	1.62	1.86
Commas, hyphens or stops	each	0.03	-	0.27	-	0.27	0.31
Direction arrows 250 mm long	each	0.40	-	3.58	0.01	3.59	4.13
Writing in two coats of oil paint on painted backgrounds							
Plain letter or numerals							
up to 50 mm high	each	0.21	-	1.88	0.01	1.89	2.17
per additional 25 mm high	each	0.09	-	0.81	0.01	0.82	0.94
Plain letters or numerals with shading							
up to 50 mm high	each	0.29	-	2.60	0.01	2.61	3.00
per additional 25 mm high	each	0.13	-	1.16	0.01	1.17	1.35
Plain letters or numerals with outlining							
up to 50 mm high	each	0.42	-	3.76	0.01	3.77	4.34
per additional 25 mm high	each	0.18	-	1.61	0.01	1.62	1.86
Ornamental letters or numerals							
up to 50 mm high	each	0.32	-	2.87	0.01	2.88	3.31
per additional 25 mm high	each	0.14	-	1.25	0.01	1.26	1.45
Ornamental letters or numerals with shading							
up to 50 mm high	each	0.44	-	3.94	0.01	3.95	4.54
per additional 25 mm high	each	0.19	-	1.70	0.01	1.71	1.97

Painting and decorating

New work	Unit	Hours C	Hours L	Labour net	Material net	Price net	Price with 15%
Signwriting				£	£	£	£
					VAT not included		

Writing in two coats of oil paint on painted backgrounds (*continued*)

	Unit	Hours C	Hours L	Labour net	Material net	Price net	Price with 15%
Ornamental letters or numerals with outlining							
up to 50 mm high	each	0.63	-	5.64	0.01	5.65	6.50
per additional 25 mm high	each	0.27	-	2.42	0.01	2.43	2.79
Commas, hyphens or stops	each	0.05	-	0.45	0.01	0.46	0.53
Direction arrows 250 mm long	each	0.60	-	5.38	0.02	5.40	6.21

Writing in two coats of oil paint in reverse on glass and one coat of protective varnish

	Unit	Hours C	Hours L	Labour net	Material net	Price net	Price with 15%
Plain letters or numerals							
up to 50 mm high	each	0.41	-	3.67	0.02	3.69	4.24
per additional 25 mm high	each	0.17	-	1.52	0.02	1.54	1.77
Plain letters or numerals with shading							
up to 50 mm high	each	0.57	-	5.11	0.02	5.13	5.90
per additional 25 mm high	each	0.24	-	2.15	0.02	2.17	2.50
Plain letters or numerals with outlining							
up to 50 mm high	each	0.81	-	7.26	0.02	7.28	8.37
per additional 25 mm high	each	0.35	-	3.14	0.02	3.16	3.63
Ornamental letters or numerals							
up to 50 mm high	each	0.61	-	5.47	0.02	5.49	6.31
per additional 25 mm high	each	0.26	-	2.33	0.02	2.35	2.70
Ornamental letters or numerals with shading							
up to 50 mm high	each	0.85	-	7.62	0.02	7.64	8.79
per additional 25 mm high	each	0.37	-	3.32	0.02	3.34	3.84
Ornamental letters or numerals with outlining							
up to 50 mm high	each	1.22	-	10.93	0.02	10.95	12.59
per additional 25 mm high	each	0.52	-	4.66	0.02	4.68	5.38
Commas, hyphens or stops	each	0.09	-	0.81	0.02	0.83	0.95
Direction arrows 250 mm long	each	1.16	-	10.39	0.03	10.42	11.98

Writing in three coats of oil paint in reverse on glass and one coat of protective varnish

	Unit	Hours C	Hours L	Labour net	Material net	Price net	Price with 15%
Plain letters or numerals							
up to 50 mm high	each	0.48	-	4.30	0.02	4.32	4.97
per additional 25 mm high	each	0.20	-	1.79	0.02	1.81	2.08
Plain letters or numerals with shading							
up to 50 mm high	each	0.67	-	6.00	0.03	6.03	6.93
per additional 25 mm high	each	0.29	-	2.60	0.03	2.63	3.02

Painting and decorating

New work	Unit	Hours C	Hours L	Labour net	Material net	Price net	Price with 15%
Signwriting				£	£	£	£
					VAT not included		

Writing in three coats of oil paint in reverse on glass and one coat of protective varnish
(*continued*)

New work	Unit	Hours C	Hours L	Labour net	Material net	Price net	Price with 15%
Plain letters or numerals with outlining							
up to 50 mm high	each	0.95	-	8.51	0.03	8.54	9.82
per additional 25 mm high	each	0.41	-	3.67	0.03	3.70	4.25
Ornamental letters or numerals							
up to 50 mm high	each	0.71	-	6.36	0.02	6.38	7.34
per additional 25 mm high	each	0.31	-	2.78	0.02	2.80	3.22
Ornamental letters or numerals with shading							
up to 50 mm high	each	1.00	-	8.96	0.03	8.99	10.34
per additional 25 mm high	each	0.43	-	3.85	0.03	3.88	4.46
Ornamental letters or numerals with outlining							
up to 50 mm high	each	1.43	-	12.81	0.03	12.84	14.77
per additional 25 mm high	each	0.61	-	5.47	0.03	5.50	6.33
Commas, hyphens or stops	each	0.10	-	0.90	0.02	0.92	1.06
Direction arrows 250 mm long	each	1.36	-	12.19	0.04	12.23	14.06

Drainage

Preamble

"Labour net" figures include allowances for all costs incidental to the employment of labour. Figures for pipework generally are based on the labour costs of three bricklayers working with two labourers as in "Brickwork and blockwork"; for cast iron pipes figures are based on the costs of an advanced plumber working with an apprentice in the third year of training.

"Plant net" figures include for all costs of plant including drivers and operators where applicable.

"Materials net" figures include for all costs of materials including an allowance for waste except where specifically stated.

"Price net" figures are the totals of the "Labour net", "Plant net" and "Materials net" figures. Prices are for a builder employing his own labour; according to the amount and nature of the work involved, it may well be possible to secure more advantageous prices from specialist sub-contractors.

Excavation prices are for work in firm soil. For other soils the following adjustments should be made:

    clay - add 25%
    hard gravel - add 50%
    chalk - add 100 to 150%
    rock - add 300 to 400%

Where excavation items are described as "including disposal of surplus on site", this has been allowed for on the basis of disposal of surplus excavated material in spoil heaps on site average 50 metres from the excavations.

Where excavation items are described as "including removal of surplus from site" this has been allowed for on the basis of removal of surplus excavated material to a tip average 15 kilometres from site.

Figures for concrete are based on the use of a hired 7/5 mixer.

Figures for formwork are based on the assumptions that timber is used and that each use of material requires the full labour content; if the work is repetitive, permitting re-use of made-up sections, some reduction of the figures could be made.

Although not in accordance with the Standard Method of Measurement, figures per square metre have been included for concrete benchings, from which figures for any particular enumerated items can be calculated.

Drainage

New work

Basic prices for materials		**£**
Granular bedding - 10 mm	m3	19.75
	tonne	13.16
Aggregates		
40 mm	m3	19.75
	tonne	13.16
20 mm	m3	19.91
	tonne	13.27
10 mm	m3	20.07
	tonne	13.38
Sand	m3	17.96
	tonne	11.22
Portland cement	tonne	89.04
Vitrified clay plain ended "Supersleve" pipes		
100 mm pipes	m	2.43
100 mm polypropylene couplings	each	1.68
150 mm pipes	m	5.56
150 mm polypropylene couplings	each	3.48
Vitrified clay socketed and flexible jointed "Hepseal" pipes		
100 mm	m	7.23
150 mm	m	9.38
225 mm	m	18.15
Vitrified clay socketed unjointed pipes		
100 mm	m	4.51
150 mm	m	7.78
225 mm	m	15.48
Flexible jointed standard concrete pipes		
150 mm	m	7.50
225 mm	m	8.75
300 mm	m	12.53
Standard concrete pipes with ogee joints		
150 mm	m	3.74
225 mm	m	4.78
300 mm	m	7.19
Timesaver cast iron pipes		
100 mm 3000 mm lengths	m	20.00
100 mm couplings	each	12.80
150 mm 3000 mm lengths	m	37.70
150 mm couplings	each	15.50
PVC-U pipes		
110 mm 3000 mm lengths	m	5.32
160 mm 3000 mm lengths	m	11.29

Prices actually to be paid for materials must be checked against the above basic prices and adjustments made as necessary.

Drainage

New work

Basic prices for materials £

Clayware field drain pipes
75 mm	m	1.04
100 mm	m	1.87
150 mm	m	3.99
225 mm	m	10.50

Flexible jointed vitrified clay "Hepline" pipes
100 mm	m	4.64
150 mm	m	8.44
225 mm	m	15.52

Concrete porous pipes
150 mm	m	3.69
225 mm	m	4.32

Plain ended PVC-U perforated pipes
110 mm pipes	m	5.62
110 mm double socket couplers	each	1.69
160 mm pipes	m	10.59
160 mm double socket couplers	each	3.09

PVC-U perforated flexible corrugated pipes
80 mm	m	1.51
100 mm	m	2.15

Prices actually to be paid for materials must be checked against the above basic prices and adjustments made as necessary.

Drainage

New work	Unit	Hours C	Hours L	Labour net	Plant net	Material net	Price net	Price with 15%

Pipe trenches
£ £ £ £ £

Excavation by machine - including disposal of surplus on site average 50 m from excavations

Excavate trenches 450 mm wide; grade bottom; fill in and compact and dispose of surplus on site - earthwork support not included - average depth

	Unit	Hours C	Hours L	Labour net	Plant net	Material net	Price net	Price with 15%
0.50 m	m	-	0.16	1.08	2.38	-	3.46	3.98
0.75 m	m	-	0.25	1.69	3.80	-	5.49	6.31
1.00 m	m	-	0.35	2.36	5.43	-	7.79	8.96
1.25 m	m	-	0.44	2.97	6.79	-	9.76	11.22
1.50 m	m	-	0.54	3.64	8.41	-	12.05	13.86

Excavate trenches 600 mm wide; grade bottom; fill in and compact and dispose of surplus on site - earthwork support not included - average depth

	Unit	Hours C	Hours L	Labour net	Plant net	Material net	Price net	Price with 15%
1.00 m	m	-	0.46	3.10	7.22	-	10.32	11.87
1.25 m	m	-	0.55	3.71	8.79	-	12.50	14.38
1.50 m	m	-	0.71	4.79	11.02	-	15.81	18.18
1.75 m	m	-	0.84	5.66	13.04	-	18.70	21.50
2.00 m	m	-	0.94	6.34	14.44	-	20.78	23.90

Excavate trenches 750 mm wide; grade bottom; fill in and compact and dispose of surplus on site - earthwork support not included - average depth

	Unit	Hours C	Hours L	Labour net	Plant net	Material net	Price net	Price with 15%
1.50 m	m	-	0.90	6.07	13.87	-	19.94	22.93
1.75 m	m	-	1.09	7.35	16.89	-	24.24	27.88
2.00 m	m	-	1.21	8.16	18.68	-	26.84	30.87
2.25 m	m	-	1.43	9.64	22.07	-	31.71	36.47
2.50 m	m	-	1.60	10.78	24.72	-	35.50	40.83
2.75 m	m	-	1.74	11.73	26.89	-	38.62	44.41
3.00 m	m	-	1.92	12.94	29.71	-	42.65	49.05

Excavate trenches 900 mm wide; grade bottom; fill in and compact and dispose of surplus on site - earthwork support not included - average depth

	Unit	Hours C	Hours L	Labour net	Plant net	Material net	Price net	Price with 15%
1.50 m	m	-	1.11	7.48	17.28	-	24.76	28.47
1.75 m	m	-	1.29	8.69	19.99	-	28.68	32.98
2.00 m	m	-	1.48	9.98	22.93	-	32.91	37.85
2.25 m	m	-	1.75	11.80	27.12	-	38.92	44.76
2.50 m	m	-	1.97	13.28	30.56	-	43.84	50.42
2.75 m	m	-	2.13	14.36	32.95	-	47.31	54.41
3.00 m	m	-	2.35	15.84	36.40	-	52.24	60.08

Drainage

New work	Unit	Hours C	Hours L	Labour net	Plant net	Material net	Price net	Price with 15%
Pipe trenches				£	£	£	£	£
					VAT not included			

Excavation by machine - including removal of surplus from site to tip average 15 km from site

Excavate trenches 450 mm wide; grade bottom; fill in and compact and remove surplus from site - earthwork support not included - average depth

0.50 m	m	-	0.16	1.08	5.37	-	6.45	7.42
0.75 m	m	-	0.25	1.69	10.69	-	12.38	14.24
1.00 m	m	-	0.35	2.36	10.92	-	13.28	15.27
1.25 m	m	-	0.44	2.97	13.57	-	16.54	19.02
1.50 m	m	-	0.54	3.64	16.89	-	20.53	23.61

Excavate trenches 600 mm wide; grade bottom; fill in and compact and remove surplus from site - earthwork support not included - average depth

1.00 m	m	-	0.46	3.10	14.30	-	17.40	20.01
1.25 m	m	-	0.55	3.71	18.17	-	21.88	25.16
1.50 m	m	-	0.71	4.79	21.99	-	26.78	30.80
1.75 m	m	-	0.84	5.66	25.61	-	31.27	35.96
2.00 m	m	-	0.94	6.34	29.11	-	35.45	40.77

Excavate trenches 750 mm wide; grade bottom; fill in and compact and remove surplus from site - earthwork support not included - average depth

1.50 m	m	-	0.90	6.07	27.84	-	33.91	39.00
1.75 m	m	-	1.09	7.35	36.15	-	43.50	50.02
2.00 m	m	-	1.21	8.16	37.24	-	45.40	52.21
2.25 m	m	-	1.43	9.64	42.42	-	52.06	59.87
2.50 m	m	-	1.60	10.78	47.37	-	58.15	66.87
2.75 m	m	-	1.74	11.73	51.83	-	63.56	73.09
3.00 m	m	-	1.92	12.94	57.15	-	70.09	80.60

Excavate trenches 900 mm wide; grade bottom; fill in and compact and remove surplus from site - earthwork support not included - average depth

1.50 m	m	-	1.11	7.48	33.74	-	41.22	47.40
1.75 m	m	-	1.29	8.69	39.45	-	48.14	55.36
2.00 m	m	-	1.48	9.98	44.88	-	54.86	63.09
2.25 m	m	-	1.75	11.80	52.06	-	63.86	73.44
2.50 m	m	-	1.97	13.28	58.00	-	71.28	81.97
2.75 m	m	-	2.13	14.36	63.18	-	77.54	89.17
3.00 m	m	-	2.35	15.84	69.12	-	84.96	97.70

Drainage

New work	Unit	Hours C	Hours L	Labour net	Plant net	Material net	Price net	Price with 15%
Pipe trenches				£	£	£	£	£
					VAT not included			

Excavation by hand - including disposal of surplus on site average 50 m from excavations

Excavate trenches 450 mm wide; grade bottom; fill in and compact and dispose of surplus on site - earthwork support not included - average depth

0.50 m	m	-	1.20	8.09	-	-	8.09	9.30
0.75 m	m	-	1.85	12.47	-	-	12.47	14.34
1.00 m	m	-	2.35	15.84	-	-	15.84	18.22
1.25 m	m	-	3.65	24.60	-	-	24.60	28.29
1.50 m	m	-	4.40	29.66	-	-	29.66	34.11

Excavate trenches 600 mm wide; grade bottom; fill in and compact and dispose of surplus on site - earthwork support not included - average depth

1.00 m	m	-	3.15	21.23	-	-	21.23	24.41
1.25 m	m	-	4.85	32.69	-	-	32.69	37.59
1.50 m	m	-	5.85	39.43	-	-	39.43	45.34
1.75 m	m	-	6.80	45.83	-	-	45.83	52.70
2.00 m	m	-	7.75	52.23	-	-	52.23	60.06

Excavate trenches 750 mm wide; grade bottom; fill in and compact and dispose of surplus on site - earthwork support not included - average depth

1.50 m	m	-	7.30	49.20	-	-	49.20	56.58
1.75 m	m	-	8.50	57.29	-	-	57.29	65.88
2.00 m	m	-	9.70	65.38	-	-	65.38	75.19
2.25 m	m	-	13.15	88.63	-	-	88.63	101.92
2.50 m	m	-	14.60	98.40	-	-	98.40	113.16
2.75 m	m	-	16.00	107.84	-	-	107.84	124.02
3.00 m	m	-	17.50	117.95	-	-	117.95	135.64

Excavate trenches 900 mm wide; grade bottom; fill in and compact and dispose of surplus on site - earthwork support not included - average depth

1.50 m	m	-	8.80	59.31	-	-	59.31	68.21
1.75 m	m	-	10.25	69.09	-	-	69.09	79.45
2.00 m	m	-	11.65	78.52	-	-	78.52	90.30
2.25 m	m	-	15.80	106.49	-	-	106.49	122.46
2.50 m	m	-	17.50	117.95	-	-	117.95	135.64
2.75 m	m	-	19.25	129.75	-	-	129.75	149.21
3.00 m	m	-	21.00	141.54	-	-	141.54	162.77

Drainage

New work	Unit	Hours C	Hours L	Labour net	Plant net	Material net	Price net	Price with 15%
Pipe trenches				£	£	£	£	£
					VAT not included			

Excavation by hand - including removal of surplus from site to tip average 15 km from site

Excavate trenches 450 mm wide; grade bottom; fill in and compact and remove surplus from site - earthwork support not included - average depth

0.50 m	m	-	1.25	8.43	4.19	-	12.62	14.51
0.75 m	m	-	1.85	12.47	6.08	-	18.55	21.33
1.00 m	m	-	2.40	16.18	8.27	-	24.45	28.12
1.25 m	m	-	3.70	24.94	10.36	-	35.30	40.59
1.50 m	m	-	4.45	29.99	12.65	-	42.64	49.04

Excavate trenches 600 mm wide; grade bottom; fill in and compact and remove surplus from site - earthwork support not included - average depth

1.00 m	m	-	3.20	21.57	10.86	-	32.43	37.29
1.25 m	m	-	4.90	33.03	13.95	-	46.98	54.03
1.50 m	m	-	5.90	39.77	16.54	-	56.31	64.76
1.75 m	m	-	6.85	46.17	19.13	-	65.30	75.09
2.00 m	m	-	7.80	52.57	22.22	-	74.79	86.01

Excavate trenches 750 mm wide; grade bottom; fill in and compact and remove surplus from site - earthwork support not included - average depth

1.50 m	m	-	7.35	49.54	20.92	-	70.46	81.03
1.75 m	m	-	8.55	57.63	24.41	-	82.04	94.35
2.00 m	m	-	9.75	65.72	27.90	-	93.62	107.66
2.25 m	m	-	13.20	88.97	30.69	-	119.66	137.61
2.50 m	m	-	14.65	98.74	34.17	-	132.91	152.85
2.75 m	m	-	16.50	111.21	37.66	-	148.87	171.20
3.00 m	m	-	17.55	118.29	41.15	-	159.44	183.36

Excavate trenches 900 mm wide; grade bottom; fill in and compact and remove surplus from site - earthwork support not included - average depth

1.50 m	m	-	8.85	59.65	24.86	-	84.51	97.19
1.75 m	m	-	10.25	69.09	28.99	-	98.08	112.79
2.00 m	m	-	11.70	78.86	32.88	-	111.74	128.50
2.25 m	m	-	15.85	106.83	37.26	-	144.09	165.70
2.50 m	m	-	17.55	118.29	41.15	-	159.44	183.36
2.75 m	m	-	19.30	130.08	45.53	-	175.61	201.95
3.00 m	m	-	21.50	144.91	49.42	-	194.33	223.48

Drainage

New work	Unit	Hours C	Hours L	Labour net	Plant net	Material net	Price net	Price with 15%
Pipe trenches				£	£	£	£	£
					VAT not included			

Breaking up by machine - excluding reinstatement

Break up surface concrete; for trenches 600 mm wide; average thickness

100 mm	m	-	0.13	0.88	0.53	-	1.41	1.62
150 mm	m	-	0.17	1.15	0.69	-	1.84	2.12
200 mm	m	-	0.25	1.69	1.02	-	2.71	3.12

Break up reinforced surface concrete; for trenches 600 mm wide; average thickness

100 mm	m	-	0.18	1.21	0.73	-	1.94	2.23
150 mm	m	-	0.24	1.62	0.98	-	2.60	2.99
200 mm	m	-	0.35	2.36	1.42	-	3.78	4.35

Break up tarmacadam paving; for trenches 600 mm wide; average thickness

100 mm	m	-	0.06	0.40	0.24	-	0.64	0.74
150 mm	m	-	0.10	0.67	0.41	-	1.08	1.24
200 mm	m	-	0.13	0.88	0.53	-	1.41	1.62

Granular beds; side filling and coverings

50 mm beds

450 mm wide	m	-	0.10	0.67	-	0.44	1.11	1.28
525 mm wide	m	-	0.11	0.74	-	0.49	1.23	1.41
600 mm wide	m	-	0.13	0.88	-	0.57	1.45	1.67
750 mm wide	m	-	0.16	1.08	-	0.72	1.80	2.07

100 mm beds

450 mm wide	m	-	0.16	1.08	-	0.86	1.94	2.23
525 mm wide	m	-	0.18	1.21	-	1.01	2.22	2.55
600 mm wide	m	-	0.20	1.35	-	1.14	2.49	2.86
750 mm wide	m	-	0.26	1.75	-	1.43	3.18	3.66

100 mm beds and side filling to half height of pipes

450 mm wide to 100 mm pipes	m	-	0.22	1.48	-	1.20	2.68	3.08
525 mm wide to 150 mm pipes	m	-	0.29	1.95	-	1.69	3.64	4.19
600 mm wide to 225 mm pipes	m	-	0.36	2.43	-	2.15	4.58	5.27
750 mm wide to 300 mm pipes	m	-	0.49	3.30	-	3.01	6.31	7.26

150 mm beds and side filling to half height of pipes

450 mm wide to 100 mm pipes	m	-	0.28	1.89	-	1.77	3.66	4.21
525 mm wide to 150 mm pipes	m	-	0.34	2.29	-	2.21	4.50	5.17
600 mm wide to 225 mm pipes	m	-	0.42	2.83	-	2.72	5.55	6.38
750 mm wide to 300 mm pipes	m	-	0.57	3.84	-	3.73	7.57	8.71

Drainage

New work	Unit	Hours C	Hours L	Labour net	Plant net	Material net	Price net	Price with 15%
Pipe trenches				£	£	£	£	£
					VAT not included			

Granular beds; side filling and coverings (*continued*)

100 mm beds and side filling to full height of pipes

450 mm wide to 100 mm pipes	m	-	0.27	1.82	-	1.81	3.63	4.17
525 mm wide to 150 mm pipes	m	-	0.37	2.49	-	2.40	4.89	5.62
600 mm wide to 225 mm pipes	m	-	0.47	3.17	-	3.14	6.31	7.26
750 mm wide to 300 mm pipes	m	-	0.66	4.45	-	4.59	9.04	10.40

150 mm beds and side filling to full heights of pipes

450 mm wide to 100 mm pipes	m	-	0.34	2.29	-	2.23	4.52	5.20
525 mm wide to 150 mm pipes	m	-	0.43	2.90	-	2.89	5.79	6.66
600 mm wide to 225 mm pipes	m	-	0.55	3.71	-	3.71	7.42	8.53
750 mm wide to 300 mm pipes	m	-	0.72	4.85	-	5.31	10.16	11.68

Beds and coverings to 100 mm pipes

450 x 350 mm	m	-	0.41	2.76	-	2.91	5.67	6.52
450 x 450 mm	m	-	0.52	3.50	-	3.81	7.31	8.41

Beds and coverings to 150 mm pipes

525 x 400 mm	m	-	0.52	3.50	-	3.62	7.12	8.19
525 x 500 mm	m	-	0.65	4.38	-	4.63	9.01	10.36

Beds and coverings to 225 mm pipes

600 x 475 mm	m	-	0.64	4.31	-	4.66	8.97	10.32
600 x 575 mm	m	-	0.79	5.32	-	5.81	11.13	12.80

Beds and coverings to 300 mm pipes

750 x 550 mm	m	-	0.89	6.00	-	6.53	12.53	14.41
750 x 650 mm	m	-	1.10	7.41	-	7.96	15.37	17.68

Concrete 1:3:6 beds; benchings and coverings

100 mm beds

450 mm wide	m	-	0.18	1.21	-	2.50	3.71	4.27
525 mm wide	m	-	0.20	1.35	-	2.95	4.30	4.95
600 mm wide	m	-	0.22	1.48	-	3.34	4.82	5.54
750 mm wide	m	-	0.25	1.69	-	4.17	5.86	6.74

150 mm beds

450 mm wide	m	-	0.25	1.69	-	3.78	5.47	6.29
525 mm wide	m	-	0.26	1.75	-	4.40	6.15	7.07
600 mm wide	m	-	0.30	2.02	-	5.01	7.03	8.08
750 mm wide	m	-	0.37	2.49	-	6.29	8.78	10.10

100 mm beds and benchings to full height of pipes

450 mm wide to 100 mm pipes	m	-	0.24	1.62	-	5.29	6.91	7.95
525 mm wide to 150 mm pipes	m	-	0.40	2.70	-	7.01	9.71	11.17
600 mm wide to 225 mm pipes	m	-	0.46	3.10	-	9.18	12.28	14.12
750 mm wide to 300 mm pipes	m	-	0.60	4.04	-	13.41	17.45	20.07

Drainage

New work	Unit	Hours C	Hours L	Labour net	Plant net	Material net	Price net	Price with 15%
Pipe trenches				£	£	£	£	£
					VAT not included			

Concrete 1:3:6 beds; benchings and coverings (*continued*)

150 mm beds and benchings to full height of pipes

450 mm wide to 100 mm pipes	m	-	0.26	1.75	-	6.51	8.26	9.50
525 mm wide to 150 mm pipes	m	-	0.44	2.97	-	8.46	11.43	13.14
600 mm wide to 225 mm pipes	m	-	0.51	3.44	-	10.85	14.29	16.43
750 mm wide to 300 mm pipes	m	-	0.66	4.45	-	15.53	19.98	22.98

Beds and coverings to 100 mm pipes

450 x 350 mm	m	-	0.50	3.37	-	8.51	11.88	13.66
450 x 450 mm	m	-	0.55	3.71	-	11.13	14.84	17.07

Beds and coverings to 150 mm pipes

525 x 400 mm	m	-	0.50	3.37	-	10.57	13.94	16.03
525 x 500 mm	m	-	0.65	4.38	-	13.52	17.90	20.59

Beds and coverings to 225 mm pipes

600 x 475 mm	m	-	0.65	4.38	-	13.63	18.01	20.71
600 x 575 mm	m	-	0.95	6.40	-	16.97	23.37	26.88

Beds and coverings to 300 mm pipes

750 x 550 mm	m	-	0.80	5.39	-	19.09	24.48	28.15
750 x 650 mm	m	-	1.00	6.74	-	23.26	30.00	34.50

Concrete 1:2:4 beds; benchings and coverings

100 mm beds

450 mm wide	m	-	0.18	1.21	-	2.65	3.86	4.44
525 mm wide	m	-	0.20	1.35	-	3.13	4.48	5.15
600 mm wide	m	-	0.22	1.48	-	3.54	5.02	5.77
750 mm wide	m	-	0.25	1.69	-	4.42	6.11	7.03

150 mm beds

450 mm wide	m	-	0.25	1.69	-	4.01	5.70	6.55
525 mm wide	m	-	0.26	1.75	-	4.66	6.41	7.37
600 mm wide	m	-	0.30	2.02	-	5.31	7.33	8.43
750 mm wide	m	-	0.37	2.49	-	6.67	9.16	10.53

100 mm beds and benchings to full height of pipes

450 mm wide to 100 mm pipes	m	-	0.24	1.62	-	5.60	7.22	8.30
525 mm wide to 150 mm pipes	m	-	0.44	2.97	-	7.43	10.40	11.96
600 mm wide to 225 mm pipes	m	-	0.46	3.10	-	9.73	12.83	14.75
750 mm wide to 300 mm pipes	m	c	0.60	4.04	-	14.22	18.26	21.00

150 mm beds and benchings to full height of pipes

450 mm wide to 100 mm pipes	m	-	0.26	1.75	-	6.90	8.65	9.95
525 mm wide to 150 mm pipes	m	-	0.42	2.83	-	8.97	11.80	13.57
600 mm wide to 225 mm pipes	m	-	0.51	3.44	-	11.50	14.94	17.18
750 mm wide to 300 mm pipes	m	-	0.66	4.45	-	16.46	20.91	24.05

Drainage

New work	Unit	Hours C	Hours L	Labour net	Plant net	Material net	Price net	Price with 15%
Pipe trenches				£	£	£	£	£
					VAT not included			
Concrete 1:2:4 beds; benchings and coverings (*continued*)								
Beds and coverings to 100 mm pipes								
450 x 350 mm	m	-	0.50	3.37	-	9.03	12.40	14.26
450 x 450 mm	m	-	0.55	3.71	-	11.80	15.51	17.84
Beds and coverings to 150 mm pipes								
525 x 400 mm	m	-	0.50	3.37	-	11.21	14.58	16.77
525 x 500 mm	m	-	0.65	4.38	-	14.33	18.71	21.52
Beds and coverings to 225 mm pipes								
600 x 475 mm	m	-	0.65	4.38	-	14.45	18.83	21.65
600 x 575 mm	m	-	0.95	6.40	-	17.99	24.39	28.05
Beds and coverings to 300 mm pipes								
750 x 550 mm	m	-	0.80	5.39	-	20.23	25.62	29.46
750 x 650 mm	m	-	1.00	6.74	-	24.66	31.40	36.11

Drainage

New work	Unit	Hours C	Hours L	Labour net	Plant net	Material net	Price net	Price with 15%
Pipework				£	£	£	£	£
						VAT not included		

Flexible jointed vitrified clay pipes and fittings to BS 65

100 mm plain ended "Supersleve" pipes jointed with polypropylene sleeve couplings and laid in trench bottom

	Unit	Hours C	Hours L	Labour net	Plant net	Material net	Price net	Price with 15%
in runs over 3.00 m long	m	0.13	0.06	1.56	-	3.24	4.80	5.52
in runs not exceeding 3.00 m long	m	0.29	0.05	2.94	-	3.24	6.18	7.11
Extra for								
bends	each	0.11	-	0.99	-	3.50	4.49	5.16
junctions	each	0.15	-	1.34	-	7.55	8.89	10.22

150 mm plain ended "Supersleve" pipes jointed with polypropylene sleeve couplings and laid in trench bottom

	Unit	Hours C	Hours L	Labour net	Plant net	Material net	Price net	Price with 15%
in runs over 3.00 m long	m	0.16	0.08	1.97	-	7.29	9.26	10.65
in runs not exceeding 3.00 m long	m	0.34	0.08	3.59	-	7.29	10.88	12.51
Extra for								
bends	each	0.10	-	0.90	-	8.14	9.04	10.40
junctions	each	0.18	-	1.61	-	10.69	12.30	14.15

100 mm socketed "Hepseal" pipes jointed with rubber sealing rings and laid in trench bottom

	Unit	Hours C	Hours L	Labour net	Plant net	Material net	Price net	Price with 15%
in runs over 3.00 m long	m	0.15	0.05	1.68	-	7.89	9.57	11.01
in runs not exceeding 3.00 m long	m	0.30	0.05	3.03	-	7.89	10.92	12.56
Extra for								
bends	each	0.18	-	1.61	-	10.85	12.46	14.33
junctions	each	0.15	-	1.34	-	15.07	16.41	18.87

150 mm socketed "Hepseal" pipes jointed with rubber sealing rings and laid in trench bottom

	Unit	Hours C	Hours L	Labour net	Plant net	Material net	Price net	Price with 15%
in runs over 3.00 m long	m	0.18	0.08	2.15	-	10.24	12.39	14.25
in runs not exceeding 3.00 m long	m	0.36	0.08	3.77	-	10.24	14.01	16.11
Extra for								
bends	each	0.20	-	1.79	-	17.90	19.69	22.64
junctions	each	0.18	-	1.61	-	23.38	24.99	28.74

225 mm socketed "Hepseal" pipes jointed with rubber sealing rings and laid in trench bottom

	Unit	Hours C	Hours L	Labour net	Plant net	Material net	Price net	Price with 15%
in runs over 3.00 m long	m	0.21	0.09	2.49	-	19.82	22.31	25.66
in runs not exceeding 3.00 m long	m	0.38	0.09	4.01	-	19.82	23.83	27.40
Extra for								
bends	each	0.13	-	1.16	-	37.46	38.62	44.41
junctions	each	0.22	-	1.97	-	56.33	58.30	67.05

Drainage

Pipework

£   £   £   £   £

VAT not included

Flexible jointed vitrified clay pipes and fittings to BS 65 (continued)

	Unit	Hours C	Hours L	Labour net	Plant net	Material net	Price net	Price with 15%
Square one-piece trapped access gullies with 150 x 110 mm vertical back inlet; rodding eye and plastics stopper and providing 100 mm outlet including polypropylene coupling to pipe and 150 x 150 mm coated cast iron grating	each	0.75	0.34	9.01	-	23.34	32.35	37.20
225 mm internal diameter x 600 mm deep three-piece inspection chambers for 100 mm pipes; comprising straight through base; 300 mm chamber raising piece with integral alloy plate and frame, and polypropylene sleeve coupling								
including two polypropylene couplings to 100 mm pipes	each	1.00	0.67	13.48	-	61.94	75.42	86.73
Extra for								
single junction bases and one additional polypropylene sleeve coupling to 100 mm pipe	each	0.15	0.10	2.01	-	8.72	10.73	12.34
double junction bases and two additional polypropylene sleeve couplings to 100 mm pipes	each	0.30	0.20	4.04	-	17.97	22.01	25.31
300 or 450 mm chamber raising pieces	each	0.35	0.25	4.83	-	13.70	18.53	21.31

Cement jointed vitrified clay pipes and fittings to BS 65

	Unit	Hours C	Hours L	Labour net	Plant net	Material net	Price net	Price with 15%
100 mm pipes jointed with tarred gaskin and cement mortar 1:3 and laid in trench bottom								
in runs over 3.00 m long	m	0.31	0.05	3.12	-	5.06	8.18	9.41
in runs not exceeding 3.00 m long	m	0.49	0.05	4.73	-	5.06	9.79	11.26
Extra for								
bends	each	0.19	-	1.70	-	3.72	5.42	6.23
junctions	each	0.19	-	1.70	-	7.61	9.31	10.71
150 mm pipes jointed with tarred gaskin and cement mortar 1:3 and laid in trench bottom								
in runs over 3.00 m long	m	0.39	0.05	3.83	-	8.67	12.50	14.38
in runs not exceeding 3.00 m long	m	0.63	0.05	5.98	-	8.67	14.65	16.85
Extra for								
bends	each	0.22	-	1.97	-	6.36	8.33	9.58
junctions	each	0.23	-	2.06	-	12.51	14.57	16.76

Drainage

New work	Unit	Hours C	Hours L	Labour net	Plant net	Material net	Price net	Price with 15%
Pipework				£	£	£	£	£
					VAT not included			

Cement jointed vitrified clay pipes and fittings to BS 65 (continued)

225 mm pipes jointed with tarred gaskin and cement mortar 1:3 and laid in trench bottom

	Unit	Hours C	Hours L	Labour net	Plant net	Material net	Price net	Price with 15%
in runs over 3.00 mm long	m	0.49	-	4.39	-	17.28	21.67	24.92
in runs not exceeding 3.00 m long	m	0.73	0.07	7.01	-	17.28	24.29	27.93
Extra for								
bends	each	0.29	-	2.60	-	19.76	22.36	25.71

Square one-piece "P" trap gullies with 100 mm outlet including cement joint to pipe; bedding in concrete 1:3:6 and providing coated cast iron grating

	Unit	Hours C	Hours L	Labour net	Plant net	Material net	Price net	Price with 15%
150 x 150 mm	each	0.60	0.10	6.05	-	24.75	30.80	35.42
Extra for								
horizontal inlets	each	-	-	-	-	14.03	14.03	16.13
vertical inlets	each	-	-	-	-	14.03	14.03	16.13
cement jointed raising pieces 150 x 150 mm	each	0.20	-	1.79	-	7.17	8.96	10.30
cement jointed raising pieces 225 x 225 mm	each	0.20	-	1.79	-	15.57	17.36	19.96
completely surrounding gullies with concrete 1:3:6 150 mm thick	each	-	0.30	2.02	-	0.83	2.85	3.28

225 x 225 mm square one-piece "P" trap gullies 585 mm deep with rubber inspection eye and 100 mm outlet including cement joint to pipe; bedding in concrete 1:3:6 and providing coated

	Unit	Hours C	Hours L	Labour net	Plant net	Material net	Price net	Price with 15%
cast iron grating	each	0.70	-	6.27	-	58.14	64.41	74.07
Extra for								
galvanised mud buckets	each	0.10	-	0.90	-	12.42	13.32	15.32
completely surrounding gullies with concrete 1:3:6 150 mm thick	each	-	0.65	4.38	-	2.07	6.45	7.42

285 x 285 mm square one-piece "P" trap gullies 585 mm deep with rubber inspection eye and 100 mm outlet including cement joint to pipe; bedding in concrete 1:3:6 and providing coated

	Unit	Hours C	Hours L	Labour net	Plant net	Material net	Price net	Price with 15%
cast iron grating	each	0.80	-	7.17	-	85.52	92.69	106.59
Extra for								
galvanised mud buckets	each	0.10	-	0.90	-	23.93	24.83	28.55
completely surrounding gullies with concrete 1:3:6 150 mm thick	each	-	0.30	2.02	-	4.82	6.84	7.87

Drainage

New work	Unit	Hours C	Hours L	Labour net	Plant net	Material net	Price net	Price with 15%
Pipework				£	£	£	£	£
						VAT not included		

Flexible jointed standard concrete pipes and fittings to BS 5911

150 mm socketed pipes jointed with rolling rubber rings and laid in trench bottom

in runs over 3.00 m long	m	0.27	0.12	3.23	-	8.27	11.50	13.23
in runs not exceeding 3.00 m long	m	0.35	0.23	4.69	-	8.27	12.96	14.90

Extra for

bends	each	0.14	0.06	1.65	-	46.61	48.26	55.50
junctions	each	0.40	0.06	3.98	-	55.32	59.30	68.19

225 mm socketed pipes jointed with rolling rubber rings and laid in trench bottom

in runs over 3.00 m long	m	0.42	0.14	4.70	-	9.65	14.35	16.50
in runs not exceeding 3.00 m long	m	0.55	0.20	6.28	-	9.65	15.93	18.32

Extra for

bends	each	0.20	0.08	2.33	-	58.14	60.47	69.54
junctions	each	0.50	0.08	5.02	-	61.09	66.11	76.03

300 mm socketed pipes jointed with rolling rubber rings and laid in trench bottom

in runs over 3.00 m long	m	0.54	0.23	6.39	-	13.81	20.20	23.23
in runs not exceeding 3.00 m long	m	0.80	0.28	9.06	-	13.81	22.87	26.30

Extra for

bends	each	0.25	0.11	2.98	-	69.74	72.72	83.63
junctions	each	0.55	0.11	5.67	-	72.14	77.81	89.48

Cement jointed standard concrete pipes and fittings to BS 5911

150 mm ogee pipes jointed with cement mortar 1:3 and laid in trench bottom

in runs over 3.00 m long	m	0.45	0.18	5.24	-	4.23	9.47	10.89
in runs not exceeding 3.00 m long	m	0.67	0.27	7.82	-	4.23	12.05	13.86

Extra for

bends	each	0.33	0.10	3.63	-	36.21	39.84	45.82
junctions	each	0.40	0.11	4.32	-	36.33	40.65	46.75

225 mm ogee pipes jointed with cement mortar 1:3 and laid in trench bottom

in runs over 3.00 m long	m	0.56	0.24	6.64	-	5.36	12.00	13.80
in runs not exceeding 3.00 m long	m	0.83	0.36	9.87	-	5.36	15.23	17.51

Extra for

bends	each	0.36	0.11	3.97	-	39.79	43.76	50.32
junctions	each	0.55	0.16	6.01	-	37.42	43.43	49.94

Drainage

New work	Unit	Hours C	Hours L	Labour net	Plant net	Material net	Price net	Price with 15%
Pipework				£	£	£	£	£
					VAT not included			

Cement jointed standard concrete pipes and fittings to BS 5911 (*continued*)

300 mm ogee pipes jointed with cement mortar 1:3 and laid in trench bottom

in runs over 3.00 m long	m	0.67	0.27	7.82	-	8.02	15.84	18.22
in runs not exceeding 3.00 m long	m	1.00	0.40	11.66	-	8.02	19.68	22.63

Extra for

bends	each	0.42	0.15	4.77	-	53.91	58.68	67.48
junctions	each	0.70	0.20	7.62	-	39.95	47.57	54.71

375 mm diameter trapped road gullies 750 mm deep with rodding eye; stopper and chain and 150 mm outlet including cement joint to pipe and bedding in concrete 1:3:6

	each	3.30	0.50	32.94	-	26.18	59.12	67.99

Extra for completely surrounding gullies in concrete 1:3:6 150 mm thick

	each	-	0.40	2.70	-	8.95	11.65	13.40

400 x 345 mm coated cast iron road gully gratings and frames bedded and flaunched in cement mortar

	each	0.80	0.22	8.65	-	105.19	113.84	130.92

Timesaver cast iron system

100 mm pipes jointed with flexible couplings and laid in trench bottom

in runs over 3.00 m long	each	0.55	0.55	9.42	-	25.46	34.88	40.11
in runs not exceeding 3.00 m long	each	0.70	0.70	11.99	-	34.46	46.45	53.42

Extra for

22.5 deg medium radius bends	each	0.62	0.62	10.62	-	32.04	42.66	49.06
35 deg medium radius bends	each	0.62	0.62	10.62	-	33.79	44.41	51.07
45 deg medium radius bends	each	0.62	0.62	10.62	-	33.79	44.41	51.07
60 deg medium radius bends	each	0.62	0.62	10.62	-	35.56	46.18	53.11
67.5 deg medium radius bends	each	0.62	0.62	10.62	-	35.56	46.18	53.11
80 deg medium radius bends	each	0.62	0.62	10.62	-	37.39	48.01	55.21
87.5 deg medium radius bends	each	0.62	0.62	10.62	-	37.39	48.01	55.21
87.5 deg medium radius bends with heel rest and bedding in concrete 1:3:6	each	0.80	0.80	13.71	-	43.47	57.18	65.76
87.5 deg long radius bends	each	0.62	0.62	10.62	-	46.91	57.53	66.16
87.5 deg long radius bends with heel rest and bedding in concrete 1:3:6	each	0.80	0.80	13.71	-	58.53	72.24	83.08
45 deg x 100 mm branches	each	0.75	0.75	12.85	-	58.66	71.51	82.24
67.5 deg x 100 mm branches	each	0.75	0.75	12.85	-	60.33	73.18	84.16
87.5 deg x 100 mm branches	each	0.75	0.75	12.85	-	58.66	71.51	82.24
45 deg x 100 mm access branches	each	0.75	0.75	12.85	-	100.17	113.02	129.97
87.5 deg x 100 mm access branches	each	0.75	0.75	12.85	-	100.17	113.02	129.97

Drainage

New work

	Unit	Hours C	Hours L	Labour net	Plant net	Material net	Price net	Price with 15%
Pipework				£	£	£	£	£
					VAT not included			

Timesaver cast iron system
(*continued*)

150 mm pipes jointed with flexible couplings and laid in trench bottom

	Unit	Hours C	Hours L	Labour net	Plant net	Material net	Price net	Price with 15%
in runs over 3.00 m long	m	0.80	0.80	13.71	-	44.96	58.67	67.47
in runs not exceeding 3.00 m long	m	1.20	1.20	20.55	-	55.86	76.41	87.87
Extra for								
10 deg medium radius bends	each	0.70	0.70	11.99	-	33.11	45.10	51.87
22.5 deg medium radius bends	each	0.70	0.70	11.99	-	34.87	46.86	53.89
35 deg medium radius bends	each	0.70	0.70	11.99	-	36.62	48.61	55.90
45 deg medium radius bends	each	0.70	0.70	11.99	-	36.62	48.61	55.90
67.5 deg medium radius bends	each	0.70	0.70	11.99	-	38.39	50.38	57.94
87.5 deg medium radius bends	each	0.70	0.70	11.99	-	40.22	52.21	60.04
87.5 deg medium radius bends with heel rest and bedding in concrete 1:3:6	each	0.90	0.90	15.41	-	103.66	119.07	136.93
87.5 deg long radius bends	each	0.70	0.70	11.99	-	87.99	99.98	114.98
87.5 deg long radius bends with heel rest and bedding in concrete 1:3:6	each	0.90	0.90	15.41	-	82.66	98.07	112.78
45 deg x 100 mm branches	each	0.85	0.85	14.56	-	101.52	116.08	133.49
87.5 deg x 100 mm branches	each	0.85	0.85	14.56	-	98.32	112.88	129.81
45 deg x 150 mm branches	each	0.85	0.85	14.56	-	94.66	109.22	125.60
87.5 deg x 150 mm branches	each	0.85	0.85	14.56	-	110.94	125.50	144.32
87.5 deg x 150 mm access branches	each	0.85	0.85	14.56	-	101.16	115.72	133.08
250 x 250 mm square one-piece "P" trap with 100 mm outlet including joint to pipe, bedding in concrete 1:3:6 and providing grating	each	1.10	1.10	18.85	-	162.48	181.33	208.53
Extra for								
galvanised sediment pans	each	-	0.10	0.59	-	22.28	22.87	26.30
completely surrounding gullies with concrete 1:3:6 150 mm thick	each	-	0.20	1.35	-	4.61	5.96	6.85

PVCU pipes and fittings to BS 4660

110 mm socketed pipes jointed with rubber sealing rings and laid in trench bottom

	Unit	Hours C	Hours L	Labour net	Plant net	Material net	Price net	Price with 15%
in runs over 3.00 m long	m	0.10	0.10	1.57	-	5.59	7.16	8.23
in runs not exceeding 3.00 m long	m	0.12	0.12	1.89	-	5.59	7.48	8.60
Extra for								
swept bends	each	0.12	0.12	1.89	-	8.48	10.37	11.93
long radius bends	each	0.10	0.10	1.57	-	17.39	18.96	21.80

Drainage

New work	Unit	Hours C	Hours L	Labour net	Plant net	Material net	Price net	Price with 15%
Pipework				£	£	£	£	£
					VAT not included			

PVCU pipes and fittings to BS 4660
(*continued*)

160 mm socketed pipes jointed with
rubber sealing rings and laid in trench
bottom

	Unit	Hours C	Hours L	Labour net	Plant net	Material net	Price net	Price with 15%
in runs over 3.00 m long	m	0.12	0.12	1.89	-	11.86	13.75	15.81
in runs not exceeding 3.00 m long	m	0.14	0.14	2.19	-	11.86	14.05	16.16

Extra for

	Unit	Hours C	Hours L	Labour net	Plant net	Material net	Price net	Price with 15%
swept bends	each	0.14	0.14	2.19	-	20.16	22.35	25.70
long radius bends	each	0.12	0.12	1.89	-	36.29	38.18	43.91
junctions	each	0.12	0.12	1.89	-	38.48	40.37	46.43

160 mm diameter one-piece "P" trap
gullies with 110 mm inlet and 110 mm
outlet including rubber sealing ring joint
to pipe, bedding in concrete 1:3:6 and

	Unit	Hours C	Hours L	Labour net	Plant net	Material net	Price net	Price with 15%
providing PVC grating	each	1.35	0.50	15.47	-	22.91	38.38	44.14

Three-piece "P" trap gullies comprising
110 mm trap, 110 mm knuckle bend and
raising piece hopper fitted with PVC
grating including rubber sealing ring
joint to pipe and bedding in concrete

	Unit	Hours C	Hours L	Labour net	Plant net	Material net	Price net	Price with 15%
1:3:6	each	1.65	0.84	20.44	-	35.59	56.03	64.43

Sundries

Precast concrete kerbs to three sides of
150 x 150 mm gullies bedded in cement
mortar including 150 mm trowelled

	Unit	Hours C	Hours L	Labour net	Plant net	Material net	Price net	Price with 15%
skirting to wall at back	each	0.42	0.28	5.65	-	3.89	9.54	10.97

Brick on edge kerbs to three sides of
gullies in cement mortar including
rendering all round and 150 mm skirting
to wall at back trowelled smooth

	Unit	Hours C	Hours L	Labour net	Plant net	Material net	Price net	Price with 15%
150 x 150 mm gullies	each	1.15	0.75	15.35	-	2.37	17.72	20.38
225 x 225 mm gullies	each	1.33	0.90	17.99	-	3.50	21.49	24.71

Drainage

New work	Unit	Hours C	Hours L	Labour net	Plant net	Material net	Price net	Price with 15%
Land drains				£	£	£	£	£
					VAT not included			

Butt jointed clayware pipes to BS 1196

75 mm pipes
laid in trench bottom	m	-	0.20	1.35	-	1.09	2.44	2.81
extra for junctions	each	-	0.16	1.08	-	5.02	6.10	7.01

100 mm pipes
laid in trench bottom	m	-	0.22	1.48	-	1.96	3.44	3.96
extra for junctions	each	-	0.15	1.01	-	5.80	6.81	7.83

150 mm pipes
laid in trench bottom	m	-	0.25	1.69	-	4.19	5.88	6.76
extra for junctions	each	-	0.20	1.35	-	7.64	8.99	10.34

225 mm pipes
laid in trench bottom	m	-	0.30	2.02	-	11.03	13.05	15.01
extra for junctions	each	0.22	-	1.97	-	9.22	11.19	12.87

Flexible jointed vitrified clay perforated pipes and fittings with integral polyethylene sleeves

100 mm "Hepline" pipes laid in trench bottom	m	-	0.22	1.48	-	4.87	6.35	7.30

Extra for
bends	each	-	0.15	1.01	-	4.67	5.68	6.53
junctions	each	-	0.18	1.21	-	9.74	10.95	12.59

150 mm "Hepline" pipes laid in trench bottom	m	-	0.25	1.69	-	8.86	10.55	12.13

Extra for
bends	each	-	0.18	1.21	-	8.07	9.28	10.67
junctions	each	-	0.20	1.35	-	10.87	12.22	14.05

225 mm "Hepline" pipes laid in trench bottom	m	-	0.30	2.02	-	16.30	18.32	21.07

Extra for
bends	each	-	0.20	1.35	-	27.09	28.44	32.71
junctions	each	-	0.22	1.48	-	40.43	41.91	48.20

Ogee jointed porous concrete pipes to BS 5911

Pipes laid in trench bottom
225 mm	m	0.11	0.22	2.47	-	4.54	7.01	8.06

Drainage

New work	Unit	Hours C	Hours L	Labour net	Plant net	Material net	Price net	Price with 15%
Land drains				£	£	£	£	£
					VAT not included			
Plain ended PVCU perforated pipes and fittings jointed with double socket couplers								
110 mm pipes laid in trench bottom	m	-	0.18	1.21	-	6.49	7.70	8.86
Extra for								
bends	each	-	0.15	1.01	-	8.69	9.70	11.15
junctions	each	-	0.18	1.21	-	12.80	14.01	16.11
160 mm pipes laid in trench bottom	m	-	0.22	1.48	-	12.19	13.67	15.72
Extra for								
bends	each	-	0.20	1.35	-	20.66	22.01	25.31
junctions	each	-	0.22	1.48	-	36.89	38.37	44.13
PVCU perforated flexible corrugated pipes and polyethylene fittings								
80 mm pipes laid in trench bottom	m	-	0.12	0.81	-	1.59	2.40	2.76
Extra for								
end caps	each	-	0.15	1.01	-	0.52	1.53	1.76
junctions	each	-	0.18	1.21	-	3.10	4.31	4.96
100 mm pipes laid in trench bottom	m	-	0.15	1.01	-	2.26	3.27	3.76
Extra for								
end caps	each	-	0.08	0.54	-	0.99	1.53	1.76
junctions	each	-	0.18	1.21	-	3.54	4.75	5.46

Drainage

New work	Unit	Hours C	Hours L	Labour net	Plant net	Material net	Price net	Price with 15%
Manholes				£	£	£	£	£
					VAT not included			

Excavation by machine - including disposal of surplus on site average 50 m from excavations

Excavate pits, part fill in and compact and dispose of surplus on site, maximum depth not exceeding

	Unit	Hours C	Hours L	Labour net	Plant net	Material net	Price net	Price with 15%
0.25 m	m3	-	0.65	4.38	4.97	-	9.35	10.75
1.00 m	m3	-	0.65	4.38	4.03	-	8.41	9.67
2.00 m	m3	-	0.65	4.38	4.41	-	8.79	10.11
4.00 m	m3	-	0.65	4.38	4.79	-	9.17	10.55

Excavate pits less than 1.25 x 1.25 m on plan, part fill in and compact and dispose of surplus on site, maximum depth not exceeding

	Unit	Hours C	Hours L	Labour net	Plant net	Material net	Price net	Price with 15%
0.25 m	m3	-	0.65	4.38	8.73	-	13.11	15.08
1.00 m	m3	-	0.65	4.38	5.91	-	10.29	11.83
2.00 m	m3	-	0.65	4.38	6.10	-	10.48	12.05

Excavation by machine - including removal of surplus from site to tip average 15 km from site

Excavate pits, part fill in and compact and remove surplus from site, maximum depth not exceeding

	Unit	Hours C	Hours L	Labour net	Plant net	Material net	Price net	Price with 15%
0.25 m	m3	-	0.42	2.83	16.94	-	19.77	22.74
1.00 m	m3	-	0.42	2.83	16.38	-	19.21	22.09
2.00 m	m3	-	0.42	2.83	16.38	-	19.21	22.09
4.00 m	m3	-	0.42	2.83	16.76	-	19.59	22.53

Excavate pits less than 1.25 x 1.25 m on plan, part fill in and compact and remove surplus from site, maximum depth not exceeding

	Unit	Hours C	Hours L	Labour net	Plant net	Material net	Price net	Price with 15%
0.25 m	m3	-	0.64	4.31	20.70	-	25.01	28.76
1.00 m	m3	-	0.46	3.10	17.88	-	20.98	24.13
2.00 m	m3	-	0.47	3.17	18.07	-	21.24	24.43

Excavation by hand - including disposal of surplus on site average 50 m from excavations

Excavate pits, part fill in and compact and dispose of surplus on site, maximum depth not exceeding

	Unit	Hours C	Hours L	Labour net	Plant net	Material net	Price net	Price with 15%
0.25 m	m3	-	5.75	38.76	-	-	38.76	44.57
1.00 m	m3	-	6.10	41.11	-	-	41.11	47.28
2.00 m	m3	-	7.40	49.88	-	-	49.88	57.36

Drainage

New work	Unit	Hours C	Hours L	Labour net	Plant net	Material net	Price net	Price with 15%
Manholes				£	£	£	£	£
					VAT not included			

Excavation by hand - including disposal of surplus on site average 50 m from excavations (*continued*)

Excavate pits less than 1.25 x 1.25 m on plan, part fill in and compact and dispose of surplus on site, maximum depth not exceeding

	Unit	Hours C	Hours L	Labour net	Plant net	Material net	Price net	Price with 15%
0.25 m	m3	-	7.15	48.19	-	-	48.19	55.42
1.00 m	m3	-	7.65	51.56	-	-	51.56	59.29
2.00 m	m3	-	8.95	60.32	-	-	60.32	69.37

Excavation by hand - including removal of surplus from site to tip average 15 km from site

Excavate pits, part fill in and compact and remove surplus from site, maximum depth not exceeding

	Unit	Hours C	Hours L	Labour net	Plant net	Material net	Price net	Price with 15%
0.25 m	m3	-	5.90	39.77	11.97	-	51.74	59.50
1.00 m	m3	-	6.25	42.13	11.97	-	54.10	62.22
2.00 m	m3	-	7.55	44.32	11.97	-	56.29	64.73

Excavate pits less than 1.25 x 1.25 m on plan, part fill in and compact and remove surplus from site, maximum depth not exceeding

	Unit	Hours C	Hours L	Labour net	Plant net	Material net	Price net	Price with 15%
0.25 m	m3	-	7.30	49.20	11.97	-	61.17	70.35
1.00 m	m3	-	7.80	52.57	11.97	-	64.54	74.22
2.00 m	m3	-	9.10	61.33	11.97	-	73.30	84.30
Level and compact bottom of excavation	m2	-	0.10	0.67	0.25	-	0.92	1.06

Granular bedding

	Unit	Hours C	Hours L	Labour net	Plant net	Material net	Price net	Price with 15%
Surrounds to PVC chambers	m3	-	2.50	16.85	-	19.03	35.88	41.26
Beds for PVC chambers								
100 mm	m2	-	0.35	2.36	-	1.90	4.26	4.90
150 mm	m2	-	0.50	3.37	-	2.86	6.23	7.16
200 mm	m2	-	0.60	4.04	-	3.81	7.85	9.03

Concrete 1:3:6

	Unit	Hours C	Hours L	Labour net	Plant net	Material net	Price net	Price with 15%
Bases, thickness								
100 - 150 mm	m3	-	6.80	45.83	-	68.88	114.71	131.92
150 - 300 mm	m3	-	5.80	39.09	-	68.88	107.97	124.17
Surrounds to manholes, thickness								
100 - 150 mm	m3	-	11.40	76.84	-	68.88	145.72	167.58
150 - 300 mm	m3	-	11.10	74.81	-	68.88	143.69	165.24

Drainage

New work	Unit	Hours C	Hours L	Labour net	Plant net	Material net	Price net	Price with 15%
Manholes				£	£	£	£	£
					VAT not included			
Concrete 1:2:4								
Bases, thickness								
100 - 150 mm	m3	-	7.55	50.89	-	73.19	124.08	142.69
150 - 300 mm	m3	-	6.55	44.15	-	73.19	117.34	134.94
Surrounds to manholes, thickness								
100 - 150 mm	m3	-	11.95	80.54	-	73.19	153.73	176.79
150 - 300 mm	m3	-	11.65	78.52	-	73.19	151.71	174.47
Reinforced suspended cover slabs, thickness								
100 - 150 mm	m3	-	12.00	80.88	-	73.19	154.07	177.18
150 - 300 mm	m3	-	11.35	76.50	-	73.19	149.69	172.14
Reinforcement								
Fabric reinforcement to BS 4483 in cover slabs								
A252 - 3.95 kg/m2	m2	-	0.25	1.69	-	1.68	3.37	3.88
B283 - 3.73 kg/m2	m2	-	0.25	1.69	-	1.42	3.11	3.58
B503 - 5.93 kg/m2	m2	-	0.30	2.02	-	2.62	4.64	5.34
Formwork								
Formwork to edges and faces of bases and surrounds (four uses)								
over 1.00 m high	m2	1.79	-	16.04	-	8.30	24.34	27.99
not exceeding 250 mm high	m	0.47	-	4.21	-	2.22	6.43	7.39
Formwork to horizontal soffits of cover slabs (four uses)								
over 1.00 m high	m2	2.90	0.16	27.06	1.00	20.73	48.79	56.11
not exceeding 1.00 m high	m2	3.20	0.16	29.75	1.00	20.73	51.48	59.20
Formwork to edges of cover slabs not exceeding 250 mm deep (four uses)	m	0.10	0.05	1.24	-	2.22	3.46	3.98
Precast concrete units (complying with BS 5911) bedded in cement mortar 1:3 and flush pointed								
Generally								
610 x 450 mm base units	each	0.75	0.95	13.12	-	31.18	44.30	50.95
610 x 450 mm chamber section units								
150 mm deep	each	0.60	0.60	9.42	-	12.38	21.80	25.07
230 mm deep	each	0.70	0.70	10.99	-	15.91	26.90	30.93
300 mm deep	each	0.75	0.75	11.77	-	20.60	32.37	37.23
Extra for step irons cast in chamber sections	each	-	-	-	-	4.32	4.32	4.97

Drainage

New work	Unit	Hours C	Hours L	Labour net	Plant net	Material net	Price net	Price with 15%
Manholes				£	£	£	£	£
					VAT not included			

Precast concrete units (complying with BS 5911) bedded in cement mortar 1:3 and flush pointed (*continued*)

	Unit	Hours C	Hours L	Labour net	Plant net	Material net	Price net	Price with 15%
Building in ends of pipes including knocking out apertures in base units								
100 mm	each	0.25	0.17	3.39	-	0.10	3.49	4.01
150 mm	each	0.33	0.22	4.44	-	0.10	4.54	5.22
610 x 450 mm concrete covers and frames	each	0.50	0.50	7.85	-	33.21	41.06	47.22
Common bricks in cement mortar 1:3								
Manhole sides								
102.5 mm	m2	1.00	0.50	12.33	-	14.28	26.61	30.60
215 mm	m2	2.00	1.00	24.66	-	28.39	53.05	61.01
Extra for fair face and flush pointing	m2	0.30	0.15	3.70	-	0.08	3.78	4.35
102.5 x 150 mm kerbs fair faced and pointed internally under manhole cover frames								
600 x 450 mm	each	0.57	0.28	7.00	-	4.67	11.67	13.42
600 x 600 mm	each	0.72	0.36	8.88	-	5.48	14.36	16.51
750 x 600 mm	each	0.80	0.40	9.87	-	5.65	15.52	17.85
215 x 150 mm kerbs fair faced and pointed internally under manhole cover frames								
600 x 450 mm	each	1.13	0.57	13.96	-	8.90	22.86	26.29
600 x 600 mm	each	1.44	0.72	17.75	-	10.32	28.07	32.28
750 x 600 mm	each	1.60	0.80	19.73	-	11.46	31.19	35.87
Building in ends of pipes								
100 mm	each	0.20	0.10	2.46	-	0.08	2.54	2.92
150 mm	each	0.26	0.13	3.21	-	0.08	3.29	3.78
225 mm	each	0.32	0.16	3.95	-	0.17	4.12	4.74
300 mm	each	0.36	0.18	4.44	-	0.25	4.69	5.39
13 mm cement and sand 1:3 steel trowelled rendering on brick sides internally	m2	1.35	0.68	16.65	-	1.09	17.74	20.40
Class B engineering bricks in cement mortar 1:3								
Manhole sides								
102.5 mm	m2	1.20	0.60	14.79	-	26.42	41.21	47.39
215 mm	m2	2.40	1.20	29.59	-	52.47	82.06	94.37
Extra for fair face and flush pointing	m2	0.40	0.20	4.93	-	0.08	5.01	5.76

Drainage

New work	Unit	Hours C	Hours L	Labour net	Plant net	Material net	Price net	Price with 15%
Manholes				£	£	£	£	£
					VAT not included			

Class B engineering bricks in cement mortar 1:3 (continued)

102.5 x 150 mm kerbs fair faced and pointed internally under manhole cover frames

	Unit	Hours C	Hours L	Labour net	Plant net	Material net	Price net	Price with 15%
600 x 450 mm	each	0.72	0.36	8.88	-	8.65	17.53	20.16
600 x 600 mm	each	0.90	0.45	11.09	-	10.25	21.34	24.54
750 x 600 mm	each	1.00	0.50	12.33	-	10.42	22.75	26.16

215 x 150 mm kerbs fair faced and pointed internally under manhole cover frames

	Unit	Hours C	Hours L	Labour net	Plant net	Material net	Price net	Price with 15%
600 x 450 mm	each	1.44	0.72	17.75	-	16.46	34.21	39.34
600 x 600 mm	each	1.80	0.90	22.20	-	19.07	41.27	47.46
750 x 600 mm	each	2.00	1.00	24.66	-	21.21	45.87	52.75

Building in ends of pipes

	Unit	Hours C	Hours L	Labour net	Plant net	Material net	Price net	Price with 15%
100 mm	each	0.20	0.10	2.46	-	0.08	2.54	2.92
150 mm	each	0.26	0.13	3.21	-	0.08	3.29	3.78
225 mm	each	0.32	0.16	3.95	-	0.17	4.12	4.74
300 mm	each	0.36	0.18	4.44	-	0.25	4.69	5.39

13 mm cement and sand 1:3 steel trowelled rendering on brick sides

	Unit	Hours C	Hours L	Labour net	Plant net	Material net	Price net	Price with 15%
internally	m2	1.35	0.68	16.65	-	1.09	17.74	20.40

Vitrified clay channels set and jointed in cement mortar 1:3

Half section straight main channels

	Unit	Hours C	Hours L	Labour net	Plant net	Material net	Price net	Price with 15%
100 x 600 mm long	each	0.35	-	3.14	-	2.47	5.61	6.45
100 x 900/1000 mm long	each	0.35	-	3.14	-	3.68	6.82	7.84
150 x 600 mm long	each	0.46	-	4.12	-	4.35	8.47	9.74
150 x 900/1000 mm long	each	0.46	-	4.12	-	6.33	10.45	12.02
225 x 600 mm long	each	0.75	-	6.72	-	12.03	18.75	21.56
225 x 900/1000 mm long	each	0.80	-	7.17	-	14.13	21.30	24.50

Half section main channel bends

	Unit	Hours C	Hours L	Labour net	Plant net	Material net	Price net	Price with 15%
100 mm	each	0.35	-	3.14	-	3.58	6.72	7.73
150 mm	each	0.50	-	4.48	-	6.13	10.61	12.20
225 mm	each	0.75	-	6.72	-	19.77	26.49	30.46

Half section straight taper main channels

	Unit	Hours C	Hours L	Labour net	Plant net	Material net	Price net	Price with 15%
150 - 100 mm	each	0.46	-	4.12	-	14.03	18.15	20.87
225 - 150 mm	each	0.75	-	6.72	-	31.34	38.06	43.77

Half section taper main channel bends

	Unit	Hours C	Hours L	Labour net	Plant net	Material net	Price net	Price with 15%
150 - 100 mm	each	0.50	-	4.48	-	21.17	25.65	29.50
225 - 150 mm	each	0.75	-	6.72	-	60.81	67.53	77.66

Drainage

New work	Unit	Hours C	Hours L	Labour net	Plant net	Material net	Price net	Price with 15%
Manholes				£	£	£	£	£
						VAT not included		

Vitrified clay channels set and jointed in cement mortar 1:3 (*continued*)

Half section branch channel bends

100 mm	each	0.35	-	3.14	-	7.04	10.18	11.71
150 mm	each	0.50	-	4.48	-	11.76	16.24	18.68
225 mm	each	0.75	-	6.72	-	38.44	45.16	51.93

Three quarter section branch channel bends

100 mm	each	0.35	-	3.14	-	7.75	10.89	12.52
150 mm	each	0.50	-	4.48	-	13.24	17.72	20.38

PVCU channels etc set in cement mortar 1:3

Pipes with main channel cut out

110 mm	each	0.20	0.20	3.14	-	20.29	23.43	26.94
160 mm	each	0.25	0.25	3.93	-	38.52	42.45	48.82

Bends with main channel cut out

110 mm	each	0.27	0.27	4.24	-	27.51	31.75	36.51
160 mm	each	0.33	0.33	5.18	-	52.67	57.85	66.53

Half section branch channel bends

110 mm	each	0.33	0.33	5.18	-	9.87	15.05	17.31
160 mm	each	0.40	0.40	6.28	-	15.44	21.72	24.98

Three quarter section branch channel bends

110 mm	each	0.33	0.33	5.18	-	11.48	16.66	19.16
160 mm	each	0.40	0.40	6.28	-	22.41	28.69	32.99

450 mm diameter inspection chambers with four integral branches, depth to invert

270 mm	each	0.75	0.75	11.77	-	76.42	88.19	101.42
500 mm	each	1.00	1.00	15.70	-	94.30	110.00	126.50
960 mm	each	1.25	1.25	19.63	-	141.14	160.77	184.89

Extra for

branch blanking-off plugs	each	0.05	0.05	0.79	-	4.26	5.05	5.81
channel covers	each	0.10	0.10	1.57	-	21.50	23.07	26.53

Drainage

New work	Unit	Hours C	Hours L	Labour net	Plant net	Material net	Price net	Price with 15%
Manholes				£	£	£	£	£
					VAT not included			

Concrete benchings

Concrete 1:3:6 benchings with steep falls to main channel, finished with cement and sand 1:3 trowelled smooth (measured overall), average thickness

150 mm	m2	-	2.00	13.48	-	8.75	22.23	25.56
225 mm	m2	-	2.80	18.87	-	13.16	32.03	36.83
300 mm	m2	-	3.75	25.28	-	17.56	42.84	49.27
450mm	m2	-	4.50	30.33	-	26.31	56.64	65.14

Concrete 1:2:4 benchings with steep falls to main channel, finished with cement and sand 1:3 trowelled smooth (measured overall), average thickness

150 mm	m2	-	2.00	13.48	-	9.30	22.78	26.20
225 mm	m2	-	2.80	18.87	-	13.98	32.85	37.78
300 mm	m2	-	3.00	20.22	-	18.66	38.88	44.71
450mm	m2	-	4.50	30.33	-	27.96	58.29	67.03

Extra for working benchings to branch channels	each	0.15	0.10	2.01	-	-	2.01	2.31

Step irons

General purpose pattern galvanised step-irons to BS 1247 built in to brick sides

115 mm tails	each	0.11	0.05	1.33	-	5.26	6.59	7.58
230 mm tails	each	0.14	0.07	1.72	-	6.72	8.44	9.71

Covers bedded and flaunched in cement mortar 1:3 and sealed in manhole grease

Grade A manhole covers and frames to BS 497

550 mm diameter - BS reference MA-55	each	1.10	1.10	17.27	-	187.92	205.19	235.97
600 mm diameter - BS reference MA-60	each	1.50	1.50	23.55	-	221.02	244.57	281.26

Grade B Class 1 manhole covers and frames to BS 497

550 mm diameter - BS reference MB1-55	each	0.75	0.75	11.77	-	124.67	136.44	156.91
600 mm diameter - BS reference MB1-60	each	0.90	0.90	14.13	-	158.23	172.36	198.21

Drainage

New work	Unit	Hours C	Hours L	Labour net	Plant net	Material net	Price net	Price with 15%
Manholes				£	£	£	£	£
					VAT not included			

Covers bedded and flaunched in cement mortar 1:3 and sealed in manhole grease (*continued*)

Grade B Class 2 manhole covers and frames to BS 497

550 diameter - BS reference MB2-55	each	2.10	2.10	32.97	-	123.65	156.62	180.11
600 mm diameter - BS reference MB2-60	each	2.50	2.50	39.25	-	177.89	217.14	249.71
600 x 450 mm - BS reference MB2-60/45	each	2.50	2.50	39.25	-	134.08	173.33	199.33
600 x 600 mm - BS reference MB2-60/60	each	3.00	3.00	47.10	-	168.50	215.60	247.94

Grade C single seal inspection covers and frames to BS 497

600 x 450 mm - BS reference MC1-60/45	each	0.90	0.90	14.13	-	37.81	51.94	59.73
600 x 600 mm - BS reference MC1-60/60	each	1.50	1.50	23.55	-	71.96	95.51	109.84

Grade C double seal inspection covers and frames to BS 497

600 x 450 mm - BS reference MC2-60/45	each	1.50	1.50	23.55	-	66.49	90.04	103.55
600 x 600 mm - BS reference MC2-60/60	each	2.00	2.00	31.40	-	102.11	133.51	153.54

Intercepting traps

Vitrified clay intercepting traps with stopper including cement joints to channel and pipe and bedding and surrounding with concrete 1:3:6 150 mm thick

100 mm	each	0.55	0.55	8.64	-	68.37	77.01	88.56
150 mm	each	0.80	0.80	12.56	-	90.99	103.55	119.08
225 mm	each	1.20	1.20	18.84	-	213.92	232.76	267.67

Testing drains

Water test

100 mm drains	m	0.70	0.05	6.61	-	-	6.61	7.60
150 mm drains	m	0.10	0.07	1.37	-	-	1.37	1.58
225 mm drains	m	0.12	0.08	1.62	-	-	1.62	1.86
300 mm drains	m	0.15	0.10	2.01	-	-	2.01	2.31

Drainage

New work	Unit	Hours C	Hours L	Labour net	Plant net	Material net	Price net	Price with 15%
Combined items				£	£	£	£	£
					VAT not included			

Notes: Figures include for excavation by machine and are based on disposal of surplus excavated material on site within 50 metres of the excavations. Drainage pipework is not included.

	Unit	Hours C	Hours L	Labour net	Plant net	Material net	Price net	Price with 15%
Concrete manholes 610 x 450 x 600mm deep to invert with 150 mm concrete 1:3:6 base, precast base unit and chamber section, 100 mm vitrified clay main channel and three, three quarter section branch channel bends, concrete benching and precast cover and frame complete	each	6.00	6.00	94.20	10.68	125.54	230.42	264.99
Extra for each 150 mm increase in depth to 900 mm maximum	each	0.50	0.75	9.54	2.52	13.32	25.38	29.19
Brick manholes 600 x 450 x 600 mm deep to invert with 150 mm concrete 1:3:6 base, sides in common bricks rendered internally, 100 mm vitrified clay main channel and three, three quarter section branch channel bends, concrete benching and Grade C single seal cast iron cover and frame complete								
102.5 mm sides	each	8.00	7.00	118.86	15.75	110.81	245.42	282.24
215 mm sides	each	11.00	10.00	165.96	23.10	144.74	333.80	383.87
Extra for each 300 mm increase in depth to 1500 mm maximum								
102.5 mm sides with step iron	each	2.00	1.50	28.03	5.97	17.56	51.56	59.30
215 mm sides with step iron	each	3.25	2.75	47.66	8.78	34.21	90.65	104.25
Brick manholes 750 x 600 x 600 mm deep to invert with 150 mm concrete 1:3:6 base, sides in common bricks rendered internally, 100 mm vitrified clay main channel and four three quarter section branch channel bends, concrete benching, 150 mm concrete 1:2:4 suspended cover slab with steel fabric reinforcement and Grade C single seal cast iron cover and frame complete								
102.5 mm sides	each	20.00	12.00	260.08	21.22	156.99	438.29	504.04
215 mm sides	each	24.00	16.00	322.88	29.42	203.40	555.70	639.06
Extra for each 300 mm increase in depth to 1500 mm maximum								
102.5 sides with step-iron	each	2.70	2.70	42.40	8.02	20.37	70.79	81.41
215 mm sides with step-iron	each	4.00	3.50	59.43	11.00	39.64	110.07	126.58

Drainage

New work	Unit	Hours C	Hours L	Labour net	Plant net	Material net	Price net	Price with 15%
Combined items				£	£	£	£	£
					VAT not included			
Extra for 100 mm vitrified clay intercepting traps with connecting pipe and fresh air inlet complete	each	1.25	1.00	17.94	0.03	73.05	91.02	104.68
450 mm diameter PVC-U manholes with four integral branches,150 mm granular bedding base and surround, 150 mm concrete 1:3:6 surround at top and Grade C single seal cast iron cover and frame complete, depth to invert								
570 mm	each	4.00	3.50	59.43	8.67	180.84	248.94	286.29
910 mm	each	4.50	4.50	70.65	12.76	222.27	305.68	351.54
2720 litre septic tanks 2000 x 1000 x 2050 mm deep internally, with 150 mm concrete 1:2:4 base, 215 mm sides and 102.5 mm baffle wall in engineering bricks rendered internally and externally, 110 mm PVC-U inlet, outlet and dip pipes, aluminium fresh air inlet, 150 mm concrete 1:2:4 suspended cover slab with steel fabric reinforcement and three 600 x 450 mm Grade C single seal cast iron covers and frames complete	each	152.00	133.00	2258.34	501.63	1587.05	4347.02	4999.08
2800 litre glass fibre spherical septic tanks 1850 mm diameter (2500 mm deep from ground level), with inlet and outlet pipes and manhole venting kit comprising cover and frame, rodding pipes and fresh air inlet vent complete, including 100 mm concrete 1:3:6 base and 150 mm filling around base of tank	each	7.00	22.00	211.00	211.02	583.26	1005.28	1156.08
Filter chambers 1500 x 1000 x 2050mm deep internally with 150 mm concrete 1:3:6 base, 215 mm sides in engineering bricks, 50 - 100 mm graded media filling, 150 mm perforated PVC-U gutter distribution channels and 100 mm precast reinforced concrete plank covers complete	each	92.00	92.00	1444.40	352.60	1152.26	2949.26	3391.65

Civil engineering

Preamble

Generally

The layout of the measured work pages in this section differs from the rest of the book due to the fact that the work priced herein involves the use of both labour and plant gangs. Although the measured rates are based on the same basic labour, plant hire and materials prices as the other chapters, the basic hourly rate calculations together with the composition of the labour and plant gangs are set out in detail so that users may make any adjustments to the rates and prices to suit their own particular circumstances.

The items in this section are intended to represent the range of civil engineering work that would be encountered on a site prior to the construction of a large building project. The work covered includes demolition, site clearance, muckshifting, filling, roads, sewers and concrete structures such as headwalls, storage and settlement tanks. The following labour and gang details set out the build-up of the hourly rates used in this section.

Labour costs

The following all-in rates are based upon the three year agreement promulgated by the CIJC on 23 July 1997. The calculations below cover the third year of the agreement with effect on and from Monday 28 June 1999. The rates have been calculated as follows:

Craftsmen and General Operatives

		Craftsman				General Operative	
			£				£
Flat time (paid hours)	1883.8 hours	6.05	11,396.99	1883.8 hours	4.55	8,571.29
Non-productive overtime	68 hours	6.05	411.40	68 hours	4.55	309.40
Public holidays	71 hours	6.05	429.55	71 hours	4.55	323.05
Sick pay	5 days	12.10	60.50	5 days	12.10	60.50
			12,298.44			9,264.24
NIC Employers' contribution	12.2%		1,500.41	12.2%		1,130.24
CITB levy	0.25%		30.75	0.25%		23.16
Holidays with pay	47 weeks	21.30	1,001.10	47 weeks	21.30	1,001.10
			14,830.70			11,418.74
Severance pay and other statutory costs	2%		296.61	2%		228.37
			15,127.31			11,647.11
Employers' liability and third party insurance	2%		302.55	2%		232.94
Total cost of 1846 Productive hours per annum		£	15,429.86		£	11,880.05
Total cost per hour		£	8.36		£	6.44

Preamble *(continued)*

Craftsmen and General Operatives *(continued)*

			General operative (skill rate 4) £
Flat time (paid hours)	1883.8 hours	4.90	9,230.62
Non-productive overtime	68 hours	4.90	333.20
Public holidays	71 hours	4.90	347.90
Sick pay	5 days	12.10	60.50
			9,972.22
NIC Employers' contribution	12.2%		1,216.61
CITB levy	0.25%		24.93
Holidays with pay	47 weeks	21.30	1,001.10
			12,214.86
Severance pay and other statutory costs	2%		244.30
			12,459.16
Employers' liability and third party insurance	2%		249.18
Total cost of 1846 Productive hours per annum		£	12,708.34
Total cost per hour		£	6.88

Bonus payments

When there is a shortage of craftsmen and general operatives, Employers sometimes make bonus payments to attract and retain staff. These payments are not related to performance and are usually expressed as weekly sums. The following all-in hourly rates reflect the effect of including the stated weekly bonus amounts into the preceding calculations.

Craftsman		General Operative		General Operative (Skill Rate 4)	
Weekly bonus £	All-in rate £	Weekly bonus £	All-in rate £	Weekly bonus £	All-in rate £
15.00	8.81	5.00	6.59	5.00	7.03
20.00	8.96	10.00	6.74	10.00	7.18
25.00	9.11	15.00	6.89	15.00	7.33
30.00	9.26	20.00	7.04	20.00	7.48

Rates used in this section

It has been assumed that a £20 weekly bonus is paid to craftsmen and £10 to general operatives. The all-in hourly rates included in this edition are:

	£
Craftsman	8.96
General operative	6.74
General operative (skill rate 4)	7.18

Preamble *(continued)*

Labour gangs

As already mentioned, the work in this section involves the use of labour gangs. There are gangs of various compositions and these are identified by a simple reference number. The composition of the gangs and the calculations of the hourly rate for each are shown below.

Ref.	Composition	Hourly Rate £
LA	1 Craftsman	8.96
LB	1 General operative	6.74
LC	1 General operative (skill rate 4)	7.18
LD	1 Ganger	8.96
	1 General operative	6.74
	1 General operative (skill rate 4)	7.18
	Total for LD	22.88
LE	2 Craftsmen	17.92
	1 General operative	6.74
	Total for LE	24.66
LF	1 Craftsman	8.96
	2 General operatives	13.48
	Total for LF	22.44
LG	1 General operative (skill rate 4)	7.18
	1 General operative	6.74
	Total for LG	13.92
LH	1 Ganger	8.96
	2 General operatives (skill rate 4)	14.36
	1 General operative	6.74
	Total for LH	30.06

Civil engineering

Preamble *(continued)*

Plant gangs

As already mentioned, the work in this section involves the use of plant gangs. There are gangs of various compositions and these are identified by a simple reference number. The composition of the gangs and the calculations of the hourly rate for each are shown below.

Ref.		Composition	Hourly Rate £
PA	General excavation	Hydraulic excavator (3.5m3)	39.38
PB	Breaking up	Compressor (375cfm)	7.00
		Drills and breakers	1.50
		Wheeled hydraulic excavator	18.95
		Dumper (1.5t)	3.85
Total for PB			31.30
PC	Trimming	Grader	14.44
		Crawler dozer	32.83
Total for PC			47.27
PD	On-site disposal	Dump truck	25.41
PE	Off-site disposal	Tipper wagon	21.38
PF	Filling	Wheeled hydraulic excavator (3.5m3)	18.95
		Dump truck	25.41
		Vibrating roller	9.98
Total for PF			54.34
PG	Concrete placing	Crawler crane (20%)	11.93
		Concrete skips (2)	1.66
		Vibrating pokers (3)	2.06
		Compressor (375cfm)	7.00
Total for PG			22.65
PH	Shuttering	Saw bench	1.00
		Crawler crane (20%)	11.93
		Small tools	1.66
Total for PH			14.59
PI	Steel fixing	Crawler crane (20%)	11.93
		Small tools	1.66
Total for PI			13.59

Preamble *(continued)*

Plant gangs *(continued)*

Ref.	Composition	Hourly Rate £
PJ Concrete jointing	Compressor (375cfm)	7.00
	Tar boiler	2.60
	Small tools	1.66
Total for PJ		11.26
PK Roads	Crawler tractor	15.78
	Motorised roller	11.99
Total for PK		26.77
PL Kerb laying	Dumper (1.5t)	3.85
PM Sewers	Crawler hydraulic excavator (3.5m3)	39.98
	Pump (275cfm m3/h)	3.08
	Trench sheets (125)	1.50
	Props (100)	2.00
	Dumper (1.5t)	3.85
	Vibrating compactor	3.87
Total for PM		54.28
PN Concrete manholes	Wheeled hydraulic excavator)	18.95
	Pump (275cfm m3/h)	3.08
	Dumper (1.5t)	3.85
	Trench sheets (50)	0.83
	Props (35)	0.50
Total for PN		27.21

As mentioned at the beginning of this preamble, the layout of the measured work pages in this section differs from the rest of the book due to the fact that the work priced herein involves the use of both labour and plant gangs. The definitions of the columns with prices are shown below.

"Labour net" figures include allowances for all costs incidental to the employment of labour.

"Plant net" figures include for all costs of plant including drivers and operators where applicable.

"Materials net" figures include for all costs of materials including an allowance for waste except where specifically stated.

"Price net" figures are the totals of the "Labour net", "Plant net", where applicable, and "Materials net" figures . Prices are for a contractor employing his own labour; according to the amount and nature of the work involved, it may well be possible to secure more advantageous prices from specialist sub-contractors.

"Price with 15%" figures include for establishment charges, overheads and profit.

Prices do not include any allowance for scaffolding, ladders or other plant necessary to reach the work. The "Preliminaries" section includes prices for scaffolding which must be considered and allowance included to suit the particular circumstances of a tender.

Civil engineering

Specialist prices

"Price with 15%" figures are all-in guide prices and include for the contractor's overheads, profit, unloading materials and general attendance (to include free use of standing scaffolding and hoists, temporary lighting and water and clearing away rubbish).

The amount of attendance required varies between the various trades and also with the circumstances of specific jobs; the percentage addition must always be considered and adjusted as necessary to suit the terms and conditions of the quotation being used.

Quantities and delivery distances are usually the most significant of the many factors which influence prices and it must be emphasised that quotations should always be obtained when preparing a tender.

Civil engineering

New work

Basic prices for materials		£
Resin impregnated sheeting, Pourform, 17.5 mm thick	Sheet	27.95
Phenolic coated sheeting good one side, 18 mm thick	Sheet	51.86
Filcrete joint filler, thickness		
12 mm	Sheet	11.16
15 mm	Sheet	16.63
19 mm	Sheet	18.98
25 mm	Sheet	23.96
PVC-U flat dumbell waterstop		
100 mm	15 m	38.93
170 mm	15 m	53.05
210 mm	15 m	68.39
250 mm	15 m	85.17
PVC-U centre bulb waterstop		
100 mm	15 m	56.80
170 mm	15 m	65.74
210 mm	15 m	82.04
250 mm	15 m	97.31
Rubber flat dumbell waterstop		
150 mm	9 m	130.43
230 mm	9 m	194.60
Rubber centre bulb waterstop		
150 mm	9 m	149.35
230 mm	9 m	224.06
Granular filling material		
DTp Spec type 1	m3	11.43
DTp Spec type 2	m3	11.09
Vitrified clayware pipes, spigot and socket, diameter		
300 mm	m	21.09
375 mm	m	40.09
400 mm	m	43.27
450 mm	m	56.20
Concrete pipes Class L, vibrated, flexible joints, diameter		
300 mm	m	13.72
375 mm	m	17.85
450 mm	m	21.34
520 mm	m	24.98
600 mm	m	29.19
Concrete pipes Class L, spun, flexible joints, diameter		
750 mm	m	48.92
900 mm	m	66.06
1200 mm	m	110.25
1500 mm	m	170.76
1800 mm	m	229.35

Civil engineering

New work

Basic prices for materials   £

Concrete pipes Class M, vibrated, flexible joints, diameter

300 mm	m	13.99
375 mm	m	18.21
450 mm	m	21.77
525 mm	m	25.48
600 mm	m	29.85

Concrete pipes Class M, spun, flexible joints, diameter

750 mm	m	49.91
900 mm	m	67.38
1200 mm	m	112.88
1500 mm	m	174.17
1800 mm	m	234.04

Concrete pipes Class H, vibrated, flexible joints, diameter

375 mm	m	18.03
450 mm	m	22.31
525 mm	m	26.11
600 mm	m	30.59

Concrete pipes Class H, spun, flexible joints, diameter

750 mm	m	49.91
900 mm	m	69.06
1200 mm	m	115.70
1500 mm	m	179.01
1800 mm	m	239.72

Vitrified clayware bends

300 mm	each	51.53
375 mm	each	98.89
400 mm	each	162.57
450 mm	each	214.07

Vitrified clayware single junctions

300 mm	each	80.95
375 mm	each	155.05
400 mm	each	165.41
450 mm	each	196.35

Vitrified clayware double junctions

300 mm	each	136.65
375 mm	each	235.90

Vitrified clayware taper (largest end stated)

300 mm	each	61.67
375 mm	each	91.72

Civil engineering

New work

Basic prices for materials   £

Concrete bends

300 mm	m	69.24
375 mm	m	85.71
450 mm	m	107.87
600 mm	m	158.20
750 mm	m	226.95
900 mm	m	317.64
1200 mm	m	519.01
1500 mm	m	785.02
1800 mm	m	1,109.20

Concrete single junctions

300 mm	each	48.89
375 mm	each	93.33
450 mm	each	100.86
600 mm	each	134.52
750 mm	each	209.40

Concrete double junctions

300 mm	each	145.40
375 mm	each	176.49

Civil engineering

New work			

Demolition

£ £

VAT not included

The following "Specialist price net" figures are guide prices only. Circumstances greatly affect the cost of this work and it is essential that quotations be obtained on every occasion.

Prices do not include for cash discount.

See the preamble notes for builder's profit and attendance.

Demolish buildings

	Unit	Specialist price net	Price with 15%
Demolish brick building to 0.5 below ground level			
not exceeding100 m3	each	500.00	575.00
101-500 m3	each	1,500.00	1,725.00
501-1000 m3	each	2,000.00	2,300.00
1001-2000 m3	each	4,000.00	4,600.00
Demolish concrete building to 0.5 below ground level			
not exceeding100 m3	each	750.00	862.50
101-500 m3	each	2,000.00	2,300.00
501-1000 m3	each	2,500.00	2,875.00
1001-2000 m3	each	5,000.00	5,750.00

Clearing the site

	Unit	Specialist price net	Price with 15%
Clear site of undergrowth and the like	ha	1,000.00	1,150.00
Cut down trees to 300mm above ground level, girth			
not exceeding 0.5 m	each	25.00	28.75
0.50-1.00 m	each	30.00	34.50
1.00-2.00 m	each	50.00	57.50
2.00-2.50 m	each	90.00	103.50
2.50-3.00 m	each	140.00	161.00
3.00-4.00 m	each	250.00	287.50
Grub up stumps and roots of trees, girth			
not exceeding 0.5m	each	40.00	46.00
0.50-1.00 m	each	50.00	57.50
1.00-2.00 m	each	60.00	69.00
2.00-2.50 m	each	110.00	126.50
2.50-3.00 m	each	175.00	201.25
3.00-4.00 m	each	300.00	345.00

Investigation and stabilisation

	Unit	Specialist price net	Price with 15%
Excavate trial holes, size			
1.00 x 2.00 x 1.00 m	each	20.00	23.00
1.00 x 2.00 x 2.00 m	each	30.00	34.50
1.00 x 2.00 x 3.00 m	each	40.00	46.00
1.00 x 2.00 x 4.00 m	each	60.00	69.00
1.00 x 3.00 x 1.00 m	each	30.00	34.50
1.00 x 3.00 x 2.00 m	each	40.00	46.00
1.00 x 3.00 x 3.00 m	each	50.00	57.50
1.00 x 3.00 x 4.00 m	each	70.00	80.50

Civil engineering

New work

Demolition

£ £

VAT not included

Investigation and stabilisation *(continued)*

	Unit	Specialist price net	Price with 15%
Excavate trial holes in rock, size			
1.00 x 2.00 x 1.00 m	each	40.00	46.00
1.00 x 2.00 x 2.00 m	each	80.00	92.00
1.00 x 2.00 x 3.00 m	each	40.00	46.00
1.00 x 2.00 x 4.00 m	each	120.00	138.00
1.00 x 3.00 x 1.00 m	each	60.00	69.00
1.00 x 3.00 x 2.00 m	each	80.00	92.00
1.00 x 3.00 x 3.00 m	each	100.00	115.00
1.00 x 3.00 x 4.00 m	each	140.00	161.00
Backfilling trial holes with excavated material including levelling and compaction			
1.00 x 2.00 x 1.00 m	each	10.00	11.50
1.00 x 2.00 x 2.00 m	each	20.00	23.00
1.00 x 2.00 x 3.00 m	each	30.00	34.50
1.00 x 2.00 x 4.00 m	each	40.00	46.00
1.00 x 3.00 x 1.00 m	each	15.00	17.25
1.00 x 3.00 x 2.00 m	each	30.00	34.50
1.00 x 3.00 x 3.00 m	each	45.00	51.75
1.00 x 3.00 x 4.00 m	each	60.00	69.00
Cable percussion boreholes 150mm diameter including erection and removal of equipment, total depth			
less than 5.00 m	m	25.00	28.75
5.00 - 10.00 m	m	30.00	34.50
10.00 - 20.00 m	m	35.00	40.25
Rotary drilled boreholes 150mm diameter including erection and removal of equipment, total depth			
less than 5.00 m	m	20.00	23.00
5.00 - 10.00 m	m	25.00	28.75
10.00 - 20.00 m	m	30.00	34.50
Backfilling boreholes 150mm diameter with concrete, 1:3:6 40mm aggregate	m	1.50	1.73

Civil engineering

New work	Unit	Labour gang hours		Labour net	Plant gang hours		Plant net	Mats net	Price net	Price with 15%
Excavation				£			£	£	£	£
								VAT not included		
Excavating										
Excavate topsoil and deposit in spoil heaps										
150 mm thick	m3	LA	0.03	0.27	PA	0.03	1.18	-	1.45	1.67
200 mm thick	m3	LA	0.04	0.36	PA	0.04	1.58	-	1.94	2.23
Excavate generally to reduce levels and deposit in spoil heaps	m3	LA	0.08	0.72	PA	0.08	3.15	-	3.87	4.45
Excavate in soft rock and deposit in spoil heaps	m3	LB	1.50	10.11	PB	1.50	46.95	-	57.06	65.62
Excavate in hard rock and deposit in spoil heaps	m3	LB	2.50	16.85	PB	2.50	78.25	-	95.10	109.36
Break up mass concrete	m3	LB	1.70	11.46	PB	1.70	53.21	-	64.67	74.37
Break up reinforced concrete	m3	LB	2.20	14.83	PB	2.20	68.86	-	83.69	96.24
Trim horizontal excavated surfaces	m2	LA	0.01	0.09	PC	0.01	0.47	-	0.56	0.64
Trim sloping excavated surfaces	m2	LA	0.02	0.18	PC	0.02	0.95	-	1.13	1.30
Load and deposit excavated material to stockpile on site, distance										
200 m	m3	LA	0.06	0.54	PD	0.06	1.52	-	2.06	2.37
400 m	m3	LA	0.10	0.90	PD	0.10	2.54	-	3.44	3.96
Load and remove excavated material from stockpile on site to tip, distance										
5 km	m3		-	-	PE	0.10	2.14	-	2.14	2.46
10 km	m3		-	-	PE	0.20	4.28	-	4.28	4.92
20 km	m3		-	-	PE	0.40	8.55	-	8.55	9.83
Excavate soft spots and backfill with										
granular fill	m3	LA	0.12	1.08	PA	0.12	4.73	19.03	24.84	28.57
concrete 1:3:6 - 40 mm aggregate	m3	LA	0.12	1.08	PA	0.12	4.73	55.65	61.46	70.68
Filling										
Excavate from stockpile, load and transport distance of 200 m and deposit										
non-selected subsoil	m3	LA	0.05	0.45	PF	0.05	2.72	-	3.17	3.65
subsoil	m3	LA	0.05	0.45	PF	0.05	2.72	-	3.17	3.65

Civil engineering

New work	Unit		Labour gang hours	Labour net		Plant gang hours	Plant net	Mats net	Price net	Price with 15%
Excavation				£			£	£	£	£
								VAT not included		
Filling *(continued)*										
Imported filling deposited and consolidated in layers										
subsoil	m3	LA	0.03	0.27	PF	0.03	1.63	5.11	7.01	8.06
granular material	m3	LA	0.03	0.27	PF	0.03	1.63	19.03	20.93	24.07
rock	m3	LA	0.03	0.27	PF	0.03	1.63	10.20	12.10	13.91
Imported filling deposited and consolidated in layer 150 mm thick										
subsoil	m2	LA	0.01	0.09	PF	0.01	0.54	0.77	1.40	1.61
granular material	m2	LA	0.01	0.09	PF	0.01	0.54	2.86	3.49	4.01
rock	m2	LA	0.01	0.09	PF	0.01	0.54	1.53	2.16	2.48
Imported filling deposited and consolidated in layer 250 mm thick										
subsoil	m2	LA	0.02	0.18	PF	0.01	0.54	1.28	2.00	2.30
granular material	m2	LA	0.02	0.18	PF	0.01	0.54	4.76	5.48	6.30
rock	m2	LA	0.02	0.18	PF	0.01	0.54	2.55	3.27	3.76
Imported filling deposited and consolidated in layer 350 mm thick										
subsoil	m2	LA	0.03	0.27	PF	0.02	1.09	1.79	3.15	3.62
granular material	m2	LA	0.03	0.27	PF	0.02	1.09	6.66	8.02	9.22
rock	m2	LA	0.03	0.27	PF	0.02	1.09	3.57	4.93	5.67
Imported filling deposited and consolidated in layer 500 mm thick										
subsoil	m2	LA	0.04	0.36	PF	0.02	1.09	2.55	4.00	4.60
granular material	m2	LA	0.04	0.36	PF	0.02	1.09	9.52	10.97	12.62
rock	m2	LA	0.04	0.36	PF	0.02	1.09	5.10	6.55	7.53
Trim horizontal filled surfaces	m2	LA	0.01	0.09	PC	0.01	0.47	-	0.56	0.64
Trim sloping filled surfaces	m2	LA	0.02	0.18	PC	0.02	0.95	-	1.13	1.30

Civil engineering

New work	Unit		Labour gang hours	Labour net	Plant gang hours	Plant net	Mats net	Price net	Price with 15%
Excavation				£		£	£	£	£
							VAT not included		
Geotextiles									
Paraweb flexible sheeting, type 100S									
Mono	m2	LB	0.05	0.34	-	-	6.62	6.96	8.01
Duplex	m2	LB	0.05	0.34	-	-	7.35	7.69	8.85
Triplex	m2	LB	0.05	0.34	-	-	8.40	8.74	10.06
Polypropylene sheeting									
G100, 0.60 mm thick	m2	LB	0.04	0.27	-	-	0.39	0.66	0.76
F2B, 0.95 mm thick	m2	LB	0.04	0.27	-	-	0.43	0.70	0.81
F32M, 2.50 mm thick	m2	LB	0.04	0.27	-	-	0.59	0.86	0.99
F3S, 1.20 mm thick	m2	LB	0.04	0.27	-	-	0.63	0.90	1.03
F33S, 1.50 mm thick	m2	LB	0.04	0.27	-	-	0.70	0.97	1.12
F45S, 1.40 mm thick	m2	LB	0.04	0.27	-	-	0.83	1.10	1.26
F4M, 3.20 mm thick	m2	LB	0.04	0.27	-	-	0.92	1.19	1.37
Polypropylene ground stabilising mats fixed with steel pins	m2	LB	0.06	0.40	-	-	-	0.40	0.46
Terram polypropylene sheeting									
Type 500, 0.40 mm thick	m2	LB	0.06	0.40	-	-	0.46	0.86	0.99
Type 700, 0.50 mm thick	m2	LB	0.06	0.40	-	-	0.54	0.94	1.08
Type 1000, 0.70 mm thick	m2	LB	0.06	0.40	-	-	0.60	1.00	1.15
Type 2000, 1.00 mm thick	m2	LB	0.06	0.40	-	-	1.07	1.47	1.69

Civil engineering

New work	Unit		Labour gang hours	Labour net		Plant gang hours	Plant net	Mats net	Price net	Price with 15%
Concrete work				£			£	£	£	£
								VAT not included		
Provide ready-mixed concrete										
1:3:6 40mm aggregate	m3		-	-		-	-	55.65	55.65	64.00
1:2:4 20mm aggregate	m3		-	-		-	-	58.99	58.99	67.84
Place un-reinforced concrete										
Blinding, thickness										
not exceeding 150 mm	m3	LD	0.20	4.58	PG	0.06	1.36	-	5.94	6.83
150-300 mm	m3	LD	0.20	4.58	PG	0.06	1.36	-	5.94	6.83
Bases, footings, pile caps and ground slabs, thickness										
not exceeding 150 mm	m3	LD	0.20	4.58	PG	0.20	4.53	-	9.11	10.48
150-300 mm	m3	LD	0.18	4.12	PG	0.18	4.08	-	8.20	9.43
300-500 mm	m3	LD	0.16	3.66	PG	0.16	3.62	-	7.28	8.37
exceeding 500 mm	m3	LD	0.14	3.20	PG	0.14	3.17	-	6.37	7.33
Walls, thickness										
not exceeding 150 mm	m3	LD	0.22	5.03	PG	0.22	4.98	-	10.01	11.51
150-300 mm	m3	LD	0.20	4.58	PG	0.20	4.53	-	9.11	10.48
300-500 mm	m3	LD	0.18	4.12	PG	0.18	4.08	-	8.20	9.43
exceeding 500 mm	m3	LD	0.16	3.66	PG	0.16	3.62	-	7.28	8.37
Place reinforced concrete										
Bases, footings, pile caps and ground slabs, thickness										
not exceeding 150 mm	m3	LD	0.24	5.49	PG	0.24	5.44	-	10.93	12.57
150-300 mm	m3	LD	0.22	5.03	PG	0.22	4.98	-	10.01	11.51
300-500 mm	m3	LD	0.20	4.58	PG	0.20	4.53	-	9.11	10.48
exceeding 500 mm	m3	LD	0.18	4.12	PG	0.18	4.08	-	8.20	9.43
Walls, thickness										
not exceeding 150 mm	m3	LD	0.28	6.41	PG	0.28	6.34	-	12.75	14.66
150-300 mm	m3	LD	0.26	5.95	PG	0.26	5.89	-	11.84	13.62
300-500 mm	m3	LD	0.24	5.49	PG	0.24	5.44	-	10.93	12.57
exceeding 500 mm	m3	LD	0.22	5.03	PG	0.22	4.98	-	10.01	11.51
Suspended slabs, thickness										
not exceeding 150 mm	m3	LD	0.28	6.41	PG	0.28	6.34	-	12.75	14.66
150-300 mm	m3	LD	0.26	5.95	PG	0.26	5.89	-	11.84	13.62
Columns and piers, cross-sectional area										
not exceeding 0.03 m2	m3	LD	0.30	6.86	PG	0.30	6.79	-	13.65	15.70
0.03-0.10 m2	m3	LD	0.28	6.41	PG	0.28	6.34	-	12.75	14.66
0.1-0.25 m2	m3	LD	0.26	5.95	PG	0.26	5.89	-	11.84	13.62
0.25-1.00 m2	m3	LD	0.24	5.49	PG	0.24	5.44	-	10.93	12.57
exceeding 1.00 m2	m3	LD	0.22	5.03	PG	0.22	4.98	-	10.01	11.51

Civil engineering

New work	Unit		Labour gang hours	Labour net		Plant gang hours	Plant net	Mats net	Price net	Price with 15%
Concrete work				£			£	£	£	£
								VAT not included		

Place reinforced concrete
(*continued*)

Beams, cross-sectional area										
not exceeding 0.03 m2	m3	LD	0.50	11.44	PG	0.50	11.32	-	22.76	26.17
0.03-0.10 m2	m3	LD	0.45	10.30	PG	0.45	10.19	-	20.49	23.56
0.1-0.25 m2	m3	LD	0.40	9.15	PG	0.40	9.06	-	18.21	20.94
0.25-1.00 m2	m3	LD	0.35	8.01	PG	0.35	7.93	-	15.94	18.33
exceeding 1.00 m2	m3	LD	0.30	6.86	PG	0.30	6.79	-	13.65	15.70
Casing to metal sections										
not exceeding 0.03 m2	m3	LD	0.55	12.58	PG	0.55	12.46	-	25.04	28.80
0.03-0.10 m2	m3	LD	0.50	11.44	PG	0.50	11.32	-	22.76	26.17
0.1-0.25 m2	m3	LD	0.45	10.30	PG	0.45	10.19	-	20.49	23.56
0.25-1.00 m2	m3	LD	0.40	9.15	PG	0.40	9.06	-	18.21	20.94
exceeding 1.00 m2	m3	LD	0.35	8.01	PG	0.35	7.93	-	15.94	18.33

Place prestressed concrete

Suspended slabs, thickness										
not exceeding 150 mm	m3	LD	0.30	6.86	PG	0.30	6.79	-	13.65	15.70
150-300 mm	m3	LD	0.28	6.41	PG	0.28	6.34	-	12.75	14.66
Beams										
not exceeding 0.03 m2	m3	LD	0.55	12.58	PG	0.55	12.46	-	25.04	28.80
0.03-0.10 m2	m3	LD	0.50	11.44	PG	0.50	11.32	-	22.76	26.17
0.1-0.25 m2	m3	LD	0.45	10.30	PG	0.45	10.19	-	20.49	23.56
0.25-1.00 m2	m3	LD	0.40	9.15	PG	0.40	9.06	-	18.21	20.94
exceeding 1.00 m2	m3	LD	0.35	8.01	PG	0.35	7.93	-	15.94	18.33

Formwork

Rough finish, plane horizontal										
not exceeding 0.10 m	m	LE	0.15	3.70	PH	0.15	2.19	0.31	6.20	7.13
0.10 -0.20 m	m	LE	0.20	4.93	PH	0.20	2.92	0.61	8.46	9.73
0.20 -0.40 m	m2	LE	0.50	12.33	PH	0.50	7.29	1.54	21.16	24.33
0.40 -1.22 m	m2	LE	0.45	11.10	PH	0.45	6.57	1.54	19.21	22.09
exceeding 1.22 m	m2	LE	0.40	9.86	PH	0.40	5.84	1.54	17.24	19.83
Rough finish, plane sloping										
not exceeding 0.10 m	m	LE	0.17	4.19	PH	0.17	2.48	0.31	6.98	8.03
0.10 -0.20 m	m	LE	0.22	5.43	PH	0.22	3.21	0.61	9.25	10.64
0.20 -0.40 m	m2	LE	0.54	13.32	PH	0.54	7.88	1.54	22.74	26.15
0.40 -1.22 m	m2	LE	0.48	11.84	PH	0.48	7.00	1.54	20.38	23.44
exceeding 1.22 m	m2	LE	0.44	10.85	PH	0.44	6.42	1.54	18.81	21.63
Rough finish, plane vertical										
not exceeding 0.10 m	m	LE	0.24	5.92	PH	0.24	3.50	0.31	9.73	11.19
0.10 -0.20 m	m	LE	0.28	6.90	PH	0.28	4.09	0.61	11.60	13.34
0.20 -0.40 m	m2	LE	0.62	15.29	PH	0.62	9.05	1.54	25.88	29.76
0.40 -1.22 m	m2	LE	0.56	13.81	PH	0.56	8.17	1.54	23.52	27.05
exceeding 1.22 m	m2	LE	0.52	12.82	PH	0.52	7.59	1.54	21.95	25.24

Civil engineering

New work	Unit	Labour gang hours	Labour net		Plant gang hours	Plant net	Mats net	Price net	Price with 15%
Concrete work			£			£	£	£	£
							VAT not included		

Formwork (*continued*)

Rough finish, curved radius in
one plane 0.5m radius

not exceeding 0.10 m	m	LE	0.28	6.90	PH	0.28	4.09	0.31	11.30	12.99
0.10 -0.20 m	m	LE	0.32	7.89	PH	0.32	4.67	0.61	13.17	15.15
0.20 -0.40 m	m2	LE	0.66	16.28	PH	0.66	9.63	1.54	27.45	31.57
0.40 -1.22 m	m2	LE	0.62	15.29	PH	0.62	9.05	1.54	25.88	29.76
exceeding 1.22 m	m2	LE	0.58	14.30	PH	0.58	8.46	1.54	24.30	27.95

Rough finish, curved radius in
one plane 1m radius

not exceeding 0.10 m	m	LE	0.26	6.41	PH	0.26	3.79	0.31	10.51	12.09
0.10 -0.20 m	m	LE	0.30	7.40	PH	0.30	4.38	0.61	12.39	14.25
0.20 -0.40 m	m2	LE	0.64	15.78	PH	0.64	9.34	1.54	26.66	30.66
0.40 -1.22 m	m2	LE	0.60	14.80	PH	0.60	8.75	1.54	25.09	28.85
exceeding 1.22 m	m2	LE	0.56	13.81	PH	0.56	8.17	1.54	23.52	27.05

Rough finish, curved radius in
one plane 2m radius

not exceeding 0.10 m	m	LE	0.24	5.92	PH	0.24	3.50	0.31	9.73	11.19
0.10 -0.20 m	m	LE	0.28	6.90	PH	0.28	4.09	0.61	11.60	13.34
0.20 -0.40 m	m2	LE	0.62	15.29	PH	0.62	9.05	1.54	25.88	29.76
0.40 -1.22 m	m2	LE	0.58	14.30	PH	0.58	8.46	1.54	24.30	27.95
exceeding 1.22 m	m2	LE	0.54	13.32	PH	0.54	7.88	1.54	22.74	26.15

Rough finish left in, for small
voids, depth

not exceeding 0.50 m	each	LE	0.35	8.63	PH	0.35	5.11	0.46	14.20	16.33
0.50 -1.00 m	each	LE	0.45	11.10	PH	0.45	6.57	0.93	18.60	21.39
1.00 -2.00 m	each	LE	0.70	17.26	PH	0.70	10.21	1.86	29.33	33.73
2.00 -5.00 m	each	LE	2.25	55.48	PH	2.25	32.83	4.64	92.95	106.89

Rough finish left in, for large
voids, depth

not exceeding 0.50 m	each	LE	0.70	17.26	PH	0.70	10.21	2.32	29.79	34.26
0.50 -1.00 m	each	LE	0.90	22.19	PH	0.90	13.13	4.64	39.96	45.95
1.00 -2.00 m	each	LE	1.40	34.52	PH	1.40	20.43	9.29	64.24	73.88
2.00 -5.00 m	each	LE	4.50	110.97	PH	4.50	65.66	23.22	199.85	229.83

New work	Unit	Labour gang hours	Labour net		Plant gang hours	Plant net	Mats net	Price net	Price with 15%
Concrete work			£			£	£	£	£
							VAT not included		

Formwork (*continued*)

Rough finish for concrete components of constant cross-section, three sides of isolated beams or columns, size

	Unit								
100 x 200 mm	each	LE	0.20	4.93	PH	0.20	2.92	0.77	8.62	9.91
100 x 250 mm	each	LE	0.25	6.17	PH	0.25	3.65	0.92	10.74	12.35
100 x 300 mm	each	LE	0.28	6.90	PH	0.28	4.09	1.08	12.07	13.88
150 x 250 mm	each	LE	0.25	6.17	PH	0.25	3.65	1.00	10.82	12.44
200 x 200 mm	each	LE	0.25	6.17	PH	0.25	3.65	0.92	10.74	12.35
200 x 250 mm	each	LE	0.28	6.90	PH	0.28	4.09	1.08	12.07	13.88
200 x 300 mm	each	LE	0.30	7.40	PH	0.30	4.38	1.24	13.02	14.97
250 x 250 mm	each	LE	0.30	7.40	PH	0.30	4.38	1.15	12.93	14.87
250 x 300 mm	each	LE	0.30	7.40	PH	0.30	4.38	1.31	13.09	15.05
250 x 350 mm	each	LE	0.30	7.40	PH	0.30	4.38	1.47	13.25	15.24
300 x 300 mm	each	LE	0.35	8.63	PH	0.35	5.11	1.38	15.12	17.39
300 x 350 mm	each	LE	0.35	8.63	PH	0.35	5.11	1.54	15.28	17.57
300 x 400 mm	each	LE	0.40	9.86	PH	0.40	5.84	1.70	17.40	20.01
300 x 500 mm	each	LE	0.40	9.86	PH	0.40	5.84	2.01	17.71	20.37
400 x 400 mm	each	LE	0.40	9.86	PH	0.40	5.84	1.85	17.55	20.18
400 x 500 mm	each	LE	0.45	11.10	PH	0.45	6.57	2.15	19.82	22.79
500 x 500 mm	each	LE	0.50	12.33	PH	0.50	7.29	2.31	21.93	25.22

Rough finish for concrete components of constant cross-section, four sides of detached columns, size

	Unit								
100 x 200 mm	each	LE	0.20	4.93	PH	0.20	2.92	0.92	8.77	10.09
100 x 250 mm	each	LE	0.25	6.17	PH	0.25	3.65	1.08	10.90	12.54
100 x 300 mm	each	LE	0.28	6.90	PH	0.28	4.09	1.23	12.22	14.05
150 x 250 mm	each	LE	0.25	6.17	PH	0.25	3.65	1.23	11.05	12.71
200 x 200 mm	each	LE	0.25	6.17	PH	0.25	3.65	1.23	11.05	12.71
200 x 250 mm	each	LE	0.28	6.90	PH	0.28	4.09	1.38	12.37	14.23
200 x 300 mm	each	LE	0.30	7.40	PH	0.30	4.38	1.38	13.16	15.13
250 x 250 mm	each	LE	0.30	7.40	PH	0.30	4.38	1.54	13.32	15.32
250 x 300 mm	each	LE	0.30	7.40	PH	0.30	4.38	1.70	13.48	15.50
250 x 350 mm	each	LE	0.30	7.40	PH	0.30	4.38	1.85	13.63	15.67
300 x 300 mm	each	LE	0.35	8.63	PH	0.35	5.11	1.85	15.59	17.93
300 x 350 mm	each	LE	0.35	8.63	PH	0.35	5.11	2.00	15.74	18.10
300 x 400 mm	each	LE	0.40	9.86	PH	0.40	5.84	2.15	17.85	20.53
300 x 500 mm	each	LE	0.40	9.86	PH	0.40	5.84	2.46	18.16	20.88
400 x 400 mm	each	LE	0.40	9.86	PH	0.40	5.84	2.46	18.16	20.88
400 x 500 mm	each	LE	0.45	11.10	PH	0.45	6.57	2.77	20.44	23.51
500 x 500 mm	each	LE	0.50	12.33	PH	0.50	7.29	3.08	22.70	26.11

Fair finish, plane horizontal

	Unit								
not exceeding 0.10 m	m	LE	0.15	3.70	PH	0.15	2.19	0.57	6.46	7.43
0.10 -0.20 m	m	LE	0.20	4.93	PH	0.20	2.92	1.14	8.99	10.34
0.20 -0.40 m	m2	LE	0.50	12.33	PH	0.50	7.29	2.86	22.48	25.85
0.40 -1.22 m	m2	LE	0.45	11.10	PH	0.45	6.57	2.86	20.53	23.61
exceeding 1.22 m	m2	LE	0.40	9.86	PH	0.40	5.84	2.86	18.56	21.34

Civil engineering

New work	Unit		Labour gang hours	Labour net		Plant gang hours	Plant net	Mats net	Price net	Price with 15%
Concrete work				£			£	£	£	£
								VAT not included		
Formwork (*continued*)										
Fair finish, plane sloping										
not exceeding 0.10 m	m	LE	0.17	4.19	PH	0.17	2.48	0.57	7.24	8.33
0.10 -0.20 m	m	LE	0.22	5.43	PH	0.22	3.21	1.14	9.78	11.25
0.20 -0.40 m	m2	LE	0.54	13.32	PH	0.54	7.88	2.86	24.06	27.67
0.40 -1.22 m	m2	LE	0.48	11.84	PH	0.48	7.00	2.86	21.70	24.95
exceeding 1.22 m	m2	LE	0.44	10.85	PH	0.44	6.42	2.86	20.13	23.15
Fair finish, plane vertical										
not exceeding 0.10 m	m	LE	0.24	5.92	PH	0.24	3.50	0.57	9.99	11.49
0.10 -0.20 m	m	LE	0.28	6.90	PH	0.28	4.09	1.14	12.13	13.95
0.20 -0.40 m	m2	LE	0.62	15.29	PH	0.62	9.05	2.86	27.20	31.28
0.40 -1.22 m	m2	LE	0.56	13.81	PH	0.56	8.17	2.86	24.84	28.57
exceeding 1.22 m	m2	LE	0.52	12.82	PH	0.52	7.59	2.86	23.27	26.76
Fair finish, curved radius in one plane 0.5m radius										
not exceeding 0.10 m	m	LE	0.28	6.90	PH	0.28	4.09	0.57	11.56	13.29
0.10 -0.20 m	m	LE	0.32	7.89	PH	0.32	4.67	1.14	13.70	15.76
0.20 -0.40 m	m2	LE	0.66	16.28	PH	0.66	9.63	2.86	28.77	33.09
0.40 -1.22 m	m2	LE	0.62	15.29	PH	0.62	9.05	2.86	27.20	31.28
exceeding 1.22 m	m2	LE	0.58	14.30	PH	0.58	8.46	2.86	25.62	29.46
Fair finish, curved radius in one plane 1m radius										
not exceeding 0.10 m	m	LE	0.26	6.41	PH	0.26	3.79	0.57	10.77	12.39
0.10 -0.20 m	m	LE	0.30	7.40	PH	0.30	4.38	1.14	12.92	14.86
0.20 -0.40 m	m2	LE	0.64	15.78	PH	0.64	9.34	2.86	27.98	32.18
0.40 -1.22 m	m2	LE	0.60	14.80	PH	0.60	8.75	2.86	26.41	30.37
exceeding 1.22 m	m2	LE	0.56	13.81	PH	0.56	8.17	2.86	24.84	28.57
Fair finish, curved radius in one plane 2m radius										
not exceeding 0.10 m	m	LE	0.24	5.92	PH	0.24	3.50	0.57	9.99	11.49
0.10 -0.20 m	m	LE	0.28	6.90	PH	0.28	4.09	1.14	12.13	13.95
0.20 -0.40 m	m2	LE	0.62	15.29	PH	0.62	9.05	2.86	27.20	31.28
0.40 -1.22 m	m2	LE	0.58	14.30	PH	0.56	8.17	2.86	25.33	29.13
exceeding 1.22 m	m2	LE	0.54	13.32	PH	0.54	7.88	2.86	24.06	27.67
Fair finish left in, for small voids, depth										
not exceeding 0.50 m	each	LE	0.35	8.63	PH	0.35	5.11	0.86	14.60	16.79
0.50 -1.00 m	each	LE	0.45	11.10	PH	0.45	6.57	1.72	19.39	22.30
1.00 -2.00 m	each	LE	0.70	17.26	PH	0.70	10.21	3.45	30.92	35.56
2.00 -5.00 m	each	LE	2.25	55.48	PH	2.25	32.83	8.61	96.92	111.46
Fair finish left in, for large voids, depth										
not exceeding 0.50 m	each	LE	0.70	17.26	PH	0.70	10.21	4.31	31.78	36.55
0.50 -1.00 m	each	LE	0.90	22.19	PH	0.90	13.13	8.61	43.93	50.52
1.00 -2.00 m	each	LE	1.40	34.52	PH	1.40	20.43	17.23	72.18	83.01
2.00 -5.00 m	each	LE	4.50	110.97	PH	4.50	65.66	43.07	219.70	252.66

Civil engineering

New work	Unit		Labour gang hours	Labour net		Plant gang hours	Plant net	Mats net	Price net	Price with 15%
				£			£	£	£	£
Concrete work										
									VAT not included	

Formwork *(continued)*

Fair finish for concrete
components of constant cross-
section, three sides of isolated
beams or columns, size

100 x 200 mm	each	LE	0.20	4.93	PH	0.20	2.92	1.43	9.28	10.67
100 x 250 mm	each	LE	0.25	6.17	PH	0.25	3.65	1.71	11.53	13.26
100 x 300 mm	each	LE	0.28	6.90	PH	0.28	4.09	2.00	12.99	14.94
150 x 250 mm	each	LE	0.25	6.17	PH	0.25	3.65	1.86	11.68	13.43
200 x 200 mm	each	LE	0.25	6.17	PH	0.25	3.65	1.71	11.53	13.26
200 x 250 mm	each	LE	0.28	6.90	PH	0.28	4.09	2.00	12.99	14.94
200 x 300 mm	each	LE	0.30	7.40	PH	0.30	4.38	2.29	14.07	16.18
250 x 250 mm	each	LE	0.30	7.40	PH	0.30	4.38	2.14	13.92	16.01
250 x 300 mm	each	LE	0.30	7.40	PH	0.30	4.38	2.43	14.21	16.34
250 x 350 mm	each	LE	0.30	7.40	PH	0.30	4.38	2.72	14.50	16.68
300 x 300 mm	each	LE	0.35	8.63	PH	0.35	5.11	2.57	16.31	18.76
300 x 350 mm	each	LE	0.35	8.63	PH	0.35	5.11	2.86	16.60	19.09
300 x 400 mm	each	LE	0.40	9.86	PH	0.40	5.84	3.15	18.85	21.68
300 x 500 mm	each	LE	0.40	9.86	PH	0.40	5.84	3.72	19.42	22.33
400 x 400 mm	each	LE	0.40	9.86	PH	0.40	5.84	3.43	19.13	22.00
400 x 500 mm	each	LE	0.45	11.10	PH	0.45	6.57	4.00	21.67	24.92
500 x 500 mm	each	LE	0.50	12.33	PH	0.50	7.29	4.29	23.91	27.50

Fair finish for concrete
components of constant cross-
section, four sides of detached
columns, size

100 x 200 mm	each	LE	0.20	4.93	PH	0.20	2.92	1.71	9.56	10.99
100 x 250 mm	each	LE	0.25	6.17	PH	0.25	3.65	2.00	11.82	13.59
100 x 300 mm	each	LE	0.28	6.90	PH	0.28	4.09	2.27	13.26	15.25
150 x 250 mm	each	LE	0.25	6.17	PH	0.25	3.65	2.27	12.09	13.90
200 x 200 mm	each	LE	0.25	6.17	PH	0.25	3.65	2.27	12.09	13.90
200 x 250 mm	each	LE	0.28	6.90	PH	0.28	4.09	2.57	13.56	15.59
200 x 300 mm	each	LE	0.30	7.40	PH	0.30	4.38	2.57	14.35	16.50
250 x 250 mm	each	LE	0.30	7.40	PH	0.30	4.38	2.86	14.64	16.84
250 x 300 mm	each	LE	0.30	7.40	PH	0.30	4.38	3.15	14.93	17.17
250 x 350 mm	each	LE	0.30	7.40	PH	0.30	4.38	3.43	15.21	17.49
300 x 300 mm	each	LE	0.35	8.63	PH	0.35	5.11	3.43	17.17	19.75
300 x 350 mm	each	LE	0.35	8.63	PH	0.35	5.11	3.70	17.44	20.06
300 x 400 mm	each	LE	0.40	9.86	PH	0.40	5.84	4.00	19.70	22.66
300 x 500 mm	each	LE	0.40	9.86	PH	0.40	5.84	4.57	20.27	23.31
400 x 400 mm	each	LE	0.40	9.86	PH	0.40	5.84	4.57	20.27	23.31
400 x 500 mm	each	LE	0.45	11.10	PH	0.45	6.57	5.13	22.80	26.22
500 x 500 mm	each	LE	0.50	12.33	PH	0.50	7.29	5.72	25.34	29.14

Civil engineering

New work	Unit	Labour gang hours	Labour net		Plant gang hours	Plant net	Mats net	Price net	Price with 15%
			£			£	£	£	£

Concrete work

VAT not included

Bar reinforcement

Mild steel round bars to BS4449
in straight lengths, nominal size

6 mm	tonne	LF	11.00	246.84	PI	11.00	149.49	491.49	887.82	1020.99
8 mm	tonne	LF	10.00	224.40	PI	10.00	135.90	449.98	810.28	931.82
10 mm	tonne	LF	9.50	213.18	PI	9.50	129.10	414.62	756.90	870.43
12 mm	tonne	LF	9.00	201.96	PI	9.00	122.31	387.44	711.71	818.47
16 mm	tonne	LF	8.00	179.52	PI	8.00	108.72	358.24	646.48	743.45
20 mm	tonne	LF	7.00	157.08	PI	7.00	95.13	350.04	602.25	692.59
25 mm	tonne	LF	6.00	134.64	PI	6.00	81.54	350.04	566.22	651.15

Deformed high yield steel bars
to BS4449 in straight lengths,
nominal size

8 mm	tonne	LF	10.00	224.40	PI	10.00	135.90	424.36	784.66	902.36
10 mm	tonne	LF	9.50	213.18	PI	9.50	129.10	395.14	737.42	848.03
12 mm	tonne	LF	9.00	201.96	PI	9.00	122.31	378.74	703.01	808.46
16 mm	tonne	LF	8.00	179.52	PI	8.00	108.72	354.14	642.38	738.74
20 mm	tonne	LF	7.00	157.08	PI	7.00	95.13	354.14	606.35	697.30
25 mm	tonne	LF	6.00	134.64	PI	6.00	81.54	354.14	570.32	655.87

Mesh reinforcement

Nominal mass not exceeding
2kg/m2

A98	m2	LF	0.06	1.35	PI	0.06	0.82	0.77	2.94	3.38

Nominal mass 2-3kg/m2

A142	m2	LF	0.07	1.57	PI	0.07	0.95	0.92	3.44	3.96
C283	m2	LF	0.07	1.57	PI	0.07	0.95	1.42	3.94	4.53

Nominal mass 3-4kg/m2

A193	m2	LF	0.08	1.80	PI	0.08	1.09	1.20	4.09	4.70
C385	m2	LF	0.08	1.80	PI	0.08	1.09	1.50	4.39	5.05
B283	m2	LF	0.08	1.80	PI	0.08	1.09	1.42	4.31	4.96
A252	m2	LF	0.08	1.80	PI	0.08	1.09	1.68	4.57	5.26

Nominal mass 4-5kg/m2

C503	m2	LF	0.10	2.24	PI	0.10	1.36	1.91	5.51	6.34
B385	m2	LF	0.10	2.24	PI	0.10	1.36	1.97	5.57	6.41

Nominal mass 5-6kg/m2

B503	m2	LF	0.12	2.69	PI	0.12	1.63	2.62	6.94	7.98

Nominal mass 6-7kg/m2

A393	m2	LF	0.15	3.37	PI	0.15	2.04	2.54	7.95	9.14

Nominal mass 8.14kg/m2

B785	m2	LF	0.18	4.04	PI	0.18	2.45	3.50	9.99	11.49

Civil engineering

New work	Unit		Labour gang hours	Labour net		Plant gang hours	Plant net	Mats net	Price net	Price with 15%
Concrete work				£			£	£	£	£
								VAT not included		
Joints										
Open surface plain joint including scabbling concrete										
not exceeding 0.50 m	m2	LG	0.08	1.11	PJ	0.08	0.90	-	2.01	2.31
0.50 - 1.00 m	m2	LG	0.07	0.97	PJ	0.07	0.79	-	1.76	2.02
1.50 m	m2	LG	0.06	0.84	PJ	0.06	0.68	-	1.52	1.75
Open surface joint with Filcrete joint filler 12 mm thick, width										
not exceeding 0.50 m	m2	LG	0.18	2.51		-	-	4.12	6.63	7.62
0.50 - 1.00 m	m2	LG	0.16	2.23		-	-	4.12	6.35	7.30
1.50 m	m2	LG	0.14	1.95		-	-	4.12	6.07	6.98
Open surface joint with Filcrete joint filler 15 mm thick, width										
not exceeding 0.50 m	m2	LG	0.20	2.78		-	-	6.14	8.92	10.26
0.50 - 1.00 m	m2	LG	0.18	2.51		-	-	6.14	8.65	9.95
1.50 m	m2	LG	0.16	2.23		-	-	6.14	8.37	9.63
Open surface joint with Filcrete joint filler 19 mm thick, width										
not exceeding 0.50 m	m2	LG	0.22	3.06		-	-	7.01	10.07	11.58
0.50 - 1.00 m	m2	LG	0.20	2.78		-	-	7.01	9.79	11.26
1.50 m	m2	LG	0.18	2.51		-	-	7.01	9.52	10.95
Open surface joint with Filcrete joint filler 25 mm thick, width										
not exceeding 0.50 m	m2	LG	0.26	3.62		-	-	8.84	12.46	14.33
0.50 - 1.00 m	m2	LG	0.24	3.34		-	-	8.84	12.18	14.01
1.50 m	m2	LG	0.22	3.06		-	-	8.84	11.90	13.69
Formed plain surface joint including formwork, width										
not exceeding 0.50 m	m2	LG	0.48	6.68	PJ	0.48	5.40	1.55	13.63	15.68
0.50 - 1.00 m	m2	LG	0.45	6.26	PJ	0.45	5.07	1.55	12.88	14.81
1.50 m	m2	LG	0.42	5.85	PJ	0.42	4.73	1.55	12.13	13.95
Formed surface joint with Filcrete joint filler 12 mm thick and formwork, width										
not exceeding 0.50 m	m2	LG	0.56	7.80	PJ	0.56	6.31	5.66	19.77	22.74
0.50 - 1.00 m	m2	LG	0.52	7.24	PJ	0.52	5.86	5.66	18.76	21.57
1.50 m	m2	LG	0.48	6.68	PJ	0.48	5.40	5.66	17.74	20.40
Formed surface joint with Filcrete joint filler 15 mm thick and formwork, width										
not exceeding 0.50 m	m2	LG	0.60	8.35	PJ	0.60	6.76	7.68	22.79	26.21
0.50 - 1.00 m	m2	LG	0.56	7.80	PJ	0.56	6.31	7.68	21.79	25.06
1.50 m	m2	LG	0.52	7.24	PJ	0.52	5.86	7.68	20.78	23.90

Civil engineering

New work	Unit		Labour gang hours	Labour net		Plant gang hours	Plant net	Mats net	Price net	Price with 15%
Concrete work				£			£	£	£	£
								VAT not included		
Joints (*continued*)										
Formed surface joint with Filcrete joint filler 19 mm thick and formwork, width										
not exceeding 0.50 m	m2	LG	0.66	9.19	PJ	0.66	7.43	8.55	25.17	28.95
0.50 - 1.00 m	m2	LG	0.62	8.63	PJ	0.62	6.98	8.55	24.16	27.78
1.50 m	m2	LG	0.58	8.07	PJ	0.58	6.53	8.55	23.15	26.62
Formed surface joint with Filcrete joint filler 25 mm thick and formwork, width										
not exceeding 0.50 m	m2	LG	0.72	10.02	PJ	0.72	8.11	10.38	28.51	32.79
0.50 - 1.00 m	m2	LG	0.68	9.47	PJ	0.68	7.66	10.38	27.51	31.64
1.50 m	m2	LG	0.64	8.91	PJ	0.64	7.21	10.38	26.50	30.48
Water stops										
PVC-U flat dumbbell, width										
100 mm	m	LG	0.14	1.95	-	-		2.72	4.67	5.37
170 mm	m	LG	0.16	2.23	-	-		3.71	5.94	6.83
210 mm	m	LG	0.18	2.51	-	-		4.79	7.30	8.39
250 mm	m	LG	0.20	2.78	-	-		5.96	8.74	10.05
PVC-U centre bulb, width										
100 mm	m	LG	0.14	1.95	-	-		3.98	5.93	6.82
170 mm	m	LG	0.16	2.23	-	-		4.60	6.83	7.85
210 mm	m	LG	0.18	2.51	-	-		5.74	8.25	9.49
250 mm	m	LG	0.20	2.78	-	-		6.53	9.31	10.71
Rubber flat dumbbell, width										
150 mm	m	LG	0.16	2.23	-	-		15.22	17.45	20.07
230 mm	m	LG	0.20	2.78	-	-		22.70	25.48	29.30
Rubber centre bulb, width										
150 mm	m	LG	0.16	2.23	-	-		17.42	19.65	22.60
230 mm	m	LG	0.20	2.78	-	-		26.14	28.92	33.26

Civil engineering

New work	Unit		Labour gang hours	Labour net		Plant gang hours	Plant net	Mats net	Price net	Price with 15%
Roads and pavings				£			£	£	£	£
									VAT not included	

Sub-bases, flexible road bases and surfaces

Granular material DTp Spec. type1, depth

100 mm	m2	LD	0.01	0.23	PK	0.01	0.27	1.23	1.73	1.99
150 mm	m2	LD	0.02	0.46	PK	0.02	0.54	1.84	2.84	3.27
200 mm	m2	LD	0.02	0.46	PK	0.02	0.54	2.46	3.46	3.98
250 mm	m2	LD	0.03	0.69	PK	0.03	0.80	3.07	4.56	5.24
300 mm	m2	LD	0.03	0.69	PK	0.03	0.80	3.68	5.17	5.95
400 mm	m2	LD	0.04	0.92	PK	0.04	1.07	4.91	6.90	7.93
500 mm	m2	LD	0.04	0.92	PK	0.04	1.07	6.14	8.13	9.35

Granular material DTp Spec type 2, depth

100 mm	m2	LD	0.01	0.23	PK	0.01	0.27	1.19	1.69	1.94
150 mm	m2	LD	0.02	0.46	PK	0.02	0.54	1.79	2.79	3.21
200 mm	m2	LD	0.02	0.46	PK	0.02	0.54	2.38	3.38	3.89
250 mm	m2	LD	0.03	0.69	PK	0.03	0.80	2.98	4.47	5.14
300 mm	m2	LD	0.03	0.69	PK	0.03	0.80	3.58	5.07	5.83
400 mm	m2	LD	0.04	0.92	PK	0.04	1.07	4.77	6.76	7.77
500 mm	m2	LD	0.04	0.92	PK	0.04	1.07	5.96	7.95	9.14

Soil/cement (100kg cement to 1m3 soil), depth

100 mm	m2	LD	0.01	0.23	PK	0.01	0.27	1.10	1.60	1.84
150 mm	m2	LD	0.02	0.46	PK	0.02	0.54	1.66	2.66	3.06
200 mm	m2	LD	0.02	0.46	PK	0.02	0.54	2.21	3.21	3.69
250 mm	m2	LD	0.03	0.69	PK	0.03	0.80	2.76	4.25	4.89
300 mm	m2	LD	0.03	0.69	PK	0.03	0.80	3.31	4.80	5.52

Hardcore filling levelled and compacted, depth

100 mm	m2	LD	0.01	0.23	PK	0.01	0.27	1.35	1.85	2.13
150 mm	m2	LD	0.02	0.46	PK	0.02	0.54	2.02	3.02	3.47
200 mm	m2	LD	0.02	0.46	PK	0.02	0.54	2.69	3.69	4.24
250 mm	m2	LD	0.03	0.69	PK	0.03	0.80	3.37	4.86	5.59
300 mm	m2	LD	0.03	0.69	PK	0.03	0.80	4.04	5.53	6.36
400 mm	m2	LD	0.04	0.92	PK	0.04	1.07	5.39	7.38	8.49
500 mm	m2	LD	0.04	0.92	PK	0.04	1.07	6.74	8.73	10.04

Civil engineering

VAT not included

New work	Unit	Specialist price net	Price with 15%

Roads and pavings

£ **£**

The following "Specialist price net" figures are guide prices only. Circumstances greatly affect the cost of this work and it is essential that quotations be obtained on every occasion.

Prices do not include for cash discount.

See the preamble notes for builder's profit and attendance.

Macadam surfaces

Wet mix macadam, single course, depth

75 mm	m2	1.40	1.61
100 mm	m2	1.95	2.24
150 mm	m2	2.85	3.28
200 mm	m2	3.75	4.31

Dry bound macadam single course, depth

75 mm	m2	1.40	1.61
100 mm	m2	1.95	2.24
150 mm	m2	2.85	3.28
200 mm	m2	3.75	4.31

Dense bitumen macadam base course, depth

40 mm	m2	2.60	2.99
50 mm	m2	3.00	3.45
75 mm	m2	4.12	4.74
100 mm	m2	5.48	6.30

Dense bitumen macadam wearing course, depth

20 mm	m2	1.95	2.24
25 mm	m2	2.15	2.47
30 mm	m2	2.40	2.76
35 mm	m2	2.75	3.16
40 mm	m2	3.05	3.51
45 mm	m2	3.35	3.85
50 mm	m2	3.60	4.14
60 mm	m2	4.25	4.89
75 mm	m2	5.00	5.75

Open textured bitumen macadam base course, depth

40 mm	m2	2.75	3.16
50 mm	m2	3.15	3.62
75 mm	m2	4.30	4.95
100 mm	m2	5.75	6.61

Open textured bitumen macadam wearing course, depth

20 mm	m2	1.55	1.78
30 mm	m2	2.05	2.36
40 mm	m2	2.25	2.59

Civil engineering

New work	Unit	Specialist price net	Price with 15%
Roads and pavings		£	£
		VAT not included	
Cold asphalt			
Cold asphalt wearing course, depth			
15 mm	m2	1.65	1.90
20 mm	m2	2.05	2.36
25 mm	m2	2.48	2.85
30 mm	m2	2.85	3.28
Coated chippings			
Coated chippings rolled into macadam course, nominal size 8mm, rate			
4kg/m2	tonne	0.40	0.46
6kg/m2	tonne	0.45	0.52
8kg/m2	tonne	0.50	0.58
Coated chippings rolled into macadam course, nominal size 10mm, rate			
6kg/m2	tonne	0.50	0.58
8kg/m2	tonne	0.55	0.63
10kg/m2	tonne	0.60	0.69
Bitumen tack coat sprayed on	m2	0.14	0.16

Civil engineering

New work	Unit		Labour gang hours	Labour net		Plant gang hours	Plant net	Mats net	Price net	Price with 15%
Roads and pavings				£			£	£	£	£
									VAT not included	

Kerbs, channels and edgings

Precast concrete kerb size 125
x 255 mm, laid on concrete
(1:3:6 40 mm aggregate)
bedded, pointed, jointed and
haunched in cement mortar
(1:3), laid

	Unit		L gang	L net		P gang	P net	Mats net	Price net	Price 15%
straight or curved to radius exceeding 12 m	m	LH	0.10	3.01	PL	0.10	0.39	10.00	13.40	15.41
curved to radius 5 m	m	LH	0.16	4.81	PL	0.16	0.62	12.21	17.64	20.29
curved to radius 7 m	m	LH	0.14	4.21	PL	0.14	0.54	12.21	16.96	19.50
curved to radius 10 m	m	LH	0.12	3.61	PL	0.12	0.46	12.21	16.28	18.72

Precast concrete kerb size 150
x 305 mm, laid on concrete
(1:3:6 40 mm aggregate)
bedded, pointed, jointed and
haunched in cement mortar
(1:3), laid

straight or curved to radius exceeding 12 m	m	LH	0.10	3.01	PL	0.10	0.39	11.30	14.70	16.91
curved to radius 5 m	m	LH	0.16	4.81	PL	0.16	0.62	11.30	16.73	19.24
curved to radius 7 m	m	LH	0.14	4.21	PL	0.14	0.54	11.30	16.05	18.46
curved to radius 10 m	m	LH	0.12	3.61	PL	0.12	0.46	11.30	15.37	17.68
Quadrants, size										
305 x 305 x 150 mm	each	LH	0.08	2.40	PL	0.08	0.31	10.01	12.72	14.63
455 x 455 x 150 mm	each	LH	0.08	2.40	PL	0.08	0.31	11.73	14.44	16.61

Precast concrete channel size
255 x 125 mm, bedded, pointed,
jointed and haunched in cement
mortar (1.3), laid

straight or curved to radius exceeding 12 m	m	LH	0.10	3.01	PL	0.10	0.39	6.22	9.62	11.06
curved to radius 5 m	m	LH	0.16	4.81	PL	0.16	0.62	8.42	13.85	15.93
curved to radius 7 m	m	LH	0.14	4.21	PL	0.14	0.54	8.42	13.17	15.15
curved to radius 10 m	m	LH	0.12	3.61	PL	0.12	0.46	8.42	12.49	14.36

Precast concrete channel size
150 x 125 mm, bedded pointed,
jointed and haunched in cement
mortar (1.3), laid

straight or curved to radius exceeding 12 m	m	LH	0.10	3.01	PL	0.10	0.39	4.77	8.17	9.40
curved to radius 5 m	m	LH	0.16	4.81	PL	0.16	0.62	6.30	11.73	13.49
curved to radius 7 m	m	LH	0.14	4.21	PL	0.14	0.54	6.30	11.05	12.71
curved to radius 10 m	m	LH	0.12	3.61	PL	0.12	0.46	6.30	10.37	11.93

Civil engineering

New work	Unit	Labour gang hours	Labour net	Plant gang hours	Plant net	Mats net	Price net	Price with 15%
Roads and pavings			£		£	£	£	£
						VAT not included		

Light duty pavements

Precast concrete plain paving
flags, bedded on dabs of
cement lime mortar

	Unit		Labour		Plant	Mats	Price	Price
900 x 600 x 50 mm	m2	LH	0.08	2.40	PL	0.08	0.31	4.89	7.60	8.74
750 x 600 x 50 mm	m2	LH	0.10	3.01	PL	0.10	0.39	5.33	8.73	10.04
600 x 600 x 50 mm	m2	LH	0.12	3.61	PL	0.12	0.46	9.06	13.13	15.10
600 x 450 x 50 mm	m2	LH	0.14	4.21	PL	0.14	0.54	6.54	11.29	12.98
450 x 450 x 50 mm	m2	LH	0.16	4.81	PL	0.16	0.62	11.03	16.46	18.93

New work	Unit	Labour gang hours		Labour net		Plant gang hours		Plant net	Mats net	Price net	Price with 15%
Sewers and manholes				£				£	£	£	£
									VAT not included		
Pipework											
Vitrified clay spigot and socket pipes, jointed in cement mortar, 300 mm diameter, depth											
not exceeding 1.50 m	m	LH	0.15	4.51	PM	0.15		8.14	20.34	32.99	37.94
1.50 - 2.00 m	m	LH	0.20	6.01	PM	0.20		10.86	20.34	37.21	42.79
2.00 - 2.50 m	m	LH	0.25	7.51	PM	0.25		13.57	20.34	41.42	47.63
2.50 - 3.00 m	m	LH	0.30	9.02	PM	0.30		16.28	20.34	45.64	52.49
3.00 - 3.50 m	m	LH	0.35	10.52	PM	0.35		19.00	20.34	49.86	57.34
3.50 - 4.00 m	m	LH	0.40	12.02	PM	0.40		21.71	20.34	54.07	62.18
4.00 - 4.50 m	m	LH	0.45	13.53	PM	0.45		24.43	20.34	58.30	67.05
4.50 - 5.00 m	m	LH	0.50	15.03	PM	0.50		27.14	20.34	62.51	71.89
Vitrified clay spigot and socket pipes, jointed in cement mortar, 375 mm diameter, depth											
not exceeding 1.50 m	m	LH	0.17	5.11	PM	0.17		9.23	38.98	53.32	61.32
1.50 - 2.00 m	m	LH	0.22	6.61	PM	0.22		11.94	38.98	57.53	66.16
2.00 - 2.50 m	m	LH	0.27	8.12	PM	0.27		14.66	38.98	61.76	71.02
2.50 - 3.00 m	m	LH	0.32	9.62	PM	0.32		17.37	38.98	65.97	75.87
3.00 - 3.50 m	m	LH	0.37	11.12	PM	0.37		20.08	38.98	70.18	80.71
3.50 - 4.00 m	m	LH	0.42	12.63	PM	0.42		22.80	38.98	74.41	85.57
4.00 - 4.50 m	m	LH	0.47	14.13	PM	0.47		25.51	38.98	78.62	90.41
4.50 - 5.00 m	m	LH	0.52	15.63	PM	0.52		28.23	38.98	82.84	95.27
Vitrified clay spigot and socket pipes, jointed in cement mortar, 400 mm diameter, depth											
not exceeding 1.50 m	m	LH	0.20	6.01	PM	0.20		10.86	39.38	56.25	64.69
1.50 - 2.00 m	m	LH	0.25	7.51	PM	0.25		13.57	39.38	60.46	69.53
2.00 - 2.50 m	m	LH	0.30	9.02	PM	0.30		16.28	39.38	64.68	74.38
2.50 - 3.00 m	m	LH	0.35	10.52	PM	0.35		19.00	39.38	68.90	79.23
3.00 - 3.50 m	m	LH	0.40	12.02	PM	0.40		21.71	39.38	73.11	84.08
3.50 - 4.00 m	m	LH	0.45	13.53	PM	0.45		24.43	39.38	77.34	88.94
4.00 - 4.50 m	m	LH	0.50	15.03	PM	0.50		27.14	39.38	81.55	93.78
4.50 - 5.00 m	m	LH	0.55	16.53	PM	0.55		29.85	39.38	85.76	98.62
Vitrified clay spigot and socket pipes, jointed in cement mortar, 450 mm diameter, depth											
not exceeding 1.50 m	m	LH	0.22	6.61	PM	0.22		11.94	52.43	70.98	81.63
1.50 - 2.00 m	m	LH	0.27	8.12	PM	0.27		14.66	52.43	75.21	86.49
2.00 - 2.50 m	m	LH	0.32	9.62	PM	0.32		17.37	52.43	79.42	91.33
2.50 - 3.00 m	m	LH	0.37	11.12	PM	0.37		20.08	52.43	83.63	96.17
3.00 - 3.50 m	m	LH	0.42	12.63	PM	0.42		22.80	52.43	87.86	101.04
3.50 - 4.00 m	m	LH	0.47	14.13	PM	0.47		25.51	52.43	92.07	105.88
4.00 - 4.50 m	m	LH	0.52	15.63	PM	0.52		28.23	52.43	96.29	110.73
4.50 - 5.00 m	m	LH	0.57	17.13	PM	0.57		30.94	52.43	100.50	115.58

Civil engineering

Sewers and manholes

£   £   £   £   £

VAT not included

Pipework (*continued*)

Concrete pipes, class L, vibrated, flexible joints, 300 mm diameter, depth

not exceeding 1.50 m	m	LH 0.20	6.01	PM 0.20	10.86	14.40	31.27	35.96
1.50 - 2.00 m	m	LH 0.25	7.51	PM 0.25	13.57	14.40	35.48	40.80
2.00 - 2.50 m	m	LH 0.30	9.02	PM 0.30	16.28	14.40	39.70	45.66
2.50 - 3.00 m	m	LH 0.35	10.52	PM 0.35	19.00	14.40	43.92	50.51
3.00 - 3.50 m	m	LH 0.40	12.02	PM 0.40	21.71	14.40	48.13	55.35
3.50 - 4.00 m	m	LH 0.45	13.53	PM 0.45	24.43	14.40	52.36	60.21
4.00 - 4.50 m	m	LH 0.50	15.03	PM 0.50	27.14	14.40	56.57	65.06
4.50 - 5.00 m	m	LH 0.55	16.53	PM 0.55	29.85	14.40	60.78	69.90

Concrete pipes, class L, vibrated, flexible joints, 375 mm diameter, depth

not exceeding 1.50 m	m	LH 0.22	6.61	PM 0.22	11.94	18.74	37.29	42.88
1.50 - 2.00 m	m	LH 0.27	8.12	PM 0.27	14.66	18.74	41.52	47.75
2.00 - 2.50 m	m	LH 0.32	9.62	PM 0.32	17.37	18.74	45.73	52.59
2.50 - 3.00 m	m	LH 0.37	11.12	PM 0.37	20.08	18.74	49.94	57.43
3.00 - 3.50 m	m	LH 0.42	12.63	PM 0.42	22.80	18.74	54.17	62.30
3.50 - 4.00 m	m	LH 0.47	14.13	PM 0.47	25.51	18.74	58.38	67.14
4.00 - 4.50 m	m	LH 0.52	15.63	PM 0.52	28.23	18.74	62.60	71.99
4.50 - 5.00 m	m	LH 0.57	17.13	PM 0.57	30.94	18.74	66.81	76.83

Concrete pipes, class L, vibrated, flexible joints, 450 mm diameter, depth

not exceeding 1.50 m	m	LH 0.25	7.51	PM 0.25	13.57	22.40	43.48	50.00
1.50 - 2.00 m	m	LH 0.30	9.02	PM 0.30	16.28	22.40	47.70	54.85
2.00 - 2.50 m	m	LH 0.35	10.52	PM 0.35	19.00	22.40	51.92	59.71
2.50 - 3.00 m	m	LH 0.40	12.02	PM 0.40	21.71	22.40	56.13	64.55
3.00 - 3.50 m	m	LH 0.45	13.53	PM 0.45	24.43	22.40	60.36	69.41
3.50 - 4.00 m	m	LH 0.50	15.03	PM 0.50	27.14	22.40	64.57	74.26
4.00 - 4.50 m	m	LH 0.55	16.53	PM 0.55	29.85	22.40	68.78	79.10
4.50 - 5.00 m	m	LH 0.60	18.04	PM 0.60	32.57	22.40	73.01	83.96

Concrete pipes, class L, vibrated, flexible joints, 525 mm diameter, depth

not exceeding 1.50 m	m	LH 0.27	8.12	PM 0.27	14.66	26.23	49.01	56.36
1.50 - 2.00 m	m	LH 0.32	9.62	PM 0.32	17.37	26.23	53.22	61.20
2.00 - 2.50 m	m	LH 0.37	11.12	PM 0.37	20.08	26.23	57.43	66.04
2.50 - 3.00 m	m	LH 0.42	12.63	PM 0.42	22.80	26.23	61.66	70.91
3.00 - 3.50 m	m	LH 0.47	14.13	PM 0.47	25.51	26.23	65.87	75.75
3.50 - 4.00 m	m	LH 0.52	15.63	PM 0.52	28.23	26.23	70.09	80.60
4.00 - 4.50 m	m	LH 0.57	17.13	PM 0.57	30.94	26.23	74.30	85.44
4.50 - 5.00 m	m	LH 0.62	18.64	PM 0.62	33.65	26.23	78.52	90.30

Civil engineering

New work	Unit	Labour gang hours	Labour net		Plant gang hours	Plant net	Mats net	Price net	Price with 15%
Sewers and manholes			£			£	£	£	£
							VAT not included		

Pipework (*continued*)

Concrete pipes, class L,
vibrated, flexible joints, 600 mm
diameter, depth

not exceeding 1.50 m	m	LH	0.30	9.02	PM	0.30	16.28	30.65	55.95	64.34
1.50 - 2.00 m	m	LH	0.35	10.52	PM	0.35	19.00	30.65	60.17	69.20
2.00 - 2.50 m	m	LH	0.40	12.02	PM	0.40	21.71	30.65	64.38	74.04
2.50 - 3.00 m	m	LH	0.45	13.53	PM	0.45	24.43	30.65	68.61	78.90
3.00 - 3.50 m	m	LH	0.50	15.03	PM	0.50	27.14	30.65	72.82	83.74
3.50 - 4.00 m	m	LH	0.55	16.53	PM	0.55	29.85	30.65	77.03	88.58
4.00 - 4.50 m	m	LH	0.60	18.04	PM	0.60	32.57	30.65	81.26	93.45
4.50 - 5.00 m	m	LH	0.65	19.54	PM	0.65	35.28	30.65	85.47	98.29

Concrete pipes, class L, spun,
flexible joints, 750 mm
diameter, depth

not exceeding 1.50 m	m	LH	0.32	9.62	PM	0.32	17.37	51.37	78.36	90.11
1.50 - 2.00 m	m	LH	0.37	11.12	PM	0.37	20.08	51.37	82.57	94.96
2.00 - 2.50 m	m	LH	0.42	12.63	PM	0.42	22.80	51.37	86.80	99.82
2.50 - 3.00 m	m	LH	0.47	14.13	PM	0.47	25.51	51.37	91.01	104.66
3.00 - 3.50 m	m	LH	0.52	15.63	PM	0.52	28.23	51.37	95.23	109.51
3.50 - 4.00 m	m	LH	0.57	17.13	PM	0.57	30.94	51.37	99.44	114.36
4.00 - 4.50 m	m	LH	0.62	18.64	PM	0.62	33.65	51.37	103.66	119.21
4.50 - 5.00 m	m	LH	0.67	20.14	PM	0.67	36.37	51.37	107.88	124.06

Concrete pipes, class L, spun,
flexible joints, 900 mm
diameter, depth

not exceeding 1.50 m	m	LH	0.35	10.52	PM	0.35	19.00	69.36	98.88	113.71
1.50 - 2.00 m	m	LH	0.40	12.02	PM	0.40	21.71	69.36	103.09	118.55
2.00 - 2.50 m	m	LH	0.45	13.53	PM	0.45	24.43	69.36	107.32	123.42
2.50 - 3.00 m	m	LH	0.50	15.03	PM	0.50	27.14	69.36	111.53	128.26
3.00 - 3.50 m	m	LH	0.55	16.53	PM	0.55	29.85	69.36	115.74	133.10
3.50 - 4.00 m	m	LH	0.60	18.04	PM	0.60	32.57	69.36	119.97	137.97
4.00 - 4.50 m	m	LH	0.65	19.54	PM	0.65	35.28	69.36	124.18	142.81
4.50 - 5.00 m	m	LH	0.70	21.04	PM	0.70	38.00	69.36	128.40	147.66

Concrete pipes, class L, spun,
flexible joints, 1200 mm
diameter, depth

not exceeding 1.50 m	m	LH	0.40	12.02	PM	0.40	21.71	115.76	149.49	171.91
1.50 - 2.00 m	m	LH	0.45	13.53	PM	0.45	24.43	115.76	153.72	176.78
2.00 - 2.50 m	m	LH	0.50	15.03	PM	0.50	27.14	115.76	157.93	181.62
2.50 - 3.00 m	m	LH	0.55	16.53	PM	0.55	29.85	115.76	162.14	186.46
3.00 - 3.50 m	m	LH	0.60	18.04	PM	0.60	32.57	115.76	166.37	191.33
3.50 - 4.00 m	m	LH	0.65	19.54	PM	0.65	35.28	115.76	170.58	196.17
4.00 - 4.50 m	m	LH	0.70	21.04	PM	0.70	38.00	115.76	174.80	201.02
4.50 - 5.00 m	m	LH	0.75	22.54	PM	0.75	40.71	115.76	179.01	205.86

Civil engineering

Sewers and manholes			£			£	£	£	£
								VAT not included	

Pipework (*continued*)

Concrete pipes, class L, spun, flexible joints, 1500 mm diameter, depth

not exceeding 1.50 m	m	LH	0.50	15.03	PM	0.50	27.14	179.29	221.46	254.68
1.50 - 2.00 m	m	LH	0.55	16.53	PM	0.55	29.85	179.29	225.67	259.52
2.00 - 2.50 m	m	LH	0.60	18.04	PM	0.60	32.57	179.29	229.90	264.38
2.50 - 3.00 m	m	LH	0.65	19.54	PM	0.65	35.28	179.29	234.11	269.23
3.00 - 3.50 m	m	LH	0.70	21.04	PM	0.70	38.00	179.29	238.33	274.08
3.50 - 4.00 m	m	LH	0.75	22.54	PM	0.75	40.71	179.29	242.54	278.92
4.00 - 4.50 m	m	LH	0.80	24.05	PM	0.80	43.42	179.29	246.76	283.77
4.50 - 5.00 m	m	LH	0.85	25.55	PM	0.85	46.14	179.29	250.98	288.63

Concrete pipes, class L, spun, flexible joints, 1800 mm diameter, depth

not exceeding 1.50 m	m	LH	0.60	18.04	PM	0.60	32.57	240.81	291.42	335.13
1.50 - 2.00 m	m	LH	0.65	19.54	PM	0.65	35.28	240.81	295.63	339.97
2.00 - 2.50 m	m	LH	0.70	21.04	PM	0.70	38.00	240.81	299.85	344.83
2.50 - 3.00 m	m	LH	0.75	22.54	PM	0.75	40.71	240.81	304.06	349.67
3.00 - 3.50 m	m	LH	0.80	24.05	PM	0.80	43.42	240.81	308.28	354.52
3.50 - 4.00 m	m	LH	0.85	25.55	PM	0.85	46.14	240.81	312.50	359.38
4.00 - 4.50 m	m	LH	0.90	27.05	PM	0.90	48.85	240.81	316.71	364.22
4.50 - 5.00 m	m	LH	0.95	28.56	PM	0.95	51.57	240.81	320.94	369.08

Concrete pipes, class M, vibrated, flexible joints, 300 mm diameter, depth

not exceeding 1.50 m	m	LH	0.20	6.01	PM	0.20	10.86	14.69	31.56	36.29
1.50 - 2.00 m	m	LH	0.25	7.51	PM	0.25	13.57	14.69	35.77	41.14
2.00 - 2.50 m	m	LH	0.30	9.02	PM	0.30	16.28	14.69	39.99	45.99
2.50 - 3.00 m	m	LH	0.35	10.52	PM	0.35	19.00	14.69	44.21	50.84
3.00 - 3.50 m	m	LH	0.40	12.02	PM	0.40	21.71	14.69	48.42	55.68
3.50 - 4.00 m	m	LH	0.45	13.53	PM	0.45	24.43	14.69	52.65	60.55
4.00 - 4.50 m	m	LH	0.50	15.03	PM	0.50	27.14	14.69	56.86	65.39
4.50 - 5.00 m	m	LH	0.55	16.53	PM	0.55	29.85	14.69	61.07	70.23

Concrete pipes, class M, vibrated, flexible joints, 375 mm diameter, depth

not exceeding 1.50 m	m	LH	0.22	6.61	PM	0.22	11.94	19.12	37.67	43.32
1.50 - 2.00 m	m	LH	0.27	8.12	PM	0.27	14.66	19.12	41.90	48.19
2.00 - 2.50 m	m	LH	0.32	9.62	PM	0.32	17.37	19.12	46.11	53.03
2.50 - 3.00 m	m	LH	0.37	11.12	PM	0.37	20.08	19.12	50.32	57.87
3.00 - 3.50 m	m	LH	0.42	12.63	PM	0.42	22.80	19.12	54.55	62.73
3.50 - 4.00 m	m	LH	0.47	14.13	PM	0.47	25.51	19.12	58.76	67.57
4.00 - 4.50 m	m	LH	0.52	15.63	PM	0.52	28.23	19.12	62.98	72.43
4.50 - 5.00 m	m	LH	0.57	17.13	PM	0.57	30.94	19.12	67.19	77.27

New work	Unit	Labour gang hours		Labour net	Plant gang hours		Plant net	Mats net	Price net	Price with 15%

Sewers and manholes £ £ £ £ £

VAT not included

Pipework (*continued*)

Concrete pipes, class M,
vibrated, flexible joints, 450 mm
diameter, depth

not exceeding 1.50 m	m	LH	0.25	7.51	PM	0.25	13.57	22.86	43.94	50.53
1.50 - 2.00 m	m	LH	0.30	9.02	PM	0.30	16.28	22.86	48.16	55.38
2.00 - 2.50 m	m	LH	0.35	10.52	PM	0.35	19.00	22.86	52.38	60.24
2.50 - 3.00 m	m	LH	0.40	12.02	PM	0.40	21.71	22.86	56.59	65.08
3.00 - 3.50 m	m	LH	0.45	13.53	PM	0.45	24.43	22.86	60.82	69.94
3.50 - 4.00 m	m	LH	0.50	15.03	PM	0.50	27.14	22.86	65.03	74.78
4.00 - 4.50 m	m	LH	0.55	16.53	PM	0.55	29.85	22.86	69.24	79.63
4.50 - 5.00 m	m	LH	0.60	18.04	PM	0.60	32.57	22.86	73.47	84.49

Concrete pipes, class M,
vibrated, flexible joints, 525 mm
diameter, depth

not exceeding 1.50 m	m	LH	0.27	8.12	PM	0.27	14.66	26.75	49.53	56.96
1.50 - 2.00 m	m	LH	0.32	9.62	PM	0.32	17.37	26.75	53.74	61.80
2.00 - 2.50 m	m	LH	0.37	11.12	PM	0.37	20.08	26.75	57.95	66.64
2.50 - 3.00 m	m	LH	0.42	12.63	PM	0.42	22.80	26.75	62.18	71.51
3.00 - 3.50 m	m	LH	0.47	14.13	PM	0.47	25.51	26.75	66.39	76.35
3.50 - 4.00 m	m	LH	0.52	15.63	PM	0.52	28.23	26.75	70.61	81.20
4.00 - 4.50 m	m	LH	0.57	17.13	PM	0.57	30.94	26.75	74.82	86.04
4.50 - 5.00 m	m	LH	0.62	18.64	PM	0.62	33.65	26.75	79.04	90.90

Concrete pipes, class M,
vibrated, flexible joints, 600 mm
diameter, depth

not exceeding 1.50 m	m	LH	0.30	9.02	PM	0.30	16.28	31.33	56.63	65.12
1.50 - 2.00 m	m	LH	0.35	10.52	PM	0.35	19.00	31.33	60.85	69.98
2.00 - 2.50 m	m	LH	0.40	12.02	PM	0.40	21.71	31.33	65.06	74.82
2.50 - 3.00 m	m	LH	0.45	13.53	PM	0.45	24.43	31.33	69.29	79.68
3.00 - 3.50 m	m	LH	0.50	15.03	PM	0.50	27.14	31.33	73.50	84.53
3.50 - 4.00 m	m	LH	0.55	16.53	PM	0.55	29.85	31.33	77.71	89.37
4.00 - 4.50 m	m	LH	0.60	18.04	PM	0.60	32.57	31.33	81.94	94.23
4.50 - 5.00 m	m	LH	0.65	19.54	PM	0.65	35.28	31.33	86.15	99.07

Concrete pipes, class M, spun,
flexible joints, 750 mm
diameter, depth

not exceeding 1.50 m	m	LH	0.32	9.62	PM	0.32	17.37	52.40	79.39	91.30
1.50 - 2.00 m	m	LH	0.37	11.12	PM	0.37	20.08	52.40	83.60	96.14
2.00 - 2.50 m	m	LH	0.42	12.63	PM	0.42	22.80	52.40	87.83	101.00
2.50 - 3.00 m	m	LH	0.47	14.13	PM	0.47	25.51	52.40	92.04	105.85
3.00 - 3.50 m	m	LH	0.52	15.63	PM	0.52	28.23	52.40	96.26	110.70
3.50 - 4.00 m	m	LH	0.57	17.13	PM	0.57	30.94	52.40	100.47	115.54
4.00 - 4.50 m	m	LH	0.62	18.64	PM	0.62	33.65	52.40	104.69	120.39
4.50 - 5.00 m	m	LH	0.67	20.14	PM	0.67	36.37	52.40	108.91	125.25

Civil engineering

Sewers and manholes

£ £ £ £ £

VAT not included

Pipework (*continued*)

Concrete pipes, class M, spun,
flexible joints, 900 mm
diameter, depth

not exceeding 1.50 m	m	LH	0.35	10.52	PM	0.35	19.00	70.75	100.27	115.31
1.50 - 2.00 m	m	LH	0.40	12.02	PM	0.40	21.71	70.75	104.48	120.15
2.00 - 2.50 m	m	LH	0.45	13.53	PM	0.45	24.43	70.75	108.71	125.02
2.50 - 3.00 m	m	LH	0.50	15.03	PM	0.50	27.14	70.75	112.92	129.86
3.00 - 3.50 m	m	LH	0.55	16.53	PM	0.55	29.85	70.75	117.13	134.70
3.50 - 4.00 m	m	LH	0.60	18.04	PM	0.60	32.57	70.75	121.36	139.56
4.00 - 4.50 m	m	LH	0.65	19.54	PM	0.65	35.28	70.75	125.57	144.41
4.50 - 5.00 m	m	LH	0.70	21.04	PM	0.70	38.00	70.75	129.79	149.26

Concrete pipes, class M, spun,
flexible joints, 1200 mm
diameter, depth

not exceeding 1.50 m	m	LH	0.40	12.02	PM	0.40	21.71	118.52	152.25	175.09
1.50 - 2.00 m	m	LH	0.45	13.53	PM	0.45	24.43	118.52	156.48	179.95
2.00 - 2.50 m	m	LH	0.50	15.03	PM	0.50	27.14	118.52	160.69	184.79
2.50 - 3.00 m	m	LH	0.55	16.53	PM	0.55	29.85	118.52	164.90	189.63
3.00 - 3.50 m	m	LH	0.60	18.04	PM	0.60	32.57	118.52	169.13	194.50
3.50 - 4.00 m	m	LH	0.65	19.54	PM	0.65	35.28	118.52	173.34	199.34
4.00 - 4.50 m	m	LH	0.70	21.04	PM	0.70	38.00	118.52	177.56	204.19
4.50 - 5.00 m	m	LH	0.75	22.54	PM	0.75	40.71	118.52	181.77	209.04

Concrete pipes, class M, spun,
flexible joints, 1500 mm
diameter, depth

not exceeding 1.50 m	m	LH	0.50	15.03	PM	0.50	27.14	182.87	225.04	258.80
1.50 - 2.00 m	m	LH	0.55	16.53	PM	0.55	29.85	182.87	229.25	263.64
2.00 - 2.50 m	m	LH	0.60	18.04	PM	0.60	32.57	182.87	233.48	268.50
2.50 - 3.00 m	m	LH	0.65	19.54	PM	0.65	35.28	182.87	237.69	273.34
3.00 - 3.50 m	m	LH	0.70	21.04	PM	0.70	38.00	182.87	241.91	278.20
3.50 - 4.00 m	m	LH	0.75	22.54	PM	0.75	40.71	182.87	246.12	283.04
4.00 - 4.50 m	m	LH	0.80	24.05	PM	0.80	43.42	182.87	250.34	287.89
4.50 - 5.00 m	m	LH	0.85	25.55	PM	0.85	46.14	182.87	254.56	292.74

Concrete pipes, class M, spun,
flexible joints, 1800 mm
diameter, depth

not exceeding 1.50 m	m	LH	0.60	18.04	PM	0.60	32.57	245.74	296.35	340.80
1.50 - 2.00 m	m	LH	0.65	19.54	PM	0.65	35.28	245.74	300.56	345.64
2.00 - 2.50 m	m	LH	0.70	21.04	PM	0.70	38.00	245.74	304.78	350.50
2.50 - 3.00 m	m	LH	0.75	22.54	PM	0.75	40.71	245.74	308.99	355.34
3.00 - 3.50 m	m	LH	0.80	24.05	PM	0.80	43.42	245.74	313.21	360.19
3.50 - 4.00 m	m	LH	0.85	25.55	PM	0.85	46.14	245.74	317.43	365.04
4.00 - 4.50 m	m	LH	0.90	27.05	PM	0.90	48.85	245.74	321.64	369.89
4.50 - 5.00 m	m	LH	0.95	28.56	PM	0.95	51.57	245.74	325.87	374.75

Civil engineering

New work	Unit	Labour gang hours	Labour net	Plant gang hours	Plant net	Mats net	Price net	Price with 15%
Sewers and manholes		£			£	£	£	£
						VAT not included		

Pipework (*continued*)

Concrete pipes, class H,
vibrated, flexible joints, 375 mm
diameter, depth

not exceeding 1.50 m	m	LH 0.22	6.61	PM 0.22	11.94	18.93	37.48	43.10
1.50 - 2.00 m	m	LH 0.27	8.12	PM 0.27	14.66	18.93	41.71	47.97
2.00 - 2.50 m	m	LH 0.32	9.62	PM 0.32	17.37	18.93	45.92	52.81
2.50 - 3.00 m	m	LH 0.37	11.12	PM 0.37	20.08	18.93	50.13	57.65
3.00 - 3.50 m	m	LH 0.42	12.63	PM 0.42	22.80	18.93	54.36	62.51
3.50 - 4.00 m	m	LH 0.47	14.13	PM 0.47	25.51	18.93	58.57	67.36
4.00 - 4.50 m	m	LH 0.52	15.63	PM 0.52	28.23	18.93	62.79	72.21
4.50 - 5.00 m	m	LH 0.57	17.13	PM 0.57	30.94	18.93	67.00	77.05

Concrete pipes, class H,
vibrated, flexible joints, 450 mm
diameter, depth

not exceeding 1.50 m	m	LH 0.25	7.51	PM 0.25	13.57	23.42	44.50	51.17
1.50 - 2.00 m	m	LH 0.30	9.02	PM 0.30	16.28	23.42	48.72	56.03
2.00 - 2.50 m	m	LH 0.35	10.52	PM 0.35	19.00	23.42	52.94	60.88
2.50 - 3.00 m	m	LH 0.40	12.02	PM 0.40	21.71	23.42	57.15	65.72
3.00 - 3.50 m	m	LH 0.45	13.53	PM 0.45	24.43	23.42	61.38	70.59
3.50 - 4.00 m	m	LH 0.50	15.03	PM 0.50	27.14	23.42	65.59	75.43
4.00 - 4.50 m	m	LH 0.55	16.53	PM 0.55	29.85	23.42	69.80	80.27
4.50 - 5.00 m	m	LH 0.60	18.04	PM 0.60	32.57	23.42	74.03	85.13

Concrete pipes, class H,
vibrated, flexible joints, 525 mm
diameter, depth

not exceeding 1.50 m	m	LH 0.27	8.12	PM 0.27	14.66	27.41	50.19	57.72
1.50 - 2.00 m	m	LH 0.32	9.62	PM 0.32	17.37	27.41	54.40	62.56
2.00 - 2.50 m	m	LH 0.37	11.12	PM 0.37	20.08	27.41	58.61	67.40
2.50 - 3.00 m	m	LH 0.42	12.63	PM 0.42	22.80	27.41	62.84	72.27
3.00 - 3.50 m	m	LH 0.47	14.13	PM 0.47	25.51	27.41	67.05	77.11
3.50 - 4.00 m	m	LH 0.52	15.63	PM 0.52	28.23	27.41	71.27	81.96
4.00 - 4.50 m	m	LH 0.57	17.13	PM 0.57	30.94	27.41	75.48	86.80
4.50 - 5.00 m	m	LH 0.62	18.64	PM 0.62	33.65	27.41	79.70	91.66

Concrete pipes, class H,
vibrated, flexible joints, 600 mm
diameter, depth

not exceeding 1.50 m	m	LH 0.30	9.02	PM 0.30	16.28	32.12	57.42	66.03
1.50 - 2.00 m	m	LH 0.35	10.52	PM 0.35	19.00	32.12	61.64	70.89
2.00 - 2.50 m	m	LH 0.40	12.02	PM 0.40	21.71	32.12	65.85	75.73
2.50 - 3.00 m	m	LH 0.45	13.53	PM 0.45	24.43	32.12	70.08	80.59
3.00 - 3.50 m	m	LH 0.50	15.03	PM 0.50	27.14	32.12	74.29	85.43
3.50 - 4.00 m	m	LH 0.55	16.53	PM 0.55	29.85	32.12	78.50	90.28
4.00 - 4.50 m	m	LH 0.60	18.04	PM 0.60	32.57	32.12	82.73	95.14
4.50 - 5.00 m	m	LH 0.65	19.54	PM 0.65	35.28	32.12	86.94	99.98

Civil engineering

New work	Unit	Labour gang hours	Labour net		Plant gang hours	Plant net	Mats net	Price net	Price with 15%
Sewers and manholes			£			£	£	£	£
							VAT not included		

Pipework (*continued*)

Concrete pipes, class H, spun,
flexible joints, 750 mm
diameter, depth

not exceeding 1.50 m	m	LH	0.32	9.62	PM	0.32	17.37	52.40	79.39	91.30
1.50 - 2.00 m	m	LH	0.37	11.12	PM	0.37	20.08	52.40	83.60	96.14
2.00 - 2.50 m	m	LH	0.42	12.63	PM	0.42	22.80	52.40	87.83	101.00
2.50 - 3.00 m	m	LH	0.47	14.13	PM	0.47	25.51	52.40	92.04	105.85
3.00 - 3.50 m	m	LH	0.52	15.63	PM	0.52	28.23	52.40	96.26	110.70
3.50 - 4.00 m	m	LH	0.57	17.13	PM	0.57	30.94	52.40	100.47	115.54
4.00 - 4.50 m	m	LH	0.62	18.64	PM	0.62	33.65	52.40	104.69	120.39
4.50 - 5.00 m	m	LH	0.67	20.14	PM	0.67	36.37	52.40	108.91	125.25

Concrete pipes, class H, spun,
flexible joints, 900 mm
diameter, depth

not exceeding 1.50 m	m	LH	0.35	10.52	PM	0.35	19.00	72.51	102.03	117.33
1.50 - 2.00 m	m	LH	0.40	12.02	PM	0.40	21.71	72.51	106.24	122.18
2.00 - 2.50 m	m	LH	0.45	13.53	PM	0.45	24.43	72.51	110.47	127.04
2.50 - 3.00 m	m	LH	0.50	15.03	PM	0.50	27.14	72.51	114.68	131.88
3.00 - 3.50 m	m	LH	0.55	16.53	PM	0.55	29.85	72.51	118.89	136.72
3.50 - 4.00 m	m	LH	0.60	18.04	PM	0.60	32.57	72.51	123.12	141.59
4.00 - 4.50 m	m	LH	0.65	19.54	PM	0.65	35.28	72.51	127.33	146.43
4.50 - 5.00 m	m	LH	0.70	21.04	PM	0.70	38.00	72.51	131.55	151.28

Concrete pipes, class H, spun,
flexible joints, 1200 mm
diameter, depth

not exceeding 1.50 m	m	LH	0.40	12.02	PM	0.40	21.71	121.49	155.22	178.50
1.50 - 2.00 m	m	LH	0.45	13.53	PM	0.45	24.43	121.49	159.45	183.37
2.00 - 2.50 m	m	LH	0.50	15.03	PM	0.50	27.14	121.49	163.66	188.21
2.50 - 3.00 m	m	LH	0.55	16.53	PM	0.55	29.85	121.49	167.87	193.05
3.00 - 3.50 m	m	LH	0.60	18.04	PM	0.60	32.57	121.49	172.10	197.91
3.50 - 4.00 m	m	LH	0.65	19.54	PM	0.65	35.28	121.49	176.31	202.76
4.00 - 4.50 m	m	LH	0.70	21.04	PM	0.70	38.00	121.49	180.53	207.61
4.50 - 5.00 m	m	LH	0.75	22.54	PM	0.75	40.71	121.49	184.74	212.45

Concrete pipes, class H, spun,
flexible joints, 1500 mm
diameter, depth

not exceeding 1.50 m	m	LH	0.50	15.03	PM	0.50	27.14	187.99	230.16	264.68
1.50 - 2.00 m	m	LH	0.55	16.53	PM	0.55	29.85	187.99	234.37	269.53
2.00 - 2.50 m	m	LH	0.60	18.04	PM	0.60	32.57	187.99	238.60	274.39
2.50 - 3.00 m	m	LH	0.65	19.54	PM	0.65	35.28	187.99	242.81	279.23
3.00 - 3.50 m	m	LH	0.70	21.04	PM	0.70	38.00	187.99	247.03	284.08
3.50 - 4.00 m	m	LH	0.75	22.54	PM	0.75	40.71	187.99	251.24	288.93
4.00 - 4.50 m	m	LH	0.80	24.05	PM	0.80	43.42	187.99	255.46	293.78
4.50 - 5.00 m	m	LH	0.85	25.55	PM	0.85	46.14	187.99	259.68	298.63

Civil engineering

New work	Unit		Labour gang hours	Labour net		Plant gang hours	Plant net	Mats net	Price net	Price with 15%
Sewers and manholes				£			£	£	£	£
								VAT not included		

Pipework (*continued*)

Concrete pipes, class H, spun,
flexible joints, 1800 mm
diameter, depth

not exceeding 1.50 m	m	LH	0.60	18.04	PM	0.60	32.57	251.76	302.37	347.73
1.50 - 2.00 m	m	LH	0.65	19.54	PM	0.65	35.28	251.76	306.58	352.57
2.00 - 2.50 m	m	LH	0.70	21.04	PM	0.70	38.00	251.76	310.80	357.42
2.50 - 3.00 m	m	LH	0.75	22.54	PM	0.75	40.71	251.76	315.01	362.26
3.00 - 3.50 m	m	LH	0.80	24.05	PM	0.80	43.42	251.76	319.23	367.11
3.50 - 4.00 m	m	LH	0.85	25.55	PM	0.85	46.14	251.76	323.45	371.97
4.00 - 4.50 m	m	LH	0.90	27.05	PM	0.90	48.85	251.76	327.66	376.81
4.50 - 5.00 m	m	LH	0.95	28.56	PM	0.95	51.57	251.76	331.89	381.67

Ductile iron spigot and socket
pipes, Tyton joints, 300 mm
diameter, depth

not exceeding 1.50 m	m	LH	0.15	4.51	PM	0.15	8.14	50.88	63.53	73.06
1.50 - 2.00 m	m	LH	0.20	6.01	PM	0.20	10.86	50.88	67.75	77.91
2.00 - 2.50 m	m	LH	0.25	7.51	PM	0.25	13.57	50.88	71.96	82.75
2.50 - 3.00 m	m	LH	0.30	9.02	PM	0.30	16.28	50.88	76.18	87.61
3.00 - 3.50 m	m	LH	0.35	10.52	PM	0.35	19.00	50.88	80.40	92.46
3.50 - 4.00 m	m	LH	0.40	12.02	PM	0.40	21.71	50.88	84.61	97.30
4.00 - 4.50 m	m	LH	0.45	13.53	PM	0.45	24.43	50.88	88.84	102.17
4.50 - 5.00 m	m	LH	0.50	15.03	PM	0.50	27.14	50.88	93.05	107.01

Ductile iron spigot and socket
pipes, Tyton joints, 450 mm
diameter, depth

not exceeding 1.50 m	m	LH	0.22	6.61	PM	0.22	11.94	103.49	122.04	140.35
1.50 - 2.00 m	m	LH	0.27	8.12	PM	0.27	14.66	103.49	126.27	145.21
2.00 - 2.50 m	m	LH	0.32	9.62	PM	0.32	17.37	103.49	130.48	150.05
2.50 - 3.00 m	m	LH	0.37	11.12	PM	0.37	20.08	103.49	134.69	154.89
3.00 - 3.50 m	m	LH	0.42	12.63	PM	0.42	22.80	103.49	138.92	159.76
3.50 - 4.00 m	m	LH	0.47	14.13	PM	0.47	25.51	103.49	143.13	164.60
4.00 - 4.50 m	m	LH	0.52	15.63	PM	0.52	28.23	103.49	147.35	169.45
4.50 - 5.00 m	m	LH	0.57	17.13	PM	0.57	30.94	103.49	151.56	174.29

Ductile iron spigot and socket
pipes, Tyton joints, 600 mm
diameter, depth

not exceeding 1.50 m	m	LH	0.30	9.02	PM	0.30	16.28	157.27	182.57	209.96
1.50 - 2.00 m	m	LH	0.35	10.52	PM	0.35	19.00	157.27	186.79	214.81
2.00 - 2.50 m	m	LH	0.40	12.02	PM	0.40	21.71	157.27	191.00	219.65
2.50 - 3.00 m	m	LH	0.45	13.53	PM	0.45	24.43	157.27	195.23	224.51
3.00 - 3.50 m	m	LH	0.50	15.03	PM	0.50	27.14	157.27	199.44	229.36
3.50 - 4.00 m	m	LH	0.55	16.53	PM	0.55	29.85	157.27	203.65	234.20
4.00 - 4.50 m	m	LH	0.60	18.04	PM	0.60	32.57	157.27	207.88	239.06
4.50 - 5.00 m	m	LH	0.65	19.54	PM	0.65	35.28	157.27	212.09	243.90

Civil engineering

New work	Unit		Labour gang hours	Labour net		Plant gang hours	Plant net	Mats net	Price net	Price with 15%
Sewers and manholes				£			£	£	£	£
								VAT not included		

Pipework (*continued*)

Ductile iron spigot and socket
pipes, Tyton joints, 700 mm
diameter, depth

not exceeding 1.50 m	m	LH	0.32	9.62	PM	0.32	17.37	184.79	211.78	243.55
1.50 - 2.00 m	m	LH	0.37	11.12	PM	0.37	20.08	184.79	215.99	248.39
2.00 - 2.50 m	m	LH	0.42	12.63	PM	0.42	22.80	184.79	220.22	253.25
2.50 - 3.00 m	m	LH	0.47	14.13	PM	0.47	25.51	184.79	224.43	258.09
3.00 - 3.50 m	m	LH	0.52	15.63	PM	0.52	28.23	184.79	228.65	262.95
3.50 - 4.00 m	m	LH	0.57	17.13	PM	0.57	30.94	184.79	232.86	267.79
4.00 - 4.50 m	m	LH	0.62	18.64	PM	0.62	33.65	184.79	237.08	272.64
4.50 - 5.00 m	m	LH	0.67	20.14	PM	0.67	36.37	184.79	241.30	277.50

Ductile iron spigot and socket
pipes, Tyton joints, 800 mm
diameter, depth

not exceeding 1.50 m	m	LH	0.35	10.52	PM	0.35	19.00	236.95	266.47	306.44
1.50 - 2.00 m	m	LH	0.40	12.02	PM	0.40	21.71	236.95	270.68	311.28
2.00 - 2.50 m	m	LH	0.45	13.53	PM	0.45	24.43	236.95	274.91	316.15
2.50 - 3.00 m	m	LH	0.50	15.03	PM	0.50	27.14	236.95	279.12	320.99
3.00 - 3.50 m	m	LH	0.55	16.53	PM	0.55	29.85	236.95	283.33	325.83
3.50 - 4.00 m	m	LH	0.60	18.04	PM	0.60	32.57	236.95	287.56	330.69
4.00 - 4.50 m	m	LH	0.65	19.54	PM	0.65	35.28	236.95	291.77	335.54
4.50 - 5.00 m	m	LH	0.70	21.04	PM	0.70	38.00	236.95	295.99	340.39

Ductile iron spigot and socket
pipes, Tyton joints, 900 mm
diameter, depth

not exceeding 1.50 m	m	LH	0.37	11.12	PM	0.37	20.08	281.05	312.25	359.09
1.50 - 2.00 m	m	LH	0.42	12.63	PM	0.42	22.80	281.05	316.48	363.95
2.00 - 2.50 m	m	LH	0.47	14.13	PM	0.47	25.51	281.05	320.69	368.79
2.50 - 3.00 m	m	LH	0.52	15.63	PM	0.52	28.23	281.05	324.91	373.65
3.00 - 3.50 m	m	LH	0.57	17.13	PM	0.57	30.94	281.05	329.12	378.49
3.50 - 4.00 m	m	LH	0.62	18.64	PM	0.62	33.65	281.05	333.34	383.34
4.00 - 4.50 m	m	LH	0.67	20.14	PM	0.67	36.37	281.05	337.56	388.19
4.50 - 5.00 m	m	LH	0.72	21.64	PM	0.72	39.08	281.05	341.77	393.04

Ductile iron spigot and socket
pipes, Tyton joints, 1000 mm
diameter, depth

not exceeding 1.50 m	m	LH	0.40	12.02	PM	0.40	21.71	360.60	394.33	453.48
1.50 - 2.00 m	m	LH	0.45	13.53	PM	0.45	24.43	360.60	398.56	458.34
2.00 - 2.50 m	m	LH	0.50	15.03	PM	0.50	27.14	360.60	402.77	463.19
2.50 - 3.00 m	m	LH	0.55	16.53	PM	0.55	29.85	360.60	406.98	468.03
3.00 - 3.50 m	m	LH	0.60	18.04	PM	0.60	32.57	360.60	411.21	472.89
3.50 - 4.00 m	m	LH	0.65	19.54	PM	0.65	35.28	360.60	415.42	477.73
4.00 - 4.50 m	m	LH	0.70	21.04	PM	0.70	38.00	360.60	419.64	482.59
4.50 - 5.00 m	m	LH	0.75	22.54	PM	0.75	40.71	360.60	423.85	487.43

Civil engineering

New work	Unit	Labour gang hours		Labour net		Plant gang hours		Plant net	Mats net	Price net	Price with 15%

Sewers and manholes — £ — — £ £ £ £

Pipework (*continued*)

Ductile iron spigot and socket
pipes, Tyton joints, 1100 mm
diameter, depth

not exceeding 1.50 m	m	LH	0.45	13.53	PM	0.45	24.43	493.31	531.27	610.96
1.50 - 2.00 m	m	LH	0.50	15.03	PM	0.50	27.14	493.31	535.48	615.80
2.00 - 2.50 m	m	LH	0.55	16.53	PM	0.55	29.85	493.31	539.69	620.64
2.50 - 3.00 m	m	LH	0.60	18.04	PM	0.60	32.57	493.31	543.92	625.51
3.00 - 3.50 m	m	LH	0.65	19.54	PM	0.65	35.28	493.31	548.13	630.35
3.50 - 4.00 m	m	LH	0.70	21.04	PM	0.70	38.00	493.31	552.35	635.20
4.00 - 4.50 m	m	LH	0.75	22.54	PM	0.75	40.71	493.31	556.56	640.04
4.50 - 5.00 m	m	LH	0.80	24.05	PM	0.80	43.42	493.31	560.78	644.90

Clay pipe fittings

Bends

300 mm	each	LH	0.15	4.51	PM	0.15	8.14	49.62	62.27	71.61
375 mm	each	LH	0.20	6.01	PM	0.20	10.86	94.39	111.26	127.95
400 mm	each	LH	0.25	7.51	PM	0.25	13.57	118.67	139.75	160.71
450 mm	each	LH	0.30	9.02	PM	0.30	16.28	157.22	182.52	209.90

Single junctions

300 mm	each	LH	0.20	6.01	PM	0.20	10.86	68.73	85.60	98.44
375 mm	each	LH	0.25	7.51	PM	0.25	13.57	136.80	157.88	181.56
400 mm	each	LH	0.30	9.02	PM	0.30	16.28	140.70	166.00	190.90
450 mm	each	LH	0.35	10.52	PM	0.35	19.00	173.25	202.77	233.19

Double junctions

300 mm	each	LH	0.25	7.51	PM	0.25	13.57	120.57	141.65	162.90
375 mm	each	LH	0.30	9.02	PM	0.30	16.28	208.14	233.44	268.46

Tapers (largest end stated)

300 mm	each	LH	0.25	7.51	PM	0.25	13.57	58.65	79.73	91.69
375 mm	each	LH	0.30	9.02	PM	0.30	16.28	88.35	113.65	130.70

Concrete pipe fittings

Bends

300 mm	each	LH	0.10	3.01	PM	0.10	5.43	72.70	81.14	93.31
375 mm	each	LH	0.15	4.51	PM	0.15	8.14	89.99	102.64	118.04
450 mm	each	LH	0.20	6.01	PM	0.20	10.86	113.26	130.13	149.65
600 mm	each	LH	0.25	7.51	PM	0.25	13.57	166.10	187.18	215.26
750 mm	each	LH	0.30	9.02	PM	0.30	16.28	238.29	263.59	303.13
900 mm	each	LH	0.35	10.52	PM	0.35	19.00	333.52	363.04	417.50
1200 mm	each	LH	0.40	12.02	PM	0.40	21.71	544.96	578.69	665.49
1550 mm	each	LH	0.45	13.53	PM	0.45	24.43	824.26	862.22	991.55
1800 mm	each	LH	0.50	15.03	PM	0.50	27.14	1,164.66	1,206.83	1,387.85

Civil engineering

VAT not included

New work	Unit	Labour gang hours	Labour net		Plant gang hours	Plant net	Mats net	Price net	Price with 15%
Sewers and manholes			£			£	£	£	£
Concrete pipe fittings *(continued)*									
Single junctions									
300 mm	each	LH	0.15	4.51	PM 0.15	8.14	51.33	63.98	73.58
375 mm	each	LH	0.20	6.01	PM 0.20	10.86	97.99	114.86	132.09
450 mm	each	LH	0.25	7.51	PM 0.25	13.57	105.90	126.98	146.03
600 mm	each	LH	0.30	9.02	PM 0.30	16.28	141.24	166.54	191.52
750 mm	each	LH	0.35	10.52	PM 0.35	19.00	219.86	249.38	286.79
Double junctions									
300 mm	each	LH	0.25	7.51	PM 0.25	13.57	152.66	173.74	199.80
375 mm	each	LH	0.30	9.02	PM 0.30	16.28	185.31	210.61	242.20
Beds and surrounds									
Sand in bed to pipe diameter									
300 mm	m	LH	0.05	1.50	PM 0.05	2.71	2.37	6.58	7.57
375 mm	m	LH	0.06	1.80	PM 0.06	3.26	2.57	7.63	8.77
400 mm	m	LH	0.06	1.80	PM 0.06	3.26	2.57	7.63	8.77
450 mm	m	LH	0.07	2.10	PM 0.07	3.80	2.77	8.67	9.97
525 mm	m	LH	0.07	2.10	PM 0.07	3.80	3.16	9.06	10.42
600 mm	m	LH	0.07	2.10	PM 0.07	3.80	3.36	9.26	10.65
700 mm	m	LH	0.08	2.40	PM 0.08	4.34	3.95	10.69	12.29
750 mm	m	LH	0.08	2.40	PM 0.08	4.34	4.15	10.89	12.52
800 mm	m	LH	0.08	2.40	PM 0.08	4.34	4.54	11.28	12.97
900 mm	m	LH	0.09	2.71	PM 0.09	4.89	5.14	12.74	14.65
1000 mm	m	LH	0.09	2.71	PM 0.09	4.89	5.33	12.93	14.87
1100 mm	m	LH	0.10	3.01	PM 0.10	5.43	5.53	13.97	16.07
1200 mm	m	LH	0.10	3.01	PM 0.10	5.43	5.73	14.17	16.30
1500 mm	m	LH	0.12	3.61	PM 0.12	6.51	6.12	16.24	18.68
1800 mm	m	LH	0.12	3.61	PM 0.12	6.51	7.11	17.23	19.81
Granular material in bed to pipe diameter									
300 mm	m	LH	0.05	1.50	PM 0.05	2.71	2.28	6.49	7.46
375 mm	m	LH	0.06	1.80	PM 0.06	3.26	2.47	7.53	8.66
400 mm	m	LH	0.06	1.80	PM 0.06	3.26	2.47	7.53	8.66
450 mm	m	LH	0.07	2.10	PM 0.07	3.80	2.66	8.56	9.84
525 mm	m	LH	0.07	2.10	PM 0.07	3.80	3.05	8.95	10.29
600 mm	m	LH	0.07	2.10	PM 0.07	3.80	3.24	9.14	10.51
700 mm	m	LH	0.08	2.40	PM 0.08	4.34	3.81	10.55	12.13
750 mm	m	LH	0.08	2.40	PM 0.08	4.34	4.00	10.74	12.35
800 mm	m	LH	0.08	2.40	PM 0.08	4.34	4.38	11.12	12.79
900 mm	m	LH	0.09	2.71	PM 0.09	4.89	4.95	12.55	14.43
1000 mm	m	LH	0.09	2.71	PM 0.09	4.89	5.14	12.74	14.65
1100 mm	m	LH	0.10	3.01	PM 0.10	5.43	5.33	13.77	15.84
1200 mm	m	LH	0.10	3.01	PM 0.10	5.43	5.52	13.96	16.05
1500 mm	m	LH	0.12	3.61	PM 0.12	6.51	5.90	16.02	18.42
1800 mm	m	LH	0.12	3.61	PM 0.12	6.51	6.85	16.97	19.52

New work	Unit	Labour gang hours	Labour net		Plant gang hours	Plant net	Mats net	Price net	Price with 15%
Sewers and manholes			£			£	£	£	£
							VAT not included		

Beds and surrounds
(continued)

Concrete 1:3:6 40 mm
aggregate in bed to pipe
diameter

300 mm	m	LH	0.05	1.50	PM	0.05	2.71	6.68	10.89	12.52
375 mm	m	LH	0.06	1.80	PM	0.06	3.26	7.23	12.29	14.13
400 mm	m	LH	0.06	1.80	PM	0.06	3.26	7.23	12.29	14.13
450 mm	m	LH	0.07	2.10	PM	0.07	3.80	7.79	13.69	15.74
525 mm	m	LH	0.07	2.10	PM	0.07	3.80	8.90	14.80	17.02
600 mm	m	LH	0.07	2.10	PM	0.07	3.80	9.46	15.36	17.66
700 mm	m	LH	0.08	2.40	PM	0.08	4.34	11.13	17.87	20.55
750 mm	m	LH	0.08	2.40	PM	0.08	4.34	11.69	18.43	21.19
800 mm	m	LH	0.08	2.40	PM	0.08	4.34	12.80	19.54	22.47
900 mm	m	LH	0.09	2.71	PM	0.09	4.89	14.47	22.07	25.38
1000 mm	m	LH	0.09	2.71	PM	0.09	4.89	15.03	22.63	26.02
1100 mm	m	LH	0.10	3.01	PM	0.10	5.43	15.58	24.02	27.62
1200 mm	m	LH	0.10	3.01	PM	0.10	5.43	16.14	24.58	28.27
1500 mm	m	LH	0.12	3.61	PM	0.12	6.51	17.25	27.37	31.48
1800 mm	m	LH	0.12	3.61	PM	0.12	6.51	20.03	30.15	34.67

Granular material in bed and
haunching to pipe diameter

300 mm	m	LH	0.05	1.50	PM	0.05	2.71	5.33	9.54	10.97
375 mm	m	LH	0.06	1.80	PM	0.06	3.26	5.33	10.39	11.95
400 mm	m	LH	0.06	1.80	PM	0.06	3.26	5.52	10.58	12.17
450 mm	m	LH	0.07	2.10	PM	0.07	3.80	6.09	11.99	13.79
525 mm	m	LH	0.07	2.10	PM	0.07	3.80	6.28	12.18	14.01
600 mm	m	LH	0.07	2.10	PM	0.07	3.80	6.66	12.56	14.44
700 mm	m	LH	0.08	2.40	PM	0.08	4.34	7.61	14.35	16.50
750 mm	m	LH	0.08	2.40	PM	0.08	4.34	8.18	14.92	17.16
800 mm	m	LH	0.08	2.40	PM	0.08	4.34	8.57	15.31	17.61
900 mm	m	LH	0.09	2.71	PM	0.09	4.89	10.09	17.69	20.34
1000 mm	m	LH	0.09	2.71	PM	0.09	4.89	11.42	19.02	21.87
1100 mm	m	LH	0.10	3.01	PM	0.10	5.43	12.75	21.19	24.37
1200 mm	m	LH	0.10	3.01	PM	0.10	5.43	13.89	22.33	25.68
1500 mm	m	LH	0.12	3.61	PM	0.12	6.51	18.08	28.20	32.43
1800 mm	m	LH	0.12	3.61	PM	0.12	6.51	22.46	32.58	37.47

New work	Unit	Labour gang hours		Labour net	Plant gang hours		Plant net	Mats net	Price net	Price with 15%
Sewers and manholes				£			£	£	£	£

Beds and surrounds
(*continued*) '

Concrete 1:3:6 40 mm
aggregate in bed and haunching
to pipe diameter

	Unit			Labour net			Plant net	Mats net	Price net	Price with 15%
300 mm	m	LH	0.07	2.10	PM	0.05	2.71	15.58	20.39	23.45
375 mm	m	LH	0.08	2.40	PM	0.06	3.26	15.58	21.24	24.43
400 mm	m	LH	0.08	2.40	PM	0.06	3.26	16.14	21.80	25.07
450 mm	m	LH	0.09	2.71	PM	0.07	3.80	17.81	24.32	27.97
525 mm	m	LH	0.09	2.71	PM	0.07	3.80	18.36	24.87	28.60
600 mm	m	LH	0.09	2.71	PM	0.07	3.80	19.48	25.99	29.89
700 mm	m	LH	0.10	3.01	PM	0.08	4.34	22.26	29.61	34.05
750 mm	m	LH	0.10	3.01	PM	0.08	4.34	23.93	31.28	35.97
800 mm	m	LH	0.10	3.01	PM	0.08	4.34	25.04	32.39	37.25
900 mm	m	LH	0.11	3.31	PM	0.09	4.89	29.49	37.69	43.34
1000 mm	m	LH	0.11	3.31	PM	0.09	4.89	33.39	41.59	47.83
1100 mm	m	LH	0.12	3.61	PM	0.10	5.43	37.29	46.33	53.28
1200 mm	m	LH	0.12	3.61	PM	0.10	5.43	40.62	49.66	57.11
1500 mm	m	LH	0.14	4.21	PM	0.12	6.51	52.87	63.59	73.13
1800 mm	m	LH	0.14	4.21	PM	0.12	6.51	65.67	76.39	87.85

Sand in bed and surround to
pipe diameter

	Unit			Labour net			Plant net	Mats net	Price net	Price with 15%
300 mm	m	LH	0.08	2.40	PM	0.08	4.34	7.70	14.44	16.61
375 mm	m	LH	0.09	2.71	PM	0.09	4.89	8.30	15.90	18.29
400 mm	m	LH	0.09	2.71	PM	0.09	4.89	8.69	16.29	18.73
450 mm	m	LH	0.10	3.01	PM	0.10	5.43	9.48	17.92	20.61
525 mm	m	LH	0.10	3.01	PM	0.10	5.43	9.50	19.50	22.43
600 mm	m	LH	0.10	3.01	PM	0.10	5.43	12.64	21.08	24.24
700 mm	m	LH	0.11	3.31	PM	0.11	5.97	13.83	23.11	26.58
750 mm	m	LH	0.11	3.31	PM	0.11	5.97	14.62	23.90	27.48
800 mm	m	LH	0.11	3.31	PM	0.11	5.97	16.59	25.87	29.75
900 mm	m	LH	0.12	3.61	PM	0.12	6.51	18.96	29.08	33.44
1000 mm	m	LH	0.12	3.61	PM	0.12	6.51	21.73	31.85	36.63
1100 mm	m	LH	0.13	3.91	PM	0.13	7.06	24.49	35.46	40.78
1200 mm	m	LH	0.13	3.91	PM	0.13	7.06	26.86	37.83	43.50
1500mm	m	LH	0.15	4.51	PM	0.13	7.06	35.56	47.13	54.20
1800 mm	m	LH	0.15	4.51	PM	0.13	7.06	44.05	55.62	63.96

Civil engineering

New work	Unit		Labour gang hours	Labour net		Plant gang hours	Plant net	Mats net	Price net	Price with 15%
Sewers and manholes				£			£	£	£	£
							VAT not included			

Beds and surrounds
(*continued*)

Granular material in bed and
surround to pipe diameter

300 mm	m	LH	0.08	2.40	PM	0.08	4.34	7.42	14.16	16.28
375 mm	m	LH	0.09	2.71	PM	0.09	4.89	7.99	15.59	17.93
400 mm	m	LH	0.09	2.71	PM	0.09	4.89	8.37	15.97	18.37
450 mm	m	LH	0.10	3.01	PM	0.10	5.43	9.14	17.58	20.22
525 mm	m	LH	0.10	3.01	PM	0.10	5.43	10.66	19.10	21.97
600 mm	m	LH	0.10	3.01	PM	0.10	5.43	12.18	20.62	23.71
700 mm	m	LH	0.11	3.31	PM	0.11	5.97	13.32	22.60	25.99
750 mm	m	LH	0.11	3.31	PM	0.11	5.97	14.09	23.37	26.88
800 mm	m	LH	0.11	3.31	PM	0.11	5.97	15.99	25.27	29.06
900 mm	m	LH	0.12	3.61	PM	0.12	6.51	18.27	28.39	32.65
1000 mm	m	LH	0.12	3.61	PM	0.12	6.51	20.94	31.06	35.72
1100 mm	m	LH	0.13	3.91	PM	0.13	7.06	23.60	34.57	39.76
1200 mm	m	LH	0.13	3.91	PM	0.13	7.06	25.89	36.86	42.39
1500 mm	m	LH	0.15	4.51	PM	0.13	7.06	34.26	45.83	52.70
1800 mm	m	LH	0.15	4.51	PM	0.13	7.06	42.45	54.02	62.12

Concrete 1:3:6 40 mm
aggregate in bed and surround
to pipe diameter

300 mm	m	LH	0.08	2.40	PM	0.08	4.34	21.70	28.44	32.71
375 mm	m	LH	0.09	2.71	PM	0.09	4.89	23.37	30.97	35.62
400 mm	m	LH	0.09	2.71	PM	0.09	4.89	24.49	32.09	36.90
450 mm	m	LH	0.10	3.01	PM	0.10	5.43	26.71	35.15	40.42
525 mm	m	LH	0.10	3.01	PM	0.10	5.43	31.16	39.60	45.54
600 mm	m	LH	0.10	3.01	PM	0.10	5.43	35.62	44.06	50.67
700 mm	m	LH	0.11	3.31	PM	0.11	5.97	38.95	48.23	55.46
750 mm	m	LH	0.11	3.31	PM	0.11	5.97	41.18	50.46	58.03
800 mm	m	LH	0.11	3.31	PM	0.11	5.97	46.75	56.03	64.43
900 mm	m	LH	0.12	3.61	PM	0.12	6.51	53.42	63.54	73.07
1000 mm	m	LH	0.12	3.61	PM	0.12	6.51	61.22	71.34	82.04
1100 mm	m	LH	0.13	3.91	PM	0.13	7.06	69.01	79.98	91.98
1200 mm	m	LH	0.13	3.91	PM	0.13	7.06	75.68	86.65	99.65
1500 mm	m	LH	0.15	4.51	PM	0.13	7.06	100.17	111.74	128.50
1800 mm	m	LH	0.15	4.51	PM	0.13	7.06	124.10	135.67	156.02

New work	Unit	Labour gang hours		Labour net	Plant gang hours		Plant net	Mats net	Price net	Price with 15%
				£			£	£	£	£

Sewers and manholes

VAT not included

Manholes

Excavate for manholes
including levelling and ramming
base, supporting sides,
backfilling, disposal of surplus
excavated material, depth

1.00 - 2.00 m	m3	LD	0.15	3.43	PN	0.15	4.08	-	7.51	8.64
2.00 - 5.00 m	m3	LD	0.20	4.58	PN	0.20	5.44	-	10.02	11.52

Provide ready mixed concrete

1:3:6 40 mm aggregate	m3		-	-		-	-	55.65	55.65	64.00
1:2:4 20 mm aggregate	m3		-	-		-	-	58.99	58.99	67.84

Place un-reinforced concrete in
manhole base and surround,
thickness

150 - 300 mm	m3	LD	0.24	5.49	PN	0.24	6.53	-	12.02	13.82
300 - 500 mm	m3	LD	0.22	5.03	PN	0.22	5.99	-	11.02	12.67
over 500 mm	m3	LD	0.20	4.58	PN	0.20	5.44	-	10.02	11.52

Precast concrete shaft rings,
diameter

625 mm	each	LD	0.35	8.01	PN	0.36	9.80	29.79	47.60	54.74
900 mm	each	LD	0.40	9.15	PN	0.40	10.88	34.83	54.86	63.09
1050 mm	each	LD	0.45	10.30	PN	0.45	12.24	41.47	64.01	73.61
1200 mm	each	LD	0.50	11.44	PN	0.50	13.61	55.37	80.42	92.48
1350 mm	each	LD	0.55	12.58	PN	0.55	14.97	80.66	108.21	124.44
1500 mm	each	LD	0.60	13.73	PN	0.60	16.33	104.58	134.64	154.84
1800 mm	each	LD	0.65	14.87	PN	0.65	17.69	121.00	153.56	176.59
2100 mm	each	LD	0.70	16.02	PN	0.70	19.05	267.39	302.46	347.83
2400 mm	each	LD	0.75	17.16	PN	0.75	20.41	304.06	341.63	392.87
2700 mm	each	LD	0.80	18.30	PN	0.80	21.77	381.98	422.05	485.36
3000 mm	each	LD	0.85	19.45	PN	0.85	23.13	544.06	586.64	674.64

Precast concrete cover slabs,
diameter

625 mm	each	LD	0.35	8.01	PN	0.36	9.80	27.77	45.58	52.42
900 mm	each	LD	0.40	9.15	PN	0.40	10.88	35.42	55.45	63.77
1050 mm	each	LD	0.45	10.30	PN	0.45	12.24	38.38	60.92	70.06
1200 mm	each	LD	0.50	11.44	PN	0.50	13.61	52.91	77.96	89.65
1350 mm	each	LD	0.55	12.58	PN	0.55	14.97	79.06	106.61	122.60
1500 mm	each	LD	0.60	13.73	PN	0.60	16.33	102.73	132.79	152.71
1800 mm	each	LD	0.65	14.87	PN	0.65	17.69	124.29	156.85	180.38
2100 mm	each	LD	0.70	16.02	PN	0.70	19.05	287.68	322.75	371.16
2400 mm	each	LD	0.75	17.16	PN	0.75	20.41	492.10	529.67	609.12
2700 mm	each	LD	0.80	18.30	PN	0.80	21.77	642.23	682.30	784.64
3000 mm	each	LD	0.85	19.45	PN	0.85	23.13	764.31	806.89	927.92

Precast concrete cover slabs,
size

1350 x 1125 mm	each	LD	0.40	9.15	PN	0.40	10.88	34.77	54.80	63.02
1650 x 1500 mm	each	LD	0.50	11.44	PN	0.50	13.61	50.97	76.02	87.42

Civil engineering

New work	Unit	Labour gang hours	Labour net		Plant gang hours	Plant net	Mats net	Price net	Price with 15%
Sewers and manholes			£			£	£	£	£
							VAT not included		

Manholes (*continued*)

New work	Unit	Labour gang hours	Labour net		Plant gang hours	Plant net	Mats net	Price net	Price with 15%
Vitrified clayware half round straight main channels, diameter									
100 mm	each	LD 0.10	2.29	PN	0.10	2.72	3.08	8.09	9.30
150 mm	each	LD 0.12	2.75	PN	0.12	3.27	5.12	11.14	12.81
225 mm	each	LD 0.14	3.20	PN	0.14	3.81	11.52	18.53	21.31
Vitrified clayware half round bends, diameter									
100 mm	each	LD 0.08	1.83	PN	0.08	2.18	3.28	7.29	8.38
150 mm	each	LD 0.10	2.29	PN	0.10	2.72	5.43	10.44	12.01
225 mm	each	LD 0.12	2.75	PN	0.12	3.27	18.26	24.28	27.92
Vitrified clayware half round taper channels, diameter									
150 - 100 mm	each	LD 0.10	2.29	PN	0.10	2.72	13.73	18.74	21.55
225 -150 mm	each	LD 0.12	2.75	PN	0.12	3.27	30.64	36.66	42.16
Vitrified clayware half round taper bends, diameter									
150 - 100 mm	each	LD 0.10	2.29	PN	0.10	2.72	20.87	25.88	29.76
225 -150 mm	each	LD 0.12	2.75	PN	0.12	3.27	60.11	66.13	76.05
Vitrified clayware three quarter section channel bends, diameter									
100 mm	each	LD 0.08	1.83	PN	0.08	2.18	7.45	11.46	13.18
150 mm	each	LD 0.10	2.29	PN	0.10	2.72	12.54	17.55	20.18
Galvanised metal step irons, size									
115 mm	each	LD 0.05	1.14	PN	0.05	1.36	5.26	7.76	8.92
230 mm	each	LD 0.06	1.37	PN	0.06	1.63	6.72	9.72	11.18
Cast iron manhole covers and frames									
grade A, 550 mm diameter	each	LD 0.30	6.86	PN	0.30	8.16	186.41	201.43	231.64
grade A, 600 mm diameter	each	LD 0.30	6.86	PN	0.30	8.16	219.31	234.33	269.48
Vitrified clayware road gullies with rodding eye and stopper									
300 x 600 mm with 150 mm outlet	each	LD 0.25	5.72	PN	0.25	6.80	42.34	54.86	63.09
450 x 900 mm with 150 mm outlet	each	LD 0.30	6.86	PN	0.30	8.16	70.43	85.45	98.27

External works

Preamble

"Labour net" figures include allowances for all costs incidental to the employment of labour.

"Plant net" figures include for all costs of plant including drivers and operators where applicable.

"Materials net" figures include for all costs of materials including an allowance for waste except where specifically stated.

"Price net" figures are the totals of the "Labour net", "Plant net" and "Materials net" figures. Prices are for a builder employing his own labour; according to the amount and nature of the work involved, it may well be possible to secure more advantageous prices from specialist sub-contractors.

Although the Standard Method of Measurement requires gates, gate-posts and post holes to be enumerated separately some combined items have been given for gates, posts or piers and the necessary excavation and concrete complete.

Specialist prices

"Price with 15%" figures are all-in guide prices and include 15% for the builder's overheads, profit, unloading materials and general attendance (to include free use of standing scaffolding and hoists, temporary lighting and water and clearing away rubbish).

The amount of attendance required varies between the various trades and also with the circumstances of specific jobs; the percentage addition must always be considered and adjusted as necessary to suit the terms and conditions of the quotation being used.

Quantities and delivery distances are usually the most significant of the many factors which influence prices and it must be emphasised that quotations should always be obtained when preparing a tender.

Specialist prices for fencing are not in accordance with the Standard Method of Measurement in so far as they include for excavating post holes and setting posts etc. in concrete.

External works

New work

Basic prices for materials		£
Ashes	m3	9.50
	tonne	7.60
Binding gravel (hoggin)	m3	15.84
	tonne	8.55
Hardcore	m3	12.25
	tonne	8.16
Sand	m3	17.96
	tonne	11.22
Timber for earthwork support	m3	244.80
Aggregates		
40 mm	m3	19.75
	tonne	13.16
20 mm	m3	19.91
	tonne	13.27
10 mm	m3	20.07
	tonne	13.38
Ready mixed concrete		
1:3:6 - 40 mm aggregate	m3	53.00
1:2:4 - 20 mm aggregate	m3	56.18
1:1½:3 - 10 mm aggregate	m3	60.42
Bituminous emulsion waterproofing liquid	5 litre	10.96
Cement		
Portland	tonne	89.04
rapid hardening	tonne	150.33
sulphate resisting	tonne	92.66
Formwork		
softwood	m3	244.80
25 mm sawn boarding	m2	7.65
25 mm wrought boarding	m2	13.77

Prices actually to be paid for materials must be checked against the above basic prices and adjustments made as necessary.

External works

New work

Basic prices for materials £

	Unit	Price
Precast concrete block paving		
65 x 200 x 100 mm plain blocks	m2	4.93
65 x 200 x 100 mm coloured blocks	m2	5.88
80 x 200 x 100 mm plain blocks	m2	5.39
80 x 200 x 100 mm coloured blocks	m2	6.09
50 mm thick broken precast concrete slabs	m2	6.76
Precast concrete slabs, 50 mm thick - plain		
450 x 450 mm	m2	9.55
600 x 450 mm	m2	5.28
600 x 600 mm	m2	7.86
600 x 750 mm	m2	4.12
600 x 900 mm	m2	3.71
Precast concrete slabs, 50 mm thick - coloured		
450 x 450 mm	m2	13.37
600 x 450 mm	m2	7.10
600 x 600 mm	m2	6.35
600 x 750 mm	m2	6.27
600 x 900 mm	m2	5.62
Brick paviours 215 x 103 x 65 mm	1,000	325.00
Vegetable soil - imported	m3	9.06
Grass seed	25 kg	48.95
Turf - imported	m2	1.63

Prices actually to be paid for materials must be checked against the above basic prices and adjustments made as necessary.

External works

New work	Unit	Hours C	Hours L	Labour net	Plant net	Material net	Price net	Price with 15%
Hard landscaping				£	£	£	£	£
					VAT not included			
Kerbs, edgings and channels								
Excavating by machine								
Excavating trenches by machine to receive kerb foundations; average size								
300 x 100 mm	m	-	0.02	0.13	0.38	-	0.51	0.59
450 x 150 mm	m	-	0.03	0.20	0.56	-	0.76	0.87
600 x 200 mm	m	-	0.04	0.27	0.75	-	1.02	1.17
Excavating curved trenches by machine to receive kerb foundation; average size								
300 x 100 mm	m	-	0.02	0.15	0.41	-	0.56	0.64
450 x 100 mm	m	-	0.03	0.22	0.62	-	0.84	0.97
600 x 200 mm	m	-	0.04	0.30	0.83	-	1.13	1.30
Excavating trenches by hand to receive kerb foundation; average size								
150 x 50mm	m	-	0.03	0.20	-	-	0.20	0.23
200 x 75mm	m	-	0.07	0.47	-	-	0.47	0.54
250 x 100mm	m	-	0.11	0.74	-	-	0.74	0.85
300 x 100 mm	m	-	0.15	1.01	-	-	1.01	1.16
Excavating curved trenches by hand to receive kerb foundation; average size								
150 x 50mm	m	-	0.04	0.27	-	-	0.27	0.31
200 x 75mm	m	-	0.08	0.54	-	-	0.54	0.62
250 x 100mm	m	-	0.12	0.81	-	-	0.81	0.93
300 x 100mm	m	-	0.15	1.01	-	-	1.01	1.16
Surface treatments								
Compact base of trench excavation by hand	m2	-	0.10	0.67	-	-	0.67	0.77
Precast concrete kerbs, edgings and channels								
Precast concrete standard units, bedding and pointing in cement mortar (1:3); bedding and haunching in concrete (1:2:4) - 20mm								
125 x 150 mm kerb - straight - 325 x 150 mm bed	m	-	0.48	3.24	-	6.84	10.08	11.59
125 x 150mm kerb - curved - 325 x 150mm bed	m	-	0.96	6.47	-	9.32	15.79	18.16
125 x 255mm kerb - straight - 325 x 150mm bed	m	-	0.51	3.44	-	8.39	11.83	13.60
125 x 255mm kerb - curved - 325 x 150mm bed	m	-	1.02	6.87	-	9.32	16.19	18.62
50 x 150mm edging - straight - 300 x 150mm bed	m	-	0.40	2.70	-	5.66	8.36	9.61
50 x 150mm edging - curved - 300 x 150mm bed	m	-	0.80	5.39	-	6.57	11.96	13.75

External works

New work	Unit	Hours C	Hours L	Labour net	Plant net	Material net	Price net	Price with 15%
Hard landscaping				£	£	£	£	£
					VAT not included			

Kerbs, edgings and channels
(*continued*)

Precast concrete kerbs, edgings and channels (*continued*)

	Unit	Hours C	Hours L	Labour net	Plant net	Material net	Price net	Price with 15%
50 x 255mm edging - straight - 300 x 150mm bed	m	-	0.45	3.03	-	6.46	9.49	10.91
50 x 255mm edging - curved - 300 x 150mm bed	m	-	0.90	6.07	-	7.84	13.91	16.00
125 x 150mm channel - straight - 450 x 150mm bed	m	-	0.40	2.70	-	8.48	11.18	12.86
125 x 150mm channel - curved - 450 x 150mm bed	m	-	0.80	5.39	-	10.01	15.40	17.71
125 x 255mm channel - straight - 450 x 150mm bed	m	-	0.50	3.37	-	9.93	13.30	15.30
125 x 255mm channel - curved - 450 x 150mm bed	m	-	1.00	6.74	-	12.13	18.87	21.70
305 x 305 x 150mm quadrant - 500 x 500 x 150mm bed	m	-	0.59	3.98	-	12.89	16.87	19.40
455 x 455 150mm quadrant - 650 x 650 x 150mm bed	m	-	0.80	5.39	-	16.51	21.90	25.18

Hardcore, granular, cement bound bases, sub-bases to roads and pavings

Filling by machine

Imported broken brick hardcore filling, to make up levels, in layers average 150 mm thick, average thickness:-

	Unit	Hours C	Hours L	Labour net	Plant net	Material net	Price net	Price with 15%
100 mm	m2	-	0.03	0.22	0.66	1.04	1.92	2.21
150 mm	m2	-	0.05	0.32	0.99	1.56	2.87	3.30
200 mm	m2	-	0.06	0.43	1.32	2.08	3.83	4.40
250 mm	m2	-	0.08	0.54	1.66	2.60	4.80	5.52
over 250 mm	m3	-	0.28	1.89	5.88	10.41	18.18	20.91

Imported ashes filling, to make up levels, in layers average 150 mm thick, average thickness:-

	Unit	Hours C	Hours L	Labour net	Plant net	Material net	Price net	Price with 15%
50 mm	m2	-	0.02	0.11	0.33	0.40	0.84	0.97
100 mm	m2	-	0.03	0.22	0.66	0.80	1.68	1.93
150 mm	m2	-	0.05	0.32	0.99	1.20	2.51	2.89
200 mm	m2	-	0.06	0.43	1.32	1.61	3.36	3.86
250 mm	m2	-	0.08	0.54	1.66	2.01	4.21	4.84
over 250 mm	m3	-	0.32	2.16	6.63	8.03	16.82	19.34

Imported sand filling, to make up levels, in layers average 150 mm thick, average thickness:-

	Unit	Hours C	Hours L	Labour net	Plant net	Material net	Price net	Price with 15%
50 mm	m2	-	0.02	0.11	0.33	0.95	1.39	1.60

External works

New work	Unit	Hours C	Hours L	Labour net	Plant net	Material net	Price net	Price with 15%
Hard landscaping				£	£	£	£	£
					VAT not included			

Hardcore, granular, cement bound bases, sub-bases to roads and pavings (*continued*)

Filling by machine (*continued*)

Imported granular fill MOT 1, filling in making up levels, in layers average 150 mm thick, average thickness:-

	Unit	Hours C	Hours L	Labour net	Plant net	Material net	Price net	Price with 15%
100 mm	m2	-	0.03	0.22	0.66	1.42	2.30	2.65
150 mm	m2	-	0.05	0.32	0.99	2.14	3.45	3.97
200 mm	m2	-	0.06	0.43	1.32	2.84	4.59	5.28
250 mm	m2	-	0.08	0.54	1.66	3.56	5.76	6.62
over 250 mm	m3	-	0.32	2.16	6.63	14.20	22.99	26.44

Filling by hand

Imported broken brick hardcore, filling in make up levels, in layers average 150 mm thick, average thickness:-

	Unit	Hours C	Hours L	Labour net	Plant net	Material net	Price net	Price with 15%
100 mm	m2	-	0.11	0.74	0.21	1.04	1.99	2.29
150 mm	m2	-	0.17	1.11	0.32	1.56	2.99	3.44
200 mm	m2	-	0.22	1.48	0.43	2.08	3.99	4.59
250 mm	m2	-	0.28	1.85	0.54	2.60	4.99	5.74
over 250 mm	m3	-	1.20	8.09	2.33	10.41	20.83	23.95

Imported ash, filling to make up levels, in layers average 150 mm thick, average thickness:-

	Unit	Hours C	Hours L	Labour net	Plant net	Material net	Price net	Price with 15%
50 mm	m2	-	0.06	0.37	0.11	0.40	0.88	1.01
100 mm	m2	-	0.11	0.74	0.21	0.80	1.75	2.01
150 mm	m2	-	0.17	1.11	0.32	1.20	2.63	3.02
200 mm	m2	-	0.21	1.43	0.33	1.67	3.43	3.94
250 mm	m2	-	0.28	1.85	0.54	2.01	4.40	5.06
over 250 mm	m2	-	1.20	8.09	2.33	8.03	18.45	21.22

Imported sand, filling to make up levels, in layers average 150 mm thick, average thickness:-

	Unit	Hours C	Hours L	Labour net	Plant net	Material net	Price net	Price with 15%
50 mm	m2	-	0.06	0.37	0.11	0.95	1.43	1.64

Imported granular fill MOT 1, filling in making up levels, in layers average 150 mm thick, average thickness:-

	Unit	Hours C	Hours L	Labour net	Plant net	Material net	Price net	Price with 15%
100 mm	m2	-	0.11	0.74	0.21	1.42	2.37	2.73
150 mm	m2	-	0.17	1.11	0.32	2.14	3.57	4.11
200 mm	m2	-	0.22	1.48	0.43	2.84	4.75	5.46
250 mm	m2	-	0.28	1.85	0.54	3.56	5.95	6.84
over 250 mm	m2	-	1.20	8.09	2.33	14.20	24.62	28.31

External works

New work	Unit	Hours C	Hours L	Labour net	Plant net	Material net	Price net	Price with 15%
Hard landscaping				£	£	£	£	£
					VAT not included			
Hardcore, granular, cement bound bases, sub-bases to roads and pavings (*continued*)								
Surface packing to filling								
To vertical or battered faces of:-								
hardcore	m2	-	0.10	0.67	-	-	0.67	0.77
ashes	m2	-	0.10	0.67	-	-	0.67	0.77
granular fill	m2	-	0.10	0.67	-	-	0.67	0.77
Surface treatments								
Compacting by hand, surfaces of:-								
hardcore, blinded by sand	m2	-	0.24	1.62	-	0.48	2.10	2.42
ashes	m2	-	0.14	0.94	-	-	0.94	1.08
sand	m2	-	0.14	0.94	-	-	0.94	1.08
granular fill	m2	-	0.19	1.28	-	-	1.28	1.47
Compacting with vibrating roller, surfaces of:-								
hardcore, blinded with sand	m2	-	0.09	0.57	0.66	0.48	1.71	1.97
ashes	m2	-	0.05	0.33	0.54	-	0.87	1.00
sand	m2	-	0.05	0.33	0.54	-	0.87	1.00
granular filling	m2	-	0.07	0.46	0.75	-	1.21	1.39
Compacting with wheeled roller, surfaces of:-								
hardcore, blinded with sand	m2	-	0.09	0.60	1.42	0.48	2.50	2.88
ashes	m2	-	0.05	0.33	0.78	-	1.11	1.28
sand	m2	-	0.05	0.33	0.78	-	1.11	1.28
granular filling	m2	-	0.07	0.47	1.10	-	1.57	1.81

External works

New work	Unit	Hours C	Hours L	Labour net	Plant net	Material net	Price net	Price with 15%
Hard landscaping				£	£	£	£	£
					VAT not included			
In situ concrete roads, pavings and bases								
Plain in situ concrete, 1:2:4 - 20mm aggregate								
Beds								
not exceeding 150 mm thick	m3	-	1.42	9.57	-	73.19	82.76	95.17
150 - 300 mm thick	m3	-	1.01	6.81	-	73.19	80.00	92.00
Beds, sloping, not exceeding 15 degrees,								
not exceeding 150 mm thick	m3	-	1.65	11.12	-	73.19	84.31	96.96
150 - 300 mm thick	m3	-	1.25	8.43	-	73.19	81.62	93.86
Reinforced in situ concrete, 1:2:4 – 20 mm aggregate								
Beds								
not exceeding 150 mm	m3	0.46	2.04	17.87	-	73.19	91.06	104.72
150 - 300 mm thick	m3	0.33	1.46	12.80	-	73.19	85.99	98.89
Beds sloping, not exceeding 15 degrees								
not exceeding 150 mm	m3	0.53	2.31	20.32	-	73.19	93.51	107.54
150 - 300 mm thick	m3	0.40	1.74	15.31	-	73.19	88.50	101.78
Formwork, basic finish								
Edges of beds								
not exceeding 250 mm high	m	0.38	-	3.36	-	3.81	7.17	8.25
250 - 500 mm	m	0.75	-	6.72	-	7.63	14.35	16.50
Fabric reinforcement								
Fabric reinforcement with one width mesh side lap and one width measured end lap, generally								
C283 (2.61kg/m2)	m2	0.08	-	0.72	-	1.42	2.14	2.46
C385 (3.41 kg/m2)	m2	0.09	-	0.81	-	1.50	2.31	2.66
C503 (4.34kg/m2)	m2	0.11	-	0.99	-	1.91	2.90	3.34
Worked finishes								
Tamped finish (by mechanical means) to								
level surfaces	m2	-	0.03	0.20	1.61	-	1.81	2.08
sloping surfaces	m2	-	0.04	0.27	2.15	-	2.42	2.78
Tamped finish manual								
tamping surface as paving	m2	-	0.03	0.20	-	-	0.20	0.23
Wood float finish								
surface of concrete	m2	0.33	-	2.96	-	-	2.96	3.40

External works

New work	Unit	Hours C	Hours L	Labour net	Plant net	Material net	Price net	Price with 15%
Hard landscaping				£	£	£	£	£
					VAT not included			

In situ concrete roads, pavings and bases (*continued*)

Worked finishes (*continued*)

New work	Unit	Hours C	Hours L	Labour net	Plant net	Material net	Price net	Price with 15%
Steel float finish								
surface of concrete	m2	-	0.33	2.22	-	-	2.22	2.55

Gravel, hoggin, roads, pavings

New work	Unit	Hours C	Hours L	Labour net	Plant net	Material net	Price net	Price with 15%
Gravel, level and to falls only, laid on sand blinded base, thickness								
50 mm thick	m2	-	0.07	0.48	0.57	0.94	1.99	2.29
75 mm thick	m2	-	0.08	0.55	0.57	1.25	2.37	2.73
100 mm thick	m2	-	0.09	0.62	0.57	1.61	2.80	3.22
Gravel, to falls and cross falls and to slopes not exceeding 15 degrees from horizontal, generally, thickness,								
50 mm thick	m2	-	0.08	0.54	0.57	0.94	2.05	2.36
75 mm thick	m2	-	0.09	0.61	0.57	1.25	2.43	2.79
100 mm thick	m2	-	0.10	0.67	0.57	1.61	2.85	3.28

Interlocking brick, block; roads, pavings

Plain precast concrete block paving

New work	Unit	Hours C	Hours L	Labour net	Plant net	Material net	Price net	Price with 15%
200 x 100 mm blocks, 50 mm sand base, fill joints with sand and brush in, compact with vibrating plate compactor, level and to falls only, thickness,								
65 mm half bond	m2	-	1.05	7.08	0.19	6.65	13.92	16.01
80 mm half bond	m2	-	1.15	7.75	0.19	7.16	15.10	17.36
200 x 100mm blocks, 50mm sand base, fill joints with sand and brush in, compact with vibrating plate compactor, level and to falls only, thickness,								
65 mm, herringbone pattern	m2	-	1.15	7.75	0.19	6.65	14.59	16.78
80mm, herringbone pattern	m2	-	1.25	8.43	0.19	7.16	15.78	18.15
200 x 100mm blocks, 50mm sand base, fill joints with sand and brush in, compact with vibrating plate compactor, to falls and crossfalls and to slopes not exceeding 15 degrees from the horizontal, thickness,								
65mm half bond	m2	-	1.07	7.21	0.19	6.65	14.05	16.16
80mm half bond	m2	-	1.17	7.89	0.19	7.16	15.24	17.53
65mm herringbone pattern	m2	-	1.17	7.89	0.19	6.65	14.73	16.94
80mm herringbone pattern	m2	-	1.27	8.56	0.19	6.65	15.40	17.71

External works

	Unit	Hours C	Hours L	Labour net	Plant net	Material net	Price net	Price with 15%
Hard landscaping				£	£	£	£	£
					VAT not included			

Interlocking brick, block; roads, pavings (*continued*)

Coloured precast concrete block paving

200 x 100mm blocks, 50mm sand base, fill joints with sand and brush in, compact with vibrating plate compactor, level and to falls only, thickness,

	Unit	Hours C	Hours L	Labour net	Plant net	Material net	Price net	Price with 15%
65mm half bond	m2	-	1.05	7.08	0.19	7.70	14.97	17.22
80mm half bond	m2	-	1.15	7.75	0.19	7.93	15.87	18.25
65mm herringbone pattern	m2	-	1.15	7.75	0.19	7.70	15.64	17.99
80mm herringbone pattern	m2	-	1.25	8.43	0.19	7.93	16.55	19.03

Slab, brick, block, sett, cobble pavings

Crazy paving

Broken precast concrete paving, laid to falls and crossfalls on a cement and sand bed (1:3) 10mm thick, pointing in cement mortar (1:3) as work proceeds,

	Unit	Hours C	Hours L	Labour net	Plant net	Material net	Price net	Price with 15%
50mm thick	m2	0.65	0.65	10.20	-	11.46	21.66	24.91

Plain PCC slab paving

50mm thick slabs, 50mm sand base, 6mm straight joints both ways, pointing in cement mortar (1:3), level and to falls only, slab size,

	Unit	Hours C	Hours L	Labour net	Plant net	Material net	Price net	Price with 15%
450 x 450mm	m2	0.28	0.28	4.40	0.19	11.79	16.38	18.84
600 x 450mm	m2	0.24	0.24	3.77	0.19	7.10	11.06	12.72
600 x 600mm	m2	0.22	0.22	3.45	0.19	9.42	13.06	15.02
600 x 750mm	m2	0.21	0.21	3.30	0.19	5.59	9.08	10.44
600 x 900mm	m2	0.20	0.20	3.14	0.19	5.15	8.48	9.75

Coloured PCC slab paving

50mm thick slabs, 50mm sand base, 6mm joints straight both ways, pointing with cement mortar (1:3), level and to falls only, slab size,

	Unit	Hours C	Hours L	Labour net	Plant net	Material net	Price net	Price with 15%
450 x 450mm	m2	0.28	0.28	4.40	0.19	15.80	20.39	23.45
450 x 600mm	m2	0.24	0.24	3.77	0.19	9.02	12.98	14.93
600 x 600mm	m2	0.22	0.22	3.45	0.19	8.03	11.67	13.42
600 x 750mm	m2	0.21	0.21	3.30	0.19	7.84	11.33	13.03
600 x 900mm	m2	0.20	0.20	3.14	0.19	7.16	10.49	12.06

External works

New work	Unit	Hours C	Hours L	Labour net	Plant net	Material net	Price net	Price with 15%
Hard landscaping				£	£	£	£	£
					VAT not included			

Slab, brick, block, sett, cobble pavings (*continued*)

Brick Paviours

215 x 103 x 65mm rough stock bricks, to falls or crossfalls, bedding in cement mortar (1:3) 10mm thick, pointing in cement mortar (1:3), as work proceeds; to hardcore base, to paved areas over 300 mm wide, straight joints both ways

	Unit	Hours C	Hours L	Labour net	Plant net	Material net	Price net	Price with 15%
laid flat	m2	1.00	0.50	12.33	-	15.79	28.12	32.34
laid on edge	m2	1.25	0.63	15.41	-	23.41	38.82	44.64
laid flat herringbone pattern	m2	1.25	0.63	15.41	-	15.79	31.20	35.88
laid on edge herringbone pattern	m2	1.50	0.75	18.49	-	23.41	41.90	48.19

External works

	Unit	Specialist price net	Price with 15%
Hard landscaping work		£	£
			VAT not included

Grass reinforcement

The following "Specialist price net" figures are guide prices provided by Grass Concrete Ltd for quantities of about 250 square metres within 150 kilometres of their works.

Prices do not include for cash discount.

See the preamble notes for builder's profit and attendance.

Grasscrete paving of concrete 28 N/mm2 at 28 days, 10 mm aggregate, spread around plastic formers on sub-base prepared by others, including burning off former tops, filling voids with fine top soil and sowing grass seed 50 g/m2

	Unit	Specialist price net	Price with 15%
76 mm paving with GC3 formers and fabric reinforcement A 193-3.02 kg/m2	m2	20.63	23.72
100 mm paving with GC1 formers and fabric reinforcement A193-3.02 kg/m2	m2	23.67	26.76
150 mm paving with GC2 formers and fabric reinforcement A252-3.95 kg/m2	m2	28.63	32.92

Grassblock paving of precast concrete blocks close butted on sub-base prepared by others, including filling voids with fine top soil and sowing grass seed 50 g/m2

	Unit	Specialist price net	Price with 15%
63 mm paving in 406 x 406 mm interlocking blocks	m2	21.50	24.72
83 mm paving in 406 x 406 mm interlocking blocks	m2	21.91	25.20
103 mm paving in 406 x 406 mm interlocking blocks	m2	22.87	26.30
125 mm paving in 406 x 406 mm interlocking blocks	m2	23.58	27.12

Grassroad paving of HDPE plastic honeycomb cell interlocking units on sub-base prepared by others, including filling voids with fine topsoil and sowing grass seed 50g/m2.

	Unit	Specialist price net	Price with 15%
	m2	20.87	24.00

Enquiries about the foregoing specialist prices should be made to Grass Concrete Ltd, Walker House, 22 Bond Street, Wakefield, West Yorkshire, WF1 2QP, tel (01924) 375997/374818, fax (01924) 290289.

External works

New work

Hard landscaping work

£   £

VAT not included

Roadwork etc

The following "Specialist price net" figures are guide prices based upon limestone aggregates unless otherwise mentioned provided by Bardon Aggregates, for work within 30 kilometres of their works.

To determine whether work can be machine laid or must be laid by hand it should be noted that a standard machine requires at least 3 m width in which to lay. Hard access must be available for machine and lorries and lorries must have sufficient room (including headroom in locations such as underground car parks) to be able to tip satisfactorily. Foundations must be strong enough to take the weight of machine and lorries.

Machine outputs vary considerably according to the shape of the area to be paved - long continuous road being easier than a square area where the machine has to stop regularly and change direction. The machine laid prices quoted take into account to some extent the latter problem.

Prices do not include for cash discount.

See preamble notes for builder's profit and attendance.

40 mm dense bituminous roadbase macadam to BS 4987, 1994 Clause 5.1

Road base in areas over 1000 m2 machine laid

	Unit	Specialist price net	Price with 15%
100 mm	m2	8.93	10.27
150 mm	m2	12.55	14.43

20 mm open-graded bituminous basecourse macadam to BS 4987, 1994 Clause 6.1

60 mm base course for carriageways in area

	Unit	Specialist price net	Price with 15%
50 - 100 m2 hand laid	m2	10.55	12.13
100 - 500 m2 hand laid	m2	7.67	8.82
500 - 1000 m2 hand laid	m2	6.09	7.00
over 1000 m2 machine laid	m2	6.14	7.74

75 mm base course for carriageways in areas

	Unit	Specialist price net	Price with 15%
50 - 100 m2 hand laid	m2	11.13	12.80
100 - 500 m2 hand laid	m2	8.56	9.84
500 - 1000 m2 hand laid	m2	6.93	7.97
over 1000 m2 machine laid	m2	7.30	8.40

40 mm base course for footpaths in areas

	Unit	Specialist price net	Price with 15%
50 - 100 m2 hand laid	m2	8.45	9.72
100 - 500 m2 hand laid	m2	5.30	6.10
500 - 1000 m2 hand laid	m2	4.36	5.01

40 mm dense bituminous basecourse macadam to BS 4987, 1994 Clause 6.3

Base course for carriageways in areas over 1000 m2 machine laid

	Unit	Specialist price net	Price with 15%
75 mm	m2	7.25	8.34
100 mm	m2	9.03	10.38

External works

New work

	Unit	Specialist price net	Price with 15%
Hard landscaping work		£	£
		VAT not included	

Roadwork etc *(continued)*

28 mm dense bituminous basecourse macadam to BS 4987, 1994 Clause 6.4

	Unit	Specialist price net	Price with 15%
50 mm base course for carriageways in areas over 1000 m2 machine laid	m2	5.36	6.16

10 mm open-graded bituminous wearing course macadam to BS 4987, 1994 Clause 7.2

25 mm wearing course for private roads, car parks for light vehicles etc in areas

	Unit	Specialist price net	Price with 15%
50 - 100 m2 hand laid	m2	7.09	8.15
100 - 500 m2 hand laid	m2	4.57	5.26
500 - 1000 m2 hand laid	m2	3.47	3.99
over 1000 m2 machine laid	m2	3.78	4.35

6 mm medium-graded bituminous wearing course macadam to BS 4987, 1994 Clause 7.6

14 mm wearing course for light traffic, footpaths, playgrounds, patching etc in areas

	Unit	Specialist price net	Price with 15%
50 - 100 m2 hand laid	m2	6.30	7.25
100 - 500 m2 hand laid	m2	3.36	3.86
500 - 1000 m2 hand laid	m2	2.42	2.78

20 mm wearing course for light traffic, footpaths, playgrounds, patching etc in areas

	Unit	Specialist price net	Price with 15%
50 - 100 m2 hand laid	m2	6.86	7.89
100 - 500 m2 hand laid	m2	4.39	5.05
500 - 1000 m2 hand laid	m2	2.87	3.61

10 mm close-graded wearing macadam to BS 4987, 1994 Clause 7.4

25 mm wearing course for roads, car parks etc in areas

	Unit	Specialist price net	Price with 15%
50 - 100 m2 hand laid	m2	7.09	8.15
100 - 500 m2 hand laid	m2	4.57	5.26
500 - 1000 m2 hand laid	m2	3.47	3.99
over 1000 m2 machine laid	m2	3.78	4.35

30 mm wearing course for roads, car parks etc for heavier traffic in areas

	Unit	Specialist price net	Price with 15%
50 - 100 m2 hand laid	m2	8.01	9.21
100 - 500 m2 hand laid	m2	4.49	5.16
500 - 1000 m2 hand laid	m2	3.34	3.84
over 1000 m2 machine laid	m2	4.13	4.75

Extra over above wearing courses for specification using hardstone or slag aggregates, aggregates with a PSV not less than 50 or rock types 1 to 7 aggregates in lieu of limestone aggregate

	Unit	Specialist price net	Price with 15%
20 mm	m2	0.32	0.37
25 mm	m2	0.42	0.48

External works

New work	Unit	Specialist price net	Price with 15%

Hard landscaping work

£   £

VAT not included

Roadwork etc *(continued)*

3 mm fine-graded bituminous wearing course to BS 4987, 1994 Clause 7.7

14 mm wearing course for footpaths in areas			
50 - 100 m2 hand laid	m2	6.46	7.43
100 - 500 m2 hand laid	m2	3.52	4.05
500 - 1000 m2 hand laid	m2	2.57	2.96
20 mm wearing course for carriageways in areas			
50 - 100 m2 hand laid	m2	7.12	8.19
100 - 500 m2 hand laid	m2	4.65	5.35
500 - 1000 m2 hand laid	m2	3.13	3.60
over 1000 m2 machine laid	m2	3.29	3 78

Sundries

Bituminous emulsion tack coat	m2	0.16	0.18
Seal bituminous macadam surfaces with bituminous grit to BS 4987, 1994 Clause 7.9			
base courses (130 m2 per tonne)	m2	0.26	0.30
wearing courses (150 m2 per tonne)	m2	0.21	0.24
Fair flush joint of new pavings to existing macadam finishings	m	0.32	0.37
Make good paving around manhole cover frames	each	1.31	1.51
Dish and make good paving around gullies and the like	each	0.63	0.79

Hot rolled asphalt paving to BS 594

Base course asphalt to BS 594/1992 Part 1 Group 1 Table 2 column 5 with 60% of 28 mm nominal size limestone coarse aggregate (asphaltic cement Table 1) in areas over 500 m2 machine laid			
60 mm	m2	9.08	10.44
75 mm	m2	10.19	11.72
40 mm wearing coarse asphalt to BS 594/1992 Part 1 Group 3 Table 5 column 21 with 30% of 14 mm nominal size hardstone coarse aggregate (asphaltic cement Table 1) with 20 mm precoated chippings rolled into the top surface, in areas over 500 m2 machine laid	m2	8.87	10.20

Note: The above examples represent typical specifications to withstand heavy wear to roads (e.g. factories). It should be noted that there are numerous variations contained in BS 594 and it is advisable that separate quotations be obtained against any deviation from the above examples.

New work	Unit	Specialist price net	Price with 15%

Hard landscaping work — £ / £

VAT not included

Roadwork etc *(continued)*

Coloured paving

20 mm red coloured bituminous macadam paving, 6 mm or 10 mm aggregate, for light traffic, footpaths, playgrounds etc, in areas

	Unit	Specialist price net	Price with 15%
100 - 500 m2 hand laid	m2	9.77	11.24
over 500 m2 machine laid	m2	11.24	12.93

Enquiries about the foregoing specialist prices should be made to Bardon Aggregates, 8 Christow Road, Marsh Barton, Exeter, EX2 8QU, tel (01392) 432281, fax (01392) 422012.

External works

New work	Unit	Hours C	Hours L	Labour net	Plant net	Material net	Price net	Price with 15%
Seeding and turfing				£	£	£	£	£
					VAT not included			

Cultivating

Spread, level and grade selected
vegetable spoil from spoil heaps for
seeding or turfing; average thickness,

	Unit	Hours C	Hours L	Labour net	Plant net	Material net	Price net	Price with 15%
75mm	m2	-	0.20	1.35	0.19	-	1.54	1.77
100mm	m2	-	0.22	1.48	0.19	-	1.67	1.92
150mm	m2	-	0.25	1.69	0.38	-	2.07	2.38

Spread, level and grade imported
vegetable soil for seeding or turfing,
average thickness,

75mm	m2	-	0.20	1.35	0.19	0.75	2.29	2.63
100mm	m2	-	0.22	1.48	0.19	1.00	2.67	3.07
150mm	m2	-	0.25	1.69	0.38	1.49	3.56	4.09

Surface applications

Treat surfaces of ground with,

fertilizer, 0.07kg/m2, rake in	m2	-	0.05	0.34	-	0.02	0.36	0.41
bone meal, 0.06kg/m2, rake in	m2	-	0.05	0.34	-	0.04	0.38	0.44
weedkiller, 0.03kg/m2	m2	-	0.05	0.34	-	0.04	0.38	0.44

Seeding

Sow grass seed at a rate of 0.07kg/m2,
rake in and roll,

general surfaces	m2	-	0.07	0.47	-	0.16	0.63	0.72

Turfing

Lay turf to general surfaces, roll and
water in,

imported turf	m2	-	0.25	1.69	-	1.71	3.40	3.91

Planting only, excluding material

Excavate or form pit, hole or trench, dig
over ground in bottom, spread and pack
around roots with finely broken soil, refill
with topsoil including growth material i.e.
manure to one third volume, water in,
remove surplus spoil, label,

small tree	each	-	0.70	4.72	-	-	4.72	5.43
medium tree	each	-	1.40	9.44	-	-	9.44	10.86
large tree	each	-	2.80	18.87	-	-	18.87	21.70
shrub	each	-	0.45	3.03	-	-	3.03	3.48
hedge plant	each	-	0.25	1.69	-	-	1.69	1.94

External works

New work	Unit	Hours C	Hours L	Labour net	Plant net	Material net	Price net	Price with 15%
Seeding and turfing				£	£	£	£	£
					VAT not included			

Planting only, excluding material
(*continued*)

Treated softwood tree stake, pointed
and driven into ground and with suitable
ties nailed to stake and secured to tree

	Unit	Hours C	Hours L	Labour net	Plant net	Material net	Price net	Price with 15%
2000mm long	each	-	0.30	2.02	-	-	2.02	2.32
3000mm long	each	-	0.30	2.02	-	-	2.02	2.32
3500mm long	each	-	0.30	2.02	-	-	2.02	2.32
4000mm long	each	-	0.30	2.02	-	-	2.02	2.32

Hedge and herbaceous plants, including
forming hole and filling after, size of
plant,

	Unit	Hours C	Hours L	Labour net	Plant net	Material net	Price net	Price with 15%
not exceeding 300mm high	each	-	0.10	0.67	-	-	0.67	0.77
300 - 450mm high	each	-	0.20	1.35	-	-	1.35	1.55
450 - 600mm	m2	-	0.30	2.02	-	-	2.02	2.32
600 - 750mm	m2	-	0.40	2.70	-	-	2.70	3.11
750 - 1000mm	m2	-	0.50	3.37	-	-	3.37	3.88

External works

New work	Unit	Specialist price net	Price with 15%

Soft landscaping work

£ £
VAT not included

Landscaping

The following "Specialist price net" figures are guide prices provided by Talastone Gardens Ltd for work to a minimum value of £200.00 within 100 kilometres of their nursery.

Prices do not include for cash discount.

Circumstances greatly affect the cost of this work, particularly that of tree felling and it is essential that quotations be obtained on every occasion.

See the preamble notes for builder's profit and attendance.

	Unit	Specialist price net	Price with 15%
Fell trees including excavating roots and removing all from site	m girth	40.00	46.00
Remove butt and roots of trees previously cut down to ground level	m girth	35.00	40.25
Move topsoil from spoil heaps not exceeding 400m, spread average 150 mm thick and roughly level	m2	0.45	0.52
Extra over last per additional 75 mm thick	m2	0.17	0.20
Supply additional topsoil from outside source and spread per 25 mm thick	m2	0.38	0.44
Dig over surface to remove debris, large stones and weeds etc, level and leave ready for planting or seeding	m2	0.60	0.69
Prepare as last and follow with pre-seeding fertilisers and best quality lawn seed, each at 68 grammes per m2, including lightly raking and rolling	m2	1.15	1.32
First cut when grass 50 mm high including making good any areas which may have failed	m2	0.03	0.03
Additional cuts	m2	0.02	0.02
Prepare surface as for seeding, but lay specially grown weed-free lawn turf	m2	4.20	4.83
Prepare surface as for seeding, but turf with best quality meadow turf, supplied and laid close butted with broken joints, cracks filled with fine soil and seeded, and the whole lightly rolled	m2	2.50	2.88
Maintain grassed areas for a period of twelve months by cutting, trimming edges and supplying and applying any necessary fertilisers and weedkillers to ensure good condition	m2	0.45	0.52
Excavate trench 500 mm wide x 500 mm deep for hedge, fill with topsoil and fertiliser and remove spoil	m	2.50	2.88
Excavate pit 600 x 600 x 600 mm deep for shrub, fill as last and remove spoil	each	3.25	3.74
Excavate pit 900 x 900 x 900 mm deep for tree, fill as last and remove spoil	each	12.00	13.80

External works

New work	Unit	Specialist price net	Price with 15%
Soft landscaping work		£	£
		VAT not included	

Schedule of prices for tree and shrub planting

Note: Prices vary with the variety of tree specified and the following prices should only be taken as an average.

These prices apply to open-ground trees, i.e. not containerised.

	Unit	Specialist price net	Price with 15%
Supply and plant an Extra Heavy Standard tree (14 - 16 cm girth) in previously dug pit, including two tanalised stakes driven in and two plastic tree ties	each	65.00	74.75
Supply and plant a Standard tree (8 - 10 cm girth) in previously dug pit, including one tanalised stake driven in and one plastic tree tie	each	25.00	28.75
Supply and plant a whip (900 - 1200 mm high) including digging a suitable hole and providing fertiliser	each	2.50	2.88
Mulching shrub beds with approved wood bark 50 mm in depth	m2	3.50	4.03

Enquiries about the foregoing specialist prices should be made to Talastone Gardens Ltd, East Penrest, Lezant, Launceston, Cornwall, PL15 9NR, tel: (01579) 370749

External works

	Unit	Hours C	Hours L	Labour net	Plant net	Material net	Price net	Price with 15%
Fencing				£	£	£	£	£
					VAT not included			

Post and wire fencing, posts set in concrete

Strained wire fencing, consisting of 3.25 mm galvanised mild steel line wire, galvanised steel components, mild steel angle posts and struts, fencing with lines of plain wire threaded through posts at 2750 mm centres let 600 mm into ground

900mm high, three lines	m	-	0.36	2.43	-	4.38	6.81	7.83
extra for end post and strut	each	-	1.72	11.59	-	21.30	32.89	37.82
extra for corner post with two struts	each	-	2.67	18.00	-	31.42	49.42	56.83
1200 mm high six lines	m	-	0.53	3.57	-	5.05	8.62	9.91
extra for end post and strut	each	-	2.03	13.68	-	25.50	39.18	45.06
extra for corner post with 2 struts	each	-	3.16	21.30	-	36.63	57.93	66.62

Chain link fencing, posts set in concrete

Chain link fencing, 36 x 51 mm plastic coated mesh, lines wires and tying wires, galvanised steel components, mild steel angle posts and struts, line wires threaded through posts and strained with eye bolts, posts at 3000mm centres let 600 mm into ground

900 mm high	m	-	0.53	3.57	-	5.87	9.44	10.86
extra for end post and strut	each	-	1.72	11.59	-	23.65	35.24	40.53
extra for corner post with 2 struts	each	-	2.67	18.00	-	33.77	51.77	59.54
1200 mm high	m	-	0.81	5.46	-	7.16	12.62	14.51
extra for end post and strut	each	-	2.03	13.68	-	30.18	43.86	50.44
extra for corner post and 2 struts	each	-	3.16	21.30	-	41.31	62.61	72.00

Chain link fencing, 36 x 51 mm plastic coated mesh, line wires and tying wires, galvanised steel components, concrete posts, line wires threaded through posts and strained with eye bolts, posts at 3000 mm centres let 600 mm into ground

900 mm high	m	-	0.57	3.84	-	5.19	9.03	10.38
extra for end post and strut	each	-	1.98	13.35	-	26.73	40.08	46.09
extra for corner post and 2 struts	each	-	3.08	20.76	-	38.89	59.65	68.60
1200 mm high	m	-	0.86	5.80	-	6.15	11.95	13.74
extra for end post and strut	each	-	2.34	15.77	-	34.27	50.04	57.55
extra for corner post and 2 struts	each	-	3.64	24.53	-	49.51	74.04	85.15

External works

New work	Unit	Hours C	Hours L	Labour net	Plant net	Material net	Price net	Price with 15%
Fencing				£	£	£	£	£
					VAT not included			

Chestnut fencing

Cleft chestnut pale fencing, consisting of
pales spaced 51 mm apart, two lines of
galvanised wire, galvanised tying wire,
64 mm diameter posts driven in at
2700 mm centres

	Unit	Hours C	Hours L	Labour net	Plant net	Material net	Price net	Price with 15%
900 mm high	m	-	0.18	1.21	-	3.41	4.62	5.31
extra for end post and strut	each	-	0.38	2.56	-	2.05	4.61	5.30
extra for corner post and 2 struts	each	-	0.48	3.24	-	2.42	5.66	6.51
1200 mm high	m	-	0.20	1.35	-	4.65	6.00	6.90
extra for end post and strut	each	-	0.45	3.03	-	2.42	5.45	6.27
extra for corner post and 2 struts	each	-	0.53	3.57	-	4.10	7.67	8.82

Panel fencing

Larch lap panel fencing, including
treated timber posts, rails, gravel board,
centre stumps, posts at 2000 mm
centres set in concrete.

	Unit	Hours C	Hours L	Labour net	Plant net	Material net	Price net	Price with 15%
900 mm high	m	-	0.89	6.00	-	10.99	16.99	19.54
1200 mm high	m	-	1.12	7.55	-	11.73	19.28	22.17

Close boarded fencing

Sawn softwood close boarded fence to
BS 1722 Part 5, consisting of 100 x 100
mm concrete tapered posts set at 2700
mm centres let into ground and set in
concrete, two arris rails (two out of 75 x
75mm) shaped and morticed into post
both ends

	Unit	Hours C	Hours L	Labour net	Plant net	Material net	Price net	Price with 15%
1000 mm high	m	0.99	0.44	11.84	-	11.26	23.10	26.57
1200 mm high	m	1.18	0.44	13.54	-	13.55	27.09	31.15
1800 mm high	m	1.78	0.44	18.92	-	18.39	37.31	42.91

External works

VAT not included

New work	Unit	Hours C	Hours L	Labour net	Plant net	Material net	Price net	Price with 15%
Fencing				£	£	£	£	£
Open type fencing								
Wrought softwood post-and-rail fencing, 25 x 150 mm horizontal nails fixed with galvanised nails to 75 x 100 mm posts at 1800 mm centres, set 1250 mm above ground and 600 mm below, including creosoting ends of posts								
1200 mm high, four rails	m	0.80	1.00	13.91	-	12.10	26.01	29.91
Wrought softwood palisade fencing, 19 x 75 mm pointed pales fixed with galvanised nails to 50 x 75 mm horizontal rails, 75 x 100 mm posts at 1800 mm centres set 900 mm above ground and 600 mm below, including creosoting ends of posts								
900 mm high, two rails	m	0.70	1.70	17.73	-	14.52	32.25	37.09
Rigid PVC ranch fencing, 20 x 125 mm horizontal rails, 81 x 81 x 750 mm posts at 1800 mm centres, set 500 mm above ground and 250 mm below ground, post caps fixed with adhesive								
450 mm high, two rails	m	0.45	1.45	13.80	-	10.53	24.33	27.98
Rigid PVC fence, 20 x 125 mm horizontal rails, 81 x 81 x 1800 mm posts at 1800 mm centres set 1250 mm above ground and 550 mm below ground, steel post inserts, post caps fixed with adhesive								
1200 mm high, four rails	m	0.50	1.50	14.59	-	19.73	34.32	39.47
Independent gate posts								
Wrought Keruing gate posts with weathered top								
175 x 175 x 2130 mm	each	0.63	1.63	16.63	-	38.44	55.07	63.33
Metal tubular gate posts with rounded tops								
50 mm diameter, 1200 mm long	each	0.50	1.50	14.59	-	16.83	31.42	36.13
50 mm diameter, 2200 mm long	each	0.63	1.63	16.63	-	20.95	37.58	43.22
80 mm diameter, 1200 mm long	each	0.50	1.50	14.59	-	29.19	43.78	50.35
80 mm diameter, 2200 mm long	each	0.63	1.63	16.63	-	34.34	50.97	58.62
Gates								
Stock joinery pattern impregnated softwood gates								
914 x 1041 mm high	each	1.00	-	8.96	-	78.25	87.21	100.29

External works

New work	Unit	Specialist price net	Price with 15%
Fencing		**£**	**£**
		VAT not included	

The following "Specialist price net" figures are guide prices provided by Darfen Durafencing, for quantities of 100 metres minimum on a clear site with ordinary easy digging where the lines are clearly pegged and within 50 kilometres of a branch depot.

Prices do not include for cash discount.

See the preamble notes for builder's profit and attendance.

In addition to general attendance the builder will be required to peg out the lines for fencing.

All posts are set in concrete in accordance with BS 1722.

Open type fencing

Post-and-wire fencing

	Unit	Specialist price net	Price with 15%
1000 mm fencing of five 4 mm line wires and 125 x 125 mm to 75 x 75 x 1670 mm reinforced concrete tapered posts at 3000 mm centres set 600 mm deep into ground in concrete	m	15.63	17.97
Extra for 125 x 125 mm end straining posts with 100 x 75 mm strut set in concrete	each	64.04	73.64
Extra for corner straining posts with two struts set in concrete	each	82.53	94.91
1400 mm fencing of seven 4 mm line wires and 125 x 125 mm to 75 x 75 x 2070 mm reinforced concrete tapered posts at 3000 mm centres set 600 mm deep into ground in concrete	m	17.61	20.25
Extra for 125 x 125 mm end straining posts with 100 x 75 mm strut set in concrete	each	68.54	78.82
Extra for corner straining posts with two struts set in concrete	each	81.80	94.07

Treated sawn softwood post-and-rail fencing

	Unit	Specialist price net	Price with 15%
1200 mm fencing of three 38 x 87 mm horizontal rails nailed to face of 75 x 150 mm posts at 1800 mm centres set 600 mm deep into ground in concrete	m	27.91	32.10
1200 mm fencing as last but with four 38 x 87 mm horizontal rails	m	29.83	34.30

Galvanised chain-link fencing

	Unit	Specialist price net	Price with 15%
900 mm fencing of 50 mm mesh x 3 mm chain link, two 3 mm line wires and 40 x 40 x 1500 mm steel angle posts at 3000 mm centres set 600 mm deep into ground in concrete	m	16.89	19.43
Extra for 50 x 50 mm angle end straining post with 40 x 40 mm steel angle strut, each bent over at bottom and set in concrete	each	60.77	69.89
Extra for corner straining posts with two struts set in concrete	each	85.97	98.87

New work	Unit	Specialist price net	Price with 15%
Fencing		£	£
		VAT not included	

Open type fencing *(continued)*

Galvanised chain-link fencing *(continued)*

	Unit	Specialist price net	Price with 15%
1400 mm fencing of 50 mm mesh x 3 mm chain link, three 3.55 mm line wires and 45 x 45 x 2000 mm steel angle posts at 3000 mm centres set 600 mm deep into ground in concrete	m	21.93	25.22
Extra for 50 x 50 mm angle end straining posts with 45 x 45 mm steel angle strut, each bent over at bottom and set in concrete	each	63.66	73.21
Extra for corner straining posts with two struts set in concrete	each	89.65	103.10
1800 mm fencing of 50 mm mesh x 3 mm chain link, three 3.55 mm line wires and 45 x 45 x 2600 mm steel angle posts at 3000 mm centres set 760 mm deep into ground in concrete	m	24.68	28.38
Extra for 60 x 60 mm angle end straining posts with 45 x 45 mm steel angle strut, each bent over at bottom and set in concrete	each	91.16	104.84
Extra for corner straining posts with two struts set in concrete	each	113.74	130.80
900 mm fencing of 50 mm mesh x 3 mm chain link, three 3 mm line wires and 120 x 120 mm to 75 x 75 x 1600 mm reinforced concrete tapered posts at 3000 mm centres set 600 mm deep into ground in concrete	m	17.50	20.12
Extra for 125 x 125 mm angle end straining posts with 100 x 75 mm strut set in concrete	each	62.11	71.42
Extra for corner straining posts with two struts set in concrete	each	75.31	86.61
1400 mm fencing of 50 mm mesh x 3 mm chain link, three 3.55 mm line wires and 125 x 125 mm to 75 x 75 x 2070 mm reinforced concrete tapered posts at 3000 mm centres set 600 mm deep into ground in concrete	m	21.57	24.81
Extra for 125 x 125 mm end straining posts with 100 x 75 mm strut set in concrete	each	68.46	78.73
Extra for corner straining posts with two struts set in concrete	each	81.80	94.07
1800 mm fencing of 50 mm mesh x 3 mm chain link, three 3.55 mm line wires and 125 x 125 mm to 75 x 75 x 2630 mm reinforced concrete tapered posts at 3000 mm centres set 760 mm deep into ground in concrete	m	24.88	28.62
Extra for 125 x 125 mm angle end straining posts with 100 x 75 mm strut set in concrete	each	87.24	100.33
Extra for corner straining posts with two struts set in concrete	each	102.94	118.38
1820 mm security fencing of 50 mm mesh x 3 mm chain link, three 3.55 mm line wires, three rows of barbed wire and 125 x 125 mm to 75 x 75 x 3000 mm reinforced concrete tapered posts, with cranked top, at 3000 mm centres set 760 mm deep into ground in concrete	m	26.41	30.37

External works

Fencing £ £

Open type fencing *(continued)*

Galvanised chain-link fencing *(continued)*

	Unit	Specialist price net	Price with 15%
Extra over last item for 125 x 125 mm end straining posts with 100 x 75 mm strut set in concrete	each	95.79	110.16
Extra for corner straining posts with two struts set in concrete	each	118.46	136.23

Cleft-pale fencing

	Unit	Specialist price net	Price with 15%
1060 mm fencing of chestnut pales 75 mm apart, two lines of binding wire and 63 mm approximate diameter x 1670 mm chestnut posts at 2280 mm centres driven 600 mm into ground	m	15.62	17.97
Extra for 75 to 85 mm approximate diameter straining posts with strut spiked to post	each	32.29	37.13
Extra for corner straining posts with two struts spiked to post	each	42.62	49.01

Galvanised steel palisade fencing

	Unit	Specialist price net	Price with 15%
Fencing of triple pointed corrugated pales (1.9 kg/m) fixed with 6 mm "Avdelok" rivets to two 50 x 40 x 6 mm horizontal angle rails, the bottom rail fitted with two support feet per bay set into ground in concrete, and with 102 x 44 mm RSJ posts at 2750 mm centres set 760 mm deep into ground in concrete			
1800 mm	m	92.35	106.20
2400 mm	m	108.95	125.30
Fencing as last but with triple pointed pales (2.42 kg/m) fixed with 8 mm "Avdelok" rivets			
1800 mm	m	107.33	123.42
2400 mm	m	124.63	143.32

Close type fencing

Treated softwood close-boarded fencing

	Unit	Specialist price net	Price with 15%
1000 mm fencing of 14 to 7 mm x 100 mm feather edged boarding, two 75 x 75 mm arris rails, 100 x 100 x 1600 mm sawn posts at 3000 mm centres set 600 mm deep into ground in concrete, 32 x 200 mm gravel board with 50 x 50 mm centre stumps driven 600 mm into ground, 25 x 65 mm counter rail and 38 x 65 mm weathered capping	m	40.98	47.13
1600 mm fencing as last but with three 75 x 75 mm arris rails and 100 x 125 x 2350 mm sawn posts at 3000 mm centres set 760 mm deep into ground in concrete	m	58.88	67.72
910 mm fencing of 1830 x 910 mm interwoven panels, of 6 x 75 mm woven slats, 19 x 38 mm framing and weathered capping, nailed between 75 x 75 x 1520 mm capped posts at 1900 mm centres set 600 mm deep into ground in concrete	m	27.78	31.95

External works

New work	Unit	Specialist price net	Price with 15%
Fencing		**£**	**£**
		VAT not included	

Close type fencing *(continued)*

Treated softwood close-boarded fencing *(continued)*

1830 mm fencing as last but with 1830 x 1830 mm interwoven panels and 75 x 75 x 2590 mm capped posts set 760 mm deep into ground in concrete	m	41.97	48.27
910 mm fencing of 1830 x 910 mm waney edged panels, of 5 x 100 to 125 mm waney edged slats, 16 x 38 mm framing and weathered capping, nailed between 75 x 75 x 1520 mm capped posts at 1900 mm centres set 600 mm deep into ground in concrete	m	30.13	34.65
1830 mm fencing as last but with 1830 x 1830 mm waney edged panels and 75 x 75 x 2590 mm capped posts set 760 mm deep into ground in concrete	m	44.27	50.91

Prices for oak fencing (subject to availability) on application.

Gates.

42 mm (outside diameter) primed tubular steel single leaf gates filled in with chain link and with two 150 x 150 mm reinforced concrete gate posts each with strut and set of fittings including setting posts and struts in concrete and hanging gates			
1000 x 900 mm	each	382.64	440.04
1000 x 1400 mm	each	411.27	472.96
1000 x 1800 mm	each	464.84	534.56
48 mm (outside diameter) primed tubular steel single leaf gates filled in with chain link and with two 150 x 150 mm reinforced concrete gate posts each with strut and set of fittings including setting posts and struts in concrete and hanging gates			
3000 x 900 mm	pair	675.48	776.81
3000 x 1400 mm	pair	730.73	840.34
3000 x 1800 mm	pair	849.66	977.11
Galvanised steel single leaf gates, to match corrugated pale (1.9 kg/m) fencing, with two 127 x 76 mm RSJ gate posts including setting posts in concrete and hanging gates			
1000 x 1800 mm	each	832.28	957.12
1000 x 2400 mm	each	979.35	1,126.26
Galvanised steel double leaf gates to match corrugated pale (1.9 kg/m) fencing, with two 125 x 125 mm SHS gate posts including setting posts in concrete and hanging gates			
3000 x 1800 mm	pair	1,658.89	1,907.72
3000 x 2400 mm	pair	1,931.12	2,220.78
1000 x 1600 mm treated sawn softwood close-boarded gates, to match close-boarded fencing, complete with all necessary hanging and closing fittings including hanging gates	each	264.33	303.98

Enquiries about the foregoing specialist prices should be made to Darfen Durafencing, Bradman Road, Knowsley Industrial Park North, Liverpool, L33 7UR, tel (0151) 547 3626, fax (0151) 549 1205.

shepherd and stone

Microcomputer Systems for the Construction Industry

We were very pleased to assist **Griffiths** in the compilation of this 45th edition of their Building Price Book.

This was achieved using the programmable features of Microsoft Word™ and Microsoft Excel™ together with a variant of our popular *Billit* for Windows™ system.

shepherd and stone specialise in the design and development of computer applications for construction professionals.

For more information please contact:

Eric Stone

shepherd and stone Consultants Ltd
9 Limeway Terrace, Dorking, Surrey, RH4 1HZ
Tel. 0181 947 8031
email = ericstone@lineone.net

Approximate estimating

Preamble

Although not recommended for preparing tenders, the use of composite prices can be a useful time-saving method of preparing approximate estimates.

The following composite items cover some of the more frequently occurring specifications for traditional domestic construction and are built up from the unit rates in the preceding trade sections of "New work".

"Price net" figures include allowances for all costs of labour, plant and materials and are based on prices for a builder employing his own labour and on specialist prices for various items as included in the preceding trade sections of "New work".

Prices do not include any allowance for preliminaries nor for scaffolding, ladders or other plant necessary to reach the work. The "Preliminaries" section deals with these items which must be considered and allowance included to suit the particular circumstances.

Basic prices

Prices for materials used in this section are as the basic prices quoted in the preceding trade sections of "New work".

Specialist prices

Prices for work more usually the province of specialist sub-contractors are as the specialist prices quoted in the preceding trade sections of "New work".

"Price with 15%" figures relative to specialist work include 15% for the builder's overheads, profit, unloading materials and general attendance (to include free use of standing scaffolding and hoists, temporary lighting and water and clearing away rubbish).

Building prices per square metre

For the first time we include a selection of typical square metre construction costs for various building types. These are for guidance only and are intended to assist at the very early stage of a development where little or no drawn or design information is available.

Elemental costs

This new section contains typical elemental cost analyses for 32 different types of buildings. The costs given reflect building costs as at the third quarter of 1999.

Approximate estimating

New work	Unit	Price net	Price with 15%
Generally		£	£
			VAT not included

Foundations

Price includes for excavation by machine, earthwork support and disposal of surplus excavated materials to tip average 15 km from site.

Hollow walls

Foundations to hollow wall, including trench excavation 750 mm deep, 600 x 225 mm ready mixed concrete 1:3:6 foundation, 675 mm high wall, concrete filling to cavity and felt damp-proof course, hollow wall of			
two 102.5 mm skins in Class B engineering bricks	m	84.54	97.22
two 102.5 mm skins in common bricks	m	64.47	74.14
two 100 mm skins in concrete dense aggregate blocks	m	37.85	43.52
Extra for 75 mm additional depth of foundation, including trench and concrete	m	4.27	4.91
Extra for 225 mm additional depth of foundation, including trench, concrete filling to cavity and hollow wall of			
102.5 mm Class B engineering bricks	m	25.30	29.09
102.5 mm common bricks	m	18.60	21.39
100 mm concrete dense aggregate blocks	m	16.13	18.55
Extra for three courses of facing bricks total 225 mm high and pointing as the work proceeds, instead of			
102.5 mm common bricks	m	4.18	4.81
100 mm skin in concrete dense aggregate blocks	m	5.42	6.23

Solid walls

Foundations to solid wall, including trench excavation 675 mm deep, 450 x 150 mm ready mixed concrete 1:3:6 foundation, 675 mm high wall and felt damp-proof course, solid wall of			
102.5 mm Class B engineering bricks	m	46.10	53.01
102.5 mm common bricks	m	31.76	36.52
100 mm concrete dense aggregate blocks	m	28.19	32.42
Extra for 75 mm additional depth of foundation, including trench and concrete	m	3.18	3.66
Extra for 225 mm additional depth of foundation, including trench and wall of			
102.5 mm Class B engineering blocks	m	13.35	15.35
102.5 mm common bricks	m	9.48	10.90
100 mm concrete dense aggregate blocks	m	8.30	9.54
Foundations to solid wall, including trench excavation 750 mm deep, 600 x 225 mm ready mixed concrete 1:3:6 foundation, 675 mm high wall and felt damp-proof course, solid wall of			
215 mm Class B engineering bricks	m	72.97	83.91
215 mm common bricks	m	49.95	57.44
215 mm concrete dense aggregate blocks (100 mm blocks laid flat)	m	42.30	48.65
Extra for 75 mm additional depth of foundation, including trench and concrete	m	3.97	4.57

Approximate estimating

	Unit	Price net	Price with 15%
Generally		£	£
		VAT not included	

Foundations *(continued)*

Solid walls *(continued)*

Extra for 225 mm additional depth of foundation, including trench and wall of			
215 mm Class B engineering bricks	m	22.85	26.28
215 mm common bricks	m	15.17	17.44
215 mm concrete dense aggregate blocks (100 mm blocks laid flat)	m	21.53	24.75

Ground floors

Prices include for excavation by machine, earthwork support and disposal of surplus excavated material to tip average 15 km from site.

Solid floors

Solid floors including surface excavation 150 mm deep, 100 mm hardcore bed, 100 mm ready mixed concrete 1:2:4 bed, 25 mm cement and sand screed 1:3 trowelled bed	m2	20.93	24.07

Extra for			
fabric reinforcement A142 - 2.22 kg/m2	m2	1.51	1.74
fabric reinforcement A193 - 3.02 kg/m2	m2	1.93	2.22
two coats liquid bituminous membrane	m2	5.89	6.77
building paper underlay	m2	1.00	1.15
heavyweight polythene sheet underlay and sand blinding	m2	1.89	2.18
self adhesive damp-proof membrane	m2	9.99	11.49
additional cement and sand bed, per 13 mm thickness	m2	2.70	3.11

Hollow floors

Hollow floors, including surface excavation 150 mm deep, 100 mm hardcore bed, 100 mm ready mixed concrete 1:2:4 bed, 102.5 mm common brick honeycombed sleeper walls 225 mm high, felt damp-proof courses, 50 x 100 mm impregnated softwood plates and joists			
25 mm softwood tongued and grooved boards in 150 mm widths	m2	108.32	124.57
25 mm tongued and grooved boards in 100 mm widths	m2	107.68	123.83
12 mm plywood in 1500 x 600 mm panels tongued and grooved on all edges	m2	110.03	126.53
18 mm chipboard tongued and grooved all edges	m2	99.26	114.15

Extra for			
two coats of polyurethane varnish on softwood boards	m2	3.71	4.27

Hollow floors including surface excavation 150 mm deep, removal from site; 50mm blinding concrete (1:2:4), 150 mm precast, prestressed beam and block floors, 75 mm cement and sand (1:3) trowelled bed; A193 mesh and thermoplastic tiling.	m2	60.00	69.00

Adjustments

Adjust for			
excavation per 100 mm thickness	m2	1.75	2.01
hardcore bed, per 25 mm thickness	m2	0.69	0.80
concrete bed, per 25 mm thickness	m2	1.73	1.99
sleeper walls, per 75 mm height	m2	1.53	1.76

Approximate estimating

New work	Unit	Price net	Price with 15%

Generally

£ £

VAT not included

Walls and partitions

Note: Prices for walls and partitions are based on ceiling heights of 2400 mm and include proportionate allowances for skirtings primed and painted three coats.

External walls

External hollow walls, including galvanised twist wall ties, two coat plastering internally, two coats of emulsion paint, and 19 x 100 mm stock pattern softwood skirting, hollow wall of

	Unit	Price net	Price with 15%
102.5 mm skin in facing bricks pointed as the work proceeds, cavity and 100 mm skin in concrete lightweight aggregate loadbearing blocks	m2	82.60	94.99
102.5 mm skin in facing bricks pointed as the work proceeds, cavity and 140 mm skin in concrete lightweight aggregate loadbearing blocks	m2	92.13	105.95
102.5 mm skin in facing bricks pointed as work proceeds, drained cavity and 25 mm insulating cavity bats and 100 mm skin in lightweight aggregate loadbearing blocks	m2	88.15	101.37

External hollow walls, including galvanised twist wall ties, two coat cement, lime and sand and Tyrolean finish externally, two coat plastering internally, two coats emulsion paint and 19 x 100 mm stock pattern softwood skirting, hollow wall of

	Unit	Price net	Price with 15%
two 100 mm skins in concrete lightweight aggregate loadbearing blocks and cavity	m2	66.00	75.90
one 100 and one 150 mm skin in concrete lightweight aggregate loadbearing blocks and cavity	m2	75.53	86.86

Extra for dry lining with flush jointed tapered edge plasterboard on 25 x 50 mm impregnated battens

	Unit	Price net	Price with 15%
9.5 mm wallboard	m2	7.41	8.53
12.5 mm wallboard	m2	8.15	9.38

Internal walls

Internal solid walls, including fair face and flush pointing one side and two coat plastering, two coats of emulsion paint and 19 x 100 mm stock pattern softwood skirting the other side, solid wall of

	Unit	Price net	Price with 15%
102.5 mm common bricks	m2	43.78	50.35
215 mm common bricks	m2	70.07	80.58

Internal solid walls, including two coat plastering, two coats of emulsion paint and 19 x 100 mm stock pattern softwood skirting both sides, solid wall of

	Unit	Price net	Price with 15%
102.5 mm common bricks	m2	50.56	58.14
215 mm common bricks	m2	76.83	88.36
75 mm concrete dense aggregate blocks	m2	41.43	47.65
100 mm concrete dense aggregate blocks	m2	45.26	52.05
140 mm concrete dense aggregate blocks	m2	52.66	60.56
215 mm concrete dense aggregate blocks (100 mm blocks laid flat)	m2	67.29	77.39
75 mm concrete lightweight aggregate blocks	m2	40.11	46.13
100 mm concrete lightweight aggregate blocks	m2	43.60	50.14
140 mm concrete lightweight blocks	m2	52.69	60.59
215 mm concrete lightweight aggregate blocks	m2	64.71	74.41

Approximate estimating

New work

Generally

£   £

VAT not included

Walls and partitions *(continued)*

Internal walls *(continued)*

Internal hollow walls, including galvanised twist wall ties, two coat plastering, two coats of emulsion paint and 19 x 100 mm stock pattern softwood skirting both sides, hollow wall of

two 102.5 mm skins in common bricks	m2	84.01	96.61
two 100 mm skins in dense concrete aggregate blocks	m2	73.05	84.01
two 100 mm skins in concrete lightweight aggregate blocks	m2	58.16	66.89

Stud partitions, including 9.5 mm flush jointed tapered edge plasterboard, two coats of emulsion paint and 19 x 100 mm stock pattern skirtings both sides, impregnated studding size

50 x 75 mm	m2	30.55	35.13
50 x 100 mm	m2	33.91	39.00
75 x 100 mm	m2	38.97	44.82

Stud partitions, including 9.5 mm plaster lath, 5 mm one coat plastering, two coats of emulsion paint and 19 x 100 mm stock pattern skirting both sides, impregnated softwood studding size

50 x 75 mm	m2	39.66	45.60
50 x 100 mm	m2	43.02	49.47
75 x100 mm	m2	48.07	55.28

Cellular core partitions, including flush jointed tapered edge panels, softwood sole plates, blockings, plugs and battens, two coats of emulsion paint and 19 x 100 mm stock pattern skirtings both sides

57 mm	m2	25.04	28.80
63 mm	m2	26.24	30.17

Laminated partitions, including flush jointed tapered edge outer layers, 19 mm square edge plank core, softwood plates and battens, two coats of emulsion paint and 19 x 100 mm stock pattern skirting both sides

50 mm - outer layers 12.5 mm wallboard and 19 mm plank core	m2	38.19	43.92
65 mm - outer layers of 19 mm plank core	m2	42.85	49.28

Extra over emulsion paint for glazed tiling on plastered walls

108 x 108 x 4 mm (Group A)	m2	29.00	33.35
108 x 108 x 4 mm (Group C)	m2	39.00	44.85
152 x 152 x 5.5 mm (Group A)	m2	28.49	32.77
152 x 152 x 5.5 mm (Group C)	m2	38.85	44.67

Approximate estimating

New work	Unit	Price net	Price with 15%

Generally

£   £

VAT not included

Upper floors

Timber floors

Timber floors, including 50 x 150 mm impregnated softwood joist, 25 x 50 mm herringbone strutting, galvanised steel straps, 9.5 mm plasterboard ceiling and textured compound finish, flooring of

	Unit	Price net	Price with 15%
25 mm softwood tongued and grooved boards in 150 mm widths	m2	37.34	42.94
25 mm softwood tongued and grooved boards in 100 mm widths	m2	37.53	43.16
12 mm plywood in 1500 x 600 mm panels tongued and grooved all edges	m2	45.43	52.24
18 mm chipboard tongued and grooved all edges	m2	28.26	32.50

Extra for

	Unit	Price net	Price with 15%
50 x 175 mm joists instead of 50 x 150 mm joists	m2	1.57	1.81
50 x 200 mm joists instead of 50 x 150 mm	m2	3.57	4.11
50 x 225 mm joists instead of 50 x 150 mm	m2	5.06	5.82
12.5 mm plasterboard instead of 9.5 mm	m2	0.63	0.73
9.5 mm plaster lath, 5 mm one coat plastering and two coats emulsion instead of 9.5 mm plasterboard and textured plastic compound finish	m2	5.93	6.82
two layers of 9.5 mm plaster lath, 5 mm one coat plastering and two coats emulsion paint instead of 9.5 mm plasterboard and textured plastic compound finish	m2	6.33	7.28
two coats of polyurethane varnish on softwood boards	m2	3.71	4.27
thermoplastic tiling on plywood or chipboard	m2	8.46	9.73

Concrete floors

In situ concrete floors, including 125 mm ready mixed concrete 1:2:4 slab, steel reinforcement, lined formwork to soffit, textured plastic compound finish, 25 mm cement and sand 1:3 trowelled bed to receive thermoplastic tiling (not included)

	Unit	Price net	Price with 15%
	m2	64.34	73.99

Extra for

	Unit	Price net	Price with 15%
150 mm slab instead of 125 mm	m2	2.21	2.54
175 mm slab instead of 125 mm	m2	3.51	4.04
200 mm slab instead of 125 mm	m2	5.59	6.43

150 mm precast, pre-stressed beam and block floors with battens and 13 plasterboards and textured plastic finish to soffit; 75mm cement and sand (1:3) trowelled bed, A193 mesh and thermoplastic tiling

	Unit	Price net	Price with 15%
	m2	61.36	70.57

	Unit	Price net	Price with 15%
Extra for 200 mm instead of 150 mm precast, pre-stressed beam and pot floors	m2	7.87	9.05

Adjustments

	Unit	Price net	Price with 15%
Extra for 10 mm two coat lightweight plastering and two coats of emulsion paint instead of textured plastic compound finish to soffits	m2	3.41	3.92
Extra for additional cement and sand bed, per 13 mm thickness	m2	2.70	3.11

Approximate estimating

New work

Generally		£	£
		VAT not included	

Roofs

Note: Prices for pitched roofs are based on gabled roofs to 30 degree pitch, steeper pitches, hips and valleys, will increase the cost.

Prices include for wrought softwood fascias, soffits, etc primed and painted three coats.

Timber roofs

Timber pitched roofs, including impregnated softwood trussed rafters at 450 mm centres, plates, noggings, binders and bracings, fascia, soffit and barge boards, galvanised steel straps, felt underlay, 100 mm glass fibre insulation, 9.5 mm plasterboard ceiling and textured plastic compound finish, roof covering of

	Unit	Price net	Price with 15%
510 x 225 mm best Welsh slating	m2	141.00	162.15
500 x 250 mm asbestos-free blue/black slating	m2	105.75	121.61
430 x 380 mm concrete interlocking slating	m2	88.13	101.34
265 x 165 mm hand made sand-faced clay plain tiling	m2	146.88	168.91
265 x 165 mm machine-made sand-faced clay plain tiling	m2	105.75	121.61
265 x 165 concrete plain tiling	m2	105.75	121.61
420 x 330 mm concrete double Roman tiling	m2	84.60	97.29

Note: all the foregoing timber pitched roof prices are for the area measured on plan

Timber flat roofs, including 38 x 150 mm impregnated softwood joists, 25 x 50 mm herringbone strutting, plates, firrings and noggings, fascia, galvanised steel straps, felt underlay, 100 mm glass fibre insulation, 9.5 mm plasterboard ceiling and textured plastic compound finish and two layer felt covering with stone chippings surfacing on decking of

	Unit	Price net	Price with 15%
25 mm sawn softwood boarding butt jointed	m2	62.69	72.09
25 mm softwood tongued and grooved boarding	m2	64.25	73.89
12 mm WBP grade plywood boarding	m2	66.89	76.93
22 mm chipboard tongued and grooved all edges	m2	57.74	66.40

Extra for

	Unit	Price net	Price with 15%
50 x 150 mm joists instead of 38 x 150 mm	m2	1.26	1.45
38 x 175 mm joists instead of 38 x 150 mm	m2	1.12	1.28
50 x 175 mm joists instead of 38 x 150 mm	m2	2.74	3.15
38 x 200 mm joists instead of 38 x 150 mm	m2	2.35	2.70
50 x 200 mm joists instead of 38 x 150 mm	m2	4.18	4.81

Adjustments

	Unit	Price net	Price with 15%
Extra for insulating grade plasterboard to ceiling	m2	0.74	0.85
Extra for 9.5 mm plaster lath, 5 mm one coat plastering and two coats of emulsion paint instead of 9.5 mm plasterboard and textured plastic compound finish	m2	5.93	6.82

Approximate estimating

New work	Unit	Price net	Price with 15%

Generally

£ £

VAT not included

Roofs *(continued)*

Concrete roofs

	Unit	Price net	Price with 15%
In situ concrete flat roofs, including ready mixed concrete 1:2:4 slab, steel reinforcement, lined formwork to soffit and textured plastic compound finish, impregnated softwood fascia, 25 mm average cement and sand 1:3 floated bed to falls and two layer felt covering with stone chippings surfacing	m2	77.24	88.83
Extra for			
125 mm slab instead of 100 mm	m2	2.75	3.16
150 mm slab instead of 100 mm	m2	4.96	5.70
175 mm slab instead of 100 mm	m2	7.92	9.11
150 mm precast, pre-stressed beam and block flat roofs with textured plastic compound finish to soffit, impregnated softwood fascia, 25 mm average cement and sand 1:3 floated bed to falls and two layer felt covering with stone chippings surfacing	m2	70.50	81.08
Extra for 200 mm instead of 150 mm precast, pre-stressed beam and pot flat roofs	m2	11.75	13.51

Adjustments

	Unit	Price net	Price with 15%
Extra for			
three layer instead of two layer felt covering	m2	4.00	4.59
25 mm limestone aggregate asphalt covering instead of two layer felt and stone chippings	m2	7.93	9.12
additional cement and sand bed, per 13 mm thickness	m2	3.49	4.01
Extra for vermiculite concrete insulating bed and 13 mm cement and sand 1:3 topping instead of 25 mm cement and sand bed, total average thickness			
63 mm	m2	4.02	4.62
88 mm	m2	6.78	7.80
113 mm	m2	16.58	19.07
Extra for 9.5 mm insulation grade plaster lath, 5 mm one coat plastering, two coats of emulsion paint and 25 x 50 mm softwood open spaced battening plugged and screwed to concrete soffit instead of textured plastic compound finish	m2	33.59	38.63

Rainwater installations

	Unit	Price net	Price with 15%
Rainwater installations, including gutter and fittings and allowance for rainwater pipe for two storeys			
aluminium - 100 mm half round gutter and 63 mm pipe	m	40.44	46.51
cast iron - 100 mm half round gutter and 65 mm pipe	m	50.84	58.47
PVC-U - 112 half round gutter and 68 mm pipe	m	19.93	22.92
PVC-U - 112 mm square section gutter and 65 mm square section pipe	m	20.56	23.65

Approximate estimating

New work	Unit	Price net	Price with 15%

Generally

£   £
VAT not included

Doors

Note: If the area of door openings has not been deducted from the area of walls and partitions then the value of walls and partitions displaced should be deducted from the following prices.

Prices include for softwood primed and painted three coats and for hardwood sealed and varnished three coats.

External doors

External doors, including butts, lock, furniture and letter plate, 63 x 75 mm softwood rebated frame, 75 x 100 mm hardwood threshold with water bar, reinforced concrete lintel, galvanised steel combined lintel and cavity tray, pitch polymer damp-proof-courses and flat arch, reveals and sills in facing bricks

838 x 1981 mm softwood stock pattern panel doors	each	562.54	646.92
838 x 1981 mm hardwood stock pattern doors	each	615.41	707.72
838 x 1981 mm softwood purpose-made panel doors	each	564.53	649.21
838 x 1981 mm hardwood purpose-made panel door	each	636.79	732.31
838 x 1981 mm stock pattern flush door for painting	each	352.92	405.86

External garage doors, including butts, bolts and cylinder latch set, 50 x 100 mm softwood rebated frame, reinforced concrete lintel, galvanised steel combined lintel and cavity tray, pitch polymer vertical damp-proof courses and flat arch and reveals in facing bricks

2135 x 1980 mm softwood stock pattern matchboarded doors	each	762.54	876.92
2134 x 1981 mm softwood stock pattern glazed and matchboarded doors	each	802.63	923.03

Internal doors

Internal doors including butts, latch and furniture, 38 x 125 mm softwood rebated lining, 19 x 50 mm stock pattern architraves both sides and galvanised steel lintel to support internal partition

762 x 1981 mm softwood stock pattern panel doors	each	241.44	277.65
762 x 1981 mm softwood purpose-made panel doors	each	282.26	324.60
762 x 1981 mm hardwood purpose-made panel doors	each	378.61	435.40
762 x 1981 mm stock pattern flush doors for painting	each	156.52	180.00
762 x 1981 mm hardwood veneered stock pattern flush doors	each	164.95	189.69

Internal fire-check doors, including butts, latch, furniture and door closer, 63 x 125 mm softwood rebated frame, 19 x 50 mm stock pattern architraves both sides and galvanised steel lintel to support internal partition

762 x 1981 mm stock pattern half-hour fire-check flush doors for painting	each	267.21	307.29
762 x 1981 mm hardwood veneered stock pattern half-hour fire-check doors	each	289.18	332.56

Internal trap doors to roof space, including 25 x 100 mm softwood lining, 13 x 50 mm stop and 19 x 50 mm architrave to ceiling and 38 x 100 mm impregnated softwood noggings

600 x 400 x 25 mm blockboard trap door laid loose	each	42.58	48.97
600 x 400 x 25 mm blockboard traps, lipped all edges, including butts, necked bolts and bow handle	each	59.54	68.47

Approximate estimating

New work

Generally

£   £

VAT not included

Windows

Note: If the area of window openings has not been deducted from the area of external wall then the value of walls displaced should be deducted from the following prices.

Prices include for softwood primed and painted three coats and for exposed galvanised metalwork primed and painted three coats.

Windows about 1.25 m2, including float glass, reinforced concrete lintel, galvanised steel combined lintel and cavity tray, pitch polymer vertical damp-proof courses, flat arch and reveals in facing bricks, external sill in roofing tiles and internal sill in quarry tiles

	Unit	Price net	Price with 15%
softwood stock pattern "non-bar" casement window	each	568.54	653.82
	m2	454.80	523.01
softwood stock pattern "all-bar" (Georgian) casement window	each	500.20	575.23
	m2	400.13	460.15
softwood purpose-made windows with 38 mm casements divided into panes 0.10 - 0.50 m2	each	221.85	255.13
	m2	192.66	221.56
softwood purpose-made windows with 50 mm casements divided into panes 0.10 - 0.50 m2	each	229.10	263.47
	m2	195.07	224.33
softwood purpose-made double-hung sash windows with 38 mm sashes divided into panes 0.10 - 0.50 m2, hung on cords and weights	each	620.39	713.45
	m2	334.89	385.12
softwood purpose-made double-hung sash windows with 50 mm sashes divided into panes 0.10 - 0.50 m2, hung on cords and weights	each	406.82	467.84
	m2	334.89	385.12
standard galvanised steel windows with weatherstripping	each	340.91	392.05
	m2	281.98	324.27
standard galvanised steel windows with weatherstripping including 50 x 75 mm softwood sub-frame with 75 x 100 mm hardwood cills	each	402.74	463.15
	m2	348.52	400.79

Windows about 1.75 m2, including float glass, reinforced concrete lintel, galvanised steel combined lintel and cavity tray, pitch polymer vertical damp-proof-course, flat arch and reveals in facing bricks, external sill in roofing tiles and internal sill in quarry tiles

	Unit	Price net	Price with 15%
softwood stock pattern "non-bar" casement windows	each	837.81	963.48
	m2	478.27	550.01
softwood stock pattern "all-bar" (Georgian) casement windows	each	1532.71	1762.61
	m2	1108.42	1274.69

Approximate estimating

New work	Unit	Price net	Price with 15%
Generally		£	£
			VAT not included

Windows *(continued)*

Windows about 1.75 m2, including float glass, reinforced concrete lintel, galvanised steel combined lintel and cavity tray, pitch polymer vertical damp-proof-course, flat arch and reveals in facing bricks, external sill in roofing tiles and internal sill in quarry tiles (Continued)

	Unit	Price net	Price with 15%
softwood purpose-made windows with 38 mm casements divided into panes 0.10 - 0.50 m2	each	278.08	319.79
	m2	158.91	182.74
softwood purpose-made windows with 50 mm casements divided into panes 0.10 - 0.50 m2	each	284.30	326.95
	m2	175.78	202.15
softwood purpose-made double-hung sash windows with 38 mm sashes divided into panes 0.10 - 0.50 m2 hung on cords and weights	each	533.71	613.76
	m2	305.46	351.28
softwood purpose-made double-hung sash windows with 50 mm sashes divided into panes 0.10 - 0.50 m2 hung on cords and weights	each	535.13	615.40
	m2	306.28	352.22
standard galvanised steel windows with weatherstripping	each	473.18	544.16
	m2	329.75	379.21
standard galvanised steel windows with weatherstripping including 50 x 75 mm softwood sub-frames with 75 x 100 mm hardwood sills	each	540.17	621.20
	m2	315.48	362.80
Pelmets including 19 x 100 mm softwood top and 13 x 150 mm front with returned ends and galvanised steel brackets			
1000 mm	each	35.05	40.31
1500 mm	each	44.52	51.20
2000 mm	each	58.08	66.79
Pelmets, including 19 x 100 mm softwood top and 13 x 150 mm front with returned ends, galvanised steel brackets and standard plastics curtain track and runners			
1000 mm	each	45.97	52.86
1500 mm	each	57.83	66.51
2000 mm	each	73.79	84.86

Approximate estimating

New work	Unit	Price net	Price with 15%
Generally		£	£
		VAT not included	

Staircases

Prices include for softwood primed and painted three coats and for hardwood sealed and varnished three coats.

Staircases 2600 mm rise, including trimming floor joists with steel joist hangers, softwood carriages, handrail balusters, newels, well-hole nosing and apron lining.

	Unit	Price net	Price with 15%
softwood stock pattern straight flights	each	1375.21	1581.49
softwood stock pattern flights with three winders	each	1611.21	1852.89
softwood purpose-made straight flights	each	1451.58	1669.32
softwood purpose-made flights with three winders	each	1628.83	1873.16
hardwood purpose-made straight flights	each	2223.89	2557.47
hardwood purpose made flights with three winders	each	2553.89	2936.97

Joinery fittings

Factory finished standard kitchen fittings, including sink unit (sink top included in "internal plumbing"), two single floor units with worktops, broom cupboard and two single wall cupboards

	Unit	Price net	Price with 15%
	set	1563.44	1797.96

Extra for additional factory finished standard kitchen units

	Unit	Price net	Price with 15%
single floor units with worktops	each	196.37	225.82
single floor drawer units with worktops	each	241.53	277.76
single floor store unit	each	502.01	577.31
single wall units	each	138.78	159.60

Purpose-made serving hatches including pair of flush doors 600 x 600 mm overall, butts, catches and handles, softwood lining and sill, primed and painted three coats and galvanised steel lintel to support internal partition

	Unit	Price net	Price with 15%
	each	79.12	90.99

Airing cupboard shelving, including three 900 x 600 mm tiers of 25 x 50 mm slats 25 mm apart and 25 x 50 mm bearers

	Unit	Price net	Price with 15%
	set	62.89	72.32

Extra for additional tiers of airing cupboard slat shelving and bearers

	Unit	Price net	Price with 15%
900 x 600 mm	set	21.01	24.16
900 x 450 mm	set	17.16	19.73
900 x 300 mm	set	14.04	16.15

Softwood store shelving, one 900 x 450 mm tier of 25 mm softwood shelving, two 900 x 300 mm tiers and 25 x 50 mm bearers

	Unit	Price net	Price with 15%
	set	56.93	65.47

Extra for additional tiers of store shelving

	Unit	Price net	Price with 15%
900 x 450 mm	set	21.27	24.46
900 x 300 mm	set	17.84	20.51
900 x 150 mm	set	10.35	11.90

Approximate estimating

	Unit	Price net	Price with 15%

Generally

		£	£
			VAT not included

Internal plumbing

Internal plumbing installations, including copper cold water and hot water services, plastic overflows, soil, waste and vent pipes, traps, plastics cistern, glass fibre filled insulating jackets, integral foam lagged copper direct cylinder (immersion heater included with electrical installation) sanitary fittings, comprising basin, bath, sinktop, WC suite, and builder's work

	Unit	Price net	Price with 15%
white sanitary fittings complete with taps	each	2505.75	2881.61
coloured sanitary fittings complete with taps	each	2564.50	2949.17
Extra for additional basin			
white complete with taps	each	207.04	238.09
coloured complete with taps	each	224.66	258.36
Extra for additional WC and basin			
coloured complete with taps	each	480.69	552.80
white complete with taps	each	434.87	500.10

For connections to public mains, see "Plumbing and engineering installations"

Heating installations

Central heating installations, including copper services, circulating pump, room thermostat and electrical connections, plastics feed and expansion cistern and overflow, extra cost of indirect cylinder (direct cylinder included with plumbing installation), six single panel steel radiators, and with air vent plugs and valves including builder's work

	Unit	Price net	Price with 15%
gas fired floor standing room sealed boiler with balanced flue and opening through external wall	each	3307.43	3803.54
gas fired floor standing boiler with conventional flue, concrete block flue, twin walled flue pipe in roof space and gas ridge terminal	each	3672.09	4222.90
gas fired wall hung room sealed boiler with balanced flue, and opening through wall	each	3169.95	3645.44
gas fired wall hung boiler with conventional flue, concrete block flue, twin walled flue pipe in roof space and gas ridge terminal	each	3612.16	4153.99
gas fired combined room heater, surround and boiler unit with conventional flue, brick chimney breast, concrete block flue, twin walled flue pipe in roof space and gas ridge terminal	each	6558.67	7542.47
oil fired standing room sealed boiler with balanced flue and opening through external wall, 1816 litre oil storage tank, stand, valves etc and connections	each	4683.68	5386.23
oil fired floor standing boiler with conventional flue, lined flue, chimney stack, 1816 litre oil storage tank, stand, valves etc and connections	each	9324.84	10723.56
solid fuel floor standing boiler, lined flue, chimney stack and fuel bunker	each	8036.51	9241.98
solid fuel free standing room heater with high output boiler, lined flue, chimney stack and fuel bunker	each	7454.88	8573.11
solid fuel open fire with high outlet back boiler, tiled surround and hearth, brick chimney breast, lined flue, chimney stack and fuel bunker	each	7184.63	8262.33

Approximate estimating

	Unit	Price net	Price with 15%

Generally

£ £
VAT not included

Heating installations *(continued)*

	Unit	Price net	Price with 15%
Extra for			
programmer unit and electrical connections	each	103.25	118.73
additional single steel radiator and air vent plug, valves, copper services and connections	each	164.34	188.99
thermostatic radiator valves	each	2.81	3.23
pumped primary circuit for hot water system cylinder, thermostat, relay, motorised three-way valve with actuator, and electrical connections	each	256.93	295.46

Electrical installations

	Unit	Price net	Price with 15%
Electrical installation, including eight single and two two-way lighting points, five single and five double socket outlet points, one cooker point and panel, one single immersion heater and point, consumer unit and builder's work	each	1265.10	1454.86
Extra for			
additional single lighting point	each	49.09	56.46
additional single socket outlets	each	44.45	51.12
additional double socket outlet	each	45.24	52.02
infra-red heaters and points (heater 1000 kW)	each	124.20	142.83
dual type shaver outlet points	each	79.11	90.98
Bell installations, including transformer and builder's work			
single bells for one door	each	87.36	100.47
bell and buzzer sets for front and back doors	each	134.40	154.56

Drainage

	Unit	Price net	Price with 15%
Drainage systems, including 25m of 100 mm flexible jointed vitrified clay pipes, excavation, granular bedding, three gullies, three brick manholes average 0.75 m deep	each	1634.81	1880.03
Extra for			
100 mm saddle connection	each	62.47	71.85
additional length of 100 mm flexible jointed vitrified clay pipes, excavation and granular bedding	m	41.91	48.20
450 x 450 mm concrete 1:3:6 bed and covering instead of granular bedding to pipes	m	6.47	7.45
additional gullies, 2 m of 100 mm branch drain, and branch channel bend etc in manhole	m	138.76	159.57
additional brick manholes average 0.75 m deep	each	195.64	224.98
100 mm intercepting traps with fresh air inlet valve	each	133.50	153.53
Grade B Class 2 instead of Grade C single seal manholes covers	each	34.85	40.08
Soakaways about 1 m3 capacity			
stone filled excavations	each	57.73	66.39
open jointed brickwork chambers	each	241.85	278.13

For septic tanks and filter chambers, see Combined items in "Drainage"

For infrastructure charges for domestic sewerage connections, see Connections to public mains in "Plumbing and engineering installations".

Approximate estimating

Building prices per square metre

The following rates are intended to provide users with typical average construction costs for a range of building types. It must be emphasised that construction prices can be affected by many factors including geographical location, site configuration, local market conditions, specification and even the season of the year.

The rates given therefore must be treated with some caution. The figure given are based on gross internal area, i.e. measured to the inside face of external walls, over all internal walls and partitions. The rates include for contingencies and external works but exclude furniture, fittings, professional fees and VAT. It is of course possible to encounter projects where the square metre construction rates are outside the ranges given below.

	Price per m2			Price per m2	
	Average £	Range £		Average £	Range £
Residential			**Medical**		
LA low rise flats	510	350-900	Medical Centres	670	420-1070
LA semi-det houses	460	340-820	Hospitals	950	590-1830
Private flats(standard)	560	380-1100	Nursing Homes	660	400-1160
Private Houses detached(std)	560	410-860	Veterinary Surgeries	700	670-790
			Dental Surgeries	700	670-780
Industrial			Hospices	880	690-990
			Ambulance Stations	690	540-950
Farm Buildings	240	200-300			
Warehouses(shells only)	320	190-720	**Education**		
Nursery Units	400	270-790			
Factories-Light Industrial	420	280-1030	Schools	730	430-1570
Factories-Heavy Industrial	560	390-1130	Universities	820	600-1800
Hi-tech Laboratories	940	600-1420			
			Sport & Leisure		
Communication					
			Sports Halls	680	490-1140
Multi Storey Car Parks	230	170-330	Sports Pavilions	730	460-1240
			Gymnasium	570	360-830
Commercial					
			Catering		
Car Showrooms	490	260-710			
Shops	630	250-1260	Hotels general	740	460-1240
Supermarkets	750	280-1340	Motels	500	420-550
Shopping Centres	990	680-1410	Tea rooms/cafes	590	560-660
Retail Warehouses	410	270-660	Restaurants	900	610-1500
Offices 1-3 storey	450	370-1130	Public Houses	800	600-1160
Offices over 3 storey	710	440-1230			
			Social		
Entertainment					
			Community Centres	680	460-1260
Cinemas	830	640-1180	Village Halls	660	440-1050
Conference/exhibition Halls	770	700-900			
Public					
Fire stations	1020	650-1510			
Law Courts	1080	730-1410			
Police Stations	980	680-1370			
Libraries	780	620-1200			

Start building the foundation of your marketing success with...

Glenigan Direct

Using our exclusive **construction database** of live and historic project information, we can make our data work for YOU.

Services available include:

- Competitor Analysis
- Market Analysis
- Ranking lists
- Company profiles
- Listings for direct mail and database building

For more information about Glenigan Direct, call us on...

01202 435990

Approximate estimating

Elemental costs

The following new section contains typical elemental cost analyses for 32 different types of buildings. Each analysis is broken down into 28 elements. The elements are tabled generally in line with the recommendations of the BCIS (Building Cost Information Service, tel: 0171-222 7000). The figures, which exclude VAT, the cost of land and professional fees, are intended to help in the early stages of the financial appraisal of a project.

Imbalances in cost allocations can easily occur unless an overall view of the approximate value of each element is known at the cost-planning and subsequent stages in the cost development of each project. If the client desires particularly expensive wall finishings, for example, it should be possible to identify areas of savings by adjusting percentages of one or more of other elements. For example, in a fire station, the percentage quoted for internal doors is 4%, i.e. the equivalent of £37 per square metre and this figure is the result of averaging the costs of internal doors of several fire stations. However, before using the 4% percentage for another project, the area of the proposed building should be calculated. The percentage may drop to 2-3% for an above-average sized fire station or increase to 5-6% for one below average size.

Where appropriate, the rates have been updated to reflect the cost of building in the third quarter of 1999. Due to rounding-up, there may be some slight discrepancies in the individual rates and the totals.

The building types included on the following pages are:

Multi-storey car park
Fire station
Ambulance station
Police station
County court building
Community centre
Sports hall
Sports pavilion
Restaurant
Public house
Gymnasium hall
Village hall
Farm building
Warehouse (shell only)
Nursery units
Factory, light industrial
Factory, heavy industrial
Hi-tech laboratory
Primary school
Middle school
Sixth form college
Special school
Training college
Library
Health centre
Welfare centre
Old persons home
Group surgery
Local authority low rise flats
Local authority semi-detached houses
Standard private flats
Luxury private flats

Approximate estimating

Elemental costs *(continued)*

	Multi-storey car park		Fire station		Ambulance station		Police station	
	%	£/m2	%	£/m2	%	£/m2	%	£/m2
PRELIMINARIES	12.00	30.56	8.00	75.92	7.00	55.50	10.00	96.33
SUBSTRUCTURES	16.00	40.75	5.00	47.45	7.00	55.50	8.00	77.07
SUPERSTRUCTURES								
Frame	9.00	22.92	4.00	37.96	6.00	47.57	5.00	48.17
Upper floors	6.00	15.28	1.00	9.49	2.00	15.86	2.00	19.27
Roof	6.00	15.28	3.00	28.47	2.00	15.86	3.00	28.90
Stairs	2.00	5.09	5.00	47.45	4.00	31.71	5.00	48.17
External walls and doors	15.00	38.20	8.00	75.92	7.00	55.50	8.00	77.07
Windows and external doors	2.00	5.09	3.00	28.47	5.00	39.64	6.00	57.80
Internal walls and partitions	1.00	2.55	4.00	37.96	5.00	39.64	4.00	38.53
Internal doors	1.00	2.55	4.00	37.96	5.00	39.64	4.00	38.53
FINISHES								
Wall finishes	1.00	2.55	4.00	37.96	1.00	7.93	1.00	9.63
Floor finishes	1.00	2.55	4.00	37.96	2.00	15.86	2.00	19.27
Ceiling finishes	1.00	2.55	1.00	9.49	2.00	15.86	2.00	19.27
FITTINGS AND FURNISHINGS	-	-	1.00	9.49	3.00	23.79	2.00	19.27
SERVICES								
Sanitary and disposal installations	-	-	2.00	18.98	2.00	15.86	2.00	19.27
Services equipment	1.00	2.55	-	-	1.00	7.93	1.00	9.63
Heat source	1.00	2.55	2.00	18.98	2.00	15.86	2.00	19.27
Hot and cold water services	-	-	-	-	1.00	7.93	1.00	9.63
Heating and air treatment	1.00	2.55	3.00	28.47	1.00	7.93	1.00	9.63
Ventilation installation	-	-	2.00	18.98	1.00	7.93	2.00	19.27
Gas services	4.00	10.19	2.00	18.98	1.00	7.93	1.00	9.63
Electrical installation	-	-	11.00	104.38	7.00	55.50	5.00	48.17
Lift and conveyor installations	6.00	15.28	-	-	-	-	-	
Protective communication	-	-	2.00	18.98	1.00	7.93	1.00	9.63
Communication installations	-	-	2.00	18.98	3.00	23.79	3.00	28.90
Special equipment	-	-	2.00	18.98	2.00	15.86	1.00	9.63
Builders' work and profit	2.00	5.09	3.00	28.47	2.00	15.86	2.00	19.27
EXTERNAL WORKS	12.00	30.56	14.00	132.85	18.00	142.71	16.00	154.13
	100.00	254.70	100.00	948.95	100.00	792.84	100.00	963.33

Approximate estimating

Elemental costs *(continued)*

	County court		Community centre		Sports hall		Sports pavilion	
	%	£/m2	%	£/m2	%	£/m2	%	£/m2
PRELIMINARIES	5.00	57.26	8.00	56.53	6.00	46.09	9.00	72.10
SUBSTRUCTURES	6.00	68.71	7.00	49.46	11.00	84.50	9.00	72.10
SUPERSTRUCTURES								
Frame	6.00	68.71	5.00	35.33	-	-	-	
Upper floors	3.00	34.35	2.00	14.13	-	-	-	
Roof	4.00	45.80	2.00	14.13	7.00	53.77	5.00	40.05
Stairs	5.00	57.26	5.00	35.33	-	-	-	
External walls and doors	9.00	103.06	8.00	56.53	10.00	76.82	9.00	72.10
Windows and external doors	5.00	57.26	5.00	35.33	7.00	53.77	8.00	64.08
Internal walls and partitions	4.00	45.80	5.00	35.33	5.00	38.41	8.00	64.08
Internal doors	5.00	57.26	5.00	35.33	6.00	46.09	8.00	64.08
FINISHES								
Wall finishes	3.00	34.35	2.00	14.13	3.00	23.05	2.00	16.02
Floor finishes	4.00	45.80	3.00	21.20	2.00	15.36	3.00	24.03
Ceiling finishes	4.00	45.80	2.00	14.13	3.00	23.05	2.00	16.02
FITTINGS AND FURNISHINGS	4.00	45.80	3.00	21.20	5.00	38.41	3.00	24.03
SERVICES								
Sanitary and disposal installations	2.00	22.90	2.00	14.13	2.00	15.36	3.00	24.03
Services equipment	-		1.00	7.07	1.00	7.68	1.00	8.01
Heat source	2.00	22.90	2.00	14.13	2.00	15.36	2.00	16.02
Hot and cold water services	1.00	11.45	1.00	7.07	2.00	15.36	2.00	16.02
Heating and air treatment	1.00	11.45	1.00	7.07	2.00	15.36	2.00	16.02
Ventilation installation	1.00	11.45	1.00	7.07	-	-	1.00	8.01
Gas services	1.00	11.45	1.00	7.07	1.00	7.68	1.00	8.01
Electrical installation	9.00	103.06	10.00	70.66	7.00	53.77	10.00	80.11
Lift and conveyor installations	2.00	22.90	-	-	-	-	-	
Protective communication	1.00	11.45	-	-	-	-	1.00	8.01
Communication installations	1.00	11.45	-	-	2.00	15.36	-	
Special equipment	1.00	11.45	-	-	2.00	15.36	1.00	8.01
Builders' work and profit	2.00	22.90	3.00	21.20	2.00	15.36	2.00	16.02
EXTERNAL WORKS	9.00	103.06	16.00	113.05	12.00	92.18	8.00	64.08
	100.00	1,145.11	100.00	706.58	100.00	768.20	100.00	801.06

Elemental costs *(continued)*

	Restaurant		Public house		Gymnasium hall		Village hall	
	%	£/m2	%	£/m2	%	£/m2	%	£/m2
PRELIMINARIES	7.00	56.07	6.00	50.04	8.00	45.60	8.00	52.75
SUBSTRUCTURES	6.00	48.06	7.00	58.37	8.00	45.60	10.00	65.93
SUPERSTRUCTURES								
Frame	-	-	-	-	-	-	-	-
Upper floors	-	-	-	-	-	-	-	-
Roof	7.00	56.07	6.00	50.04	8.00	45.60	7.00	46.15
Stairs	5.00	40.05	1.00	8.34	-	-	-	-
External walls and doors	9.00	72.10	8.00	66.71	9.00	51.30	8.00	52.75
Windows and external doors	6.00	48.06	7.00	58.37	7.00	39.90	7.00	46.15
Internal walls and partitions	8.00	64.08	6.00	50.04	6.00	34.20	5.00	32.97
Internal doors	7.00	56.07	6.00	50.04	6.00	34.20	7.00	46.15
FINISHES								
Wall finishes	3.00	24.03	3.00	25.02	2.00	11.40	2.00	13.19
Floor finishes	3.00	24.03	3.00	25.02	2.00	11.40	3.00	19.78
Ceiling finishes	2.00	16.02	2.00	16.68	1.00	5.70	2.00	13.19
FITTINGS AND FURNISHINGS	4.00	32.04	4.00	33.36	3.00	17.10	3.00	19.78
SERVICES								
Sanitary and disposal installations	3.00	24.03	4.00	33.36	4.00	22.80	2.00	13.19
Services equipment	1.00	8.01	1.00	8.34	1.00	5.70	1.00	6.59
Heat source	2.00	16.02	2.00	16.68	2.00	11.40	2.00	13.19
Hot and cold water services	1.00	8.01	1.00	8.34	1.00	5.70	1.00	6.59
Heating and air treatment	2.00	16.02	2.00	16.68	2.00	11.40	2.00	13.19
Ventilation installation	3.00	24.03	2.00	16.68	2.00	11.40	1.00	6.59
Gas services	1.00	8.01	1.00	8.34	1.00	5.70	1.00	6.59
Electrical installation	11.00	88.12	14.00	116.75	10.00	57.00	14.00	92.31
Lift and conveyor installations	-	-	-	-	-	-	-	-
Protective communication	1.00	8.01	1.00	8.34	1.00	5.70	1.00	6.59
Communication installations	1.00	8.01	1.00	8.34	2.00	11.40	-	-
Special equipment	2.00	16.02	1.00	8.34	2.00	11.40	1.00	6.59
Builders' work and profit	2.00	16.02	2.00	16.68	2.00	11.40	1.00	6.59
EXTERNAL WORKS	3.00	24.03	9.00	75.05	10.00	57.00	11.00	72.53
	100.00	801.06	100.00	833.92	100.00	569.99	100.00	659.33

Approximate estimating

Elemental costs *(continued)*

	Farm building		Warehouse (shell only)		Nursery units		Factory, light industrial	
	%	£/m2	%	£/m2	%	£/m2	%	£/m2
PRELIMINARIES	6.00	19.72	8.00	23.66	7.00	30.63	6.00	28.35
SUBSTRUCTURES	7.00	23.00	6.00	17.75	6.00	26.25	5.00	23.62
SUPERSTRUCTURES								
Frame	-	-	8.00	23.66	-	-	7.00	33.07
Upper floors	-	-	-	-	-	-	-	-
Roof	14.00	46.01	10.00	29.58	11.00	48.13	9.00	42.52
Stairs	-	-	-	-	-	-	-	-
External walls and doors	24.00	78.87	12.00	35.49	12.00	52.50	9.00	42.52
Windows and external doors	2.00	6.57	4.00	11.83	7.00	30.63	5.00	23.62
Internal walls and partitions	6.00	19.72	-	-	-	-	2.00	9.45
Internal doors	-	-	-	-	1.00	4.38	2.00	9.45
FINISHES								
Wall finishes	-	-	2.00	5.92	1.00	4.38	1.00	4.72
Floor finishes	-	-	2.00	5.92	2.00	8.75	2.00	9.45
Ceiling finishes	-	-	-	-	1.00	4.38	1.00	4.72
FITTINGS AND FURNISHINGS	-	-	-	-	-	-	2.00	9.45
SERVICES								
Sanitary and disposal installations	-	-	-	-	1.00	4.38	2.00	9.45
Services equipment	-	-	-	-	-	-	1.00	4.72
Heat source	-	-	-	-	-	-	2.00	9.45
Hot and cold water services	-	-	2.00	5.92	2.00	8.75	2.00	9.45
Heating and air treatment	-	-	-	-	-	-	2.00	9.45
Ventilation installation	-	-	-	-	-	-	2.00	9.45
Gas services	-	-	1.00	2.96	1.00	4.38	1.00	4.72
Electrical installation	26.00	85.45	35.00	103.52	37.00	161.88	25.00	118.11
Lift and conveyor installations	-	-	-	-	-	-	-	-
Protective communication	-	-	-	-	-	-	-	-
Communication installations	-	-	-	-	-	-	2.00	9.45
Special equipment	7.00	23.00	-	-	-	-	2.00	9.45
Builders' work and profit	4.00	13.15	3.00	8.87	2.00	8.75	3.00	14.17
EXTERNAL WORKS	4.00	13.15	7.00	20.70	9.00	39.38	5.00	23.62
	100.00	328.64	100.00	295.78	100.00	437.50	100.00	472.42

Elemental costs *(continued)*

	Factory, heavy industrial		Hi-tech laboratory		Primary school		Middle school	
	%	£/m2	%	£/m2	%	£/m2	%	£/m2
PRELIMINARIES	7.00	47.88	5.00	45.96	8.00	67.37	10.00	89.86
SUBSTRUCTURES	7.00	47.88	5.00	45.96	6.00	50.53	7.00	62.90
SUPERSTRUCTURES								
Frame	8.00	54.72	6.00	55.15	-	-	-	-
Upper floors	-	-	4.00	36.77	-	-	-	-
Roof	11.00	75.24	7.00	64.34	6.00	50.53	6.00	53.92
Stairs	-	-	3.00	27.57	-	-	3.00	26.96
External walls and doors	10.00	68.40	5.00	45.96	10.00	84.21	13.00	116.82
Windows and external doors	4.00	27.36	8.00	73.53	7.00	58.95	8.00	71.89
Internal walls and partitions	3.00	20.52	3.00	27.57	8.00	67.37	7.00	62.90
Internal doors	1.00	6.84	2.00	18.38	7.00	58.95	4.00	35.95
FINISHES								
Wall finishes	2.00	13.68	2.00	18.38	2.00	16.84	2.00	17.97
Floor finishes	2.00	13.68	2.00	18.38	2.00	16.84	2.00	17.97
Ceiling finishes	1.00	6.84	2.00	18.38	1.00	8.42	2.00	17.97
FITTINGS AND FURNISHINGS	2.00	13.68	3.00	27.57	3.00	25.26	3.00	26.96
SERVICES								
Sanitary and disposal installations	2.00	13.68	1.00	9.19	2.00	16.84	1.00	8.99
Services equipment	1.00	6.84	1.00	9.19	1.00	8.42	1.00	8.99
Heat source	2.00	13.68	2.00	18.38	2.00	16.84	2.00	17.97
Hot and cold water services	2.00	13.68	2.00	18.38	2.00	16.84	2.00	17.97
Heating and air treatment	3.00	20.52	2.00	18.38	2.00	16.84	2.00	17.97
Ventilation installation	3.00	20.52	3.00	27.57	-	-	-	-
Gas services	1.00	6.84	2.00	18.38	1.00	8.42	1.00	8.99
Electrical installation	13.00	88.92	7.00	64.34	13.00	109.48	5.00	44.93
Lift and conveyor installations	-	-	4.00	36.77	-	-	2.00	17.97
Protective communication	1.00	6.84	1.00	9.19	-	-	-	-
Communication installations	2.00	13.68	2.00	18.38	1.00	8.42	1.00	8.99
Special equipment	3.00	20.52	3.00	27.57	-	-	-	-
Builders' work and profit	4.00	27.36	4.00	36.77	5.00	42.11	6.00	53.92
EXTERNAL WORKS	5.00	34.20	9.00	82.72	11.00	92.64	10.00	89.86
	100.00	683.98	100.00	919.17	100.00	842.14	100.00	898.63

Approximate estimating

Elemental costs (continued)

	Sixth form college		Special school		Training college		Library	
	%	£/m2	%	£/m2	%	£/m2	%	£/m2
PRELIMINARIES	9.00	65.44	10.00	77.03	8.00	59.16	10.00	87.91
SUBSTRUCTURES	6.00	43.63	5.00	38.51	7.00	51.76	7.00	61.54
SUPERSTRUCTURES								
Frame	-	-	-	-	-	-	-	-
Upper floors	-	-	-	-	-	-	-	-
Roof	4.00	29.08	5.00	38.51	7.00	51.76	6.00	52.75
Stairs	4.00	29.08	3.00	23.11	4.00	29.58	3.00	26.37
External walls and doors	11.00	79.98	9.00	69.32	9.00	66.55	10.00	87.91
Windows and external doors	9.00	65.44	10.00	77.03	8.00	59.16	9.00	79.12
Internal walls and partitions	5.00	36.36	5.00	38.51	6.00	44.37	5.00	43.96
Internal doors	3.00	21.81	4.00	30.81	4.00	29.58	4.00	35.16
FINISHES								
Wall finishes	3.00	21.81	2.00	15.41	2.00	14.79	2.00	17.58
Floor finishes	3.00	21.81	2.00	15.41	2.00	14.79	2.00	17.58
Ceiling finishes	2.00	14.54	1.00	7.70	2.00	14.79	2.00	17.58
FITTINGS AND FURNISHINGS	4.00	29.08	3.00	23.11	3.00	22.18	3.00	26.37
SERVICES								
Sanitary and disposal installations	1.00	7.27	2.00	15.41	2.00	14.79	1.00	8.79
Services equipment	1.00	7.27	1.00	7.70	1.00	7.39	1.00	8.79
Heat source	2.00	14.54	2.00	15.41	2.00	14.79	2.00	17.58
Hot and cold water services	1.00	7.27	1.00	7.70	2.00	14.79	1.00	8.79
Heating and air treatment	1.00	7.27	2.00	15.41	1.00	7.39	2.00	17.58
Ventilation installation	-	-	-	-	-	-	-	-
Gas services	1.00	7.27	1.00	7.70	1.00	7.39	1.00	8.79
Electrical installation	10.00	72.71	12.00	92.43	8.00	59.16	13.00	114.28
Lift and conveyor installations	3.00	21.81	3.00	23.11	3.00	22.18	-	-
Protective communication	-	-	-	-	-	-	1.00	8.79
Communication installations	1.00	7.27	1.00	7.70	1.00	7.39	1.00	8.79
Special equipment	-	-	-	-	-	-	-	-
Builders' work and profit	5.00	36.36	7.00	53.92	5.00	36.97	7.00	61.54
EXTERNAL WORKS	11.00	79.98	9.00	69.32	12.00	88.73	7.00	61.54
	100.00	727.12	100.00	770.25	100.00	739.44	100.00	879.11

Approximate estimating

Elemental costs *(continued)*

	Health centre		Welfare centre		Old persons' home		Group surgery	
	%	£/m2	%	£/m2	%	£/m2	%	£/m2
PRELIMINARIES	7.00	60.96	12.00	101.92	11.00	80.77	10.00	70.35
SUBSTRUCTURES	10.00	87.09	6.00	50.96	7.00	51.40	6.00	42.21
SUPERSTRUCTURES								
Frame	-	-	-	-	-	-	-	-
Upper floors	-	-	-	-	-	-	-	-
Roof	9.00	78.38	10.00	84.93	7.00	51.40	6.00	42.21
Stairs	-	-	-	-	-	-	-	-
External walls and doors	10.00	87.09	8.00	67.95	8.00	58.74	6.00	42.21
Windows and external doors	6.00	52.25	5.00	42.47	6.00	44.06	5.00	35.17
Internal walls and partitions	5.00	43.54	5.00	42.47	6.00	44.06	6.00	42.21
Internal doors	4.00	34.84	3.00	25.48	5.00	36.72	4.00	28.14
FINISHES								
Wall finishes	3.00	26.13	2.00	16.99	3.00	22.03	2.00	14.07
Floor finishes	2.00	17.42	2.00	16.99	2.00	14.69	2.00	14.07
Ceiling finishes	2.00	17.42	2.00	16.99	2.00	14.69	2.00	14.07
FITTINGS AND FURNISHINGS	3.00	26.13	3.00	25.48	3.00	22.03	3.00	21.10
SERVICES								
Sanitary and disposal installations	2.00	17.42	2.00	16.99	2.00	14.69	2.00	14.07
Services equipment	2.00	17.42	1.00	8.49	2.00	14.69	3.00	21.10
Heat source	2.00	17.42	2.00	16.99	2.00	14.69	2.00	14.07
Hot and cold water services	2.00	17.42	2.00	16.99	1.00	7.34	2.00	14.07
Heating and air treatment	3.00	26.13	2.00	16.99	2.00	14.69	2.00	14.07
Ventilation installation	-	-	-	-	-	-	-	-
Gas services	1.00	8.71	1.00	8.49	1.00	7.34	1.00	7.03
Electrical installation	5.00	43.54	11.00	93.43	11.00	80.77	17.00	119.59
Lift and conveyor installations	-	-	-	-	-	-	-	-
Protective communication	1.00	8.71	1.00	8.49	1.00	7.34	2.00	14.07
Communication installations	2.00	17.42	2.00	16.99	2.00	14.69	2.00	14.07
Special equipment	2.00	17.42	1.00	8.49	2.00	14.69	2.00	14.07
Builders' work and profit	6.00	52.25	5.00	42.47	4.00	29.37	8.00	56.28
EXTERNAL WORKS	11.00	95.80	12.00	101.92	10.00	73.43	5.00	35.17
	100.00	870.90	100.00	849.33	100.00	734.31	100.00	703.50

Approximate estimating

Elemental costs *(continued)*

	Local authority low rise flats		Local authority semi-detached houses		Standard private apartments		Luxury private apartments	
	%	£/m2	%	£/m2	%	£/m2	%	£/m2
PRELIMINARIES	10.00	53.92	6.00	30.50	9.00	52.22	7.00	53.92
SUBSTRUCTURES	7.00	37.74	6.00	30.50	10.00	58.03	6.00	46.22
SUPERSTRUCTURES								
Frame	-	-	-	-	-	-	-	-
Upper floors	-	-	-	-	-	-	-	-
Roof	10.00	53.92	10.00	50.84	8.00	46.42	9.00	69.32
Stairs	2.00	10.78	3.00	15.25	3.00	17.41	3.00	23.11
External walls and doors	6.00	32.35	8.00	40.67	11.00	63.83	10.00	77.03
Windows and external doors	6.00	32.35	7.00	35.59	5.00	29.01	6.00	46.22
Internal walls and partitions	6.00	32.35	5.00	25.42	5.00	29.01	6.00	46.22
Internal doors	4.00	21.57	4.00	20.33	4.00	23.21	6.00	46.22
FINISHES								
Wall finishes	2.00	10.78	2.00	10.17	2.00	11.61	2.00	15.41
Floor finishes	2.00	10.78	1.00	5.08	2.00	11.61	2.00	15.41
Ceiling finishes	1.00	5.39	2.00	10.17	2.00	11.61	2.00	15.41
FITTINGS AND FURNISHINGS	2.00	10.78	3.00	15.25	2.00	11.61	3.00	23.11
SERVICES								
Sanitary and disposal installations	3.00	16.18	2.00	10.17	2.00	11.61	3.00	23.11
Services equipment	2.00	10.78	2.00	10.17	2.00	11.61	2.00	15.41
Heat source	2.00	10.78	2.00	10.17	3.00	17.41	3.00	23.11
Hot and cold water services	2.00	10.78	1.00	5.08	2.00	11.61	2.00	15.41
Heating and air treatment	-	-	-	-	-	-	1.00	7.70
Ventilation installation	-	-	-	-	1.00	5.80	1.00	7.70
Gas services	1.00	5.39	1.00	5.08	1.00	5.80	1.00	7.70
Electrical installation	16.00	86.27	18.00	91.51	13.00	75.43	12.00	92.43
Lift and conveyor installations	-	-	-	-	-	-	-	-
Protective communication	-	-	-	-	1.00	5.80	1.00	7.70
Communication installations	2.00	10.78	-	-	1.00	5.80	1.00	7.70
Special equipment	-	-	-	-	-	-	-	-
Builders' work and profit	9.00	48.53	8.00	40.67	7.00	40.62	6.00	46.22
EXTERNAL WORKS	5.00	26.96	9.00	45.75	4.00	23.21	5.00	38.51
	100.00	539.18	100.00	508.37	100.00	580.26	100.00	770.25

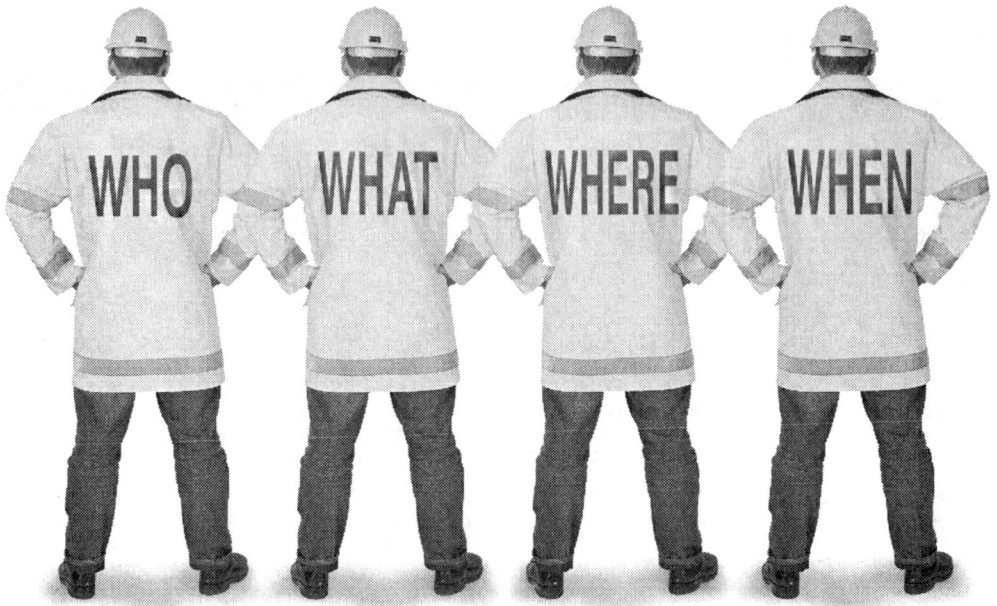

You need to know **Who** is building **What** and **Where**,
you need to know **When** they are going to do it
and you need to know **NOW...**

Glenigan is the solution!

Saving you time and money in the search for new business, our sales leads
provide you with comprehensive, fully researched information about who is doing
what in *your* marketplace right now!

Timely, accurate and reliable, Glenigan leads offer real profit potential.

Phone us now for more details on: **0800 373771**

41-47 Seabourne Road, Bournemouth, Dorset BH5 2HU
Tel: 01202 432121   Fax: 01202 431204
e mail: info@glenigan.emap.co.uk

Alterations and repairs

Figures in this section are intended as a guide to pricing typical mid to small scale contracts for alterations, extensions, conversions, repairs and re-decorations. Many items may be found useful for jobbing work but only after suitable addition has been made to cover travelling time, transport, extra supervision, etc. In the preparation of figures the contract envisaged involves a reasonably substantial amount of work in a situation where space is limited so that, for example, cement could not be taken at more than a tonne at a time and aggregates would have to be wheeled in from the road.

As noted for "New work", it must be emphasised that the locality should be thoroughly investigated and quotations for materials should be obtained (with particular attention to the extra cost resulting from delivery in small lots) so as to ensure that a tender takes full account of local market conditions.

Throughout this section, in addition to net prices, there are shown prices with 25% added to cover general establishment charges, attendance and profit.

Demolition

Preamble

"Labour net" figures include allowances for all costs incidental to the employment of labour.

"Plant net" figures include for all costs of plant including drivers and operators where applicable.

"Materials net" figures include for all costs of materials including an allowance for waste except where specifically stated.

"Price net" figures are the totals of the "Labour net", "Plant net" and "Materials net" figures. Prices are for a builder employing his own labour; according to the amount and nature of the work involved, it may well be possible to secure more advantageous prices from specialist sub-contractors.

Prices do not include any allowance for scaffolding, ladders or other plant necessary to reach the work. The "Preliminaries" section includes prices for general scaffolding which must be considered and allowance included to suit the particular circumstances of a tender. Items for scaffolding incidental to demolitions and alterations are included in this section.

Where items are described as "including removal of debris", this has been allowed for on the basis of removal to a tip average 15 kilometres from site.

Demolition

Alterations and repairs

	Unit	Price
Basic prices for materials		£
Building paper - standard grade	m2	0.62
3.2 mm tempered hardboard	m2	2.05
13 mm ivory faced insulation board	m2	3.09
Roofing felt	m2	1.00
Sawn softwood		
smaller scantlings, ungraded	m3	244.80
larger scantlings, SC3 grade	m3	244.80
Structural steel	tonne	986.65

Prices actually to be paid for materials must be checked against the above basic prices and adjustments made as necessary.

Demolition

Alterations and repairs	Unit	Hours C	Hours L	Labour net	Plant net	Material net	Price net	Price with 25%
Generally				£	£	£	£	£
					VAT not included			
Pulling down - excluding removal of debris								
Demolish brickwork in lime mortar and drop to ground level								
102.5 mm walls	m2	-	0.35	2.36	-	-	2.36	2.95
215 mm walls	m2	-	0.75	5.05	-	-	5.05	6.31
327.5 mm walls	m2	-	1.15	7.75	-	-	7.75	9.69
Demolish brickwork in cement mortar and drop to ground level								
102.5 mm walls	m2	-	0.55	3.71	-	-	3.71	4.64
215 mm walls	m2	-	1.15	7.75	-	-	7.75	9.69
327.5 mm walls	m2	-	1.75	11.80	-	-	11.80	14.75
Demolish blockwork in cement mortar and drop to ground level								
75 mm walls	m2	-	0.40	2.70	-	-	2.70	3.38
100 mm walls	m2	-	0.50	3.37	-	-	3.37	4.21
Demolish stonework in lime mortar and drop to ground level								
300 mm walls	m2	-	1.05	7.08	-	-	7.08	8.85
450 mm walls	m2	-	1.60	10.78	-	-	10.78	13.48
Pulling down - including removal of debris								
Demolish brickwork in lime mortar and drop to ground level								
102.5 mm walls	m2	-	0.60	4.04	2.19	-	6.23	7.79
215 mm walls	m2	-	1.30	8.76	3.69	-	12.45	15.56
327.5 mm walls	m2	-	2.00	13.48	5.88	-	19.36	24.20
Demolish brickwork in cement mortar and drop to ground level								
102.5 mm	m2	-	0.80	5.39	1.99	-	7.38	9.22
215mm walls	m2	-	1.70	11.46	3.69	-	15.15	18.94
327.5 mm walls	m2	-	2.60	17.52	5.88	-	23.40	29.25
Demolish blockwork in cement mortar and drop to ground level								
75 mm walls	m2	-	0.60	4.04	1.29	-	5.33	6.66
100 mm walls	m2	-	0.75	5.05	1.99	-	7.04	8.80
Demolish stonework in lime mortar and drop to ground level								
300 mm walls	m2	-	1.80	12.13	5.68	-	17.81	22.26
450 mm walls	m2	-	2.75	18.54	8.27	-	26.81	33.51

Demolition

Alterations and repairs	Unit	Hours C	Hours L	Labour net	Plant net	Material net	Price net	Price with 25%
Generally				£	£	£	£	£
					VAT not included			

Pulling down with care - excluding removal of debris

Carefully take down brickwork in lime mortar and lower average one storey

	Unit	Hours C	Hours L	Labour net	Plant net	Material net	Price net	Price with 25%
102.5 mm walls	m2	-	0.42	2.83	-	-	2.83	3.54
215 mm walls	m2	-	0.90	6.07	-	-	6.07	7.59
327.5 mm	m2	-	1.38	9.30	-	-	9.30	11.63

Carefully take down brickwork in cement mortar and lower average one storey

102.5 mm walls	m2	-	0.66	4.45	-	-	4.45	5.56
215 mm walls	m2	-	1.38	9.30	-	-	9.30	11.63
327.5 mm walls	m2	-	2.10	14.15	-	-	14.15	17.69

Carefully take down blockwork in cement mortar and lower average one storey

75 mm walls	m2	-	0.48	3.24	-	-	3.24	4.05
100 mm walls	m2	-	0.60	4.04	-	-	4.04	5.05

Carefully take down stonework in lime mortar and lower average one storey

300 mm walls	m2	-	1.26	8.49	-	-	8.49	10.61
450 mm walls	m2	-	1.92	12.94	-	-	12.94	16.18

Pulling down with care - including removal of debris

Carefully take down brickwork in lime mortar and lower average one storey

102.5 mm walls	m2	-	0.67	4.52	1.99	-	6.51	8.14
215 mm walls	m2	-	1.45	9.77	3.69	-	13.46	16.82
327.5 mm walls	m2	-	2.23	15.03	5.88	-	20.91	26.14

Carefully take down brickwork in cement mortar and lower average one storey

102.5 mm	m2	-	1.46	9.84	1.99	-	11.83	14.79
215 mm walls	m2	-	1.93	13.01	3.69	-	16.70	20.88
327.5 mm walls	m2	-	2.95	19.88	5.88	-	25.76	32.20

Carefully take down blockwork in cement mortar and lower average one storey

75 mm walls	m2	-	0.68	4.58	1.29	-	5.87	7.34
100 mm walls	m2	-	0.85	5.73	1.99	-	7.72	9.65

Carefully take down stonework in lime mortar and lower average one storey

300 mm walls	m2	-	2.01	13.55	5.68	-	19.23	24.04
450 mm walls	m2	-	3.07	20.69	8.27	-	28.96	36.20

Demolition

Alterations and repairs	Unit	Hours C	Hours L	Labour net	Plant net	Material net	Price net	Price with 25%
Generally				£	£	£	£	£
					VAT not included			

Cutting openings - excluding removal of debris

Cut away brickwork in lime mortar to form door or window openings and lower average one storey

	Unit	Hours C	Hours L	Labour net	Plant net	Material net	Price net	Price with 25%
102.5 mm walls	m2	-	0.85	5.73	-	-	5.73	7.16
215 mm walls	m2	-	1.28	8.63	-	-	8.63	10.79
327.5 mm walls	m2	-	1.91	12.87	-	-	12.87	16.09

Cut away brickwork in cement mortar to form door or window openings and lower average one storey

	Unit	Hours C	Hours L	Labour net	Plant net	Material net	Price net	Price with 25%
102.5 mm walls	m2	-	1.00	6.74	-	-	6.74	8.43
215 mm walls	m2	-	1.50	10.11	-	-	10.11	12.64
327.5 mm walls	m2	-	2.00	13.48	-	-	13.48	16.85

Cut away blockwork in cement mortar to form door or window openings and lower average one storey

	Unit	Hours C	Hours L	Labour net	Plant net	Material net	Price net	Price with 25%
75 mm walls	m2	-	0.80	5.39	-	-	5.39	6.74
100 mm walls	m2	-	1.00	6.74	-	-	6.74	8.43

Cut away stonework in lime mortar to form door or window openings and lower average one storey

	Unit	Hours C	Hours L	Labour net	Plant net	Material net	Price net	Price with 25%
300 mm walls	m2	2.40	2.40	37.68	-	-	37.68	47.10
450 mm walls	m2	3.60	3.60	56.52	-	-	56.52	70.65

Cut away suspended reinforced concrete floor or roof to form openings and lower average one storey

	Unit	Hours C	Hours L	Labour net	Plant net	Material net	Price net	Price with 25%
100 mm walls	m2	2.50	2.50	39.25	-	-	39.25	49.06
150 mm walls	m2	4.00	4.00	62.80	-	-	62.80	78.50
200 mm walls	m2	5.50	5.50	86.35	-	-	86.35	107.94

Cutting openings - including removal of debris

Cut away brickwork in lime mortar to form door or window openings and lower average one storey

	Unit	Hours C	Hours L	Labour net	Plant net	Material net	Price net	Price with 25%
102.5 mm walls	m2	-	0.91	6.13	1.99	-	8.12	10.15
215 mm walls	m2	-	1.42	9.57	3.19	-	12.76	15.95
327.5 mm walls	m2	-	2.11	14.22	5.88	-	20.10	25.13

Cut away brickwork in cement mortar to form door or window openings and lower average one storey

	Unit	Hours C	Hours L	Labour net	Plant net	Material net	Price net	Price with 25%
102.5 mm walls	m2	-	1.06	7.14	1.99	-	9.13	11.41
215 mm walls	m2	-	1.64	11.05	3.69	-	14.74	18.43
327.5 mm walls	m2	-	2.20	14.83	5.88	-	20.71	25.89

Demolition

Alterations and repairs	Unit	Hours C	Hours L	Labour net	Plant net	Material net	Price net	Price with 25%
Generally				£	£	£	£	£
					VAT not included			

Cutting openings - including removal of debris *(continued)*

Cut away blockwork to form door or window openings and lower average one storey

75 mm walls	m2	-	0.85	5.73	1.29	-	7.02	8.78
100 mm walls	m2	-	1.06	7.14	1.99	-	9.13	11.41

Cut away stonework in lime mortar to form door or window openings and lower average one storey

300 mm walls	m2	2.40	3.15	42.73	5.68	-	48.41	60.51
450 mm walls	m2	3.60	4.75	64.28	8.27	-	72.55	90.69

Cut away suspended reinforced concrete floor or roof to form openings and lower average one storey

100 mm slabs	m2	2.50	2.75	40.94	1.99	-	42.93	53.66
150 mm slabs	m2	4.00	4.40	65.50	2.59	-	68.09	85.11
200 mm slabs	m2	5.50	6.00	89.72	3.69	-	93.41	116.76

General Demolition

Remove average fireplaces, surround and backing etc. complete and remove from site	each	-	2.50	16.85	1.84	-	18.69	23.36
Carefully remove average fireplaces, surround and backing etc complete aside and store surround and remove debris from site	each	-	2.75	18.54	0.84	-	19.38	24.23
Carefully strip medium size roof slates, lower average two storeys, set aside and store salvaged slates	m2	0.25	0.25	3.93	-	-	3.93	4.91
Carefully strip plain roof tiles, lower average two storeys, set aside and store salvaged tiles	m2	0.20	0.20	3.14	-	-	3.14	3.92
Carefully take up floor boarding, draw nails, set aside and store salvaged boards	m2	0.13	0.13	2.04	-	-	2.04	2.55

Carefully remove timber joists or rafters, draw nails, set aside and store salvaged timbers - average sectional area

0.005 m2 (50 x 100 mm)	m2	0.15	0.15	2.35	-	-	2.35	2.94
0.010 m2 (50 x 200 mm)	m2	0.23	0.23	3.61	-	-	3.61	4.51

Demolition

Alterations and repairs	Unit	Hours C	Hours L	Labour net	Plant net	Material net	Price net	Price with 25%
Generally				£	£	£	£	£
					VAT not included			
General Demolition (*continued*)								
Carefully remove doors and frames, set aside and store								
single doors up to 2 m2	each	0.50	0.50	7.85	-	-	7.85	9.81
double doors up to 4 m2	each	0.70	0.70	10.99	-	-	10.99	13.74
Carefully remove windows, set aside and store								
up to 1 m2	each	0.33	0.33	5.18	-	-	5.18	6.47
up to 2 m2	each	0.50	0.50	7.85	-	-	7.85	9.81
up to 4 m2	each	0.85	0.85	13.35	-	-	13.35	16.69
Hack off lath and plaster ceiling, lower average one storey, draw nails from joists and remove debris from site	m2	-	0.85	5.73	0.04	-	5.77	7.21
Hack off lath and plaster from both sides of partition, lower average one storey and remove debris from site, remove studding, draw nails, set aside and store salvaged timbers	m2	-	1.75	11.80	0.04	-	11.84	14.80
Hack off wall finishings, lower average one storey and remove debris from site								
Plastering	m2	-	0.61	4.13	0.04	-	4.17	5.21
Tiling and cement backing	m2	-	0.80	5.39	0.07	-	5.46	6.83
Rendering, roughcast or pebbledash	m2	-	0.85	5.73	0.07	-	5.80	7.25
Sort and clean lime mortar from brick, set aside and stack salvaged bricks	1000	-	8.00	53.92	-	-	53.92	67.40
Load general rubbish and debris and remove to tip average 15 km from site	m3	-	2.00	13.48	18.53	-	32.01	40.01
Add or deduct for every 1 km difference in distance	m3	-	-	-	0.20	-	0.20	0.25

Demolition

Alterations and repairs	Unit	Hours C	Hours L	Labour net	Plant net	Material net	Price net	Price with 25%
Generally				£	£	£	£	£
					VAT not included			

Temporary screens

For temporary fencing and hoardings, see "Preliminaries".

Alterations and repairs	Unit	Hours C	Hours L	Labour net	Plant net	Material net	Price net	Price with 25%
Screens of 50 x 50 mm sawn softwood framing (two uses) covered one side with								
building paper (one use)	m2	0.77	-	6.90	-	2.66	9.56	11.95
polythene sheet (one use)	m2	0.77	-	6.90	-	1.71	8.61	10.76
13 mm insulation board (two uses)	m2	0.80	-	7.17	-	2.55	9.72	12.15
3.2 mm hardboard (two uses)	m2	0.80	-	7.17	-	2.26	9.43	11.79
Dustproof screens of 50 x 50 mm sawn softwood framing (two uses) covered one side with								
building paper (one use) taped at joints and edges	m2	0.78	-	6.99	-	2.72	9.71	12.14
polythene sheet (one use) taped at joints and edges	m2	0.78	-	6.99	-	1.85	8.84	11.05
13 mm insulation board (two uses) taped at joints and edges	m2	0.55	-	4.93	-	2.69	7.62	9.53
3.2 mm hardboard (two uses) taped at joints and edges	m2	0.55	-	4.93	-	2.42	7.35	9.19
Doors of 38 x 75 mm sawn softwood framing covered both sides with 3.2 mm hardboard including additional framing to screen and fastenings (two uses)								
900 x 2000 mm single doors	each	4.50	-	40.32	-	28.38	68.70	85.88
1800 x 2000 mm pair of doors	pair	7.50	-	67.20	-	44.98	112.18	140.22
Weatherproof screens of 50 x 75 mm sawn softwood framing (two uses) covered with roofing felt (one use) and lined with								
building paper (one use) taped at joints and edges	m2	0.87	-	7.80	-	2.72	10.52	13.15
13 mm insulation board (two uses) taped at joints and edges	m2	0.90	-	8.06	-	2.69	10.75	13.44
3.2 mm hardboard (two uses) taped at joints and edges	m2	0.90	-	8.06	-	2.42	10.48	13.10
Doors of 38 x 75 mm sawn softwood framing covered both sides with 3.2 mm hardboard including additional framing to screen and fastenings (two uses) and covered with roofing felt (one use) externally								
900 x 2000 mm single doors	each	4.50	-	40.32	-	33.05	73.37	91.71
1800 x 2000 mm pairs of doors	pair	7.50	-	67.20	-	54.32	121.52	151.90

Demolition

Alterations and repairs	Unit	Hours C	Hours L	Labour net	Plant net	Material net	Price net	Price with 25%
Generally				£	£	£	£	£
					VAT not included			

Shoring

Prices for temporary shores and strutting are based on six uses of the materials of the materials described and include for providing and erecting the shoring and strutting and for clearing away.

Prices for permancent shores and strutting include the full cost of new materials as described but include only for providing and erecting the shoring and strutting.

Dead shores to support 80 kn. (8 tons), comprising two 150 x 150 mm softwood legs and one 203 x 133 mm x 30 kg/m steel UB needle, complete with sole pieces, wedges and braces including cutting away through wall, floor and ceiling and making good

temporary shores	each	14.00	8.00	179.36	-	36.83	216.19	270.24
permanent shores	each	9.50	6.00	125.56	-	168.28	293.84	267.30

Dead shores to support 140 kn. (14 tons), comprising two 200 x 200 mm softwood legs and one 203 x 133 mm x 30 kg/m steel UB needle, complete with sole pieces, wedges and braces including cutting away through wall, floor and ceiling and making good

temporary shores	each	15.00	8.50	191.69	-	43.50	235.19	293.99
permanent shores	each	10.50	6.50	137.89	-	208.19	346.08	432.60

Flying shores in softwood comprising 100 x 150 mm shore, 50 x 175 mm wall pieces and 50 x 100 mm struts and straining pieces - distance between walls 4.50 m

temporary shores	each	12.00	12.00	188.40	-	20.20	208.60	260.75
temporary shores	m3	52.40	52.40	822.68	-	59.51	882.19	1102.74
permanent shores	each	10.00	10.00	157.00	-	101.70	258.70	323.38
permanent shores	m3	43.67	43.67	685.62	-	376.27	1061.89	1327.36

Raking shores in softwood comprising two 150 x 150 mm rakers (total length 11 m), 50 x 175 mm wall piece complete with sole pieces, needles, cleats and 25 mm board bracing

temporary shores	each	20.00	20.00	314.00	-	32.52	346.52	433.15
temporary shores	m3	48.80	48.80	766.16	-	79.37	845.53	1056.91
permanent shores	each	15.00	15.00	235.50	-	172.57	408.07	510.09
permanent shores	m3	36.60	36.60	574.62	-	417.24	991.86	1239.83

Demolition

Alterations and repairs	Unit	Hours C	Hours L	Labour net	Plant net	Material net	Price net	Price with 25%
Generally				£	£	£	£	£
						VAT not included		

Shoring (continued)

Raking shores in softwood comprising two 200 x 200 mm rakers and rider (total length 21 m), 50 x 225 mm wall piece complete with sole pieces, needles, cleats and 25 mm board bracing

	Unit	Hours C	Hours L	Labour net	Plant net	Material net	Price net	Price with 25%
temporary shores	each	28.50	28.50	447.45	-	89.46	536.91	671.14
temporary shores	m3	26.50	26.50	416.05	-	85.97	502.02	627.52
permanent shores	each	21.50	21.50	337.55	-	342.04	679.59	849.49
permanent shores	m3	20.00	20.00	314.00	-	319.37	633.37	791.71

Strutting floors 2.75 m high to underside in softwood comprising 50 x 150 mm head and cill pieces and 75 x 100 mm legs at one metre centres with two 25 x 75 mm braces

temporary strutting (six uses)	m	0.67	0.67	10.52	-	2.37	12.89	16.11
temporary strutting (six uses)	m3	16.75	16.75	262.98	-	55.44	318.42	398.02
permanent strutting	m	0.50	0.50	7.85	-	13.91	21.76	27.20
permanent strutting	m3	12.50	12.50	196.25	-	323.46	519.71	649.64

Strutting window openings in softwood comprising 50 x 100 wall pieces and struts (total length 6 m)

temporary strutting (six uses)	each	0.50	-	4.48	-	1.50	5.98	7.47
temporary strutting (six uses)	m3	16.67	-	149.36	-	50.25	199.61	249.51
permanent strutting	each	0.38	-	3.40	-	9.04	12.44	15.55
permanent strutting	m3	12.67	-	113.52	-	301.62	415.14	518.92

Incidental Scaffolding

Putlog scaffolding in connection with demolitions and alterations, erected, standing four weeks and dismantled

builder's own plant	m2	0.67	-	6.00	-	0.65	6.65	8.31
each additional four weeks of builder's own plant	m2	-	-	-	-	0.65	0.65	0.81
hired plant	m2	0.67	-	6.00	-	0.84	6.84	8.55
each additional four week period of hired plant	m2	-	-	-	-	0.84	0.84	1.05

Independent scaffolding in connection with demolitions and alterations, erected, standing four weeks and dismantled

builder's own plant	m2	0.90	-	8.06	-	0.60	8.66	10.82
each additional four weeks of builder's own plant	m2	-	-	-	-	0.60	0.60	0.75
hired plant	m2	0.90	-	8.06	-	0.79	8.85	11.06
each additional four week of hired plant	m2	-	-	-	-	0.79	0.79	0.99

Excavation and earthwork

Preamble

"Labour net" figures include allowances for all costs incidental to the employment of labour.

"Plant net" figures include for all costs of plant including drivers and operators where applicable.

"Materials net" figures include for all costs of materials including an allowance for waste except where specifically stated.

"Price net" figures are the totals of the "Labour net", "Plant net" and "Materials net" figures. Prices are for a builder employing his own labour; according to the amount and nature of the work involved, it may well be possible to secure more advantageous prices from specialist sub-contractors.

Excavation prices are for work in firm soil. For other soils the following adjustments should be made:

    clay - add 25%
    hard gravel - add 50%
    chalk - add 100 to 150%
    rock - add 300 to 400%

Earthwork support is to a large extent a risk item. The poling boards may need to be close together in poor ground, they may be well apart in good ground or they may have to be driven as the depth increases. It may be possible for timbering as a whole to be put in afterwards or at wide intervals or not at all; the cost can vary widely. Figures given are for semi-close boarding with conditions assumed as poor to moderate. Material for earthwork support is assumed as being timber used ten times.

Figures for disposal of excavated material include allowance for 25% increase in bulk after excavation.

Excavation and earthwork

Alterations and repairs

	Unit	Price
Basic prices for materials		**£**
Ashes	m3	9.50
	tonne	7.60
Binding gravel (hoggin)	m3	15.84
	tonne	8.55
Hardcore	m3	12.25
	tonne	8.16
Sand	m3	17.96
	tonne	11.22
Timber for earthwork support	m3	253.37

Prices actually to be paid for materials must be checked against the above basic prices and adjustments made as necessary.

Excavation and earthwork

Alterations and repairs	Unit	Hours C	Hours L	Labour net	Plant net	Material net	Price net	Price with 25%
Generally				£	£	£	£	£
					VAT not included			
Site preparation								
Cut down hedge and grub up roots								
privet hedge 1 m high	m	-	1.50	10.11	-	-	10.11	12.64
field hedge 2 m height including								
small trees	m	-	6.25	42.13	-	-	42.13	52.66
Clear site of bushes, scrub and undergrowth and grub up roots	m2	-	0.13	0.88	-	-	0.88	1.10
Lift and roll turf to be preserved, remove 25 m and stack	m2	-	0.55	3.71	-	-	3.71	4.64
Excavate by machine topsoil to be preserved, average depth								
0.150 m	m2	-	-	-	0.94	-	0.94	1.18
0.225 m	m2	-	-	-	1.31	-	1.31	1.64
Excavate by hand topsoil to be preserved, average depth								
0.150 m	m2	-	0.34	2.29	-	-	2.29	2.86
0.225 m	m2	-	0.55	3.71	-	-	3.71	4.64
Excavation by machine								
Excavate to reduce levels, maximum depth not exceeding								
0.25 m	m3	-	-	-	1.27	-	1.27	1.59
1.00 m	m3	-	-	-	1.52	-	1.52	1.90
2.00 m	m3	-	-	-	1.52	-	1.52	1.90
4.00 m	m3	-	-	-	1.78	-	1.78	2.23
Excavate basements, maximum depth not exceeding								
1.00 m	m3	-	0.07	0.44	1.31	-	1.75	2.19
2.00 m	m3	-	0.13	0.89	2.44	-	3.33	4.16
4.00 m	m3	-	0.25	1.70	4.69	-	6.39	7.99
Excavate pits to receive bases of stanchions etc, maximum depth not exceeding								
0.25 m	m3	-	0.30	2.02	4.69	-	6.71	8.39
1.00 m	m3	-	0.20	1.35	3.75	-	5.10	6.38
2.00 m	m3	-	0.22	1.46	4.13	-	5.59	6.99
4.00 m	m3	-	0.24	1.62	4.51	-	6.13	7.66
Excavate pits less than 1.25 x 1.25 m on plan to receive bases of stanchions etc, maximum depth not exceeding								
0.25 m	m3	-	0.45	3.03	8.45	-	11.48	14.35
1.00 m	m3	-	0.30	2.02	5.63	-	7.65	9.56
2.00 m	m3	-	0.31	2.06	5.82	-	7.88	9.85

Excavation and earthwork

Alterations and repairs	Unit	Hours C	Hours L	Labour net	Plant net	Material net	Price net	Price with 25%
Generally				£	£	£	£	£
					VAT not included			

Excavate by machine (*continued*)

Excavate trenches not exceeding 0.30 m in width to receive foundations, average depth

0.25 m	m	-	0.30	2.02	0.56	-	2.58	3.23
0.50 m	m	-	0.06	0.40	1.13	-	1.53	1.91
0.75 m	m	-	0.08	0.53	1.50	-	2.03	2.54
1.00 m	m	-	0.10	0.65	1.88	-	2.53	3.16

Excavate trenches over 0.30 m in width to receive foundations, maximum depth not exceeding

0.25 m	m3	-	0.34	2.26	6.38	-	8.64	10.80
1.00 m	m3	-	0.29	1.98	5.44	-	7.42	9.28
2.00 m	m3	-	0.31	2.10	5.82	-	7.92	9.90
4.00 m	m3	-	0.34	2.31	6.38	-	8.69	10.86

Excavate trenches to receive service pipes, cables etc, grade bottom, fill in and compact and remove surplus – earthwork support not included – average depth not exceeding

0.25 m	m	-	0.09	0.58	0.83	-	1.41	1.76
0.50 m	m	-	0.13	0.89	1.41	-	2.30	2.88
0.75 m	m	-	0.20	1.33	2.22	-	3.55	4.44
1.00 m	m	-	0.26	1.78	2.48	-	4.26	5.33

Excavate and fill basement working space, maximum depth not exceeding

1.00 m	m3	-	1.41	9.52	1.77	-	11.29	14.11
2.00 m	m3	-	1.69	11.42	3.09	-	14.51	18.14
4.00 m	m3	-	2.04	13.75	4.97	-	18.72	23.40

Excavate and fill pit working space, maximum depth not exceeding

0.25 m	m3	-	1.18	7.94	4.97	-	12.91	16.14
1.00 m	m3	-	1.44	9.68	4.59	-	14.27	17.84
2.00 m	m3	-	1.71	11.50	4.59	-	16.09	20.11
4.00 m	m3	-	2.04	13.75	4.97	-	18.72	23.40

Excavate and fill trench working space, maximum depth not exceeding

0.25 m	m3	-	1.18	7.95	5.15	-	13.10	16.38
1.00 m	m3	-	1.44	9.68	4.59	-	14.27	17.84
2.00 m	m3	-	1.71	11.50	4.59	-	16.09	20.11
4.00 m	m3	-	2.04	13.75	4.21	-	17.96	22.45

Excavation and earthwork

Alterations and repairs	Unit	Hours C	Hours L	Labour net	Plant net	Material net	Price net	Price with 25%
Generally				£	£	£	£	£
					VAT not included			
Excavation by hand								
Excavate to reduce levels, maximum depth not exceeding								
0.25 m	m3	-	2.76	18.60	-	-	18.60	23.25
1.00 m	m3	-	2.93	19.75	-	-	19.75	24.69
Excavate basements, maximum depth not exceeding								
0.25 m	m3	-	2.76	18.60	-	-	18.60	23.25
1.00 m	m3	-	2.93	19.75	-	-	19.75	24.69
2.00 m	m3	-	3.91	26.35	-	-	26.35	32.94
Excavate pits to receive bases of stanchions etc, maximum depth not exceeding								
0.25 m	m3	-	3.91	26.35	-	-	26.35	32.94
1.00 m	m3	-	4.08	27.50	-	-	27.50	34.38
2.00 m	m3	-	4.66	31.41	-	-	31.41	39.26
Excavate pits less than 1.25 x 1.25 m on plan to receive bases of stanchions etc, maximum depth not exceeding								
0.25 m	m3	-	5.08	34.24	-	-	34.24	42.80
1.00 m	m3	-	5.31	35.79	-	-	35.79	44.74
2.00 m	m3	-	6.32	42.60	-	-	42.60	53.25
Excavate trenches not exceeding 0.30 m width to receive foundations, average depth								
0.25 m	m	-	0.29	1.95	-	-	1.95	2.44
0.50 m	m	-	0.61	4.11	-	-	4.11	5.14
0.75 m	m	-	0.91	6.13	-	-	6.13	7.66
1.00 m	m	-	1.21	8.16	-	-	8.16	10.20
Excavate trenches over 0.30 m width to receive foundations, maximum depth not exceeding								
0.25 m	m3	-	3.62	24.40	-	-	24.40	30.50
1.00 m	m3	-	3.82	25.75	-	-	25.75	32.19
2.00 m	m3	-	4.77	32.15	-	-	32.15	40.19
4.00 m	m3	-	7.61	51.29	-	-	51.29	64.11
Excavate short trenches not exceeding 0.30 m width to receive foundations, average depth								
0.25 m	m	-	0.30	2.02	-	-	2.02	2.52
0.50 m	m	-	0.64	4.31	-	-	4.31	5.39
0.75 m	m	-	0.96	6.47	-	-	6.47	8.09
1.00 m	m	-	1.27	8.56	-	-	8.56	10.70

Excavation and earthwork

Alterations and repairs	Unit	Hours C	Hours L	Labour net	Plant net	Material net	Price net	Price with 25%
Generally				£	£	£	£	£
					VAT not included			

Excavation by hand (*continued*)

Excavate short trenches over 0.30 m in width to receive foundations, maximum depth not exceeding

0.25 m	m3	-	3.80	25.61	-	-	25.61	32.01
1.00 m	m3	-	4.01	27.03	-	-	27.03	33.79
2.00 m	m3	-	5.00	33.70	-	-	33.70	42.13
4.00 m	m3	-	7.99	53.85	-	-	53.85	67.31

Excavate trenches to receive service pipes, cables etc, grade bottom, fill in and compact and remove surplus – earthwork support not included – average depth not exceeding

0.25 m	m	-	0.63	4.25	0.38	-	4.63	5.79
0.50 m	m	-	1.27	8.56	0.38	-	8.94	11.18
0.75 m	m	-	1.84	12.40	0.38	-	12.78	15.98
1.00 m	m	-	2.47	16.65	0.38	-	17.03	21.29

Excavate and fill basement working space, maximum depth not exceeding

0.25 m	m3	-	4.88	32.89	-	-	32.89	41.11
1.00 m	m3	-	5.70	38.42	-	-	38.42	48.02
2.00 m	m3	-	7.25	48.87	-	-	48.87	61.09

Excavate and fill pit working space, maximum depth not exceeding

0.25 m	m3	-	6.30	42.46	-	-	42.46	53.08
1.00 m	m3	-	6.72	45.29	-	-	45.29	56.61
2.00 m	m3	-	8.51	57.36	-	-	57.36	71.70

Excavate and fill trench working space, maximum depth not exceeding

0.25 m	m3	-	5.29	35.65	-	-	35.65	44.56
1.00 m	m3	-	5.92	39.90	-	-	39.90	49.88
2.00 m	m3	-	7.70	51.90	-	-	51.90	64.88

Breaking up by machine - excluding reinstatement

Break up surface concrete, average thickness

100 mm	m2	-	0.24	1.62	0.98	-	2.60	3.25
150 mm	m2	-	0.31	2.09	1.26	-	3.35	4.19
200 mm	m2	-	0.46	3.10	1.87	-	4.97	6.21

Break up reinforced surface concrete, average thickness

100 mm	m2	-	0.34	2.29	1.38	-	3.67	4.59
150 mm	m2	-	0.43	2.90	1.75	-	4.65	5.81
200 mm	m2	-	0.65	4.38	2.64	-	7.02	8.78

Excavation and earthwork

Alterations and repairs	Unit	Hours C	Hours L	Labour net	Plant net	Material net	Price net	Price with 25%
Generally				£	£	£	£	£
					VAT not included			

Breaking up by machine - excluding reinstatement (*continued*)

Alterations and repairs	Unit	Hours C	Hours L	Labour net	Plant net	Material net	Price net	Price with 25%
Break up tarmacadam paving, average thickness								
100 mm	m2	-	0.11	0.74	0.45	-	1.19	1.49
150 mm	m2	-	0.18	1.21	0.73	-	1.94	2.42
200 mm	m2	-	0.24	1.62	0.98	-	2.60	3.25
Break up obstructions in excavations								
concrete	m3	-	5.32	35.86	21.64	-	57.50	71.88
reinforced concrete	m3	-	7.44	50.15	30.26	-	80.41	100.51
brick, block or stonework in lime mortar	m3	-	1.50	10.11	6.10	-	16.21	20.26
brick, block or stonework in cement mortar	m3	-	1.75	11.80	7.12	-	18.92	23.65
Break up surface concrete for service trenches 300 mm wide, average thickness								
100 mm	m	-	0.08	0.52	0.33	-	0.85	1.06
150 mm	m	-	0.09	0.59	0.37	-	0.96	1.20
200 mm	m	-	0.14	0.94	0.57	-	1.51	1.89
Break up reinforced surface concrete for service trenches 300 mm wide, average thickness								
100 mm	m	-	0.10	0.67	0.41	-	1.08	1.35
150 mm	m	-	0.13	0.89	0.53	-	1.42	1.77
200 mm	m	-	0.20	1.35	0.81	-	2.16	2.70
Break up tarmacadam paving for service trenches 300 mm wide, average thickness								
100 mm	m	-	0.03	0.22	0.12	-	0.34	0.42
150 mm	m	-	0.06	0.37	0.24	-	0.61	0.76
200 mm	m	-	0.08	0.52	0.33	-	0.85	1.06
Break up hardcore								
hardcore	m3	-	0.33	2.22	7.54	-	9.76	12.20

Breaking up by machine - including reinstatement

Alterations and repairs	Unit	Hours C	Hours L	Labour net	Plant net	Material net	Price net	Price with 25%
Breaking up surface concrete for service trenches 300 mm wide and reinstate concrete paving and 150 mm hardcore bed, average concrete thickness								
100mm	m	-	6.52	43.93	9.22	70.90	124.05	155.06
150mm	m	-	6.09	41.03	9.26	70.90	121.19	151.49
200mm	m	-	5.15	34.71	9.46	70.90	115.07	143.84

Excavation and earthwork

Alterations and repairs	Unit	Hours C	Hours L	Labour net	Plant net	Material net	Price net	Price with 25%
Generally				£	£	£	£	£
					VAT not included			

Breaking up by machine - including reinstatement *(continued)*

Breaking up reinforced concrete for service trenches 300 mm wide and reinstate concrete paving and 150 mm hardcore bed, average concrete thickness

	Unit	Hours C	Hours L	Labour net	Plant net	Material net	Price net	Price with 25%
100mm	m	-	3.47	23.37	9.22	57.67	90.26	112.83
150mm	m	-	3.48	23.44	9.26	57.67	90.37	112.96
200mm	m	-	2.74	18.46	9.46	57.67	85.59	106.99

Breaking up tarmacadam paving for service trenches 300 mm wide and reinstate concrete paving and 150 mm hardcore bed average tarmacadam thickness

	Unit	Hours C	Hours L	Labour net	Plant net	Material net	Price net	Price with 25%
100 mm	m	-	0.57	3.84	3.07	23.59	30.50	38.13
150 mm	m	-	0.57	3.84	3.19	35.16	42.19	52.74
200 mm	m	-	0.62	4.18	3.35	44.49	52.02	65.03

Breaking up by hand - excluding reinstatement

Break up surface concrete, average thickness

	Unit	Hours C	Hours L	Labour net	Plant net	Material net	Price net	Price with 25%
100 mm	m2	-	1.08	7.28	-	-	7.28	9.10
150 mm	m2	-	1.92	12.94	-	-	12.94	16.18
200 mm	m2	-	3.04	20.49	-	-	20.49	25.61

Break up reinforced concrete, average thickness

	Unit	Hours C	Hours L	Labour net	Plant net	Material net	Price net	Price with 25%
100 mm	m2	-	1.51	10.18	-	-	10.18	12.73
150 mm	m2	-	2.69	18.13	-	-	18.13	22.66
200 mm	m2	-	4.26	28.71	-	-	28.71	35.89

Break up tarmacadam paving, average thickness

	Unit	Hours C	Hours L	Labour net	Plant net	Material net	Price net	Price with 25%
100 mm	m2	-	0.90	6.07	-	-	6.07	7.59
150 mm	m2	-	1.35	9.10	-	-	9.10	11.38
200 mm	m2	-	1.80	12.13	-	-	12.13	15.16

Break up obstructions in excavations

	Unit	Hours C	Hours L	Labour net	Plant net	Material net	Price net	Price with 25%
concrete	m3	-	6.50	43.81	-	-	43.81	54.76
reinforced concrete	m3	-	8.00	53.92	-	-	53.92	67.40
brick, block or stonework in lime mortar	m3	-	3.33	22.44	-	-	22.44	28.05
brick, block or stonework in cement mortar	m3	-	5.00	33.70	-	-	33.70	42.13

Excavation and earthwork

Alterations and repairs	Unit	Hours C	Hours L	Labour net	Plant net	Material net	Price net	Price with 25%
Generally				£	£	£	£	£
					VAT not included			

Breaking up by hand - excluding reinstatement (*continued*)

Break up surface concrete for service trenches 300 mm wide, average thickness

100 mm	m	-	0.32	2.16	-	-	2.16	2.70
150 mm	m	-	0.58	3.91	-	-	3.91	4.89
200 mm	m	-	0.91	6.13	-	-	6.13	7.66

Break up reinforced surface concrete for service trenches 300 mm wide, average thickness

100 mm	m	-	0.45	3.03	-	-	3.03	3.79
150 mm	m	-	0.81	5.46	-	-	5.46	6.83
200 mm	m	-	1.28	8.63	-	-	8.63	10.79

Break up tarmacadam paving for service trenches 300 mm wide, average thickness

100 mm	m	-	0.27	1.82	-	-	1.82	2.27
150 mm	m	-	0.41	2.76	-	-	2.76	3.45
200 mm	m	-	0.54	3.64	-	-	3.64	4.55

Break up hardcore

hardcore	m3	-	4.00	26.96	-	-	26.96	33.70

Breaking up by hand - including reinstatement

Breaking up surface concrete for service trenches 300 mm wide and reinstate concrete paving and 150 mm hardcore bed, average concrete thickness

100mm	m	-	10.43	70.30	1.35	70.90	142.55	178.19
150mm	m	-	10.25	69.08	1.35	70.90	141.33	176.66
200mm	m	-	9.59	64.64	1.35	70.90	136.89	171.11

Breaking up reinforced concrete for service trenches 300 mm wide and reinstate concrete paving and 150 mm hardcore bed, average concrete thickness

100mm	m	-	10.56	71.18	1.35	70.90	143.43	179.29
150mm	m	-	10.48	70.63	1.35	70.90	142.88	178.60
200mm	m	-	9.96	67.13	1.35	70.90	139.38	174.22

Alterations and repairs	Unit	Hours C	Hours L	Labour net	Plant net	Material net	Price net	Price with 25%
Generally				£	£	£	£	£
					VAT not included			

Breaking up by hand - including reinstatement (*continued*)

Breaking up tarmacadam paving for service trenches 300 mm wide and reinstate concrete paving and 150 mm hardcore bed, average tarmacadam thickness

	Unit	Hours C	Hours L	Labour net	Plant net	Material net	Price net	Price with 25%
100 mm	m	-	1.10	7.41	0.53	23.59	31.53	39.41
150 mm	m	-	1.40	9.44	0.69	34.13	44.26	55.33
200 mm	m	-	1.70	11.46	0.85	44.49	56.80	71.00

Earthwork support

Earthwork support to sides of trenches not exceeding 2.00 m between opposing faces, in firm soil, maximum depth not exceeding

	Unit	Hours C	Hours L	Labour net	Plant net	Material net	Price net	Price with 25%
1.00 m	m2	-	0.12	0.81	-	0.87	1.68	2.10
2.00 m	m2	-	0.14	0.94	-	0.87	1.81	2.26
4.00 m	m2	-	0.21	1.42	-	0.87	2.29	2.86

Earthwork support to sides of pits not exceeding 2.00 m between opposing faces, in firm soil, maximum depth not exceeding

1.00 m	m2	-	0.12	0.81	-	0.87	1.68	2.10
2.00 m	m2	-	0.14	0.94	-	0.87	1.81	2.26
4.00 m	m2	-	0.21	1.42	-	0.87	2.29	2.86

Earthwork support to sides of trenches not exceeding 2.00 m between opposing faces, in moderately firm soil, maximum depth not exceeding,

1.00 m	m2	-	0.48	3.24	-	2.32	5.56	6.95
2.00 m	m2	-	0.63	4.25	-	2.32	6.57	8.21
4.00 m	m2	-	0.77	5.19	-	2.32	7.51	9.39

Earthwork support to sides of pits not exceeding 2.00 m between opposing faces in moderately firm soil, maximum depth not exceeding,

1.00 m	m2	-	0.48	3.24	-	2.32	5.56	6.95
2.00 m	m2	-	0.63	4.25	-	2.32	6.57	8.21
4.00 m	m2	-	0.77	5.19	-	2.32	7.51	9.39

Earthwork support to sides of trenches not exceeding 2.00 m between opposing faces in loose soil, maximum depth not exceeding,

1.00 m	m2	-	0.97	6.54	-	3.19	9.73	12.16
2.00 m	m2	-	1.13	7.62	-	3.19	10.81	13.51
4.00 m	m2	-	1.56	10.51	-	3.19	13.70	17.13

Excavation and earthwork

Alterations and repairs	Unit	Hours C	Hours L	Labour net	Plant net	Material net	Price net	Price with 25%
Generally				£	£	£	£	£
					VAT not included			

Earthwork support (*continued*)

Earthwork support to sides of pits not exceeding 2.00 m between opposing faces, in loose soil, maximum depth not exceeding,

	Unit	Hours C	Hours L	Labour net	Plant net	Material net	Price net	Price with 25%
1.00 m	m2	-	0.97	6.54	-	3.19	9.73	12.16
2.00 m	m2	-	1.13	7.62	-	3.19	10.81	13.51
4.00 m	m2	-	1.56	10.51	-	3.19	13.70	17.13

Disposal of excavated material

	Unit	Hours C	Hours L	Labour net	Plant net	Material net	Price net	Price with 25%
Excavated material moved by machine and deposited on site in spoil heaps average 100 m from excavation	m3	-	-	-	2.03	-	2.03	2.54
Add or deduct for every 25 m difference in distance	m3	-	-	-	0.51	-	0.51	0.64
Excavated material moved by hand and deposited on site in spoil heaps average 50 m from excavation	m3	-	1.65	11.12	-	-	11.12	13.90
Add or deduct for every 10 m difference in distance	m3	-	0.33	2.22	-	-	2.22	2.77
Excavated material moved by machine and spread on site average 100 m from excavation	m3	-	1.49	10.04	1.52	-	11.56	14.45
Add or deduct for every 25 m difference in distance	m3	-	-	-	0.51	-	0.51	0.64
Excavated material moved by hand and spread on site average 50 m from excavation	m3	-	2.75	18.54	-	-	18.54	23.18
Add or deduct for every 10 m difference in distance	m3	-	0.33	2.22	-	-	2.22	2.77
Excavated material removed from site to tip average 15 km from site								
loaded by machine direct from excavations	m3	-	-	-	14.00	-	14.00	17.50
loaded by machine from spoil heaps	m3	-	-	-	14.91	-	14.91	18.64
loaded by hand from spoil heaps	m3	-	1.76	11.86	18.53	-	30.39	37.99
Add or deduct for every 1 km difference in distance	m3	-	-	-	0.20	-	0.20	0.25

Excavation and earthwork

Alterations and repairs	Unit	Hours C	Hours L	Labour net	Plant net	Material net	Price net	Price with 25%
Generally				£	£	£	£	£
					VAT not included			

Filling

Excavated material filling to excavations deposited and compacted in 225 mm layers	m3	-	1.38	9.30	1.75	-	11.05	13.81
Excavated material filling to make up levels, deposited and compacted in 225 mm layers								
over 250 mm thick	m3	-	1.76	11.86	2.80	-	14.66	18.32
average 100 mm thick	m2	-	0.18	1.19	0.23	-	1.42	1.77
average 150 mm thick	m2	-	0.26	1.78	0.33	-	2.11	2.64
average 225 mm thick	m2	-	0.40	2.67	0.51	-	3.18	3.98
63 mm layer or binding gravel (hogging) rolled and consolidated	m2	-	0.28	1.85	2.92	1.10	5.87	7.34
Hardcore filling to make up levels, deposited and compacted in 150 mm layers								
over 250 mm thick	m3	-	2.20	14.83	2.80	13.47	31.10	38.88
average 75 mm thick	m2	-	0.28	1.85	0.33	1.01	3.19	3.99
average 100 mm thick	m2	-	0.33	2.22	0.43	1.35	4.00	5.00
average 150 mm thick	m2	-	0.44	2.97	1.35	2.02	6.34	7.92
average 225 mm thick	m2	-	0.60	4.08	0.79	3.10	7.97	9.96

Surface treatments

Level and compact surface of filling or bottom of excavation	m2	-	0.06	0.37	0.15	-	0.52	0.65
Grade and compact surface of filling or bottom of excavation								
to falls	m2	-	0.11	0.74	0.28	-	1.02	1.27
to cross-falls	m2	-	0.22	1.48	0.56	-	2.04	2.55
50 mm blinding on hardcore								
sand	m2	-	0.08	0.52	-	0.99	1.51	1.89
ashes	m2	-	0.09	0.59	-	0.52	1.11	1.39

Concrete work

Preamble

"Labour net" figures include allowances for all costs incidental to the employment of labour.

"Plant net" figures include for all costs of plant including drivers and operators where applicable.

"Materials net" figures include for all costs of materials including an allowance for waste except where specifically stated.

"Price net" figures are the totals of the "Labour net", "Plant net" where applicable, and "Materials net" figures. Prices are for a builder employing his own labour; according to the amount and nature of the work involved, it may well be possible to secure more advantageous prices from specialist sub-contractors.

Figures for site mixed concrete are based on the use of a hired 7/5 mixer.

Figures for formwork are based on the assumptions that timber is used and that each use of material requires the full labour content; if the work is repetitive, permitting the re-use of made up sections, some reduction of the figures could be made.

Prices do not include any allowance for scaffolding, ladders or other plant necessary to reach the work. The "Preliminaries" section includes prices for scaffolding which must be considered and allowance included to suit the particular circumstances of a tender.

Specialist prices

"Price with 25%" figures are all-in guide prices and include 25% for the builder's overheads, profit, unloading materials and general attendance (to include free use of standing scaffolding and hoists, temporary lighting and water and clearing away rubbish).

The amount of attendance required varies between the various trades and also with the circumstances of specific jobs; the percentage addition must always be considered and adjusted as necessary to suit the terms and conditions of the quotation being used.

Quantities and delivery distances are usually the most significant of the many factors which influence prices and it must be emphasised that quotations should always be obtained when preparing a tender.

Piling

Specialist prices for piling have been included in the "New work" section.

Concrete work

	Unit	Price
Basic prices for materials		**£**
Aggregates		
40 mm	m3	19.75
	tonne	13.16
20 mm	m3	19.91
	tonne	13.27
10 mm	m3	20.07
	tonne	13.38
Ready mixed concrete		
1:3:6 - 40 mm aggregate	m3	53.00
1:2:4 - 20 mm aggregate	m3	56.18
1:1½:3 - 10 mm aggregate	m3	60.42
Bituminous emulsion waterproofing liquid	5 litre	11.16
Cement		
Portland	tonne	89.04
rapid hardening	tonne	150.33
sulphate resisting	tonne	92.66
Formwork		
softwood	m3	244.80
25 mm sawn boarding	m2	7.65
25 mm wrought boarding	m2	13.77
18 mm plywood	m2	10.20
3.2 mm tempered hardboard	m2	1.50
Liquid surface hardener	5 litre	8.95
Polythene building sheet		
medium	m2	0.14
heavy	m2	0.42
Retarder liquid	5 litres	16.60
Sand	m3	17.96
	tonne	11.22
Self-adhesive damp-proof membrane	m2	6.50
Sub-soil building paper	m2	0.62
Waterproofer		
liquid	5 litre	16.79
powder	5 kg	12.42

Prices actually to be paid for materials must be checked against the above basic prices and adjustments made as necessary.

Concrete work

Alterations and repairs	Unit	Hours C	Hours L	Labour net	Plant net	Material net	Price net	Price with 25%
In-situ concrete				£	£	£	£	£
					VAT not included			
Ready mixed								
Concrete 1:3:6 - 40 mm aggregate								
Foundations in trenches, thickness								
150 - 300 mm	m3	-	2.09	14.09	-	55.65	69.74	87.17
over 300 mm	m3	-	1.87	12.60	-	55.65	68.25	85.31
Isolated foundation bases to columns and piers; thickness								
150 - 300 mm	m3	-	2.37	15.97	-	55.65	71.62	89.53
over 300 mm	m3	-	2.20	14.83	-	55.65	70.48	88.10
Beds; thickness								
not exceeding 100 mm	m3	-	2.62	17.66	-	55.65	73.31	91.64
100 - 150 mm	m3	-	2.62	17.66	-	55.65	73.31	91.64
150 - 300 mm	m3	-	1.83	12.33	-	55.65	67.98	84.97
Concrete 1:2:4 - 20 mm aggregate								
Foundations in trenches; thickness								
150 - 300 mm	m3	-	2.09	14.09	-	58.99	73.08	91.35
over 300 mm	m3	-	1.87	12.60	-	58.99	71.59	89.49
Isolated foundation bases to columns and piers; thickness								
150 - 300 mm	m3	-	2.37	15.97	-	58.99	74.96	93.70
over 300 mm	m3	-	2.20	14.83	-	58.99	73.82	92.28
Foundations and haunching to kerbs; sectional area								
not exceeding 0.03 m2	m3	-	5.17	34.85	-	58.99	93.84	117.30
0.03 - 0.10 m2	m3	-	4.73	31.88	-	58.99	90.87	113.59
Ground beams; sectional area								
0.10 - 0.25 m2	m3	-	2.31	15.57	-	58.99	74.56	93.20
Pile caps	m3	-	2.20	14.83	-	58.99	73.82	92.28
Beds; thickness								
not exceeding 100 mm	m3	-	2.62	17.66	-	58.99	76.65	95.81
100 - 150 mm	m3	-	2.62	17.66	-	58.99	76.65	95.81
150 - 300 mm	m3	-	1.83	12.33	-	58.99	71.32	89.15
Suspended slabs and attached beams; thickness								
100 - 150 mm	m3	-	4.97	33.50	-	58.99	92.49	115.61
150 - 300 mm	m3	-	4.13	27.84	-	58.99	86.83	108.54
Upstands and kerbs; sectional area not exceeding 0.03 m2	m3	-	4.95	33.36	-	58.99	92.35	115.44

Concrete work

Alterations and repairs	Unit	Hours C	Hours L	Labour net	Plant net	Material net	Price net	Price with 25%
In-situ concrete				£	£	£	£	£
					VAT not included			

Ready mixed (*continued*)

Concrete 1:2:4 - 20mm aggregate
(*continued*)

	Unit	Hours C	Hours L	Labour net	Plant net	Material net	Price net	Price with 25%
Walls thickness								
100 - 150 mm	m3	-	4.90	33.03	-	58.99	92.02	115.03
150 - 300 mm	m3	-	4.73	31.88	-	58.99	90.87	113.59
Isolated beams and casings to isolated steel beams; sectional area								
not exceeding 0.03 m2	m3	-	5.30	35.72	-	58.99	94.71	118.39
0.03 - 0.10 m2	m3	-	4.95	33.36	-	58.99	92.35	115.44
Isolated columns and isolated casings to steel column; sectional area								
not exceeding 0.03 m2	m3	-	7.26	48.93	-	58.99	107.92	134.90
0.03 - 0.10 m2	m3	-	6.56	44.21	-	58.99	103.20	129.00
Steps; staircases and landings	m3	-	7.98	53.79	-	58.99	112.78	140.97
Filling to hollow walls; thickness not exceeding 100 mm	m3	-	4.79	32.28	-	58.99	91.27	114.09
Concrete 1:1½:3 - 10 mm aggregate								
Ground beams; sectional area 0.10 - 0.25 m2	m3	-	2.31	15.57	-	56.76	72.33	90.41
Pile caps	m3	-	2.20	14.83	-	56.76	71.59	89.49
Suspended slabs and attached beams; thickness								
100 - 150 mm	m3	-	4.97	33.50	-	56.76	90.26	112.83
150 - 300 mm	m3	-	4.13	27.80	-	56.76	84.56	105.70
Upstands and kerbs; sectional area not exceeding 0.03 m2	m3	-	4.95	33.36	-	56.76	90.12	112.65
Walls; thickness								
100 - 150 mm	m3	-	4.90	33.03	-	56.76	89.79	112.24
150 - 300 mm	m3	-	4.73	31.88	-	56.76	88.64	110.80
Isolated beams and casings to isolated steel beams; sectional area								
not exceeding 0.03 m2	m3	-	5.30	35.72	-	56.76	92.48	115.60
0.03 - 0.10 m2	m3	-	4.95	33.36	-	56.76	90.12	112.65
Isolated columns and isolated casings to steel columns; sectional area								
not exceeding 0.03 m2	m3	-	7.26	48.93	-	56.76	105.69	132.11
0.03 - 0.10 m2	m3	-	6.56	44.21	-	56.76	100.97	126.21

Concrete work

Alterations and repairs	Unit	Hours C	Hours L	Labour net	Plant net	Material net	Price net	Price with 25%
In-situ concrete				£	£	£	£	£
					VAT not included			
Ready mixed *(continued)*								
Concrete 1:1½:3 - 10 mm aggregate *(continued)*								
Steps; staircases and landings	m3	-	7.98	53.79	-	56.76	110.55	138.19
Site mixed								
Concrete 1:3:6 - 40 mm aggregate								
Foundations in trenches; thickness								
150 - 300 mm	m3	-	3.47	23.39	-	68.88	92.27	115.34
over 300 mm	m3	-	3.25	21.91	-	68.88	90.79	113.49
Isolated foundation bases to columns and piers, thickness								
150 - 300 mm	m3	-	3.74	25.21	-	68.88	94.09	117.61
over 300 mm	m3	-	3.58	24.13	-	68.88	93.01	116.26
Beds; thickness								
not exceeding 100 mm	m3	-	5.67	38.22	-	68.88	107.10	133.88
100 - 150 mm	m3	-	5.23	35.25	-	68.88	104.13	130.16
150 - 300 mm	m3	-	4.24	28.58	-	68.88	97.46	121.83
Concrete 1:2:4 - 20 mm aggregate								
Foundations in trenches; thickness								
150 - 300 mm	m3	-	4.07	27.43	-	73.19	100.62	125.78
over 300 mm	m3	-	3.85	25.95	-	73.19	99.14	123.93
Isolated foundation bases to columns and piers; thickness								
150 - 300 mm	m3	-	4.35	29.32	-	73.19	102.51	128.14
over 300 mm	m3	-	4.18	28.17	-	73.19	101.36	126.70
Foundations and haunching to kerbs; sectional area								
not exceeding 0.03 m2	m3	-	7.15	48.19	-	73.19	121.38	151.72
0.03 - 0.10 m2	m3	-	6.71	45.23	-	73.19	118.42	148.03
Ground beams; sectional area								
0.10 - 0.25 m2	m3	-	4.29	28.91	-	75.47	104.38	130.47
Pile caps	m3	-	4.18	28.17	-	73.19	101.36	126.70
Beds; thickness								
not exceeding 100 mm	m3	-	6.27	42.26	-	73.19	115.45	144.31
100 - 150 mm	m3	-	5.83	39.29	-	73.19	112.48	140.60
150 - 300 mm	m3	-	4.84	32.62	-	73.19	105.81	132.26

Concrete work

Alterations and repairs	Unit	Hours C	Hours L	Labour net	Plant net	Material net	Price net	Price with 25%
In-situ concrete				£	£	£	£	£
					VAT not included			

Site mixed (*continued*)

Concrete 1:2:4 - 20 mm aggregate (*continued*)

	Unit	Hours C	Hours L	Labour net	Plant net	Material net	Price net	Price with 25%
Suspended slabs and attached beams; thickness								
100 - 150 mm	m3	-	11.22	75.62	-	73.19	148.81	186.01
150 - 300 mm	m3	-	10.23	68.95	-	73.19	142.14	177.68
Upstands and kerbs; sectional area not exceeding 0.03 m2	m3	-	12.54	84.52	-	73.19	157.71	197.14
Walls; thickness								
100 - 150 mm	m3	-	11.77	79.33	-	73.19	152.52	190.65
150 - 300 mm	m3	-	11.44	77.11	-	73.19	150.30	187.88
Isolated beams and casings to isolated steel beams; sectional area								
not exceeding 0.03 m2	m3	-	13.20	88.97	-	73.19	162.16	202.70
0.03 - 0.10 m2	m3	-	11.88	80.07	-	73.19	153.26	191.57
Isolated columns and isolated casings to steel columns; sectional area								
not exceeding 0.03 m2	m3	-	13.86	93.42	-	73.19	166.61	208.26
0.03 - 0.10 m2	m3	-	12.32	83.04	-	73.19	156.23	195.29
Steps; staircases and landings	m3	-	11.55	77.85	-	73.19	151.04	188.80
Filling to hollow walls; thickness not exceeding 100 mm	m3	-	6.77	45.63	-	73.19	118.82	148.53
Concrete 1:1½:3 - 10 mm aggregate								
Ground beams; sectional area 0.10 - 0.25 m2	m3	-	4.68	31.54	-	78.55	110.09	137.61
Pile caps	m3	-	4.57	30.80	-	78.55	109.35	136.69
Suspend slabs and attached beams; thickness								
100 - 150 mm	m3	-	11.61	78.25	-	78.55	156.80	196.00
150 - 300 mm	m3	-	10.62	71.58	-	78.55	150.13	187.66
Upstands and kerbs, sectional area not exceeding 0.03 m2	m3	-	12.93	87.15	-	78.55	165.70	207.13
Walls; thickness								
100 - 150 mm	m3	-	12.16	81.96	-	78.55	160.51	200.64
150 - 300 mm	m3	-	11.83	79.73	-	78.55	158.28	197.85

Concrete work

Alterations and repairs	Unit	Hours C	Hours L	Labour net	Plant net	Material net	Price net	Price with 25%
In-situ concrete				£	£	£	£	£
					VAT not included			
Site mixed (continued)								
Concrete 1:1½:3 - 10 mm aggregate *(continued)*								
Isolated beams and casings to isolated steel beams; sectional area								
not exceeding 0.03 m2	m3	-	13.59	91.60	-	78.55	170.15	212.69
0.03 - 0.10 m2	m3	-	12.27	82.70	-	78.55	161.25	201.56
Isolated columns and isolated casings to steel columns; sectional area								
not exceeding 0.03 m2	m3	-	14.25	96.05	-	78.55	174.60	218.25
0.03 - 0.10 m2	m3	-	12.71	85.67	-	78.55	164.22	205.28
Steps; staircases and landings	m3	-	11.94	80.48	-	78.55	159.03	198.79
No-fines concrete 1:8 - 10 mm aggregate								
Walls; thickness 150 - 300 mm	m3	-	10.62	71.58	-	85.34	156.92	196.15

Concrete work

Alterations and repairs	Unit	Hours C	Hours L	Labour net	Plant net	Material net	Price net	Price with 25%
Generally				£	£	£	£	£
					VAT not included			
Joints								
Impregnated fibreboard expansion joints in concrete including formwork (four uses)								
12 mm in 100 mm slab	m	0.28	-	2.51	-	2.15	4.66	5.83
12 mm in 125 mm slab	m	0.28	-	2.51	-	2.32	4.83	6.04
12 mm in 150 mm slab	m	0.28	-	2.51	-	2.46	4.97	6.21
Prime and seal top of expansion joint with bituminous compound	m	-	0.20	1.35	-	0.01	1.36	1.70
Labours on concrete								
Extra over beds etc. for								
levelling surface	m2	-	0.03	0.21	-	-	0.21	0.27
tamping surface as paving	m2	-	0.03	0.21	-	-	0.21	0.27
trowelling	m2	-	0.36	2.43	-	-	2.43	3.04
grading to falls	m2	-	0.40	2.70	-	-	2.70	3.38
grading to cross-falls	m2	-	0.06	0.41	-	-	0.41	0.52
laying in bays including temporary fillets and jointing and pointing	m2	0.08	0.08	1.21	-	0.58	1.79	2.24
Mortices								
Cut mortices 100 mm deep in concrete for rag bolts and grout in	m2	-	0.33	2.22	1.14	0.01	3.37	4.21
Sundries								
Grout under steel stanchion bases including grouting in four anchor bolts; base size								
300 x 300 mm	each	0.55	0.55	8.64	-	0.59	9.23	11.54
450 x 300 mm	each	0.66	0.72	10.76	-	0.79	11.55	14.44
450 x 450 mm	each	0.83	0.94	13.78	-	0.99	14.77	18.46
Fix only anchor bolts including template and								
temporary boxing	each	0.88	-	7.88	-	2.96	10.84	13.55
expanded metal boxing	each	0.42	-	3.76	-	2.56	6.32	7.90
Form 150 x 150 mm holes for pipes through 150 mm concrete floor and make good	each	0.66	0.36	8.34	-	0.59	8.93	11.16
Abrasive grain applied at 0.75 kg per m2 and trowelled in to surface	m2	0.11	0.11	1.73	-	1.59	3.32	4.15
Three coats of liquid surface hardener to concrete floors	m2	-	0.11	0.74	-	1.41	2.15	2.69

Concrete work

Alterations and repairs	Unit	Hours C	Hours L	Labour net	Plant net	Material net	Price net	Price with 25%
Generally				£	£	£	£	£
					VAT not included			
Sundries (continued)								
Liquid bituminous membrane to concrete surfaces and blinded with sand								
two coats	m2	-	0.44	2.97	-	3.98	6.95	8.69
three coats	m2	-	0.66	4.45	-	5.79	10.24	12.80
Building paper underlay lapped 150 mm at joints	m2	-	0.03	0.22	-	0.68	0.90	1.13
Polythene sheet underlay lapped 150 mm at joints								
medium weight	m2	-	0.03	0.22	-	0.15	0.37	0.46
heavy weight	m2	-	0.04	0.30	-	0.46	0.76	0.95
Self-adhesive damp-proof membrane on concrete lapped 50 mm at joints	m2	0.17	0.17	2.67	-	7.15	9.82	12.28

Concrete work

Alterations and repairs	Unit	Hours C	Hours L	Labour net	Plant net	Material net	Price net	Price with 25%
Reinforcement				£	£	£	£	£
					VAT not included			

Mild steel

Plain round mild steel bar reinforcement to BS 4449 delivered cut, bent and bundled and fixed including tying wire, distance blocks and ordinary spacers

	Unit	Hours C	Hours L	Labour net	Plant net	Material net	Price net	Price with 25%
6 mm	tonne	66.00	2.00	604.84	-	499.38	1104.22	1380.28
6 mm	m	0.01	-	0.13	-	0.15	0.28	0.35
8 mm	tonne	55.00	2.00	506.28	-	457.87	964.15	1205.19
8 mm	m	0.02	-	0.20	-	0.24	0.44	0.55
10 mm	tonne	46.00	2.00	425.64	-	422.51	848.15	1060.19
10 mm	m	0.03	-	0.25	-	0.32	0.57	0.71
12 mm	tonne	40.00	2.00	371.88	-	395.33	767.21	959.01
12 mm	m	0.04	-	0.31	-	0.38	0.69	0.86
16 mm	tonne	33.00	2.00	309.16	-	366.13	675.29	844.11
16 mm	m	0.05	-	0.47	-	0.57	1.04	1.30
20 mm	tonne	28.00	2.00	264.36	-	357.93	622.29	777.86
20 mm	m	0.07	-	0.59	-	0.89	1.48	1.85
25 mm	tonne	22.00	2.00	210.60	-	357.93	568.53	710.66
25 mm	m	0.09	-	0.76	-	1.30	2.06	2.58

High yield

Deformed high yield steel bar reinforcement to BS 4449 delivered cut, bent and bundled and fixed including tying wire, distance blocks and ordinary spacers

	Unit	Hours C	Hours L	Labour net	Plant net	Material net	Price net	Price with 25%
8 mm	tonne	53.00	2.20	489.71	-	432.25	921.96	1152.45
8 mm	m	0.02	-	0.20	-	0.15	0.35	0.44
10 mm	tonne	44.20	2.20	410.86	-	395.14	806.00	1007.50
10 mm	m	0.03	-	0.25	-	0.32	0.57	0.71
12 mm	tonne	37.60	2.20	351.73	-	378.74	730.47	913.09
12 mm	m	0.04	-	0.31	-	0.38	0.69	0.86
16 mm	tonne	31.00	2.20	292.59	-	354.14	646.73	808.41
16 mm	m	0.05	-	0.47	-	0.57	1.04	1.30
20 mm	tonne	25.50	2.20	243.31	-	362.03	605.34	756.67
20 mm	m	0.07	-	0.59	-	0.79	1.38	1.73
25 mm	tonne	20.00	2.20	194.03	-	362.03	556.06	695.08
25 mm	m	0.09	-	0.76	-	1.09	1.85	2.31

Links, stirrups and binders

Mild steel bar links, stirrups, binders and special spacers delivered cut, bent and bundled and fixed including tying wire

	Unit	Hours C	Hours L	Labour net	Plant net	Material net	Price net	Price with 25%
6 mm	tonne	80.00	2.20	731.63	-	676.30	1407.93	1759.91
6 mm	m	0.02	-	0.18	-	0.12	0.30	0.38
8 mm	tonne	70.00	2.20	642.03	-	614.28	1256.31	1570.39
8 mm	m	0.03	-	0.28	-	0.20	0.48	0.60

Concrete work

Alterations and repairs	Unit	Hours C	Hours L	Labour net	Plant net	Material net	Price net	Price with 25%
Reinforcement				£	£	£	£	£
					VAT not included			
Fabric								
Fabric reinforcement to BS 4483 in slabs including tying wire and distance blocks/chairs and allowance for 200 mm laps								
A98 - 1.54 kg/m2	m2	0.07	-	0.59	-	0.91	1.50	1.88
A142 - 2.22 kg/m2	m2	0.08	-	0.69	-	1.06	1.75	2.19
A193 - 3.02 kg/m2	m2	0.09	-	0.79	-	1.34	2.13	2.66
A252 - 3.95 kg/m2	m2	0.12	-	1.08	-	1.82	2.90	3.63
A393 - 6.16 kg/m2	m2	0.15	-	1.34	-	2.68	4.02	5.03
B283 - 3.73 kg/m2	m2	0.11	-	0.99	-	1.56	2.55	3.19
B385 - 4.53 kg/m2	m2	0.13	-	1.16	-	2.11	3.27	4.09
B503 - 5.93 kg/m2	m2	0.15	-	1.34	-	2.76	4.10	5.13
B785 - 8.14 kg/m2	m2	0.18	-	1.61	-	3.64	5.25	6.56
C283 - 2.61 kg/m2	m2	0.09	-	0.79	-	1.56	2.35	2.94
C385 - 3.41 kg/m2	m2	0.10	-	0.89	-	1.64	2.53	3.16
C503 - 4.34 kg/m2	m2	0.12	-	1.08	-	2.05	3.13	3.91
Fabric reinforcement to BS 4483 in casings to steel columns and beams including bending, tying wire and distance blocks and allowance for 200 mm laps								
D49 - 0.77 kg/m2	m2	0.17	-	1.48	-	1.53	3.01	3.76
D98 - 1.54 kg/m2	m2	0.22	-	1.97	-	2.20	4.17	5.21
Self Centering								
Self-centering combined reinforcement and formwork including temporary supports and allowance for laps but excluding bar reinforcement								
21 mm rib x 0.575 mm	m2	1.00	-	8.96	1.00	14.35	24.31	30.39
21 mm rib x 0.750 mm	m2	1.05	-	9.41	1.00	17.43	27.84	34.80

Concrete work

Alterations and repairs	Unit	Hours C	Hours L	Labour net	Plant net	Material net	Price net	Price with 25%
Generally				£	£	£	£	£
					VAT not included			

Formwork

Figures are based on the use of sawn softwood unless otherwise described.

Formwork to edges and faces of foundations; ground beams and beds (four uses)								
over 1.00 m high	m2	1.79	-	16.04	-	8.31	24.35	30.44
not exceeding 250 mm high	m	0.48	-	4.30	-	2.22	6.52	8.15
250 - 500 mm high	m	0.93	-	8.33	-	4.12	12.45	15.56
500 mm - 1.00 m high	m	1.60	-	14.34	-	8.26	22.60	28.25
Formwork to horizontal soffits of slabs (six uses)	m2	2.92	0.17	27.23	1.00	13.71	41.94	52.42
Formwork to sloping soffits of staircases over 15 deg from horizontal (six uses)	m2	3.63	0.20	33.85	1.00	12.24	47.09	58.86
Formwork to attached beams (four uses)								
400 mm girth	m	1.65	-	14.78	-	4.72	19.50	24.38
600 mm girth	m	1.93	-	17.29	-	5.31	22.60	28.25
Formwork to soffits of projecting eaves and edges of slabs (six uses)								
250 mm girth	m	0.99	0.06	9.24	-	1.36	10.60	13.25
325 mm girth	m	1.21	0.07	11.28	-	1.76	13.04	16.30
Formwork not exceeding 250 mm high to both sides of kickers to walls (four uses)	m	1.65	-	14.78	-	7.83	22.61	28.26
Formwork to edges of slabs (six uses)								
not exceeding 250 mm deep	m	0.99	0.06	9.24	-	1.36	10.60	13.25
250 - 500 mm deep	m	1.82	0.11	17.05	-	2.71	19.76	24.70
Formwork to risers of staircases not exceeding 250 mm deep (six uses)	m	0.55	0.03	5.15	-	1.36	6.51	8.14
Formwork to edges of staircase flights (three uses); maximum width								
200 mm	m	0.94	-	8.42	-	1.67	10.09	12.61
300 mm	m	1.38	-	12.36	-	2.35	14.71	18.39
Formwork to wall faces (six uses)								
vertical	m2	3.19	0.18	29.77	-	7.80	37.57	46.96
battering	m2	3.52	0.22	33.02	-	9.28	42.30	52.88
Formwork to pilasters or attached columns (four uses)								
400 mm girth	m	1.65	-	14.78	-	2.55	17.33	21.66
500 mm girth	m	1.76	-	15.77	-	3.53	19.30	24.13
600 mm girth	m	1.87	-	16.76	-	4.12	20.88	26.10

Concrete work

Alterations and repairs	Unit	Hours C	Hours L	Labour net	Plant net	Material net	Price net	Price with 25%
Generally				£	£	£	£	£
					VAT not included			

Formwork (*continued*)

Alterations and repairs	Unit	Hours C	Hours L	Labour net	Plant net	Material net	Price net	Price with 25%
Formwork to isolated beams (four uses)								
400 mm girth	m	1.65	-	14.78	-	3.53	18.31	22.89
500 mm girth	m	1.82	-	16.31	-	4.42	20.73	25.91
600 mm girth	m	1.93	-	17.29	-	5.31	22.60	28.25
700 mm girth	m	2.04	-	18.28	-	6.21	24.49	30.61
800 mm girth	m	2.20	-	19.71	-	6.25	25.96	32.45
Formwork to isolated columns (four uses)								
600 mm girth	m	1.85	-	16.58	-	5.31	21.89	27.36
750 mm girth	m	2.04	-	18.28	-	6.65	24.93	31.16
900 mm girth	m	2.20	-	19.71	-	7.97	27.68	34.60
1050 mm girth	m	2.37	-	21.24	-	9.30	30.54	38.17
1200 mm girth	m	2.59	-	23.21	-	10.62	33.83	42.29
Cutting formwork								
raking	m	0.17	-	1.48	-	0.59	2.07	2.59
curved	m	0.40	-	3.58	-	0.59	4.17	5.21
Extra over formwork for								
wrought face timber (four uses)	m2	0.45	-	4.03	-	0.83	4.86	6.08
hardboard lining (two uses)	m2	0.40	-	3.58	-	0.89	4.47	5.59
Saving on formwork by using shuttering quality plywood (four uses)	m2	-	-	-	-	1.35	1.35	1.69
Retarder liquid applied to formwork and brushing surface of concrete								
as key for plastering	m2	-	0.33	2.22	-	3.65	5.87	7.34
to provide exposed aggregate finish	m2	-	1.21	8.16	-	4.57	12.73	15.91
Precast concrete								
300 x 75 mm weathered and throated copings bedded in gauged mortar and pointed								
straight	m	0.25	0.25	3.93	-	12.39	16.32	20.40
Extra for								
fair ends	each	0.10	0.10	1.57	-	2.54	4.11	5.14
angles	each	0.10	0.10	1.57	-	9.18	10.75	13.44
intersections	each	0.10	0.10	1.57	-	12.24	13.81	17.26
356 x 75 mm weathered and throated copings bedded in gauged mortar and pointed								
straight	m	0.39	0.39	6.12	-	15.28	21.40	26.75

Concrete work

Alterations and repairs	Unit	Hours C	Hours L	Labour net	Plant net	Material net	Price net	Price with 25%
Generally				£	£	£	£	£
					VAT not included			
Precast concrete (continued)								
Extra for								
fair ends	each	0.10	0.10	1.57	-	3.04	4.61	5.76
angles	each	0.10	0.10	1.57	-	11.82	13.39	16.74
intersections	each	0.10	0.10	1.57	-	15.76	17.33	21.66
450 x 100 mm weathered and throated copings bedded in gauged mortar and pointed								
straight	m	0.50	0.50	7.85	-	20.25	28.10	35.13
Extra for								
fair ends	each	0.10	0.10	1.57	-	3.91	5.48	6.85
angles	each	0.10	0.10	1.57	-	12.02	13.59	16.99
intersections	each	0.10	0.10	1.57	-	16.02	17.59	21.99
Weathered and throated pier caps bedded in gauged mortar								
406 x 406 x 75 mm	each	0.44	0.44	6.91	-	9.95	16.86	21.07
450 x 450 x 75 mm	each	0.44	0.44	6.91	-	12.56	19.47	24.34
530 x 530 x 75 mm	each	0.60	0.60	9.42	-	16.39	25.81	32.26
Weathered; throated and grooved sills with stooled ends bedded in gauged mortar and pointed								
150 x 67 x 900 mm	each	0.22	0.22	3.45	-	13.41	16.86	21.07
150 x 67 x 1200 mm	each	0.27	0.27	4.24	-	17.75	21.99	27.49
150 x 67 x 1500 mm	each	0.36	0.36	5.66	-	22.12	27.78	34.73
203 x 67 x 900 mm	each	0.25	0.25	3.93	-	16.23	20.16	25.20
203 x 67 x 1200 mm	each	0.35	0.35	5.50	-	21.28	26.78	33.48
203 x 67 x 1500 mm	each	0.42	0.42	6.59	-	26.71	33.30	41.63
Padstones bedded in gauged mortar								
215 x 225 x 150 mm	each	0.39	0.39	6.12	-	4.96	11.08	13.85
328 x 225 x 150 mm	each	0.50	0.50	7.85	-	6.36	14.21	17.76
Reinforced lintels bedded in gauged mortar								
75 x 150 x 900 mm	each	0.25	0.25	3.93	-	6.90	10.83	13.54
100 x 150 x 900 mm	each	0.25	0.25	3.93	-	8.14	12.07	15.09
100 x 150 x 1200 mm	each	0.31	0.31	4.87	-	10.22	15.09	18.86
215 x 150 x 900 mm	each	0.29	0.29	4.55	-	13.28	17.83	22.29
215 x 150 x 1200 mm	each	0.39	0.39	6.12	-	17.45	23.57	29.46
100 x 225 x 900 mm	each	0.30	0.30	4.71	-	10.70	15.41	19.26
100 x 225 x 1200 mm	each	0.39	0.39	6.12	-	13.60	19.72	24.65
100 x 225 x 1500 mm	each	0.48	0.48	7.54	-	17.94	25.48	31.85
215 x 225 x 900 mm	each	0.55	0.55	8.64	-	18.33	26.97	33.71
215 x 225 x 1200 mm	each	0.73	0.73	11.46	-	24.90	36.36	45.45
215 x 225 x 1500 mm	each	0.91	0.91	14.28	-	31.14	45.42	56.77

Concrete work

	Unit	Hours C	Hours L	Labour net	Plant net	Material net	Price net	Price with 25%
Generally				£	£	£	£	£
					VAT not included			

Precast concrete (*continued*)

Reinforced boot lintels bedded in gauged mortar

250 x 225 x 900 mm	each	0.66	0.66	10.36	-	19.22	29.58	36.98
250 x 225 x 1200 mm	each	0.88	0.88	13.81	-	26.08	39.89	49.86
275 x 225 x 1200 mm	each	0.66	0.66	10.36	-	20.10	30.46	38.08
275 x 225 x 1200 mm	each	0.88	0.88	13.81	-	27.35	41.16	51.45

Prestressed lintels bedded in gauged mortar; including strutting

100 x 66 x 900 mm	each	0.25	0.25	3.93	-	3.96	7.89	9.86
100 x 66 x 1200 mm	each	0.31	0.31	4.87	-	5.16	10.03	12.54
150 x 66 x 900 mm	each	0.25	0.25	3.93	-	5.16	9.09	11.36
150 x 66 x 1200 mm	each	0.31	0.31	4.87	-	6.62	11.49	14.36
215 x 66 x 900 mm	each	0.30	0.30	4.71	-	8.00	12.71	15.89
215 x 66 x 1200 mm	each	0.39	0.39	6.12	-	10.61	16.73	20.91
250 x 66 x 900 mm	each	0.30	0.30	4.71	-	9.39	14.10	17.63
250 x 66 x 1200 mm	each	0.40	0.40	6.28	-	12.35	18.63	23.29
105 x 145 x 1200 mm	each	0.33	0.33	5.18	-	10.15	15.33	19.16
105 x 145 x 1500 mm	each	0.42	0.42	6.59	-	12.68	19.27	24.09
105 x 220 x 1500 mm	each	0.48	0.48	7.54	-	18.43	25.97	32.46
105 x 220 x 1800 mm	each	0.58	0.58	9.11	-	22.13	31.24	39.05

Brickwork and blockwork

Preamble

"Labour net" figures include allowances for all costs incidental to the employment of labour.

"Materials net" figures include for all costs of materials including an allowance for waste except where specifically stated.

"Price net" figures are the totals of the "Labour net" and "Materials net" figures. Prices are for a builder employing his own labour; according to the amount and nature of the work involved, it may well be possible to secure more advantageous prices from specialist sub-contractors.

Prices do not include any allowance for scaffolding, ladders or other plant necessary to reach the work. The "Preliminaries" section includes prices for scaffolding which must be considered and allowance included to suit the particular circumstances of a tender.

Specialist prices

"Price with 25%" figures are all-in guide prices and include 25% for the builder's overheads, profit, unloading materials and general attendance (to include free use of standing scaffolding and hoists, temporary lighting and water and clearing away rubbish).

The amount of attendance required varies between the various trades and also with the circumstances of specific jobs; the percentage addition must always be considered and adjusted as necessary to suit the terms and conditions of the quotation being used.

Quantities and delivery distances are usually the most significant of the many factors which influence prices and it must be emphasised that quotations should always be obtained when preparing a tender.

Composite walls

Although not in accordance with the Standard Method of Measurement, figures for a range of composite walls have been included at the end of this section.

Brickwork and blockwork

Alterations and repairs

	Unit	Price

Basic prices for materials		£

Bricks

class B engineering	1,000	360.00
commons	1,000	170.00
refractory	1,000	920.00
facings	1,000	325.00

Bituminous emulsion waterproofing liquid	5 litre	10.94

Cavity insulation retaining clips	1,000	55.61

Clay flue linings

185 mm square x 300 mm - straight	each	5.51
185 mm square - curved	each	14.60

Clay tiles - 265 x 165 mm

creasing	10	4.27
roofing	10	4.37

Damp-proof courses

fibre base

112.5 mm wide	8 m roll	4.88
150 mm wide	8 m roll	6.50
225 mm wide	8 m roll	9.76

fibre base lead lined

112.5 mm wide	8 m roll	17.34
225 mm wide	8 m roll	34.63

hessian base

112.5 mm wide	8 m roll	6.65
150 mm wide	8 m roll	8.86
225 mm wide	8 m roll	13.30

hessian base lead lined

112.5 mm wide	8 m roll	18.25
225 mm wide	8 m roll	36.50

pitch polymer

112.5 mm wide	20 m roll	18.48
150 mm wide	20 m roll	24.63
225 mm wide	20 m roll	36.95

polyethylene

112.5 mm wide	30 m roll	4.78
225 mm wide	30 m roll	9.56

slates

350 x 112.5 mm	10	5.61
350 x 225 mm	10	12.00

Fireclay	tonne	255.00

Glass blocks

190 x 190 x 80 mm	each	3.92
240 x 240 x 80 mm	each	6.75

Hydrated lime	tonne	139.84

Prices actually to be paid for materials must be checked against the above basic prices and adjustments made as necessary.

Brickwork and blockwork

	Unit	Price
Basic prices for materials		**£**
Mesh reinforcement		
64 mm wide	25 m coil	9.32
178 mm wide	25 m coil	25.16
Portland cement	tonne	89.04
Typex Twin Wall metal flue pipes	m	26.32
Sand	m3	17.96
	tonne	11.22
Wall-ties - 200 mm long		
galvanised butterfly	1,000	101.40
stainless steel butterfly	1,000	121.68
3 mm galvanised vertical-twist	400	75.36
0.6 mm stainless steel - pressed	250	62.34

Prices actually to be paid for materials must be checked against the above basic prices and adjustments made as necessary.

Brickwork and blockwork

Alterations and repairs	Unit	Hours C	Hours L	Labour net	Material net	Price net	Price with 25%
Brickwork				£	£	£	£
					VAT not included		
Class B engineering bricks in cement mortar 1:3							
Walls							
102.5 mm	m2	1.70	0.85	20.96	27.08	48.04	60.05
215 mm	m2	2.64	1.32	32.55	53.78	86.33	107.91
327.5 mm	m2	3.63	1.82	44.79	79.85	124.64	155.80
Skins of hollow walls							
102.5 mm	m2	1.32	0.66	16.28	27.08	43.36	54.20
215 mm	m2	3.74	1.87	46.11	53.78	99.89	124.86
327.5 mm	m2	5.67	2.84	69.94	79.85	149.79	187.24
Common bricks in gauged mortar 1:1:6							
Walls							
102.5 mm	m2	1.10	0.55	13.57	14.28	27.85	34.81
215 mm	m2	2.20	1.10	27.12	28.39	55.51	69.39
327.5 mm	m2	2.97	1.49	36.65	41.83	78.48	98.10
Walls - curved to 3.00 m radius							
102.5 mm	m2	1.38	0.69	17.01	14.28	31.29	39.11
215 mm	m2	2.75	1.38	33.94	28.39	62.33	77.91
327.5 mm	m2	3.72	1.86	45.87	41.83	87.70	109.63
Honeycomb walls							
102.5 mm	m2	1.00	0.50	12.33	9.02	21.35	26.69
Filling existing openings							
102.5 mm	m2	1.21	0.61	14.95	14.28	29.23	36.54
215 mm	m2	2.42	1.21	29.84	28.39	58.23	72.79
327.5 mm	m2	3.30	1.65	40.69	41.83	82.52	103.15
Skins of hollow walls							
102.5 mm	m2	1.32	0.66	16.28	14.28	30.56	38.20
215 mm	m2	2.42	1.21	29.84	28.39	58.23	72.79
327.5 mm	m2	3.56	1.78	43.90	41.83	85.73	107.16
Projections of footings and chimney-breasts							
102.5 mm	m2	1.60	0.80	19.73	14.28	34.01	42.51
215 mm	m2	3.19	1.60	39.36	28.39	67.75	84.69
Isolated piers and chimney-stacks							
215 mm	m2	2.86	1.43	35.27	28.39	63.66	79.58
327.5 mm	m2	3.86	1.93	47.60	41.83	89.43	111.79
Backing to masonry including cutting and bonding to masonry							
102.5 mm	m2	1.60	0.80	19.73	16.49	36.22	45.27
215 mm	m2	2.92	1.46	36.00	30.60	66.60	83.25
327.5 mm	m2	4.29	2.15	52.93	44.04	96.97	121.21

Brickwork and blockwork

	Unit	Hours C	Hours L	Labour net	Material net	Price net	Price with 25%
Brickwork				£	£	£	£
					VAT not included		

Common bricks in gauged mortar 1:1:6
(continued)

Projections of attached piers; plinths; bands;
oversailing courses and the like

	Unit	Hours C	Hours L	Labour net	Material net	Price net	Price with 25%
215 x 102.5 mm	m	0.44	0.22	5.42	3.09	8.51	10.64
215 x 215 mm	m	0.88	0.44	10.85	6.16	17.01	21.26
327.5 x 102.5 mm	m	0.66	0.33	8.13	4.67	12.80	16.00
327.5 x 215 mm	m	1.21	0.60	14.88	9.35	24.23	30.29

Thickening existing walls including cutting and
bonding new to existing and extra material for
bonding

102.5 mm	m2	2.75	1.35	33.74	16.49	50.23	62.79
215 mm	m2	4.02	2.01	49.57	30.60	80.17	100.21

Projections on existing walls of attached piers;
chimney-breasts and the like including cutting and
bonding new to existing and extra material for
bonding

215 x 102.5 mm	m	0.74	0.37	9.12	3.09	12.21	15.26
215 x 215 mm	m	1.32	0.66	16.28	6.16	22.44	28.05
327.5 x 102.5 mm	m	1.05	0.53	12.95	4.67	17.62	22.02
327.5 x 215 mm	m	1.93	0.96	23.79	9.35	33.14	41.42

Form 50 mm cavities in hollow walls with 200 mm
wall-ties at 5 per m2 using

galvanised butterfly ties	m2	0.13	0.07	1.62	0.56	2.18	2.73
stainless steel butterfly ties	m2	0.13	0.07	1.62	0.67	2.29	2.86
3 mm galvanised vertical-twist ties	m2	0.13	0.07	1.62	1.03	2.65	3.31
0.6 mm stainless steel pressed ties	m2	0.13	0.07	1.62	1.37	2.99	3.74
0.6 mm stainless steel pressed ties with insulation retaining clips	m2	0.17	0.08	2.04	1.66	3.70	4.63

Close 50 mm cavities at ends of hollow walls and
at jambs or sills of openings with

102.5 mm brickwork and additional ties	m	0.44	0.22	5.42	3.06	8.48	10.60
102.5 mm brickwork; 150 mm fibre base bitumen damp-proof course bedded in gauged mortar 1:1:6 and additional ties	m	0.55	0.28	6.78	2.48	9.26	11.57

Extra for rough arches 215 mm thick

in one half brick ring	m	0.66	0.33	8.13	3.10	11.23	14.04
in two half brick rings	m	1.00	0.50	12.33	6.19	18.52	23.15

Block bond ends of new walls to existing including
cutting pockets in existing work and extra material
for bonding

102.5 mm	m	0.46	0.23	5.67	0.89	6.56	8.20
215 mm	m	0.88	0.44	10.85	1.77	12.62	15.78

Brickwork and blockwork

Alterations and repairs	Unit	Hours C	Hours L	Labour net	Material net	Price net	Price with 25%

Brickwork

				£	£	£	£
					VAT not included		

Common bricks in gauged mortar 1:1:6
(*continued*)

Quoin up jambs of new openings in existing walls

102.5 mm	m	0.54	0.27	6.66	5.26	11.92	14.90
215 mm	m	0.82	0.41	10.11	13.59	23.70	29.63

Quoin up jambs of new openings in existing walls
including raking out joints and flush pointing

102.5 mm	m	1.91	0.27	18.93	5.43	24.36	30.45
215 mm	m	2.19	0.41	22.38	4.80	27.18	33.98

Refractory bricks in fireclay mortar

102.5 mm linings to flues flush pointed one side and bonded to surrounding brickwork	m2	2.26	1.13	27.87	128.71	156.58	195.72
215 mm boiler flues flush pointed one side and fair faced and flush pointed the other side	m2	3.66	1.83	45.12	256.45	301.57	376.96

Brick facework

Common bricks in gauged mortar 1:1:6

Extra over common brickwork for fair face and
flush pointing as the work proceeds

stretcher bond	m2	0.50	0.25	6.17	0.08	6.25	7.81
Flemish bond	m2	0.55	0.28	6.78	0.08	6.86	8.57
margins	m	0.09	0.04	1.09	0.08	1.17	1.46
flat arch 215 mm on face and 50 mm on exposed soffit	m	0.44	0.22	5.42	0.08	5.50	6.88

Extra over common brickwork for fair face; raking
out joints and flush pointing in cement mortar 1:3

stretcher bond	m2	0.33	0.17	4.07	0.10	4.17	5.21
Flemish bond	m2	0.20	0.10	2.46	0.10	2.56	3.20
margins	m	0.10	0.05	1.24	0.10	1.34	1.68
flat arch 215 mm on face and 50 mm on exposed soffit	m	0.44	0.22	5.42	0.10	5.52	6.90

102.5 mm walls in stretcher bond fair faced and flush pointed both sides as the work proceeds	m2	1.76	0.88	21.70	14.28	35.98	44.98
102.5 mm walls in stretcher bond fair faced and flush pointed both sides in cement mortar 1:3 including raking out joints	m2	2.70	1.35	33.29	14.48	47.77	59.71
215 mm walls in Flemish bond fair faced and flush pointed both sides in cement mortar 1:3 including raking out joints	m2	4.18	2.09	51.54	28.59	80.13	100.16
Fair squint angles in purpose made squint bricks	m	0.28	0.14	3.45	3.61	7.06	8.82

Brickwork and blockwork

Alterations and repairs	Unit	Hours C	Hours L	Labour net	Material net	Price net	Price with 25%
Brickwork				£	£	£	£
					VAT not included		

Brick facework (*continued*)

Common bricks in gauged mortar 1:1:6 (*continued*)

	Unit	Hours C	Hours L	Labour net	Material net	Price net	Price with 25%
Fair rounded angles in purpose made bullnose bricks	m	0.28	0.14	3.45	3.61	7.06	8.82
Extra over 215 mm wall for projecting double course tile creasings 20 mm set-forward and with cement fillets both sides	m	0.72	0.36	8.88	7.47	16.35	20.44
215 x 102.5 mm brick on edge copings fair faced and flush pointed as the work proceeds	m	0.88	0.44	10.85	3.18	14.03	17.54

Facing bricks in gauged mortar 1:1:6

	Unit	Hours C	Hours L	Labour net	Material net	Price net	Price with 25%
Extra over common brickwork for facework and flush pointing as the work proceeds							
Flemish bond	m2	0.73	0.36	9.00	13.19	22.19	27.74
margins	m	0.09	0.04	1.09	0.08	1.17	1.46
flat arch 215 mm on face and 50 mm on exposed soffit	m	0.45	0.45	7.06	1.74	8.80	11.00
segmental arch 215 mm on face and 50 mm on exposed soffit	m	1.65	1.65	25.90	1.74	27.64	34.55
semicircular arch 215 mm on face and 50 mm on exposed soffit	m	1.65	1.65	25.90	1.74	27.64	34.55
Extra over common brickwork for facework; raking out joints and flush pointing in cement mortar 1:3							
Flemish bond	m2	0.90	0.90	14.13	13.21	27.34	34.17
margins	m	0.15	0.15	2.35	0.10	2.45	3.06
flat arch 215 mm on face and 50 mm on exposed soffit	m	0.51	0.26	6.29	1.76	8.05	10.06
segmental arch 215 mm on face and on exposed soffit	m	1.65	0.82	20.34	1.76	22.10	27.63
semicircular arch 215 mm on face and on exposed soffit	m	1.65	0.82	20.34	1.76	22.10	27.63
102.5 mm walls in stretcher bond faced and flush pointed as the work proceeds							
one side	m2	1.98	0.99	24.41	24.26	48.67	60.84
both sides	m2	2.64	1.32	32.55	24.36	56.91	71.14
102.5 mm walls in stretcher bond faced and flush pointed in cement mortar 1:3 including raking out joints							
one side	m2	2.28	1.14	28.11	24.26	52.37	65.46
both sides	m2	2.94	1.47	36.25	24.36	60.61	75.76
215 mm walls in Flemish bond faced and flush pointed both sides as the work proceeds	m2	3.52	1.76	43.40	48.16	91.56	114.45

Brickwork and blockwork

Alterations and repairs	Unit	Hours C	Hours L	Labour net	Material net	Price net	Price with 25%
Brickwork				£	£	£	£
					VAT not included		
Brickwork facework (*continued*)							
Facing bricks in gauged mortar 1:1:6 (*continued*)							
215 mm walls in Flemish bond faced and flush pointed both sides in cement mortar 1:3 including raking out joints	m2	4.31	2.16	53.18	48.19	101.37	126.71
102.5 mm skins of hollow walls in stretcher bond faced and flush pointed one side as the work proceeds	m2	1.98	0.99	24.41	24.16	48.57	60.71
102.5 mm skins of hollow walls in stretcher bond faced and flush pointed one side in cement mortar 1:3 including raking out joints	m2	2.37	1.19	29.26	24.26	53.52	66.90
Fair squint angles in purpose made squint bricks	m	0.28	0.14	3.45	6.75	10.20	12.75
Fair rounded angles in purpose made bullnose bricks	m	0.28	0.14	3.45	6.01	9.46	11.82
Extra for facework to							
65 mm recessed bands 25 mm set-back	m	0.28	0.14	3.45	-	3.45	4.31
65 mm projecting bands 25 mm set-forward	m	0.24	0.12	2.96	-	2.96	3.70
Extra over 215 mm wall for projecting double course tile creasings 20 mm set-forward and with cement fillets both sides	m	0.72	0.36	8.88	7.47	16.35	20.44
215 x 102.5 mm brick on edge copings faced and flush pointed in cement mortar 1:3 including raking out joints	m	0.88	0.44	10.85	5.46	16.31	20.39
215 x 102.5 mm brick on edge sills faced and flush pointed in cement mortar 1:3 including raking out joints	m	0.82	0.41	10.11	5.46	15.57	19.46
Engineering bricks in cement mortar 1:3							
Brick on edge steps fair faced and flush pointed as the work proceeds							
215 x 102.5 mm	m	0.91	0.46	11.22	6.09	17.31	21.64
327.5 x 102.5 mm	m	1.64	0.82	20.22	9.15	29.37	36.71
Extra for rounded edges formed with							
bullnosed bricks	m	0.28	0.14	3.45	2.40	5.85	7.31
angles	each	0.10	0.05	1.24	0.66	1.90	2.38

Brickwork and blockwork

Alterations and repairs	Unit	Hours C	Hours L	Labour net	Material net	Price net	Price with 25%
Blockwork				£	£	£	£
					VAT not included		

Precast concrete dense aggregate blocks to BS 6073 in gauged mortar 1:1:6

	Unit	Hours C	Hours L	Labour net	Material net	Price net	Price with 25%
Walls and partitions							
75 mm solid	m2	0.72	0.53	10.02	5.97	15.99	19.99
100 mm solid	m2	0.88	0.66	12.33	6.67	19.00	23.75
140 mm solid	m2	1.10	0.83	15.45	12.77	28.22	35.27
215 mm hollow	m2	1.32	0.99	18.50	15.47	33.97	42.46
Honeycomb walls							
100 mm solid	m2	0.77	0.58	10.80	4.41	15.21	19.01
Filling existing openings							
75 mm solid	m2	0.92	0.69	12.89	5.97	18.86	23.57
100 mm solid	m2	1.14	0.86	16.01	6.67	22.68	28.35
140 mm solid	m2	1.43	1.07	20.02	12.77	32.79	40.99
215 mm hollow	m2	1.43	1.07	20.02	15.30	35.32	44.15
Skins of hollow walls							
75 mm solid	m2	0.86	0.64	12.02	5.97	17.99	22.49
100 mm solid	m2	1.06	0.79	14.82	6.67	21.49	26.86
140 mm solid	m2	1.32	0.99	18.50	12.77	31.27	39.09
215 mm hollow	m2	1.34	1.00	18.75	15.47	34.22	42.77
Walls and partitions fair faced and flush pointed one side as the work proceeds							
75 mm solid	m2	0.91	0.69	12.80	5.97	18.77	23.46
100 mm solid	m2	1.08	0.81	15.14	6.67	21.81	27.26
140 mm solid	m2	1.30	0.97	18.19	12.77	30.96	38.70
Walls and partitions fair faced and flush pointed both sides as the work proceeds							
75 mm solid	m2	1.11	0.83	15.54	5.97	21.51	26.89
100 mm solid	m2	1.28	0.96	17.94	6.67	24.61	30.76
140 mm solid	m2	1.50	1.12	20.99	12.77	33.76	42.20
Extra for filling every fourth void of 215 mm hollow blocks with concrete 1:2:4 and reinforcing with one 6 mm bar	m2	0.09	0.32	2.95	1.01	3.96	4.95
Form 50 mm cavities in hollow walls with 200 mm wall-ties at 5 per m2							
galvanised butterfly ties	m2	0.13	0.07	1.62	0.56	2.18	2.73
stainless steel butterfly ties	m2	0.13	0.07	1.62	0.67	2.29	2.86
3 mm galvanised vertical-twist ties	m2	0.13	0.07	1.62	1.03	2.65	3.31
0.6 mm stainless steel pressed ties	m2	0.13	0.07	1.62	1.37	2.99	3.74
0.6 mm stainless steel pressed ties with insulation retaining clips	m2	0.13	0.09	1.77	1.66	3.43	4.29

Brickwork and blockwork

Alterations and repairs	Unit	Hours C	Hours L	Labour net	Material net	Price net	Price with 25%
Blockwork				£	£	£	£
					VAT not included		

Precast concrete dense aggregate blocks to BS 6073 in gauged mortar 1:1:6 *(continued)*

Close 50 mm cavities at ends of hollow walls and at jambs or sills of openings with

100 mm blockwork and additional ties	m	0.11	0.08	1.55	0.91	2.46	3.08
100 mm blockwork; 150 mm fibre base bitumen damp-proof course bedded in gauged mortar 1:1:6 and additional ties	m	0.22	0.17	3.08	1.76	4.84	6.05

Close 50 - 100 mm cavities at tops of hollow walls with single course of blocks laid flat in gauged mortar 1:1:6

100 mm blocks	m	0.22	0.17	3.08	0.66	3.74	4.67
140 mm blocks	m	0.20	0.15	2.80	1.04	3.84	4.80

Bond ends of new walls to other types of construction including forming pockets in new construction and extra material for bonding

75 mm	m	0.28	0.20	3.86	0.37	4.23	5.29
100 mm	m	0.33	0.25	4.65	0.41	5.06	6.33
140 mm	m	0.39	0.29	5.44	0.72	6.16	7.70
215 mm	m	0.44	0.33	6.16	0.92	7.08	8.85

Bond ends of new walls to other types of construction including cutting pockets in existing construction and extra material for bonding

75 mm	m	0.44	0.33	6.16	0.43	6.59	8.24
100 mm	m	0.50	0.37	6.97	0.47	7.44	9.30
140 mm	m	0.61	0.45	8.50	0.72	9.22	11.53
215 mm	m	0.66	0.50	9.28	0.98	10.26	12.82

Precast concrete clinker aggregate blocks to BS 6073 in gauged mortar 1:1:6

Walls and partitions

60 mm	m2	0.50	0.37	6.97	6.50	13.47	16.84
75 mm	m2	0.55	0.41	7.69	6.50	14.19	17.74
100 mm	m2	0.61	0.45	8.50	7.71	16.21	20.26
140 mm	m2	0.97	0.72	13.54	9.40	22.94	28.68
215 mm	m2	1.05	0.78	14.67	17.35	32.02	40.02

Filling existing openings

60 mm	m2	1.00	0.75	14.01	6.42	20.43	25.54
75 mm	m2	1.10	0.83	15.45	6.50	21.95	27.44
100 mm	m2	1.20	0.90	16.82	7.71	24.53	30.66
140 mm	m2	1.40	1.05	19.62	9.40	29.02	36.27
215 mm	m2	1.55	1.16	21.71	17.35	39.06	48.83

Brickwork and blockwork

Alterations and repairs	Unit	Hours C	Hours L	Labour net	Material net	Price net	Price with 25%
Blockwork				£	£	£	£
					VAT not included		

Precast concrete clinker aggregate blocks to BS 6073 in gauged mortar 1:1:6 (*continued*)

	Unit	Hours C	Hours L	Labour net	Material net	Price net	Price with 25%
Skins of hollow walls							
60 mm	m2	0.59	0.45	8.32	6.50	14.82	18.52
75 mm	m2	0.66	0.50	9.28	6.50	15.78	19.73
100 mm	m2	0.73	0.54	10.18	7.71	17.89	22.36
140 mm	m2	0.92	0.69	12.89	9.40	22.29	27.86
215 mm	m2	1.25	0.94	17.54	17.35	34.89	43.61
Form 50 mm cavities in hollow walls with 200 mm wall-ties at 5 per m2 using							
galvanised butterfly ties	m2	0.13	0.07	1.62	0.56	2.18	2.73
stainless steel butterfly ties	m2	0.13	0.07	1.62	0.67	2.29	2.86
3 mm galvanised vertical-twist ties	m2	0.13	0.07	1.62	1.03	2.65	3.31
0.6 mm stainless steel pressed ties	m2	0.13	0.07	1.62	1.37	2.99	3.74
0.6 mm stainless steel pressed ties with insulation retaining clips	m2	0.13	0.09	1.77	1.66	3.43	4.29
Close 50 mm cavities at ends of hollow walls and at jambs or sills of openings with							
100 mm blockwork and additional ties	m	0.11	0.06	1.36	0.99	2.35	2.94
100 mm blockwork; 150 mm fibre base bitumen damp-proof course bedded in gauged mortar 1:1:6 and additional ties	m	0.22	0.11	2.71	1.84	4.55	5.69
Close 50 - 100 mm cavities at tops of hollow walls with single course of blocks laid flat in gauged mortar 1:1:6							
100 mm blocks	m	0.22	0.11	2.71	0.65	3.36	4.20
140 mm blocks	m	0.25	0.13	3.08	0.77	3.85	4.81
Bond ends of new walls to other types of construction including forming pockets in new construction and extra material for bonding							
60 mm	m	0.22	0.11	2.71	0.40	3.11	3.89
75 mm	m	0.28	0.14	3.45	0.40	3.85	4.81
100 mm	m	0.33	0.17	4.07	0.46	4.53	5.66
140 mm	m	0.36	0.18	4.44	0.54	4.98	6.22
215 mm	m	0.44	0.22	5.42	0.95	6.37	7.96
Bond ends of new walls to other types of construction including cutting pockets in existing construction and extra material for bonding							
60 mm	m	0.36	0.18	4.44	0.40	4.84	6.05
75 mm	m	0.44	0.33	6.16	0.40	6.56	8.20
100 mm	m	0.50	0.36	6.91	0.46	7.37	9.21
140 mm	m	0.61	0.45	8.50	0.45	8.95	11.19
215 mm	m	0.72	0.54	10.09	0.95	11.04	13.80

Brickwork and blockwork

Alterations and repairs	Unit	Hours C	Hours L	Labour net	Material net	Price net	Price with 25%
Blockwork				£	£	£	£
					VAT not included		

Precast concrete lightweight aggregate loadbearing blocks to BS 6073 in gauged mortar 1:1:6

Walls and partitions

75 mm	m2	0.50	0.37	6.97	9.15	16.12	20.15
100 mm	m2	0.55	0.42	7.76	10.98	18.74	23.43
140 mm	m2	0.72	0.54	10.09	18.25	28.34	35.42
215 mm	m2	0.99	0.75	13.92	26.02	39.94	49.92

Honeycomb walls

100 mm	m2	0.66	0.50	9.28	7.25	16.53	20.66

Filling existing openings

75 mm	m2	0.80	0.60	11.21	6.50	17.71	22.14
100 mm	m2	0.95	0.71	13.30	7.71	21.01	26.26
140 mm	m2	1.20	0.90	16.82	9.40	26.22	32.77
215 mm	m2	1.35	1.00	18.84	17.35	36.19	45.24

Skins of hollow walls

75 mm	m2	0.59	0.45	8.32	9.15	17.47	21.84
100 mm	m2	0.66	0.50	9.28	10.98	20.26	25.32
150 mm	m2	0.86	0.65	12.09	18.25	30.34	37.92
215 mm	m2	1.19	0.89	16.66	26.02	42.68	53.35

Form 50 mm cavities in hollow walls with 200 mm wall-ties at 5 per m2 using

galvanised butterfly ties	m2	0.13	0.07	1.62	0.56	2.18	2.73
stainless steel butterfly ties	m2	0.13	0.07	1.62	0.67	2.29	2.86
3 mm galvanised vertical-twist ties	m2	0.13	0.07	1.62	1.03	2.65	3.31
0.6 mm stainless steel pressed ties	m2	0.13	0.07	1.62	1.37	2.99	3.74
0.6 mm stainless steel pressed ties with insulation retaining clips	m2	0.13	0.09	1.77	1.66	3.43	4.29

Close 50 mm cavities at ends of hollow walls and at jambs or sills of openings with

100 mm blockwork and additional ties	m	0.09	0.07	1.23	1.25	2.48	3.10
100 mm blockwork; 150 mm fibre base bitumen damp-proof course bedded in gauged mortar 1:1:6 and additional ties	m	0.20	0.15	2.80	2.10	4.90	6.13

Close 50 - 100 mm cavities at tops of hollow walls with single course of blocks laid flat in gauged mortar 1:1:6

100 mm blocks	m	0.17	0.12	2.33	0.91	3.24	4.05
150 mm blocks	m	0.22	0.17	3.08	1.46	4.54	5.67

Brickwork and blockwork

Alterations and repairs	Unit	Hours C	Hours L	Labour net	Material net	Price net	Price with 25%
Blockwork				£	£	£	£
					VAT not included		

Precast concrete lightweight aggregate loadbearing blocks to BS 6073 in gauged mortar 1:1:6 (continued)

Bond ends of new walls to other types of construction including forming pockets in new construction and extra material for bonding

	Unit	Hours C	Hours L	Labour net	Material net	Price net	Price with 25%
75 mm	m	0.28	0.20	3.86	0.54	4.40	5.50
100 mm	m	0.33	0.25	4.65	0.64	5.29	6.61
150 mm	m	0.36	0.28	5.12	1.01	6.13	7.66
215 mm	m	0.44	0.33	6.16	1.41	7.57	9.46

Bond ends of new walls to other types of construction including cutting pockets in existing construction and extra material for bonding

	Unit	Hours C	Hours L	Labour net	Material net	Price net	Price with 25%
75 mm	m	0.44	0.33	6.16	0.54	6.70	8.38
100 mm	m	0.50	0.37	6.97	0.64	7.61	9.51
150 mm	m	0.61	0.45	8.50	1.01	9.51	11.89
215 mm	m	0.72	0.54	10.09	1.41	11.50	14.38

290 x 290 mm perforated precast concrete screen walling blocks in gauged mortar 1:1:6

90 mm walls flush pointed both sides as the work proceeds

	Unit	Hours C	Hours L	Labour net	Material net	Price net	Price with 25%
	m2	1.16	0.87	16.25	22.13	38.38	47.98

Tie ends of 90 mm walls to brickwork with galvanised butterfly ties at 300 mm centres vertically

	Unit	Hours C	Hours L	Labour net	Material net	Price net	Price with 25%
	m	0.12	0.09	1.67	0.34	2.01	2.51

Glass blocks

80 mm partitions in gauged mortar 1:1:6 flush pointed both sides as the work proceeds

	Unit	Hours C	Hours L	Labour net	Material net	Price net	Price with 25%
in 190 x 190 mm blocks	m2	2.64	1.98	37.00	180.26	217.26	271.57
in 240 x 240 mm blocks	m2	1.98	1.49	27.75	153.78	181.53	226.91

Brickwork and blockwork

Alterations and repairs	Unit	Hours C	Hours L	Labour net	Material net	Price net	Price with 25%
Damp proof courses				£	£	£	£
					VAT not included		

Fibre based bitumen

Horizontal damp-proof courses bedded in cement mortar 1:3

over 225 mm wide	m2	0.28	0.14	3.39	5.76	9.15	11.44
112.5 mm wide	m	0.03	0.01	0.38	0.64	1.02	1.27
225 mm wide	m	0.06	0.03	0.77	1.28	2.05	2.56

150 mm fibre based bitumen vertical damp-proof courses bedded in gauged mortar 1:1:6

	m	0.07	0.04	0.92	0.85	1.77	2.21

Lead lined fibre base bitumen

Horizontal damp-proof courses bedded in cement mortar 1:3

over 225 mm wide	m2	0.36	0.18	4.44	20.45	24.89	31.11
112.5 mm wide	m	0.04	0.02	0.54	2.28	2.82	3.52
225 mm wide	m	0.08	0.04	1.01	4.54	5.55	6.94

Hessian base bitumen

Horizontal damp-proof courses bedded in cement mortar 1:3

over 225 mm wide	m2	0.28	0.14	3.39	7.86	11.25	14.06
112.5 mm wide	m	0.03	0.01	0.38	0.87	1.25	1.56
225 mm wide	m	0.06	0.03	0.77	1.75	2.52	3.15

150 mm hessian based bitumen vertical damp-proof courses bedded in gauged mortar 1:1:6

	m	0.07	0.04	0.91	1.16	2.07	2.59

Lead lined hessian base bitumen

Horizontal damp-proof courses bedded in cement mortar 1:3

over 225 mm wide	m2	0.36	0.18	4.44	21.56	26.00	32.50
112.5 mm wide	m	0.04	0.02	0.54	2.40	2.94	3.67
225 mm wide	m	0.08	0.04	1.01	4.79	5.80	7.25

Pitch polymer

Horizontal damp-proof courses bedded in cement mortar 1:3

over 225 mm wide	m2	0.28	0.14	3.45	8.73	12.18	15.23
112.5 mm wide	m	0.03	0.01	0.38	0.97	1.35	1.69
225 mm wide	m	0.06	0.03	0.77	1.94	2.71	3.39

150 mm vertical damp-proof courses bedded in gauged mortar 1:1:6

	m	0.07	0.04	0.91	1.29	2.20	2.75

Brickwork and blockwork

Alterations and repairs	Unit	Hours C	Hours L	Labour net	Material net	Price net	Price with 25%
Damp proof courses				£	£	£	£
					VAT not included		
Polythene							
Horizontal damp-proof courses bedded in cement mortar 1:3							
over 225 mm wide	m2	0.28	0.14	3.45	1.51	4.96	6.20
112.5 mm wide	m	0.03	0.01	0.38	0.17	0.55	0.69
225 mm wide	m	0.06	0.03	0.77	0.34	1.11	1.39
Slate							
Double course horizontal damp-proof courses bedded in cement mortar 1:3							
over 225 mm wide	m2	1.21	0.66	15.29	22.20	37.49	46.86
112.5 mm wide	m	0.17	0.08	2.04	2.81	4.85	6.06
225 mm wide	m	0.28	0.14	3.45	5.04	8.49	10.61

Brickwork and blockwork

Alterations and repairs	Unit	Hours C	Hours L	Labour net	Material net	Price net	Price with 25%
Sundries				£	£	£	£
					VAT not included		

Cavities

Form 50 mm cavities in hollow walls with wall-ties at 5 per m2 using

	Unit	Hours C	Hours L	Labour net	Material net	Price net	Price with 25%
galvanised butterfly ties	m2	0.13	0.07	1.62	0.56	2.18	2.73
stainless steel butterfly ties	m2	0.13	0.07	1.62	0.67	2.29	2.86
3 mm galvanised vertical-twist ties	m2	0.13	0.07	1.62	1.03	2.65	3.31
0.6 mm stainless steel pressed ties	m2	0.13	0.07	1.62	1.37	2.99	3.74
0.6 mm stainless steel pressed ties with insulation retaining clips	m2	0.13	0.09	1.77	1.66	3.43	4.29

Seal 50 - 100 mm cavities at eaves and sills in hollow walls with single course of slates in gauged mortar 1:1:6

	Unit	Hours C	Hours L	Labour net	Material net	Price net	Price with 25%
mortar 1:1:6	m	0.17	0.08	2.04	4.68	6.72	8.40

Reinforcement

Mesh reinforcement in walls

	Unit	Hours C	Hours L	Labour net	Material net	Price net	Price with 25%
64 mm	m	0.03	0.02	0.41	0.41	0.82	1.02
178 mm	m	0.03	0.02	0.41	1.11	1.52	1.90

Joints

	Unit	Hours C	Hours L	Labour net	Material net	Price net	Price with 25%
Rake out joints of brickwork to form key for plastering etc	m2	0.22	0.11	2.71	-	2.71	3.39

Chases

Horizontal rough chases

	Unit	Hours C	Hours L	Labour net	Material net	Price net	Price with 25%
25 x 25 mm	m	0.22	0.11	2.71	-	2.71	3.39
50 x 50 mm	m	0.33	0.17	4.11	-	4.11	5.14
75 x 75 mm	m	0.44	0.22	5.42	-	5.42	6.78
100 x 100 mm	m	0.55	0.28	6.78	-	6.78	8.47

Vertical rough chases

	Unit	Hours C	Hours L	Labour net	Material net	Price net	Price with 25%
25 x 25 mm	m	0.33	0.17	4.07	-	4.07	5.09
50 x 50 mm	m	0.50	0.25	6.17	-	6.17	7.71
75 x 75 mm	m	0.66	0.33	8.13	-	8.13	10.16
100 x 100 mm	m	0.83	0.41	10.24	-	10.24	12.80

Waterproofing

Three coats of bituminous emulsion water-proofing liquid on brick or block walls including blinding final coat with sand

	Unit	Hours C	Hours L	Labour net	Material net	Price net	Price with 25%
blinding final coat with sand	m2	-	0.66	4.45	3.80	8.25	10.31

Angle fillets

Mortar angle-fillets

	Unit	Hours C	Hours L	Labour net	Material net	Price net	Price with 25%
50 mm	m	0.28	0.14	3.45	0.30	3.75	4.69
75 mm	m	0.50	0.25	6.17	0.60	6.77	8.46

Brickwork and blockwork

Alterations and repairs	Unit	Hours C	Hours L	Labour net	Material net	Price net	Price with 25%
Sundries				£	£	£	£
					VAT not included		

Weatherings

150 mm sills or weatherings to projections of two courses of plain clay roofing tiles set weathering and breaking joint and bedded; jointed and pointed in cement mortar 1:3

	Unit	Hours C	Hours L	Labour net	Material net	Price net	Price with 25%
straight lengths	m	0.61	0.30	7.49	5.04	12.53	15.66
cut and fitted ends	each	0.11	0.06	1.36	0.69	2.05	2.56

Bedding

	Unit	Hours C	Hours L	Labour net	Material net	Price net	Price with 25%
Bed plates 100 mm wide in mortar	m	0.06	0.03	0.68	0.10	0.78	0.97
Bed wood frames and sills in mortar and point one side (includes allowance for unloading and hoisting)	m	0.10	0.05	1.23	0.50	1.73	2.16

Wedge and pin

Wedge and pin up to underside of existing construction with slates in cement mortar 1:3

	Unit	Hours C	Hours L	Labour net	Material net	Price net	Price with 25%
102.5 mm walls	m	0.94	0.47	11.59	2.53	14.12	17.65
215 mm walls	m	1.54	0.77	18.99	4.82	23.81	29.76
327.5 mm walls	m	2.15	1.08	26.54	6.13	32.67	40.84

Rake out

Rake out joints for turned-in edges of flashings and point

	Unit	Hours C	Hours L	Labour net	Material net	Price net	Price with 25%
horizontal	m	0.28	0.14	3.45	0.10	3.55	4.44
stepped	m	0.39	0.14	4.47	0.20	4.67	5.84

Build in

Build in metal windows including building in lugs at jambs; plugging and screwing frames at head and sill; filling backs of frames with cement mortar 1:3 and pointing one side; window area approximately

	Unit	Hours C	Hours L	Labour net	Material net	Price net	Price with 25%
0.50 m2	each	0.74	0.37	9.12	1.03	10.15	12.69
1.00 m2	each	1.21	0.60	14.92	1.51	16.43	20.54
1.50 m2	each	1.60	0.80	19.73	1.98	21.71	27.14
2.00 m2	each	1.82	0.91	22.44	2.45	24.89	31.11
2.50 m2	each	2.37	1.19	29.23	2.93	32.16	40.20
3.00 m2	each	2.56	1.28	31.57	3.42	34.99	43.74

	Unit	Hours C	Hours L	Labour net	Material net	Price net	Price with 25%
Build in 800 x 2100 mm metal door frames including building in lugs and filling backs of frames with cement mortar	each	0.95	0.47	11.71	0.80	12.51	15.64

Brickwork and blockwork

Alterations and repairs	Unit	Hours C	Hours L	Labour net	Material net	Price net	Price with 25%
Sundries				£	£	£	£
					VAT not included		

Cut and pin

Cut and pin ends of steel sections and make good

small (not exceeding 250 mm deep)	each	0.66	0.33	8.13	0.10	8.23	10.29
large (250 - 500 mm deep)	each	0.99	0.50	12.21	0.20	12.41	15.51
extra large (over 500 mm deep)	each	1.32	0.66	16.28	0.20	16.48	20.60

Holes

Holes for small pipes (not exceeding 55 mm) through walls and make good

75 mm	each	0.19	0.09	2.31	0.10	2.41	3.01
102.5 mm	each	0.33	0.17	4.07	0.20	4.27	5.34
215 mm	each	0.50	0.25	6.11	0.30	6.41	8.01

Holes for large pipes (55 - 110 mm) through walls and make good

75 mm	each	0.30	0.15	3.66	0.10	3.76	4.70
102.5 mm	each	0.41	0.20	5.05	0.20	5.25	6.56
215 mm	each	0.55	0.28	6.78	0.30	7.08	8.85

Holes for extra large pipes (over 110 mm) through walls and make good

75 mm	each	0.42	0.21	5.18	0.20	5.38	6.72
102.5 mm	each	0.50	0.25	6.17	0.30	6.47	8.09
215 mm	each	0.83	0.41	10.24	0.40	10.64	13.30

Holes for small pipes (not exceeding 55 mm) through walls and make good facings

102.5 mm	each	0.44	0.22	5.42	0.10	5.52	6.90
215 mm	each	0.61	0.30	7.53	0.10	7.63	9.54

Holes for large pipes (55 - 110 mm) through walls and make good facings

102.5 mm	each	0.61	0.30	7.53	0.20	7.73	9.66
215 mm	each	0.88	0.44	10.85	0.20	11.05	13.81

Holes for extra large pipes (over 110 mm) through walls and make good facings

102.5 mm	each	0.74	0.37	9.12	0.20	9.32	11.65
215 mm	each	1.10	0.55	13.57	0.30	13.87	17.34

Mortices

Mortices for bolts and run with mortar	each	0.39	0.19	4.75	0.10	4.85	6.06

Openings

225 x 75 mm openings through walls with slate lintels and make good

102.5 mm	each	0.13	0.07	1.60	0.78	2.38	2.98
215 mm	each	0.22	0.11	2.71	0.88	3.59	4.49

Brickwork and blockwork

Alterations and repairs	Unit	Hours C	Hours L	Labour net	Material net	Price net	Price with 25%
Sundries				£	£	£	£
					VAT not included		
Openings (*continued*)							
225 x 150 mm openings through walls with slate lintels and make good							
102.5 mm	each	0.17	0.09	2.09	0.78	2.87	3.59
215 mm	each	0.28	0.14	3.45	0.88	4.33	5.41
225 x 225 mm openings through walls with slate lintels and make good							
102.5 mm	each	0.20	0.10	2.44	0.78	3.22	4.03
215 mm	each	0.33	0.17	4.07	0.88	4.95	6.19
250 x 250 mm openings through walls with slate lintels and make good							
102.5 mm	each	0.33	0.17	4.07	0.78	4.85	6.06
215 mm	each	0.55	0.28	6.78	0.88	7.66	9.57
225 x 75 mm openings through walls with slate lintels and make good facings							
102.5 mm	each	0.20	0.10	2.46	0.78	3.24	4.05
215 mm	each	0.30	0.15	3.70	0.88	4.58	5.72
225 x 150 mm openings through walls with slate lintels and make good facings							
102.5 mm	each	0.25	0.13	3.08	0.78	3.86	4.83
215 mm	each	0.36	0.18	4.44	0.88	5.32	6.65
225 x 225 mm openings through walls with slate lintels and make good facings							
102.5 mm	each	0.30	0.15	3.70	0.78	4.48	5.60
215 mm	each	0.44	0.22	5.42	0.88	6.30	7.88
250 x 250 mm openings through walls with slate lintels and make good facings							
102.5 mm	each	0.46	0.23	5.67	0.78	6.45	8.06
215 mm	each	0.69	0.34	8.51	0.88	9.39	11.74
Air bricks and soot doors							
Provide and build in air-bricks - nominal size							
229 x 76 mm galvanised	each	0.09	0.04	1.09	2.77	3.86	4.83
229 x 152 mm galvanised	each	0.09	0.04	1.09	5.10	6.19	7.74
229 x 229 mm galvanised	each	0.09	0.04	1.09	7.47	8.56	10.70
229 x 76 mm terra-cotta	each	0.09	0.04	1.09	1.60	2.69	3.36
229 x 152 mm terra-cotta	each	0.09	0.04	1.09	2.22	3.31	4.14
229 x 229 mm terra-cotta	each	0.09	0.04	1.09	6.10	7.19	8.99

Brickwork and blockwork

Alterations and repairs	Unit	Hours C	Hours L	Labour net	Material net	Price net	Price with 25%
Sundries				£	£	£	£
					VAT not included		
Air bricks and soot doors (*continued*)							
Provide and build in terra-cotta air brick extension cavity liners 300 mm long - nominal size							
229 x 76 mm horizontal	each	0.11	0.06	1.36	3.24	4.60	5.75
229 x 152 mm horizontal	each	0.11	0.06	1.36	3.51	4.87	6.09
229 x 229 mm horizontal	each	0.11	0.06	1.36	9.37	10.73	13.41
229 x 76 mm inclined	each	0.11	0.06	1.36	21.79	23.15	28.94
229 x 152 mm inclined	each	0.11	0.06	1.36	25.98	27.34	34.17
229 x 229 mm inclined	each	0.11	0.06	1.36	51.14	52.50	65.63
Provide and build in 250 x 250 mm (nominal) cast iron soot-doors with double covers	each	0.30	0.25	4.38	19.96	24.34	30.43
Clay linings and pots							
Linings							
185 x 185 mm clay flue linings to BS 1181with rebated joints	m	0.44	0.22	5.42	19.77	25.19	31.49
extra for bends	each	0.12	0.06	1.49	15.33	16.82	21.02
Clay chimney-pots set and flaunched in mortar							
300 mm	each	0.55	0.28	6.78	18.69	25.47	31.84
450 mm	each	0.55	0.28	6.78	22.93	29.71	37.14
600 mm	each	0.61	0.30	7.53	33.71	41.24	51.55
Fix only							
Fix only in fireplace openings							
continuous burning fires	each	0.83	0.41	10.24	16.07	26.31	32.89
back boilers	each	1.65	0.82	20.34	0.20	20.54	25.68
Fix only average fireplace surrounds and hearths including assembling and jointing and setting in mortar	each	4.40	2.20	54.25	16.07	70.32	87.90
Gas flue blocks							
Typex HP gas flue blocks bedded and jointed in "Fluejoint" mortar							
Gas fire recess set comprising three 405 x 147 x 222 mm recess blocks HP1	set	0.33	0.17	4.07	27.75	31.82	39.77
385 x 140 x 222 mm cover block HP2	each	0.13	0.07	1.62	10.57	12.19	15.24
280 x 140 x 222 mm standard block HP3	each	0.12	0.06	1.48	9.01	10.49	13.11
380 x 140 x 222 mm closer block HP4	each	0.13	0.07	1.60	8.92	10.52	13.15
400 x 140 x 222 mm side offset block HP5	each	0.13	0.07	1.60	9.61	11.21	14.01
280 x 210 x 222 mm back offset block HP6	each	0.19	0.10	2.34	14.86	17.20	21.50
280 x 181 x 222 mm vertical exit block HP7	each	0.17	0.09	2.09	11.96	14.05	17.56
280 x 240 x 230 mm angled entry/exit block HP8	each	0.17	0.09	2.09	11.96	14.05	17.56
280 x 140 x 222 mm double rebate block HP9	each	0.12	0.06	1.48	10.50	11.98	14.98
280 x 262 x 222 mm corbel block HP10	each	0.17	0.09	2.09	11.65	13.74	17.18

Brickwork and blockwork

Alterations and repairs	Unit	Hours C	Hours L	Labour net	Material net	Price net	Price with 25%
Sundries				£	£	£	£
					VAT not included		
Metal flue pipes							
Typex Twin Wall metal flue pipes							
125 mm pipes in roof space	m	0.28	0.14	3.45	23.35	26.80	33.50
Extra for adjustable bends	each	0.36	0.18	4.44	12.14	16.58	20.73
Joints of 125 mm pipe to terminal block	each	0.19	0.10	2.34	0.10	2.44	3.05
Type ridge adapters bolted to ridge terminal and jointed to 125 mm pipe	each	0.28	0.14	3.45	12.94	16.39	20.49
Gas ridge terminal bedded and pointed in cement mortar 1:3	each	0.14	0.07	1.72	63.10	64.82	81.03
Centering							
Centering for flat brick arches not exceeding 2.00 m span; 102.5 mm wide							
first use	m	0.33	0.33	5.18	18.05	23.23	29.04
subsequent uses	m	0.22	0.22	3.45	0.02	3.47	4.34
Centering for segmental brick arches 1.00 m span; 102.5 mm wide and 25 mm rise							
first use	each	0.78	0.78	12.25	5.06	17.31	21.64
subsequent uses	each	0.55	0.55	8.64	0.33	8.97	11.21
Centering for segmental brick arches 2.00 m span; 102.5 mm wide and 25 mm rise							
first use	each	1.21	1.21	19.00	6.28	25.28	31.60
subsequent uses	each	0.66	0.66	10.36	0.66	11.02	13.78
Centering for semicircular brick arches 1.00 m span and 215 mm wide							
first use	each	3.63	3.63	56.99	8.32	65.31	81.64
subsequent uses	each	1.98	1.98	31.09	0.34	31.43	39.29
Centering for semicircular brick arches 2.00 m span and 215 mm wide							
first use	each	4.95	4.95	77.71	13.70	91.41	114.26
subsequent uses	each	2.97	2.97	46.63	-	46.63	58.29
Cavity insulation							
Expanded polystyrene cavity batts in cavities of hollow walls - retaining ties included with forming cavities							
25 mm	m2	0.28	0.14	3.45	1.62	5.07	6.34
40 mm	m2	0.28	0.14	3.45	2.60	6.05	7.56
50 mm	m2	0.28	0.14	3.45	3.38	6.83	8.54
Glass fibre cavity batts filling cavities of hollow walls							
50 mm	m2	0.28	0.14	3.45	3.21	6.66	8.32
75 mm	m2	0.28	0.14	3.45	4.27	7.72	9.65
100 mm	m2	0.28	0.14	3.45	5.68	9.13	11.41

Brickwork and blockwork

Alterations and repairs	Unit	Specialist price net	Price with 25%
Damp-proof-courses		£	£
		VAT not included	

The following "Specialist price net" figures are guide prices provided by Cavity Trays Ltd.

Prices do not include for cash discount.

Supply and deliver the following:

	Unit	Specialist price net	Price with 25%
Cavitrays Type C for building into hollow walls to form cavity gutters over openings	m from	7.42	9.28
Cavitrays Type G for building into hollow walls to form cavity gutters at abutments of flat roofs	m from	6.00	7.50
Preformed injection moulded polypropylene standard Cavitrays Type E 457 mm long for insertion into existing hollow walls to form cavity gutters	m	7.42	9.28
Standard 90 deg angles for "Cavitrays Type E"	each	5.30	6.63
Non-standard angles for "Cavitrays Type E"	each	7.95	9.94
Special polypropylene "Cavitrays Type X" complete with ready shaped lead flashing (BS Code 4) for building into hollow walls to form combined stepped cavity gutter and flashing to roof pitches of 26 deg to 45 deg (measured vertically from plate to ridge)	m rise from	49.27	61.59
Undercill Trays Type U in various profiles for building into hollow walls under cills and with removable front edge guide for alignment of brick or tile cills	m from	7.42	9.28
Cavity Closure/Water Check Trays Type R for building into hollow walls to form water check groove at underside of concrete floors	m	7.42	9.28
Lintel Stopends Type L for fixing at ends of steel lintels to prevent discharge of water into cavity	pair	1.06	1.33
450 x 450 x 1.80 mm (BS Code 4) lead slates, with collar for large pipe, for pitched roofs	each from	8.00	10.00
Caulking Mastic silicone formula, for positive adhesion in flashings caulking etc	310 ml tube	5.41	6.76
Cut away outer skin in short lengths, supply and insert "Cavitrays Type E" and rebuild outer skin	m from	24.00	30.00

Prices on request for stainless steel "Cavitrays" and "Cavitrays Type A" for arched openings.

	Unit	Specialist price net	Price with 25%
Universal Cavicloser Type H, for closing, damp coursing and insulating openings in cavity walls. Suits cavities from 50 mm to 100 mm	m	4.51	5.64
Combined "caviweep" and "cavivent" dual-function Type W offers unequalled bi-pressure performance. Suits perpendicular joints and satisfies all NHBC and Building Regulation requirements	each	0.41	0.51
Small Adjustable Telescopic Weep. Discreet, small-sized outlet for discharging water from dpc's, trays, etc.	each	0.41	0.51

Brickwork and blockwork

Alterations and repairs	Unit	Specialist price net	Price with 25%

Damp-proof-courses

		£	£
		VAT not included	
Type H cavicloser. Closes cavity reveal, acts as dpc, insulates and creates thermal zoning.	m	4.51	5.64

Enquiries about the foregoing specialist prices, free design and advisory service (purpose-made designs and specials available within ten days), should be made to Cavity Trays Ltd, Administration Centre, Lufton Trading Estate, Yeovil, Somerset, BA22 8HU, tel (01935) 474769 fax (01935) 428223

The following "Specialist price net" figures are minimum guide prices provided by Dampcoursing Ltd, for work within 50 kilometres of their works.

Prices do not include for cash discount.

See the preamble notes for builder's profit and attendance.

Chase out mortar joint in brickwork with mechanical saw, insert flexible membrane of lead, zinc, copper, bituminous material or low density polythene and force in new mortar (material cost of membrane not included)			
102.5 mm walls	m from	26.00	32.50
215 mm walls	m from	52.00	65.00
327.5 mm walls	m from	78.00	97.50

Quotations for other widths and other materials will be given on request.

Drill brickwork and provide "Vandex" damp-proof course			
102.5 mm walls	m from	4.50	5.63
215 mm walls	m from	9.00	11.25
327.5 mm walls	m from	11.00	13.75

Drill brickwork and provide chemical damp-proof course injected under pressure - aqueous			
102.5 mm walls	m from	4.50	5.63
215 mm walls	m from	9.00	11.25
327.5 mm walls	m from	11.00	13.75

Prices for other thickness walls on request.

To meet guarantee requirements, where a chemical damp-proof course is to be infused or injected, existing plastering or rendering to a height of one metre above the damp-proof course is to be hacked off and later replaced with two coats of Premix DR 5 and a "Thistle Multicoat" setting coat. Prices on request.

Five coat "Cementitious" tanking including preparation of exposed brickwork. Prices on request.

Inspection and quotations are recommended especially in the case of stone or very thick walls. Prices for " Newton 500" dry line tanking on request.

Enquiries about the foregoing specialist prices should be made to Dampcoursing Ltd, 10 - 12 Dorset Road, Tottenham, London, N15 5AJ, tel (0181) 802 2233/2333, fax (0181) 809 1839.

Brickwork and blockwork

Alterations and repairs	Unit	Specialist price net	Price with 25%

Damp-proof-courses

£   £

VAT not included

The following "Specialist price net" figures are minimum guide prices provided by Kiltox Chemicals Ltd, for work by one of their approved contractors within 200 kilometres of a branch depot.

Prices do not include for cash discount.

See the preamble notes for builder's profit and attendance.

Chemical damp-proofing using "Kiltox" aqueous-based siliconate DPC pressure injection system

Horizontal or vertical damp-proof courses to brickwork

	Unit	Specialist price net	Price with 25%
102.5 mm walls	m from	6.16	7.70
215 mm walls	m from	11.09	13.86
327.5 mm walls	m from	14.16	17.71
440 mm walls	m from	18.47	23.09
For each additional 102.5 mm thickness of wall	m from	6.16	7.70

Prices for damp-proof courses to other wall constructions and thicknesses available on request.

Enquiries about the foregoing specialist prices should be made to Kiltox Chemicals Ltd, Kiltox House, Park Row, Greenwich, London, SE10 9NL, tel (0208) 858 6277, fax (0208) 853 3572.

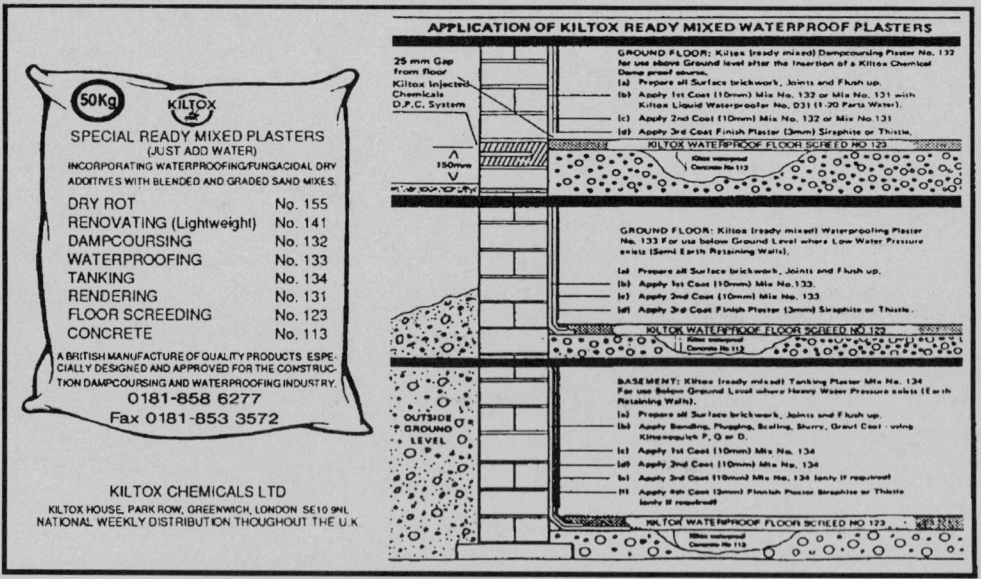

Brickwork and blockwork

	Unit	Specialist price net	Price with 25%
Alterations and repairs			

UF foam cavity wall insulation

		£	£
		VAT not included	

The foam is installed by specialist sub-contractors who should be registered with the British Standards Institution. For existing buildings the work is carried out from the outside. The small injection holes in mortar joints or rendering are made good with matching mortar before the sub-contractor leaves the site

The specialist sub-contractor supplies all equipment and materials required for the installation. Prices depend partly on quantity of work per contract; quotations should be obtained for each contract.

The following "Specialist price net" figures are guide prices provided by the Cavity Foam Bureau.

Prices do not include for cash discount.

	Unit	Specialist price net	Price with 25%
UF foam insulation to hollow walls in areas 50 - 100 m2, cavity widths			
up to 65 mm	m2	3.80	4.75
65 to 75 mm	m2	4.05	5.06
UF foam insulation to hollow walls in areas 100 - 500 m2, cavity widths			
up to 65 mm	m2	3.60	4.50
65 to 75 mm	m2	3.85	4.81
UF foam insulation to hollow walls in areas over 500 m2, cavity widths			
up to 65 mm	m2	3.30	4.13
65 to 75 mm	m2	3.55	4.44

Enquiries about the foregoing specialist prices and technical information about UF foam should be made to The Cavity Foam Bureau, PO Box 79, Oldbury, Warley, West Midlands, B69 4PW, tel (0121) 544 4949.

Brickwork and blockwork

Alterations and repairs	Unit	Hours C	Hours L	Labour net	Material net	Price net	Price with 25%
Composite walls				£	£	£	£
					VAT not included		
Facing bricks external skin							
Hollow walls in gauged mortar 1:1:6 of 102.5 mm facing brick external skin pointed as the work proceeds; 50 - 100 mm cavity with galvanised butterfly wall-ties and internal skin of							
100 mm dense aggregate blocks	m2	3.17	1.85	40.86	31.40	72.26	90.33
140 mm dense aggregate blocks	m2	3.43	2.05	44.54	37.50	82.04	102.55
100 mm clinker aggregate blocks	m2	2.84	1.60	36.21	32.44	68.65	85.81
140 mm clinker aggregate blocks	m2	3.03	1.75	38.94	34.13	73.07	91.34
100 mm lightweight aggregate blocks	m2	2.77	1.56	35.32	35.71	71.03	88.79
150 mm lightweight aggregate blocks	m2	2.97	1.71	38.13	42.97	81.10	101.38
Dense aggregate block external skin							
Hollow walls in gauged mortar 1:1:6 of 100 mm dense aggregate block external skin; 50 - 100 mm cavity with galvanised butterfly wall-ties and internal skin of							
100 mm dense aggregate blocks	m2	2.15	1.56	29.77	13.91	43.68	54.60
140 mm dense aggregate blocks	m2	2.51	1.85	34.95	20.01	54.96	68.70
100 mm clinker aggregate blocks	m2	1.92	1.40	26.63	14.95	41.58	51.98
140 mm clinker aggregate blocks	m2	2.11	1.55	29.34	16.64	45.98	57.48
100 mm lightweight aggregate blocks	m2	1.85	1.36	25.73	18.22	43.95	54.94
150 mm lightweight aggregate blocks	m2	2.05	1.51	28.54	25.48	54.02	67.53
Lightweight aggregate block external skin							
Hollow walls in gauged mortar 1:1:6 of 100 mm lightweight aggregate block external skin; 50 - 100 mm cavity with galvanised butterfly wall-ties and internal skin of							
100 mm dense aggregate blocks	m2	1.85	1.36	25.73	18.22	43.95	54.94
140 mm dense aggregate blocks	m2	2.11	1.56	29.41	24.32	53.73	67.16
100 mm clinker aggregate blocks	m2	1.52	1.11	21.09	19.26	40.35	50.44
140 mm clinker aggregate blocks	m2	1.71	1.26	23.81	20.95	44.76	55.95
100 mm lightweight aggregate blocks	m2	1.45	1.07	20.19	22.52	42.71	53.39
150 mm lightweight aggregate blocks	m2	1.65	1.22	23.00	29.79	52.79	65.99
Insulation							
Extra for expanded polystyrene cavity batts in cavities and 0.6 mm stainless steel pressed ties with insulation retaining clips (ties and clips measured in form cavity)							
25 mm batts	m2	0.28	0.14	3.45	1.62	5.07	6.34
40 mm batts	m2	0.28	0.14	3.45	2.60	6.05	7.56
50 mm batts	m2	0.28	0.14	3.45	3.38	6.83	8.54
Extra for glass fibre cavity batts filling cavities							
50 mm batts	m2	0.22	0.11	2.71	3.21	5.92	7.40
75 mm batts	m2	0.22	0.11	2.71	4.27	6.98	8.72
100 mm batts	m2	0.22	0.11	2.71	5.68	8.39	10.49

Brickwork and blockwork

Alterations and repairs	Unit	Hours C	Hours L	Labour net	Material net	Price net	Price with 25%
Repairs and maintenance				£	£	£	£
					VAT not included		
Rake out old mortar pointing to brickwork and repoint in cement mortar	m2	0.65	0.65	10.20	0.20	10.40	13.00
Rake out joints, rewedge and repoint flashings	m	0.55	-	4.93	0.30	5.23	6.54
Cut out cracks in 102.5 mm brick walls and make good with common bricks	m	2.00	2.00	31.40	4.22	35.62	44.52
Cut out cracks in 102.5 mm brick walls and make good with facing bricks and point to match existing	m	2.50	2.50	39.25	7.14	46.39	57.99
Cut out and remove defective facing bricks, renew with facing bricks and point to match existing	each	0.33	0.33	5.18	0.44	5.62	7.03
Take down defective brick on end camber arches 1000 mm span, 225 mm on face and 112.5 mm on soffit, provide centering, rebuild arch in new bricks pointed to match existing and make good all works disturbed	each	8.30	4.40	104.03	15.83	119.86	149.82
Cut away for and build in 75 x 12 mm galvanised steel arch bars to support defective or dropped camber arch 1000 mm span and make good all work disturbed	each	1.65	1.35	23.88	1.00	24.88	31.10
Cut out 200 x 90 mm defective stone or concrete sills 1000 mm long and cast new in-situ concrete sills including formwork and make good all work disturbed	each	4.00	2.50	52.69	5.90	58.59	73.24
Cut out and build in ends of joists and make good brickwork	each	0.13	0.13	2.04	0.73	2.77	3.46
Cut out wall plates in short lengths and make good brickwork	m	0.50	0.50	7.85	2.09	9.94	12.43
Cut away and renew mortar angle fillets							
50 mm	m	0.50	0.50	7.85	0.10	7.95	9.94
75 mm	m	0.75	0.75	11.77	0.30	12.07	15.09
Take down brick chimney-stacks and rebuild in common bricks fair faced and flush pointed as the work proceeds including parging and coring flues							
440 x 440 mm	m	8.00	8.35	127.96	27.27	155.23	194.04
777.5 mm x 440 mm	m	13.00	13.55	207.81	36.44	244.25	305.31
777.5 x 777.5 mm	m	22.00	22.90	351.47	57.05	408.52	510.65
Remove defective chimney-pots and provide and fix new set and flaunched in mortar							
450 mm	each	1.25	1.75	23.00	20.22	43.22	54.02
900 mm	each	1.50	2.00	26.92	38.62	65.54	81.92

Alterations and repairs	Unit	Hours C	Hours L	Labour net	Material net	Price net	Price with 25%
Repairs and maintenance				£	£	£	£
					VAT not included		
Take out average fireplaces, unload and fix only new fireplace with fireback and hearth and make good brickwork and plaster	each	7.00	8.00	116.64	1.03	117.67	147.09
Cut away 215 mm brick wall in short lengths and insert double slate horizontal damp-proof course bedded in cement mortar, make good to brickwork over with brick on edge course and point to match existing (making good skirtings, plaster and decorations	m	4.40	4.40	69.08	5.03	74.11	92.64

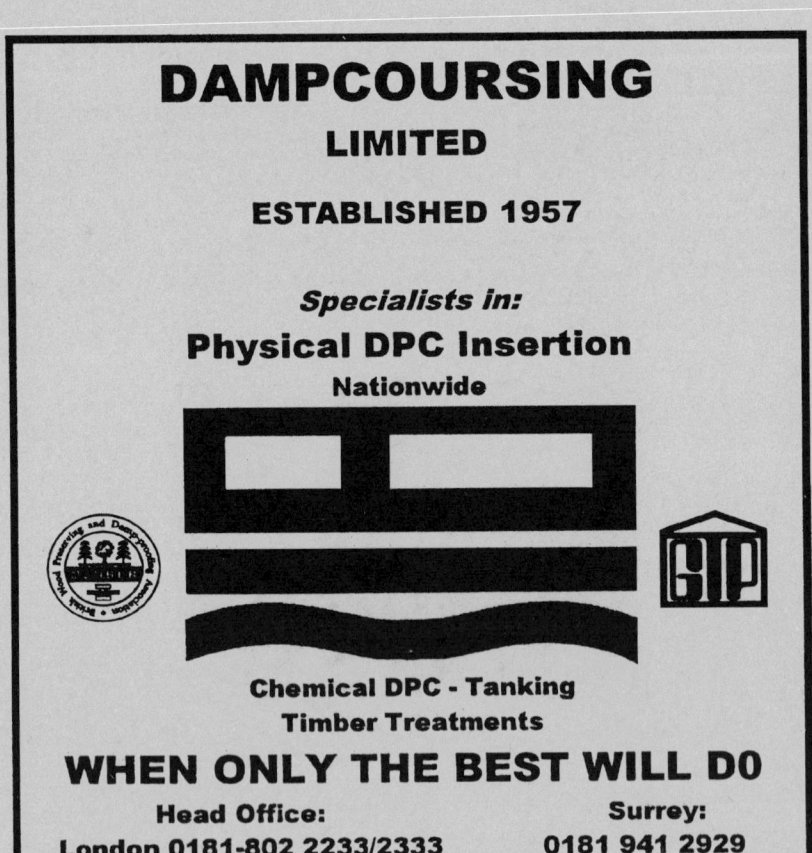

Underpinning

Preamble

"Labour net" figures include allowances for all costs incidental to the employment of labour.

"Plant net" figures include for all costs of plant including drivers and operators applicable.

"Materials net" figures include for all costs of materials including an allowance for waste except where specifically stated.

"Price net" figures are the totals of the "Labour net", "Plant net" and "Materials net" figures. Prices are for a builder employing his own labour; according to the amount and nature of the work involved, it may well be possible to secure more advantageous prices from specialist sub-contractors.

Prices do not include any allowance for scaffolding, ladders or other plant necessary to reach the work. The "Preliminaries" section includes prices for scaffolding which must be considered and allowance included to suit the particular circumstances of a tender.

Underpinning

Alterations and repairs

	Unit	Price
Basic prices for materials		**£**
Aggregates		
40 mm	m3	19.75
	tonne	13.16
20 mm	m3	19.91
	tonne	13.27
Damp-proof courses		
fibre base - 225 mm wide	8 m roll	9.76
pitch polymer - 225 mm wide	20 m roll	36.95
slates - 350 x 225 mm	10	12.00
Portland cement	tonne	89.04
Sand	m3	17.96
	tonne	11.22

Prices actually to be paid for materials must be checked against the above basic prices and adjustments made as necessary.

Underpinning

Alterations and repairs	Unit	Hours C	Hours L	Labour net	Plant net	Material net	Price net	Price with 25%
Work in all trades				£	£	£	£	£
					VAT not included			

For temporary supports to work to be underpinned, see "Demolition" section.

Preliminary trenches

Excavate preliminary trenches down to the level of the base of the existing foundation, maximum depth not exceeding

1.00 m	m3	-	5.98	40.31	-	-	40.31	50.39
2.00 m	m3	-	8.05	54.26	-	-	54.26	67.83

Excavate below foundations

Excavate below the level of the base of the existing foundation, maximum depth not exceeding

1.00 m	m3	-	6.78	45.70	-	-	45.70	57.13
2.00 m	m3	-	8.97	60.46	-	-	60.46	75.58

Working space

Excavate and fill working space, maximum depth not exceeding

1.00 m	m3	-	7.94	53.52	-	-	53.52	66.90
2.00 m	m3	-	9.89	66.66	-	-	66.66	83.33

Projecting foundations

Cut away projecting foundations two courses of footings and

600 x 225 mm concrete	m	-	2.99	20.15	-	-	20.15	25.19

three courses of footings and

825 x 300 mm concrete	m	-	5.64	38.01	-	-	38.01	47.51

Prepare underside

Prepare the underside of the existing work to receive the pinning up of the new

300 mm wide	m	-	0.35	2.36	-	-	2.36	2.95
450 mm wide	m	-	0.50	3.37	-	-	3.37	4.21
600 mm wide	m	-	0.75	5.05	-	-	5.05	6.31
900 mm wide	m	-	0.98	6.61	-	-	6.61	8.26
1200 mm wide	m	-	1.32	8.90	-	-	8.90	11.13

Underpinning

Alterations and repairs	Unit	Hours C	Hours L	Labour net	Plant net	Material net	Price net	Price with 25%
Work in all trades				£	£	£	£	£
					VAT not included			
Earthwork support								
Earthwork support to sides of preliminary trenches not exceeding 2.00 m between opposing faces, in stiff soil, maximum depth not exceeding								
1.00 m	m2	-	0.17	1.15	-	0.87	2.02	2.52
2.00 m	m2	-	0.19	1.28	-	0.87	2.15	2.69
Earthwork support to sides of excavation below the level of the base of existing foundation not exceeding 2.00 m between opposing faces, in stiff soil, maximum depth not exceeding								
1.00 m	m2	-	0.18	1.21	-	0.87	2.08	2.60
2.00 m	m2	-	0.21	1.42	-	1.74	3.16	3.95
Earthwork support to sides of preliminary trenches, not exceeding 2.00 m between opposing faces, in moderately firm soil, maximum depth not exceeding,								
1.00 m	m2	-	0.65	4.38	-	2.32	6.70	8.38
2.00 m	m2	-	0.84	5.66	-	2.32	7.98	9.97
Earthwork support to sides of excavation below the level of the base of existing foundation, not exceeding 2.00 m between opposing faces, in moderately firm soil, maximum depth not exceeding,								
1.00 m	m2	-	0.71	4.79	-	2.32	7.11	8.89
2.00 m	m2	-	0.92	6.20	-	2.32	8.52	10.65
Earthwork support to sides of preliminary trenches not exceeding 2.00 m between opposing faces, in soft soil, maximum depth not exceeding,								
1.00 m	m2	-	1.29	8.69	-	3.19	11.88	14.85
2.00 m	m2	-	1.51	10.18	-	3.19	13.37	16.71
Earthwork support to sides of excavation below the level of existing foundation, not exceeding 2.00 m between opposing faces, in soft soil, maximum depth not exceeding,								
1.00 m	m2	-	1.42	9.57	-	3.19	12.76	15.95
2.00 m	m2	-	1.66	11.19	-	3.19	14.38	17.98

Underpinning

Alterations and repairs	Unit	Hours C	Hours L	Labour net	Plant net	Material net	Price net	Price with 25%
Work in all trades				£	£	£	£	£
					VAT not included			

Earthwork support (*continued*)

Earthwork support left in below level of base

1.00 m deep - stiff soil	m2	-	0.12	0.81	-	5.21	6.02	7.53
2.00 m deep - stiff soil	m2	-	0.14	0.94	-	5.21	6.15	7.69
1.00 m deep - moderately soft soil	m2	-	0.47	3.17	-	13.90	17.07	21.34
2.00 m deep - moderately stiff soil	m2	-	0.61	4.11	-	13.90	18.01	22.51
1.00 m deep - soft soil	m2	-	0.95	6.40	-	19.12	25.52	31.90
2.00 m deep - soft soil	m2	-	1.11	7.48	-	19.12	26.60	33.25

Disposal

Excavated material removed from site to tip average 15 km from site	m3	-	2.02	13.61	19.92	-	33.53	41.91
Add or deduct for every 1 km difference in distance	m3	-	-	-	0.20	-	0.20	0.25

Filling

Excavated material filling to excavations deposited and compacted in 225 mm layers	m3	-	1.87	12.60	3.64	-	16.24	20.30

Trench bottoms

Level and compact bottom of excavation	m2	-	0.29	1.95	0.74	-	2.69	3.36

Concrete work

Foundations in trenches over 300 mm thick

1:3:6 concrete	m3	-	6.03	40.64	-	75.47	116.11	145.14
1:2:4 concrete	m3	-	6.03	40.64	-	73.19	113.83	142.29

Brickwork

Walls in common bricks in cement mortar 1:3

215 mm	m2	4.60	2.30	56.72	-	29.70	86.42	108.03
327.5	m2	6.21	3.11	76.57	-	43.63	120.20	150.25

Underpinning

Alterations and repairs	Unit	Hours C	Hours L	Labour net	Plant net	Material net	Price net	Price with 25%
Work in all trades				£	£	£	£	£
					VAT not included			

Damp proof courses

Fibre base bitumen horizontal damp-proof courses bedded in cement mortar 1:3

Alterations and repairs	Unit	Hours C	Hours L	Labour net	Plant net	Material net	Price net	Price with 25%
over 225 mm wide	m2	0.29	0.14	3.58	-	5.76	9.34	11.68
225 mm wide	m	0.64	0.32	7.89	-	1.28	9.17	11.46

Pitch polymer horizontal damp-proof courses bedded in cement mortar 1:3

over 225 mm wide	m2	0.29	0.14	3.58	-	8.73	12.31	15.39
225 mm wide	m	0.06	0.03	0.79	-	1.94	2.73	3.41

Double course slate horizontal damp-proof courses bedded in cement mortar 1:3

over 225 mm wide	m2	1.90	0.95	23.42	-	22.20	45.62	57.02
225 mm wide	m	0.44	0.22	5.42	-	5.04	10.46	13.07

Wedge and pin

Wedge and pin up to underside of existing construction with slates in cement mortar 1:3

215 mm walls	m	0.55	0.28	6.82	-	4.36	11.18	13.98
327.5 mm walls	m	0.85	0.42	10.48	-	6.39	16.87	21.09

Rubble walling

Preamble

"Labour net" figures include allowances for all costs incidental to the employment of labour. The labour for rubble walling has been based generally on a team of three masons to two labourers as for brickwork.

"Materials net" figures include for all costs of materials including an allowance for waste except where specifically stated.

"Price net" figures are the totals of the "Labour net" and "Materials net" figures. Prices are for a builder employing his own labour; according to the amount and nature of the work involved, it may well be possible to secure more advantageous prices from specialist sub-contractors.

Prices do not include any allowance for scaffolding, ladders or other plant necessary to reach the work. The "Preliminaries" section includes prices for scaffolding which must be considered and allowance included to suit the particular circumstances of a tender.

Rubble walling

Alterations and repairs

	Unit	Price
Basic prices for materials		**£**
Galvanised butterfly wall-ties	1000	101.40
Hydrated lime	tonne	139.84
Portland cement	tonne	89.04
Sand	m3	17.96
	tonne	11.22
Walling stone		
150 mm on bed	tonne	94.50

Prices actually to be paid for materials must be checked against the above basic prices and adjustments made as necessary.

Rubble walling

Alterations and repairs	Unit	Hours C	Hours L	Labour net	Material net	Price net	Price with 25%
Stone rubble work				£	£	£	£
					VAT not included		
Walling laid in mortar							
150 mm random rubble filling to existing openings in gauged mortar 1:1:6 coursed average 450 mm high and flush pointed one side	m2	5.15	3.45	69.39	45.15	114.54	143.18
Extra for dressed face to rubble work	m2	2.10	-	18.82	-	18.82	23.52
Bond 150 mm walls to existing including cutting pockets in existing work and extra material for bonding	m	0.75	0.75	11.77	2.34	14.11	17.64
Work to new openings							
Quoin up jambs of new openings in existing 450 mm walls including dressing quoin stones one side and flush pointing	m	2.30	2.30	36.11	22.07	58.18	72.72
Quoin up jambs of new openings in existing 450 mm walls including dressing up quoin stones both sides and flush pointing	m	2.70	2.70	42.39	22.57	64.96	81.20
Cut holes through rubble work and make good for							
small pipes (not exceeding 55 mm)	each	1.75	1.17	23.57	0.40	23.97	29.96
large pipes (55 - 110 mm)	each	3.25	2.17	43.75	2.01	45.76	57.20
extra large pipes (over 110 mm)	each	6.25	4.17	84.11	4.02	88.13	110.16

Masonry

Preamble

Prices do not include any allowance for scaffolding, ladders or other plant necessary to reach the work. The "Preliminaries" section includes prices for scaffolding which must be considered and allowance included to suit the particular circumstances of a tender.

Specialist prices

"Price with 25%" figures are all-in guide prices and include 25% for the builder's overheads, profit, unloading materials and general attendance (to include free use of standing scaffolding and hoists, temporary lighting and water and clearing away rubbish).

The amount of attendance required varies between the various trades and also with the circumstances of specific jobs; the percentage addition must always be considered and adjusted as necessary to suit the terms and conditions of the quotation being used.

Quantities and delivery distances are usually the most significant of the many factors which influence prices and it must be emphasised that quotations should always be obtained when preparing a tender.

Masonry

Alterations and repairs	Unit	Specialist price net	Price with 25%

Stonework

		£	£
			VAT not included

The following "Specialist price net" figures are guide prices for work on sites within about 60 kilometres of a specialist depot.

Prices for fixing stonework allow for all bedding and pointing materials and for cleaning down on completion but do not include for cramps or dowels.

Prices do not include for cash discount.

See the preamble notes for builder's profit and attendance.

The builder would be required to provide a mechanical hoist, all necessary scaffolding, power etc for the use of the masons.

Fixing stonework

	Unit	Specialist price net	Price with 25%
Fix only the following cast stone:			
50mm	m2	82.50	103.12
75mm	m2	110.00	137.50
Cut out and piece in to existing natural or reconstructed stonework	m3	5,891.61	7,364.51

Cleaning stonework or brickwork

	Unit	Specialist price net	Price with 25%
Clean by nebulous cold water spray process assisted by suitable graded brushes	m2	16.65	20.82
Clean by chemical and high pressure water process	m2	14.28	17.85
Dry clean by use of silica-free abrasive grit under regulated air pressure	m2	17.04	21.30
Dry clean by use of spinning carborundum pads	m2	28.87	36.09

Repointing stonework

	Unit	Specialist price net	Price with 25%
Rake out and repoint			
ashlar facing	m2	24.14	30.17
flint walling	m2	110.42	138.02
rubble walling	m2	52.74	65.93

Masonry

Alterations and repairs	Unit	Specialist price net	Price with 25%

Stonework

£   £

VAT not included

Repairing stonework

Bath or similar soft stone

Cut out decayed or defective stonework (normally found in small isolated areas or individual stones) to a minimum depth of 25 mm or to sound stone, whichever the greater, properly key, dowel and reinforce as necessary with non-ferrous metal and make good in "Plastic Artificial Stone" to match existing to the following:

	Unit	Specialist price net	Price with 25%
ashlar	m2	441.68	552.10
mouldings 75 mm girth	m	138.02	172.52
per 25 mm increase in girth of mouldings	m	46.01	57.52
circular columns, plain	m2	572.00	715.00
circular columns, fluted	m2	1,309.00	1,636.25
mullion fronts from glazing 150 mm wide	m	195.54	244.42
stoolings for jambs or mullions, plain	each	60.96	76.20
stoolings for jambs or mullions, moulded	each	103.52	129.40
plain copings 300 mm wide	m	210.10	262.62
per 25 mm increase in width of copings	m	46.01	57.52
circular labels or hoods to 225 mm girth	m	276.06	345.07
tracery	m2	1,309.00	1,636.25

Portland or similar stone

Cut out decayed or defective stonework (normally found in small isolated areas or individual stones) to a minimum depth of 25 mm or to sound stone, whichever the greater, properly key, dowel and reinforce as necessary with non-ferrous metal and make good in "Plastic Artificial Stone" to match existing to the following:

	Unit	Specialist price net	Price with 25%
ashlar	m2	529.10	661.37
mouldings 75 mm girth	m	165.64	207.05
per 25 mm increase in girth of mouldings	m	55.21	69.01
circular columns, plain	m2	657.80	822.25
circular columns, fluted	m2	1,504.80	1,881.00
mullion fronts from glazing 150 mm wide	m	234.64	293.30
stoolings for jambs or mullions, plain	each	73.15	91.44
stoolings for jambs or mullions, moulded	each	124.22	155.28
plain copings 300 mm wide	m	242.00	302.50
per 25 mm increase in width of copings	m	55.21	69.01
circular labels or hoods to 225 mm girth	m	331.26	414.08
tracery	m2	1,504.80	1,881.00

Masonry

	Unit	Specialist price net	Price with 25%

Stonework

		£	£
		VAT not included	

Repairing stonework *(continued)*

York or similar stone

Cut out decayed or defective stonework (normally found in small isolated areas or individual stones) to a minimum depth of 25 mm or to sound stone, which ever the greater, properly key, dowel and reinforce as necessary with non-ferrous metal and make good in "Plastic Artificial Stone" to match existing to the following:

	Unit	Specialist price net	Price with 25%
ashlar	m2	582.01	727.51
mouldings 75 mm girth	m	182.19	227.74
per 25 mm increase in girth of mouldings	m	60.74	75.93
circular columns, plain	m2	674.24	842.81
circular columns, fluted	m2	1,542.20	1,927.75
mullion fronts from glazing 150 mm wide	m	258.10	322.63
stoolings for jambs or mullions, plain	each	79.71	99.63
stoolings for jambs or mullions, moulded	each	136.65	170.82
plain copings 300 mm wide	m	248.05	310.06
per 25 mm increase in width of copings	m	60.74	75.93
circular labels or hoods to 225 mm girth	m	364.40	455.50
tracery	m2	1,542.20	1,927.75

Although prices for repairing stonework allow for cutting out to a depth of 25 mm, it is essential always to cut back to sound stone and it may be necessary to increase the depth depending on the extent to which the stone has decayed.

Asphalt work

Preamble

Asphalt work is the province of specialist sub-contractors; consequently prices in this section are based on those of specialists.

Prices do not include any allowance for scaffolding, ladders or other plant necessary to reach the work. The "Preliminaries" section includes prices for scaffolding which must be considered and allowance included to suit the particular circumstances of a tender.

Specialist prices

"Price with 25%" figures are all-in guide prices and include 25% for the builder's overheads, profit, unloading materials and general attendance (to include free use of standing scaffolding and hoists, temporary lighting and water and clearing away rubbish).

The amount of attendance required varies between the various trades and also with the circumstances of specific jobs; the percentage addition must always be considered and adjusted as necessary to suit the terms and conditions of the quotation being used.

Quantities and delivery distances are usually the most significant of the many factors which influence prices and it must be emphasised that quotations should always be obtained when preparing a tender.

Asphalt work

Alterations and repairs

Generally

£ £

VAT not included

The following "Specialist price net" figures are guide prices provided by Asphaltic Contracts for quantities of about 50 square metres within 25 kilometres of a branch depot.

Prices do not include for cash discount.

See the preamble notes for builder's profit and attendance.

Damp-proofing and tanking

Mastic asphalt (limestone aggregate) to BS 6925 Type T1097

	Unit	Specialist price net	Price with 25%
13 mm one coat horizontal coverings on concrete			
over 300 mm wide	m2	13.87	17.34
not exceeding 150 mm wide	m	4.37	5.46
150 - 300 mm wide	m	6.21	7.76
20 mm two coat horizontal coverings on concrete			
over 300 mm wide	m2	17.07	21.34
not exceeding 150 mm wide	m	5.05	6.31
150 - 300 mm wide	m	7.97	9.96
30 mm three coat horizontal coverings on concrete			
over 300 mm wide	m2	24.68	30.85
not exceeding 150 mm wide	m	6.57	8.21
150 - 300 mm wide	m	10.68	13.35
13 mm two coat vertical coverings on brickwork			
over 300 mm wide	m2	38.67	48.34
not exceeding 150 mm wide	m	8.24	10.30
150 - 300 mm wide	m	13.17	16.46
20 mm three coat vertical coverings on brickwork			
over 300 mm wide	m2	48.05	60.06
not exceeding 150 mm wide	m	10.49	13.11
150 - 300 mm wide	m	12.32	15.40
Internal angle fillets	m	4.05	5.06
Turning nibs into grooves	m	2.67	3.34
Working into outlets	each	20.50	25.63
Collars and internal angle fillets around			
small pipes (not exceeding 55 mm)	each	14.35	17.94
large pipes (55 - 110 mm)	each	16.40	20.50

Asphalt work

Alterations and repairs	Unit	Specialist price net	Price with 25%
Generally		£	£
		VAT not included	

Flooring

Mastic asphalt (limestone aggregate) to BS 6925 Type F1076

15 mm one coat light duty flooring and isolating membrane			
over 300 mm wide	m2	16.26	20.33
not exceeding 150 mm wide	m	4.56	5.70
150 - 300 mm wide	m	7.72	9.65
20 mm one coat medium duty flooring and isolating membrane			
over 300 mm wide	m2	18.25	22.81
not exceeding 150 mm wide	m	5.20	6.50
150 - 300 mm wide	m	8.64	10.80
30 mm one coat heavy duty flooring and isolating membrane			
over 300 mm wide	m2	26.25	32.81
not exceeding 150 mm wide	m	7.50	9.38
150 - 300 mm wide	m	10.84	13.55
Working against metal frames	m	1.85	2.31
Extra for working flooring into recessed covers not exceeding 1.00 m2	each	18.45	23.06
13 x 150 mm two coat skirtings with fair edge, coved angle fillet and nib turned into groove - including angles	m	8.21	10.26

Coloured mastic asphalt (limestone aggregate) to BS 6925 Type F1451

15 mm one coat light duty brown flooring and isolating membrane			
over 300 mm wide	m2	20.24	25.30
not exceeding 150 mm wide	m	6.32	7.90
150 - 300 mm wide	m	10.77	13.46
Working against metal frames	m	1.95	2.44
Extra for working flooring into recessed covers not exceeding 1.00 m2	each	18.45	23.06
13 x 150 mm two coat brown skirtings with fair edge, coved angle fillet and nib turned into groove - including angles	m	8.66	10.83

Roofing

Mastic asphalt (limestone aggregate)to BS 6925 Type R988

20 mm two coat flat coverings and isolating membrane			
over 300 mm wide	m2	16.37	20.46
not exceeding 150 mm wide	m	6.00	7.50
150 - 300 mm wide	m	9.00	11.25
25 mm two coat flat coverings and isolating membrane			
over 300 mm wide	m2	20.77	25.96
not exceeding 150 mm wide	m	5.73	7.16
150 - 300 mm wide	m	9.51	11.89

Asphalt work

Alterations and repairs

	Unit	Specialist price net	Price with 25%
Generally		£	£
		VAT not included	

Roofing *(continued)*

Mastic asphalt (limestone aggregate)to BS 6925 Type R988 *(continued)*

	Unit	Specialist price net	Price with 25%
Turning nibs into grooves	m	1.85	2.31
Working to metal flashings	m	1.95	2.44
Working into outlets	each	20.50	25.63
Two coat skirtings with fair edge, internal angle fillet and nib turned into groove - including angles			
13 x 150 mm	m	8.22	10.28
13 x 250 mm	m	9.52	11.90
Two coat aprons with undercut drip edge and rounded arris - including angles			
13 x 75 mm	m	5.36	6.70
13 x 100 mm	m	5.85	7.31
13 x 300 mm two coat linings to gutter with two rounded arrises and two internal angle fillets - including angles and intersections	m	17.75	22.19
ends	each	4.72	5.90
outlets	each	20.50	25.63
Collars and internal angle fillets around			
small pipes (not exceeding 55 mm)	each	15.38	19.23
large pipes (55 - 110 mm)	each	17.43	21.79
Accessories			
50 x 65 mm aluminium edge trims including butt straps and working asphalt to trim	m	14.44	18.05
Extra for right angle corner pieces			
internal	each	13.12	16.40
external	each	13.12	16.40
75 x 65 mm aluminium edge trims including butt straps and working asphalt to trim	m	15.06	18.83
Extra for right angle corner pieces			
internal	each	13.57	16.96
external	each	13.57	16.96
Pressure release breather ventilators including asphalt collars	each	30.75	38.44

Enquiries about the foregoing specialist prices should be made to Asphaltic Contracts Ltd, Meesons Wharf, 1-15 High Street, Stratford, London E15 2QQ, tel (0181) 519 9555, fax (0181) 519 9666.

Roofing

Preamble

"Labour net" figures include allowances for all costs incidental to the employment of labour. "Labour net" figures for sheet lead work are based on the labour costs of an advanced plumber working with an apprentice in the third year of training and include allowances for all costs incidental to the employment of labour.

"Materials net" figures include for all costs of materials including an allowance for waste except where specifically stated.

"Price net" figures are the totals of the "Labour net" and "Materials net" figures.

Prices do not include any allowance for scaffolding, ladders or other plant necessary to reach the work. The "Preliminaries" section includes prices for scaffolding which must be considered and allowance included to suit the particular circumstances of a tender.

Specialist prices

"Price with 25%" figures are all-in guide prices and include 25% for the builder's overheads, profit, unloading materials and general attendance (to include free use of standing scaffolding and hoists, temporary lighting and water and clearing away rubbish).

The amount of attendance required varies between the various trades and also with the circumstances of specific jobs; the percentage addition must always be considered and adjusted as necessary to suit the terms and conditions of the quotation being used.

Quantities and delivery distances are usually the most significant of the many factors which influence prices and it must be emphasised that quotations should always be obtained when preparing a tender.

Roofing

Alterations and repairs

	Unit	Price
Basic prices for materials		**£**
508 x 254 mm slates	100	308.27
265 x 165 mm machine made ornamental clay tiles	100	63.00
Impregnated softwood battens		
19 x 38 mm	100 m	19.00
19 x 50 mm	100 m	35.15
Reinforced bituminous felt - 15 x 1 m rolls	each	26.00
Bitumen emulsion	25 litre	35.00
Milled sheet lead - Code 5, 2.24mm	m2	24.92
Zinc sheet 0.65mm	m2	10.23

Prices actually to be paid for materials must be checked against the above basic prices and adjustments made as necessary.

Roofing

Alterations and repairs	Unit	Hours C	Hours L	Labour net	Material net	Price net	Price with 25%
Corrugated/troughed sheeting				£	£	£	£
					VAT not included		
Roof decking							
0.7 x 35 mm galvanised steel troughed decking							
decking bolted to steelwork	m2	0.26	0.26	4.08	12.87	16.95	21.19
square cutting to large openings	m	0.17	0.08	2.06	1.11	3.17	3.96
holes for pipes	each	0.80	0.38	9.73	-	9.73	12.16
0.7 x 63 mm galvanised steel troughed decking							
decking bolted to steelwork	m2	0.26	0.26	4.08	13.21	17.29	21.61
square cutting to large openings	m	0.17	0.08	2.06	1.14	3.20	4.00
holes for pipes	each	0.80	0.38	9.73	-	9.73	12.16
50 mm wood wool slabs							
decking in 1800 x 600 mm standard slabs							
nailed to timber joists	m2	0.17	0.08	2.06	8.18	10.24	12.80
square cutting to large openings	m	0.22	0.11	2.71	0.81	3.52	4.40
raking cutting	m	0.26	0.13	3.21	4.05	7.26	9.07
holes for pipes	each	0.77	0.38	9.46	-	9.46	11.82
75 mm woodwool slabs							
decking in 2700 x 600 mm standard slabs							
nailed to timber joists	m2	0.22	0.11	2.71	13.64	16.35	20.44
square cutting to large openings	m	0.22	0.11	2.71	1.29	4.00	5.00
raking cutting	m	0.26	0.13	3.21	6.45	9.66	12.07
holes for pipes	each	0.77	0.38	9.46	-	9.46	11.82

Roofing

Alterations and repairs	Unit	Hours C	Hours L	Labour net	Material net	Price net	Price with 25%
Flexible sheet finishings				£	£	£	£
					VAT not included		
Bitumen-felt roofing							
Two layer glass fibre felt coverings and stone chippings surfacing							
Flat coverings over 300 mm wide	m2	0.31	0.16	3.86	5.35	9.21	11.51
Working outlets	each	0.72	0.36	8.88	2.10	10.98	13.73
Three layer glass fibre felt coverings and stone chippings surfacing							
Flat coverings over 300 mm wide	m2	0.35	0.17	4.29	7.53	11.82	14.78
Working into outlets	each	0.77	0.38	9.46	2.43	11.89	14.86
Two layer "Asbex" glass fibre felt mineral-surfaced coverings							
Sloping coverings over 300 mm wide	m2	0.38	0.19	4.68	5.05	9.73	12.16
Working into outlets	each	0.71	0.37	8.85	2.07	10.92	13.65
Three layer "Asbex" glass fibre felt mineral-surfaced coverings							
Sloping coverings over 300 mm wide	m2	0.52	0.26	6.41	7.85	14.26	17.82
Working into outlets	each	0.77	0.38	9.46	2.91	12.37	15.46
Mineral-surfaced "Asbex" glass fibre felt finishes							
Aprons with fair drip edge at eaves or verges							
75 mm	m	0.20	0.10	2.46	0.92	3.38	4.22
150 mm	m	0.24	0.12	2.96	1.33	4.29	5.36
Skirtings dressed over angle fillet							
150 mm girth	m	0.13	0.07	1.63	1.23	2.86	3.58
300 mm girth	m	0.26	0.13	3.21	2.45	5.66	7.08
200 mm three coat linings to gutters dressed over two angle fillets							
straight	m	0.33	0.16	4.04	1.78	5.82	7.28
ends	each	0.44	0.22	5.42	1.71	7.13	8.91
outlets	each	0.55	0.27	6.75	2.54	9.29	11.61
300 mm three coat linings to gutters dressed over two angle fillets							
straight	m	0.49	0.25	6.08	2.66	8.74	10.93
ends	each	0.57	0.28	7.00	2.73	9.73	12.16
outlets	each	0.66	0.33	8.13	3.55	11.68	14.60
Collars around							
small pipes (not exceeding 55 mm)	each	1.38	0.69	17.01	1.43	18.44	23.05
large pipes (55 - 110 mm)	each	1.76	0.88	21.70	1.99	23.69	29.61

Roofing

Alterations and repairs	Unit	Hours C	Hours L	Labour net	Material net	Price net	Price with 25%

Flexible sheet finishings

				£	£	£	£
					VAT not included		

Underlays

19 mm fibre insulation board bedded in hot bitumen	m2	0.11	0.06	1.39	4.06	5.45	6.81
Holes for pipes	each	0.33	0.16	4.04	-	4.04	5.05
25 mm resin-bonded glass fibre slabs bedded in hot bitumen	m2	0.13	0.07	1.63	4.90	6.53	8.16
Holes for pipes	each	0.34	0.17	4.20	-	4.20	5.25
Felt vapour barrier bedded in hot bitumen	m2	0.08	0.04	0.99	1.78	2.77	3.46

Accessories

40 x 65 mm aluminium edge trims including butt straps and working feltwork to trim	m	0.33	0.17	4.11	5.22	9.33	11.66
Extra for right angle corner pieces							
internal	each	0.38	0.19	4.68	4.50	9.18	11.48
external angle	each	0.38	0.19	4.68	5.04	9.72	12.15
75 x 65 mm aluminium edge trims including butt straps and working feltwork to trim	m	0.36	0.18	4.44	6.10	10.54	13.18
Extra for right angle corner pieces							
internal angle	each	0.38	0.19	4.68	4.88	9.56	11.95
external angle	each	0.38	0.19	4.68	4.88	9.56	11.95

High performance polyester based roofing

Two layer coverings and stone chippings surfacing							
Flat coverings over 300 mm wide	m2	0.31	0.16	3.86	12.43	16.29	20.36
Working into outlets	each	0.72	0.36	8.88	3.13	12.01	15.01
Flat coverings over 300 mm wide							
with Type 2B (BS 747) base layer	m2	0.35	0.17	4.29	14.46	18.75	23.44
with Type 3B (BS 747) base layer	m2	0.35	0.17	4.29	15.61	19.90	24.88
Working into outlets	each	0.77	0.38	9.46	4.38	13.84	17.30
Two layer mineral surfaced coverings							
Flat coverings over 300 mm wide	m2	0.31	0.16	3.86	14.58	18.44	23.05
Working into outlets	each	0.71	0.36	8.79	4.94	13.73	17.16
Three layer mineral surfaced coverings							
Flat coverings over 300 mm wide with Type 3B (BS 747) base layer	m2	0.35	0.17	4.29	16.05	20.34	25.43
Working into outlets	each	0.77	0.38	9.46	5.54	15.00	18.75

Roofing

Alterations and repairs	Unit	Hours C	Hours L	Labour net	Material net	Price net	Price with 25%
Flexible sheet finishings				£	£	£	£
					VAT not included		
Mineral surfaced finishes							
Aprons with fair drip edge at eaves or verges							
75 mm wide	m	0.20	0.10	2.46	2.41	4.87	6.09
150 mm	m	0.24	0.12	2.96	3.55	6.51	8.14
Skirtings dressed over angle fillet							
150 mm	m	0.13	0.06	1.56	4.44	6.00	7.50
300 mm girth	m	0.26	0.13	3.21	6.90	10.11	12.64
200 mm three coat linings to gutters dressed over two angle fillets							
straight	m	0.33	0.16	4.04	3.62	7.66	9.57
ends	each	0.38	0.19	4.68	3.70	8.38	10.48
outlets	each	0.55	0.27	6.75	5.51	12.26	15.32
300 mm three coat linings to gutters dressed over two angle fillets							
straight	m	0.50	0.25	6.17	5.46	11.63	14.54
ends	each	0.57	0.28	7.00	5.53	12.53	15.66
outlets	each	0.66	0.33	8.13	7.29	15.42	19.27
Collars around							
small pipes (not exceeding 55 mm)	each	1.38	0.69	17.01	5.94	22.95	28.69
large pipes (55 - 110 mm)	each	1.76	0.88	21.70	2.79	24.49	30.61

Roofing

Alterations and repairs	Unit	Hours C	Hours L	Labour net	Material net	Price net	Price with 25%
Sheet metal roofing and flashings				£	£	£	£
					VAT not included		
Milled sheet lead							
2.24 mm (BS Code 5) flat roof and gutter coverings	m2	3.45	3.45	59.10	36.12	95.22	119.03
2.24 mm (BS Code 5) sloping roof coverings							
over 50 deg	m2	4.15	4.15	71.09	29.25	100.34	125.43
bossed ends to rolls	each	0.50	0.50	8.57	0.29	8.86	11.07
bossed intersections to rolls	each	1.00	1.00	17.13	0.51	17.64	22.05
copper nailing at 50 mm centres	m	0.25	0.25	4.28	0.42	4.70	5.88
soldered dots with brass screws	each	0.50	0.50	8.57	0.74	9.31	11.64
1.80 mm (BS Code 4) flashings lapped 100 mm and lead wedged							
150 mm girth	m	0.75	0.75	12.85	3.45	16.30	20.38
210 mm girth	m	0.95	0.95	16.28	4.68	20.96	26.20
1.80 mm (BS Code 4) stepped flashings lapped 100 mm and lead wedged							
180 mm girth	m	1.15	1.15	19.70	4.06	23.76	29.70
240 mm girth	m	1.30	1.30	22.27	5.38	27.65	34.56
1.80 mm (BS Code 4) sloping gutters lapped 100 mm lead wedged and dressed over tilting fillet							
240 mm girth	m	1.45	1.45	24.84	5.66	30.50	38.13
300 mm girth	m	1.60	1.60	27.41	6.89	34.30	42.88
1.80 mm (Code 4) soakers (for fixing by slater or tiler)							
175 x 165 mm	each	0.04	0.04	0.68	0.66	1.34	1.68
300 x 165 mm	each	0.04	0.04	0.68	1.10	1.78	2.23
1.80 mm (BS Code 4) slates (for fixing by slater or tiler)							
600 x 450 mm	each	2.00	2.00	34.26	5.95	40.21	50.26
600 x 600 mm	each	2.00	2.00	34.26	7.93	42.19	52.74
Pipe flashings							
Aluminium slates (for fixing by asphalter or roofer to flat roofs) with synthetic rubber cone and elastomeric seal around							
75 mm pipes	each	0.28	0.28	4.79	15.38	20.17	25.21
100 mm pipes	each	0.28	0.28	4.79	20.07	24.86	31.07
Aluminium slates (for fixing by slater or tiler to sloping roofs) with synthetic rubber cone and elastomeric seal around							
75 mm pipes	each	0.28	0.28	4.79	15.38	20.17	25.21
100 mm pipes	each	0.28	0.28	4.79	20.80	25.59	31.99

Roofing

Alterations and repairs	Unit	Hours C	Hours L	Labour net	Material net	Price net	Price with 25%
Sheet metal roofing and flashings				£	£	£	£
				VAT not included			
Sheet zinc/titanium alloy							
Flat roof coverings 0.80 mm	m2	3.00	3.00	51.39	13.29	64.68	80.85
Sloping roof coverings over 50 deg 0.80 mm	m2	3.20	3.20	54.81	13.29	68.10	85.13
Capped ends or saddles to rolls	each	0.25	0.25	4.28	3.99	8.27	10.34
100 mm flashings lapped 100 mm and wedged one edge 0.80 mm	m	0.80	0.80	13.71	1.43	15.14	18.93
0.80 mm soakers (for fixing by slater or tiler)							
175 x 165 mm	each	0.20	0.20	3.42	0.32	3.74	4.67
300 x 165 mm	each	0.20	0.20	3.42	0.54	3.96	4.95
Sheet aluminium (SIC) (commercial purity)							
0.90 mm flat roof coverings	m2	3.00	3.00	51.39	7.84	59.23	74.04
0.90 mm sloping roof coverings over 50 deg	m2	3.20	3.20	54.81	7.84	62.65	78.31
Capped ends or saddles to rolls	each	0.25	0.25	4.28	1.90	6.18	7.72
Forming standing seams	m	0.30	0.30	5.14	0.63	5.77	7.21
0.90 x 100 mm flashings lapped 100 mm and wedged one edge	m	0.80	0.80	13.71	0.73	14.44	18.05
0.90 mm soakers (for fixing by slater or tiler)							
175 x 165 mm	each	0.20	0.20	3.42	0.19	3.61	4.51
300 x 165 mm	each	0.20	0.20	3.42	0.32	3.74	4.67
Sheet soft copper							
0.55 mm flat roof coverings	m2	3.66	3.66	62.69	15.21	77.90	97.38
0.55 mm sloping roof coverings over 50 deg	m2	3.90	3.90	66.80	15.21	82.01	102.51
Capped ends or saddles to rolls	each	0.25	0.25	4.28	4.56	8.84	11.05
Forming standing seams	m	0.30	0.30	5.14	1.52	6.66	8.32
0.55 x 100 mm flashings lapped 100 mm and wedged one edge	m	0.80	0.80	13.71	2.38	16.09	20.11
0.55 mm soakers (for fixing by slater or tiler)							
175 x 165 mm	each	0.20	0.20	3.42	0.46	3.88	4.85
300 x 165 mm	each	0.20	0.20	3.42	0.76	4.18	5.22

Roofing

Alterations and repairs	Unit	Hours C	Hours L	Labour net	Material net	Price net	Price with 25%
Sheet metal roofing and flashings				£	£	£	£
					VAT not included		
Stainless steel (roofing quality)							
0.38 mm flat roof coverings	m2	3.66	3.66	62.69	22.20	84.89	106.11
0.38 mm sloping roof coverings over 50 deg	m2	3.90	3.90	66.80	22.20	89.00	111.25
Capped ends or saddles to rolls	each	0.25	0.25	4.28	6.66	10.94	13.68
0.38 mm x 100 mm flashings lapped 100 mm and wedged one edge	m	0.80	0.80	13.71	2.32	16.03	20.04
0.38 mm soakers (for fixing by slater or tiler)							
175 x 165 mm	each	0.20	0.20	3.42	0.67	4.09	5.11
300 x 165 mm	each	0.20	0.20	3.42	1.11	4.53	5.66
Terne coated stainless steel							
0.38 mm flat roof coverings	m2	4.00	4.00	68.52	26.34	94.86	118.58
0.38 mm sloping roof coverings over 50 deg							
	m2	4.25	4.25	72.80	26.34	99.14	123.93
Capped ends or saddles to rolls	each	0.28	0.28	4.79	7.90	12.69	15.86
0.38 x 100 mm flashings lapped and wedged one edge	m	1.20	1.20	20.55	4.05	24.60	30.75
0.38 mm soakers (for fixing by slater or tiler)							
175 x 165 mm	each	0.22	0.22	3.77	0.79	4.56	5.70
300 x 165 mm	each	0.22	0.22	3.77	1.32	5.09	6.36
Brown sheathing felt to BS 747							
Underlay close butted at joints and fixed with galvanised felt nails	m2	0.06	0.03	0.74	3.00	3.74	4.67

Alterations and repairs	Unit	Specialist price net	Price with 25%

Thatching

		£	£
		VAT not included	

The following "Specialist price net" figures are guide prices for average size jobs provided by the National Council of Master Thatchers Associations. Quantities and locations may affect prices, therefore it is advisable to submit details and request a firm quotation for any proposed work.

Prices do not include for cash discount.

See the preamble notes for builder's profit and attendance.

Thatching to a thickness of about 300 mm, fixed with iron hooks and finished with a block cut, patterned and saddled ridge

	Unit	Specialist price net	Price with 25%
with best quality water reed	m2	92.64	115.80
with best quality combed wheat reed	m2	84.32	105.40
with best quality long straw	m2	77.27	96.59
Wiring over thatching with 1200 x 19 mm x 20G galvanised wired netting	m2	6.04	7.55

Repairs

	Unit	Specialist price net	Price with 25%
Renewing ridging with a block cut, patterned and saddled ridge	m	201.66	252.07
Wiring over thatching with 1200 x 19 mm x 20G galvanised wired netting	m2	6.04	7.55

Enquiries about the foregoing specialist prices should be made to the National Council of Master Thatchers Associations, tel 07000 781909.

Roofing

Alterations and repairs	Unit	Hours C	Hours L	Labour net	Material net	Price net	Price with 25%
Roofing repairs				£	£	£	£
					VAT not included		
Carefully strip 510 x 255 mm slates, supply 25% new slates and re-slate with galvanised nails including all labours	m2	0.75	0.75	11.77	17.94	29.71	37.14
Examine old slating battens, re-nail as necessary and supply and fix 25% new 19 x 50 mm impregnated softwood battens	m2	0.10	0.10	1.57	0.49	2.06	2.58
Remove old slating battens completely and supply and fix new 19 x 50 mm impregnated softwood battens throughout	m2	0.20	0.20	3.14	1.93	5.07	6.34
Carefully strip 265 x 165 mm tiles, supply 25% new machine made sand-faced tiles and re-tile with galvanised nails including all labours	m2	1.10	1.10	17.27	9.02	26.29	32.86
Examine old tiling battens, re-nail as necessary and supply and fix 25% new 19 x 38 mm impregnated softwood battens throughout	m2	0.15	0.15	2.35	0.70	3.05	3.81
Remove old tiling battens completely and supply and fix 19 x 38 mm impregnated softwood battens throughout	m2	0.30	0.30	4.71	2.74	7.45	9.31
Remove old underfelting and supply and fix new reinforced bituminous felt with galvanised clout nails	m2	0.13	0.13	2.04	2.91	4.95	6.19
Sweep clean flat and shallow pitched roofs and apply two coats bitumen emulsion at the rate of 1.5 m2 per litre per coat	m2	-	0.33	2.22	2.12	4.34	5.42
Strip two slates wide on both sides of valley, remove defective valley gutter & replace with 2.24 mm (BS Code 5) sheet lead dressed into gutter sole and sides and over tilting fillets, supply 25% new 510 x 255 mm slates & re-slate with galvanised nails							
450 mm girth lead	m	3.25	3.25	55.67	23.08	78.75	98.44
600 mm girth lead	m	3.50	3.50	59.96	27.19	87.15	108.94
750 mm girth lead	m	3.75	3.75	64.24	31.30	95.54	119.43
Strip two slates wide on both sides of valley, remove defective valley gutter and replace with 0.65 mm zinc dressed into gutter sole and sides and over tilting fillets, supply 25% new 510 x 255 mm slates and re-slate with galvanised nails							
450 mm girth zinc	m	3.00	3.00	51.39	15.80	67.19	83.99
600 mm girth zinc	m	3.13	3.13	53.61	17.49	71.10	88.88
750 mm girth zinc	m	3.25	3.25	55.67	19.18	74.85	93.56

Woodwork

Preamble

"Labour net" figures refer to site labour only and include allowances for all costs incidental to the employment of labour. Except for unloading and helping with heavy lifting, no labourer assistance has been allowed in compiling this section, carpenters and joiners usually being able to work together and assist each other more economically.

The cost of workshop labour, manufacturing the joinery etc, has been included in figures entered under the "Materials net" heading.

"Materials net" figures include for all costs of materials including workshop labour and an allowance for waste except where specifically stated.

"Price net" figures are the totals of the "Labour net" and "Materials net" figures. Prices are for a builder employing his own labour; according to the amount and nature of the work involved, it may well be possible to secure more advantageous prices from specialist sub-contractors.

Stated sizes of woodwork sections are basic (nominal) sizes before planing and prices include for fixing with nails except where another method of fixing has been described.

Prices for boarding and flooring include allowance for reduced coverage resulting from machined or tongued and grooved edges to boards.

For curved work, add from 100% to the stated figures according to the types of curves required.

Prices do not include any allowance for scaffolding, ladders or other plant necessary to reach the work. The "Preliminaries" section includes prices for scaffolding which must be considered and allowance included to suit the particular circumstances of a tender.

Specialist prices

"Price with 25%" figures are all-in guide prices and include 25% for the builder's overheads, profit, unloading materials and general attendance (to include free use of standing scaffolding and hoists, temporary lighting and water and clearing away rubbish).

The amount of attendance required varies between the various trades and also with the circumstances of specific jobs; the percentage addition must always be considered and adjusted as necessary to suit the terms and conditions of the quotation being used.

Quantities and delivery distances are usually the most significant of the many factors which influence prices and it must be emphasised that quotations should always be obtained when preparing a tender.

Woodwork

Alterations and repairs

Basic prices for materials £

Carcassing softwood

	Unit	Price
Smaller scantlings	m3	244.80
Larger scantlings	m3	244.80
Extra for stress grading		
SC3 grade	m3	10.20
SC4 grade	m3	20.40

Joinery softwood

	Unit	Price
Smaller scantlings	m3	494.00
Larger scantlings	m3	428.40

Boarding

	Unit	Price
Softwood sawn boarding		
19 mm	m2	5.87
25 mm	m2	7.65
19 mm softwood plain edged and tongued and grooved flooring		
in 150 mm widths	m2	12.75
in 100 mm widths	m2	12.75
25 mm softwood plain edged and tongued and grooved flooring		
in 150 mm widths	m2	13.77
in 100 mm widths	m2	13.77
Softwood tongued and grooved and V jointed boarding		
19 mm	m2	13.72
25 mm	m2	15.45
19 mm sawn softwood feather edged boarding	m2	5.41
Softwood matchboarding or shiplap boarding		
19 mm	m2	13.12
25 mm	m2	14.34
Western red cedar matchboarding or shiplap boarding		
19 mm	m2	34.88
25 mm	m2	46.67
19 mm waney edge larch boarding	m2	23.28

Prices actually to be paid for materials must be checked against the above basic prices and adjustments made as necessary.

Woodwork

Alterations and repairs

Basic prices for materials   £

Wrought softwood battens and mouldings

Battens

	Unit	Price
13 x 38 mm	m	0.42
13 x 50 mm	m	0.58
13 x 75 mm	m	0.76
16 x 50 mm	m	0.55
19 x 25 mm	m	0.51
19 x 38 mm	m	0.56
19 x 50 mm	m	0.72
19 x 75 mm	m	1.14
25 x 38 mm	m	0.59
25 x 50 mm	m	0.65
25 x 75 mm	m	1.31
25 x 100 mm	m	1.80
25 x 125 mm	m	2.28
25 x 175 mm	m	3.24
25 x 225 mm	m	4.55
32 x 75 mm	m	1.53
32 x 100 mm	m	1.86
32 x 150 mm	m	2.80
38 x 50 mm	m	1.09
50 x 50 mm	m	1.09
50 x 75 mm	m	1.59
50 x 100 mm	m	2.11

Stock pattern skirtings

	Unit	Price
19 x 75 mm	m	0.83
19 x 100 mm	m	1.71
25 x 150 mm	m	3.28

Stock pattern architraves

	Unit	Price
19 x 50 mm	m	1.06
25 x 25 mm	m	0.64
25 x 75 mm	m	1.86

Quadrants

	Unit	Price
19 mm	m	0.48
25 mm	m	0.65

	Unit	Price
16 x 50 mm stock pattern moulded stop	m	0.86

Blockboard, plywood and sundries

Blockboard

	Unit	Price
18 mm	m2	19.11
25 mm	m2	24.89

Prices actually to be paid for materials must be checked against the above basic prices and adjustments made as necessary.

Woodwork

Alterations and repairs

Basic prices for materials £

Blockboard, plywood and sundries *(continued)*

	Unit	Price
Blockboard veneered both sides		
18 mm	m2	19.11
25 mm	m2	24.89
Plywood WBP grade		
4 mm	m2	4.98
6 mm	m2	7.57
12 mm	m2	12.51
Plywood veneered both sides		
6 mm	m2	27.73
12 mm	m2	33.67
Plywood in 1500 x 600 mm tongued and grooved panels		
12 mm	m2	13.70
15 mm	m2	17.12
18 mm	m2	18.44
25 mm chipboard standard grade	m2	6.22
Chipboard flooring grade Type C4		
18 mm	m2	4.22
22 mm	m2	4.62
Chipboard tongued and grooved all edges Type C4		
18 mm	m2	4.25
22 mm	m2	4.84
18 mm melamine faced chipboard	m2	6.36
Medium density fibreboard		
6 mm	m2	2.76
12 mm	m2	4.64
18 mm	m2	6.10
25 mm	m2	7.38
3.2 mm hardboard	m2	1.31
3.2 mm decorated hardboard	m2	1.99
Plastic laminate sheet		
covering	m2	18.02
balancing	m2	2.83
13 mm sanded finish fibre insulation boards	m2	1.66

Prices actually to be paid for materials must be checked against the above basic prices and adjustments made as necessary.

Woodwork

Alterations and repairs

	Unit	Price
Basic prices for materials		£
Insulating materials		
Expanded polystyrene ISD grade		
25 mm	m2	1.41
40 mm	m2	2.26
13 mm sound deadening quilt	m2	1.66
Thermal insulating quilt		
80 mm	m2	2.91
100 mm	m2	3.46
150 mm	m2	5.38
Granular loose fill	110 litre bag	7.57
Building paper		
standard grade	m2	1.07
reflective grade	m2	1.61
Sundries		
Anti-woodworm fluid	25 litre	104.13

Prices actually to be paid for materials must be checked against the above basic prices and adjustments made as necessary.

Woodwork

Alterations and repairs	Unit	Hours C	Hours L	Labour net	Material net	Price net	Price with 25%
Carcassing				£	£	£	£
					VAT not included		
Impregnated sawn softwood - SC3 Grade							
Floors							
38 x 75mm	m	0.08	-	0.69	1.15	1.84	2.30
38 x 100mm	m	0.08	-	0.69	1.50	2.19	2.74
38 x 125mm	m	0.09	-	0.79	1.84	2.63	3.29
38 x 150mm	m	0.10	-	0.85	2.17	3.02	3.78
38 x 175mm	m	0.11	-	0.99	2.52	3.51	4.39
38 x 200mm	m	0.12	-	1.08	2.90	3.98	4.98
38 x 225mm	m	0.14	-	1.25	3.32	4.57	5.71
50 x 75mm	m	0.09	-	0.79	1.16	1.95	2.44
50 x 100mm	m	0.09	-	0.84	1.54	2.38	2.98
50 x 125mm	m	0.10	-	0.89	1.92	2.81	3.51
50 x 150mm	m	0.12	-	1.08	2.31	3.39	4.24
50 x 175mm	m	0.14	-	1.25	2.69	3.94	4.93
50 x 200mm	m	0.17	-	1.52	3.09	4.61	5.76
50 x 225mm	m	0.17	-	1.52	3.48	5.00	6.25
75 x 150mm	m	0.17	-	1.52	3.68	5.20	6.50
75 x 175mm	m	0.19	-	1.70	4.36	6.06	7.58
75 x 200mm	m	0.22	-	1.97	4.92	6.89	8.61
75 x 225mm	m	0.24	-	2.15	5.53	7.68	9.60
Partitions							
38 x 75mm	m	0.14	-	1.25	1.15	2.40	3.00
38 x 100mm	m	0.19	-	1.70	1.50	3.20	4.00
50 x 75mm	m	0.19	-	1.70	1.16	2.86	3.58
50 x 100mm	m	0.25	-	2.24	1.54	3.78	4.73
75 x 75mm	m	0.24	-	2.15	1.92	4.07	5.09
75 x 100mm	m	0.32	-	2.87	2.45	5.32	6.65
100 x 100mm	m	0.43	-	3.85	3.52	7.37	9.21
Flat roofs							
38 x 75mm	m	0.08	-	0.69	1.15	1.84	2.30
38 x 100mm	m	0.11	-	0.99	1.50	2.49	3.11
38 x 125mm	m	0.13	-	1.16	1.84	3.00	3.75
38 x 150mm	m	0.15	-	1.34	2.17	3.51	4.39
38 x 175mm	m	0.18	-	1.61	2.52	4.13	5.16
38 x 200mm	m	0.20	-	1.79	2.90	4.69	5.86
50 x 75mm	m	0.11	-	0.99	1.16	2.15	2.69
50 x 100mm	m	0.14	-	1.25	1.54	2.79	3.49
50 x 125mm	m	0.17	-	1.52	1.76	3.28	4.10
50 x 150mm	m	0.20	-	1.79	2.31	4.10	5.13
50 x 175mm	m	0.23	-	2.06	2.44	4.50	5.63
50 x 200mm	m	0.26	-	2.33	3.09	5.42	6.78
50 x 225mm	m	0.28	-	2.51	3.48	5.99	7.49
75 x 150mm	m	0.28	-	2.51	3.68	6.19	7.74
75 x 175mm	m	0.32	-	2.87	4.36	7.23	9.04
75 x 200mm	m	0.36	-	3.23	4.92	8.15	10.19
75 x 225mm	m	0.40	-	3.58	5.53	9.11	11.39

Woodwork

Alterations and repairs	Unit	Hours C	Hours L	Labour net	Material net	Price net	Price with 25%

Carcassing

				£	£	£	£
					VAT not included		

Impregnated sawn softwood - SC3 Grade
(continued)

Pitched roofs including ceiling joists

	Unit	Hours C	Hours L	Labour net	Material net	Price net	Price with 25%
25 x 150mm	m	0.15	-	1.34	1.43	2.77	3.46
32 x 175mm	m	0.21	-	1.88	2.64	4.52	5.65
38 x 75mm	m	0.11	-	0.99	1.16	2.15	2.69
38 x 100mm	m	0.15	-	1.34	1.54	2.88	3.60
38 x 125mm	m	0.19	-	1.70	1.87	3.57	4.46
38 x 150mm	m	0.21	-	1.88	2.21	4.09	5.11
38 x 175mm	m	0.25	-	2.24	2.58	4.82	6.03
38 x 200mm	m	0.29	-	2.60	2.95	5.55	6.94
38 x 225mm	m	0.32	-	2.87	3.38	6.25	7.81
50 x 75mm	m	0.15	-	1.34	1.18	2.52	3.15
50 x 100mm	m	0.20	-	1.79	1.57	3.36	4.20
50 x 125mm	m	0.23	-	2.06	1.96	4.02	5.03
75 x 75mm	m	0.21	-	1.88	1.94	3.82	4.78
75 x 100mm	m	0.29	-	2.60	2.48	5.08	6.35
75 x 125mm	m	0.35	-	3.14	3.10	6.24	7.80
75 x 150mm	m	0.37	-	3.32	3.72	7.04	8.80
100 x 100mm	m	0.37	-	3.32	3.55	6.87	8.59
100 x 150mm	m	0.50	-	4.48	5.16	9.64	12.05
100 x 175mm	m	0.58	-	5.20	6.03	11.23	14.04

Kerbs, bearers etc.

	Unit	Hours C	Hours L	Labour net	Material net	Price net	Price with 25%
38 x 75mm	m	0.04	-	0.39	1.12	1.51	1.89
38 x 100mm	m	0.06	-	0.49	1.48	1.97	2.46
50 x 100mm	m	0.08	-	0.69	1.51	2.20	2.75
75 x 100mm	m	0.10	-	0.89	2.41	3.30	4.13
50 x 100mm - bolted	m	0.14	-	1.25	1.71	2.96	3.70

Noggings between joists

	Unit	Hours C	Hours L	Labour net	Material net	Price net	Price with 25%
38 x 50mm	m	0.19	-	1.70	0.80	2.50	3.13
50 x 50mm	m	0.20	-	1.79	0.80	2.59	3.24
50 x 75mm	m	0.22	-	1.97	1.21	3.18	3.98

Herringbone strutting (size 38 x 38 mm) to joists (measured over the joists)

	Unit	Hours C	Hours L	Labour net	Material net	Price net	Price with 25%
100 mm deep	m	0.33	-	2.96	1.19	4.15	5.19
150 mm deep	m	0.33	-	2.96	1.17	4.13	5.16
175 mm deep	m	0.36	-	3.23	1.24	4.47	5.59
200 mm deep	m	0.40	-	3.58	1.30	4.88	6.10
225 mm deep	m	0.43	-	3.85	1.35	5.20	6.50

Herringbone strutting (size 50 x 50 mm) to joists, (measured over joists)

	Unit	Hours C	Hours L	Labour net	Material net	Price net	Price with 25%
100 mm deep	m	0.33	-	2.96	1.56	4.52	5.65
150 mm deep	m	0.33	-	2.96	1.56	4.52	5.65
175 mm deep	m	0.36	-	3.23	1.63	4.86	6.08
200 mm deep	m	0.40	-	3.58	1.71	5.29	6.61
225 mm deep	m	0.43	-	3.85	1.79	5.64	7.05

Woodwork

Alterations and repairs	Unit	Hours C	Hours L	Labour net	Material net	Price net	Price with 25%

Carcassing

				£	£	£	£
					VAT not included		

Impregnated sawn softwood - SC3 Grade
(*continued*)

Solid bridging between joists (measured over the joists)

	Unit	Hours C	Hours L	Labour net	Material net	Price net	Price with 25%
38 x 100mm	m	0.22	-	1.97	1.56	3.53	4.41
50 x 100mm	m	0.22	-	1.97	1.60	3.57	4.46
50 x 150mm	m	0.22	-	1.97	2.30	4.27	5.34
50 x 175 mm	m	0.24	-	2.15	2.67	4.82	6.03
50 x 200 mm	m	0.26	-	2.33	3.04	5.37	6.71
50 x 225 mm	m	0.29	-	2.60	3.42	6.02	7.53
50 x 75 x 450mm sprockets	each	0.17	-	1.52	0.54	2.06	2.58
Notch and fit ends of members to metal	each	0.28	-	2.51	-	2.51	3.14

Trim members around openings with four skew nailed butt joints

	Unit	Hours C	Hours L	Labour net	Material net	Price net	Price with 25%
50 x 75mm	each	0.74	-	6.63	0.36	6.99	8.74
50 x 100mm	each	0.83	-	7.44	0.36	7.80	9.75

Trim members around openings with six skew nailed butt joints

	Unit	Hours C	Hours L	Labour net	Material net	Price net	Price with 25%
50 x 75mm	each	1.10	-	9.86	0.53	10.39	12.99
50 x 100mm	each	1.27	-	11.38	0.53	11.91	14.89

Trim members around openings with four framed joints

	Unit	Hours C	Hours L	Labour net	Material net	Price net	Price with 25%
50 x 150mm	each	4.40	-	39.42	0.36	39.78	49.73
50 x 175mm	each	4.51	-	40.41	0.36	40.77	50.96
50 x 200mm	each	4.62	-	41.40	0.54	41.94	52.42
50 x 225mm	each	4.62	-	41.40	0.54	41.94	52.42

Trim members around openings with five framed joints

	Unit	Hours C	Hours L	Labour net	Material net	Price net	Price with 25%
50 x 150mm	each	4.95	-	44.35	0.44	44.79	55.99
50 x 175mm	each	5.09	-	45.61	0.44	46.05	57.56
50 x 200mm	each	5.23	-	46.86	0.68	47.54	59.42
50 x 225mm	each	5.36	-	48.03	0.68	48.71	60.89

Trim members around openings with six framed joints

	Unit	Hours C	Hours L	Labour net	Material net	Price net	Price with 25%
50 x 150mm	each	5.50	-	49.28	0.53	49.81	62.26
50 x 175mm	each	5.67	-	50.80	0.53	51.33	64.16
50 x 200mm	each	5.83	-	52.24	0.81	53.05	66.31
50 x 225mm	each	6.00	-	53.76	0.81	54.57	68.21

Trim members around openings with seven framed joints

	Unit	Hours C	Hours L	Labour net	Material net	Price net	Price with 25%
50 x 150mm	each	6.05	-	54.21	0.62	54.83	68.54
50 x 175mm	each	6.25	-	56.00	0.62	56.62	70.78
50 x 200mm	each	6.44	-	57.70	0.95	58.65	73.31
50 x 225mm	each	6.63	-	59.40	0.95	60.35	75.44

Woodwork

Alterations and repairs	Unit	Hours C	Hours L	Labour net	Material net	Price net	Price with 25%

Carcassing

				£	£	£	£
					VAT not included		

Impregnated sawn softwood - SC3 Grade
(*continued*)

Trim members around openings with eight framed joints

50 x 150mm	each	6.60	-	59.14	0.71	59.85	74.81
50 x 175mm	each	6.82	-	61.11	0.71	61.82	77.28
50 x 200mm	each	7.04	-	63.08	1.08	64.16	80.20
50 x 225mm	each	7.26	-	65.05	1.08	66.13	82.66

Trim members around openings with four steel joist hangers

50 x 150mm	each	2.75	-	24.64	6.16	30.80	38.50
50 x 175mm	each	2.86	-	25.63	6.57	32.20	40.25
50 x 200mm	each	2.97	-	26.61	7.50	34.11	42.64
50 x 225mm	each	3.08	-	27.60	7.50	35.10	43.88

Trim members around openings with five steel joist hangers

50 x 150mm	each	3.44	-	30.82	7.70	38.52	48.15
50 x 175mm	each	3.58	-	32.08	8.21	40.29	50.36
50 x 200mm	each	3.72	-	33.33	9.38	42.71	53.39
50 x 225mm	each	3.85	-	34.50	9.38	43.88	54.85

Trim members around openings with six steel joist hangers

50 x 150mm	each	4.13	-	37.00	9.23	46.23	57.79
50 x 175mm	each	4.29	-	38.44	9.85	48.29	60.36
50 x 200mm	each	4.46	-	39.96	11.26	51.22	64.03
50 x 225mm	each	4.62	-	41.40	11.26	52.66	65.83

Trim members around openings with seven steel joist hangers

50 x 150mm	each	4.82	-	43.19	10.78	53.97	67.46
50 x 175mm	each	5.00	-	44.80	11.50	56.30	70.38
50 x 200mm	each	5.20	-	46.59	13.13	59.72	74.65
50 x 225mm	each	5.39	-	48.29	13.13	61.42	76.78

Trim members around openings with eight steel joist hangers

50 x 150mm	each	5.50	-	49.28	12.31	61.59	76.99
50 x 175mm	each	5.72	-	51.25	13.14	64.39	80.49
50 x 200mm	each	5.94	-	53.22	15.01	68.23	85.29
50 x 225mm	each	6.16	-	55.19	15.01	70.20	87.75

Woodwork

Alterations and repairs	Unit	Hours C	Hours L	Labour net	Material net	Price net	Price with 25%
First fixings				£	£	£	£
					VAT not included		
Floors							
25mm softwood plain edged board flooring							
in 150mm widths	m2	0.61	-	5.47	12.81	18.28	22.85
in 125mm widths	m2	0.67	-	6.00	12.01	18.01	22.51
25mm softwood tongued and grooved board flooring							
in 150mm widths	m2	0.73	-	6.54	13.32	19.86	24.82
in 100mm widths	m2	0.85	-	7.62	11.06	18.68	23.35
Plywood flooring in 1500 x 600mm panels tongued and grooved all edges							
12mm	m2	0.33	-	2.96	16.52	19.48	24.35
15mm	m2	0.36	-	3.23	20.59	23.82	29.77
18mm	m2	0.39	-	3.49	22.17	25.66	32.08
Chipboard type C4 flooring							
18mm	m2	0.28	-	2.51	4.70	7.21	9.01
22mm	m2	0.33	-	2.96	5.12	8.08	10.10
Chipboard type C4 flooring tongued and grooved all edges							
18mm	m2	0.36	-	3.23	4.73	7.96	9.95
22mm	m2	0.44	-	3.94	5.36	9.30	11.63
Make out 25 mm board flooring where old partition or wall removed							
boards parallel with wall	m	0.10	-	0.90	5.80	6.70	8.38
boards at right angles to wall	m	0.30	-	2.69	7.20	9.89	12.36
External walls							
19mm impregnated softwood feather edged boarding							
over 1.00m2	m2	0.66	-	5.91	4.45	10.36	12.95
not exceeding 1.00m2	each	0.83	-	7.44	4.45	11.89	14.86
raking cutting	m	0.08	-	0.69	0.44	1.13	1.41
19mm impregnated softwood matchboarding or shiplap boarding							
over 1.00m2	m2	0.72	-	6.45	10.80	17.25	21.56
not exceeding 1.00m2	each	0.88	-	7.88	10.80	18.68	23.35
raking cutting	m	0.08	-	0.69	1.07	1.76	2.20
25mm impregnated softwood matchboarding or shiplap boarding							
over 1.00m2	m2	0.77	-	6.90	12.28	19.18	23.98
not exceeding 1.00m2	each	0.97	-	8.69	12.28	20.97	26.21
raking cutting	m	0.09	-	0.79	1.22	2.01	2.51

Woodwork

Alterations and repairs	Unit	Hours C	Hours L	Labour net	Material net	Price net	Price with 25%
First fixings				£	£	£	£
					VAT not included		
External walls (*continued*)							
19mm western red cedar matchboarding or shiplap boarding, fixed with galvanised nails							
over 1.00m2	m2	0.77	-	6.90	26.23	33.13	41.41
not exceeding 1.00m2	each	0.97	-	8.69	26.23	34.92	43.65
raking cutting	m	0.09	-	0.79	2.61	3.40	4.25
25mm western red cedar matchboarding or shiplap boarding, fixed with galvanised nails							
over 1.00m2	m2	0.83	-	7.44	33.92	41.36	51.70
not exceeding 1.00m2	each	1.05	-	9.41	33.92	43.33	54.16
raking cutting	m	0.09	-	0.79	3.38	4.17	5.21
19mm waney edge elm boarding lapped average 50mm, fixed with galvanised nails							
over 1.00m2	m2	0.88	-	7.88	16.98	24.86	31.07
not exceeding 1.00m2	each	1.10	-	9.86	16.98	26.84	33.55
raking cutting	m	0.10	-	0.89	1.69	2.58	3.23
Cellular PVCU cladding of 150mm (cover width) shiplap planks , fixed with galvanised nails							
over 1.00m2	m2	0.77	-	6.90	23.28	30.18	37.73
not exceeding 1.00m2	each	1.16	-	10.39	23.28	33.67	42.09
raking cutting	m	0.08	-	0.69	0.35	1.04	1.30
Cellular PVCU cladding of 100mm (cover width) shiplap planks , fixed with galvanised nails							
over 1.00m2	m2	1.10	-	9.86	25.24	35.10	43.88
not exceeding 1.00m2	each	1.65	-	14.78	25.24	40.02	50.02
raking cutting	m	0.11	-	0.99	0.25	1.24	1.55
PVCU cladding of 150mm (cover width) shiplap planks , fixed with clips and aluminium nails							
over 1.00m2	m2	0.66	-	5.91	14.00	19.91	24.89
not exceeding 1.00m2	each	0.99	-	8.87	14.00	22.87	28.59
raking cutting	m	0.09	-	0.79	0.21	1.00	1.25
PVCU cladding of 100mm (cover width) shiplap planks , fixed with clips and aluminium nails							
over 1.00m2	m2	0.99	-	8.87	14.82	23.69	29.61
not exceeding 1.00m2	each	1.49	-	13.35	14.82	28.17	35.21
raking cutting	m	0.11	-	0.99	0.25	1.24	1.55
Ancillaries for PVCU cladding							
starter strip fixed with aluminium nails	m	0.08	-	0.69	0.65	1.34	1.68
top channel fixed with aluminium nails	m	0.11	-	0.99	0.55	1.54	1.93
side channel fixed with aluminium nails	m	0.13	-	1.16	1.12	2.28	2.85
top cover plank snap fixed to top channel	m	0.08	-	0.69	1.70	2.39	2.99

Woodwork

	Unit	Hours C	Hours L	Labour net	Material net	Price net	Price with 25%
First fixings				£	£	£	£
					VAT not included		
Roofs							
19mm impregnated sawn softwood boarding							
flat	m2	0.55	-	4.93	11.95	16.88	21.10
sloping	m2	0.83	-	7.44	11.95	19.39	24.24
raking cutting	m	0.08	-	0.69	0.18	0.87	1.09
25mm impregnated sawn softwood boarding							
flat	m2	0.62	-	5.56	13.62	19.18	23.98
sloping	m2	0.67	-	6.00	13.62	19.62	24.52
raking cutting	m	0.08	-	0.69	0.20	0.89	1.11
19mm impregnated softwood tongued and grooved boarding							
flat	m2	0.66	-	5.91	6.00	11.91	14.89
sloping	m2	0.99	-	8.87	6.00	14.87	18.59
raking cutting	m	0.08	-	0.69	0.09	0.78	0.97
25mm impregnated softwood tongued and grooved boarding							
flat	m2	0.73	-	6.54	6.74	13.28	16.60
sloping	m2	1.09	-	9.77	6.74	16.51	20.64
raking cutting	m	0.09	-	0.79	0.10	0.89	1.11
12mm WBP grade plywood boarding							
flat	m2	0.77	-	6.90	13.53	20.43	25.54
sloping	m2	0.95	-	8.51	13.60	22.11	27.64
raking cutting	m	0.03	-	0.27	1.35	1.62	2.02
18mm chipboard boarding tongued and grooved all edges							
flat	m2	0.29	-	2.60	4.62	7.22	9.03
sloping	m2	0.33	-	2.96	4.72	7.68	9.60
raking cutting	m	0.08	-	0.69	0.46	1.15	1.44
22mm chipboard boarding tongued and grooved all edges							
flat	m2	0.33	-	2.96	5.35	8.31	10.39
sloping	m2	0.42	-	3.76	5.35	9.11	11.39
raking cutting	m	0.09	-	0.79	0.52	1.31	1.64
Labour chamfered, rebated or rounded edge to boarding	m	0.12	-	1.08	-	1.08	1.35
Impregnated sawn softwood gutter boarding etc							
25mm gutter and lay boarding	m2	1.71	-	15.32	13.53	28.85	36.06
Splayed chimney gutter or valley boards							
25 x 150mm	m	0.28	-	2.51	2.05	4.56	5.70
25 x 225mm	m	0.42	-	3.76	3.17	6.93	8.66
25mm boxed cesspools 225 x 225 x 100mm	each	2.75	-	24.64	3.31	27.95	34.94

Woodwork

Alterations and repairs	Unit	Hours C	Hours L	Labour net	Material net	Price net	Price with 25%
First fixings				£	£	£	£
					VAT not included		
Eaves and verge boarding							
19mm impregnated softwood matchboarded soffits							
over 300mm wide	m2	1.54	-	13.80	8.91	22.71	28.39
225mm wide	m	0.43	-	3.85	2.45	6.30	7.88
19mm Western red cedar matchboarded soffits							
over 300mm wide	m2	1.32	-	11.83	26.25	38.08	47.60
225mm wide	m	0.31	-	2.78	5.92	8.70	10.88
6mm sanded finish asbestos-free insulation board soffits							
over 300mm wide	m2	0.88	-	7.88	10.98	18.86	23.57
225mm wide	m	0.24	-	2.15	2.54	4.69	5.86
Impregnated wrought softwood eaves and verge boarding							
25 x 150mm square							
fascias	m	0.28	-	2.51	1.92	4.43	5.54
returned ends	each	0.06	-	0.49	-	0.49	0.61
mitres	each	0.22	-	1.97	-	1.97	2.46
32 x 175mm moulded							
fascias	m	0.33	-	2.96	2.47	5.43	6.79
returned ends	each	0.28	-	2.51	-	2.51	3.14
mitres	each	0.36	-	3.23	-	3.23	4.04
25 x 150mm square							
barge boards	m	0.28	-	2.51	1.92	4.43	5.54
returned ends with sprocket piece	each	0.42	-	3.76	0.68	4.44	5.55
mitres	each	0.28	-	2.51	-	2.51	3.14
25 x 175mm moulded							
barge boards	m	0.32	-	2.87	3.93	6.80	8.50
returned end with sprocket piece	each	1.10	-	9.86	1.29	11.15	13.94
mitres	each	0.36	-	3.23	-	3.23	4.04
32 x 175mm moulded							
barge boards	m	0.32	-	2.87	2.47	5.34	6.67
returned ends with sprocket piece	each	1.14	-	10.21	0.84	11.05	13.81
mitres	each	0.36	-	3.23	-	3.23	4.04
32 x 225mm moulded							
barge boards	m	0.50	-	4.48	3.12	7.60	9.50
returned ends with sprocket piece	each	1.17	-	10.48	1.05	11.53	14.41
mitres	each	0.36	-	3.23	-	3.23	4.04
25 x 63mm bed moulding including ends and mitres	m	0.14	-	1.25	3.79	5.04	6.30

Woodwork

Alterations and repairs	Unit	Hours C	Hours L	Labour net	Material net	Price net	Price with 25%
First fixings				£	£	£	£
					VAT not included		
Impregnated sawn softwood firrings, bearers etc							
38mm firrings, average depth							
38mm	m	0.07	-	0.59	0.85	1.44	1.80
50mm	m	0.07	-	0.59	1.29	1.88	2.35
63mm	m	0.08	-	0.69	1.52	2.21	2.76
50mm firrings, average depth							
38mm	m	0.07	-	0.59	1.26	1.85	2.31
50mm	m	0.08	-	0.69	1.93	2.62	3.28
63mm	m	0.09	-	0.79	2.26	3.05	3.81
Bearers							
25 x 50mm	m	0.13	-	1.16	0.57	1.73	2.16
32 x 50mm	m	0.13	-	1.16	0.73	1.89	2.36
38 X 50mm	m	0.13	-	1.16	0.85	2.01	2.51
50 x 50mm	m	0.14	-	1.25	1.11	2.36	2.95
38 x 63mm	m	0.14	-	1.25	0.88	2.13	2.66
50 x 63mm	m	0.14	-	1.25	1.26	2.51	3.14
50 x 75mm	m	0.16	-	1.43	1.43	2.86	3.58
Bearers not exceeding 300mm long							
25 x 50mm	each	0.06	-	0.49	0.17	0.66	0.82
38 x 50mm	each	0.06	-	0.49	0.25	0.74	0.93
50 x 50mm	each	0.07	-	0.59	0.33	0.92	1.15
38 x 63mm	each	0.07	-	0.59	0.26	0.85	1.06
50 x 63mm	each	0.07	-	0.59	0.38	0.97	1.21
50 x 75mm	each	0.08	-	0.69	0.43	1.12	1.40
Angle fillets							
38 x 38mm	m	0.14	-	1.25	0.50	1.75	2.19
50 x 50mm	m	0.15	-	1.34	0.72	2.06	2.58
75 x 75mm	m	0.16	-	1.43	1.42	2.85	3.56
Tilting fillets							
25 x 50mm	m	0.08	-	0.69	0.83	1.52	1.90
38 x 75mm	m	0.11	-	0.99	1.07	2.06	2.58
50 x 50mm rolls for metal roofing	m	0.16	-	1.43	1.25	2.68	3.35

Woodwork

Alterations and repairs	Unit	Hours C	Hours L	Labour net	Material net	Price net	Price with 25%
First fixings				£	£	£	£
					VAT not included		
Impregnated sawn softwood grounds and battens							
Open spaced battening at 400mm centres one way							
13 x 50mm	m2	0.28	-	2.51	0.86	3.37	4.21
25 x 50mm	m2	0.31	-	2.78	1.34	4.12	5.15
Open spaced battening at 400mm centres both ways							
13 x 50mm	m2	0.68	-	6.09	1.71	7.80	9.75
25 x 50mm	m2	0.77	-	6.90	2.69	9.59	11.99
Individual grounds and battens							
13 x 25mm	m	0.11	-	0.99	0.17	1.16	1.45
13 x 38mm	m	0.11	-	0.99	0.26	1.25	1.56
13 x 50mm	m	0.11	-	0.99	0.35	1.34	1.68
25 x 50mm	m	0.12	-	1.08	0.54	1.62	2.02
Impregnated floor fillets fixed to clips in concrete							
50 x 50mm	m	0.08	-	0.69	0.88	1.57	1.96
50 x 75mm	m	0.10	-	0.89	1.32	2.21	2.76
Fixing blocks wedged into steelwork							
50 x 50 x 150mm long	each	0.11	-	0.99	0.13	1.12	1.40
50 x 75 x 300mm long	each	0.11	-	0.99	0.40	1.39	1.74
Impregnated sawn softwood framework							
Cradling to steel beams with 50 x 75mm wedged blocks and 25 x 50mm members at 400mm centres	m2	2.37	-	21.24	4.10	25.34	31.68
50 x 50mm bracketing to false ceilings etc	m	0.34	-	3.05	0.92	3.97	4.96
38 x 50mm bath panel framing							
to front only	set	1.36	-	12.19	4.74	16.93	21.16
to front and one end	set	2.13	-	19.08	6.58	25.66	32.08

Woodwork

Alterations and repairs	Unit	Hours C	Hours L	Labour net	Material net	Price net	Price with 25%
Second fixings				£	£	£	£
					VAT not included		
Softwood							
Stock pattern skirtings including ends and mitres							
19 x 75mm	m	0.11	-	0.99	0.93	1.92	2.40
19 x 100mm	m	0.12	-	1.08	1.90	2.98	3.73
25 x 150mm							
stock pattern skirtings	m	0.13	-	1.16	3.63	4.79	5.99
returned ends	each	0.11	-	0.99	-	0.99	1.24
mitres	each	0.17	-	1.52	-	1.52	1.90
25 x 175mm							
skirtings moulded to detail	m	0.15	-	1.34	4.20	5.54	6.92
returned ends	each	0.11	-	0.99	-	0.99	1.24
mitres	each	0.19	-	1.70	-	1.70	2.13
25 x 125mm							
skirtings moulded to match existing	m	0.13	-	1.16	3.53	4.69	5.86
returned ends	each	0.11	-	0.99	-	0.99	1.24
mitres	each	0.17	-	1.52	-	1.52	1.90
mitres with existing	each	0.23	-	2.06	-	2.06	2.58
25 x 175mm							
skirtings moulded to match existing	each	0.15	-	1.34	4.58	5.92	7.40
returned ends	each	0.11	-	0.99	-	0.99	1.24
mitres	each	0.19	-	1.70	-	1.70	2.13
mitres with existing	each	0.26	-	2.33	-	2.33	2.91
25 x 225							
skirtings moulded to match existing	m	0.15	-	1.34	6.02	7.36	9.20
returned ends	each	0.13	-	1.16	-	1.16	1.45
mitres	each	0.21	-	1.88	-	1.88	2.35
mitres with existing	each	0.28	-	2.51	-	2.51	3.14
25 x 50mm picture rails moulded to match existing including mitres and ends	m	0.22	-	1.97	1.93	3.90	4.88
25 x 50mm dado rails moulded to detail including ends and mitres	m	0.22	-	1.97	1.21	3.18	3.98
32 x 75mm dado rails moulded to detail	m	0.11	-	0.99	2.22	3.21	4.01
returned ends	each	0.11	-	0.99	-	0.99	1.24
mitres	each	0.16	-	1.43	-	1.43	1.79
Stock pattern architraves including ends and mitres							
19 x 50mm	m	0.14	-	1.25	1.19	2.44	3.05
25 x 75mm	m	0.16	-	1.43	2.07	3.50	4.38
Architraves moulded to detail including ends and mitres							
19 x 50mm	m	0.14	-	1.25	1.19	2.44	3.05
25 x 75mm	m	0.16	-	1.43	2.10	3.53	4.41

Woodwork

Alterations and repairs	Unit	Hours C	Hours L	Labour net	Material net	Price net	Price with 25%

Second fixings

				£	£	£	£
				VAT not included			

Softwood (*continued*)

Architraves moulded to match existing including ends, mitres and joints of new to existing							
19 x 50 mm	m	0.17	-	1.52	2.14	3.66	4.58
25 x 75 mm	m	0.20	-	1.79	2.79	4.58	5.72
Architraves moulded to match existing							
32 x 75mm	m	0.17	-	1.52	3.03	4.55	5.69
mitres	each	0.12	-	1.08	-	1.08	1.35
mitres with existing	each	0.15	-	1.34	-	1.34	1.68
32 x 100mm architraves moulded to match existing	m	0.17	-	1.52	3.40	4.92	6.15
mitres	each	0.12	-	1.08	-	1.08	1.35
mitres with existing	each	0.15	-	1.34	-	1.34	1.68
Quadrant cover fillets including ends and mitres							
19mm	m	0.08	-	0.69	0.54	1.23	1.54
25mm	m	0.08	-	0.69	0.72	1.41	1.76
13 x 50mm square edged cover fillets including ends and mitres	m	0.08	-	0.69	1.03	1.72	2.15
13 x 50mm twice moulded cover fillets including ends and mitres	m	0.08	-	0.69	1.70	2.39	2.99
19 x 75mm chamfered cover fillets including ends and mitres	m	0.08	-	0.69	2.29	2.98	3.73
25 x 25mm stock pattern cover fillets to match existing including ends and mitres	m	0.17	-	1.52	1.02	2.54	3.17
25 x 32mm cover fillets moulded to match existing including returned ends and mitres	m	0.18	-	1.61	1.29	2.90	3.63
Stops including ends and mitres							
13 x 50mm	m	0.14	-	1.25	0.89	2.14	2.67
13 x 75mm	m	0.14	-	1.25	1.11	2.36	2.95
16 x 50mm	m	0.14	-	1.25	0.96	2.21	2.76
25 x 50mm	m	0.14	-	1.25	1.06	2.31	2.89
Glazing beads including mitres							
13 x 19mm	m	0.11	-	0.99	0.24	1.23	1.54
13 x 25mm	m	0.11	-	0.99	0.28	1.27	1.59
13 x 32mm	m	0.13	-	1.16	0.31	1.47	1.84
19 x 32mm	m	0.13	-	1.16	0.41	1.57	1.96
19 x 38mm	m	0.13	-	1.16	0.46	1.62	2.02

Woodwork

Alterations and repairs	Unit	Hours C	Hours L	Labour net	Material net	Price net	Price with 25%
Second fixings				£	£	£	£
					VAT not included		
Softwood (*continued*)							
Glazing beads fixed with brass screws and cups including mitres							
13 x 19mm	m	0.27	-	2.42	1.89	4.31	5.39
13 x 25mm	m	0.27	-	2.42	1.93	4.35	5.44
13 x 32mm	m	0.29	-	2.60	1.95	4.55	5.69
Slat shelves of 25 x 50mm slats 25mm apart	m2	1.49	-	13.35	8.83	22.18	27.73
19mm shelves							
150mm wide	m	0.17	-	1.52	1.61	3.13	3.91
200mm wide	m	0.18	-	1.61	3.87	5.48	6.85
19mm crosstongued shelves							
over 300mm wide	m2	2.48	-	22.22	20.30	42.52	53.15
250mm wide	m	0.19	-	1.70	3.58	5.28	6.60
25mm shelves							
150mm wide	m	0.20	-	1.79	1.75	3.54	4.42
200m wide	m	0.21	-	1.88	4.26	6.14	7.67
25mm crosstongued shelves							
over 300mm wide	m2	2.59	-	23.21	22.22	45.43	56.79
250mm wide	m	0.22	-	1.97	5.39	7.36	9.20
Crosstongued worktops button blocked to framing							
25mm	m2	4.24	-	37.99	22.87	60.86	76.08
32mm	m2	4.40	-	39.42	38.93	78.35	97.94
38mm	m2	4.62	-	41.40	29.81	71.21	89.01
Bearers							
19 x 38mm	m	0.17	-	1.52	0.64	2.16	2.70
25 x 50mm	m	0.19	-	1.70	0.73	2.43	3.04
38 x 50mm	m	0.19	-	1.70	1.22	2.92	3.65
50 x 50mm	m	0.20	-	1.79	1.27	3.06	3.83
50 x 75mm	m	0.22	-	1.97	1.82	3.79	4.74
Bearers not exceeding 300mm long							
19 x 38mm	each	0.08	-	0.69	0.19	0.88	1.10
25 x 50mm	each	0.09	-	0.79	0.23	1.02	1.27
38 x 50mm	each	0.09	-	0.79	0.38	1.17	1.46
50 x 50mm	each	0.09	-	0.79	0.43	1.22	1.52
50 x 75mm	each	0.10	-	0.89	0.59	1.48	1.85
Legs and bearers framed on site							
50 x 50mm	m	0.74	-	6.63	1.27	7.90	9.88
50 x 75mm	m	0.91	-	8.15	1.82	9.97	12.46
32 x 50mm window nosings tongued on including ends	m	0.31	-	2.78	1.83	4.61	5.76

Woodwork

Alterations and repairs	Unit	Hours C	Hours L	Labour net	Material net	Price net	Price with 25%
Second fixings				£	£	£	£
					VAT not included		
Softwood (*continued*)							
25 x 150mm							
rounded window boards tongued on	m	0.33	-	2.96	2.94	5.90	7.38
returned ends	each	0.13	-	1.16	-	1.16	1.45
32 x 200mm							
rounded window boards tongued on	m	0.36	-	3.23	7.58	10.81	13.51
returned ends	each	0.19	-	1.70	-	1.70	2.13
63 x 75mm							
moulded handrails screwed on	m	0.23	-	2.06	2.97	5.03	6.29
returned ends	each	0.66	-	5.91	-	5.91	7.39
Hardwood							
Skirtings moulded to detail including ends and mitres							
19 x 75mm	m	0.15	-	1.34	6.00	7.34	9.18
19 x 100mm	m	0.17	-	1.52	6.56	8.08	10.10
25 x 150mm							
skirtings moulded to detail	m	0.18	-	1.61	16.04	17.65	22.06
returned ends	each	0.28	-	2.51	-	2.51	3.14
mitres	each	0.28	-	2.51	-	2.51	3.14
25 x 175mm							
skirtings moulded to detail	m	0.20	-	1.79	21.40	23.19	28.99
returned ends	each	0.33	-	2.96	-	2.96	3.70
mitres	each	0.44	-	3.94	-	3.94	4.92
25 x 50mm dado rails moulded to detail including ends and mitres	m	0.16	-	1.43	5.41	6.84	8.55
32 x 75mm							
dado rails moulded to detail	m	0.17	-	1.52	10.17	11.69	14.61
returned ends	each	0.33	-	2.96	-	2.96	3.70
mitres	each	0.38	-	3.40	-	3.40	4.25
25 x 75mm architraves moulded to detail including ends and mitres	m	0.20	-	1.79	8.03	9.82	12.28
32 x100mm							
architraves moulded to detail	m	0.22	-	1.97	11.16	13.13	16.41
returned ends	each	0.33	-	2.96	-	2.96	3.70
mitres	each	0.38	-	3.40	-	3.40	4.25
Plinth blocks							
32 x 88mm x 175mm high	each	0.28	-	2.51	1.26	3.77	4.71
38 x 113mm x 225mm high	each	0.39	-	3.49	2.51	6.00	7.50

Woodwork

Alterations and repairs	Unit	Hours C	Hours L	Labour net	Material net	Price net	Price with 25%
Second fixings				£	£	£	£
					VAT not included		
Hardwood (continued)							
Quadrant cover fillets including ends and mitres							
19mm	m	0.11	-	0.99	0.80	1.79	2.24
25mm	m	0.12	-	1.08	1.82	2.90	3.63
16 x 50mm square edged cover fillets including ends and mitres	m	0.11	-	0.99	2.03	3.02	3.77
13 x 50 twice moulded cover fillets including ends and mitres	m	0.11	-	0.99	2.61	3.60	4.50
19 x 75mm chamfered cover fillets including ends and mitres	m	0.11	-	0.99	3.75	4.74	5.92
25 x 32mm scotia moulds including ends	m	0.20	-	1.79	2.40	4.19	5.24
Stops including ends and mitres							
13 x 50mm	m	0.19	-	1.70	1.90	3.60	4.50
16 x 50mm	m	0.20	-	1.79	2.03	3.82	4.78
25 x 50mm	m	0.22	-	1.97	2.98	4.95	6.19
Glazing beads including mitres							
13 x 19mm	m	0.14	-	1.25	1.14	2.39	2.99
13 x 25mm	m	0.14	-	1.25	1.43	2.68	3.35
13 x 32mm	m	0.16	-	1.43	1.74	3.17	3.96
19 x 32mm	m	0.18	-	1.61	2.02	3.63	4.54
19 x 38mm	m	0.18	-	1.61	2.12	3.73	4.66
Glazing beads fixed with brass screws and cups including mitres							
13 x 19mm	m	0.35	-	3.14	1.33	4.47	5.59
13 x 25mm	m	0.35	-	3.14	1.62	4.76	5.95
13 x 32mm	m	0.37	-	3.32	1.93	5.25	6.56
19mm shelves							
150mm wide	m	0.23	-	2.06	6.94	9.00	11.25
200mm wide	m	0.24	-	2.15	9.26	11.41	14.26
19mm crosstongued shelves							
over 300mm wide	m2	3.25	-	29.12	48.18	77.30	96.63
250mm wide	m	0.25	-	2.24	11.94	14.18	17.73
25mm shelves							
150mm wide	m	0.28	-	2.51	8.89	11.40	14.25
200 mm wide	m	0.29	-	2.60	11.84	14.44	18.05
25mm crosstongued shelves							
over 300 mm wide	m2	3.36	-	30.11	61.05	91.16	113.95
250 mm wide	m	0.31	-	2.78	15.15	17.93	22.41

Woodwork

Alterations and repairs	Unit	Hours C	Hours L	Labour net	Material net	Price net	Price with 25%
Second fixings				£	£	£	£
					VAT not included		
Hardwood (*continued*)							
Crosstongued worktops button blocked to framing							
25mm	m2	6.60	-	59.14	63.39	122.53	153.16
32mm	m2	6.60	-	59.14	85.89	145.03	181.29
38mm	m2	6.60	-	59.14	102.46	161.60	202.00
Bearers							
19 x 38mm	m	0.22	-	1.97	1.78	3.75	4.69
25 x 50mm	m	0.25	-	2.24	2.14	4.38	5.47
Bearers not exceeding 300mm long							
19 x 38mm	each	0.10	-	0.90	0.54	1.44	1.80
25 x 50mm	each	0.11	-	0.99	0.90	1.89	2.36
25 x 150mm							
rounded window boards tongued on	m	0.43	-	3.85	4.99	8.84	11.05
returned ends	each	0.28	-	2.51	-	2.51	3.14
25 x 200mm							
rounded window boards tongued on	m	0.54	-	4.84	6.53	11.37	14.21
returned ends	each	0.33	-	2.96	-	2.96	3.70
50 x 75mm							
moulded handrails screwed on	m	0.32	-	2.87	7.42	10.29	12.86
returned ends	each	0.99	-	8.87	-	8.87	11.09
Sundry worktops							
25mm blockboard worktops screwed on	m2	0.99	-	8.87	26.83	35.70	44.63
25mm chipboard worktops screwed on	m2	0.70	-	6.27	6.75	13.02	16.27
Extra for lippings to 25mm worktop							
softwood	m	0.22	-	1.97	0.23	2.20	2.75
hardwood	m	0.26	-	2.33	0.59	2.92	3.65
Laminated plastic sheet fixed to worktop with adhesive							
balancers	m2	0.99	-	8.87	4.58	13.45	16.81
coverings	m2	0.22	-	1.97	18.96	20.93	26.16
edgings 25mm	m	0.40	-	3.58	0.52	4.10	5.13
edgings 32mm	m	0.22	-	1.97	0.65	2.62	3.27
edgings 38mm	m	0.22	-	1.97	0.78	2.75	3.44

Woodwork

Alterations and repairs	Unit	Hours C	Hours L	Labour net	Material net	Price net	Price with 25%
Second fixings				£	£	£	£
					VAT not included		
Sheet linings and casings							
7mm pre-finished veneered plywood, grooved to resemble plank panelling, pinned to studding or battens to walls							
over 300mm wide	m2	0.83	-	7.44	7.04	14.48	18.10
not exceeding 100mm wide	m	0.21	-	1.88	0.71	2.59	3.24
100 - 200mm wide	m	0.28	-	2.51	1.42	3.93	4.91
200 - 300mm wide	m	0.33	-	2.96	2.13	5.09	6.36
3.2mm hardboard pinned to studding or battens to walls							
over 300mm wide	m2	0.30	-	2.69	1.71	4.40	5.50
not exceeding 100mm wide	m	0.08	-	0.69	0.15	0.84	1.05
100 - 200mm wide	m	0.10	-	0.89	0.30	1.19	1.49
200 - 300mm wide	m	0.12	-	1.08	0.45	1.53	1.91
3.2mm hardboard bath tile (white painted), pinned to studding or battens to walls							
over 300mm wide	m2	0.55	-	4.93	2.19	7.12	8.90
not exceeding 100mm wide	m	0.14	-	1.25	0.22	1.47	1.84
100 - 200mm wide	m	0.19	-	1.70	0.45	2.15	2.69
200 - 300mm wide	m	0.22	-	1.97	0.67	2.64	3.30
3.2mm hardboard bath panels with melamine surface fixed with chromium plated screws to framing							
to front only	each	0.33	-	2.96	9.60	12.56	15.70
to front and one end including stainless steel angle strip	set	0.50	-	4.48	16.51	20.99	26.24
Moulded acrylic bath panels fixed with chromium plated screws to framing							
to front only	each	0.33	-	2.96	24.30	27.26	34.08
to front and one end	set	0.50	-	4.48	38.05	42.53	53.16
18mm melamine faced chipboard pinned to studding or battens to walls							
over 300mm wide	m2	0.24	-	2.15	6.89	9.04	11.30
not exceeding 100mm wide	m	0.17	-	1.52	0.69	2.21	2.76
100 - 200mm wide	m	0.21	-	1.88	1.39	3.27	4.09
200 - 300mm wide	m	0.25	-	2.24	2.08	4.32	5.40
13mm sanded finish fibre insulation board pinned to studding or battens to walls							
over 300mm wide	m2	0.30	-	2.69	15.10	17.79	22.24
not exceeding 100mm wide	m	0.08	-	0.72	1.51	2.23	2.79
100 - 200mm wide	m	0.10	-	0.90	3.02	3.92	4.90
200 - 300mm wide	m	0.12	-	1.08	4.52	5.60	7.00

Woodwork

Alterations and repairs	Unit	Hours C	Hours L	Labour net	Material net	Price net	Price with 25%
Second fixings				£	£	£	£
					VAT not included		
Sheet linings and casings (*continued*)							
6mm sanded finish asbestos free insulation board pinned to studding or battens to walls							
over 300mm wide	m2	0.51	-	4.57	10.92	15.49	19.36
not exceeding 100mm wide	m	0.13	-	1.16	1.10	2.26	2.83
100 - 200mm wide	m	0.17	-	1.52	2.18	3.70	4.63
200 - 300mm wide	m	0.20	-	1.79	3.27	5.06	6.33
9mm sanded finish asbestos free insulation board pinned to studding or battens to walls							
over 300mm wide	m2	0.55	-	4.93	22.13	27.06	33.83
not exceeding 100mm wide	m	0.14	-	1.25	2.22	3.47	4.34
100 - 200mm wide	m	0.17	-	1.52	4.43	5.95	7.44
200 - 300mm wide	m	0.22	-	1.97	6.63	8.60	10.75
25mm sterlingboard nailed to studding or battens to walls							
over 300mm wide	m2	0.46	-	4.12	10.96	15.08	18.85
not exceeding 100mm wide	m	0.12	-	1.08	1.10	2.18	2.73
100 - 200mm wide	m	0.15	-	1.34	1.86	3.20	4.00
200 - 300mm wide	m	0.19	-	1.70	3.29	4.99	6.24
Extra for fixing sheet linings with adhesive							
to battens	m2	0.11	-	0.99	0.32	1.31	1.64
to flat wall surfaces	m2	0.20	-	1.79	1.61	3.40	4.25
50mm paper faced cotton tape pasted over joints	m	0.06	-	0.54	0.05	0.59	0.74
19mm plastic cover strips fixed with adhesive	m	0.05	-	0.45	0.54	0.99	1.24
19 x 19mm white plastic angle cover strips fixed with adhesive	m	0.08	-	0.72	0.74	1.46	1.83
Black or white plastic joint holder for 3mm sheets							
to intermediate joints	m	0.36	-	3.23	0.84	4.07	5.09
to internal angles	m	0.36	-	3.23	0.84	4.07	5.09
to external angles	m	0.36	-	3.23	0.84	4.07	5.09
Angle pipe casings 225mm girth with 25mm softwood side, 19 x 25mm backings and front fixed with brass screws and cups							
with 19mm softwood beaded front	m	1.21	-	10.84	8.29	19.13	23.91
with 6mm plywood front	m	1.21	-	10.84	7.73	18.57	23.21
with 3.2mm hardboard front	m	1.21	-	10.84	6.15	16.99	21.24
Boxed pipe casings 300mm girth with 25mm softwood sides, 19 x 25mm backings and front fixed with brass screws and cups							
with 19mm softwood beaded front	m	1.29	-	11.56	12.53	24.09	30.11
with 6mm plywood front	m	1.29	-	11.56	8.35	19.91	24.89
with 3.2mm hardboard front	m	1.29	-	11.56	6.96	18.52	23.15

Woodwork

Alterations and repairs	Unit	Hours C	Hours L	Labour net	Material net	Price net	Price with 25%
Purpose made composite items				£	£	£	£
					VAT not included		
Hardwood panelling							
19mm panelling with 63mm square stiles and rails and 6mm veneered plywood panels	m2	0.89	-	7.97	18.73	26.70	33.38
25mm panelling with 63mm moulded stiles and rails and 6mm veneered plywood panels	m2	0.89	-	7.97	20.43	28.40	35.50
32 x 63mm							
moulded cappings	m	0.28	-	2.51	6.43	8.94	11.18
mitres	each	0.28	-	2.51	-	2.51	3.14
32 x 75mm							
moulded cappings	m	0.28	-	2.51	7.52	10.03	12.54
mitres	each	0.36	-	3.23	-	3.23	4.04
Softwood doors							
762 x 1981mm ledged and braced doors of 25mm matchboarding and 25mm ledges and braces	each	0.94	-	8.42	70.90	79.32	99.15
Framed, ledged and braced doors of 50mm framing, 25mm matchboarding and 25mm ledges and braces							
762 x 1981mm	each	1.21	-	10.84	87.21	98.05	122.56
1067 x 2134mm	each	1.32	-	11.83	89.00	100.83	126.04
762 x 1981 x 50mm casement doors open for glass and divided into eight panes	each	1.21	-	10.84	113.12	123.96	154.95
838 x 2057 x 50mm one panel doors moulded one side, bolection moulded the other side and with 12mm plywood panel	each	1.38	-	12.36	172.46	184.82	231.02
762 x 1981 x 50mm three panel doors, the lower panels bead butt and square and the upper panel with diminished stiles, open for glass and divided into six panes	each	1.27	-	11.38	130.49	141.87	177.34
Four panel doors square both sides with 6mm plywood panels							
762 x 1981 x 38mm	each	1.21	-	10.84	232.66	243.50	304.38
762 x 1981 x 50mm	each	1.46	-	13.08	232.66	245.74	307.18
Four panel doors moulded on solid both sides and with 6mm plywood panels							
762 x 1981 x 38mm	each	1.21	-	10.84	116.74	127.58	159.48
762 x 1981 x 50mm	each	1.46	-	13.08	122.25	135.33	169.16
762 x 1981 x 50mm four panel solid doors bead butt both sides	each	1.54	-	13.80	214.25	228.05	285.06

Woodwork

Alterations and repairs	Unit	Hours C	Hours L	Labour net	Material net	Price net	Price with 25%
Purpose made composite items				£	£	£	£
					VAT not included		

Softwood doors (*continued*)

762 x 1981 x 50mm three panel doors, the lower panels solid bead butt both sides and the upper panel open for glass and divided into four panes	each	1.54	-	13.80	147.05	160.85	201.06
Extra for							
rebated meeting stiles (both sides measured)	m	-	-	-	0.57	0.57	0.71
rebated and beaded meeting stiles (both sides measured)	m	-	-	-	3.36	3.36	4.20
rounded heels or stiles	m	-	-	-	2.04	2.04	2.55
32 x 100mm twice splayed weatherboards not exceeding 900mm long screwed on including notching frames	each	0.33	-	2.96	2.00	4.96	6.20
50 x 75mm moulded and throated weatherboards not exceeding 900mm long screwed on including notching frames	each	0.33	-	2.96	2.03	4.99	6.24

Hardwood doors

50mm casement doors open for glass in one pane							
762 x 1981mm	each	2.86	-	25.63	126.49	152.12	190.15
838 x 2057mm	each	3.03	-	27.15	126.49	153.64	192.05
762 x 1981 x 50mm one panel doors moulded one side, bolection moulded the other side and with 12mm plywood panel	each	2.75	-	24.64	186.74	211.38	264.23
838 x 2057 x 50mm one panel doors moulded one side, bolection moulded the other side and with 12mm plywood panel	each	3.30	-	29.57	327.08	356.65	445.81
762 x 1981 x 50mm four panel doors moulded both sides with 6mm veneered plywood panels	each	2.75	-	24.64	260.89	285.53	356.91
762 x 1981 x 50mm four panel solid doors bead butt both sides	each	2.75	-	24.64	354.81	379.45	474.31
762 x 1981 x 50mm three panel doors, the lower panels solid bead butt both sides and the upper panel open for glass and divided into four panes	each	2.75	-	24.64	311.70	336.34	420.43
Extra for							
rebated meeting stiles (both sides measured)	m	-	-	-	0.57	0.57	0.71
rebated and beaded meeting stiles (both sides measured)	m	-	-	-	4.03	4.03	5.04
rounded heels or stiles	m	-	-	-	2.29	2.29	2.86
50 x 75mm moulded and grooved weatherboards not exceeding 900mm long screwed on and pelleted including notching frames	each	0.33	-	2.96	12.53	15.49	19.36

Woodwork

Alterations and repairs	Unit	Hours C	Hours L	Labour net	Material net	Price net	Price with 25%
Purpose made composite items				£	£	£	£
					VAT not included		
Flush doors							
762 x 1981 x 35mm flush doors of 32mm softwood skeleton framed core covered on both sides with 3.2mm hardboard							
unlipped	each	1.10	-	9.86	22.49	32.35	40.44
lipped on long edges	each	1.10	-	9.86	22.49	32.35	40.44
726 x 2040 x 40mm flush doors of 32mm softwood skeleton framed core covered on both sides with 4mm plywood							
unlipped	each	1.10	-	9.86	24.47	34.33	42.91
lipped on long edges	each	1.10	-	9.86	24.47	34.33	42.91
726 x 2040 x 40mm flush doors of 32mm softwood skeleton framed core covered on both sides with 4mm hardwood veneered plywood and hardwood lipped on all edges	each	1.10	-	9.86	33.02	42.88	53.60
Softwood door frames and linings							
Frames							
50 x 75mm	m	0.24	-	2.15	1.75	3.90	4.88
50 x 100mm	m	0.24	-	2.15	2.32	4.47	5.59
50 x 150mm	m	0.24	-	2.15	4.51	6.66	8.32
Rebated frames							
50 x 75mm	m	0.26	-	2.33	2.05	4.38	5.47
50 x 100mm	m	0.26	-	2.33	2.62	4.95	6.19
50 x 125mm	m	0.26	-	2.33	3.96	6.29	7.86
50 x 150mm	m	0.28	-	2.51	4.81	7.32	9.15
63 x 75mm	m	0.26	-	2.33	2.93	5.26	6.58
63 x 100mm	m	0.26	-	2.33	3.78	6.11	7.64
63 x 125 mm	m	0.26	-	2.33	5.80	8.13	10.16
63 x 150mm	m	0.28	-	2.51	7.06	9.57	11.96
Rebated and moulded frames							
50 x 100mm	m	0.26	-	2.33	2.92	5.25	6.56
50 x 150mm	m	0.28	-	2.51	5.05	7.56	9.45
63 x 100 mm	m	0.26	-	2.33	3.78	6.11	7.64
63 x 125 mm	m	0.26	-	2.33	6.10	8.43	10.54
63 x 150 mm	m	0.28	-	2.51	7.36	9.87	12.34
75 x 100 mm	m	0.26	-	2.33	4.94	7.27	9.09
75 x 125 mm	m	0.26	-	2.33	6.10	8.43	10.54
75 x 150 mm	m	0.28	-	2.51	7.27	9.78	12.23

Woodwork

Alterations and repairs	Unit	Hours C	Hours L	Labour net	Material net	Price net	Price with 25%
Purpose made composite items				£	£	£	£
					VAT not included		
Softwood door frames and linings (*continued*)							
Linings tongued at angles							
25 x 75mm	m	0.22	-	1.97	1.45	3.42	4.28
25 x 100mm	m	0.24	-	2.15	1.99	4.14	5.17
25 x 125mm	m	0.24	-	2.15	2.52	4.67	5.84
25 x 150mm	m	0.25	-	2.24	1.76	4.00	5.00
32 x 100mm	m	0.24	-	2.15	2.06	4.21	5.26
32 x 125mm	m	0.24	-	2.15	3.09	5.24	6.55
32 x 150mm	m	0.25	-	2.24	3.09	5.33	6.66
Rebated linings tongued at angles							
38 x 100mm	m	0.24	-	2.15	2.36	4.51	5.64
38 x 125mm	m	0.24	-	2.15	3.39	5.54	6.92
38 x 150mm	m	0.26	-	2.33	3.39	5.72	7.15
Extra for hollow groove in frames for heel of swing door	m	-	-	-	0.33	0.33	0.41
Hardwood door frames and linings							
Rebated and moulded frames							
63 x 75mm	m	0.40	-	3.58	23.37	26.95	33.69
75 x 100mm	m	0.45	-	4.03	26.91	30.94	38.67
75 x 100mm twice moulded frames	m	0.46	-	4.12	15.85	19.97	24.96
Rebated linings tongued at angles							
38 x 125mm	m	0.43	-	3.85	15.44	19.29	24.11
38 x 150mm	m	0.43	-	3.85	18.52	22.37	27.96
Extra for hollow groove in frames for heel of swing door	m	-	-	-	0.47	0.47	0.59
Sunk weathered and grooved thresholds							
75 x 100mm	m	0.31	-	2.78	27.00	29.78	37.23
75 x 175mm	m	0.31	-	2.78	40.03	42.81	53.51
Softwood casements							
Moulded casements in one pane 0.50 - 1.00m2							
38mm	m2	-	-	-	38.95	38.95	48.69
50mm	m2	-	-	-	41.00	41.00	51.25
Moulded casements divided into panes 0.10 - 0.50m2							
38mm	m2	-	-	-	59.45	59.45	74.31
50mm	m2	-	-	-	61.50	61.50	76.88
Moulded casements divided into panes not exceeding 0.10m2							
38mm	m2	-	-	-	88.15	88.15	110.19
50mm	m2	-	-	-	92.25	92.25	115.31
Fitting and fixing casements	each	0.36	-	3.23	0.07	3.30	4.13

Woodwork

Alterations and repairs	Unit	Hours C	Hours L	Labour net	Material net	Price net	Price with 25%

Purpose made composite items

				£	£	£	£
					VAT not included		

Softwood casements (*continued*)

Fitting and hanging casements on 63mm light steel butts	each	0.74	-	6.63	0.63	7.26	9.07
Fitting and hanging casements on stove enamelled pivots	each	1.27	-	11.38	7.91	19.29	24.11

Hardwood casements

Moulded casements in one pane 0.50 - 1.00m2							
38mm	m2	-	-	-	75.51	75.51	94.39
50mm	m2	-	-	-	81.31	81.31	101.64
Moulded casements divided into panes 0.10 - 0.50m2							
38mm	m2	-	-	-	110.36	110.36	137.95
50mm	m2	-	-	-	116.16	116.16	145.20
Moulded casements divided into panes not exceeding 0.10m2							
38mm	m2	-	-	-	162.63	162.63	203.29
50mm	m2	-	-	-	174.25	174.25	217.81
Fitting and fixing casements	each	0.55	-	4.93	0.07	5.00	6.25
Fitting and hanging casements on 63mm brass butts	each	1.24	-	11.11	5.10	16.21	20.26
Fitting and hanging casements on brass pivots	each	1.93	-	17.29	25.09	42.38	52.98

Softwood frames

Rebated casement frames							
50 x 63mm	m	0.24	-	2.15	2.09	4.24	5.30
63 x 75mm	m	0.24	-	2.15	2.63	4.78	5.97
75 x 100mm	m	0.24	-	2.15	4.64	6.79	8.49
Rebated and moulded casement frames							
50 x 63mm	m	0.24	-	2.15	2.39	4.54	5.67
63 x 75mm	m	0.24	-	2.15	3.23	5.38	6.72
75 x 100mm	m	0.24	-	2.15	4.94	7.09	8.86
Twice rebated mullions or transoms							
50 x 63mm	m	0.24	-	2.15	2.39	4.54	5.67
63 x 75mm	m	0.24	-	2.15	3.23	5.38	6.72
75 x 100mm	m	0.24	-	2.15	4.94	7.09	8.86

Woodwork

Alterations and repairs	Unit	Hours C	Hours L	Labour net	Material net	Price net	Price with 25%
Purpose made composite items				£	£	£	£
					VAT not included		
Softwood frames (continued)							
Twice rebated and twice moulded mullions or transoms							
50 x 63mm	m	0.24	-	2.15	2.99	5.14	6.42
63 x 75mm	m	0.24	-	2.15	3.83	5.98	7.47
75 x 100mm	m	0.22	-	1.97	5.54	7.51	9.39
16 x 22mm cut and mitred beads to pivot hung casement and frame	m	0.28	-	2.51	0.28	2.79	3.49
Hardwood frames							
Rebated and grooved cills							
63 x 63mm	m	0.31	-	2.78	21.38	24.16	30.20
63 x 75mm	m	0.31	-	2.78	23.37	26.15	32.69
Rebated, sunk weathered, throated, check throated and grooved cills							
63 x 150mm	m	0.31	-	2.78	31.71	34.49	43.11
75 x 150mm	m	0.31	-	2.78	41.23	44.01	55.01
75 x 175mm	m	0.31	-	2.78	60.77	63.55	79.44
Rebated, moulded, sunk weathered, throated, check throated and grooved cills							
63 x 150mm	m	0.31	-	2.78	32.14	34.92	43.65
75 x 150 mm	m	0.31	-	2.78	41.66	44.44	55.55
75 x 175mm	m	0.31	-	2.78	48.17	50.95	63.69
Mitres to cill including handrail bolt	each	0.38	-	3.40	1.68	5.08	6.35
Softwood window surrounds							
19 x 125mm window linings tongued at angles and tongued on	m	0.39	-	3.49	1.63	5.12	6.40
Pelmets of 19 x 100mm top, 13 x 150mm front							
straight	m	0.66	-	5.91	3.27	9.18	11.48
returned ends	each	0.22	-	1.97	0.17	2.14	2.67
Hardwood window surrounds							
19 x 125mm window linings tongued at angles and tongued on	m	0.52	-	4.66	8.87	13.53	16.91
38 x 75mm rebated and moulded shop window framing	m	0.77	-	6.90	10.10	17.00	21.25
50 x 75mm twice rebated and twice moulded mullion or transom in shop window framing	m	0.88	-	7.88	17.59	25.47	31.84
50 x 100 mm rebated, weathered and moulded cill in shop window framing	m	0.36	-	3.23	21.76	24.99	31.24

Woodwork

Alterations and repairs	Unit	Hours C	Hours L	Labour net	Material net	Price net	Price with 25%
Purpose made composite items				£	£	£	£
					VAT not included		

Sash Windows

Softwood double hung sash windows with sashes
divided into panes 0.10 - 0.50 m2, 32 mm pulley
stiles, 25 mm inside and 19 mm outside linings,
13 mm beads and back linings and hardwood cills
- cord and weight not included

	Unit	Hours C	Hours L	Labour net	Material net	Price net	Price with 25%
38 mm sashes and 75 x 125 mm cills – based on 1.75 m2 overall	m2	19.25	-	172.48	45.64	218.12	272.65
38 mm sashes and 75 x 125 mm cills – based on 1.25 m2 overall	m2	13.75	-	123.20	49.97	173.17	216.46
50 mm sashes and 75 x 150 mm cills - based on 1.75 m2 overall	m2	19.25	-	172.48	52.11	224.59	280.74
50 mm sashes and 75 x 150 mm cills - based on 1.25 m2 overall	m2	13.75	-	123.20	61.21	184.41	230.51
Extra for moulded horns							
to 38 mm sashes	each	0.20	-	1.79	0.19	1.98	2.48
to 50 mm sashes	each	0.20	-	1.79	0.23	2.02	2.52
Fitting and hanging sashes on cords and weights							
600 x 600 mm sashes	each	2.00	-	17.92	18.70	36.62	45.77
900 x 900 mm sashes	m2	2.00	-	17.92	28.22	46.14	57.67
Fitting and hanging sashes on chains and weights							
600 x 600 mm sashes	each	2.00	-	17.92	44.00	61.92	77.40
900 x 900 mm sashes	each	2.00	-	17.92	59.86	77.78	97.22

Sash windows combined items

	Unit	Hours C	Hours L	Labour net	Material net	Price net	Price with 25%
Softwood double hung sash windows 1.75 m2 overall complete with 38 mm sashes including fitting and hanging sashes with							
cords and weights	each	21.45	-	192.19	73.78	265.97	332.46
chains and weights	each	21.45	-	192.19	105.69	297.88	372.35
Softwood double hung sash windows 1.25 m2 overall complete with 38 mm sashes including fitting and hanging sashes with							
cords and weights	each	15.95	-	142.91	78.38	221.29	276.61
chains and weights	each	15.95	-	142.91	110.02	252.93	316.16
Softwood double hung sash windows 1.75 m2 overall complete with 50 mm sashes including fitting and hanging sashes with							
cords and weights	each	21.45	-	192.19	80.52	272.71	340.89
chains and weights	each	21.45	-	192.19	112.16	304.35	380.44
Softwood double hung sash windows 1.25 m2 overall complete with 50 mm sashes including fitting and hanging with							
cords and weights	each	15.95	-	142.91	89.62	232.53	290.66
chains and weights	each	15.95	-	142.91	121.26	264.17	330.21

Woodwork

Alterations and repairs	Unit	Hours C	Hours L	Labour net	Material net	Price net	Price with 25%
Purpose made composite items				£	£	£	£
					VAT not included		

Sash windows combined items (*continued*)

Softwood double hung sash window with sashes divided into panes 0.10 - 0.50 m2, 38 mm stiles, 25 mm inside and 19 mm outside linings, 13 mm beads and oak cills including fitting and hanging sashes on spring balances

38 mm sashes and 75 x 125 mm cills – based on 1.75 m2 overall	m2	21.39	-	191.65	67.27	258.92	323.65
38 mm sashes and 75 x 125 mm cills - based on 1.25 m2 overall	m2	16.35	-	146.50	80.19	226.69	283.36
50 mm sashes and 75 x 150 mm cills - based on 1.75 m2 overall	m2	21.39	-	191.65	73.74	265.39	331.74
50 mm sashes and 75 x 150 mm cills - based on 1.25 m2 overall	m2	16.35	-	146.50	91.43	237.93	297.41

Softwood screens and borrowed lights

Square framing

25 x 100mm	m	0.15	-	1.34	1.98	3.32	4.15
50 x 50mm	m	0.15	-	1.34	1.20	2.54	3.17
50 x 75mm	m	0.15	-	1.34	1.75	3.09	3.86
50 x 100mm	m	0.17	-	1.52	2.32	3.84	4.80

Rebated and moulded framing

50 x 75mm	m	0.15	-	1.34	2.35	3.69	4.61
50 x 100mm	m	0.17	-	1.52	2.92	4.44	5.55
75 x 100mm	m	0.18	-	1.61	2.20	3.81	4.76

Twice rebated and twice moulded framing

50 x 75mm	m	0.15	-	1.34	2.95	4.29	5.36
50 x 100mm	m	0.17	-	1.52	3.52	5.04	6.30
75 x 100mm	m	0.18	-	1.61	5.54	7.15	8.94

Square glazing bars

25 x 50mm	m	0.11	-	0.99	0.71	1.70	2.13
25 x 75mm	m	0.11	-	0.99	1.44	2.43	3.04

Twice rebated and twice moulded glazing bars

32 x 75mm	m	0.11	-	0.99	2.88	3.87	4.84
32 x 100mm	m	0.11	-	0.99	3.25	4.24	5.30
38 x 75mm	m	0.11	-	0.99	2.93	3.92	4.90
38 x 100mm	m	0.11	-	0.99	3.44	4.43	5.54

16 x 50mm stock pattern moulded stop	m	0.14	-	1.25	0.97	2.22	2.77

Twice rebated feature rail

50 x 100mm	m	0.17	-	1.52	2.92	4.44	5.55
50 x 150mm	m	0.18	-	1.61	5.11	6.72	8.40

Woodwork

Alterations and repairs	Unit	Hours C	Hours L	Labour net	Material net	Price net	Price with 25%
Purpose made composite items				£	£	£	£
					VAT not included		

Softwood closed tread staircases

	Unit	Hours C	Hours L	Labour net	Material net	Price net	Price with 25%
32 mm treads with rounded nosings and scotia moulding under and 19 mm risers, glued and blocked together including carriages							
flights	m2	5.80	5.80	91.06	42.27	133.33	166.66
ends of tread and riser housed and wedged	each	0.35	-	3.14	0.07	3.21	4.01
Extra for bullnose quadrant steps with pre formed riser	each	2.00	-	17.92	48.66	66.58	83.22
32 mm winders with rounded nosings and scotia moulding under and 19 mm risers, glued and blocked together including carriages							
flights	m2	6.50	6.50	102.05	70.40	172.45	215.56
narrow ends of winder and riser housed and wedged	each	0.30	-	2.69	0.07	2.76	3.45
wide ends of winder and riser housed and wedged	each	0.52	-	4.66	0.07	4.73	5.91
Landings of 25mm tongued and grooved flooring with rounded nosing, including framed bearers	m2	1.74	0.66	20.04	15.50	35.54	44.42
Solid balustrade fillings of 25 x 50mm uprights and 19 x 38mm fillets covered on both sides with 3mm hardboard							
square	m2	2.11	-	18.91	6.58	25.49	31.86
raking	m2	2.69	-	24.10	6.58	30.68	38.35
19 x 200mm beaded apron linings	m	0.33	-	2.96	1.82	4.78	5.97
25 x 50mm chamfered floor fillets	m	0.11	-	0.99	0.98	1.97	2.46
32 x 32mm							
balusters	m	0.06	-	0.54	3.30	3.84	4.80
fitted ends	each	0.02	-	0.20	-	0.20	0.25
fitted ends on rake	each	0.03	-	0.30	-	0.30	0.38
32 x 50mm moulded cappings	m	0.19	-	1.70	1.71	3.41	4.26
25 x 100mm rebated nosings to flooring including scotia moulding under	m	0.33	-	2.96	4.08	7.04	8.80
32 x 225mm							
wall strings	m	0.66	-	5.91	4.20	10.11	12.64
fitted ends	each	0.33	-	2.96	-	2.96	3.70
extra for ramps	each	0.38	-	3.40	-	3.40	4.25
38 x 225mm							
outer strings	m	0.66	-	5.91	6.55	12.46	15.57
ends framed to newel	each	0.44	-	3.94	-	3.94	4.92
50 x 100mm half newels	m	0.72	-	6.45	8.33	14.78	18.48

Woodwork

	Unit	Hours C	Hours L	Labour net	Material net	Price net	Price with 25%

Purpose made composite items

				£	£	£	£
					VAT not included		

Softwood closed tread staircases (*continued*)

	Unit	Hours C	Hours L	Labour net	Material net	Price net	Price with 25%
100 x 100mm							
newels	m	0.99	-	8.87	11.29	20.16	25.20
shaped or rounded ends	each	0.83	-	7.44	-	7.44	9.30
63 x 150 x 150mm moulded newel caps	each	0.55	-	4.93	12.27	17.20	21.50
63 x 75mm							
moulded handrails	m	0.22	-	1.97	2.92	4.89	6.11
ends framed to newel	each	0.36	-	3.23	-	3.23	4.04
ends framed on rake to newel	each	0.55	-	4.93	-	4.93	6.16
heading joint including handrail screw	each	1.67	-	14.96	2.50	17.46	21.82
mitres including handrail screw	each	2.00	-	17.92	2.50	20.42	25.52
extra for level bend 350 mm girth and two heading joints including handrail screws	each	2.57	-	23.03	19.67	42.70	53.38
extra for ramps 225 mm girth and two heading joints including handrail screws	each	2.90	-	25.98	23.09	49.07	61.34
extra for wreaths 450 mm girth and two heading joints including handrail screws	each	2.90	-	25.98	26.03	52.01	65.01

Softwood closed tread staircases, combined items

	Unit	Hours C	Hours L	Labour net	Material net	Price net	Price with 25%
Straight flight staircases, with 32 x 225 mm wall string, 38 x 225 mm outer string, 32 mm treads, 19 mm risers, 100 x 100 mm newels and quadrant step at bottom							
900 mm wide x 2650 mm rise	each	12.16	5.80	148.04	320.00	468.04	585.05
fixing staircase on site	each	5.14	5.14	80.69	1.02	81.71	102.14
timber carriages and fixing	each	2.57	-	23.03	7.19	30.22	37.77
balustrades of 32 x 32 mm balusters, 63 x 75 mm moulded handrail, moulded string capping and chamfered floor fillet	each	1.16	-	10.42	8.91	19.33	24.16
half newels and newel caps	each	2.15	-	19.26	18.99	38.25	47.81
32 x 100 mm well-hole nosings with scotia moulding under, 19 x 200 mm apron linings	each	2.40	-	21.50	27.70	49.20	61.50
Straight staircases complete							
per flight	each	12.26	5.14	144.49	605.00	749.49	936.86

Woodwork

Alterations and repairs	Unit	Hours C	Hours L	Labour net	Material net	Price net	Price with 25%
Purpose made composite items				£	£	£	£
					VAT not included		

Softwood closed tread staircases, combined items (*continued*)

Straight flight staircases, with 32 x 225 mm wall string, 38 x 225 mm outer string, 32 mm tread and risers and three winders at top, 19 mm risers, 100 x 100 mm newels and quadrant step at bottom

Alterations and repairs	Unit	Hours C	Hours L	Labour net	Material net	Price net	Price with 25%
900 mm wide and 2650 mm rise	each	19.48	18.80	301.25	460.00	761.25	951.56
fixing staircases on site	each	9.14	9.14	143.49	2.04	145.53	181.91
timber carriages and fixing	each	4.57	-	40.95	23.68	64.63	80.79
balustrades of 32 x 32 mm balusters, 63 x 75 mm moulded handrail, moulded sting capping and chamfered floor fillet	each	4.17	-	37.39	11.41	48.80	61.00
half newels and newel caps	each	2.15	-	19.26	37.99	57.25	71.56
32 x 100 mm well-hole nosings with scotia moulding under, 19 x 200 mm apron linings	each	2.07	-	18.55	55.40	73.95	92.44

Straight staircases complete

	Unit	Hours C	Hours L	Labour net	Material net	Price net	Price with 25%
per flight	each	17.93	9.14	222.25	790.00	1012.25	1265.31

Softwood open tread staircases

Staircases

	Unit	Hours C	Hours L	Labour net	Material net	Price net	Price with 25%
50 x 200 mm treads with rounded arrises	each	0.40	0.20	4.93	7.53	12.46	15.57
ends housed and nailed to string	each	0.35	-	3.14	0.01	3.15	3.94
50 x 225 mm strings	m	0.28	-	2.51	7.39	9.90	12.38
fitted ends	each	0.33	-	2.96	-	2.96	3.70
50 x 75 mm newels bolted to string	m	0.60	-	5.38	1.75	7.13	8.91
M10 x 110 mm steel carriage bolts and washer including boring softwood	each	0.28	-	2.51	0.47	2.98	3.73
50 x 75 mm handrails with rounded tops	m	0.28	-	2.51	2.33	4.84	6.05
Returned ends	each	0.33	-	2.96	-	2.96	3.70
Ends housed to newels	each	0.38	-	3.40	-	3.40	4.25
32 x 125 mm rails screwed on	each	0.28	-	2.51	3.12	5.63	7.04
Returned ends	each	0.38	-	3.40	-	3.40	4.25
extra for quadrant steps with preformed riser	each	2.00	-	17.92	48.66	66.58	83.22

Woodwork

Alterations and repairs	Unit	Hours C	Hours L	Labour net	Material net	Price net	Price with 25%
Purpose made composite items				£	£	£	£
					VAT not included		
Softwood open tread staircases, combined items							
Straight flight staircases 900 mm wide and 3000 mm rise with 50 x 225 mm strings and 50 x 200 mm treads, no risers	each	1.36	0.20	13.54	122.00	135.54	169.42
Fixing staircases on site	each	2.29	-	20.52	0.51	21.03	26.29
50 x 75 mm rounded handrail to one side with three 50 x 75 mm newel bolted to string and 32 x 125 mm rails	each	2.15	-	19.26	43.00	62.26	77.82
Straight staircases per flight	each	11.20	5.80	139.44	585.00	724.44	905.55
Hardwood closed tread staircase							
32 mm treads with rounded nosings and scotia moulding under and 19 mm risers, glued and blocked together including carriages							
flights	m2	7.54	7.54	118.38	150.03	268.41	335.51
ends of tread and riser housed, wedged	each	0.46	-	4.12	0.20	4.32	5.40
Extra for bullnose quadrant steps with preformed riser	each	3.00	-	26.88	68.50	95.38	119.22
32 mm winders with rounded nosings and scotia moulding under and 19 mm risers, glued and blocked together including carriages							
per flight	each	8.45	8.45	132.66	217.92	350.58	438.23
narrow ends of winder and riser housed and wedged	each	0.40	-	3.58	0.20	3.78	4.72
wide ends of winder and riser housed and wedged	each	0.68	-	6.09	0.20	6.29	7.86
Landings of 25 mm tongued and grooved flooring with rounded nosing, including framed bearers	m2	2.26	0.86	26.05	63.47	89.52	111.90
19 x 225 mm beaded apron linings	m	0.61	-	5.47	2.96	8.43	10.54
25 x 50 mm chamfered floor fillets	m	0.21	-	1.88	3.70	5.58	6.97
32 x 32 mm							
balusters	m	0.09	-	0.81	9.58	10.39	12.99
fitted ends	each	0.05	-	0.45	-	0.45	0.56
fitted ends on the rake	each	0.06	-	0.54	-	0.54	0.68
32 x 50 mm moulded cappings	m	0.33	-	2.96	5.14	8.10	10.13
25 x 100 mm rebated nosings to flooring including scotia moulding under	m	0.61	-	5.47	9.81	15.28	19.10

Woodwork

Alterations and repairs	Unit	Hours C	Hours L	Labour net	Material net	Price net	Price with 25%

Purpose made composite items

				£	£	£	£
				VAT not included			

Hardwood closed tread staircases (*continued*)

	Unit	Hours C	Hours L	Labour net	Material net	Price net	Price with 25%
32 x 225 mm							
wall strings	m	0.55	-	4.93	19.87	24.80	31.00
fitted ends	each	0.55	-	4.93	-	4.93	6.16
extra for ramps	each	0.35	-	3.14	-	3.14	3.92
38 x 225 mm							
outer strings	m	0.55	-	4.93	27.91	32.84	41.05
ends framed to newel	each	0.98	-	8.78	-	8.78	10.98
50 x 100 mm half newels	m	1.29	-	11.56	0.70	12.26	15.32
100 x 100 mm							
newels	m	1.98	-	17.74	0.84	18.58	23.23
shaped or rounded ends	each	1.52	-	13.62	-	13.62	17.02
63 x 150 x 150 mm moulded newel cap	each	0.57	-	5.11	16.72	21.83	27.29
63 x 75 mm							
moulded handrail	each	0.32	-	2.87	11.73	14.60	18.25
moulded handrail, grooved for core rail	each	0.40	-	3.58	11.70	15.28	19.10
ends framed to newel	each	0.57	-	5.11	-	5.11	6.39
ends framed on rake to newel	each	0.86	-	7.71	-	7.71	9.64
heading joint including handrail screw	each	2.15	-	19.26	2.50	21.76	27.20
mitres including handrail screw	each	2.00	-	17.92	2.50	20.42	25.52
extra for level bends 350 mm girth and two heading joints including handrail screws	each	4.57	-	40.95	19.67	60.62	75.78
extra for ramps 225 mm girth and two heading joints including handrail screws	each	5.47	-	49.01	23.09	72.10	90.13
extra for wreaths 450 mm girth and two heading joints including handrail screws	each	4.57	-	40.95	26.03	66.98	83.72

Hardwood closed tread staircases, combined items

	Unit	Hours C	Hours L	Labour net	Material net	Price net	Price with 25%
Straight flight staircases, with 32 x 225 mm wall string, 38 x 225 mm outer string, 32 mm treads, 19 mm risers, 100 x 100 mm newels and quadrant step at bottom							
900 mm wide and 2650 mm rise	each	16.53	7.54	198.93	890.00	1088.93	1361.16
fixing staircases on site	each	6.86	6.86	107.71	2.04	109.75	137.19
timber carriages and fixing	each	2.57	-	23.03	41.81	64.84	81.05
balustrades of 32 x 32 mm balusters, 63 x 75 mm moulded handrail, moulded string capping and chamfered floor fillet	each	1.87	-	16.76	30.15	46.91	58.64
half newels and newel caps	each	2.72	-	24.37	17.78	42.15	52.69
32 x 100 mm well-hole nosings with scotia moulding under and 19 x 225 mm apron linings	each	2.40	-	21.50	53.01	74.51	93.14
Straight staircases							
per flight	each	30.31	14.40	368.64	1400.00	1768.64	2210.80

Woodwork

Alterations and repairs	Unit	Hours C	Hours L	Labour net	Material net	Price net	Price with 25%
Purpose made composite items				£	£	£	£
					VAT not included		

Hardwood closed tread staircases, combined items (continued)

Straight staircases with 32 x 225 mm string, 38 x 225 mm outer string, 32 mm treads and three winders at top, 19 mm risers, 100 x 100 mm newels and quadrant step at bottom

	Unit	Hours C	Hours L	Labour net	Material net	Price net	Price with 25%
900 mm wide and 2650 mm rise	each	26.06	15.99	341.27	1350.00	1691.27	2114.09
fixing staircases on site	each	11.43	11.43	179.45	1.02	180.47	225.59
timber carriages and fixing	each	4.57	-	40.95	69.53	110.48	138.10
balustrades of 32 x 32 mm balusters, 63 x 75 mm moulded handrail, moulded string capping and chamfered floor fillet	each	11.91	-	106.71	79.27	185.98	232.47
half newel and newel caps	each	2.72	-	24.37	17.98	42.35	52.94
32 x 100 mm well-hole nosing with scotia moulding under, 19 x 225 mm apron lining	each	3.63	-	32.52	56.37	88.89	111.11

Staircases with winders

	Unit	Hours C	Hours L	Labour net	Material net	Price net	Price with 25%
per flight	each	49.80	22.85	600.22	1800.00	2400.22	3000.27

Hardwood open tread staircases

Staircases

	Unit	Hours C	Hours L	Labour net	Material net	Price net	Price with 25%
38 x 250 mm treads with rounded arises, ends housed to string, screwed and pellated	m	0.90	0.45	1.09	25.00	36.09	45.11
ends housed to string, screwed and pellated	each	0.70	-	6.27	3.04	9.31	11.64
38 x 250 mm strings	m	1.00	-	8.96	32.37	41.33	51.66
fair splay cut ends	each	0.33	-	2.96	-	2.96	3.70
splay cut and fitted ends	each	0.30	-	2.69	-	2.69	3.36
ends housed to newel on rake	each	0.95	-	8.51	-	8.51	10.64
ends slotted through newel on rake	each	1.05	-	9.41	-	9.41	11.76
19 x 38 mm riser slats	m	0.10	-	0.90	2.31	3.21	4.01
ends housed to string	m	0.12	-	1.08	-	1.08	1.35
19 x 200 mm apron lining	m	0.61	-	5.47	11.24	16.71	20.89
32 x 100 mm nosings and scotia	m	0.61	-	5.47	19.90	25.37	31.71
50 x 100 mm half newels	m	1.29	-	11.56	0.70	12.26	15.32
100 x 100 mm newels	m	1.98	-	17.74	0.84	18.58	23.23
shaped or rounded ends	each	1.52	-	13.62	-	13.62	17.02
63 x 150 x 150 mm moulded newel caps	each	0.57	-	5.11	16.72	21.83	27.29

38 x 125 mm

	Unit	Hours C	Hours L	Labour net	Material net	Price net	Price with 25%
handrails and intermediate rails	m	0.35	-	3.14	15.42	18.56	23.20
ends framed to newel	each	0.57	-	5.11	-	5.11	6.39
ends framed on rake to newel	each	0.86	-	7.71	-	7.71	9.64

25 x 38 mm twice rounded balustrade stiffeners screwed and pellated

	Unit	Hours C	Hours L	Labour net	Material net	Price net	Price with 25%
	each	0.20	-	1.79	3.35	5.14	6.42

Woodwork

Alterations and repairs	Unit	Hours C	Hours L	Labour net	Material net	Price net	Price with 25%
Purpose made composite items				£	£	£	£
					VAT not included		

Hardwood open tread staircases combined items

Staircase with 38 x 250 mm treads slightly rounded on all edges and 19 x 38 mm riser slats, all housed and screwed to 38 x 250 mm strings.

Description	Unit	Hours C	Hours L	Labour net	Material net	Price net	Price with 25%
900 mm wide and 2650 mm rise	each	3.12	0.45	30.99	500.00	530.99	663.74
Fixing staircases on site	each	5.71	-	51.16	3.03	54.19	67.74
100 x 100 mm newels bolted to carriages and floor joist, 38 x 125 mm handrail slightly rounded on top edges, 38 x 125 mm infill rails and 50 x 100 mm half newel	each	5.82	-	52.15	37.03	89.18	111.47
32 x 100 mm landing nosing, 19 x 200 mm apron lining and scotia moulding	each	1.22	-	10.93	31.14	42.07	52.59
Straight staircases per flight	each	15.87	0.45	145.23	850.00	995.23	1244.04

Softwood built-in cupboards

Description	Unit	Hours C	Hours L	Labour net	Material net	Price net	Price with 25%
Cupboard fronts moulded one side, with 4mm plywood panels; doors hung on pairs 63mm pressed steel butts							
full height	m2	0.94	-	8.42	13.87	22.29	27.86
dwarf	m2	1.21	-	10.84	15.35	26.19	32.74
Cupboard ends moulded one side, with 4mm plywood panels							
full height	m2	1.47	-	13.17	11.16	24.33	30.41
dwarf	m2	2.20	-	19.71	11.64	31.35	39.19
Flush cupboard fronts of 32mm outer frame and 25mm door framing, covered one side with 3.2mm hardboard; doors hung on pairs 63mm pressed steel butts							
full height	m2	0.94	-	8.42	8.98	17.40	21.75
dwarf	m2	1.22	-	10.93	10.08	21.01	26.26
Flush cupboard ends of 25mm framing covered on one side with 3.2mm hardboard							
full height	m2	1.47	-	13.17	4.49	17.66	22.07
dwarf	m2	2.20	-	19.71	6.27	25.98	32.48
Flush cupboard fronts with blockboard doors lipped on long edges and hung on pairs 63mm pressed steel butts							
full height with 32mm outer frame and 25mm doors	m2	0.94	-	8.42	32.46	40.88	51.10
dwarf with 25mm outer frame and 18mm doors	m2	1.22	-	10.93	27.21	38.14	47.67

Woodwork

Alterations and repairs	Unit	Hours C	Hours L	Labour net	Material net	Price net	Price with 25%
Purpose made composite items				£	£	£	£
					VAT not included		
Softwood built-in cupboards (*continued*)							
Flush cupboard ends of 18mm blockboard							
full height	m2	1.47	-	13.17	23.70	36.87	46.09
dwarf	m2	2.20	-	19.71	23.43	43.14	53.92
Crosstongued shelves or divisions							
19mm	m2	3.08	-	27.60	20.30	47.90	59.88
25mm	m2	3.30	-	29.57	22.22	51.79	64.74
Crosstongued tops to dwarf cupboards, fixed with buttons							
25mm	m2	2.38	-	21.32	22.29	43.61	54.51
32mm	m2	2.53	-	22.67	23.38	46.05	57.56
Extra for moulded edges to tops	m	-	-	-	0.60	0.60	0.75
Blockboard tops to dwarf cupboards							
18mm	m2	2.38	-	21.32	20.55	41.87	52.34
25mm	m2	2.53	-	22.67	26.76	49.43	61.79
Extra for softwood lippings to tops							
18mm	m2	0.22	-	1.97	0.17	2.14	2.67
25mm	m2	0.22	-	1.97	0.23	2.20	2.75
Hardwood built-in cupboards							
Cupboard fronts moulded one side, with 6mm veneered plywood panels; doors hung on pairs 63mm brass butts							
full height	m2	1.24	-	11.11	38.75	49.86	62.33
dwarf	m2	1.24	-	11.11	43.40	54.51	68.14
Cupboard ends moulded one side, with 6mm veneered plywood panels							
full height	m2	1.84	-	16.49	21.73	38.22	47.77
dwarf	m2	2.75	-	24.64	14.49	39.13	48.91
Flush cupboard fronts with veneered blockboard doors lipped on all edges and doors hung on pairs 63mm brass butts							
full height with 32mm outer frame and 25mm doors	m2	1.24	-	11.11	39.33	50.44	63.05
dwarf with 25mm outer frame and 18mm doors	m2	1.40	-	12.55	44.05	56.60	70.75
Flush cupboard ends of 18mm veneered blockboard							
full height	m2	1.84	-	16.49	33.17	49.66	62.08
dwarf	m2	2.50	-	22.40	39.75	62.15	77.69

Woodwork

Alterations and repairs	Unit	Hours C	Hours L	Labour net	Material net	Price net	Price with 25%
Purpose made composite items				£	£	£	£
					VAT not included		
Hardwood built-in cupboards (*continued*)							
Crosstongued tops to dwarf cupboards, fixed with buttons							
25mm	m2	2.48	-	22.22	68.94	91.16	113.95
32mm	m2	2.48	-	22.22	89.89	112.11	140.14
Extra for moulded edges to tops	m	-	-	-	0.86	0.86	1.08

Woodwork

Alterations and repairs	Unit	Hours C	Hours L	Labour net	Material net	Price net	Price with 25%
Standard composite items				£	£	£	£
					VAT not included		

Redwood overhead garage doors

Vertically boarded garage doors with galvanised
steel gear screwed to timber frame

2135 x 1980mm	each	4.40	-	39.42	424.62	464.04	580.05
2135 x 2135mm	each	4.40	-	39.42	444.00	483.42	604.27
4270 x 1980mm	each	6.60	-	59.14	1028.77	1087.91	1359.89
4270 x 2135mm	each	6.60	-	59.14	1028.77	1087.91	1359.89

Vertically boarded garage doors with galvanised
steel gear screwed and integral frame plugged
and screwed

2135 x 1980mm	each	4.40	-	39.42	477.66	517.08	646.35
2135 x 2135mm	each	4.40	-	39.42	511.32	550.74	688.42

Cedar overhead garage doors

Vertically boarded garage doors with galvanised
steel gear screwed to timber frame

2135 x 1980mm	each	4.40	-	39.42	464.40	503.82	629.77
2135 x 2135mm	each	4.40	-	39.42	490.92	530.34	662.92
4270 x 1980mm	each	6.60	-	59.14	1098.13	1157.27	1446.59
4270 x 2135mm	each	6.60	-	59.14	1098.13	1157.27	1446.59

Horizontally boarded garage doors with
galvanised steel gear screwed to timber frame

2135 x 1980mm	each	4.40	-	39.42	464.40	503.82	629.77
2135 x 2135mm	each	4.40	-	39.42	490.92	530.34	662.92

Diagonally boarded garage doors with galvanised
steel gear screwed to timber frame

2135 x 1980mm	each	4.40	-	39.42	589.85	629.27	786.59
2135 x 2135mm	each	4.40	-	39.42	607.19	646.61	808.26

Vertically boarded garage doors with galvanised
steel gear screwed and integral frame plugged
and screwed

2135 x 1980mm	each	4.40	-	39.42	511.32	550.74	688.42
2135 x 2135mm	each	4.40	-	39.42	543.95	583.37	729.21

Horizontally boarded garage doors with
galvanised steel gear screwed and integral frame
plugged and screwed

2135 x 1980mm	each	4.40	-	39.42	511.32	550.74	688.42
2135 x 2135mm	each	4.40	-	39.42	543.95	583.37	729.21

Diagonally boarded garage doors with galvanised
steel gear screwed and integral frame plugged
and screwed

2135 x 1980mm	each	4.40	-	39.42	642.89	682.31	852.89
2135 x 2135mm	each	4.40	-	39.42	660.23	699.65	874.56

Woodwork

Alterations and repairs	Unit	Hours C	Hours L	Labour net	Material net	Price net	Price with 25%
Standard composite items				£	£	£	£
					VAT not included		

Roof windows

Velux roof windows in treated Nordic red pine; exterior aluminium cladding; factory double glazed sealed unit with 3 mm clear float glass panes; screwed to softwood

	Unit	Hours C	Hours L	Labour net	Material net	Price net	Price with 25%
550 x 780 mm GGL 102	each	0.88	-	7.88	145.69	153.57	191.96
550 x 980 mm GGL 104	each	0.88	-	7.88	162.10	169.98	212.47
780 x 980 mm GGL 304	each	0.99	-	8.87	185.07	193.94	242.43
940 x 1600 mm GGL 410	each	1.10	-	9.86	273.66	283.52	354.40
1340 x 980 mm GGL 804	each	1.10	-	9.86	257.25	267.11	333.89
550 x 980 mm GHL 104	each	0.88	-	7.88	210.32	218.20	272.75
780 x 980 mm GHL 304	each	0.99	-	8.87	236.34	245.21	306.51
1340 x 980 mm GHL 804	each	1.10	-	9.86	302.93	312.79	390.99

Velux roof windows in treated Nordic red pine; exterior aluminium cladding; factory double glazed sealed unit with 4 mm toughened outer pane and 3 mm clear float glass inner pane; screwed to softwood

	Unit	Hours C	Hours L	Labour net	Material net	Price net	Price with 25%
550 x 780 mm GGL 102	each	0.88	-	7.88	190.54	198.42	248.03
550 x 980 mm GGL 104	each	0.88	-	7.88	210.22	218.10	272.63
780 x 980 mm GGL 304	each	0.99	-	8.87	209.72	218.59	273.24
940 x 1600 mm GGL 410	each	1.10	-	9.86	329.81	339.67	424.59
1340 x 980 mm GGL 804	each	1.10	-	9.86	309.90	319.76	399.70
550 x 980 mm GHL 104	each	0.88	-	7.88	252.15	260.03	325.04
780 x 980 mm GHL 304	each	0.99	-	8.87	286.59	295.46	369.32
1340 x 980 mm GHL 804	each	1.10	-	9.86	377.61	387.47	484.34

Velux flashings for roof windows

	Unit	Hours C	Hours L	Labour net	Material net	Price net	Price with 25%
EDZ 102	each	0.88	-	7.88	31.72	39.60	49.50
EDZ 104	each	0.94	-	8.42	33.90	42.32	52.90
EDZ 304	each	0.99	-	8.87	37.19	46.06	57.58
EDZ 410	each	1.10	-	9.86	47.04	56.90	71.13
EDZ 804	each	1.10	-	9.86	48.13	57.99	72.49
EDH 102	each	0.88	-	7.88	38.29	46.17	57.71
EDH 104	each	0.94	-	8.42	39.38	47.80	59.75
EDH 304	each	0.99	-	8.87	44.84	53.71	67.14
EDH 410	each	1.10	-	9.86	49.79	59.65	74.56
EDH 804	each	1.10	-	9.86	54.77	64.63	80.79
EDL 102	each	0.88	-	7.88	27.35	35.23	44.04
EDL 104	each	0.94	-	8.42	29.53	37.95	47.44
EDL 304	each	0.99	-	8.87	32.81	41.68	52.10
EDL 410	each	1.10	-	9.86	41.56	51.42	64.28
EDL 804	each	1.10	-	9.86	39.38	49.24	61.55

Woodwork

Alterations and repairs	Unit	Hours C	Hours L	Labour net	Material net	Price net	Price with 25%
Standard composite items				£	£	£	£
					VAT not included		

Roof windows (*continued*)

Velux roller blinds to suit roof windows

GGL 102	each	0.28	-	2.51	35.05	37.56	46.95
GGL 104	each	0.28	-	2.51	35.05	37.56	46.95
GGL 304	each	0.33	-	2.96	47.92	50.88	63.60
GGL 410	each	0.40	-	3.58	54.94	58.52	73.15
GGL 804	each	0.40	-	3.58	72.49	76.07	95.09
GHL104	each	0.28	-	2.51	35.05	37.56	46.95
GHL304	each	0.33	-	2.96	47.92	50.88	63.60
GHL804	each	0.40	-	3.58	72.49	76.07	95.09

Velux portable rods for roof windows

800 mm long ZCZ 080	each	0.17	-	1.52	14.04	15.56	19.45
1000 - 1800 mm long telescopic ZCT 200	each	0.07	-	0.59	26.89	27.48	34.35
1000 mm long extension ZCT 100	each	0.07	-	0.59	5.98	6.57	8.21

Fixing fittings

Fixing floor units including allowance for scribing to floor and/or walls and plugging and screwing units

600 x 500 x 900mm high	each	0.83	-	7.44	0.37	7.81	9.76
1200 x 500 x 900mm high	each	1.37	-	12.28	0.37	12.65	15.81
1500 x 500 x 900mm high	each	1.65	-	14.78	0.37	15.15	18.94

Fixing wall units including allowance for scribing to walls and plugging and screwing units

600 x 300 x 900mm high	each	1.10	-	9.86	0.37	10.23	12.79
1200 x 300 x 900mm high	each	1.93	-	17.29	0.37	17.66	22.07

Woodwork

Alterations and repairs	Unit	Hours C	Hours L	Labour net	Material net	Price net	Price with 25%
Sundries				£	£	£	£
					VAT not included		
Extra over fixing with nails for fixing with							
steel screws	m	0.17	-	1.52	0.07	1.59	1.99
steel screws including sinking heads and							
pellating	m	0.60	-	5.38	0.91	6.29	7.86
brass screws	m	0.33	-	2.96	0.88	3.84	4.80
"Rawlnuts" to plasterboard	each	0.11	-	0.99	0.39	1.38	1.73
Plugging etc							
Plugging brickwork or concrete							
for open spaced members at 400mm centres							
one way	m2	0.92	-	8.24	0.17	8.41	10.51
for open spaced members at 400mm centres							
both ways	m2	2.10	-	18.82	0.34	19.16	23.95
for individual members	m	0.36	-	3.23	0.09	3.32	4.15
Extra over fixing with nails for fixing with							
hardened masonry pins to brickwork or concrete							
open spaced members at 400mm centres one							
way	m2	0.28	-	2.51	0.01	2.52	3.15
open spaced members at 400mm centres							
both ways	m2	0.66	-	5.91	0.02	5.93	7.41
individual members	m	0.11	-	0.99	0.01	1.00	1.25
Extra over fixing with nails for shot fixing to							
brickwork or concrete							
open spaced members at 400 mm centres							
one way	m2	0.11	-	0.99	0.45	1.44	1.80
open spaced members at 400 mm centres							
both ways	m2	0.29	-	2.60	0.90	3.50	4.38
individual members	m	0.06	-	0.54	0.23	0.77	0.96
Holes in timber							
Holes for bolts through timber							
25mm	each	0.09	-	0.79	-	0.79	0.99
50mm	each	0.13	-	1.16	-	1.16	1.45
75mm	each	0.20	-	1.79	-	1.79	2.24
100mm	each	0.26	-	2.33	-	2.33	2.91
Holes for pipes not exceeding 110mm diameter							
through timber							
25mm	each	0.36	-	3.23	-	3.23	4.04
50mm	each	0.55	-	4.93	-	4.93	6.16
75mm	each	0.83	-	7.44	-	7.44	9.30
100mm	each	1.10	-	9.86	-	9.86	12.32
Holes for ducting etc not exceeding 0.025m2							
through timber							
25mm	each	0.48	-	4.30	-	4.30	5.38
50mm	each	0.73	-	6.54	-	6.54	8.18
75mm	each	1.10	-	9.86	-	9.86	12.32
100mm	each	1.43	-	12.81	-	12.81	16.01

Woodwork

Alterations and repairs	Unit	Hours C	Hours L	Labour net	Material net	Price net	Price with 25%

Sundries

£ £ £ £
VAT not included

Insulating materials

	Unit	Hours C	Hours L	Labour net	Material net	Price net	Price with 25%
13mm sound deadening quilts cut and laid under flooring between joists at 400mm centres (measured overall)	m2	0.11	-	0.99	1.70	2.69	3.36
Thermal insulating quilts cut and laid in roof between joists at 400mm centres (measured overall)							
80mm	m2	0.13	-	1.16	2.98	4.14	5.17
100mm	m2	0.13	-	1.16	3.55	4.71	5.89
150mm	m2	0.13	-	1.16	5.51	6.67	8.34
Expanded polystyrene ISD grade laid with close butted joints taped with waterproof tape on concrete							
25mm	m2	0.13	-	1.16	1.50	2.66	3.33
40mm	m2	0.13	-	1.16	2.39	3.55	4.44
25 x 75 mm expanded polystyrene ISD grade upstand fixed with adhesive to walls.	m	0.06	-	0.54	1.28	1.82	2.27
Granular loose fill spread 50mm thick between joists at 400mm centres (measured overall)	m2	0.18	-	1.61	7.76	9.37	11.71
Building paper lapped 150mm and fixed with galvanised clout nails							
standard grade	m2	0.22	-	1.97	1.27	3.24	4.05
reflective grade	m2	0.22	-	1.97	1.91	3.88	4.85

Metalwork

	Unit	Hours C	Hours L	Labour net	Material net	Price net	Price with 25%
Galvanised steel dowels							
10 x 50mm	each	0.09	-	0.79	0.20	0.99	1.24
12 x 50mm	each	0.09	-	0.79	0.31	1.10	1.38
10 x 50mm non-ferrous dowels	each	0.09	-	0.79	0.20	0.99	1.24
Steel carriage bolts with nut and two washers							
M10 x 75mm	each	0.11	-	0.99	0.32	1.31	1.64
M12 x 75mm	each	0.11	-	0.99	0.52	1.51	1.89
M12 x 100mm	each	0.13	-	1.16	0.64	1.80	2.25
M12 x 150mm	each	0.17	-	1.52	0.89	2.41	3.01
Galvanised steel water bars including grooves in timber							
3 x 25mm	each	0.28	-	2.51	2.06	4.57	5.71
5 x 30mm	each	0.29	-	2.60	3.02	5.62	7.03
6 x 40mm	each	0.30	-	2.69	4.71	7.40	9.25
25 x 3.2 x 250mm girth galvanised steel cramps, the other end built in	each	0.11	-	0.99	0.31	1.30	1.63

Woodwork

Alterations and repairs	Unit	Hours C	Hours L	Labour net	Material net	Price net	Price with 25%
Sundries				£	£	£	£
					VAT not included		

Metalwork (*continued*)

30 x 2.5mm galvanised steel holding down straps, twice bent, one end drilled and screwed to timber, the other end built in

	Unit	Hours C	Hours L	Labour net	Material net	Price net	Price with 25%
400mm girth	each	0.33	-	2.96	0.74	3.70	4.63
600mm girth	each	0.36	-	3.23	1.12	4.35	5.44
900mm girth	each	0.44	-	3.94	1.65	5.59	6.99

Galvanised steel joist hangers - built in

	Unit	Hours C	Hours L	Labour net	Material net	Price net	Price with 25%
38 x 100mm	each	0.11	-	0.99	2.13	3.12	3.90
38 x 125mm	each	0.11	-	0.99	2.15	3.14	3.92
38 x 150mm	each	0.13	-	1.16	2.23	3.39	4.24
38 x 175mm	each	0.13	-	1.16	2.24	3.40	4.25
38 x 200mm	each	0.14	-	1.25	2.37	3.62	4.53
38 x 225mm	each	0.14	-	1.25	2.72	3.97	4.96
50 x 100mm	each	0.11	-	0.99	2.13	3.12	3.90
50 x 125mm	each	0.13	-	1.16	2.15	3.31	4.14
50 x 150mm	each	0.13	-	1.16	2.23	3.39	4.24
50 x 175mm	each	0.13	-	1.16	2.24	3.40	4.25
50 x 200mm	each	0.14	-	1.25	2.48	3.73	4.66
50 x 225mm	each	0.14	-	1.25	2.76	4.01	5.01
75 x 150mm	each	0.14	-	1.25	3.05	4.30	5.38
75 x 175mm	each	0.14	-	1.25	3.11	4.36	5.45
75 x 200mm	each	0.16	-	1.43	3.17	4.60	5.75
75 x 225mm	each	0.17	-	1.52	3.26	4.78	5.97

Galvanised steel truss clips, fixed with nails to suit

	Unit	Hours C	Hours L	Labour net	Material net	Price net	Price with 25%
38 mm thick members	each	0.22	-	1.97	0.33	2.30	2.88
50 mm thick members	each	0.22	-	1.97	0.35	2.32	2.90

Galvanised steel square toothed plate timber connectors to BS 1579 Table 4

	Unit	Hours C	Hours L	Labour net	Material net	Price net	Price with 25%
38 mm diameter, single sided	each	0.03	-	0.30	0.16	0.46	0.57
38 mm diameter, double sided	each	0.03	-	0.30	0.25	0.55	0.69
50 mm diameter, single sided	each	0.03	-	0.30	0.21	0.51	0.64
50 mm diameter, double sided	each	0.03	-	0.30	0.26	0.56	0.70
63 mm diameter, single sided	each	0.03	-	0.30	0.30	0.60	0.75
63 mm diameter, double sided	each	0.03	-	0.30	0.35	0.65	0.81
75 mm diameter, single sided	each	0.03	-	0.30	0.32	0.62	0.78
75 mm diameter, double sided	each	0.03	-	0.30	0.40	0.70	0.88

Galvanised steel herringbone joist struts, fixed with nails, to suit joist centres of

	Unit	Hours C	Hours L	Labour net	Material net	Price net	Price with 25%
400 mm	each	0.30	-	2.69	1.05	3.74	4.67
450 mm	each	0.29	-	2.60	1.03	3.63	4.54
600 mm	each	0.24	-	2.15	0.86	3.01	3.76

Galvanised steel pelmet brackets

	Unit	Hours C	Hours L	Labour net	Material net	Price net	Price with 25%
75 x 100mm, screwed to timber	each	0.17	-	1.52	0.96	2.48	3.10
plugged and screwed to brickwork or concrete	each	0.28	-	2.51	1.09	3.60	4.50

Woodwork

Alterations and repairs	Unit	Hours C	Hours L	Labour net	Material net	Price net	Price with 25%
Sundries				£	£	£	£
					VAT not included		
Metalwork (*continued*)							
Enamelled shelf brackets							
150 x 200mm grey, screwed to timber	each	0.17	-	1.52	0.38	1.90	2.38
plugged and screwed to brickwork or concrete	each	0.28	-	2.51	0.55	3.06	3.83

Woodwork

Alterations and repairs	Unit	Hours C	Hours L	Labour net	Material net	Price net	Price with 25%
Ironmongery				£	£	£	£
					VAT not included		
Butts etc, excluding hanging doors							
Light steel butts							
50 mm	pair	-	-	-	0.39	0.39	0.49
75 mm	pair	-	-	-	0.67	0.67	0.84
100 mm	pair	-	-	-	1.43	1.43	1.79
Light steel sheradised butts							
50 mm	pair	-	-	-	0.39	0.39	0.49
75 mm	pair	-	-	-	0.67	0.67	0.84
100 mm	pair	-	-	-	1.43	1.43	1.79
Cast iron butts							
75 mm	pair	-	-	-	2.82	2.82	3.52
100 mm	pair	-	-	-	3.27	3.27	4.09
100 mm steel rising butts	pair	-	-	-	5.14	5.14	6.42
Steel washered brass butts							
75 mm	pair	-	-	-	6.52	6.52	8.15
100 mm	pair	-	-	-	11.95	11.95	14.94
450 mm japanned tee hinges	pair	-	-	-	4.04	4.04	5.05
450 mm light reversible hinges	pair	-	-	-	9.58	9.58	11.98
600 mm heavy cast iron reversible hinges with cast or malleable cups and bolts	pair	-	-	-	16.31	16.31	20.39
125 mm double action regulating spring hinges	pair	-	-	-	38.14	38.14	47.67
Check action floor springs with all fittings							
single action	each	-	-	-	211.66	211.66	264.57
double action	each	-	-	-	227.88	227.88	284.85
Overhead sliding door tracks with hangers and steel pelmet	set	-	-	-	20.72	20.72	25.90
Overhead garage door gear	set	-	-	-	109.23	109.23	136.54
Butts etc, including hanging doors							
Light steel butts							
50mm to small cupboard doors	pair	0.55	-	4.93	0.39	5.32	6.65
75mm to 38 mm softwood doors	pair	1.21	-	10.84	0.67	11.51	14.39
100mm to 50 mm softwood doors	pair	1.43	-	12.81	1.43	14.24	17.80
Light steel sheradised butts and labour hanging doors							
50mm to small cupboard doors	pair	0.55	-	4.93	0.39	5.32	6.65
75mm to 38 mm softwood doors	pair	1.21	-	10.84	0.67	11.51	14.39
100mm to 50 mm softwood doors	pair	1.43	-	12.81	1.43	14.24	17.80

Woodwork

Alterations and repairs	Unit	Hours C	Hours L	Labour net	Material net	Price net	Price with 25%
Ironmongery				£	£	£	£
					VAT not included		
Butts etc, including hanging doors (*continued*)							
Cast iron butts and labour hanging doors							
75mm to 38 mm softwood doors	pair	1.21	-	10.84	2.00	12.84	16.05
100mm to 50 mm softwood doors	pair	1.43	-	12.81	2.77	15.58	19.48
100mm steel rising butts	pair	1.76	-	15.77	5.14	20.91	26.14
Steel washered brass butts and labour hanging doors							
75mm to 38 mm softwood doors	pair	2.20	-	19.71	3.31	23.02	28.77
100mm to 50 mm softwood doors	pair	2.75	-	24.64	6.96	31.60	39.50
450mm japanned tee hinges and labour hanging softwood doors	pair	0.94	-	8.42	4.04	12.46	15.57
450mm light reversible hinges and labour hanging softwood doors	pair	1.21	-	10.84	9.58	20.42	25.52
600mm heavy cast iron reversible hinges with cast or malleable cups and bolts and labour hanging garage doors	pair	1.65	-	14.78	16.31	31.09	38.86
125mm double action regulating spring hinges and labour hanging 50mm softwood doors	pair	3.30	-	29.57	38.14	67.71	84.64
Check action floor springs with all fittings and loose box and labour hanging 50mm hardwood door including setting box in floor							
single action	each	6.05	-	54.21	211.66	265.87	332.34
double action	each	6.60	-	59.14	227.88	287.02	358.77
Overhead sliding door tracks with hangers and steel pelmet etc and labour hanging 38mm or 50mm single softwood door (side fixing to door head)	set	3.58	-	32.08	20.72	52.80	66.00
Overhead garage door gear and fixing to softwood frame and labour hanging 50mm softwood door	set	3.30	-	29.57	109.23	138.80	173.50
For gate hanging see fencing							
Fixing only to softwood							
Barrel or tower bolts							
small	each	0.28	-	2.51	-	2.51	3.14
medium	each	0.36	-	3.23	-	3.23	4.04
large	each	0.46	-	4.12	-	4.12	5.15
Necked bolts							
small	each	0.36	-	3.23	-	3.23	4.04
medium	each	0.46	-	4.12	-	4.12	5.15

Woodwork

Alterations and repairs	Unit	Hours C	Hours L	Labour net	Material net	Price net	Price with 25%
Ironmongery				£	£	£	£
					VAT not included		
Fixing only to softwood (continued)							
Flush bolts							
small	each	0.83	-	7.44	-	7.44	9.30
medium	each	1.10	-	9.86	-	9.86	12.32
large	each	1.46	-	13.08	-	13.08	16.35
Espagnolette bolt sets	each	1.37	-	12.28	-	12.28	15.35
Panic bolt sets	each	1.65	-	14.78	-	14.78	18.48
Ball catches							
small	each	0.36	-	3.23	-	3.23	4.04
medium	each	0.55	-	4.93	-	4.93	6.16
Cupboard catches	each	0.19	-	1.70	-	1.70	2.13
Roller bolt catches	each	0.91	-	8.15	-	8.15	10.19
Fanlight catches	each	0.19	-	1.70	-	1.70	2.13
Casement fasteners							
with hook plate	each	0.19	-	1.70	-	1.70	2.13
with mortice plate	each	0.55	-	4.93	-	4.93	6.16
Casement stays	each	0.28	-	2.51	-	2.51	3.14
Sash fasteners	each	0.36	-	3.23	-	3.23	4.04
Quadrant stays	each	0.36	-	3.23	-	3.23	4.04
Thumb latches	each	0.66	-	5.91	-	5.91	7.39
Rim night latches	each	1.10	-	9.86	-	9.86	12.32
Cupboard and drawer locks							
surface pattern	each	0.55	-	4.93	-	4.93	6.16
mortice pattern	each	1.10	-	9.86	-	9.86	12.32
Rim locks and furniture	each	0.83	-	7.44	-	7.44	9.30
Mortice locks and furniture							
shallow pattern	each	1.65	-	14.78	-	14.78	18.48
deep pattern	each	2.20	-	19.71	-	19.71	24.64
Locking bars	each	0.55	-	4.93	-	4.93	6.16
Bow handles							
medium	each	0.14	-	1.25	-	1.25	1.56
large	each	0.36	-	3.23	-	3.23	4.04
Drawer pulls	each	0.11	-	0.99	-	0.99	1.24

Woodwork

Alterations and repairs	Unit	Hours C	Hours L	Labour net	Material net	Price net	Price with 25%
Ironmongery				£	£	£	£
					VAT not included		
Fixing only to softwood (*continued*)							
Flush pulls	each	0.66	-	5.91	-	5.91	7.39
Finger plates	each	0.14	-	1.25	-	1.25	1.56
Helical door springs	each	0.28	-	2.51	-	2.51	3.14
Overhead door closers							
surface fixing	each	0.83	-	7.44	-	7.44	9.30
mortice fixing	each	1.93	-	17.29	-	17.29	21.61
Letter plates in 50mm doors	each	1.65	-	14.78	-	14.78	18.48
Coat hooks	each	0.11	-	0.99	-	0.99	1.24
Flap table brackets	each	0.66	-	5.91	-	5.91	7.39
Fixing only to hardwood							
Barrel or tower bolts							
small	each	0.42	-	3.76	-	3.76	4.70
medium	each	0.55	-	4.93	-	4.93	6.16
large	each	0.69	-	6.18	-	6.18	7.72
Necked bolts							
small	each	0.55	-	4.93	-	4.93	6.16
medium	each	0.69	-	6.18	-	6.18	7.72
Flush bolts							
small	each	1.24	-	11.11	-	11.11	13.89
medium	each	1.65	-	14.78	-	14.78	18.48
large	each	2.20	-	19.71	-	19.71	24.64
Espagnolette bolt sets	each	2.07	-	18.55	-	18.55	23.19
Panic bolt sets	each	2.48	-	22.22	-	22.22	27.77
Ball catches							
small	each	0.55	-	4.93	-	4.93	6.16
medium	each	0.83	-	7.44	-	7.44	9.30
Cupboard catches	each	0.29	-	2.60	-	2.60	3.25
Roller bolt catches	each	1.38	-	12.36	-	12.36	15.45
Fanlight catches	each	0.29	-	2.60	-	2.60	3.25
Casement fasteners							
with hook plate	each	0.29	-	2.60	-	2.60	3.25
with mortice plate	each	0.83	-	7.44	-	7.44	9.30
Casement stays	each	0.42	-	3.76	-	3.76	4.70

Woodwork

Woodwork

Alterations and repairs	Unit	Hours C	Hours L	Labour net	Material net	Price net	Price with 25%
Ironmongery				£	£	£	£
					VAT not included		
Fixing only to hardwood (*continued*)							
Sash fasteners	each	0.55	-	4.93	-	4.93	6.16
Quadrant stays	each	0.55	-	4.93	-	4.93	6.16
Thumb latches	each	0.99	-	8.87	-	8.87	11.09
Rim night latches	each	1.65	-	14.78	-	14.78	18.48
Cupboard and drawer locks							
surface pattern	each	0.83	-	7.44	-	7.44	9.30
mortice pattern	each	1.65	-	14.78	-	14.78	18.48
Rim locks and furniture	each	1.24	-	11.11	-	11.11	13.89
Mortice locks and furniture							
shallow pattern	each	2.48	-	22.22	-	22.22	27.77
deep pattern	each	3.30	-	29.57	-	29.57	36.96
Locking bars	each	0.83	-	7.44	-	7.44	9.30
Bow handles							
medium	each	0.22	-	1.97	-	1.97	2.46
large	each	0.55	-	4.93	-	4.93	6.16
Drawer pulls	each	0.17	-	1.52	-	1.52	1.90
Flush pulls	each	0.99	-	8.87	-	8.87	11.09
Finger plates	each	0.22	-	1.97	-	1.97	2.46
Helical door springs	each	0.38	-	3.40	-	3.40	4.25
Overhead door closers							
surface fixing	each	1.24	-	11.11	-	11.11	13.89
mortice fixing	each	2.89	-	25.89	-	25.89	32.36
Letter plates in 50mm doors	each	2.48	-	22.22	-	22.22	27.77
Coat hooks	each	0.17	-	1.52	-	1.52	1.90
Flap table brackets	each	0.99	-	8.87	-	8.87	11.09

Woodwork

Alterations and repairs	Unit	Hours C	Hours L	Labour net	Material net	Price net	Price with 25%
Ironmongery				£	£	£	£
					VAT not included		
Sundries							
Fibre sliding cupboard door tracks including groove	m	0.11	-	0.99	3.24	4.23	5.29
Nylon sliders including fitting doors	pair	0.83	-	7.44	0.43	7.87	9.84
Aluminium bookcase							
strips including groove	m	0.31	-	2.78	2.43	5.21	6.51
set of four adjustable studs	set	0.11	-	0.99	0.52	1.51	1.89
Standard curtain tracks and runners							
anodised aluminium	m	0.74	-	6.63	3.73	10.36	12.95
plastics	m	0.74	-	6.63	4.41	11.04	13.80
Rubber door stops							
screwed to timber	each	0.14	-	1.25	0.36	1.61	2.01
plugged and screwed to concrete	each	0.28	-	2.51	0.40	2.91	3.64

Woodwork

Alterations and repairs	Unit	Hours C	Hours L	Labour net	Material net	Price net	Price with 25%
Repairs and maintenance				£	£	£	£
					VAT not included		

Generally

	Unit	Hours C	Hours L	Labour net	Material net	Price net	Price with 25%
Treat infested timbers with anti-woodworm fluid							
by spray	m2	0.10	-	0.90	0.38	1.28	1.60
by brush	m2	0.25	-	2.24	0.66	2.90	3.63
Remove dust and cobwebs and spray roof timbers with anti-woodworm fluid, including ceiling joists, rafters, purlins, struts and roof boarding, per m3 of roof voids	m3	0.35	-	3.14	-	3.14	3.92

Flooring repairs

	Unit	Hours C	Hours L	Labour net	Material net	Price net	Price with 25%
Take up damaged or worn flooring, renew in 25 mm plain edged board flooring and make good to existing in areas							
not exceeding 0.50 m2	m2	2.75	-	24.64	13.54	38.18	47.73
0.05 - 2.00 m2	m2	1.75	-	15.68	14.31	29.99	37.49

Door and window repairs

	Unit	Hours C	Hours L	Labour net	Material net	Price net	Price with 25%
Take down average doors							
ease, rehang on existing butts and adjust	each	1.00	-	8.96	0.11	9.07	11.34
make up width of door by planting on 13 mm strip	each	1.00	-	8.96	3.35	12.31	15.39
take down average doors and rehang on existing butts to opposite hand, and make out old butt sinkings	each	2.00	-	17.92	0.25	18.17	22.71
Take down average doors and rehang on existing butts to opposite side of frame							
in frame with loose stops	each	3.00	-	26.88	0.25	27.13	33.91
in frame solid moulded, including cutting back and providing loose stops	each	4.50	-	40.32	2.06	42.38	52.98
Take out doors, frames/linings and architraves complete							
door 762 x 1981 mm	each	1.25	0.06	11.60	1.68	13.28	16.60
Extra for							
fanlight over	each	1.50	0.04	13.71	1.68	15.39	19.24
Cut out 550 x 550 mm top panel of old doors, cut away moulding on stiles and rails one side, glaze with obscured glass and fit and fix beads	each	2.25	-	20.16	8.52	28.68	35.85
Take down 50 mm doors, remove damaged stile, fit new moulded stile to match and rehang doors	each	5.00	-	44.80	10.48	55.28	69.10
32 x 100 mm twice splayed weatherboards not exceeding 900 mm long screwed on including notching frames	each	1.10	-	9.86	1.55	11.41	14.26

Woodwork

Alterations and repairs	Unit	Hours C	Hours L	Labour net	Material net	Price net	Price with 25%
Repairs and maintenance				£	£	£	£
					VAT not included		
Door and window repairs (*continued*)							
Cut away defective foot of 75 x 100 mm door frames, provide and scarf in new piece about 300 mm long and secure frame	each	3.00	-	26.88	1.34	28.22	35.27
Cut away defective sill and jambs to 75 x 100 mm door frames, provide and fix new 75 x 175 mm hardwood sill and new 300 mm lengths of frame, scarf in new lengths of frame and secure frame and sill	each	8.50	-	76.16	51.55	127.71	159.64
Take down average casement sashes, ease, rehang on existing butts and adjust	each	0.75	-	6.72	0.11	6.83	8.54
Take down sashes, cramp up, supply and fix 100 x 16 mm angle repair plate, and rehang on existing butts	each	1.25	-	11.20	3.77	14.97	18.71
Take down double hung sashes and replace broken line with new							
one line	each	1.00	-	8.96	0.54	9.50	11.88
two lines	set	1.25	-	11.20	1.19	12.39	15.49
four lines	set	2.00	-	17.92	2.39	20.31	25.39
Take out existing casement/sash windows	each	0.75	0.10	7.39	1.68	9.07	11.34
Cut away defective sill to double hung sash window frames, supply and fit new hardwood cill 1060 mm long and short lengths of new pulley stiles, inside and outside linings, scarf in new ends and fit new parting beads	each	9.00	-	80.64	71.53	152.17	190.21
Remove locks and make good door and frame							
rim locks	each	1.00	-	8.96	0.53	9.49	11.86
mortice locks	each	1.25	-	11.20	0.53	11.73	14.66

Woodwork

Alterations and repairs	Unit	Specialist price net	Price with 25%

Wood boring insect and general timber treatment

		£	£
			VAT not included

The following "Specialist price net" figures are guide prices supplied by Rentokil Ltd for work in the normal dwelling of conventional construction.

Prices do not include for cash discount.

See the notes below for treatment against wood-rotting fungi and site conditions.

Clean and spray all accessible timber surfaces

	Unit	Specialist price net	Price with 25%
Ground and first floors - joist depth			
100 mm	m2	3.66	4.58
150 mm	m2	4.05	5.06
200 mm	m2	4.31	5.39
Extra for lifting and reinstating sufficient floorboards to carry out treatment (eg approximately 1 in 5)			
straight edged boards	m2	4.49	5.61
tongued and grooved boards	m2	5.59	6.99
steel tongued	m2	6.68	8.35
Staircases	per riser	3.60	4.50
Extra for drilling risers	per riser	1.95	2.44
Roof voids with			
rafters with tile/slate battens exposed, ceiling joists and lath and plaster ceilings	m2	4.02	5.02
rafters with underfelting, ceiling joists and plasterboard ceilings	m2	2.92	3.65
Extra for lifting insulation material between joists	m2	2.27	2.84

Notes

The foregoing prices are for treatment against wood-boring insects only and extra amounts would be charged where:

1. replacement or reinforcement of any structurally weakened timbers is necessary;
2. dirt and debris is excessive;
3. other preparatory works are necessary (eg removing and replacing of furniture, furnishings and floor coverings);
4. treatment of dry or wet rot attacks is necessary.

In all instances estimates are subject to a site inspection in order that a fixed price quotation can be submitted. All treatments carry a 30 year guarantee.

Woodwork

Alterations and repairs	Unit	Specialist price net	Price with 25%
Wood boring insect and general timber treatment		£	£
		VAT not included	
Treat all accessible timber surfaces using "Rentokil Fogging Technique"			
Ground and first floors - joist depths up to 200 mm	m2	5.40	6.75
Staircases	m2	2.79	3.49
Roof voids with rafters with tile/slate battens exposed or underfelting and ceiling joists with plaster or plasterboard ceilings	m2	2.93	3.66

Note

It is not normally necessary to lift floorboards, remove carpets or furnishings when using "Rentokil Fogging Technique". All treatments carry a 30 year guarantee.

Enquiries about the above specialist prices should be made to Rentokil Initial UK Ltd, Property Care Division, Garland Road, East Grinstead, West Sussex, RH19 1DY, tel (01342) 327171, fax (01342) 318298.

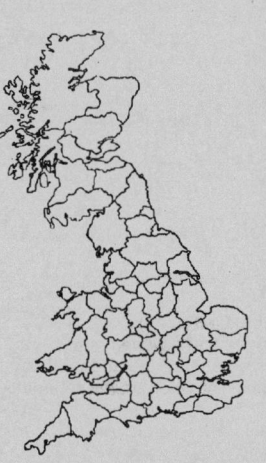

Woodwork

Alterations and repairs	Unit	Specialist price net	Price with 25%

Wood boring insect and general timber treatment

£   £
VAT not included

The following "Specialist price net" figures are minimum guide prices provided by Kiltox Chemicals Ltd, for work by one of their approved contractors within 200 kilometres of a branch depot.

Prices do not include for cash discount.

Kiltox products used in accordance with the manufacturer's instructions carry a 30 year guarantee.

See the preamble notes for builder's profit and attendance.

Kiltox timber preservative treatment

	Unit	Specialist price net	Price with 25%
Roof spaces (any pitch and spacing of rafters etc)	m2 on plan	3.27	4.08
Suspended intermediate floors	m2 on plan	3.27	4.08
Suspended ground floors	m2 on plan	3.27	4.08
Stud partitions	m2	2.61	3.26
Staircases including balustrades, strings etc	m2 on plan	11.41	14.26
Skirtings, linings, window boards and architraves	m	0.68	0.85
Removal of loft insulation (excluding granules)	m	3.03	3.78
Doors, both sides not exceeding 4 m2 total	each	11.49	14.37
Additional "Timber Paste Preservative" application to timbers			
over 300 mm girth	m	7.82	9.78
not exceeding 300 mm girth	m	5.19	6.49

Sundries

	Unit	Specialist price net	Price with 25%
Extra for lifting and reinstating sufficient floorboards to carry out treatment (about 1 in 6 boards)			
plain edged boards	m2	5.28	6.60
tongued and grooved boards	m2	7.30	9.13
Extra for lifting insulation material between joists	m2	6.15	7.68
Kiltox Dry Rot Wall Irrigation/Sterilisation to brickwork			
102.5 mm walls	m2	7.18	8.97
215 mm walls	m2	14.34	17.93
327.5 mm walls	m2	21.52	26.90
440 mm walls	m2	28.69	35.87
Kiltox Dry Rot Sterilisation by surface spray treatment to masonry	m2	3.84	4.80

Enquiries about the foregoing specialist prices should be made to Kiltox Chemicals Ltd, Kiltox House, Park Row, Greenwich, London, SE10 9NL, tel (0208) 858 6277, fax (0208) 853 3572.

Plumbing and engineering installations

Preamble

With the exception of lead pipework and connections in alterations, figures in this section are limited to repair and maintenance items. For other plumbing and engineering items, see prices in the "New work" section and add from 10% upwards according to circumstances.

"Labour net" figures for plumbing and engineering work are based on the labour cost of an advanced plumber working with an apprentice in the third year of training and include allowances for all costs incidental to the employment of labour.

"Materials net" figures include for all costs of materials including an allowance for waste except where specifically stated.

"Price net" figures are the totals of the "Labour net" and "Materials net" figures. Prices are for a builder employing his own labour; according to the amount and nature of the work involved, it may well be possible to secure more advantageous prices from specialist sub-contractors.

Prices do not include any allowance for scaffolding, ladders or other plant necessary to reach the work. The "Preliminaries" section includes prices for scaffolding which must be considered and allowance included to suit the particular circumstances of a tender.

Specialist prices

"Price with 25% figures are all-in guide prices and include 25% for the builder's overheads, profit, unloading materials and general attendance (to include free use of standing scaffolding and hoists, temporary lighting and water and clearing away rubbish).

The amount of attendance required varies between the various trades and also with the circumstances of specific jobs; the percentage addition must always be considered and adjusted as necessary to suit the terms and conditions of the quotation being used.

Quantities and delivery distances are usually the most significant of the many factors which influence prices and it must be emphasised that quotations should always be obtained when preparing a tender.

Plumbing and engineering installations

Alterations and repairs

	Unit	Price

Basic prices for materials £

Cast iron soil and waste branches		
50 mm	each	19.23
100 mm	each	27.36
100 mm with access door	each	47.11
100 mm cast iron bends	each	17.69
100 mm cast iron WC connecting pipes	each	19.24
PVC-U self-locking boss connectors		
32 mm	each	3.41
38 mm	each	3.56
100 mm PVC-U WC connecting bends	each	6.35
100 mm PVC-U branches with access door	each	15.29
Capillary straight couplings for copper		
15 mm	each	0.13
22 mm	each	0.29
28 mm	each	0.69
35 mm	each	1.76
42 mm	each	2.61
54 mm	each	5.70
Capillary tees for copper		
15 mm	each	0.43
22 mm	each	0.98
28 mm	each	2.38
35 mm	each	6.27
42 mm	each	9.26
54 mm	each	18.03
Slip pattern capillary tees for copper		
15 mm	each	1.76
22 mm	each	4.21
Slip pattern capillary adaptor tees for existing imperial size copper		
½"	each	0.85
3/4"	each	0.94

Prices actually to be paid for materials must be checked against the above basic prices and adjustments made as necessary.

Plumbing and engineering installations

Alterations and repairs

	Unit	Price
Basic prices for materials		**£**
Compression tees for copper		
15 mm	each	1.73
22 mm	each	2.92
28 mm	each	6.19
35 mm	each	11.33
42 mm	each	18.85
54 mm	each	29.56
Galvanised malleable tees for steel		
15 mm	each	0.72
20 mm	each	1.04
25 mm	each	1.51
32 mm	each	2.49
40 mm	each	3.42
50 mm	each	4.92
Galvanised steel connectors		
15 mm	each	2.18
20 mm	each	2.58
25 mm	each	3.59
32 mm	each	4.64
40 mm	each	5.61
50 mm	each	8.34
100 mm cast iron half round gutters		
1830 mm lengths	each	12.34
115 mm cast iron ogee gutters		
1830 mm lengths	each	15.14
112 mm PVC-U half round gutters		
4000 mm lengths	each	7.76
112 mm PVC-U square section gutters		
4000 mm lengths	each	7.83
65 mm cast iron rainwater pipes		
1830 mm lengths	each	22.56
68 mm PVC-U rainwater pipes		
2000 mm lengths	each	4.39
65 mm PVC-U square section rainwater pipes		
2500 mm lengths	each	5.50

Prices actually to be paid for materials must be checked against the above basic prices and adjustments made as necessary.

Plumbing and engineering installations

Alterations and repairs	Unit	Hours C	Hours L	Labour net	Material net	Price net	Price with 25%

Pipework

£ £ £ £

VAT not included

Soil and waste pipe connections

Connect branch waste pipes to existing 50 mm cast iron stack including cutting and joining cast iron branch

	Unit	Hours C	Hours L	Labour net	Material net	Price net	Price with 25%
32 mm PVCU branches	each	4.00	4.00	68.52	52.47	120.99	151.24
38 mm PVCU branches	each	4.10	4.10	70.24	53.33	123.57	154.46
50 mm cast iron branches	each	4.50	4.50	77.08	121.23	198.31	247.89

Connect branch soil or waste pipes to existing 100 mm cast iron stack including cutting in and jointing cast iron branch

	Unit	Hours C	Hours L	Labour net	Material net	Price net	Price with 25%
32 mm PVCU branches	each	6.50	6.50	111.35	86.75	198.10	247.63
38 mm PVCU branches	each	6.60	6.60	113.06	96.77	209.83	262.29
50 mm cast iron branches	each	7.00	7.00	119.91	127.49	247.40	309.25
100 mm cast iron branches	each	7.25	7.25	124.20	160.83	285.03	356.29

Connect WC outlets to existing 100 mm cast iron drain stack including cast iron WC connecting pipe and bend and cutting in and joining cast iron branch with access door

	Unit	Hours C	Hours L	Labour net	Material net	Price net	Price with 25%
	each	8.00	8.00	137.04	128.71	265.75	332.19

Connect branch waste pipe to existing 100 mm PVCU stack including drilling and fixing PVCU self locking boss connector

	Unit	Hours C	Hours L	Labour net	Material net	Price net	Price with 25%
32 mm PVCU branches	each	0.75	0.75	12.85	9.44	22.29	27.86
38 mm PVCU branches	each	0.85	0.85	14.56	9.51	24.07	30.09

Connect WC outlets to existing 100 mm PVCU WC connecting and cutting in and jointing PVCU branch with access door

	Unit	Hours C	Hours L	Labour net	Material net	Price net	Price with 25%
	each	4.00	4.00	68.52	25.16	93.68	117.10

Service pipe connections

Connect branch to existing copper service pipe including draining down, cutting in and jointing capillary tee and straight coupling; existing service pipe size

	Unit	Hours C	Hours L	Labour net	Material net	Price net	Price with 25%
15 mm	each	1.00	1.00	17.13	1.63	18.76	23.45
22 mm	each	1.10	1.10	18.85	3.43	22.28	27.85
28 mm	each	1.20	1.20	20.55	6.07	26.62	33.27
35 mm	each	1.30	1.30	22.27	14.82	37.09	46.36
42 mm	each	1.45	1.45	24.84	20.21	45.05	56.31
54 mm	each	1.60	1.60	27.41	34.70	62.11	77.64

Connect branch to existing copper service pipe including draining down, cutting in and jointing slip pattern capillary tee; existing service pipes size

	Unit	Hours C	Hours L	Labour net	Material net	Price net	Price with 25%
15 mm	each	0.75	0.75	12.85	2.95	15.80	19.75
22 mm	each	0.80	0.80	13.71	6.64	20.35	25.44

Plumbing and engineering installations

Alterations and repairs	Unit	Hours C	Hours L	Labour net	Material net	Price net	Price with 25%
Pipework				£	£	£	£
					VAT not included		

Service pipe connections (*continued*)

Connect branch to existing imperial size copper service pipe including draining down, cutting in and jointing slip pattern capillary adapter tee for metric branch; existing service pipes imperial size

	Unit	Hours C	Hours L	Labour net	Material net	Price net	Price with 25%
½"	each	0.75	0.75	12.85	3.80	16.65	20.81
¾"	each	0.80	0.80	13.71	7.58	21.29	26.61

Connect branch to existing copper service pipe including draining down, cutting in and jointing compression tee; existing pipe size

15 mm	each	0.75	0.75	12.85	3.93	16.78	20.98
22 mm	each	0.80	0.80	13.71	6.83	20.54	25.68
28 mm	each	0.85	0.85	14.56	9.18	23.74	29.68
35 mm	each	0.90	0.90	15.41	20.00	35.41	44.26
42 mm	each	1.00	1.00	17.13	30.04	47.17	58.96
54 mm	each	1.10	1.10	18.85	46.52	65.37	81.71

Connect branch to existing imperial size copper service pipe including draining down, cutting in and jointing compression tee with two adapter rings for metric fitting; existing service pipes imperial size

¾"	each	0.90	0.90	15.41	10.27	25.68	32.10
1¼"	each	1.00	1.00	17.13	25.01	42.14	52.67
1½"	each	1.10	1.10	18.85	36.00	54.85	68.56

Connect branch to existing galvanised steel service pipe including draining down, cutting in threading and jointing tee and connector; existing service pipes size

15 mm	each	1.50	1.50	25.69	7.70	33.39	41.74
20 mm	each	1.75	1.75	29.97	9.27	39.24	49.05
25 mm	each	2.00	2.00	34.26	9.46	43.72	54.65
32 mm	each	2.25	2.25	38.55	17.30	55.85	69.81
40 mm	each	2.50	2.50	42.83	21.41	64.24	80.30
50 mm	each	3.00	3.00	51.39	30.94	82.33	102.91

Plumbing and engineering installations

Alterations and repairs	Unit	Hours C	Hours L	Labour net	Material net	Price net	Price with 25%
Repairs and maintenance				£	£	£	£
					VAT not included		

Gutterwork

Take down defective 100 mm cast iron half round
gutters and brackets and supply and fix new
gutters and brackets to match

continuous runs	m	0.65	0.65	11.14	11.25	22.39	27.99
1830 mm isolated lengths	each	1.67	1.67	28.60	16.14	44.74	55.92

Take down defective 115 mm cast iron ogee
gutters and supply and fix new gutters to match

continuous runs	m	0.67	0.67	11.47	13.26	24.73	30.91
1830 mm isolated lengths	each	1.75	1.75	29.97	21.79	51.76	64.70

Take down defective gutters and supply and fix
new

112 mm half round PVCU gutters and brackets	m	0.55	0.55	9.42	4.93	14.35	17.94
112 mm square section PVCU gutters and brackets	m	0.60	0.60	10.28	5.50	15.78	19.73

Rainwater pipework

Take down defective 65 mm cast iron rainwater
pipes and supply and fix new pipes top match

continuous runs	each	0.60	0.60	10.28	18.60	28.88	36.10
1830 mm continuous lengths	each	1.50	1.50	25.69	24.13	49.82	62.27

Take down defective 65 mm cast iron rainwater
pipes and supply and fix new

68 mm PVCU pipes and brackets	each	0.52	0.52	8.91	6.26	15.17	18.96
68 mm square section PVCU pipes and brackets	each	0.54	0.54	9.25	4.97	14.22	17.77

Ancillaries

Arrange with the local water authority and re-washer main stopcocks	each	1.50	1.50	25.69	0.22	25.91	32.39
Re-washer readily accessible stopcocks bib valves or ball valves	each	0.50	0.50	8.57	0.22	8.79	10.99

Remove defective bib valves and supply and fix
new bib valves

13 mm brass	each	0.65	0.65	11.14	6.23	17.37	21.71
19 mm brass	each	0.70	0.70	11.99	9.02	21.01	26.26
13 mm chromium plated	each	0.65	0.65	11.14	9.16	20.30	25.38
19 mm chromium plated	each	0.70	0.70	11.99	12.25	24.24	30.30

Plumbing and engineering installations

Alterations and repairs	Unit	Hours C	Hours L	Labour net	Material net	Price net	Price with 25%
Repairs and maintenance				£	£	£	£
					VAT not included		

Ancillaries (*continued*)

Remove defective pillar valves and supply and fix chromium plated pillar valves

13 mm	each	1.00	1.00	17.13	9.16	26.29	32.86
19 mm	each	1.10	1.10	18.85	12.25	31.10	38.88
19 mm, including removing and refixing bath panel	each	1.50	1.50	25.69	12.25	37.94	47.42

Remove defective ball valves and supply and fix brass high pressure ball valves with copper float

13 mm	each	1.00	1.00	17.13	5.57	22.70	28.38
19 mm	each	1.10	1.10	18.85	14.81	33.66	42.08
25 mm	each	1.25	1.25	21.41	32.28	53.69	67.11

Remove defective ball valves and supply and fix brass low pressure ball valves with plastics float

13 mm	each	1.00	1.00	17.13	5.44	22.57	28.21
19 mm	each	1.10	1.10	18.85	12.11	30.96	38.70
25 mm	each	1.25	1.25	21.41	29.21	50.62	63.27

Equipment

Remove defective cisterns and supply and fix galvanised steel cisterns including drilling and re-connecting 4 pipes

191 litre (42 gallon) BS 417 Type SCM 270	each	6.00	6.00	102.78	99.95	202.73	253.41
327 litre (72 gallon) BS417 Type SCM 450/1	each	6.50	6.50	111.35	125.64	236.99	296.24

Remove defective cisterns and supply and fix plastic cisterns including drilling and re-connecting 4 pipes

114 litre (40 gallon) BS 4213 Type PC 25	each	4.50	4.50	77.08	60.51	137.59	171.99
227 litre (50 gallon) BS 4213 Type PC 50	each	4.75	4.75	81.36	113.99	195.35	244.19

Remove defective tanks and supply and fix galvanised steel tanks including drilling and re-connecting 4 pipes, bolting hand hole cover and removing and replacing slat shelving

95 litre (21 gallon) BS 417 Type T25/1	each	8.00	8.00	137.04	116.40	253.44	316.80
114 litre (25 gallon) BS 417 Type T30/1	each	8.50	8.50	145.61	123.97	269.58	336.98

Remove defective cylinders, supply and fix galv. steel direct cylinders with 5 bosses and immersion heater boss, re-connect 5 pipes and immersion heater (electrical work not included), bolting hand hole cover and removing and replacing slatted shelving

100 litre (22 gallon) BS417 Type Y25	each	7.00	7.00	119.91	191.65	311.56	389.45
123 litre (27 gallon) BS 417 Type Y31	each	7.50	7.50	128.47	200.55	329.02	411.27

Plumbing and engineering installations

Alterations and repairs	Unit	Hours C	Hours L	Labour net	Material net	Price net	Price with 25%

Repairs and maintenance

£ £ £ £

VAT not included

Equipment (*continued*)

Remove defective cylinders and supply and fix
galvanised steel indirect cylinders with 5 bosses
and immersion heater boss including re-
connecting 5 pipes and immersion heater
(electrical work not included), and removing and
replacing slatted shelving

109 litre (24 gallon) BS 1565 size BSG 1M	each	5.00	5.00	85.65	278.62	364.27	455.34
136 litre (30 gallon) BS 1565 size BSG 2M	each	5.50	5.50	94.22	295.78	390.00	487.50

Remove defective cylinders and supply and fix
copper direct cylinders with 4 bosses and
immersion heater boss including re-connecting
4 pipes and immersion heater (electrical work not
included), and removing and replacing slat
shelving

116 litre (25 gallon) BS 699 Type 3	each	4.75	4.75	81.36	89.98	171.34	214.18
144 litre (32 gallon) BS 699 Type 8	each	5.00	5.00	85.65	93.77	179.42	224.28

Remove defective cylinders and supply and fix
copper indirect cylinders with 4 bosses and
immersion heater boss including re-connecting
4 pipes and immersion heater (electrical work not
included), and removing and replacing slat
shelving

108 litre (24 gallon) BS 1566 Type 7	each	4.75	4.75	81.36	100.73	182.09	227.61
152 litre (33 gallon) BS 1566 Type 9	each	5.25	5.25	89.94	143.49	233.43	291.79

Remove defective lavatory basins and supply and
fix new 560 x 405 mm vitreous china basins to
existing brackets including re-fixing waste and
valves and re-connecting service pipes

white basins	each	4.00	4.00	68.52	28.38	96.90	121.13
coloured basins	each	4.00	4.00	68.52	38.38	106.90	133.63

Remove defective lavatory basins, supply and fix
new 560 x 405 mm vitreous china basins, pair of
13 mm chromium plated pillar taps waste, plug,
chain & stay and pair of towel rail brackets
plugged and screwed to wall, re-connect waste
and service pipes

white basin sets complete	each	4.50	4.50	77.08	55.66	132.74	165.93
coloured basin sets complete	each	4.50	4.50	77.08	65.66	142.74	178.43

Plumbing and engineering installations

Alterations and repairs	Unit	Hours C	Hours L	Labour net	Material net	Price net	Price with 25%
Repairs and maintenance				£	£	£	£
					VAT not included		

Equipment (*continued*)

Remove defective baths and supply and fix new 1700 mm baths with pair of 19 mm pillar taps, chromium plated overflow, waste, plug, chain and stay complete including re-connecting overflow, waste and service pipes (bath panels not included)

	Unit	Hours C	Hours L	Labour net	Material net	Price net	Price with 25%
white acrylic bath sets complete with cradle and brackets	each	6.00	6.00	102.78	139.54	242.32	302.90
coloured acrylic bath sets complete with cradle and brackets	each	6.00	6.00	102.78	144.55	247.33	309.16
white porcelain enamelled pressed steel bath sets complete with cradle and brackets	each	6.00	6.00	102.78	120.55	223.33	279.16
coloured porcelain enamelled pressed steel bath sets complete with cradle and brackets	each	6.00	6.00	102.78	131.55	234.33	292.91

Remove defective WC seats and supply and fix new BS pattern black seats with plastics fittings

	Unit	Hours C	Hours L	Labour net	Material net	Price net	Price with 25%
single seats	each	0.67	0.67	11.47	18.12	29.59	36.99
seats with cover	each	0.67	0.67	11.47	25.87	37.34	46.67

Remove defective WC pan and supply and fix new vitreous china pedestal pan including re-connecting flush pipe, plastics connector to soil pipe and removing and refixing existing seat

	Unit	Hours C	Hours L	Labour net	Material net	Price net	Price with 25%
white pans	each	2.75	2.75	47.11	44.87	91.98	114.97
coloured pans	each	2.75	2.75	47.11	58.87	105.98	132.47

Remove defective high level WC flushing cistern and supply and fix new 9 litre (2 gallon) standard finish black plastics flushing cisterns complete with all fittings including re-connecting flush pipe, overflow and service pipes

	Unit	Hours C	Hours L	Labour net	Material net	Price net	Price with 25%
cisterns with plastic ball valves	each	3.50	3.50	59.96	34.17	94.13	117.66
cisterns with BS ball valves	each	3.50	3.50	59.96	39.10	99.06	123.83

Remove defective low level WC flushing cistern and supply and fix new 9 litre (2 gallon) standard finish black plastics flushing cisterns complete with all fittings including connecting flush bend and re-connecting overflow and service pipes

	Unit	Hours C	Hours L	Labour net	Material net	Price net	Price with 25%
cisterns with plastic ball valves	each	3.50	3.50	59.96	34.17	94.13	117.66
cisterns with BS ball valves	each	3.50	3.50	59.96	39.10	99.06	123.83

Remove defective high level WC suites, supply and fix new white vitreous china pan, single seat, 9 litre (2 gallon) standard finish. black plastics flushing cistern complete with BS ball valve and all fittings, including plastics connector to soil pipe connecting new flush pipe and re-connecting overflow and service pipes

	Unit	Hours C	Hours L	Labour net	Material net	Price net	Price with 25%
	each	6.25	6.25	107.07	121.53	228.60	285.75

Plumbing and engineering installations

Alterations and repairs	Unit	Hours C	Hours L	Labour net	Material net	Price net	Price with 25%
Repairs and maintenance				£	£	£	£
					VAT not included		

Equipment (*continued*)

Remove defective low level WC suites, supply and fix new white suites with vitreous china pan, single seat, 9 litre (2 gallon) streamlined finish plastics flushing cistern complete with BS ball valve connector and all fittings plastics connection to soil pipe, connecting flush bend and re-connecting overflow and service pipes	each	6.25	6.25	107.07	125.53	232.60	290.75
Remove defective low level WC suites, supply and fix new coloured suites with vitreous china pan, single seat, 9 litre (2 gallon) streamlined finish plastics flushing cistern complete with BS ball valve connector and all fittings plastics connection to soil pipe, connecting flush bend and re-connecting overflow and service pipes	each	6.25	6.25	107.07	149.53	256.60	320.75
Remove defective high level WC suites and supply and fix new white low level WC suites with vitreous china pan, single seat, 9 litre (2 gallon) streamlined finish plastics flushing cistern complete with BS ball valve and all fittings plastics connection to soil pipe, connecting flush bend, removing old and providing new overflow pipe and connecting old service pipes	each	10.00	10.00	171.30	143.61	314.91	393.64
Remove defective high level WC suites and supply and fix new coloured low level WC suites with vitreous china pan, single seat, 9 litre (2 gallon) streamlined finish plastics flushing cistern complete with BS ball valve and all fittings plastics connection to soil pipe, connecting flush bend, removing old and providing new overflow pipe and connecting old service pipes	each	10.00	10.00	171.30	154.68	325.98	407.48

Plumbing and engineering installations

Repairs and maintenance

£   £

VAT not included

Replacement rainwater systems

Replacement seamless systems are available as a complete installed service through specialist installers.

The system consists of continuous lengths of gutter, roll formed on site through specialist machinery to a 125 mm ogee shape from factory coated aluminium alloy sheet in coil form, joints occur only at corners. Gutters are fitted with concealed internal aluminium hanger brackets and are fixed by stainless steel fixings. The system is complemented with extruded aluminium downpipes to BS2997.

The following "Specialist price net" figures are guide prices provided by Kingswood Guttering Services for installed quantities of around 50 metres of gutter and downpipes with good ladder access or existing scaffolding.

Prices include for polyester powder coating in standard colours.

Prices include for removing and disposing of existing rainwater goods, for travelling to site and a ten year guarantee.

Prices do not include for repair or replacing defective woodwork or for cash discount.

Replacement aluminium seamless gutters

125 x 95 mm ogee coated aluminium gutters complete with all angles and outlets, fitted with internal brackets fixed at maximum 600 mm centres by stainless steel fixings

	Unit	Specialist price net	Price with 25%
to timber	m	15.05	18.81
to walls	m	16.20	20.25

125 x 95 mm ogee coated aluminium gutters complete with all angles and outlets

	Unit	Specialist price net	Price with 25%
fitted with strap hangers fixed to timber	m	17.94	22.43
fitted with 350 mm projection rise and fall brackets fixed to walls	m	23.15	28.94

Replacement aluminium downpipes

68 mm diameter coated aluminium pipes complete with all bends, etc and aluminium or stainless steel fixings to walls; swaged joints in running length

	Unit	Specialist price net	Price with 25%
	m	18.34	22.93
cast-eared collars	each	10.41	13.01
non-eared collars	each	10.41	13.01
cast-eared shoes	each	10.64	13.30

75 x 75 mm square section coated aluminium pipes with all bends, etc and aluminium or stainless steel fixings to walls; swaged joints in running length

	Unit	Specialist price net	Price with 25%
	m	24.08	30.10
cast-eared connecting collars	each	11.46	14.32
non-eared collars	each	11.46	14.32
cast-eared shoes	each	11.69	14.61

Enquiries about the foregoing specialist prices should be made to Kingswood, Kingswood House, Exeter Road, South Brent, Devon, TQ10 9DF, tel & fax(01364) 73631.

Kingswood
EST 1980
GUTTERING SERVICES

ALL GUTTERING SYSTEMS INCLUDING
ALUMINIUM-SEAMLESS &
SECTION SYSTEMS, CAST IRON, ETC
PVCu
WINDOWS - FASCIAS - BARGES - CONSERVATORIES

TEL OR FAX
**SOUTH BRENT (O1364) 73631
KINGSWOOD HOUSE, EXETER ROAD, SOUTH BRENT
DEVON TQ10 9DF**

Electrical installations

Preamble

Electrical work is now almost exclusively the province of specialist sub-contractors, consequently, with the exception of builder's work, prices in this section are based on those of specialists.

Figures in this section are limited to work in existing dwellings. For alterations involving new building and electrical work, it is considered more appropriate to refer to figures provided in "Electrical installations" in the "New work" section and adjust according to the circumstances.

"Labour net" figures include allowances for all costs incidental to the employment of labour.

"Materials net" figures for builder's work include for all costs of materials including an allowance for waste except where specifically stated.

"Price net" figures for builder's work are the totals of the "Labour net" and "Materials net" figures. Prices are for a builder employing his own labour; according to the amount and nature of the work involved, it may well be possible to secure more advantageous prices from specialist sub-contractors.

Specialist prices

"Price with 25%" figures are all-in guide prices and include 25% for the builder's overheads, profit, unloading materials and general attendance (to include free use of standing scaffolding and hoists, temporary lighting and water and clearing away rubbish).

The amount of attendance required varies between the various trades and also with the circumstances of specific jobs; the percentage addition must always be considered and adjusted as necessary to suit the terms and conditions of the quotation being used.

Quantities and delivery distances are usually the most significant of the many factors which influence prices and it must be emphasised that quotations should always be obtained when preparing a tender.

Electrical installations

	Unit	Specialist price net	Price with 25%
Alterations and repairs			

Points in existing dwellings

£ £
VAT not included

The following "Specialist price net" figures are guide prices provided by Buckman & Hayward Ltd for an installation minimum of about 20 points within 75 kilometres of a specialist sub-contractor's work.

Prices do not include for lamps, Electricity Board charges or cash discount.

See the preamble notes for builder's profit and attendance.

PVC-U insulated and sheathed cables in existing houses or bungalows, installed in concealed positions where possible

	Unit	Specialist price net	Price with 25%
single lighting points controlled by one switch	each	45.68	57.11
two lighting points controlled by one switch	each	65.14	81.43
single lighting points controlled by two switches	each	75.18	93.98
single lighting points controlled by three switches	each	85.98	107.47
13 amp single switched socket outlets	each	59.07	73.83
13 amp double switched socket outlets	each	64.37	80.46
30 amp cooker points with cooker panel	each	127.89	159.87
45 amp cooker points with cooker panel	each	165.60	207.00
immersion heater points with control switch (excluding heater)	each	61.26	76.58
infra-red heater points with control switch (excluding heater and earth bonding)	each	62.98	78.72
dual type shaver socket outlets	each	90.31	112.89
single extractor units with control switch for bathroom or kitchen (excluding ducting)	each	180.30	225.37
bells controlled by front door push including transformer	each	92.50	115.62
Hi-lo chimes controlled by front door push and back door push including transformer	each	100.66	125.83
3 kW storage heater points controlled by switch adjacent to heater position	each	71.93	89.92

Enquiries about the foregoing specialist prices should be made in writing to Buckman & Hayward Ltd, 145a Ashford Road, Eastbourne, East Sussex, BN21 3UA, tel (01323) 642815, fax (01323) 410225.

Electrical installations

Alterations and repairs	Unit	Hours C	Hours L	Labour net	Material net	Price net	Price with 25%
Builder's work				£	£	£	£
					VAT not included		

Cutting away for and making good after the electrician in existing structures

Holes for cables or conduits through brick or block walls and make good, wall thickness

	Unit	Hours C	Hours L	Labour net	Material net	Price net	Price with 25%
75 mm	each	0.20	0.13	2.68	-	2.68	3.35
102.5 mm	each	0.35	0.25	4.83	-	4.83	6.04
215 mm	each	0.55	0.35	7.29	-	7.29	9.11

Mortices for boxes in brick or block walls and make good

single switches or sockets	each	0.15	0.10	2.03	0.16	2.19	2.74
double sockets	each	0.20	0.13	2.68	0.21	2.89	3.61

Horizontal rough chases in brickwork

25 x 25 mm	m	0.25	0.20	3.59	-	3.59	4.49
50 x 50 mm	m	0.35	0.25	4.83	-	4.83	6.04

Vertical rough chases in blockwork

25 x 25 mm	m	0.35	0.25	4.83	-	4.83	6.04
50 x 50 mm	m	0.55	0.35	7.29	-	7.29	9.11

For specialist prices for chasing, cutting and drilling concrete and brickwork, see "Concrete work" in the New work section

Lifting and replacing existing floorboards including notching joists for groups of not exceeding 3

cables or conduits	m	0.35	-	3.14	0.21	3.35	4.19

Make good plastering around switch socket boxes etc

not exceeding 0.30 m girth	each	0.20	0.12	2.61	0.20	2.81	3.51
0.30 - 1.00 m girth	each	0.30	0.18	3.91	0.31	4.22	5.28

Floor, wall and ceiling finishings

Preamble

Plastering differs little whether on new work or on alterations and repairs except that quantities are likely to be smaller and more making good will probably be involved. Materials bought in small lots will cost more than the larger lots allowed for in the "New work" section, therefore much of the "New work" section has been repeated but calculated on the higher basic prices for materials and a generally increased labour allowance.

"Labour net" figures include allowances for all costs incidental to the employment of labour.

"Plant net" figures include for all costs of plant including drivers and operators where applicable.

"Materials net" figures include for all costs of materials including an allowance for waste except where specifically stated.

"Price net" figures are the totals of the "Labour net", "Plant net", where applicable, and "Materials net" figures. Prices are for a builder employing his own labour; according to the amount and nature of the work involved, it may well be possible to secure more advantageous prices from specialist sub-contractors.

Although not in accordance with the Standard Method of Measurement, additional figures have been included for repair work in patches between one and two square metres.

Prices do not include any allowance for scaffolding, ladders or other plant necessary to reach the work. The "Preliminaries" section includes prices for scaffolding which must be considered and allowance included to suit the particular circumstances of a tender.

Specialist prices

"Price with 25%" figures are all-in guide prices and include 25% for the builder's overheads, profit, unloading materials and general attendance (to include free use of standing scaffolding and hoists, temporary lighting and water and clearing away rubbish).

The amount of attendance required varies between the various trades and also with the circumstances of specific jobs; the percentage addition must always be considered and adjusted as necessary to suit the terms and conditions of the quotation being used.

Quantities and delivery distances are usually the most significant of the many factors which influence prices and it must be emphasised that quotations should always be obtained when preparing a tender.

Floor, wall and ceiling finishings

Alterations and repairs

	Unit	Price
Basic prices for materials		**£**
Abrasive grain	kg	2.25
Bonding agent	litre	5.47
Cement		
Portland	tonne	89.04
white	tonne	281.96
Expanded metal lathing		
0.500 mm	m2	4.36
0.725 mm	m2	5.08
Floor repairing compound	25 kg	42.90
Glazed ceramic tiles		
108 x 108 x 4 mm	100	15.46
152 x 152 x 5.5 mm	100	49.77
Granite chippings	tonne	29.00
Hydrated lime	tonne	139.84
90 mm jute scrim	100 m roll	3.69
Liquid colouring agent	litre	9.12
Liquid surface hardener	1 litre	1.79
Liquid waterproofer and hardener	1 litre	1.83
9.5 mm plasterboard baseboard	m2	1.59
Plasterboard wallboard		
9.5 mm	m2	1.57
12.5 mm	m2	1.82
9.5 mm insulating grade	m2	2.14
12.5 mm insulating grade	m2	2.39
Premixed lightweight ("Carlite") plasters		
browning grade	tonne	175.12
bonding grade	tonne	172.64
finish grade	tonne	136.00
Retarded hemihydrate ("Thistle") plasters		
board finish grade	tonne	128.52
hardwall grade	tonne	175.12
multi-finish grade	tonne	128.52

Prices actually to be paid for materials must be checked against the above basic prices and adjustments made as necessary.

Floor, wall and ceiling finishings

	Unit	Price
Alterations and repairs		
Basic prices for materials		£
Sand	m3	17.96
	tonne	11.22
Tyrolean finish	tonne	330.00

Prices actually to be paid for materials must be checked against the above basic prices and adjustments made as necessary.

Floor, wall and ceiling finishings

Alterations and repairs	Unit	Hours C	Hours L	Labour net	Material net	Price net	Price with 25%
Internal in situ finishings				£	£	£	£
					VAT not included		
Retarded hemihydrate gypsum ("Thistle") plasters to BS 1191, Class B							
5 mm one coat plastering on plasterboard walls including scrimming joints							
over 300 mm wide	m2	0.42	0.21	5.18	1.25	6.43	8.04
not exceeding 300 mm wide	m2	0.63	0.32	7.80	1.25	9.05	11.31
repairs not exceeding 1.00 m2	m2	0.67	0.34	8.29	1.25	9.54	11.93
repairs 1.00 - 2.00 m2	m2	0.55	0.28	6.82	1.25	8.07	10.09
5 mm one coat plastering on plasterboard ceilings including scrimming joints							
over 300 mm wide	m2	0.44	0.22	5.42	1.02	6.44	8.05
not exceeding 300 mm wide	m2	0.66	0.33	8.13	1.02	9.15	11.44
repairs not exceeding 1.00 m2	m2	0.70	0.35	8.63	1.25	9.88	12.35
repairs 1.00 - 2.00 m2	m2	0.55	0.27	6.75	1.25	8.00	10.00
13 mm two coat plastering on brick or block walls							
over 300 mm wide	m2	0.60	0.30	7.40	2.68	10.08	12.60
not exceeding 300 mm wide	m2	0.91	0.45	11.18	2.68	13.86	17.32
repairs not exceeding 1.00 m2	m2	0.95	0.47	11.68	2.68	14.36	17.95
repairs 1.00 - 2.00 m2	m2	0.75	0.37	9.21	2.68	11.89	14.86
16 mm three coat plastering on brick or block walls							
over 300 mm wide	m2	0.82	0.41	10.11	3.28	13.39	16.74
not exceeding 300 mm wide	m2	1.23	0.62	15.20	3.28	18.48	23.10
repairs not exceeding 1.00 m2	m2	1.30	0.65	16.03	3.28	19.31	24.14
repairs 1.00 - 2.00 m2	m2	1.00	0.50	12.33	3.28	15.61	19.51
13 mm three coat plastering on metal lathing to ceilings							
over 300 mm wide	m2	0.82	0.41	10.11	6.54	16.65	20.81
not exceeding 300 mm wide	m2	1.23	0.61	15.13	6.54	21.67	27.09
repairs not exceeding 1.00 m2	m2	1.35	0.68	16.68	6.54	23.22	29.02
repairs 1.00 - 2.00 m2	m2	1.05	0.53	12.98	6.54	19.52	24.40
Retarded hemihydrate gypsum ("Thistle") plasters to BS 1191, Class B and cement and sand 1:3 undercoats							
13 mm two coat plastering on brick or block walls							
over 300 mm wide	m2	0.66	0.33	8.13	1.86	9.99	12.49
not exceeding 300 mm wide	m2	0.99	0.49	12.17	1.86	14.03	17.54
repairs not exceeding 1.00 m2	m2	1.05	0.63	13.66	1.75	15.41	19.26
repairs 1.00 - 2.00 m2	m2	0.85	0.43	10.52	1.75	12.27	15.34

Floor, wall and ceiling finishings

Alterations and repairs	Unit	Hours C	Hours L	Labour net	Material net	Price net	Price with 25%
Internal in situ finishings				£	£	£	£
					VAT not included		
Premixed lightweight ("Carlite") plaster to BS 1191							
13 mm two coat plastering on brick or block walls							
over 300 mm wide	m2	0.60	0.30	7.40	2.00	9.40	11.75
not exceeding 300 mm wide	m2	0.90	0.45	11.09	2.37	13.46	16.82
repairs not exceeding 1.00 m2	m2	0.95	0.48	11.75	2.37	14.12	17.65
repairs 1.00 - 2.00 m2	m2	0.75	0.38	9.28	2.37	11.65	14.56
13 mm two coat plastering on "Newtonite" lathing to walls							
over 300 mm wide	m2	0.66	0.33	8.13	4.46	12.59	15.74
not exceeding 300 mm wide	m2	1.30	0.65	16.03	4.46	20.49	25.61
10 mm two coat plastering on plasterboard ceilings including scrimming joints							
over 300 mm wide	m2	0.71	0.35	8.72	2.23	10.95	13.69
not exceeding 300 mm wide	m2	1.07	0.54	13.23	2.23	15.46	19.32
repairs not exceeding 1.00 m2	m2	1.15	0.67	14.82	2.32	17.14	21.43
repairs 1.00 - 2.00 m2	m2	0.90	0.45	11.09	2.32	13.41	16.76
13 mm three coat plastering on metal lathing to ceilings							
over 300 mm wide	m2	0.83	0.42	10.27	6.55	16.82	21.02
not exceeding 300 mm wide	m2	1.24	0.62	15.29	6.55	21.84	27.30
repairs not exceeding 1.00 m2	m2	1.30	0.65	16.03	6.55	22.58	28.23
repairs 1.00 - 2.00 m2	m2	1.05	0.63	13.66	6.55	20.21	25.26
Plastering sundries							
Extra for dubbing out walls per 6 mm thickness							
over 300 mm wide	m2	0.13	0.07	1.60	1.49	3.09	3.86
not exceeding 300 mm wide	m2	0.25	0.13	3.08	1.49	4.57	5.71
repairs not exceeding 1.00 m2	m2	0.20	0.10	2.46	1.49	3.95	4.94
repairs 1.00 - 2.00 m2	m2	0.15	0.07	1.85	1.49	3.34	4.17
Treatment to edges							
Arrises	m	0.12	0.06	1.48	-	1.48	1.85
Rounded external angles not exceeding 10 mm radius	m	0.17	0.09	2.09	-	2.09	2.61
Fair joints of plastering to flush edge of existing finishes	each	0.12	0.06	1.48	-	1.48	1.85
Make good plastering around pipes							
not exceeding 0.30 m girth	each	0.19	0.10	2.37	0.37	2.74	3.42
0.30 - 1.00 m girth	each	0.28	0.14	3.45	0.75	4.20	5.25
Galvanised steel angle beads fixed with plaster dabs	m	0.09	-	0.79	0.64	1.43	1.79

Floor, wall and ceiling finishings

Alterations and repairs	Unit	Hours C	Hours L	Labour net	Material net	Price net	Price with 25%
Internal in situ finishings				£	£	£	£
					VAT not included		
Plastering sundries (*continued*)							
100 mm girth plaster core coved cornices fixed with adhesive							
straight	m	0.17	0.08	2.06	1.42	3.48	4.35
angles	each	0.26	-	2.33	0.38	2.71	3.39
127 mm girth plaster core coved cornices fixed with adhesive							
straight	m	0.17	0.09	2.13	1.62	3.75	4.69
angles	each	0.26	-	2.33	0.45	2.78	3.48
225 mm girth moulded plaster cornices to match existing (based on minimum 6.00 m)							
lengths	m	1.90	0.95	23.42	3.23	26.65	33.31
stopped ends	each	0.30	0.15	3.70	3.23	6.93	8.66
returned ends	each	0.95	0.47	11.68	3.23	14.91	18.64
angles	each	0.65	0.33	8.04	3.23	11.27	14.09
Moulded plaster cornice to match existing per 25 mm girth (based on a minimum of 6.00 m)	each	0.21	0.12	2.69	0.35	3.04	3.80
Galvanised metal lathing to BS 1369							
0.500 mm lathing stapled to softwood to ceilings							
over 300 mm wide	m2	0.38	-	3.40	4.94	8.34	10.43
not exceeding 300 mm wide	m2	0.47	-	4.21	4.94	9.15	11.44
repairs not exceeding 1.00 m2	m2	0.60	-	5.38	4.94	10.32	12.90
repairs 1.00 - 2.00 m2	m2	0.50	-	4.48	4.94	9.42	11.78
0.725 mm lathing stapled to softwood to ceilings							
over 300 mm wide	m2	0.38	-	3.40	5.74	9.14	11.43
not exceeding 300 mm wide	m2	0.46	-	4.12	5.74	9.86	12.32
repairs not exceeding 1.00 m2	m2	0.60	-	5.38	5.74	11.12	13.90
repairs 1.00 - 2.00 m2	m2	0.50	-	4.48	5.74	10.22	12.78
Gypsum plasterboard to BS 1230							
Lathing fixed with galvanised nails to softwood to ceilings over 300 mm wide							
9.5 mm	m2	0.18	0.08	2.15	2.60	4.75	5.94
12.5 mm	m2	0.20	0.10	2.46	2.95	5.41	6.76
9.5 mm baseboard fixed with galvanised nails to softwood to ceilings							
over 300 mm wide	m2	0.18	0.09	2.22	2.24	4.46	5.58
not exceeding 300 mm wide	m2	0.20	0.10	2.46	2.24	4.70	5.88
repairs not exceeding 1.00 m2	m2	0.30	0.15	3.70	2.24	5.94	7.42
repairs 1.00 - 2.00 m2	m2	0.25	0.15	3.25	2.24	5.49	6.86

Floor, wall and ceiling finishings

Alterations and repairs	Unit	Hours C	Hours L	Labour net	Material net	Price net	Price with 25%
Internal in situ finishings				£	£	£	£
					VAT not included		

Newtonite fibre based corrugated lath with butt jointed horizontal and lapped vertical joints and with 100 mm bitumen felt strip behind joints

Lathing fixed with galvanised clout nails to existing plastered walls.

over 300 mm wide	m2	0.40	-	3.58	8.40	11.98	14.98
not exceeding 300 mm wide	m2	0.80	-	7.17	8.40	15.57	19.46

Lathing fixed with galvanised clout nails including 25 x 50 mm impregnated softwood battens plugged to existing walls.

over 300 mm wide	m2	1.80	-	16.13	8.78	24.91	31.14
not exceeding 300 mm wide	m2	3.60	-	32.26	8.78	41.04	51.30

Cement and sand 1:3

Steel trowelled pavings, level and to falls, on concrete over 300 mm wide including thoroughly cleaning concrete base

25 mm	m2	0.43	0.22	5.33	2.51	7.84	9.80
32 mm	m2	0.46	0.23	5.67	3.21	8.88	11.10
38 mm	m2	0.48	0.24	5.92	3.82	9.74	12.18
50 mm	m2	0.54	0.27	6.66	5.02	11.68	14.60
65 mm	m2	0.57	0.28	7.00	6.53	13.53	16.91
75 mm	m2	0.60	0.30	7.40	7.53	14.93	18.66

Steel trowelled pavings, level and to falls, on concrete not exceeding 300 mm wide including thoroughly cleaning concrete base

25 mm	m2	0.66	0.33	8.13	2.51	10.64	13.30
32 mm	m2	0.69	0.35	8.54	3.21	11.75	14.69
38 mm	m2	0.73	0.37	9.03	3.82	12.85	16.06
50 mm	m2	0.81	0.40	9.96	5.02	14.98	18.73
65 mm	m2	0.86	0.43	10.61	6.53	17.14	21.43
75 mm	m2	0.89	0.45	11.00	7.53	18.53	23.16

19 x 150 mm steel trowelled skirtings with fair edge on brick or block walls

straight	m	0.30	0.15	3.70	0.10	3.80	4.75
angles	each	0.10	0.05	1.23	-	1.23	1.54

19 x 225 mm steel trowelled skirtings with fair edge on brick or block walls

straight	m	0.33	0.16	4.04	0.20	4.24	5.30
angles	each	0.11	0.06	1.36	-	1.36	1.70

Floor, wall and ceiling finishings

Alterations and repairs	Unit	Hours C	Hours L	Labour net	Material net	Price net	Price with 25%
Internal in situ finishings				£	£	£	£
					VAT not included		
Granolithic 1:2½							
Steel trowelled pavings, level and to falls, on concrete over 300 mm wide including thoroughly cleaning concrete base							
25 mm	m2	0.49	0.24	6.01	3.50	9.51	11.89
32 mm	m2	0.55	0.27	6.75	4.48	11.23	14.04
38 mm	m2	0.60	0.30	7.40	5.32	12.72	15.90
Steel trowelled pavings, level and to falls, on concrete not exceeding 300 mm wide including thoroughly cleaning concrete base							
25 mm	m2	0.75	0.37	9.21	3.50	12.71	15.89
32 mm	m2	0.83	0.42	10.27	4.48	14.75	18.44
38 mm	m2	0.91	0.45	11.18	5.32	16.50	20.63
19 x 150 mm steel trowelled skirtings on brick or block walls							
with fair edge	m	0.33	0.16	4.04	0.42	4.46	5.58
angles on skirting with fair edge	each	0.13	0.07	1.60	-	1.60	2.00
19 x 225 mm steel trowelled skirtings on brick or block walls							
with fair edge	m	0.40	0.20	4.93	0.56	5.49	6.86
angles on skirting with fair edge	each	0.20	0.10	2.46	-	2.46	3.08
Ancillary work							
Extra for liquid waterproofer and hardener - per 25 mm thickness	m2	0.02	0.01	0.27	0.11	0.38	0.47
Extra for liquid colouring agent at the rate of 0.25 packs per m2 per 25 mm thickness in pavings - per 25 mm thickness	m2	0.04	0.02	0.54	0.31	0.85	1.06
Three coats of liquid surface hardener on pavings	m2	0.06	0.03	0.67	1.97	2.64	3.30
Abrasive grain sprinkled on pavings at the rate of 0.80 kg per m2 and trowelled in	m2	0.11	0.06	1.36	1.79	3.15	3.94
Solution 1:2 of bonding agent on concrete before laying pavings or screeds	m2	-	0.11	0.74	1.20	1.94	2.42

Floor, wall and ceiling finishings

Alterations and repairs	Unit	Hours C	Hours L	Labour net	Material net	Price net	Price with 25%
External in situ finishings				£	£	£	£
					VAT not included		
16 mm cement and sand 1:3 two coat wood floated finish on brick or block walls							
over 300 mm wide	m2	0.72	0.36	8.88	1.61	10.49	13.11
not exceeding 300 mm wide	m2	1.07	0.54	13.23	1.61	14.84	18.55
repairs not exceeding 1.00 m2	m2	1.15	0.58	14.21	1.61	15.82	19.77
repairs 1.00 - 2.00 m2	m2	0.90	0.45	11.09	1.61	12.70	15.88
Extra for waterproofer in backing coat	m2	0.01	0.06	0.47	0.35	0.82	1.02
19 mm two coat finish on brick or block walls of 13 mm cement, lime and sand 1:1:5 rendering with waterproofer and 6 mm white cement, lime and silver sand 1:1:5 lightly scraped finish							
over 300 mm wide	m2	0.77	0.38	9.46	1.67	11.13	13.91
not exceeding 300 mm wide	m2	1.16	0.58	14.30	1.64	15.94	19.93
repairs not exceeding 1.00 m2	m2	1.25	0.63	15.45	1.64	17.09	21.36
repairs 1.00 - 2.00 m2	m2	0.95	0.47	11.68	1.64	13.32	16.65
22 mm three coat finish on brick or block walls of 10 mm cement and sand 1:3 rendering with waterpoofer, 6 mm cement, lime and sand 1:1:5 floating and 6 mm white cement, lime and silver sand 1:1:5 lightly scraped finish							
over 300 mm wide	m2	0.99	0.50	12.24	1.87	14.11	17.64
not exceeding 300 mm wide	m2	1.50	0.75	18.49	1.87	20.36	25.45
repairs not exceeding 1.00 m2	m2	1.60	0.80	19.73	1.87	21.60	27.00
repairs 1.00 - 2.00 m2	m2	1.25	0.63	15.45	1.87	17.32	21.65
15 mm two coat finish on brick or block walls of 10 mm cement, lime and sand 1:1:6 rendering and Tyrolean finish							
over 300 mm wide	m2	0.66	0.33	8.13	2.76	10.89	13.61
not exceeding 300 mm wide	m2	0.99	0.50	12.24	2.76	15.00	18.75
repairs not exceeding 1.00 m2	m2	1.05	0.63	13.66	2.76	16.42	20.52
repairs 1.00 - 2.00 m2	m2	0.85	0.43	10.52	2.76	13.28	16.60
25 mm three coat finish on brick or block walls of 10 mm cement and sand 1:3 rendering with waterproofer, 10 mm cement, lime and sand 1:1:6 floating and Tyrolean finish							
over 300 mm wide	m2	0.93	0.47	11.50	3.76	15.26	19.07
not exceeding 300 mm wide	m2	1.40	0.70	17.26	3.76	21.02	26.27
repairs not exceeding 1.00 m2	m2	1.50	0.75	18.49	3.76	22.25	27.81
repairs 1.00 - 2.00 m2	m2	1.20	0.60	14.79	3.76	18.55	23.19
20 mm three coat finish on brick or block walls of 8 mm cement and sand 1:3 rendering and floating coats and pebbledash finish							
over 300 mm wide	m2	0.77	0.38	9.46	2.56	12.02	15.03
not exceeding 300 mm wide	m2	1.16	0.58	14.30	2.96	17.26	21.57
repairs not exceeding 1.00 m2	m2	1.25	0.63	15.45	2.56	18.01	22.51
repairs 1.00 - 2.00 m2	m2	0.95	0.48	11.75	2.56	14.31	17.89

Floor, wall and ceiling finishings

Alterations and repairs	Unit	Hours C	Hours L	Labour net	Material net	Price net	Price with 25%
External in situ finishings				£	£	£	£
					VAT not included		
Extra for waterproofer in backing coat	m2	0.01	0.01	0.16	0.04	0.20	0.25
Arrises on external finishings	m	0.13	0.07	1.60	-	1.60	2.00
Fair joints of external finishings to flush edges of existing finishes	each	0.12	0.06	1.48	-	1.48	1.85

Floor, wall and ceiling finishings

Alterations and repairs	Unit	Hours C	Hours L	Labour net	Material net	Price net	Price with 25%
Beds and backings				£	£	£	£
					VAT not included		

Cement and sand 1:3

Screeded beds, level and to falls, on concrete
over 300 mm wide including thoroughly cleaning
concrete base

	Unit	Hours C	Hours L	Labour net	Material net	Price net	Price with 25%
19 mm	m2	0.28	0.14	3.45	1.91	5.36	6.70
25 mm	m2	0.31	0.16	3.86	2.51	6.37	7.96
32 mm	m2	0.34	0.17	4.20	3.21	7.41	9.26
38 mm	m2	0.36	0.18	4.44	3.82	8.26	10.32
50 mm	m2	0.42	0.21	5.18	5.02	10.20	12.75
65 mm	m2	0.45	0.24	5.65	6.53	12.18	15.23
75 mm	m2	0.48	0.24	5.92	7.53	13.45	16.81

Screeded beds, level and to falls, on concrete not
exceeding 300 mm wide including thoroughly
cleaning concrete base

	Unit	Hours C	Hours L	Labour net	Material net	Price net	Price with 25%
19 mm	m2	0.42	0.21	5.18	1.91	7.09	8.86
25 mm	m2	0.46	0.23	5.67	2.51	8.18	10.23
32 mm	m2	0.52	0.26	6.41	3.21	9.62	12.03
38 mm	m2	0.55	0.28	6.82	3.82	10.64	13.30
50 mm	m2	0.63	0.32	7.80	5.02	12.82	16.02
65 mm	m2	0.68	0.34	8.38	6.53	14.91	18.64
75 mm	m2	0.73	0.37	9.03	7.53	16.56	20.70

Floated beds, level and to falls, on concrete over
300 mm wide including thoroughly cleaning
concrete base

	Unit	Hours C	Hours L	Labour net	Material net	Price net	Price with 25%
19 mm	m2	0.33	0.16	4.04	1.91	5.95	7.44
25 mm	m2	0.38	0.19	4.68	2.51	7.19	8.99
32 mm	m2	0.44	0.22	5.42	3.21	8.63	10.79
38 mm	m2	0.44	0.22	5.42	3.82	9.24	11.55
50 mm	m2	0.51	0.26	6.32	5.02	11.34	14.18
65 mm	m2	0.53	0.26	6.50	6.53	13.03	16.29
75 mm	m2	0.57	0.28	7.00	7.53	14.53	18.16

Floated beds, level and to falls, on concrete not
exceeding 300 mm wide including thoroughly
cleaning concrete base

	Unit	Hours C	Hours L	Labour net	Material net	Price net	Price with 25%
19 mm	m2	0.50	0.25	6.17	1.91	8.08	10.10
25 mm	m2	0.58	0.29	7.15	2.51	9.66	12.07
32 mm	m2	0.68	0.34	8.38	3.21	11.59	14.49
38 mm	m2	0.68	0.34	8.38	3.82	12.20	15.25
50 mm	m2	0.78	0.39	9.62	5.02	14.64	18.30
65 mm	m2	0.79	0.39	9.71	6.53	16.24	20.30
75 mm	m2	0.86	0.43	10.61	7.53	18.14	22.68

Floor, wall and ceiling finishings

Alterations and repairs	Unit	Hours C	Hours L	Labour net	Material net	Price net	Price with 25%
Beds and backings				£	£	£	£
					VAT not included		

Cement and sand 1:3 (*continued*)

Trowelled beds, level and to falls, on concrete over 300 mm wide including thoroughly cleaning concrete base

	Unit	Hours C	Hours L	Labour net	Material net	Price net	Price with 25%
19 mm	m2	0.38	0.19	4.68	1.91	6.59	8.24
25 mm	m2	0.43	0.22	5.33	2.51	7.84	9.80
32 mm	m2	0.46	0.23	5.67	3.21	8.88	11.10
38 mm	m2	0.48	0.24	5.92	3.82	9.74	12.18
50 mm	m2	0.54	0.27	6.66	5.02	11.68	14.60
65 mm	m2	0.57	0.28	7.00	6.53	13.53	16.91
75 mm	m2	0.60	0.30	7.40	7.53	14.93	18.66

Trowelled beds, level and to falls, on concrete not exceeding 300 mm wide including thoroughly cleaning concrete base

	Unit	Hours C	Hours L	Labour net	Material net	Price net	Price with 25%
19 mm	m2	0.58	0.29	7.15	1.91	9.06	11.32
25 mm	m2	0.65	0.37	8.31	2.51	10.82	13.53
32 mm	m2	0.69	0.34	8.47	3.21	11.68	14.60
38 mm	m2	0.73	0.36	8.97	3.82	12.79	15.99
50 mm	m2	0.81	0.40	9.96	5.02	14.98	18.73
65 mm	m2	0.86	0.43	10.61	6.53	17.14	21.43
75 mm	m2	0.91	0.45	11.18	7.53	18.71	23.39

13 mm screeded backings on brick or block walls

	Unit	Hours C	Hours L	Labour net	Material net	Price net	Price with 25%
over 300 mm wide	m2	0.33	0.16	4.04	1.31	5.35	6.69
not exceeding 300 mm wide	m2	0.49	0.24	6.01	1.31	7.32	9.15

13 mm floated backings on brick or block walls

	Unit	Hours C	Hours L	Labour net	Material net	Price net	Price with 25%
over 300 mm wide	m2	0.38	0.19	4.68	1.31	5.99	7.49
not exceeding 300 mm wide	m2	0.57	0.28	7.00	1.31	8.31	10.39

13 mm trowelled backings on brick or block walls

	Unit	Hours C	Hours L	Labour net	Material net	Price net	Price with 25%
over 300 mm wide	m2	0.44	0.22	5.42	1.31	6.73	8.41
not exceeding 300 mm wide	m2	0.66	0.33	8.13	1.31	9.44	11.80

Tile, slab and block finishings

Glazed ceramic tiles fixed with adhesive to plastered backings and grouted

108 x 108 x 4 mm wall tiling (Group A)

	Unit	Hours C	Hours L	Labour net	Material net	Price net	Price with 25%
over 300 mm wide	m2	0.99	0.50	12.24	20.26	32.50	40.63
not exceeding 300 mm wide	m2	1.49	0.75	18.40	20.26	38.66	48.33
Extra for rounded edge	m	-	-	-	0.69	0.69	0.86

152 x 152 x 5.5 mm wall tiling (Group A)

	Unit	Hours C	Hours L	Labour net	Material net	Price net	Price with 25%
over 300 mm wide	m2	0.93	0.46	11.43	24.19	35.62	44.52
not exceeding 300 mm wide	m2	1.40	0.70	17.26	23.86	41.12	51.40
Extra for rounded edge tiles	m	-	-	-	0.69	0.69	0.86

Floor, wall and ceiling finishings

Alterations and repairs	Unit	Hours C	Hours L	Labour net	Material net	Price net	Price with 25%
Sheet finishings				£	£	£	£
					VAT not included		

Gypsum plasterboard to BS 1230

Square edge wallboard fixed to softwood to walls with galvanised nails including covering joints with 50 mm paper backed cotton scrim and filling nail holes with joint filler

	Unit	Hours C	Hours L	Labour net	Material net	Price net	Price with 25%
9.5 mm over 300 mm wide	m2	0.29	0.15	3.61	2.39	6.00	7.50
9.5 mm not exceeding 300 mm wide	m2	0.43	0.22	5.33	2.39	7.72	9.65
12.5 mm over 300 mm wide	m2	0.33	0.16	4.04	2.65	6.69	8.36
12.5 mm not exceeding 300 mm wide	m2	0.50	0.25	6.17	2.65	8.82	11.03

Square edge insulating grade wallboard fixed to softwood to walls with galvanised nails including covering joints with 50 mm paper backed cotton scrim and filling nail holes with joint filler

	Unit	Hours C	Hours L	Labour net	Material net	Price net	Price with 25%
9.5 mm over 300 mm wide	m2	0.29	0.15	3.61	2.95	6.56	8.20
9.5 mm not exceeding 300 mm wide	m2	0.50	0.25	6.17	2.95	9.12	11.40
12.5 mm over 300 mm wide	m2	0.33	0.16	4.04	3.22	7.26	9.07
12.5 mm not exceeding 300 mm wide	m2	0.50	0.25	6.17	3.22	9.39	11.74

Tapered edge wallboard fixed to softwood to walls with galvanised nails including flush jointing with joint filler, joint tape and joint finish and filling nail holes with joint filler

	Unit	Hours C	Hours L	Labour net	Material net	Price net	Price with 25%
9.5 mm over 300 mm wide	m2	0.34	0.17	4.20	2.60	6.80	8.50
9.5 mm not exceeding 300 mm wide	m2	0.52	0.26	6.41	2.60	9.01	11.26
12.5 mm over 300 mm wide	m2	0.40	0.20	4.93	2.75	7.68	9.60
12.5 mm not exceeding 300 mm wide	m2	0.59	0.30	7.31	2.75	10.06	12.57

	Unit	Hours C	Hours L	Labour net	Material net	Price net	Price with 25%
54 mm reinforced paper corner tape bedded in joint filler and flushed over with joint filler and joint finish	m	0.11	0.06	1.36	0.16	1.52	1.90

Plaster core coved cornices fixed with adhesive

100 mm girth cornices

	Unit	Hours C	Hours L	Labour net	Material net	Price net	Price with 25%
straight	m	0.17	0.09	2.09	1.42	3.51	4.39
angles	each	0.28	-	2.51	0.38	2.89	3.61

127 mm girth cornices

	Unit	Hours C	Hours L	Labour net	Material net	Price net	Price with 25%
straight	m	0.18	0.09	2.22	1.62	3.84	4.80
angles	each	0.28	-	2.51	0.45	2.96	3.70

Floor, wall and ceiling finishings

Alterations and repairs	Unit	Hours C	Hours L	Labour net	Plant net	Material net	Price net	Price with 25%
Repairs and maintenance				£	£	£	£	£
					VAT not included			
Hack off wall finishings, lower average one storey and remove debris from site								
plastering	m2	-	0.60	4.04	0.63	-	4.67	5.84
tiling and cement backing	m2	-	0.80	5.39	1.26	-	6.65	8.31
rendering, roughcast or pebbledash	m2	-	0.85	5.73	0.63	-	6.36	7.95
Rake out joints and hack face of existing walls to form key for plastering	m2	-	0.60	4.04	-	-	4.04	5.05
Brush down and apply two coats of bonding agent (diluted 1:6 and 1:3) on sound wall surfaces to form key for plastering	m2	-	0.25	1.69	-	0.16	1.85	2.31
Cut out cracks for full depth of plasterwork 13 mm wide and make good	m	0.35	0.25	4.83	-	0.04	4.87	6.09
Cut 32 mm wide chase for conduit for the full depth of the plaster work and make good	m	0.40	0.30	5.60	-	0.07	5.67	7.09
Hack off lath and plaster ceiling, lower average one storey, draw nails from joists and remove debris from site, provide and fix 9.5 mm gypsum baseboard, scrim joints, finish with 5 mm one coat plastering and make good walls								
over 300 mm wide	m2	0.75	1.33	15.68	0.35	4.10	20.13	25.16
not exceeding 300 mm wide	m2	1.50	1.85	25.91	0.35	4.10	30.36	37.95
Hack off lath and plaster ceiling, lower one storey, draw nails from joists, remove debris from site, provide and fix two layers of 9.5 mm gypsum baseboard, cut away setting coat around edges, reinforce joints with 90 mm jute scrim								
finish with 5 mm one coat plaster in repairs not exceeding 1.00 m2	m2	2.85	1.90	38.35	0.81	5.60	44.76	55.95
finish with 5 mm one coat plaster in repairs 1.00 - 2.00 m2	m2	2.40	1.60	32.28	0.81	5.60	38.69	48.36
Cut back edges of wall or ceiling plastering where partition removed and make out plastering average 175 mm wide	m	0.65	0.40	8.52	-	0.42	8.94	11.18
Extra for filling toothing in walls	m	0.18	0.13	2.49	-	0.98	3.47	4.34
Cut out cracks in cement and sand paving 25 mm wide and 25 mm deep and make good	m	0.50	0.35	6.84	-	0.10	6.94	8.68

Floor, wall and ceiling finishings

Alterations and repairs	Unit	Hours C	Hours L	Labour net	Plant net	Material net	Price net	Price with 25%
Repairs and maintenance				£	£	£	£	£
					VAT not included			
Hack surface of concrete floor or paving with electric hammer and clear away, apply bonding agent and finish with 32 mm average trowelled paving								
cement and sand 1:3	m2	1.15	1.25	18.73	1.19	5.13	25.05	31.31
granolithic 1:2½	m2	1.25	1.35	20.30	1.33	4.78	26.41	33.01
Hack surface of concrete treads and risers with electric hammer and clear away, apply bonding agent and finish with 19 mm cement and sand 1:3 trowelled paving with rounded arrises								
150 mm wide	m	0.50	0.50	7.85	0.21	0.31	8.37	10.46
225 wide	m	0.70	0.70	10.99	0.25	0.41	11.65	14.56
300 mm wide	m	0.85	0.85	13.35	0.28	1.01	14.64	18.30
Hack surface of concrete treads and risers with electric hammer and clear away, apply bonding agent and finish with 19 mm granolithic 1:2½ trowelled paving with rounded arrises								
150 mm wide	m	0.52	0.52	8.16	0.21	0.43	8.80	11.00
225 mm wide	m	0.72	0.72	11.30	0.25	0.57	12.12	15.15
300 mm wide	m	0.88	0.88	13.81	0.28	0.85	14.94	18.68
6 mm average quick drying floor repairing compound trowelled over worn concrete or granolithic paving over 300 mm wide including thoroughly cleaning base	m2	0.40	0.40	6.28	-	3.43	9.71	12.14

Floor, wall and ceiling finishings

Alterations and repairs	Unit	Specialist price net	Price with 25%
Remedial re-plastering systems		£	£
			VAT not included

The following "Specialist price net" figures are minimum guide prices provided by Kiltox Chemicals Ltd, for work by one of their approved contractors within 200 kilometres of a branch depot.

Prices do not include for cash discount.

See the preambles notes for builder's profit and attendance.

Kiltox products used in accordance with the manufacturer's instructions carry a 30 year insured guarantee.

	Unit	Specialist price net	Price with 25%
Hack off existing wall plaster and prepare surfaces for re-plastering	m2	7.09	8.87

Kiltox plastering and rendering systems to walls

	Unit	Specialist price net	Price with 25%
Dampcoursing Plaster No 132 after DPC treatment (2 backing coats and 1 finishing coat to hold residual dampness)	m2	38.90	48.62
Waterproofing Plaster No 133 (for retaining walls in relatively dry situations)	m2	48.48	60.61
Tanking Plaster No 134 (for retaining walls in wet situations, vaults etc)	m2	58.13	72.66
Fungicidal Render No 155 (to form dry rot isolating barrier)	m2	61.81	77.26
One Coat Render No OCR after DPC treatment (1 OCR coat and 1 finishing coat) for all-purpose use	m2	45.25	56.56
Flexible K liquid applied, flexible cementitious waterproof coating for walls, floors and water-retaining structures. For exterior or internal use			
Formula 1	m2	41.46	51.83
Formula 2	m2	42.76	53.46

Enquiries about the foregoing specialist prices should be made to Kiltox Chemicals Ltd, Kiltox House, Park Row, Greenwich, London, SE10 9NI, tel (0208) 858 6277, instant estimating (0208) 853 5567, fax (0208) 853 3572.

Glazing

Preamble

"Labour net" figures include allowances for all costs incidental to the employment of labour.

"Materials net" figures include for all costs of materials including an allowance for waste except where specifically stated.

"Price net" figures are the totals of the "Labour net" and "Materials net" figures. Prices are for a builder employing his own labour; according to the amount and nature of the work involved, it may well be possible to secure more advantageous prices from specialist sub-contractors.

Although not in accordance with the Standard Method of Measurement, composite prices per square metre have been included for hacking out and re-glazing.

Prices do not include any allowance for scaffolding, ladders or other plant necessary to reach the work. The "Preliminaries" section includes prices for scaffolding which must be considered and allowance included to suit the particular circumstances of a tender.

Glazing

	Unit	Price
Alterations and repairs		

		£
Basic prices for materials		
Float glass		
3 mm - not exceeding 2400 x 1300 mm	m2	21.91
4 mm - not exceeding 2400 x 1300 mm	m2	21.91
White patterned glass		
4 mm - not exceeding 2100 x 1300 mm	m2	21.76
6 mm - not exceeding 2100 x 1300 mm	m2	38.01
Georgian wired glass		
7 mm cast - not exceeding 3450 x 1900 mm	m2	33.73
6 mm polished - not exceeding 3250 x 1900 mm	m2	71.61
Putty		
linseed oil	25 kg	14.08
metal casement	25 kg	14.80

Prices actually to be paid for materials must be checked against the above basic prices and adjustments made as necessary.

Glazing

Alterations and repairs	Unit	Hours C	Hours L	Labour net	Material net	Price net	Price with 25%
Glass pre-cut to size in openings				£	£	£	£
					VAT not included		
Hack out existing glass from wood and prepare rebates	m	0.20	-	1.79	0.09	1.88	2.35
Hack out existing glass from metal and prepare rebates	m	0.20	-	1.79	0.62	2.41	3.01
Hack out existing, prepare rebates and reglaze with 3m mm float glass.							
To wood with putty in panes							
not exceeding 0.10 m2	m2	5.32	-	47.67	30.06	77.73	97.16
0.10 - 0.50 m2	m2	1.70	-	15.23	26.97	42.20	52.75
0.50 - 1.00 m2	m2	1.24	-	11.11	24.72	35.83	44.79
To metal with metal casement putty in panes							
not exceeding 0.10 m2	m2	5.25	-	47.04	41.33	88.37	110.46
0.10 - 0.50 m2	m2	1.66	-	14.87	29.97	44.84	56.05
0.50 - 1.00 m2	m2	1.21	-	10.84	27.47	38.31	47.89
Hack out existing, prepare rebates and reglaze with 4mm float glass							
To wood with putty in panes							
not exceeding 0.10 m2	m2	5.32	-	47.67	30.06	77.73	97.16
0.10 - 0.50 m2	m2	1.70	-	15.23	26.97	42.20	52.75
0.50 - 1.00 m2	m2	1.24	-	11.11	25.45	36.56	45.70
over 1.00 m2	m2	1.12	-	10.04	24.79	34.83	43.54
To metal with metal casement putty in panes							
not exceeding 0.10 m2	m2	5.25	-	47.04	39.82	86.86	108.58
0.10 - 0.50 m2	m2	1.66	-	14.87	29.97	44.84	56.05
0.50 - 1.00 m2	m2	1.21	-	10.84	27.47	38.31	47.89
over 1.00 m2	m2	1.17	-	10.48	26.84	37.32	46.65
Hack out existing, prepare rebates and reglaze with 4 mm white patterned glass							
To wood with putty in panes							
not exceeding 0.10 m2	m2	5.50	-	49.28	29.90	79.18	98.97
0.10 - 0.50 m2	m2	1.79	-	16.04	26.81	42.85	53.56
0.50 - 1.00 m2	m2	1.33	-	11.92	25.29	37.21	46.51
To metal with metal casement putty in panes							
not exceeding 0.10 m2	m2	5.41	-	48.47	41.17	89.64	112.05
0.10 - 0.50 m2	m2	1.75	-	15.68	29.81	45.49	56.86
0.50 - 1.00 m2	m2	1.28	-	11.47	27.31	38.78	48.48

Glazing

Alterations and repairs	Unit	Hours C	Hours L	Labour net	Material net	Price net	Price with 25%
Glass pre-cut to size in openings				£	£	£	£
					VAT not included		
Hack out existing, prepare rebates and reglaze with 6 mm white patterned glass							
To wood with putty in panes							
not exceeding 0.10 m2	m2	5.58	-	50.00	47.37	97.37	121.71
0.10 - 0.50 m2	m2	1.85	-	16.58	44.28	60.86	76.08
0.50 - 1.00 m2	m2	1.37	-	12.28	42.76	55.04	68.80
over 1.00 m2	m2	1.22	-	10.93	42.10	53.03	66.29
To metal with metal casement putty in panes							
not exceeding 0.10 m2	m2	5.50	-	49.28	58.64	107.92	134.90
0.10 - 0.50 m2	m2	1.79	-	16.04	46.00	62.04	77.55
0.50 - 1.00 m2	m2	1.33	-	11.92	44.78	56.70	70.88
over 1.00 m2	m2	1.19	-	10.66	44.15	54.81	68.51
Hack out existing, prepare rebates and reglaze with 7 mm Georgian wired cast glass							
To wood with putty in panes							
not exceeding 0.10 m2	m2	5.54	-	49.64	40.20	89.84	112.30
0.10 - 0.50 m2	m2	2.08	-	18.64	38.16	56.80	71.00
0.50 - 1.00 m2	m2	1.57	-	14.07	37.58	51.65	64.56
over 1.00 m2	m2	1.35	-	12.10	37.44	49.54	61.92
To metal with metal casement putty in panes							
not exceeding 0.10 m2	m2	5.45	-	48.83	51.61	100.44	125.55
0.10 - 0.50 m2	m2	2.01	-	18.01	41.26	59.27	74.09
0.50 - 1.00 m2	m2	1.50	-	13.44	39.64	53.08	66.35
over 1.00 m2	m2	1.31	-	11.74	39.51	51.25	64.06
Hack out existing, prepare rebates and reglaze with 6 mm Georgian wire polished glass							
To wood with putty in panes							
not exceeding 0.10 m2	m2	6.09	-	54.57	80.06	134.63	168.29
0.10 - 0.50 m2	m2	2.52	-	22.58	78.31	100.89	126.11
0.50 - 1.00 m2	m2	1.96	-	17.56	78.17	95.73	119.66
over 1.00 m2	m2	1.63	-	14.60	78.03	92.63	115.79
To metal with metal casement putty in panes							
not exceeding 0.10 m2	m2	5.96	-	53.40	92.20	145.60	182.00
0.10 - 0.50 m2	m2	2.41	-	21.59	81.85	103.44	129.30
0.50 - 1.00 m2	m2	1.87	-	16.76	80.23	96.99	121.24
over 1.00 m2	m2	1.56	-	13.98	79.97	93.95	117.44

Repainting and redecorating

Preamble

"Labour net" figures include allowances for all costs incidental to the employment of labour.

"Materials net" figures include for all costs of materials including an allowance for waste except where specifically stated and an extra allowance to cover the cost of wear on brushes.

"Price net" figures are the totals of the "Labour net" and "Materials net" figures. Prices are for a builder employing his own labour; according to the amount and nature of the work involved, it may well be possible to secure more advantageous prices from specialist sub-contractors.

Although not specifically mentioned in the various descriptions, allowance has been made for rubbing down, stopping and general preparation to permit high class work.

Where estimates are prepared without bills of quantities, it is essential that painting work be fully measured, extra care should therefore be taken to ensure that full account is taken of narrow widths and of the extra girth of moulded work, edges and returns.

The term "prepare" has been used in this section to cover stopping holes and cracks, rubbing down and making good surfaces ready for redecoration.

Prices do not include any allowance for scaffolding, ladders or other plant necessary to reach the work. The "Preliminaries" section includes prices for scaffolding which must be considered and allowance included to suit the particular circumstances of a tender.

Repainting and redecorating

Alterations and repairs	Unit	Price	Cover allowed per coat Smooth Surfaces	Textured Surfaces	Rough Surfaces
Basic prices for materials		£	m2	m2	m2
Bituminous paint	25 litre	38.86	250	-	-
Cement paint	5 litre	13.90	-	200	100
Eggshell paint	5 litre	22.40	65	50	-
Emulsion paint - vinyl matt	5 litre	14.94	60	45	20
Emulsion paint - vinyl silk	5 litre	15.45	60	45	20
Fungicidal solution	5 litre	11.45	150	-	75
Gloss finish	5 litre	19.25	60	45	-
Linseed oil	5 litre	13.46	80	-	-
Masonry oil paint	5 litre	20.80	-	45	20
Oil varnish stain floor seal	5 litre	19.12	60	-	-
Plaster stabilising solution	5 litre	13.05	30	-	-
Polyurethane					
floor seal	5 litre	30.17	65	-	-
varnish	5 litre	24.59	65	-	-
Silicone waterproofing solution	5 litre	32.10	-	25	10
Stabilising solution for cement or stone paints	5 litre	13.05	-	25	10
Stone paint					
finish	5 litre	13.73	-	10	5
emulsion	5 litre	17.00	-	45	20
Textured cement paint	5 litre	17.17	-	30	15
Undercoat	5 litre	19.25	60	45	-
Water repellent decorative timber dressing	5 litre	11.65	75	-	-
Wood primer	5 litre	23.11	50	-	-

Prices actually to be paid for materials must be checked against the above basic prices and
adjustments made as necessary.

Repainting and redecorating

Alterations and repairs	Unit	Hours C	Hours L	Labour net	Material net	Price net	Price with 25%
Generally				£	£	£	£
					VAT not included		
Walls and ceilings internally							
Wash off ceiling distemper and prepare for redecoration	m2	0.25	-	2.24	0.06	2.30	2.88
Wash walls or ceilings previously painted with washable distemper, emulsion paint or oil paint and prepare for redecoration	m2	0.13	-	1.16	0.06	1.22	1.52
Wash and leather down painted walls as finished work	m2	0.10	-	0.90	-	0.90	1.13
Strip vinyl surface from wallpaper (no making good or rubbing down)							
areas	m2	0.07	-	0.63	-	0.63	0.79
per roll	roll	0.33	-	2.96	-	2.96	3.70
Strip one layer of average quality wallpaper and prepare for decoration							
areas	m2	0.25	-	2.24	0.06	2.30	2.88
per roll	roll	1.13	-	10.12	0.31	10.43	13.04
Strip one layer of glazed wallpaper and prepare for redecoration							
areas	m2	0.38	-	3.40	0.06	3.46	4.33
per roll	roll	1.70	-	15.23	0.31	15.54	19.43
Make good setting or skimming coat of plaster in small areas pulled away in stripping paper	m2	1.75	-	15.68	0.48	16.16	20.20
Brush down and apply one coat of stabilising solution on flaking or friable plastered surfaces	m2	0.25	-	2.24	0.13	2.37	2.96
Two coats of vinyl matt emulsion paint on walls and ceilings							
Surfaces of							
smooth plaster	m2	0.26	-	2.33	0.50	2.83	3.54
rendering, fair face or similar textured surface	m2	0.29	-	2.60	0.53	3.13	3.91
Two coats of vinyl silk emulsion paint on walls and ceilings							
Surfaces of							
smooth plaster	m2	0.26	-	2.33	0.50	2.83	3.54
rendering, fair face or similar textured surface	m2	0.28	-	2.51	0.85	3.36	4.20

Repainting and redecorating

Alterations and repairs	Unit	Hours C	Hours L	Labour net	Material net	Price net	Price with 25%
Generally				£	£	£	£
					VAT not included		
Walls and ceilings internally (*continued*)							
Two coats of eggshell paint on walls and ceilings							
Surfaces of							
smooth plaster	m2	0.40	-	3.58	0.83	4.41	5.51
rendering, fair face or similar textured surface	m2	0.46	-	4.12	1.08	5.20	6.50
One undercoat and one coat of gloss finish paint on walls and ceilings							
Surfaces of							
smooth plaster	m2	0.50	-	4.48	0.62	5.10	6.38
rendering, fair face or similar textured surface	m2	0.58	-	5.20	0.78	5.98	7.47
Two undercoats and one coat of gloss finish paint on walls and ceilings							
Surfaces of							
smooth plaster	m2	0.67	-	6.00	0.93	6.93	8.66
rendering, fair face or similar textured surface	m2	0.77	-	6.90	1.17	8.07	10.09
Paperhanging							
For Paperhanging, see "Painting and decorating" in "New work"							
Floors internally							
Wash and apply polyurethane floor seal							
Previously sealed floors							
one coat	m2	0.23	-	2.06	0.65	2.71	3.39
two coats	m2	0.41	-	3.67	1.30	4.97	6.21
Wash and apply oil varnish stain floor seal							
Floors							
one coat	m2	0.23	-	2.06	0.42	2.48	3.10
two coats	m2	0.40	-	3.58	0.84	4.42	5.53

Repainting and redecorating

Alterations and repairs	Unit	Hours C	Hours L	Labour net	Material net	Price net	Price with 25%
Generally				£	£	£	£
					VAT not included		
Woodwork internally							
Wash and leather down previously painted joinery as finished work	m2	0.12	-	1.08	-	1.08	1.35
Take off set of average door furniture for redecoration and refix	each	0.20	-	1.79	-	1.79	2.24
Wash prepare and paint one undercoat and one coat of gloss finish paint on previously painted joinery							
General surfaces							
over 300 mm girth	m2	0.70	-	6.27	0.55	6.82	8.53
not exceeding 150 mm girth	m	0.18	-	1.61	0.09	1.70	2.13
150 - 300 mm girth	m	0.28	-	2.51	0.20	2.71	3.39
not exceeding 0.50 m2	each	0.55	-	4.93	0.28	5.21	6.51
Glazed doors and screens in panes							
small - not exceeding 0.10m2	m2	1.55	-	13.89	0.44	14.33	17.91
medium - 0.10 - 0.50 m2	m2	0.95	-	8.51	0.36	8.87	11.09
large - 0.50 - 1.00 m2	m2	0.85	-	7.62	0.31	7.93	9.91
extra large - over 1.00 m2	m2	0.77	-	6.90	0.22	7.12	8.90
Windows (measured flat overall) in panes							
small - not exceeding 0.10 m2	m2	1.70	-	15.23	0.44	15.67	19.59
medium - 0.10 - 0.50 m2	m2	1.10	-	9.86	0.37	10.23	12.79
large - 0.50 - 1.00 m2	m2	0.75	-	6.72	0.31	7.03	8.79
extra large - over 1.00 m2	m2	0.69	-	6.18	0.22	6.40	8.00
Edges of opening casements	m	0.10	-	0.90	0.04	0.94	1.18
Wash, prepare and paint two undercoats and one coat of gloss finish paint on previously painted joinery							
General surfaces							
over 300 mm girth	m2	0.95	-	8.51	0.82	9.33	11.66
not exceeding 150 mm girth	m	0.24	-	2.15	0.13	2.28	2.85
150 - 300 mm girth	m	0.38	-	3.40	0.25	3.65	4.56
not exceeding 0.50 m2	each	0.70	-	6.27	0.41	6.68	8.35
Glazed doors and screens in panes							
small - not exceeding 0.10 m2	m2	2.10	-	18.82	0.67	19.49	24.36
medium - 0.10 - 0.50 m2	m2	1.25	-	11.20	0.55	11.75	14.69
large - 0.50 - 1.00 m2	m2	1.12	-	10.04	0.46	10.50	13.13
extra large - over 1.00 m2	m2	1.00	-	8.96	0.33	9.29	11.61
Windows (measured flat overall) in panes							
small - not exceeding 0.10 m2	m2	2.30	-	20.61	0.67	21.28	26.60
medium - 0.10 - 0.50 m2	m2	1.45	-	12.99	0.55	13.54	16.93
large - 0.50 - 1.00 m2	m2	1.05	-	9.41	0.46	9.87	12.34
extra large - over 1.00 m2	m2	0.85	-	7.62	0.33	7.95	9.94

Repainting and redecorating

Alterations and repairs	Unit	Hours C	Hours L	Labour net	Material net	Price net	Price with 25%
Generally				£	£	£	£
					VAT not included		
Woodwork internally (continued)							
Wash, prepare and paint two undercoats and one coat of gloss finish paint on previously painted joinery (continued)							
Edges of opening casements	m	0.13	-	1.16	0.05	1.21	1.51
Wash, prepare and apply two coats of polyurethane varnish on previously varnished joinery							
General surfaces							
over 300 mm girth	m2	0.65	-	5.82	1.32	7.14	8.93
not exceeding 150 mm girth	m	0.16	-	1.43	0.20	1.63	2.04
150 - 300 mm girth	m	0.26	-	2.33	0.38	2.71	3.39
not exceeding 0.50 m2	m2	0.50	-	4.48	0.66	5.14	6.42
Metalwork internally							
Wirebrush, prepare and paint one undercoat and one coat of gloss finish on previously painted metalwork							
General surfaces							
over 300 mm girth	m2	0.67	-	6.00	0.57	6.57	8.21
not exceeding 150 mm girth	m	0.17	-	1.52	0.09	1.61	2.01
150 - 300 mm girth	m	0.27	-	2.42	0.27	2.69	3.36
not exceeding 0.50 m2	each	0.50	-	4.48	0.29	4.77	5.96
Glazed doors, screens and windows (measured flat overall) in panes							
small - not exceeding 0.10 m2	m2	1.65	-	14.78	0.41	15.19	18.99
medium - 0.10 - 0.50 m2	m2	1.05	-	9.41	0.40	9.81	12.26
large - 0.50 - 1.00 m2	m	0.75	-	6.72	0.34	7.06	8.82
extra large - over 1.00 m2	m2	0.60	-	5.38	0.25	5.63	7.04
Edges of opening casements	m	0.09	-	0.79	0.04	0.83	1.04
Structural members and pipes							
over 300 mm girth	m2	0.75	-	6.72	0.57	7.29	9.11
not exceeding 150 mm girth	m	0.19	-	1.70	0.09	1.79	2.24
150 - 300 mm girth	m	0.30	-	2.69	0.18	2.87	3.59
Wirebrush, prepare and paint two undercoats and one coat of gloss finish on previously painted metalwork							
General surfaces							
over 300 mm girth	m2	0.90	-	8.06	0.85	8.91	11.14
not exceeding 150 mm girth	m	0.23	-	2.06	0.14	2.20	2.75
150 - 300 mm girth	m	0.36	-	3.23	0.27	3.50	4.38
not exceeding 0.50 m2	each	0.70	-	6.27	0.43	6.70	8.38

Repainting and redecorating

Alterations and repairs	Unit	Hours C	Hours L	Labour net	Material net	Price net	Price with 25%
Generally				£	£	£	£
					VAT not included		
Metalwork internally (*continued*)							
Wirebrush, prepare and paint two undercoats and one coat of gloss finish on previously painted metalwork (*continued*)							
Glazed doors, screens and windows (measured flat overall) in panes							
small - not exceeding 0.10 m2	m2	2.20	-	19.71	0.63	20.34	25.43
medium - 0.10 - 0.50 m2	m2	1.40	-	12.54	0.59	13.13	16.41
large - 0.50 - 1.00 m2	m2	1.00	-	8.96	0.49	9.45	11.81
extra large - over 1.00 m2	m2	0.80	-	7.17	0.37	7.54	9.43
Edges of opening casements	m	0.14	-	1.25	0.06	1.31	1.64
Structural members and pipes							
over 300 mm girth	m2	1.00	-	8.96	0.60	9.56	11.95
not exceeding 150 mm girth	m	0.25	-	2.24	0.31	2.55	3.19
150 - 300 mm girth	m	0.40	-	3.58	0.19	3.77	4.71
Walls externally							
Wash, prepare and paint two coats of emulsion paint on previously painted walls							
Rendering, fair face or similar textured surfaces	m2	0.50	-	4.48	0.34	4.82	6.03
Roughcast surfaces	m2	0.70	-	6.27	0.89	7.16	8.95
Wash, prepare and two coats of masonry oil paint on previously painted walls							
Rendering, fair face or similar textured surfaces	m2	0.60	-	5.38	0.98	6.36	7.95
Roughcast surfaces	m2	0.85	-	7.62	1.40	9.02	11.28
Brush down, prepare and paint one coat of cement paint on previously painted walls							
Rendering, fair face or similar textured surfaces	m2	0.30	-	2.69	0.18	2.87	3.59
Roughcast surfaces	m2	0.40	-	3.58	0.48	4.06	5.08
Brush down, prepare and paint two coats of cement paint on previously painted walls							
Rendering, fair face or similar textured surfaces	m2	0.50	-	4.48	0.33	4.81	6.01
Roughcast surfaces	m2	0.70	-	6.27	0.83	7.10	8.88

Repainting and redecorating

Alterations and repairs	Unit	Hours C	Hours L	Labour net	Material net	Price net	Price with 25%
Generally				£	£	£	£
					VAT not included		
Walls externally (continued)							
Brush down, prepare and paint one coat of textured cement paint on previously painted walls							
Rendering, fair face or similar textured surfaces	m2	0.45	-	4.03	0.21	4.24	5.30
Roughcast surfaces	m2	0.55	-	4.93	0.95	5.88	7.35
Brush down, prepare and paint two coats of textured cement paint on previously painted walls							
Rendering, fair face or similar textured surfaces	m2	0.55	-	4.93	0.66	5.59	6.99
Roughcast surfaces	m2	0.80	-	7.17	0.95	8.12	10.15
Brush down, prepare and paint one coat of stone paint on previously painted walls							
Rendering, fair face or similar textured surfaces	m2	0.35	-	3.14	0.33	3.47	4.34
Roughcast surfaces	m2	0.45	-	4.03	0.82	4.85	6.06
One coat of emulsion paint to redecorated stone paint finish on walls							
Rendering, fair face or similar textured surfaces	m2	0.20	-	1.79	0.31	2.10	2.63
Roughcast surfaces	m2	0.28	-	2.51	0.85	3.36	4.20
One coat of fungicidal solution scrubbed into surfaces affected by algae etc and, after 24 hours, washed off							
Rendering, fair face or similar textured surfaces	m2	0.35	-	3.14	0.14	3.28	4.10
Roughcast surfaces	m2	0.45	-	4.03	0.19	4.22	5.28
One coat stabilising solution on flaking or friable surfaces							
Rendering, fair face or similar textured surfaces	m2	0.20	-	1.79	0.16	1.95	2.44
Roughcast surfaces	m2	0.28	-	2.51	0.22	2.73	3.41
Brush down and apply one coat of silicone waterproofing solution on							
Rendering, fair face or similar textured surfaces	m2	0.25	-	2.24	0.35	2.59	3.24
Roughcast surfaces	m2	0.33	-	2.96	0.49	3.45	4.31

Repainting and redecorating

Alterations and repairs	Unit	Hours C	Hours L	Labour net	Material net	Price net	Price with 25%
Generally				£	£	£	£
					VAT not included		

Woodwork externally

Wash prepare and paint one undercoat and one coat of gloss finish paint on previously painted joinery

	Unit	Hours C	Hours L	Labour net	Material net	Price net	Price with 25%
General surfaces							
over 300 mm girth	m2	0.80	-	7.17	0.78	7.95	9.94
not exceeding 150 mm girth	m	0.20	-	1.79	0.13	1.92	2.40
150 - 300 mm girth	m	0.32	-	2.87	0.24	3.11	3.89
not exceeding 0.50 m2	each	0.60	-	5.38	0.39	5.77	7.21
Glazed doors and screens in panes							
small - not exceeding 0.10 m2	m2	1.80	-	16.13	0.78	16.91	21.14
medium - 0.10 - 0.50 m2	m2	1.10	-	9.86	0.81	10.67	13.34
large - 0.50 - 1.00 m2	m	0.70	-	6.27	0.78	7.05	8.81
extra large - over 1.00 m2	m2	0.50	-	4.48	0.78	5.26	6.58
Windows (measured flat overall) in panes							
small - not exceeding 0.10 m2	m2	1.95	-	17.47	0.78	18.25	22.81
medium - 0.10 - 0.50 m2	m2	1.25	-	11.20	0.78	11.98	14.98
large - 0.50 - 1.00 m2	m2	0.85	-	7.62	0.81	8.43	10.54
extra large - over 1.00 m2	m2	0.70	-	6.27	0.78	7.05	8.81
Edges of opening casements	m	0.08	-	0.72	0.09	0.81	1.01

Wash, prepare and paint two undercoats and one coat of gloss finish paint on previously painted joinery

	Unit	Hours C	Hours L	Labour net	Material net	Price net	Price with 25%
General surfaces							
over 300 mm girth	m2	1.10	-	9.86	1.16	11.02	13.78
not exceeding 150 mm girth	m	0.28	-	2.51	0.17	2.68	3.35
150 - 300 mm girth	m	0.44	-	3.94	0.39	4.33	5.41
not exceeding 0.50 m2	each	0.85	-	7.62	0.56	8.18	10.23
Glazed doors and screens in panes							
small - not exceeding 0.10 m2	m2	2.45	-	21.95	1.16	23.11	28.89
medium - 0.10 - 0.50 m2	m2	1.45	-	12.99	1.16	14.15	17.69
large - 0.50 - 1.00 m2	m2	1.00	-	8.96	1.16	10.12	12.65
extra large - over 1.00 m2	m2	0.75	-	6.72	1.16	7.88	9.85
Windows (measured flat overall) in panes							
small - not exceeding 0.10 m2	m2	2.70	-	24.19	1.16	25.35	31.69
medium - 0.10 - 0.50 m2	m2	1.70	-	15.23	1.16	16.39	20.49
large - 0.50 - 1.00 m2	m2	1.25	-	11.20	1.16	12.36	15.45
extra large - over 1.00 m2	m2	1.00	-	8.96	0.56	9.52	11.90
Edges of opening casements	m	0.11	-	0.99	0.17	1.16	1.45

Repainting and redecorating

Alterations and repairs	Unit	Hours C	Hours L	Labour net	Material net	Price net	Price with 25%
Generally				£	£	£	£
					VAT not included		

Woodwork externally (*continued*)

Burn off, rub down, knot, prime, stop and paint one undercoat and one coat of gloss finish paint on previously painted joinery

Surfaces

over 300 mm girth	m2	2.00	-	17.92	1.38	19.30	24.13
not exceeding 150 mm girth	m	0.50	-	4.48	0.19	4.67	5.84
150 - 300 mm girth	m	0.80	-	7.17	0.42	7.59	9.49
not exceeding 0.50 m2	each	1.50	-	13.44	0.68	14.12	17.65

Burn off, rub down, knot, stop and paint two undercoats and one coat of gloss finish paint on previously painted joinery

over 300 mm girth	m2	2.25	-	20.16	2.44	22.60	28.25
not exceeding 150 mm girth	m	0.55	-	4.93	0.23	5.16	6.45
150 - 300 mm girth	m	0.90	-	8.06	0.55	8.61	10.76
not exceeding 0.50 m2	each	1.75	-	15.68	0.85	16.53	20.66

Clean and apply two coats of boiled linseed oil rubbed in to hardwood joinery

Surfaces

over 300 mm girth	m2	0.45	-	4.03	1.03	5.06	6.33
not exceeding 150 mm girth	m	0.11	-	0.99	0.15	1.14	1.43
150 - 300 mm girth	m	0.18	-	1.61	0.30	1.91	2.39
not exceeding 0.50 m2	each	0.35	-	3.14	0.50	3.64	4.55

Clean and apply two coats of water repellent decorative timber dressing on cedar boarding etc

Surfaces

over 300 mm girth	m2	0.55	-	4.93	0.32	5.25	6.56
not exceeding 150 mm girth	m	0.14	-	1.25	0.03	1.28	1.60
150 - 300 mm girth	m	0.22	-	1.97	0.09	2.06	2.58
not exceeding 0.50 m2	each	0.45	-	4.03	0.16	4.19	5.24

Clean, prepare and apply two coats of polyurethane varnish on previously painted joinery

General surfaces

over 300 mm girth	m2	0.75	-	6.72	1.32	8.04	10.05
not exceeding 150 mm girth	m	0.19	-	1.70	0.21	1.91	2.39
150 - 300 mm girth	m	0.30	-	2.69	0.42	3.11	3.89
not exceeding 0.50 m2	each	0.60	-	5.38	0.63	6.01	7.51

Repainting and redecorating

Alterations and repairs	Unit	Hours C	Hours L	Labour net	Material net	Price net	Price with 25%
Generally				£	£	£	£
					VAT not included		
Metalwork externally							
Clean out average 125 mm eaves gutters	m	0.15	-	1.34	-	1.34	1.68
Wirebrush, prepare and paint one undercoat and one coat of gloss finish on previously painted metalwork							
General surfaces							
over 300 mm girth	m2	0.77	-	6.90	0.87	7.77	9.71
not exceeding 150 mm girth	m	0.19	-	1.70	0.09	1.79	2.24
150 - 300 mm girth	m	0.31	-	2.78	0.27	3.05	3.81
not exceeding 0.50 m2	each	0.60	-	5.38	0.44	5.82	7.28
Glazed doors, screens and windows (measured flat overall) in panes							
small - not exceeding 0.10 m2	m2	1.90	-	17.02	0.87	17.89	22.36
medium - 0.10 - 0.50 m2	m2	1.20	-	10.75	0.87	11.62	14.53
large - 0.50 - 1.00 m2	m2	0.85	-	7.62	0.87	8.49	10.61
extra large - over 1.00 m2	m2	0.70	-	6.27	0.87	7.14	8.93
Edges of opening casements	m	0.08	-	0.72	0.09	0.81	1.01
Structural members, gutters and pipes							
over 300 mm girth	m2	0.85	-	7.62	0.87	8.49	10.61
not exceeding 150 mm girth	m	0.21	-	1.88	0.09	1.97	2.46
150 - 300 mm girth	m	0.34	-	3.05	0.27	3.32	4.15
Perforated plate landings and treads - each side	m2	0.95	-	8.51	0.87	9.38	11.73
Ornamental railings and gates - each side	m2	1.30	-	11.65	0.87	12.52	15.65
Wirebrush, prepare and paint two undercoats and one coat of gloss finish on previously painted metalwork							
General surfaces							
over 300 mm girth	m2	1.05	-	9.41	1.30	10.71	13.39
not exceeding 150 mm girth	m	0.26	-	2.33	0.18	2.51	3.14
150 - 300 mm girth	m	0.42	-	3.76	0.40	4.16	5.20
not exceeding 0.50 m2	each	0.80	-	7.17	0.65	7.82	9.78
Glazed doors, screens and windows (measured flat overall) in panes							
small - not exceeding 0.10 m2	m2	2.60	-	23.30	1.30	24.60	30.75
medium - 0.10 - 0.50 m2	m2	1.65	-	14.78	1.30	16.08	20.10
large - 0.50 - 1.00 m2	m2	1.20	-	10.75	1.30	12.05	15.06
extra large - over 1.00 m2	m2	0.95	-	8.51	1.30	9.81	12.26
Edges of opening casements	m	0.11	-	0.99	0.18	1.17	1.46

Repainting and redecorating

Alterations and repairs	Unit	Hours C	Hours L	Labour net	Material net	Price net	Price with 25%
Generally				£	£	£	£
					VAT not included		
Metalwork externally (*continued*)							
Wirebrush, prepare and paint two undercoats and one coat of gloss finish on previously painted metalwork (*continued*)							
Structural members, gutters and pipes							
over 300 mm girth	m2	1.15	-	10.30	1.30	11.60	14.50
not exceeding 150 mm girth	m	0.29	-	2.60	0.18	2.78	3.48
150 - 300 mm girth	m	0.46	-	4.12	0.40	4.52	5.65
Perforated plate landings and treads - each side	m2	1.30	-	11.65	1.30	12.95	16.19
Ornamental railings and gates - each side	m2	1.75	-	15.68	1.30	16.98	21.23
Clean and apply one coat of bituminous paint on previously painted metalwork							
General surfaces							
over 300 mm girth	m2	0.35	-	3.14	0.43	3.57	4.46
not exceeding 150 mm girth	m	0.09	-	0.81	0.06	0.87	1.09
150 - 300 mm girth	m	0.14	-	1.25	0.14	1.39	1.74
not exceeding 0.50 m2	each	0.30	-	2.69	0.20	2.89	3.61
Clean and apply two coats of bituminous paint on metalwork							
General surfaces							
over 300 mm girth	m2	0.55	-	4.93	0.86	5.79	7.24
not exceeding 150 mm girth	m	0.14	-	1.25	0.14	1.39	1.74
150 - 300 mm girth	m	0.22	-	1.97	0.26	2.23	2.79
not exceeding 0.50 m2	each	0.45	-	4.03	0.43	4.46	5.58

Repainting and redecorating

Alterations and repairs	Unit	Hours C	Hours L	Labour net	Material net	Price net	Price with 25%
French polishing				£	£	£	£
					VAT not included		
Wash, prepare and repolish previously polished hardwood joinery							
General surfaces							
over 300 mm girth	m2	1.20	-	10.75	1.11	11.86	14.82
not exceeding 150 mm girth	m	0.27	-	2.42	0.17	2.59	3.24
150 - 300 mm girth	m	0.45	-	4.03	0.33	4.36	5.45
not exceeding 0.50 m2	each	0.90	-	8.06	0.57	8.63	10.79
Glazed doors and screens in panes							
small - not exceeding 0.10 m2	m2	2.67	-	23.92	0.85	24.77	30.96
medium - 0.10 - 0.50 m2	m2	1.60	-	14.34	0.70	15.04	18.80
large - 0.50 - 1.00 m2	m2	1.07	-	9.59	0.57	10.16	12.70
extra large - over 1.00 m2	m2	0.80	-	7.17	0.39	7.56	9.45
Windows (measured flat overall) in panes							
small - not exceeding 0.10 m2	m2	2.94	-	26.34	0.85	27.19	33.99
medium - 0.10 - 0.50 m2	m2	1.87	-	16.76	0.70	17.46	21.82
large - 0.50 - 1.00 m2	m2	1.34	-	12.01	0.57	12.58	15.73
extra large - over 1.00 m2	m2	1.07	-	9.59	0.39	9.98	12.48
Edges of opening casements	m	0.14	-	1.25	0.06	1.31	1.64
Strip, prepare, seal, body in and fully French polish previously polished hardwood joinery							
General surfaces							
over 300 mm girth	m2	3.60	-	32.26	1.87	34.13	42.66
not exceeding 150 mm girth	m	0.81	-	7.26	0.31	7.57	9.46
150 - 300 mm girth	m	1.35	-	12.10	0.58	12.68	15.85
not exceeding 0.50 m2	each	2.70	-	24.19	0.52	24.71	30.89
Glazed doors and screens in panes							
small - not exceeding 0.10 m2	m2	8.00	-	71.68	1.47	73.15	91.44
medium - 0.10 - 0.50 m2	m2	4.80	-	43.01	1.19	44.20	55.25
large - 0.50 - 1.00 m2	m2	3.20	-	28.67	0.98	29.65	37.06
extra large - over 1.00 m2	m2	2.40	-	21.50	0.69	22.19	27.74
Windows (measured flat overall) in panes							
small - not exceeding 0.10 m2	m2	8.80	-	78.85	1.47	80.32	100.40
medium - 0.10 - 0.50 m2	m2	5.60	-	50.18	1.19	51.37	64.21
large - 0.50 - 1.00 m2	m2	4.00	-	35.84	0.98	36.82	46.02
extra large - over 1.00 m2	m2	3.20	-	28.67	0.69	29.36	36.70
Edges of opening casements	m	0.40	-	3.58	0.10	3.68	4.60

Repainting and redecorating

Alterations and repairs	Unit	Hours C	Hours L	Labour net	Material net	Price net	Price with 25%
French polishing				£	£	£	£
					VAT not included		
Strip, prepare, stain, seal, body in and fully French polish previously polished hardwood joinery							
General surfaces							
over 300 mm girth	m2	3.75	-	33.60	2.52	36.12	45.15
not exceeding 150 mm girth	m	0.84	-	7.53	0.42	7.95	9.94
150 - 300 mm girth	m	1.40	-	12.54	0.78	13.32	16.65
not exceeding 0.50 m2	each	2.80	-	25.09	1.26	26.35	32.94
Glazed doors and screens in panes							
small - not exceeding 0.10 m2	m2	8.33	-	74.64	1.98	76.62	95.78
medium - 0.10 - 0.50 m2	m2	5.00	-	44.80	1.61	46.41	58.01
large - 0.50 - 1.00 m2	m2	3.33	-	29.84	1.32	31.16	38.95
extra large - over 1.00 m2	m2	2.50	-	22.40	0.93	23.33	29.16
Windows (measured flat overall) in panes							
small - not exceeding 0.10 m2	m2	9.16	-	82.07	1.98	84.05	105.06
medium - 0.10 - 0.50 m2	m2	5.83	-	52.24	1.61	53.85	67.31
large - 0.50 - 1.00 m2	m2	4.16	-	37.27	1.32	38.59	48.24
extra large - over 1.00 m2	m2	3.33	-	29.84	0.93	30.77	38.46
Edges of opening casements	m	0.42	-	3.76	0.14	3.90	4.88

Repainting and redecorating

Alterations and repairs	Unit	Hours C	Hours L	Labour net	Material net	Price net	Price with 25%
Signwriting				£	£	£	£
					VAT not included		
Writing in one coat of oil paint on painted backgrounds							
Plain letters or numerals							
up to 50 mm high	each	0.14	-	1.25	-	1.25	1.56
per additional 25 mm high	each	0.06	-	0.54	-	0.54	0.68
Plain letters or numerals with shading							
up to 50 mm high	each	0.20	-	1.79	0.01	1.80	2.25
per additional 25 mm high	each	0.08	-	0.72	0.01	0.73	0.91
Plain letters or numerals with outlining							
up to 50 mm high	each	0.28	-	2.51	0.01	2.52	3.15
per additional 25 mm high	each	0.12	-	1.08	0.01	1.09	1.36
Ornamental letters or numerals							
up to 50 mm high	each	0.21	-	1.88	-	1.88	2.35
per additional 25 mm high	each	0.09	-	0.81	-	0.81	1.01
Ornamental letters or numerals with shading							
up to 50 mm high	each	0.29	-	2.60	0.01	2.61	3.26
per additional 25 mm high	each	0.13	-	1.16	0.01	1.17	1.46
Ornamental letters or numerals with outlining							
up to 50 mm high	each	0.42	-	3.76	0.01	3.77	4.71
per additional 25 mm high	each	0.18	-	1.61	0.01	1.62	2.02
Commas, hyphens or stops	each	0.03	-	0.27	-	0.27	0.34
Direction arrows 250 mm long	each	0.40	-	3.58	0.01	3.59	4.49
Writing in two coats of oil paint on painted backgrounds							
Plain letter or numerals							
up to 50 mm high	each	0.21	-	1.88	0.01	1.89	2.36
per additional 25 mm high	each	0.09	-	0.81	0.01	0.82	1.02
Plain letters or numerals with shading							
up to 50 mm high	each	0.29	-	2.60	0.01	2.61	3.26
per additional 25 mm high	each	0.13	-	1.16	0.01	1.17	1.46
Plain letters or numerals with outlining							
up to 50 mm high	each	0.42	-	3.76	0.01	3.77	4.71
per additional 25 mm high	each	0.18	-	1.61	0.01	1.62	2.02
Ornamental letters or numerals							
up to 50 mm high	each	0.32	-	2.87	0.01	2.88	3.60
per additional 25 mm high	each	0.14	-	1.25	0.01	1.26	1.58
Ornamental letters or numerals with shading							
up to 50 mm high	each	0.44	-	3.94	0.01	3.95	4.94
per additional 25 mm high	each	0.19	-	1.70	0.01	1.71	2.14

Repainting and redecorating

Alterations and repairs	Unit	Hours C	Hours L	Labour net	Material net	Price net	Price with 25%
Signwriting				£	£	£	£
					VAT not included		
Writing in two coats of oil paint on painted backgrounds (*continued*)							
Ornamental letters or numerals with outlining							
up to 50 mm high	each	0.63	-	5.64	0.01	5.65	7.06
per additional 25 mm high	each	0.27	-	2.42	0.01	2.43	3.04
Commas, hyphens or stops	each	0.05	-	0.45	0.01	0.46	0.57
Direction arrows 250 mm long	each	0.60	-	5.38	0.02	5.40	6.75
Writing in two coats of oil paint in reverse on glass and one coat of protective varnish							
Plain letters or numerals							
up to 50 mm high	each	0.41	-	3.67	0.02	3.69	4.61
per additional 25 mm high	each	0.17	-	1.52	0.02	1.54	1.93
Plain letters or numerals with shading							
up to 50 mm high	each	0.57	-	5.11	0.02	5.13	6.41
per additional 25 mm high	each	0.24	-	2.15	0.02	2.17	2.71
Plain letters or numerals with outlining							
up to 50 mm high	each	0.81	-	7.26	0.02	7.28	9.10
per additional 25 mm high	each	0.35	-	3.14	0.02	3.16	3.95
Ornamental letters or numerals							
up to 50 mm high	each	0.61	-	5.47	0.02	5.49	6.86
per additional 25 mm high	each	0.26	-	2.33	0.02	2.35	2.94
Ornamental letters or numerals with shading							
up to 50 mm high	each	0.85	-	7.62	0.02	7.64	9.55
per additional 25 mm high	each	0.37	-	3.32	0.02	3.34	4.17
Ornamental letters or numerals with outlining							
up to 50 mm high	each	1.22	-	10.93	0.02	10.95	13.69
per additional 25 mm high	each	0.52	-	4.66	0.02	4.68	5.85
Commas, hyphens or stops	each	0.09	-	0.81	0.02	0.83	1.04
Direction arrows 250 mm long	each	1.16	-	10.39	0.03	10.42	13.03
Writing in three coats of oil paint in reverse on glass and one coat of protective varnish							
Plain letters or numerals							
up to 50 mm high	each	0.48	-	4.30	0.02	4.32	5.40
per additional 25 mm high	each	0.20	-	1.79	0.02	1.81	2.26
Plain letters or numerals with shading							
up to 50 mm high	each	0.67	-	6.00	0.03	6.03	7.54
per additional 25 mm high	each	0.29	-	2.60	0.03	2.63	3.29

Repainting and redecorating

Alterations and repairs	Unit	Hours C	Hours L	Labour net	Material net	Price net	Price with 25%
Signwriting				£	£	£	£
					VAT not included		
Writing in three coats of oil paint in reverse on glass and one coat of protective varnish (*continued*)							
Plain letters or numerals with outlining							
up to 50 mm high	each	0.95	-	8.51	0.03	8.54	10.68
per additional 25 mm high	each	0.41	-	3.67	0.03	3.70	4.63
Ornamental letters or numerals							
up to 50 mm high	each	0.71	-	6.36	0.02	6.38	7.97
per additional 25 mm high	each	0.31	-	2.78	0.02	2.80	3.50
Ornamental letters or numerals with shading							
up to 50 mm high	each	1.00	-	8.96	0.03	8.99	11.24
per additional 25 mm high	each	0.43	-	3.85	0.03	3.88	4.85
Ornamental letters or numerals with outlining							
up to 50 mm high	each	1.43	-	12.81	0.03	12.84	16.05
per additional 25 mm high	each	0.61	-	5.47	0.03	5.50	6.88
Commas, hyphens or stops	each	0.10	-	0.90	0.02	0.92	1.15
Direction arrows 250 mm long	each	1.36	-	12.19	0.06	12.25	15.31

Drainage

Preamble

"Labour net" figures include allowances for all costs incidental to the employment of labour. Figures for pipework generally are based on the labour costs of three bricklayers working with two labourers as in "Brickwork and blockwork"; for cast iron pipes figures are based on the costs of an advanced plumber working with an apprentice in the third year of training.

"Plant net" figures include for all costs of plant including drivers and operators where applicable.

"Materials net" figures include for all costs of materials including an allowance for waste except where specifically stated.

"Price net" figures are the totals of the "Labour net", "Plant net" and "Materials net" figures. Prices are for a builder employing his own labour; according to the amount and nature of the work involved, it may well be possible to secure more advantageous prices from specialist sub-contractors.

Excavation prices are for work in firm soil. For other soils the following adjustments should be made:

      clay - add 25%
      hard gravel - add 50%
      chalk - add 100 to 150%
      rock - add 300 to 400%

Where excavation items are described as "including disposal of surplus on site", this has been allowed for on the basis of disposal of surplus excavated material in spoil heaps on site average 50 metres from the excavations.

Where excavation items are described as "including removal of surplus from site" this has been allowed for on the basis of removal of surplus excavated material to a tip average 15 kilometres from site.

Figures for concrete are based on the use of a hired 7/5 mixer.

Figures for formwork are based on the assumptions that timber is used and that each use of material requires the full labour content; if the work is repetitive, permitting re-use of made-up sections, some reduction of the figures could be made.

Although not in accordance with the Standard Method of Measurement, figures per square metre have been included for concrete benchings, from which figures for any particular enumerated items can be calculated.

Drainage

Alterations and repairs

	Unit	Price
Basic prices for materials		£
Granular bedding - 10 mm	m3	19.75
Aggregates		
40 mm	m3	19.75
	tonne	13.16
20 mm	m3	19.91
	tonne	13.27
10 mm	m3	20.07
	tonne	13.38
Sand	m3	17.96
	tonne	11.22
Portland cement	tonne	89.04
Vitrified clay plain ended "Supersleve" pipes		
100 mm pipes	m	2.43
100 mm polypropylene couplings	each	1.68
150 mm pipes	m	5.56
150 mm polypropylene couplings	each	3.48
Vitrified clay socketed and flexible jointed "Hepseal" pipes		
100 mm	m	7.23
150 mm	m	9.38
225 mm	m	18.15
Vitrified clay socketed unjointed pipes		
100 mm	m	4.51
150 mm	m	7.78
225 mm	m	15.48
Flexible jointed standard concrete pipes		
150 mm	m	7.50
225 mm	m	8.75
300 mm	m	12.53
Standard concrete pipes with ogee joints		
150 mm	m	3.74
225 mm	m	4.78
300 mm	m	7.19
Timesaver cast iron pipes		
100 mm 3000 mm lengths	m	20.00
100 mm couplings	each	12.80
150 mm 3000 mm lengths	m	37.70
150 mm couplings	each	15.50
PVC-U pipes		
110 mm 3000 mm lengths	m	5.32
160 mm 3000 mm lengths	m	11.29

Prices actually to be paid for materials must be checked against the above basic prices and adjustments made as necessary

Drainage

Alterations and repairs

	Unit	Price
Basic prices for materials		**£**
Clayware field drain pipes		
75 mm	m	1.04
100 mm	m	1.87
150 mm	m	3.99
225 mm	m	10.50
Flexible jointed vitrified clay "Hepline" pipes		
100 mm	m	4.64
150 mm	m	8.44
225 mm	m	15.52
Concrete porous pipes		
150 mm	m	3.69
225 mm	m	4.32
Plain ended unplasticised PVC perforated pipes		
110 mm pipes	m	5.62
110 mm double socket couplers	m	1.69
160 mm pipes	m	10.59
160 mm double socket couplers	m	3.09
Unplasticised PVC perforated flexible corrugated pipes		
80 mm	m	1.51
100 mm	m	2.15

Prices actually to be paid for materials must be checked against the above basic prices and adjustments made as necessary.

Drainage

Alterations and repairs

Pipe trenches

				£	£	£	£	£

VAT not included

Excavation by machine - including disposal of surplus on site average 50 m from excavations

Excavate trenches 450 mm wide; grade bottom; fill in and compact and dispose of surplus on site - earthwork support not included - average depth

	Unit	Hours C	Hours L	Labour net	Plant net	Material net	Price net	Price with 25%
0.50 m	m	-	0.18	1.21	2.59	-	3.80	4.75
0.75 m	m	-	0.28	1.89	4.19	-	6.08	7.60
1.00 m	m	-	0.39	2.63	6.01	-	8.64	10.80
1.25 m	m	-	0.48	3.24	7.41	-	10.65	13.31
1.50 m	m	-	0.59	3.98	9.22	-	13.20	16.50

Excavate trenches 600 mm wide; grade bottom; fill in and compact and dispose of surplus on site - earthwork support not included - average depth

	Unit	Hours C	Hours L	Labour net	Plant net	Material net	Price net	Price with 25%
1.00 m	m	-	0.51	3.44	8.02	-	11.46	14.32
1.25 m	m	-	0.61	4.11	9.59	-	13.70	17.13
1.50 m	m	-	0.78	5.26	12.21	-	17.47	21.84
1.75 m	m	-	0.92	6.20	14.44	-	20.64	25.80
2.00 m	m	-	1.03	6.94	15.85	-	22.79	28.49

Excavate trenches 750 mm wide; grade bottom; fill in and compact and dispose of surplus on site - earthwork support not included - average depth

	Unit	Hours C	Hours L	Labour net	Plant net	Material net	Price net	Price with 25%
1.50 m	m	-	0.99	6.67	15.29	-	21.96	27.45
1.75 m	m	-	1.20	8.09	18.49	-	26.58	33.23
2.00 m	m	-	1.33	8.96	20.50	-	29.46	36.83
2.25 m	m	-	1.57	10.58	24.30	-	34.88	43.60
2.50 m	m	-	1.76	11.86	27.12	-	38.98	48.73
2.75 m	m	-	1.91	12.87	29.51	-	42.38	52.98
3.00 m	m	-	2.11	14.22	32.71	-	46.93	58.66

Excavate trenches 900 mm wide; grade bottom; fill in and compact and dispose of surplus on site - earthwork support not included - average depth

	Unit	Hours C	Hours L	Labour net	Plant net	Material net	Price net	Price with 25%
1.50 m	m	-	1.22	8.22	19.10	-	27.32	34.15
1.75 m	m	-	1.42	9.57	22.01	-	31.58	39.48
2.00 m	m	-	1.63	10.99	25.15	-	36.14	45.17
2.25 m	m	-	1.93	13.01	29.93	-	42.94	53.67
2.50 m	m	-	2.17	14.63	33.59	-	48.22	60.27
2.75 m	m	-	2.34	15.77	36.18	-	51.95	64.94
3.00 m	m	-	2.58	17.39	40.03	-	57.42	71.78

Drainage

Alterations and repairs	Unit	Hours C	Hours L	Labour net	Plant net	Material net	Price net	Price with 25%
Pipe trenches				£	£	£	£	£
					VAT not included			

Excavation by machine - including removal of surplus from site to tip average 15 km from site

Excavate trenches 450 mm wide; grade bottom; fill in and compact and remove surplus from site - earthwork support not included - average depth

0.50 m	m	-	0.18	1.21	5.78	-	6.99	8.74
0.75 m	m	-	0.28	1.89	12.16	-	14.05	17.56
1.00 m	m	-	0.39	2.63	12.20	-	14.83	18.54
1.25 m	m	-	0.48	3.24	13.67	-	16.91	21.14
1.50 m	m	-	0.59	3.98	18.60	-	22.58	28.23

Excavate trenches 600 mm wide; grade bottom; fill in and compact and remove surplus from site - earthwork support not included - average depth

1.00 m	m	-	0.51	3.44	15.30	-	18.74	23.43
1.25 m	m	-	0.61	4.11	19.37	-	23.48	29.35
1.50 m	m	-	0.78	5.26	23.58	-	28.84	36.05
1.75 m	m	-	0.92	6.20	27.41	-	33.61	42.01
2.00 m	m	-	1.03	6.94	31.11	-	38.05	47.56

Excavate trenches 750 mm wide; grade bottom; fill in and compact and remove surplus from site - earthwork support not included - average depth

1.50 m	m	-	0.99	6.67	29.85	-	36.52	45.65
1.75 m	m	-	1.20	8.09	39.85	-	47.94	59.92
2.00 m	m	-	1.33	8.96	39.65	-	48.61	60.76
2.25 m	m	-	1.57	10.58	45.45	-	56.03	70.04
2.50 m	m	-	1.76	11.86	50.56	-	62.42	78.03
2.75 m	m	-	1.91	12.87	55.45	-	68.32	85.40
3.00 m	m	-	2.11	14.22	61.17	-	75.39	94.24

Excavate trenches 900 mm wide; grade bottom; fill in and compact and remove surplus from site - earthwork support not included - average depth

1.50 m	m	-	1.22	8.22	36.16	-	44.38	55.48
1.75 m	m	-	1.42	9.57	42.26	-	51.83	64.79
2.00 m	m	-	1.63	10.99	47.89	-	58.88	73.60
2.25 m	m	-	1.93	13.01	55.87	-	68.88	86.10
2.50 m	m	-	2.17	14.63	62.02	-	76.65	95.81
2.75 m	m	-	2.34	15.77	67.40	-	83.17	103.96
3.00 m	m	-	2.59	17.46	73.95	-	91.41	114.26

Drainage

Alterations and repairs	Unit	Hours C	Hours L	Labour net	Plant net	Material net	Price net	Price with 25%
Pipe trenches				£	£	£	£	£
					VAT not included			

Excavation by hand - including disposal of surplus on site average 50 m from excavations

Excavate trenches 450 mm wide; grade bottom; fill in and compact and dispose of surplus on site - earthwork support not included - average depth

0.50 m	m	-	1.32	8.90	-	-	8.90	11.13
0.75 m	m	-	2.04	13.75	-	-	13.75	17.19
1.00 m	m	-	2.59	17.46	-	-	17.46	21.82
1.25 m	m	-	4.02	27.09	-	-	27.09	33.86
1.50 m	m	-	4.84	32.62	-	-	32.62	40.77

Excavate trenches 600 mm wide; grade bottom; fill in and compact and dispose of surplus on site - earthwork support not included - average depth

1.00 m	m	-	3.47	23.39	-	-	23.39	29.24
1.25 m	m	-	5.36	36.13	-	-	36.13	45.16
1.50 m	m	-	6.44	43.41	-	-	43.41	54.26
1.75 m	m	-	7.48	50.42	-	-	50.42	63.02
2.00 m	m	-	8.53	57.49	-	-	57.49	71.86

Excavate trenches 750 mm wide; grade bottom; fill in and compact and dispose of surplus on site - earthwork support not included - average depth

1.50 m	m	-	8.03	54.12	-	-	54.12	67.65
1.75 m	m	-	9.35	63.02	-	-	63.02	78.78
2.00 m	m	-	10.67	71.92	-	-	71.92	89.90
2.25 m	m	-	14.47	97.53	-	-	97.53	121.91
2.50 m	m	-	16.06	108.24	-	-	108.24	135.30
2.75 m	m	-	17.60	118.62	-	-	118.62	148.28
3.00 m	m	-	19.25	129.75	-	-	129.75	162.19

Excavate trenches 900 mm wide; grade bottom; fill in and compact and dispose of surplus on site - earthwork support not included - average depth

1.50 m	m	-	9.68	65.24	-	-	65.24	81.55
1.75 m	m	-	11.28	76.03	-	-	76.03	95.04
2.00 m	m	-	12.82	86.41	-	-	86.41	108.01
2.25 m	m	-	17.38	117.14	-	-	117.14	146.43
2.50 m	m	-	19.25	129.75	-	-	129.75	162.19
2.75 m	m	-	21.18	142.75	-	-	142.75	178.44
3.00 m	m	-	23.10	155.69	-	-	155.69	194.61

Drainage

Alterations and repairs	Unit	Hours C	Hours L	Labour net	Plant net	Material net	Price net	Price with 25%
Pipe trenches				£	£	£	£	£
					VAT not included			

Excavation by hand - including removal of surplus from site to tip average 15 km from site

Excavate trenches 450 mm wide; grade bottom; fill in and compact and remove surplus from site - earthwork support not included - average depth

	Unit	Hours C	Hours L	Labour net	Plant net	Material net	Price net	Price with 25%
0.50 m	m	-	1.38	9.30	4.38	-	13.68	17.10
0.75 m	m	-	2.04	13.75	6.47	-	20.22	25.27
1.00 m	m	-	2.64	17.79	8.67	-	26.46	33.08
1.25 m	m	-	4.07	27.43	11.06	-	38.49	48.11
1.50 m	m	-	4.90	33.03	13.45	-	46.48	58.10

Excavate trenches 600 mm wide; grade bottom; fill in and compact and remove surplus from site - earthwork support not included - average depth

	Unit	Hours C	Hours L	Labour net	Plant net	Material net	Price net	Price with 25%
1.00 m	m	-	3.52	23.72	11.46	-	35.18	43.98
1.25 m	m	-	5.39	36.33	14.74	-	51.07	63.84
1.50 m	m	-	6.49	43.74	17.53	-	61.27	76.59
1.75 m	m	-	7.54	50.82	20.32	-	71.14	88.92
2.00 m	m	-	8.58	57.83	23.41	-	81.24	101.55

Excavate trenches 750 mm wide; grade bottom; fill in and compact and remove surplus from site - earthwork support not included - average depth

	Unit	Hours C	Hours L	Labour net	Plant net	Material net	Price net	Price with 25%
1.50 m	m	-	8.09	54.53	22.12	-	76.65	95.81
1.75 m	m	-	9.41	63.42	25.80	-	89.22	111.53
2.00 m	m	-	10.73	72.32	29.49	-	101.81	127.26
2.25 m	m	-	14.52	97.86	32.47	-	130.33	162.91
2.50 m	m	-	16.12	108.65	36.16	-	144.81	181.01
2.75 m	m	-	18.15	122.33	39.85	-	162.18	202.72
3.00 m	m	-	19.31	130.15	43.53	-	173.68	217.10

Excavate trenches 900 mm wide; grade bottom; fill in and compact and remove surplus from site - earthwork support not included - average depth

	Unit	Hours C	Hours L	Labour net	Plant net	Material net	Price net	Price with 25%
1.50 m	m	-	9.74	65.65	26.20	-	91.85	114.81
1.75 m	m	-	11.28	76.03	30.58	-	106.61	133.26
2.00 m	m	-	12.87	86.74	34.87	-	121.61	152.01
2.25 m	m	-	17.44	117.55	39.45	-	157.00	196.25
2.50 m	m	-	19.31	130.15	43.53	-	173.68	217.10
2.75 m	m	-	21.23	143.09	48.12	-	191.21	239.01
3.00 m	m	-	23.65	159.40	52.20	-	211.60	264.50

Drainage

Alterations and repairs	Unit	Hours C	Hours L	Labour net	Plant net	Material net	Price net	Price with 25%
Pipe trenches				£	£	£	£	£
					VAT not included			
Breaking up by machine - excluding reinstatement								
Break up surface concrete; for trenches 600 mm wide; average thickness								
100 mm	m	-	0.14	0.94	0.57	-	1.51	1.89
150 mm	m	-	0.19	1.28	0.77	-	2.05	2.56
200 mm	m	-	0.28	1.89	1.14	-	3.03	3.79
Break up reinforced surface concrete; for trenches 600 mm wide; average thickness								
100 mm	m	-	0.20	1.35	0.81	-	2.16	2.70
150 mm	m	-	0.26	1.75	1.06	-	2.81	3.51
200 mm	m	-	0.39	2.63	1.59	-	4.22	5.28
Break up tarmacadam paving; for trenches 600 mm wide; average thickness								
100 mm	m	-	0.07	0.44	0.28	-	0.72	0.90
150 mm	m	-	0.11	0.74	0.45	-	1.19	1.49
200 mm	m	-	0.14	0.94	0.57	-	1.51	1.89
Breaking up by machine - including reinstatement								
Break up surface concrete, for trenches 600 mm wide and reinstate concrete paving and 150 mm hardcore bed, average concrete thickness								
100 mm	m	-	1.30	8.76	0.57	8.00	17.33	21.66
150 mm	m	-	2.37	15.97	1.30	7.41	24.68	30.85
200 mm	m	-	2.86	19.28	1.71	9.48	30.47	38.09
Break up reinforced surface concrete, for trenches 600 mm wide and reinstate concrete paving fabric reinforcement and 150 mm hardcore bed, average concrete thickness								
100 mm	m	0.28	2.03	16.19	1.30	6.06	23.55	29.44
150 mm	m	0.28	2.49	19.29	1.91	8.13	29.33	36.66
200 mm	m	0.28	3.02	22.86	1.59	10.20	34.65	43.31
Break up tarmacadam paving, for trenches 600 mm wide and reinstate tarmacadam and 100 mm hardcore bed, tarmacadam thickness								
100 mm	m	-	1.21	8.16	1.71	71.09	80.96	101.20
150 mm	m	-	1.46	9.84	2.20	105.94	117.98	147.47
200 mm	m	-	1.68	11.32	3.56	140.78	155.66	194.57

Drainage

Alterations and repairs	Unit	Hours C	Hours L	Labour net	Plant net	Material net	Price net	Price with 25%
Pipe trenches				£	£	£	£	£
					VAT not included			

Breaking up by hand - excluding reinstatement

Break up surface concrete, for trenches 600 mm wide, average thickness

	Unit	Hours C	Hours L	Labour net	Plant net	Material net	Price net	Price with 25%
100 mm	m	-	1.30	8.76	-	-	8.76	10.95
150 mm	m	-	1.95	13.14	-	-	13.14	16.43
200 mm	m	-	2.60	17.52	-	-	17.52	21.90

Break up reinforced surface concrete, for trenches 600 mm wide, average thickness

100 mm	m	-	1.95	13.14	-	-	13.14	16.43
150 mm	m	-	2.93	19.75	-	-	19.75	24.69
200 mm	m	-	3.90	26.29	-	-	26.29	32.86

Break up tarmacadam paving, for trenches 600 mm wide, average thickness

100 mm	m	-	0.65	4.38	-	-	4.38	5.47
150 mm	m	-	1.00	6.74	-	-	6.74	8.43
200 mm	m	-	1.30	8.76	-	-	8.76	10.95

Break up surface concrete, for trenches 600 mm wide and reinstate concrete paving and 150 mm thick hardcore bed, average concrete thickness

100 mm	m	-	3.10	20.89	-	5.34	26.23	32.79
150 mm	m	-	4.15	27.97	-	7.41	35.38	44.23
200 mm	m	-	5.25	35.38	-	9.48	44.86	56.08

Break up reinforced surface concrete, for trenches 600 mm wide and reinstate concrete paving fabric reinforcement and 150 mm hardcore bed, average concrete thickness

100 mm	m	0.28	3.75	27.79	-	6.06	33.85	42.31
150 mm	m	0.28	5.15	37.22	-	8.13	45.35	56.69
200 mm	m	0.28	6.55	46.66	-	10.20	56.86	71.08

Break up tarmacadam paving, for trenches 600 mm wide and reinstate tarmacadam and 100 mm hardcore bed, average tarmacadam thickness

100 mm	m	-	1.75	11.80	1.43	37.76	50.99	63.74
150 mm	m	-	2.25	15.17	1.75	105.94	122.86	153.57
200 mm	m	-	2.75	18.54	2.71	140.78	162.03	202.54

Drainage

Alterations and repairs	Unit	Hours C	Hours L	Labour net	Plant net	Material net	Price net	Price with 25%
Pipe trenches				£	£	£	£	£
					VAT not included			
Granular beds; side filling and coverings								
50 mm beds								
450 mm wide	m	-	0.11	0.74	-	0.44	1.18	1.48
525 mm wide	m	-	0.12	0.81	-	0.49	1.30	1.63
600 mm wide	m	-	0.14	0.94	-	0.57	1.51	1.89
750 mm wide	m	-	0.18	1.21	-	0.72	1.93	2.41
100 mm beds								
450 mm wide	m	-	0.18	1.21	-	0.86	2.07	2.59
525 mm wide	m	-	0.20	1.35	-	1.01	2.36	2.95
600 mm wide	m	-	0.22	1.48	-	1.14	2.62	3.27
750 mm wide	m	-	0.29	1.95	-	1.43	3.38	4.22
100 mm beds and side filling to half height of pipes								
450 mm wide to 100 mm pipes	m	-	0.24	1.62	-	1.20	2.82	3.52
525 mm wide to 150 mm pipes	m	-	0.35	2.36	-	1.69	4.05	5.06
600 mm wide to 225 mm pipes	m	-	0.40	2.70	-	2.15	4.85	6.06
750 mm wide to 300 mm pipes	m	-	0.54	3.64	-	3.01	6.65	8.31
150 mm beds and side filling to half height of pipes								
450 mm wide to 100 mm pipes	m	-	0.31	2.09	-	1.77	3.86	4.83
525 mm wide to 150 mm pipes	m	-	0.37	2.49	-	2.21	4.70	5.88
600 mm wide to 225 mm pipes	m	-	0.46	3.10	-	2.72	5.82	7.28
750 mm wide to 300 mm pipes	m	-	0.63	4.25	-	3.73	7.98	9.97
100 mm beds and side filling to full height of pipes								
450 mm wide to 100 mm pipes	m	-	0.30	2.02	-	1.81	3.83	4.79
525 mm wide to 150 mm pipes	m	-	0.41	2.76	-	2.40	5.16	6.45
600 mm wide to 225 mm pipes	m	-	0.52	3.50	-	3.14	6.64	8.30
750 mm wide to 300 mm pipes	m	-	0.73	4.92	-	4.59	9.51	11.89
150 mm beds and side filling to full heights of pipes								
450 mm wide to 100 mm pipes	m	-	0.37	2.49	-	2.23	4.72	5.90
525 mm wide to 150 mm pipes	m	-	0.47	3.17	-	2.89	6.06	7.58
600 mm wide to 225 mm pipes	m	-	0.61	4.11	-	3.71	7.82	9.78
750 mm wide to 300 mm pipes	m	-	0.79	5.32	-	5.31	10.63	13.29
Beds and coverings to 100 mm pipes								
450 x 350 mm	m	-	0.45	3.03	-	2.91	5.94	7.42
450 x 450 mm	m	-	0.57	3.84	-	3.81	7.65	9.56
Beds and coverings to 150 mm pipes								
525 x 400 mm	m	-	0.57	3.84	-	3.62	7.46	9.32
525 x 500 mm	m	-	0.72	4.85	-	4.63	9.48	11.85

Drainage

Alterations and repairs	Unit	Hours C	Hours L	Labour net	Plant net	Material net	Price net	Price with 25%
Pipe trenches				£	£	£	£	£
					VAT not included			

Granular beds; side filling and coverings (*continued*)

Beds and coverings to 225 mm pipes

600 x 475 mm	m	-	0.70	4.72	-	4.66	9.38	11.73
600 x 575 mm	m	-	0.87	5.86	-	5.81	11.67	14.59

Beds and coverings to 300 mm pipes

750 x 550 mm	m	-	0.98	6.61	-	6.53	13.14	16.43
750 x 650 mm	m	-	1.21	8.16	-	7.96	16.12	20.15

Concrete 1:3:6 beds; benchings and coverings

100 mm beds

450 mm wide	m	-	0.22	1.48	-	2.50	3.98	4.97
525 mm wide	m	-	0.24	1.62	-	3.65	5.27	6.59
600 mm wide	m	-	0.28	1.89	-	4.13	6.02	7.53
750 mm wide	m	-	0.35	2.36	-	5.17	7.53	9.41

150 mm beds

450 mm wide	m	-	0.28	1.89	-	3.78	5.67	7.09
525 mm wide	m	-	0.29	1.95	-	4.40	6.35	7.94
600 mm wide	m	-	0.33	2.22	-	5.01	7.23	9.04
750 mm wide	m	-	0.41	2.76	-	6.29	9.05	11.31

100 mm beds and benchings to full height of pipes

450 mm wide to 100 mm pipes	m	-	0.26	1.75	-	5.29	7.04	8.80
525 mm wide to 150 mm pipes	m	-	0.44	2.97	-	7.01	9.98	12.48
600 mm wide to 225 mm pipes	m	-	0.51	3.44	-	9.18	12.62	15.78
750 mm wide to 300 mm pipes	m	-	0.66	4.45	-	13.41	17.86	22.32

150 mm beds and benchings to full height of pipes

450 mm wide to 100 mm pipes	m	-	0.29	1.95	-	6.51	8.46	10.57
525 mm wide to 150 mm pipes	m	-	0.48	3.24	-	8.46	11.70	14.63
600 mm wide to 225 mm pipes	m	-	0.56	3.77	-	10.85	14.62	18.27
750 mm wide to 300 mm pipes	m	-	0.73	4.92	-	15.53	20.45	25.56

Beds and coverings to 100 mm pipes

450 x 350 mm	m	-	0.55	3.71	-	8.51	12.22	15.28
450 x 450 mm	m	-	0.61	4.11	-	11.13	15.24	19.05

Beds and coverings to 150 mm pipes

525 x 400 mm	m	-	0.55	3.71	-	10.57	14.28	17.85
525 x 500 mm	m	-	0.72	4.85	-	13.52	18.37	22.96

Beds and coverings to 225 mm pipes

600 x 475 mm	m	-	0.72	4.85	-	13.63	18.48	23.10
600 x 575 mm	m	-	1.05	7.08	-	16.97	24.05	30.06

Beds and coverings to 300 mm pipes

750 x 550 mm	m	-	0.88	5.93	-	19.09	25.02	31.27
750 x 650 mm	m	-	1.10	7.41	-	23.26	30.67	38.34

Drainage

Alterations and repairs	Unit	Hours C	Hours L	Labour net	Plant net	Material net	Price net	Price with 25%
Pipe trenches				£	£	£	£	£
					VAT not included			

Alterations and repairs	Unit	Hours C	Hours L	Labour net	Plant net	Material net	Price net	Price with 25%
Concrete 1:2:4 beds; benchings and coverings								
100 mm beds								
450 mm wide	m	-	0.20	1.35	.-	2.65	4.00	5.00
525 mm wide	m	-	0.22	1.48	-	3.13	4.61	5.76
600 mm wide	m	-	0.24	1.62	-	3.54	5.16	6.45
750 mm wide	m	-	0.28	1.89	-	4.42	6.31	7.89
150 mm beds								
450 mm wide	m	-	0.28	1.89	-	4.01	5.90	7.38
525 mm wide	m	-	0.29	1.95	-	4.66	6.61	8.26
600 mm wide	m	-	0.33	2.22	-	5.31	7.53	9.41
750 mm wide	m	-	0.41	2.76	-	6.67	9.43	11.79
100 mm beds and benchings to full height of pipes								
450 mm wide to 100 mm pipes	m	-	0.26	1.75	-	5.60	7.35	9.19
525 mm wide to 150 mm pipes	m	-	0.48	3.24	-	7.43	10.67	13.34
600 mm wide to 225 mm pipes	m	-	0.51	3.44	-	9.73	13.17	16.46
750 mm wide to 300 mm pipes	m	-	0.66	4.45	-	14.22	18.67	23.34
150 mm beds and benchings to full height of pipes								
450 mm wide to 100 mm pipes	m	-	0.29	1.95	-	6.90	8.85	11.06
525 mm wide to 150 mm pipes	m	-	0.46	3.10	-	8.97	12.07	15.09
600 mm wide to 225 mm pipes	m	-	0.56	3.77	-	11.50	15.27	19.09
750 mm wide to 300 mm pipes	m	-	0.73	4.92	-	16.46	21.38	26.73
Beds and coverings to 100 mm pipes								
450 x 350 mm	m	-	0.55	3.71	-	9.03	12.74	15.93
450 x 450 mm	m	-	0.61	4.11	-	11.80	15.91	19.89
Beds and coverings to 150 mm pipes								
525 x 400 mm	m	-	0.55	3.71	-	11.21	14.92	18.65
525 x 500 mm	m	-	0.72	4.85	-	14.33	19.18	23.98
Beds and coverings to 225 mm pipes								
600 x 475 mm	m	-	0.72	4.85	-	14.45	19.30	24.13
600 x 575 mm	m	-	1.05	7.08	-	17.99	25.07	31.34
Beds and coverings to 300 mm pipes								
750 x 550 mm	m	-	0.88	5.93	-	20.23	26.16	32.70
750 x 650 mm	m	-	1.10	7.41	-	24.66	32.07	40.09

Drainage

Alterations and repairs	Unit	Hours C	Hours L	Labour net	Plant net	Material net	Price net	Price with 25%
Pipework				£	£	£	£	£
					VAT not included			

Flexible jointed vitrified clay pipes and fittings to BS 65

100 mm plain ended "Supersleve" pipes jointed with polypropylene sleeve couplings and laid in trench bottom

	Unit	Hours C	Hours L	Labour net	Plant net	Material net	Price net	Price with 25%
in runs over 3.00 m long	m	0.14	0.07	1.69	-	3.24	4.93	6.16
in runs not exceeding 3.00 m long	m	0.32	0.06	3.24	-	3.24	6.48	8.10

Extra for

bends	each	0.12	-	1.08	-	3.50	4.58	5.72
junctions	each	0.17	-	1.52	-	7.55	9.07	11.34
socket adaptors	each	0.42	0.11	4.50	-	3.00	7.50	9.38

150 mm plain ended "Supersleve" pipes jointed with polypropylene sleeve couplings and laid in trench bottom

in runs over 3.00 m long	m	0.18	0.09	2.20	-	7.29	9.49	11.86
in runs not exceeding 3.00 m long	m	0.37	0.09	3.91	-	7.29	11.20	14.00

Extra for

bends	each	0.11	-	0.99	-	8.14	9.13	11.41
junctions	each	0.20	-	1.79	-	10.69	12.48	15.60
socket adaptors	each	0.52	0.11	5.40	-	7.06	12.46	15.57

100 mm socketed "Hepseal" pipes jointed with rubber sealing rings and laid in trench bottom

in runs over 3.00 m long	m	0.17	0.06	1.89	-	7.89	9.78	12.23
in runs not exceeding 3.00 m long	m	0.33	0.06	3.33	-	7.89	11.22	14.03

Extra for

bends	each	0.20	-	1.79	-	10.85	12.64	15.80
junctions	each	0.17	-	1.52	-	15.07	16.59	20.74
double collars	each	0.33	0.11	3.70	-	11.22	14.92	18.65

150 mm socketed "Hepseal" pipes jointed with rubber sealing rings and laid in trench bottom

in runs over 3.00 m long	m	0.20	0.09	2.38	-	10.24	12.62	15.78
in runs not exceeding 3.00 m long	m	0.40	0.09	4.17	-	10.24	14.41	18.01

Extra for

bends	each	0.22	-	1.97	-	17.90	19.87	24.84
junctions	each	0.20	-	1.79	-	23.38	25.17	31.46
double collar	each	0.42	0.11	4.50	-	23.38	27.88	34.85

225 mm socketed "Hepseal" pipes jointed with rubber sealing rings and laid in trench bottom

in runs over 3.00 m long	m	0.23	0.10	2.73	-	19.82	22.55	28.19
in runs not exceeding 3.00 m long	m	0.42	0.10	4.43	-	19.82	24.25	30.31

Drainage

Alterations and repairs	Unit	Hours C	Hours L	Labour net	Plant net	Material net	Price net	Price with 25%
Pipework				£	£	£	£	£
					VAT not included			

Flexible jointed vitrified clay pipes and fittings to BS 65 (*continued*)

	Unit	Hours C	Hours L	Labour net	Plant net	Material net	Price net	Price with 25%
Extra for								
bends	each	0.14	-	1.25	-	37.46	38.71	48.39
junctions	each	0.24	-	2.15	-	56.33	58.48	73.10
Square one-piece trapped access gullies with 150 x 110 mm vertical back inlet, rodding eye and plastic stopper and 100 mm outlet including polypropylene sleeve coupling to pipe and providing 150 x 150 mm coated cast iron grating	each	0.82	-	7.35	-	23.34	30.69	38.36
225 mm int. dia. x 600 mm deep 3 piece insp. chambers for 100 mm pipes; straight through base; 300 mm chamber raising piece with integral alloy plate and frame and polyprop, sleeve coupling; including two polypropylene sleeve couplings to 100 mm pipes	each	1.10	0.74	14.85	-	61.94	76.79	95.99
Extra for								
single junction bases and one additional polypropylene sleeve coupling to 100 mm pipe	each	0.17	0.11	2.26	-	8.72	10.98	13.73
double junction bases and two additional polypropylene sleeve couplings to 100 mm pipes	each	0.33	0.22	4.44	-	17.97	22.41	28.01
300 or 450 mm chamber raising pieces	each	0.38	0.28	5.29	-	13.70	18.99	23.74

Cement jointed vitrified clay pipes and fittings to BS 65

	Unit	Hours C	Hours L	Labour net	Plant net	Material net	Price net	Price with 25%
100 mm pipes jointed with tarred gaskin and cement mortar 1:3 and laid in trench bottom								
in runs over 3.00 m long	m	0.34	0.06	3.42	-	5.06	8.48	10.60
in runs not exceeding 3.00 m long	m	0.54	0.06	5.21	-	5.06	10.27	12.84
Extra for								
bends	each	0.22	-	1.97	-	3.72	5.69	7.11
junctions	each	0.21	-	1.88	-	7.51	9.39	11.74
double collar	each	0.45	0.30	6.05	-	7.94	13.99	17.49
150 mm pipes jointed with tarred gaskin and cement mortar 1:3 and laid in trench bottom								
in runs over 3.00 m long	m	0.43	0.06	4.25	-	8.67	12.92	16.15
in runs not exceeding 3.00 m long	m	0.69	0.06	6.55	-	8.67	15.22	19.02

Drainage

Alterations and repairs	Unit	Hours C	Hours L	Labour net	Plant net	Material net	Price net	Price with 25%
Pipework				£	£	£	£	£
					VAT not included			

Cement jointed vitrified clay pipes and fittings to BS 65 (*continued*)

Alterations and repairs	Unit	Hours C	Hours L	Labour net	Plant net	Material net	Price net	Price with 25%
Extra for								
bends	each	0.24	-	2.15	-	6.36	8.51	10.64
junctions	each	0.25	-	2.24	-	12.51	14.75	18.44
double collar	each	0.60	0.40	8.08	-	12.61	20.69	25.86
225 mm pipes jointed with tarred gaskin and cement mortar 1:3 and laid in trench bottom								
in runs over 3.00 mm long	m	0.54	0.08	5.38	-	17.28	22.66	28.32
in runs not exceeding 3.00 m long	m	0.80	0.08	7.69	-	17.28	24.97	31.21
Extra for								
bends	each	0.32	-	2.87	-	19.76	22.63	28.29
Square one-piece "P" trap gullies with 100 mm outlet including cement joint to pipe; bedding in concrete 1:3:6 and providing coated cast iron grating								
150 x 150 mm	each	0.66	0.11	6.65	-	24.84	31.49	39.36
Extra for								
horizontal inlets	each	-	-	-	-	14.03	14.03	17.54
vertical inlets	each	-	-	-	-	14.03	14.03	17.54
cement jointed raising pieces 150 x 150 mm	each	0.22	-	1.97	-	7.17	9.14	11.43
cement jointed raising pieces 225 x 225 mm	each	0.22	-	1.97	-	15.57	17.54	21.93
completely surrounding gullies with concrete 1:3:6 150 mm thick	each	-	0.30	2.02	-	0.83	2.85	3.56
225 x 225 mm square one-piece "P" trap gullies 585 mm deep with rubber inspection eye and 100 mm outlet including cement joint to pipe; bedding in concrete 1:3:6 and providing coated cast iron grating	each	0.77	-	6.90	-	58.14	65.04	81.30
Extra for								
galvanised mud buckets	each	0.11	-	0.99	-	12.42	13.41	16.76
completely surrounding gullies with concrete 1:3:6 150 mm thick	each	-	0.72	4.85	-	2.07	6.92	8.65
285 x 285 mm square one-piece "P" trap gullies 585 mm deep with rubber inspection eye and 100 mm outlet including cement joint to pipe; bedding in concrete 1:3:6 and providing coated cast iron grating	each	0.88	-	7.88	-	85.52	93.40	116.75

Drainage

Alterations and repairs	Unit	Hours C	Hours L	Labour net	Plant net	Material net	Price net	Price with 25%
Pipework				£	£	£	£	£
					VAT not included			

Cement jointed vitrified clay pipes and fittings to BS 65 (continued)

	Unit	Hours C	Hours L	Labour net	Plant net	Material net	Price net	Price with 25%
Extra for								
galvanised mud buckets	each	0.11	-	0.99	-	23.93	24.92	31.15
completely surrounding gullies with concrete 1:3:6 150 mm thick	each	-	0.30	2.02	-	4.82	6.84	8.55

Flexible jointed standard concrete pipes and fittings to BS 5911

	Unit	Hours C	Hours L	Labour net	Plant net	Material net	Price net	Price with 25%
150 mm socketed pipes jointed with rolling rubber rings and laid in trench bottom								
in runs over 3.00 m long	m	0.30	0.13	3.57	-	8.27	11.84	14.80
in runs not exceeding 3.00 m long	m	0.39	0.25	5.18	-	8.27	13.45	16.81
Extra for								
bends	each	0.15	0.07	1.78	-	46.61	48.39	60.49
junctions	each	0.44	0.07	4.38	-	55.32	59.70	74.63
225 mm socketed pipes jointed with rolling rubber rings and laid in trench bottom								
in runs over 3.00 m long	m	0.46	0.15	5.13	-	9.65	14.78	18.48
in runs not exceeding 3.00 m long	m	0.61	0.22	6.95	-	9.65	16.60	20.75
Extra for								
bends	each	0.22	0.09	2.56	-	58.14	60.70	75.88
junctions	each	0.55	0.09	5.52	-	61.09	66.61	83.26
300 mm socketed pipes jointed with rolling rubber rings and laid in trench bottom								
in runs over 3.00 m long	m	0.59	0.25	6.98	-	13.81	20.79	25.99
in runs not exceeding 3.00 m long	m	0.88	0.31	9.97	-	13.81	23.78	29.73
Extra for								
bends	each	0.28	0.12	3.32	-	69.74	73.06	91.33
junctions	each	0.61	0.12	6.28	-	72.14	78.42	98.03

Cement jointed standard concrete pipes and fittings to BS 5911

	Unit	Hours C	Hours L	Labour net	Plant net	Material net	Price net	Price with 25%
150 mm ogee pipes jointed with cement mortar 1:3 and laid in trench bottom								
in runs over 3.00 m long	m	0.50	0.20	5.83	-	4.23	10.06	12.57
in runs not exceeding 3.00 m long	m	0.74	0.30	8.65	-	4.23	12.88	16.10
Extra for								
bends	each	0.36	0.11	3.97	-	36.21	40.18	50.23
junctions	each	0.44	0.12	4.75	-	36.33	41.08	51.35

Drainage

Alterations and repairs	Unit	Hours C	Hours L	Labour net	Plant net	Material net	Price net	Price with 25%
Pipework				£	£	£	£	£
					VAT not included			
Cement jointed standard concrete pipes and fittings to BS 5911								
225 mm ogee pipes jointed with cement mortar 1:3 and laid in trench bottom								
in runs over 3.00 m long	m	0.62	0.28	7.45	-	5.36	12.81	16.01
in runs not exceeding 3.00 m long	m	0.91	0.40	10.85	-	5.36	16.21	20.26
Extra for								
bends	each	0.40	0.12	4.39	-	39.79	44.18	55.23
junctions	each	0.60	0.18	6.59	-	37.42	44.01	55.01
300 mm ogee pipes jointed with cement mortar 1:3 and laid in trench bottom								
in runs over 3.00 m long	m	0.74	0.30	8.65	-	8.02	16.67	20.84
in runs not exceeding 3.00 m long	m	1.10	0.44	12.83	-	8.02	20.85	26.06
Extra for								
bends	each	0.46	0.17	5.27	-	53.91	59.18	73.97
junctions	each	0.77	0.22	8.38	-	40.99	49.37	61.71
375 mm diameter trapped road gullies 750 mm deep with rodding eye; stopper and chain and 150 mm outlet including cement joint to pipe and bedding in concrete 1:3:6	each	3.63	0.55	36.23	-	26.38	62.61	78.26
Extra for completely surrounding gullies in concrete 1:3:6 150 mm thick	each	-	0.44	2.97	-	8.95	11.92	14.90
400 x 345 mm coated cast iron road gully gratings and frames bedded and flaunched in cement mortar	each	0.88	0.24	9.50	-	105.19	114.69	143.36

Drainage

Alterations and repairs	Unit	Hours C	Hours L	Labour net	Plant net	Material net	Price net	Price with 25%
Pipework				£	£	£	£	£
						VAT not included		

Timesaver cast iron system

100 mm pipes jointed with flexible couplings and laid in trench bottom

in runs over 3.00 m long	m	0.61	0.61	10.45	-	25.46	35.91	44.89
in runs not exceeding 3.00 m long	m	0.77	0.77	13.19	-	34.46	47.65	59.56

Extra for

10 deg medium radius bends	each	0.68	0.68	11.65	-	30.28	41.93	52.41
22.5 deg medium radius bends	each	0.68	0.68	11.65	-	32.04	43.69	54.61
35 deg medium radius bends	each	0.68	0.68	11.65	-	33.79	45.44	56.80
45 deg medium radius bends	each	0.68	0.68	11.65	-	33.79	45.44	56.80
60 deg medium radius bends	each	0.68	0.68	11.65	-	35.56	47.21	59.01
67.5 deg medium radius bends	each	0.68	0.68	11.65	-	35.56	47.21	59.01
80 deg medium radius bends	each	0.68	0.68	11.65	-	37.39	49.04	61.30
87.5 deg medium radius bends	each	0.68	0.68	11.65	-	37.39	49.04	61.30
87.5 deg medium radius bends with heel rest and bedding in concrete 1:3:6	each	0.88	0.88	15.08	-	43.47	58.55	73.19
87.5 deg long radius bends	each	0.68	0.68	11.65	-	46.91	58.56	73.20
87.5 deg long radius bends with heel rest and bedding in concrete 1:3:6	each	0.88	0.88	15.08	-	58.53	73.61	92.01
45 deg x 100 mm branches	each	0.82	0.82	14.04	-	58.66	72.70	90.88
67.5 deg x 100 mm branches	each	0.82	0.82	14.04	-	60.33	74.37	92.96
87.5 deg x 100 mm branches	each	0.82	0.82	14.04	-	58.66	72.70	90.88
45 deg x 100 mm access branches	each	0.82	0.82	14.04	-	100.17	114.21	142.76
87.5 deg x 100 mm access branches	each	0.82	0.82	14.04	-	100.17	114.21	142.76

150 mm pipes jointed with flexible couplings and laid in trench bottom

in runs over 3.00 m long	m	0.88	0.88	15.08	-	44.96	60.04	75.05
in runs not exceeding 3.00 m long	m	1.32	1.32	22.61	-	55.86	78.47	98.09

Extra for

10 deg medium radius bends	each	0.77	0.77	13.19	-	33.11	46.30	57.88
22.5 deg medium radius bends	each	0.77	0.77	13.19	-	34.87	48.06	60.08
35 deg medium radius bends	each	0.77	0.77	13.19	-	36.62	49.81	62.26
45 deg medium radius bends	each	0.77	0.77	13.19	-	36.62	49.81	62.26
67.5 deg medium radius bends	each	0.77	0.77	13.19	-	38.39	51.58	64.47
87.5 deg medium radius bends	each	0.77	0.77	13.19	-	40.22	53.41	66.76
87.5 deg medium radius bends with heel rest and bedding in concrete 1:3:6	each	0.99	0.99	16.96	-	80.72	97.68	122.10
87.5 deg long radius bends	each	0.77	0.77	13.19	-	87.99	101.18	126.47
87.5 deg long radius bends with heel rest and bedding in concrete 1:3:6	each	0.99	0.99	16.96	-	82.66	99.62	124.53
45 deg x 100 mm branches	each	0.94	0.94	16.10	-	101.52	117.62	147.03
87.5 deg x 100 mm branches	each	0.94	0.94	16.10	-	98.32	114.42	143.03
45 deg x 150 mm branches	each	0.94	0.94	16.10	-	110.94	127.04	158.80
87.5 deg x 150 mm branches	each	0.94	0.94	16.10	-	110.94	127.04	158.80
87.5 deg x 150 mm access branches	each	0.94	0.94	16.10	-	101.16	117.26	146.57

Drainage

Alterations and repairs	Unit	Hours C	Hours L	Labour net	Plant net	Material net	Price net	Price with 25%
Pipework				£	£	£	£	£
					VAT not included			

Timesaver cast iron system
(*continued*)

Alterations and repairs	Unit	Hours C	Hours L	Labour net	Plant net	Material net	Price net	Price with 25%
250 x 250 mm square one-piece "P" trap with 100 mm outlet including joint to pipe, bedding in concrete 1:3:6 and providing grating	each	1.21	1.21	20.72	-	162.48	183.20	229.00
Extra for								
galvanised sediment pans	each	-	0.11	0.65	-	22.28	22.93	28.66
completely surrounding gullies with concrete 1:3:6 150 mm thick	each	-	0.22	1.29	-	4.61	5.90	7.38

PVCU pipes and fittings to BS 4660

Alterations and repairs	Unit	Hours C	Hours L	Labour net	Plant net	Material net	Price net	Price with 25%
110 mm socketed pipes jointed with rubber sealing rings and laid in trench bottom								
in runs over 3.00 m long	m	0.11	0.11	1.73	-	5.59	7.32	9.15
in runs not exceeding 3.00 m long	m	0.13	0.13	2.04	-	5.59	7.63	9.54
Extra for								
swept bends	each	0.13	0.13	2.04	-	8.48	10.52	13.15
long radius bends	each	0.11	0.11	1.73	-	17.39	19.12	23.90
junctions	each	0.13	0.13	2.04	-	13.52	15.56	19.45
160 mm socketed pipes jointed with rubber sealing rings and laid in trench bottom								
in runs over 3.00 m long	m	0.13	0.13	2.04	-	11.86	13.90	17.38
in runs not exceeding 3.00 m long	m	0.15	0.15	2.35	-	11.86	14.21	17.76
Extra for								
swept bends	each	0.15	0.15	2.35	-	20.16	22.51	28.14
long radius bends	each	0.13	0.13	2.04	-	36.29	38.33	47.91
junctions	each	0.13	0.13	2.04	-	38.48	40.52	50.65
160 mm diameter one-piece "P" trap gullies with 110 mm inlet and 110 mm outlet including rubber sealing ring joint to pipe, bedding in concrete 1:3:6 and providing PVC grating	each	1.49	-	13.35	-	22.91	36.26	45.33
Three-piece "P" trap gullies comprising 110 mm trap, 110 mm knuckle bend and raising piece hopper fitted with PVC grating including rubber sealing ring joint to pipe and bedding in concrete 1:3:6	each	1.82	-	16.31	-	35.59	51.90	64.88

Drainage

Alterations and repairs	Unit	Hours C	Hours L	Labour net	Plant net	Material net	Price net	Price with 25%

Pipework

£   £   £   £   £

VAT not included

Sundries

Precast concrete kerbs to three sides of150 x 150 mm gullies bedded in cement mortar including 150 mm trowelled skirting to wall at back	each	0.46	0.31	6.21	-	3.89	10.10	12.63
Brick on edge kerbs to three sides of gullies in cement mortar including rendering all round and 150 mm skirting to wall at back trowelled smooth								
150 x 150 mm gullies	each	1.27	0.83	16.97	-	2.37	19.34	24.18
225 x 225 mm gullies	each	1.46	0.99	19.75	-	3.50	23.25	29.06

Drainage

Alterations and repairs	Unit	Hours C	Hours L	Labour net	Plant net	Material net	Price net	Price with 25%
Land drains				£	£	£	£	£
					VAT not included			
Butt jointed clayware pipes to BS 1196								
75 mm pipes								
laid in trench bottom	m	-	0.22	1.48	-	1.09	2.57	3.21
extra for junctions	each	-	0.18	1.21	-	5.02	6.23	7.79
100 mm pipes								
laid in trench bottom	m	-	0.24	1.62	-	1.96	3.58	4.47
extra for junctions	each	-	0.17	1.15	-	5.80	6.95	8.69
150 mm pipes								
laid in trench bottom	m	-	0.28	1.89	-	4.19	6.08	7.60
extra for junctions	each	-	0.22	1.48	-	7.64	9.12	11.40
225 mm pipes								
laid in trench bottom	m	-	0.33	2.22	-	11.03	13.25	16.56
extra for junctions	each	-	0.24	1.62	-	9.22	10.84	13.55
Flexible jointed vitrified clay perforated pipes and fittings with integral polyethylene sleeves								
100 mm "Hepline" pipes laid in trench bottom	m	-	0.24	1.62	-	4.87	6.49	8.11
Extra for								
bends	each	-	0.18	1.21	-	4.67	5.88	7.35
junctions	each	-	0.20	1.35	-	9.74	11.09	13.86
150 mm "Hepline" pipes laid in trench bottom	m	-	0.28	1.89	-	8.86	10.75	13.44
Extra for								
bends	each	-	0.20	1.35	-	8.07	9.42	11.78
junctions	each	-	0.22	1.48	-	10.87	12.35	15.44
225 mm "Hepline" pipes laid in trench bottom	m	-	0.33	2.22	-	16.30	18.52	23.15
Extra for								
bends	each	-	0.22	1.48	-	27.09	28.57	35.71
junctions	each	-	0.24	1.62	-	40.43	42.05	52.56
Ogee jointed porous concrete pipes to BS 5911								
Pipes laid in trench bottom								
150 mm	m	0.09	0.18	2.02	-	3.88	5.90	7.38
225 mm	m	0.12	0.24	2.70	-	4.54	7.24	9.05

Drainage

Alterations and repairs	Unit	Hours C	Hours L	Labour net	Plant net	Material net	Price net	Price with 25%
Land drains				£	£	£	£	£
					VAT not included			
Plain ended PVCU perforated pipes and fittings jointed with double socket couplers								
110 mm pipes laid in trench bottom	m	-	0.20	1.35	-	6.49	7.84	9.80
Extra for								
bends	each	-	0.18	1.21	-	8.69	9.90	12.38
junctions	each	-	0.20	1.35	-	12.80	14.15	17.69
160 mm pipes laid in trench bottom	m	-	0.24	1.62	-	12.19	13.81	17.26
Extra for								
bends	each	-	0.22	1.48	-	20.66	22.14	27.68
junctions	each	-	0.24	1.62	-	36.89	38.51	48.14
PVCU perforated flexible corrugated pipes and polyethylene fittings								
80 mm pipes laid in trench bottom	m	-	0.13	0.88	-	1.59	2.47	3.09
Extra for								
end caps	each	-	0.09	0.61	-	0.52	1.13	1.41
junctions	each	-	0.20	1.35	-	3.10	4.45	5.56
100 mm pipes laid in trench bottom	m	-	0.17	1.15	-	2.26	3.41	4.26
Extra for								
end caps	each	-	0.09	0.61	-	0.99	1.60	2.00
junctions	each	-	0.20	1.35	-	3.54	4.89	6.11

Drainage

Alterations and repairs	Unit	Hours C	Hours L	Labour net	Plant net	Material net	Price net	Price with 25%
Manholes				£	£	£	£	£
					VAT not included			

Excavation by machine - including disposal of surplus on site average 50 m from excavations

Excavate pits, part fill in and compact and dispose of surplus on site, maximum depth not exceeding

0.25 m	m3	-	0.72	4.85	5.57	-	10.42	13.03
1.00 m	m3	-	0.71	4.79	4.44	-	9.23	11.54
2.00 m	m3	-	0.72	4.85	4.82	-	9.67	12.09
4.00 m	m3	-	0.72	4.85	5.57	-	10.42	13.03

Excavate pits less than 1.25 x 1.25 m on plan, part fill in and compact and disposal of surplus on site, maximum depth not exceeding

0.25 m	m3	-	0.72	4.85	9.51	-	14.36	17.95
1.00 m	m3	-	0.72	4.85	6.51	-	11.36	14.20
2.00 m	m3	-	0.72	4.85	6.69	-	11.54	14.43

Excavation by machine - including removal of surplus from site to tip average 15 km from site

Excavate pits, part fill in and compact and remove surplus from site, maximum depth not exceeding

0.25 m	m3	-	0.46	3.10	17.54	-	20.64	25.80
1.00 m	m3	-	0.39	2.63	16.38	-	19.01	23.76
2.00 m	m3	-	0.41	2.76	16.76	-	19.52	24.40
4.00 m	m3	-	0.45	3.03	18.51	-	21.54	26.93

Excavate pits less than 1.25 x 1.25 m on plan, part fill in and compact and remove surplus from site, maximum depth not exceeding

0.25 m	m3	-	0.70	4.72	22.48	-	27.20	34.00
1.00 m	m3	-	0.51	3.44	18.48	-	21.92	27.40
2.00 m	m3	-	0.52	3.50	18.66	-	22.16	27.70

Excavation by hand - including disposal of surplus on site average 50 m from excavations

Excavate pits, part fill in and compact and dispose of surplus on site, maximum depth not exceeding

0.25 m	m3	-	6.60	44.48	-	-	44.48	55.60
1.00 m	m3	-	7.00	47.18	-	-	47.18	58.98
2.00 m	m3	-	8.50	57.29	-	-	57.29	71.61

Drainage

Alterations and repairs	Unit	Hours C	Hours L	Labour net	Plant net	Material net	Price net	Price with 25%
Manholes				£	£	£	£	£
					VAT not included			

Excavation by hand - including disposal of surplus on site average 50 m from excavations (*continued*)

Excavate pits less than 1.25 x 1.25 m on plan, part fill in and compact and dispose of surplus on site, maximum depth not exceeding

0.25 m	m3	-	8.20	55.27	-	-	55.27	69.09
1.00 m	m3	-	8.75	58.98	-	-	58.98	73.72
2.00 m	m3	-	10.25	69.09	-	-	69.09	86.36

Excavation by hand - including removal of surplus from site to tip average 15 km from site

Excavate pits, part fill in and compact and remove surplus from site, maximum depth not exceeding

0.25 m	m3	-	6.75	45.50	11.97	-	57.47	71.84
1.00 m	m3	-	7.15	48.19	11.97	-	60.16	75.20
2.00 m	m3	-	8.65	58.30	11.97	-	70.27	87.84

Excavate pits less than 1.25 x 1.25 m on plan, part fill in and compact and remove surplus from site, maximum depth not exceeding

0.25 m	m3	-	8.35	56.28	11.97	-	68.25	85.31
1.00 m	m3	-	8.90	59.99	11.97	-	71.96	89.95
2.00 m	m3	-	10.80	72.79	11.97	-	84.76	105.95

Level and compact bottom of excavation	each	-	0.10	0.67	0.25	-	0.92	1.15

Granular bedding

Surrounds to PVC chambers	m3	-	2.75	18.54	-	19.03	37.57	46.96

Beds for PVC chambers
100 mm	m2	-	0.38	2.56	-	1.90	4.46	5.58
150 mm	m2	-	0.55	3.71	-	2.86	6.57	8.21
200 mm	m2	-	0.66	4.45	-	3.81	8.26	10.32

Concrete 1:3:6

Bases, thickness
100 - 150 mm	m3	-	7.50	50.55	-	68.88	119.43	149.29
150 - 300 mm	m3	-	6.50	43.81	-	68.88	112.69	140.86

Surrounds to manholes, thickness
100 - 150 mm	m3	-	12.54	84.52	-	68.88	153.40	191.75
150 - 300 mm	m3	-	12.20	82.23	-	68.88	151.11	188.89

Drainage

Alterations and repairs	Unit	Hours C	Hours L	Labour net	Plant net	Material net	Price net	Price with 25%
Manholes				£	£	£	£	£
					VAT not included			
Concrete 1:2:4								
Bases, thickness								
100 - 150 mm	m3	-	7.50	50.55	-	75.47	126.02	157.53
150 - 300 mm	m3	-	6.50	43.81	-	75.47	119.28	149.10
Surrounds to manholes, thickness								
100 - 150 mm	m3	-	13.15	88.63	-	75.47	164.10	205.13
150 - 300 mm	m3	-	12.82	86.41	-	75.47	161.88	202.35
Reinforced suspended cover slabs, thickness								
100 - 150 mm	m3	-	13.20	88.97	-	75.47	164.44	205.55
150 - 300 mm	m3	-	12.50	84.25	-	75.47	159.72	199.65
Reinforcement								
Fabric reinforcement to BS 4483 in cover slabs								
A252 - 3.95 kg/m2	m2	-	0.28	1.89	-	1.68	3.57	4.46
B283 - 3.73 kg/m2	m2	-	0.28	1.89	-	1.42	3.31	4.14
B503 - 5.93 kg/m2	m2	-	0.33	2.22	-	2.62	4.84	6.05
Formwork								
Formwork to edges and faces of bases and surrounds (four uses)								
over 1.00 m high	m2	1.97	-	17.65	-	8.31	25.96	32.45
not exceeding 250 mm high	m	0.52	-	4.66	-	2.22	6.88	8.60
Formwork to horizontal soffits of cover slabs (four uses)								
over 1.00 m high	m2	3.20	0.18	29.88	1.00	20.73	51.61	64.51
not exceeding 1.00 m high	m2	3.52	0.18	32.75	1.00	20.73	54.48	68.10
Formwork to edges of cover slabs not exceeding 250 mm deep (four uses)	m	0.11	0.06	1.35	-	2.22	3.57	4.46
Precast concrete units (complying with BS 5911) bedded in cement mortar 1:3 and flush pointed								
610 x 450 mm base units	each	0.83	1.05	14.52	-	31.18	45.70	57.13
610 x 450 mm chamber section units								
150 mm deep	each	0.66	0.66	10.36	-	12.38	22.74	28.43
230 mm deep	each	0.77	0.77	12.09	-	15.91	28.00	35.00
300 mm deep	each	0.83	0.83	13.03	-	20.60	33.63	42.04
Extra for step irons cast in chamber sections	each	-	-	-	-	4.32	4.32	5.40

Drainage

Alterations and repairs	Unit	Hours C	Hours L	Labour net	Plant net	Material net	Price net	Price with 25%
Manholes				£	£	£	£	£
					VAT not included			
Precast concrete units (complying with BS 5911) bedded in cement mortar 1:3 and flush pointed (*continued*)								
Building in ends of pipes including knocking out apertures in base units								
100 mm	each	0.28	0.19	3.79	-	0.10	3.89	4.86
150 mm	each	0.36	0.24	4.85	-	0.10	4.95	6.19
610 x 450 mm concrete covers and frames	each	0.55	0.55	8.64	-	33.21	41.85	52.31
Common bricks in cement mortar 1:3								
Manhole sides								
102.5 mm	m2	1.10	0.55	13.57	-	14.28	27.85	34.81
215 mm	m2	2.20	1.10	27.12	-	28.39	55.51	69.39
Extra for fair face and flush pointing	m2	0.33	0.17	4.11	-	0.08	4.19	5.24
102.5 x 150 mm kerbs fair faced and pointed internally under manhole cover frames								
600 x 450 mm	each	0.63	0.32	7.80	-	4.67	12.47	15.59
600 x 600 mm	each	0.79	0.39	9.71	-	5.48	15.19	18.99
750 x 600 mm	each	0.88	0.44	10.85	-	3.50	14.35	17.94
215 x 150 mm kerbs fair faced and pointed internally under manhole cover frames								
600 x 450 mm	each	1.24	0.62	15.29	-	8.90	24.19	30.24
600 x 600 mm	each	1.58	0.79	19.48	-	10.32	29.80	37.25
750 x 600 mm	each	1.76	0.83	21.36	-	11.46	32.82	41.02
Building in ends of pipes								
100 mm	each	0.22	0.11	2.71	-	0.08	2.79	3.49
150 mm	each	0.29	0.15	3.61	-	0.08	3.69	4.61
225 mm	each	0.36	0.18	4.44	-	0.17	4.61	5.76
300 mm	each	0.40	0.20	4.93	-	0.25	5.18	6.47
13 mm cement and sand 1:3 steel trowelled rendering on brick sides internally	each	1.50	0.75	18.49	-	1.09	19.58	24.48
Class B engineering bricks in cement mortar 1:3								
Manhole sides								
102.5 mm	m2	1.32	0.66	16.28	-	26.42	42.70	53.38
215 mm	m2	3.60	1.80	44.39	-	52.47	96.86	121.08
Extra for fair face and flush pointing	m2	0.44	0.22	5.42	-	0.08	5.50	6.88

Alterations and repairs	Unit	Hours C	Hours L	Labour net	Plant net	Material net	Price net	Price with 25%
Manholes				£	£	£	£	£
						VAT not included		

Class B engineering bricks in cement mortar 1:3 (continued)

102.5 x 150 mm kerbs fair faced and pointed internally under manhole cover frames

600 x 450 mm	each	0.80	0.40	9.87	-	8.65	18.52	23.15
600 x 600 mm	each	0.99	0.49	12.17	-	10.25	22.42	28.02
750 x 600 mm	each	1.10	0.55	13.57	-	10.42	23.99	29.99

215 x 150 mm kerbs fair faced and pointed internally under manhole cover frames

600 x 450 mm	each	1.58	0.79	19.48	-	16.46	35.94	44.92
600 x 600 mm	each	1.98	0.99	24.41	-	19.07	43.48	54.35
750 x 600 mm	each	2.20	1.10	27.12	-	21.21	48.33	60.41

Building in ends of pipes

100 mm	each	0.22	0.11	2.71	-	0.08	2.79	3.49
150 mm	each	0.29	0.15	3.61	-	0.08	3.69	4.61
225 mm	each	0.35	0.17	4.29	-	0.17	4.46	5.58
300 mm	each	0.40	0.20	4.93	-	0.25	5.18	6.47

13 mm cement and sand 1:3 steel trowelled rendering on brick sides internally

	m2	1.49	0.75	18.40	-	1.09	19.49	24.36

Vitrified clay channels set and jointed in cement mortar 1:3

Half section straight main channels

100 x 600 mm long	each	0.20	-	1.79	-	2.47	4.26	5.33
100 x 900/1000 mm long	each	0.22	-	1.97	-	3.68	5.65	7.06
150 x 600 mm long	each	0.28	-	2.51	-	4.35	6.86	8.57
150 x 900/1000 mm long	each	0.30	-	2.69	-	6.33	9.02	11.28
225 x 600 mm long	each	0.41	-	3.67	-	12.03	15.70	19.63
225 x 900/1000 mm long	each	0.44	-	3.94	-	14.13	18.07	22.59

Half section main channel bends

100 mm	each	0.20	-	1.79	-	3.58	5.37	6.71
150 mm	each	0.28	-	2.51	-	6.13	8.64	10.80
225 mm	each	0.41	0.30	5.69	-	19.77	25.46	31.82

Half section straight taper main channels

150 - 100 mm	each	0.28	-	2.51	-	14.03	16.54	20.68
225 - 150 mm	each	0.41	0.30	5.69	-	31.34	37.03	46.29

Half section taper main channel bends

150 - 100 mm	each	0.28	-	2.51	-	21.17	23.68	29.60
225 - 150 mm	each	0.41	-	3.67	-	60.81	64.48	80.60

Drainage

Alterations and repairs	Unit	Hours C	Hours L	Labour net	Plant net	Material net	Price net	Price with 25%
Manholes				£	£	£	£	£
						VAT not included		
Vitrified clay channels set and jointed in cement mortar 1:3 *(continued)*								
Half section branch channel bends								
100 mm	each	0.20	-	1.79	-	7.04	8.83	11.04
150 mm	each	0.28	-	2.51	-	11.76	14.27	17.84
225 mm	each	0.41	-	3.67	-	38.44	42.11	52.64
Three quarter section branch channel bends								
100 mm	each	0.20	0.14	2.73	-	7.75	10.48	13.10
150 mm	each	0.28	0.20	3.86	-	13.24	17.10	21.38
PVCU channels etc set in cement mortar 1:3								
Pipes with main channel cut out								
110 mm	each	0.22	0.22	3.45	-	20.29	23.74	29.68
160 mm	each	0.28	0.28	4.40	-	38.52	42.92	53.65
Bends with main channel cut out								
110 mm	each	0.31	0.31	4.87	-	27.51	32.38	40.48
160 mm	each	0.36	0.36	5.66	-	52.67	58.33	72.91
Half section branch channel bends								
110 mm	each	0.36	0.36	5.66	-	9.87	15.53	19.41
160 mm	each	0.44	0.44	6.91	-	15.44	22.35	27.94
Three quarter section branch channel bends								
110 mm	each	0.36	0.36	5.66	-	11.48	17.14	21.43
160 mm	each	0.44	0.44	6.91	-	22.41	29.32	36.65
450 mm diameter inspection chambers with four integral branches, depth to invert								
270 mm	each	0.82	0.82	12.88	-	76.42	89.30	111.63
500 mm	each	1.10	1.10	17.27	-	94.30	111.57	139.46
960 mm	each	1.38	1.38	21.66	-	141.14	162.80	203.50
Extra for								
branch blanking-off plugs	each	0.06	0.06	0.94	-	4.26	5.20	6.50
channel covers	each	0.11	0.11	1.73	-	21.50	23.23	29.04

Drainage

Alterations and repairs	Unit	Hours C	Hours L	Labour net	Plant net	Material net	Price net	Price with 25%
Manholes				£	£	£	£	£
					VAT not included			
Concrete benchings								
Concrete 1:3:6 benchings with steep falls to main channel, finished with cement and sand 1:3 trowelled smooth (measured overall), average thickness								
150 mm	m2	-	2.20	14.83	-	8.75	23.58	29.48
225 mm	m2	-	3.08	20.76	-	13.16	33.92	42.40
300 mm	m2	-	4.13	27.84	-	17.56	45.40	56.75
450 mm	m2	-	4.95	33.36	-	26.31	59.67	74.59
Concrete 1:2:4 benchings with steep falls to main channel, finished with cement and sand 1:3 trowelled smooth (measured overall), average thickness								
150 mm	m2	-	2.20	14.83	-	9.30	24.13	30.16
225 mm	m2	-	3.08	20.76	-	14.41	35.17	43.96
300 mm	m2	-	3.30	22.24	-	18.66	40.90	51.13
450 mm	m2	-	4.95	33.36	-	27.96	61.32	76.65
Extra for working benchings to branch channels	each	0.16	0.11	2.17	-	-	2.17	2.71
Step irons								
General purpose pattern galvanised step-irons to BS 1247 built in to brick sides								
115 mm tails	each	0.12	0.06	1.45	-	5.26	6.71	8.39
230 mm tails	each	0.15	0.08	1.86	-	6.72	8.58	10.73
Covers bedded and flaunched in cement mortar 1:3 and sealed in manhole grease								
Grade A manhole covers and frames to BS 497								
550 mm diameter - BS reference MA-55	each	1.21	1.21	19.00	-	187.92	206.92	258.65
600 mm diameter - BS reference MA-60	each	1.65	1.65	25.90	-	221.02	246.92	308.65
Grade B Class 1 manhole covers and frames to BS 497								
550 mm diameter - BS reference MB1-55	each	0.83	0.83	13.03	-	124.67	137.70	172.13
600 mm diameter - BS reference MB1-60	each	0.99	0.99	15.54	-	158.23	173.77	217.21

Drainage

Manholes

				£	£	£	£	£
					VAT not included			

Covers bedded and flaunched in cement mortar 1:3 and sealed in manhole grease (*continued*)

Grade B Class 2 manhole covers and frames to BS 497

550 diameter - BS reference								
MB2-55	each	2.30	2.30	36.11	-	123.14	159.25	199.06
600 mm diameter - BS reference								
MB2-60	each	2.75	2.75	43.18	-	177.89	221.07	276.34
600 x 450 mm - BS reference								
MB2-60/45	each	2.75	2.75	43.18	-	134.08	177.26	221.57
600 x 600 mm - BS reference								
MB2-60/60	each	3.30	3.30	51.81	-	168.50	220.31	275.39

Grade C single seal inspection covers and frames to BS 497

600 x 450 mm - BS reference								
MC1-60/45	each	0.99	0.99	15.54	-	41.43	56.97	71.21
600 x 600 mm - BS reference								
MC1-60/60	each	1.65	1.65	25.90	-	71.96	97.86	122.33

Grade C double seal inspection covers and frames to BS 497

600 x 450 mm - BS reference								
MC2-60/45	each	1.65	1.65	25.90	-	66.49	92.39	115.49
600 x 600 mm - BS reference								
MC2-60/60	each	2.20	2.20	34.54	-	102.11	136.65	170.81

Intercepting traps

Vitrified clay intercepting traps with stopper including cement joints to channel and pipe and bedding and surrounding with concrete 1:3:6 150 mm thick

100 mm	each	0.61	0.61	9.58	-	68.37	77.95	97.44
150 mm	each	0.88	0.88	13.81	-	90.99	104.80	131.00
225 mm	each	1.32	1.32	20.73	-	213.92	234.65	293.31

Alterations and repairs	Unit	Hours C	Hours L	Labour net	Plant net	Material net	Price net	Price with 25%
Combined items				£	£	£	£	£
					VAT not included			

Including excavation by hand and removal of surplus excavated material to a tip average 15 kilometres from site, excluding drainage pipework

Concrete manholes size 610 x 450 x 600 mm deep to invert with 150 mm concrete 1:3:6 base, precast base unit and chamber sections, 100 mm vitrified clay main channel and three three-quarter section branch bends, concrete benching and precast cover and frame

	Unit	Hours C	Hours L	Labour net	Plant net	Material net	Price net	Price with 25%
Concrete manholes (above)	each	8.24	21.62	219.55	12.22	293.04	524.81	656.01

Extra for each 150 mm increase in depth to 900 mm maximum

	each	0.81	9.64	72.21	11.97	19.10	103.28	129.10

Brick manholes with 150 mm concrete 1:3:6 base, sides in common bricks rendered internally, 100 mm vitrified clay main channel and three three-quarter section branch channel bends, concrete benching and Grade C single seal cast iron cover and frame

	Unit	Hours C	Hours L	Labour net	Plant net	Material net	Price net	Price with 25%
manholes size 600 x 450 x 600 mm deep to invert with 102.5 mm sides	each	4.97	22.29	194.77	12.22	156.85	363.84	454.80
manholes size 600 x 450 x 600 mm deep to invert with 215 mm sides	each	6.07	22.84	208.32	12.22	170.96	391.50	489.38

Extra for each 300 mm increase in depth to 1500 mm maximum

	Unit	Hours C	Hours L	Labour net	Plant net	Material net	Price net	Price with 25%
102.5 mm sides with step iron	each	2.75	12.18	106.71	11.97	22.10	140.78	175.97
215 mm sides with step iron	each	3.85	12.73	120.28	11.97	36.20	168.45	210.56

Brick manhole with 150 mm concrete 1:3:6 base, sides in common bricks rendered internally, 100 mm vitrified clay main channel and four three-quarter section branch channel bends, concrete benching, 150 mm reinforced concrete 1:2:4 suspended cover slab

	Unit	Hours C	Hours L	Labour net	Plant net	Material net	Price net	Price with 25%
manhole size 750 x 600 x 600 mm deep to invert with 102.5 mm sides, grade C single seal cast iron cover and frame	each	5.00	35.77	285.86	12.22	230.73	528.81	661.01
manhole size 750 x 600 x 600 mm deep to invert with 215 mm sides, grade C single seal cast iron cover and frame	each	6.10	36.31	299.41	12.22	244.85	556.48	695.60

Extra for each 300 mm increase in depth to 1500 mm maximum

	Unit	Hours C	Hours L	Labour net	Plant net	Material net	Price net	Price with 25%
102.5 mm sides with step iron	each	2.75	12.18	106.71	11.97	22.09	140.77	175.96
215 mm sides with step iron	each	3.85	12.73	120.28	11.97	36.20	168.45	210.56

Drainage

Alterations and repairs	Unit	Hours C	Hours L	Labour net	Plant net	Material net	Price net	Price with 25%
Combined items				£	£	£	£	£
					VAT not included			
Extra for 100 mm vitrified clay intercepting traps with connecting pipe and fresh air inlet complete	each	1.30	1.50	21.76	-	61.37	83.13	103.91
450 mm diameter PVCU manholes with four integral branches, 150 mm granular bedding base and surround, 150 mm concrete 1:3:6 surround at top and Grade C single seal cast iron cover and frame complete, depth to invert								
570 mm	each	3.40	28.24	220.81	12.22	308.22	541.25	676.56
910 mm	each	3.68	28.52	225.20	12.22	355.05	592.47	740.59
Work to existing drains								
Break into existing 100 mm vitrified clay pipe, cut away concrete bed and benching, provide and insert branch, joint to existing with double collar and renew one length of pipe and make out concrete bed and benching	each	4.00	4.30	64.82	-	34.91	99.73	124.66
Extra for constructing manhole on existing drain run including breaking out 100 mm vitrified clay pipe and concrete bed and benching, building in ends of pipes and making out concrete bed and benching	each	2.00	2.00	31.40	-	3.77	35.17	43.96
Open existing manhole, cut away side and build in end of 100 mm vitrified clay drain pipe, cut away benching and provide and set 100 mm vitrified clay three-quarter section branch channel bend, make good benching and other work disturbed and re-seal	each	5.00	5.00	78.50	-	7.02	85.52	106.90
Remove defective manhole cover and frame and provide and fix new 600 x 450 mm Grade C single seal cast iron cover and frame bedded and flaunched in cement mortar 1:3 and sealed in manhole grease	each	1.25	1.25	19.63	-	38.21	57.84	72.30
Take up defective gully kerbs and provide new brick on edge kerbs to three sides in cement mortar including rendering all round and 150 mm high skirting to wall at back trowelled smooth								
150 x 150 mm gullies	each	1.55	1.55	24.34	-	0.84	25.18	31.48
225 x 225 mm	each	1.77	1.77	27.79	-	1.02	28.81	36.01

Drainage

Alterations and repairs	Unit	Hours C	Hours L	Labour net	Plant net	Material net	Price net	Price with 25%
Combined items				£	£	£	£	£
					VAT not included			
Testing drains								
Water test								
100 mm drains	m	0.08	0.06	1.06	-	-	1.06	1.33
150 mm drains	m	0.11	0.08	1.51	-	-	1.51	1.89
225 mm drains	m	0.13	0.09	1.75	-	-	1.75	2.19
300 mm drains	m	0.17	0.11	2.26	-	-	2.26	2.83

Combined trades spot items

Preamble

"Labour net" figures include allowances for all costs incidental to the employment of labour.

"Plant net" figures include for all costs of plant including drivers and operation where applicable.

"Materials net" figures include for all costs of materials including an allowance for waste except where specifically stated.

"Price net" figures are the totals of the "Labour net", "Plant net", where applicable, and "Materials net" figures. Prices are for a builder employing his own labour.

Prices do not include any allowance for scaffolding, ladders or other plant necessary to reach the work. The "Preliminaries" section includes prices for scaffolding which must be considered and allowance included to suit the particular circumstances of a tender.

Figures for items in this section include allowance for cutting away and making good but exclude decorating in all cases.

Basic prices

Prices for materials used in this section are as the basic prices quoted in the preceding trade sections of "Alterations and repairs".

Combined trades spot items

Alterations and repairs	Unit	Hours C	Hours L	Labour net	Plant net	Material net	Price net	Price with 25%

Walls

				£	£	£	£	£

VAT not included

Filling existing door openings

Take out 762 x 1981 mm doors complete with frame, fanlight over, and architraves from 60 mm block partition 2590 mm high, cut away and take up flooring, fill opening with 60 mm clinker aggregate blockwork in gauged mortar 1:1:6 cut and bonded to existing and make out plastering and 25 x 175 mm softwood moulded skirting both sides to match existing

complete	each	12.75	8.25	169.85	-	43.66	213.51	266.89

Take out 762 x 1981 mm doors complete with linings and architraves from 102.5 mm internal brick wall, cut away and take up flooring, fill opening with 102.5 mm brickwork in gauge mortar 1:1:6 cut and bonded to existing and make out plastering and 25 x 175 mm softwood moulded skirting both sides to match existing

complete	each	15.00	11.50	211.91	-	66.31	278.22	347.77

Take out 762 x 1981 mm doors complete with linings or frame and architraves from 215 mm internal brick wall, cut away and take up flooring, fill opening with 215 mm brickwork in gauged mortar 1:1:6 cut and bonded to existing and make out plastering and 25 x 175 mm softwood moulded skirting both sides to match existing

complete	each	20.50	16.75	296.57	-	110.08	406.65	508.31

Take out 762 x 1981 mm doors complete with frame from 215 mm external faced brick wall, fill opening with 215 mm brickwork in gauged mortar 1:1:6 cut and bonded to existing, faced one side with facing bricks and make out plastering and 25 x 175 mm softwood moulded skirting the other side to match existing

complete	each	18.00	15.75	267.43	-	127.63	395.06	493.82

Take out 762 x 1981 mm doors complete with linings and architraves from 100 mm stud partition, fill opening with 50 x 100 mm softwood studding, line both sides with 9.5 mm plasterboard and make out plastering and 25 x 175 mm softwood moulded skirting both sides to match existing

complete	each	10.00	4.75	121.61	-	40.29	161.90	202.37

Filling existing window openings

Take out windows complete with sill from 215 mm faced brick external wall, fill 1000 x 2000 mm opening with 215 mm brickwork in gauged mortar 1:1:6 cut and bonded to existing, faced one side with facing bricks and make out plastering the other side

complete	each	19.50	17.00	289.30	-	136.69	425.99	532.49

Take out windows complete with sill from 215 mm rendered external wall, fill 1000 x 2000 mm opening with 215 mm brickwork in gauged mortar 1:1:6 cut and bonded to existing, make out cement and sand 1:3 wood floated finish one side and plastering the other side

complete	each	20.50	17.75	303.31	-	109.49	412.80	516.00

Combined trades spot items

Alterations and repairs	Unit	Hours C	Hours L	Labour net	Plant net	Material net	Price net	Price with 25%
Walls				£	£	£	£	£
						VAT not included		

Cutting clear openings in walls

Cut away skirtings both sides and cut 850 x 2050 mm openings through brick wall, remove debris, cut away for and build in precast concrete reinforced lintels, pin up to work over, quoin up jambs with old bricks, plaster around opening and make good to existing, fix old skirting to jambs and mitre to existing and make out softwood boarded flooring - as shown on drawing A on page 715

in 102.5 mm wall, with one 100 x 150 mm lintel	each	16.75	9.75	215.79	-	41.43	257.22	321.52
in 215 mm wall, with two 100 x 150 mm lintels	each	22.75	15.25	306.62	-	68.17	374.79	468.49

Cut away skirtings both sides and cut 850 x 2050 mm openings through brick wall, remove debris, cut away for and build in precast concrete reinforced lintels, pin up to work over, quoin up jambs with old bricks, plaster around opening and make good to existing, fix old skirting to jambs and mitre to existing make out softwood boarded flooring - as shown on drawing B on page 716

in 250 mm hollow wall, with two 100 x 150 mm prestressed concrete lintels	each	24.75	17.25	338.02	-	75.97	413.99	517.49
in 327.5 mm wall, with three 100 x 150 mm pre-stressed concrete lintels	each	28.50	20.75	395.22	-	100.85	496.07	620.09

Shore up floor and wall over and cut 3350 x 2500 mm openings through 102.5 mm loadbearing brick wall, remove debris, cut away boarded flooring and extend foundation to take the 215 x 102.5 mm brick piers bonded to existing walls, cut away for and cast in 215 x 215 x 150 mm concrete padstones, fix 178 x 102 mm rolled steel joist and pin up to work over, make out softwood boarded flooring where wall removed, form softwood cradling to steel joist and cover with two layers of gypsum plasterboard, plaster piers and make good to existing, fix old skirting around piers and mitre to existing - as shown on drawing C on page 717

complete	each	57.00	34.50	743.25	-	233.51	976.76	1220.95

Cutting openings and providing and fitting doors

Cut away skirtings both sides and cut openings through 100 mm block partition, remove debris, fix 32 x 125 mm softwood door frame with 13 x 38 mm stops, fixed light over extended to ceiling, make good blockwork and make out plastering both sides, fix 19 x 50 mm softwood stock pattern architraves both sides, make good skirtings and make out boarded flooring, glaze fixed light with 4 mm white patterned glass secured with softwood beads, hang 762 x 1981 x 38 mm plywood faced flush door on pair 75 mm steel butts and fit mortice lock and furniture - as shown on drawing D on page 718

complete	each	12.50	3.00	132.22	-	138.60	270.82	338.52

Combined trades spot items

Alterations and repairs	Unit	Hours C	Hours L	Labour net	Plant net	Material net	Price net	Price with 25%
Walls		£	£	£	£	£	£	£
						VAT not included		

Cutting openings and providing and fitting doors *(continued)*

Cut away skirtings both sides and cut openings through brick wall, remove debris, cut away for and build in precast concrete reinforced lintels, pin up to work over, quoin up jambs with old bricks, make out plastering both sides, fix 19 x 50 mm softwood stock pattern architraves, make good skirtings and make out board flooring, hang 762 x 1981 x 38 mm plywood faced flush door on pair 75 mm steel butts and fit mortice lock and furniture - as shown on drawing E on page 719

	Unit	Hours C	Hours L	Labour net	Plant net	Material net	Price net	Price with 25%
in 102.5 mm wall, with one 100 x 150 mm lintel, 32 x 150 mm softwood linings with 13 x 38 mm stop and architraves both sides	each	22.50	9.75	267.31	-	137.56	404.87	506.09
in 215 mm wall, with two 100 x 150 mm lintels, 50 x 100 mm softwood frame, 13 x 38 mm stop and architraves one side	each	29.00	16.00	367.68	-	168.99	536.67	670.84

Cut away skirting one side and cut openings through faced brick external wall, remove debris, cut away for and build in steel box and precast concrete lintels internally and pin up to work over, form 102.5 x 215 mm flat arch externally and quoin up jambs with old bricks faced and pointed externally, fix 75 x 100 mm softwood rebated and moulded door frame with oak sill and 6 x 40 mm galvanised steel water bar, make out plastering and skirting, hang 838 x 1981 x 44 mm plywood faced flush door on 1½ pairs 100 mm steel butts, glaze panel with 6 mm white patterned glass secured with softwood beads and fit cylinder rim night latch and 32 x 113 mm softwood weather board - as shown on drawing F on page 720

	Unit	Hours C	Hours L	Labour net	Plant net	Material net	Price net	Price with 25%
in 215 mm wall, with one steel box lintel	each	23.00	14.75	305.49	-	239.04	544.53	680.66
in 327.5 mm wall, with one steel box lintel, one precast concrete 100 x 150 mm lintel	each	30.75	21.00	417.06	-	280.64	697.70	872.12

Form 1200 x 327.5 x 102.5 mm brick on edge steps in engineering bricks in cement mortar 1:3 fair face and flush pointed on 150 mm concrete base and hardcore filling - as shown on drawing F on page 720

	Unit	Hours C	Hours L	Labour net	Plant net	Material net	Price net	Price with 25%
complete	each	1.75	3.25	37.58	-	22.72	60.30	75.37

Cut away skirting one side and cut openings through 250 mm faced brick external hollow wall, remove debris, cut away for and build in proprietary galvanised steel combined lintel and cavity tray to support inner and outer skins and pin up to work over, build in vertical damp-proof courses, quoin up jambs with old bricks faced and pointed externally, fix 50 x 100 mm softwood rebated and moulded door frame with hardwood sill and 6 x 40 mm galvanised steel water bar, make out plastering and skirting, hang 838 x 1981 x 44 mm plywood faced flush door on 1½ pairs 100 mm steel butts, glaze panel with 4 mm white patterned glass secured with softwood beads and fit cylinder rim night latch and 32 x 113 mm softwood weather board - as shown on drawing G on page 721

	Unit	Hours C	Hours L	Labour net	Plant net	Material net	Price net	Price with 25%
complete	each	25.25	15.75	332.39	-	273.41	605.80	757.25

Form 1200 x 327.5 x 102.5 mm brick on edge steps in engineering bricks in cement mortar 1:3 fair faced and flush pointed on 150 mm concrete base and hardcore filling - as shown on drawing G on page 721

	Unit	Hours C	Hours L	Labour net	Plant net	Material net	Price net	Price with 25%
complete	each	1.75	3.25	37.58	0.09	22.72	60.39	75.49

Combined trades spot items

Alterations and repairs	Unit	Hours C	Hours L	Labour net	Plant net	Material net	Price net	Price with 25%
Walls				£	£	£	£	£
						VAT not included		

Cutting openings and providing and fitting doors *(continued)*

Cut away skirtings both sides and cut openings through 100 mm stud partition, remove debris, trim opening with additional 50 x 100 mm softwood stud and head piece, fix 32 x 125 mm softwood linings with 13 x 38 mm softwood stop, make good plasterboard both sides, fix 19 x 50 mm softwood stock pattern architraves both sides, make good skirtings and make out boarded flooring, hang 762 x 1981 x 38 mm plywood faced flush door on pair 75 mm steel butts and fit mortice lock and furniture - as shown on drawing H on page 722

complete	each	17.00	5.00	186.02	-	232.01	418.03	522.54

Cutting openings and providing and fitting windows

Cut openings through faced brick wall, remove debris, cut away for and build in precast concrete reinforced lintels internally and pin up to work over, form 102.5 x 215 mm flat arch externally using proprietary galvanised steel box lintel with lip to support flat arch, and quoin up jambs with old bricks faced and pointed externally, fix 1200 x 1350mm softwood stock pattern casement windows with 50 x 150 mm softwood sill and complete with fasteners and stays, make out plastering and glaze window with sealed double glazed units secured with softwood beads - as shown on drawings J1 and J2 on page 723

in 215 mm wall, with one 100 x 150 mm lintel, 32 x 175 mm softwood window board and plastered reveals internally	each	20.00	14.75	278.61	-	228.49	507.10	633.87
in 327.5 mm wall, with two 100 x 150 mm lintels, 32 x 275 mm softwood window board, and plastered reveals internally	each	28.75	21.25	400.82	-	286.40	687.22	859.02

Cut openings through 250 mm faced brick hollow wall, remove debris, cut away for and build in proprietary galvanised steel combined lintel and cavity tray to support inner and outer skins and pin up to work over, build in vertical damp-proof courses, quoin up jambs with old bricks faced and pointed externally, fix 1770 x 1200 mm softwood stock pattern casement window with flush sill and complete with fasteners and stays, fix two course roofing tile sill externally, fix 32 x 200 mm softwood window board, make out plastering and glaze window with sealed double glazed units secured with softwood beads - as shown on drawing K on page 724

complete	each	24.00	16.75	327.93	-	336.62	664.55	830.69

Cut openings through 250 mm faced brick hollow wall, remove debris, cut away for and build in proprietary galvanised steel combined lintel and cavity tray to support inner and outer skins and pin up to work over, build in vertical damp-proof courses, quoin up jambs with old bricks faced and pointed externally, fix 1486 x 1067 mm standard metal window with 75 x 150 mm hardwood sill, fix 200 x 19 mm quarry tile sill internally, make out plastering and glaze window with sealed double glazed units - as shown on drawing K on page 724

complete	each	21.25	17.25	306.66	-	432.43	730.09	912.61

Combined trades spot items

Alterations and repairs	Unit	Hours C	Hours L	Labour net	Plant net	Material net	Price net	Price with 25%
Walls				£	£	£	£	£
						VAT not included		

Cutting openings and providing and fitting hatches

Cut openings through 102.5 mm brick wall, remove debris, cut away for and build in 100 x 150 mm precast concrete reinforced lintel and pin up to work over, quoin up jambs with old bricks, fix 600 x 550 mm serving hatch constructed in softwood with 32 x 150 mm linings, 32 x 225 mm sill, 13 x 38 mm stops, hardboard faced flush doors lipped on all edges, hung on two pairs 50 mm steel butts and fitted with four "D" handles and four ball catches and 19 x 50 mm stock pattern architraves both sides - as shown on drawing L on page 724

	Unit	Hours C	Hours L	Labour net	Plant net	Material net	Price net	Price with 25%
complete	each	9.50	4.25	113.76	-	118.39	232.15	290.19

Combined trades spot items

Drawing A : Cutting clear opening in wall

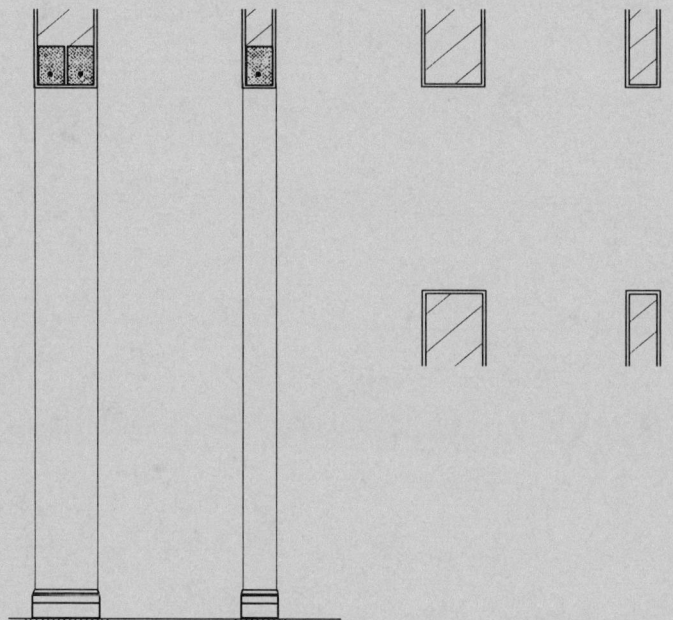

Combined trades spot items

Drawing B : Cutting clear opening in wall

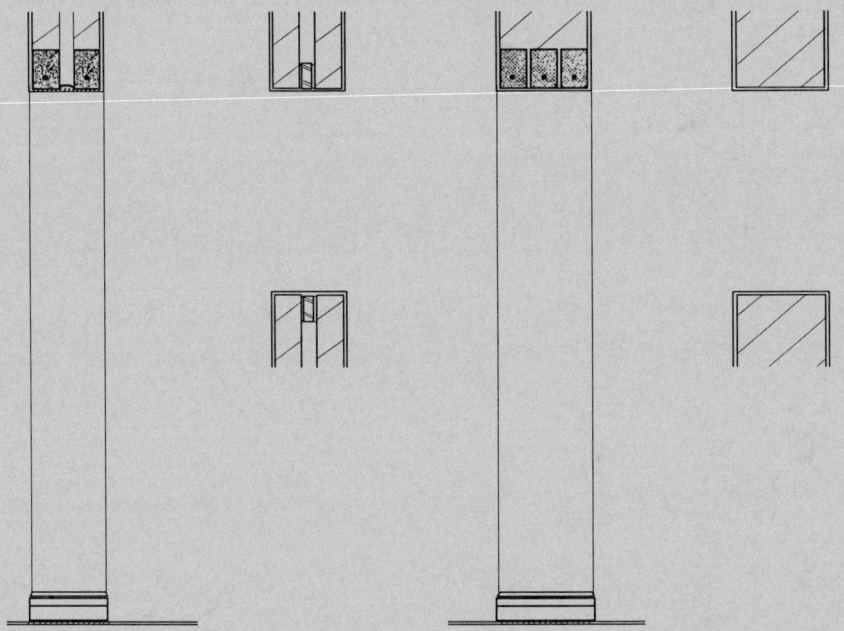

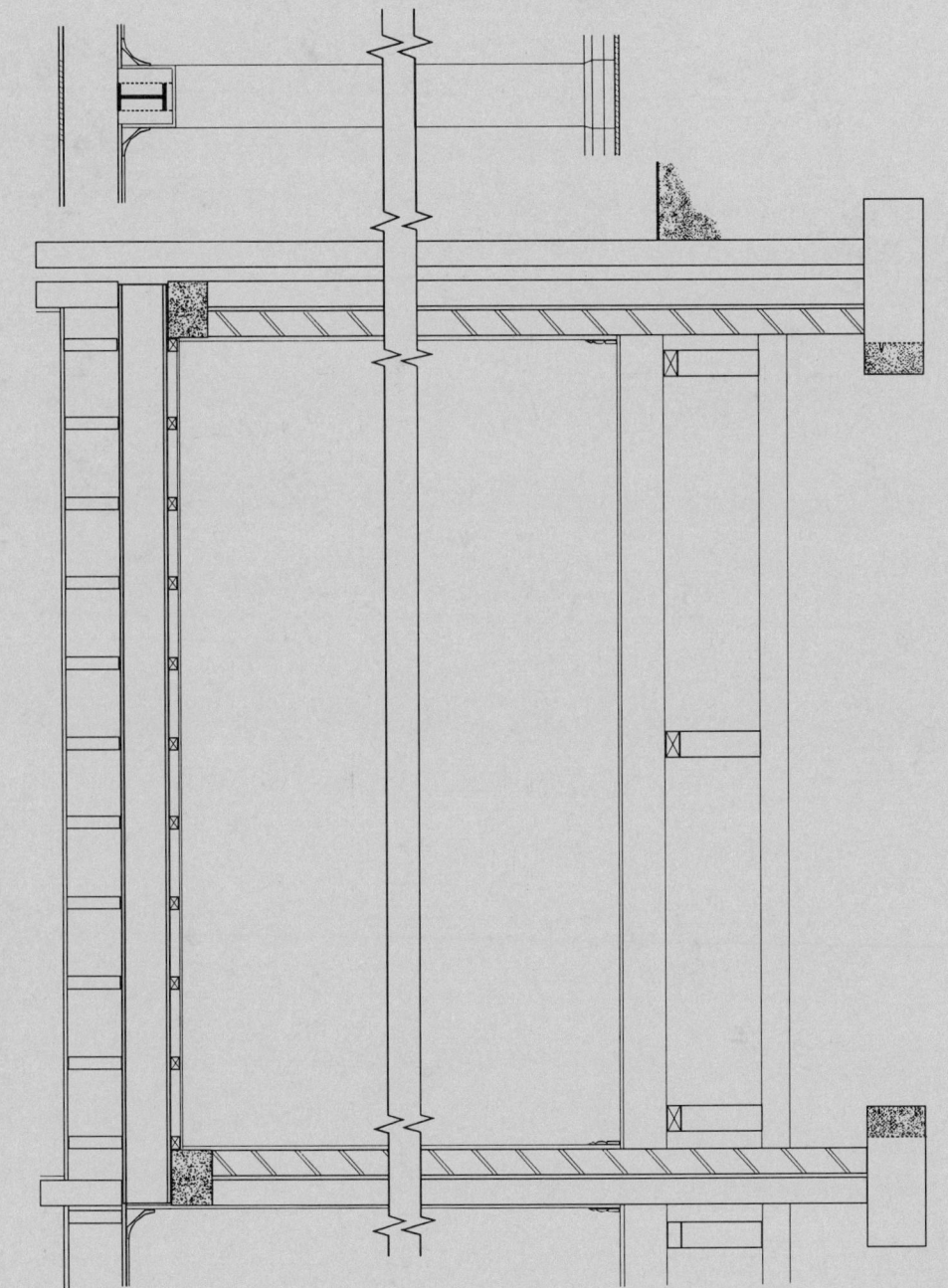

Drawing C : Cutting clear opening in wall

Combined trades spot items

Drawing D : Cutting opening and providing and fitting door

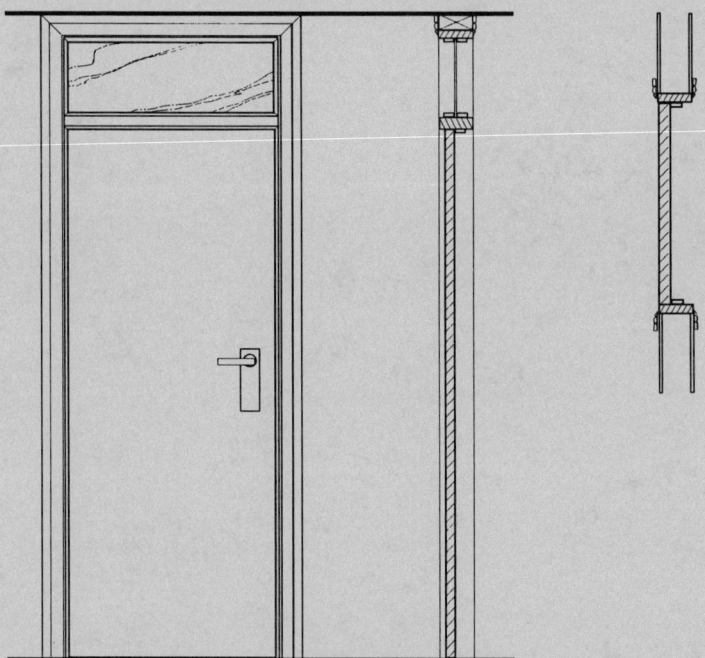

Combined trades spot items

Drawing E : Cutting opening and providing and fitting door

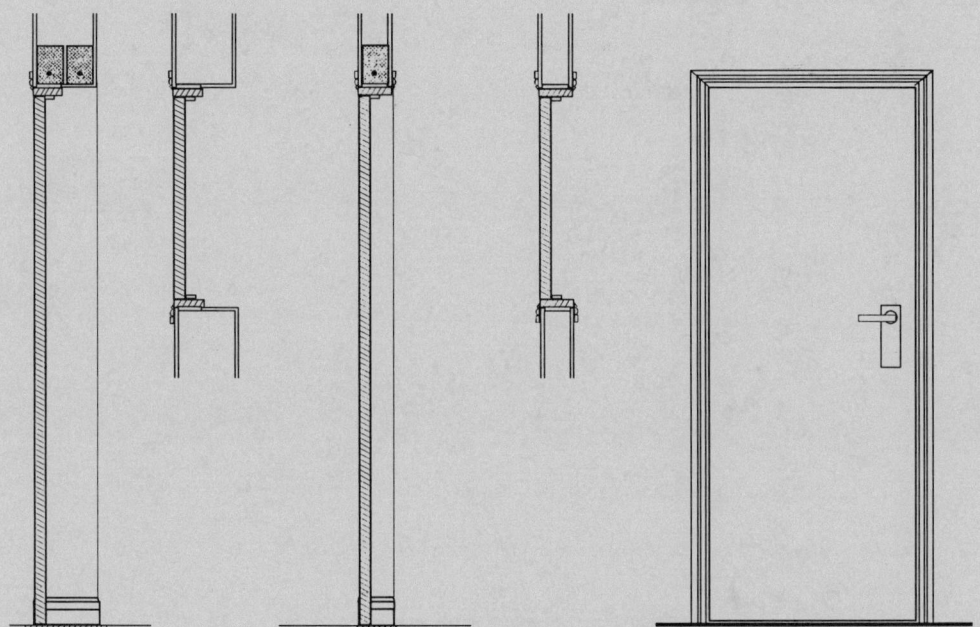

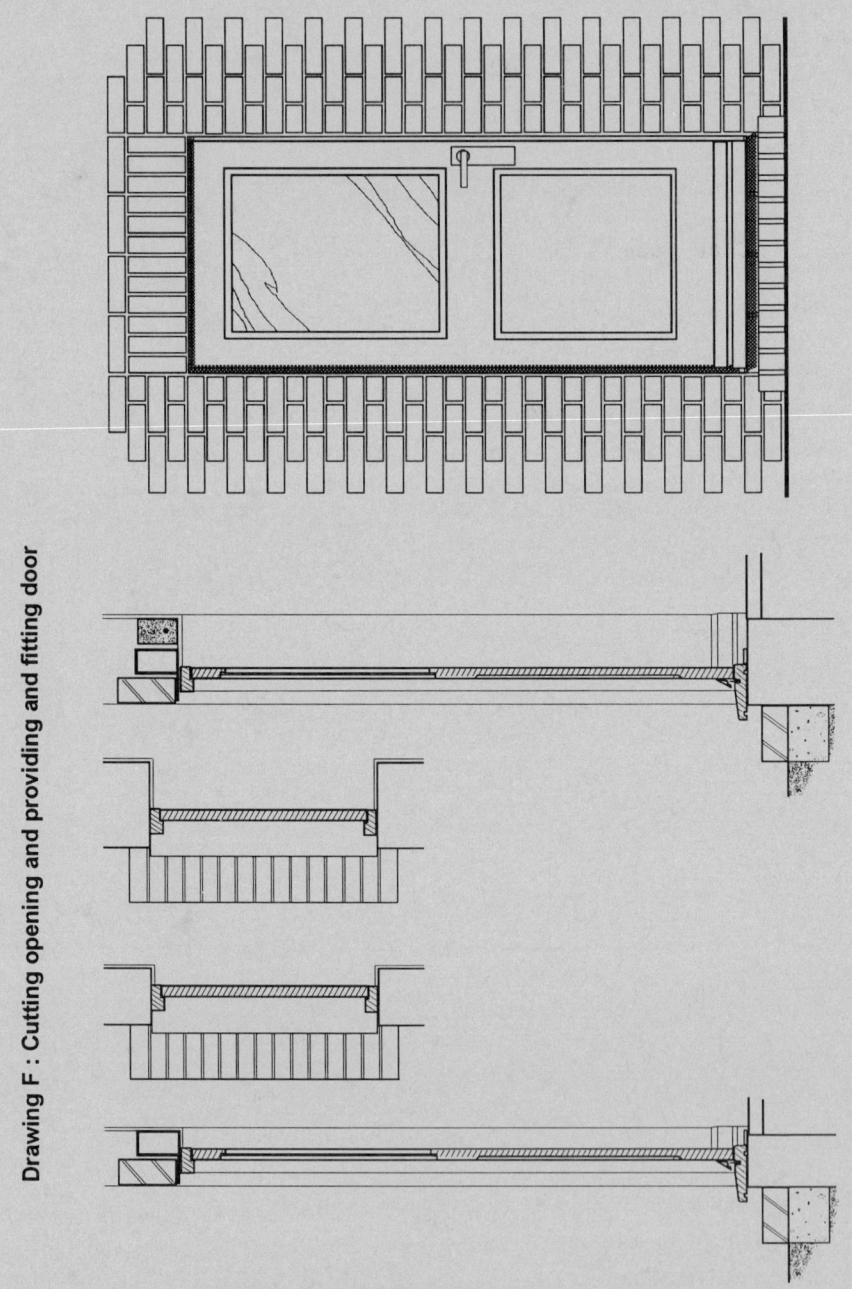

Drawing F : Cutting opening and providing and fitting door

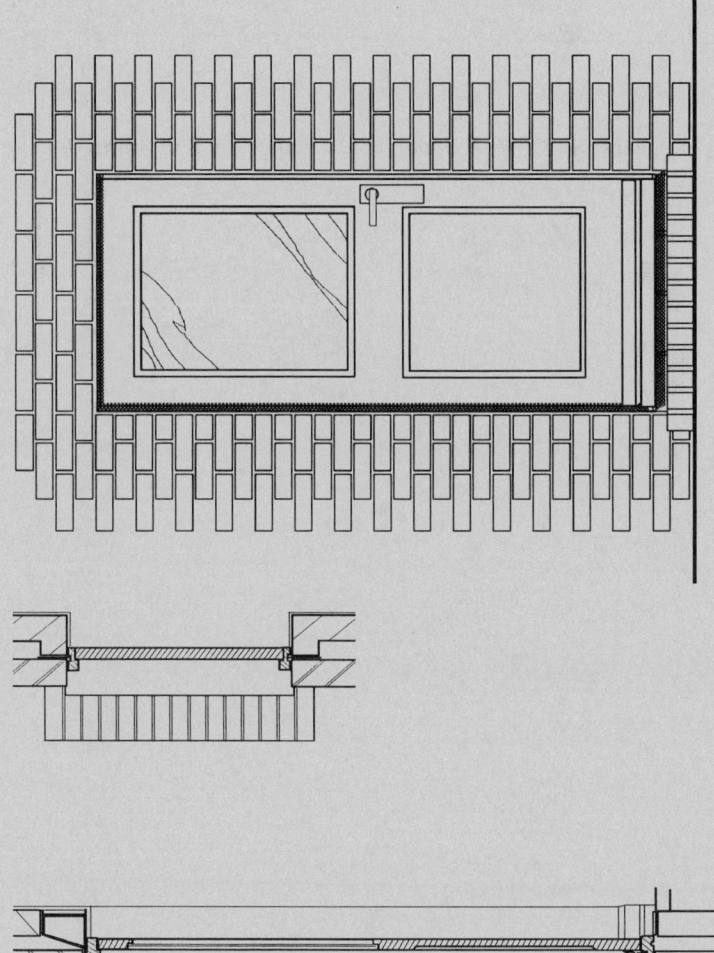

Drawing G : Cutting opening and providing and fitting door

Drawing H : Cutting opening and providing and fitting door

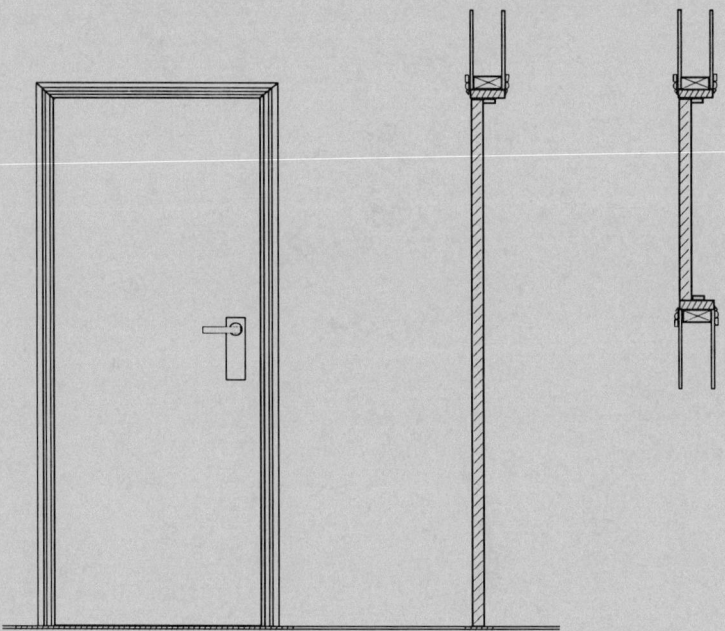

Combined trades spot items

Drawing J1 : Cutting opening and providing and fitting window

Drawing J2 : Cutting opening and providing and fitting window

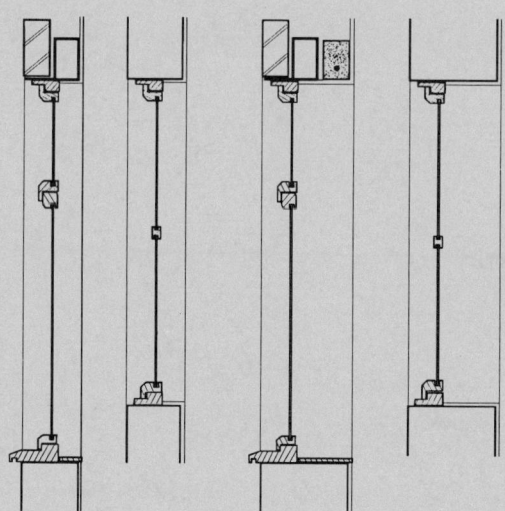

Combined trades spot items

Drawing K : Cutting opening and providing and fitting window

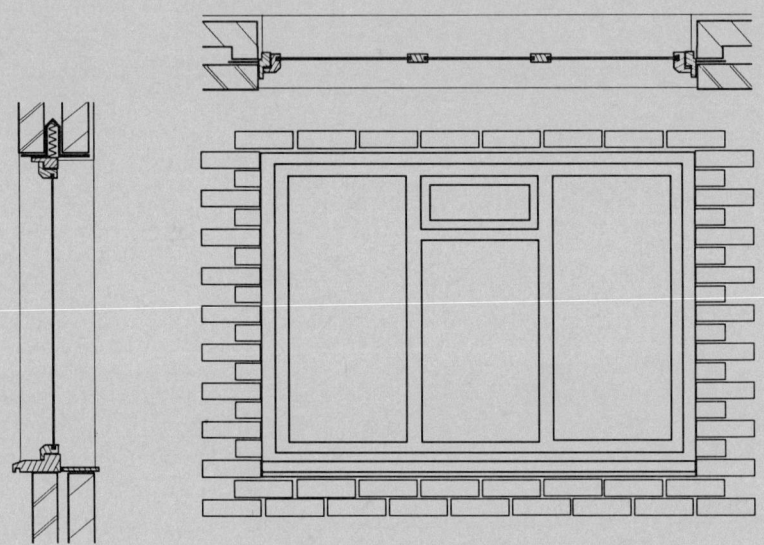

Drawing L : Cutting opening and providing and fitting hatch

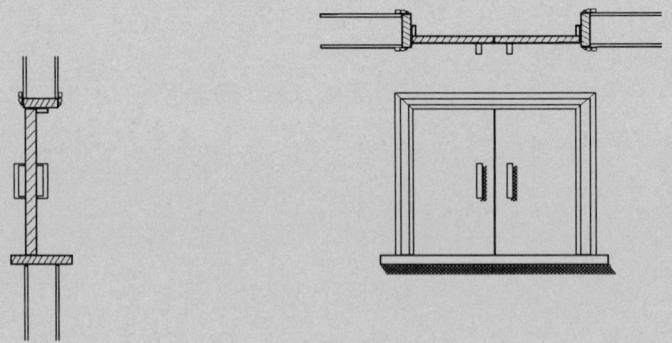

Combined trades spot items

Alterations and repairs	Unit	Hours C	Hours L	Labour net	Plant net	Material net	Price net	Price with 25%

Flues

£ £ £ £ £

VAT not included

Installing flexible flue linings

Remove chimney-pot and register plate, sweep flue and fix 100 mm flexible flue lining in existing flue using 7.5 m stainless steel lining pack comprising lining, nose cone, draw cord, sealing plate and clamp and GC1 terminal complete, fix ceiling plate, cut lining to length, fix loose collar to smoke pipe, seal in lining, fix new two piece register plate seal all around and fix GC1 terminal

	Unit	Hours C	Hours L	Labour net	Plant net	Material net	Price net	Price with 25%
complete	each	11.00	11.00	172.70	-	81.93	254.63	318.29
For every metre variation in the length of flue lining (within a limit of 3 metres either way) add or deduct	m	0.17	0.17	2.67	-	4.41	7.08	8.85
Vermiculite filling between flexible flue lining and existing flue per 7.5 metre length	each	1.25	1.25	19.62	-	33.08	52.70	65.87
For every metre variation in the length of flue add or deduct	m	0.17	0.17	2.67	-	4.41	7.08	8.85

Removing and filling in old fireplaces

Cut away and remove fireplaces, fill in opening with 60 mm clinker aggregate blockwork cut and bonded to brickwork, make out plastering and 25 x 175 mm softwood moulded skirting to match existing, make out flooring in cement and sand where hearth removed, fix chimney-pot hood to existing pot and cut away into flue externally at low level and fix 225 x 225 mm air brick to ventilate flue

	Unit	Hours C	Hours L	Labour net	Plant net	Material net	Price net	Price with 25%
complete	each	9.25	7.75	135.11	-	49.69	184.80	231.00

Removing chimney-breasts

Cut away and remove 1350 x 327.5 (projection) x 2400 mm (storey height) brick chimney-breasts and concrete hearths, fill opening in floor with 50 x 100 mm softwood bearers and 25 mm softwood tongued and grooved flooring, face up wall and make out wall plastering where breast removed, fix 25 x 175 mm softwood moulded skirting to match existing and fill opening in ceiling with 50 x 50 mm softwood bearers and two layers of 9.5 mm plasterboard and skim coat made good to existing

	Unit	Hours C	Hours L	Labour net	Plant net	Material net	Price net	Price with 25%
	each per storey	20.50	20.00	318.48	24.44	52.53	395.45	494.31

Taking down chimney-stacks

Carefully take down 777.5 x 440 mm brick chimney-stacks to below line of roof (a height of 1000 mm), remove debris, seal top of flues and fill opening in roof with 50 x 100 mm softwood rafters, reinforced bituminous felt underlay, softwood battens and

	Unit	Hours C	Hours L	Labour net	Plant net	Material net	Price net	Price with 25%
508 x 254 mm slating	each	8.00	9.00	132.34	3.32	62.76	198.42	248.02
215 x 165 mm machine-made sand face clay tiling	each	10.00	10.50	160.37	3.32	44.58	208.27	260.34

Alterations and repairs	Unit	Hours C	Hours L	Labour net	Plant net	Material net	Price net	Price with 25%
Roofs				£	£	£	£	£
					VAT not included			

Access to roofs

Cutting openings and providing and fitting trap doors

Cut openings through plasterboard ceiling, remove debris, trim opening with 50 x 100 mm softwood, fix 25 x 125 mm softwood linings with planted stops, make good plasterboard, fix 19 x 50 mm softwood stock moulded architrave and fit 900 x 600 x 19 MDF loose trap

	Unit	Hours C	Hours L	Labour net	Plant net	Material net	Price net	Price with 25%
complete	each	8.00	2.50	88.53	-	29.14	117.67	147.09

Cutting openings and providing and fitting loft ladders

Cut openings through plasterboard ceiling, remove debris, trim opening with 50 x 100 mm softwood, fix 25 x 125 mm softwood linings with planted stops, make good plasterboard, fix 19 x 50 mm softwood stock moulded architrave, hang 900 x 600 x 19 mm MDF trap door on pair 75 mm steel butts and fit aluminium loft stairway with handrail

	Unit	Hours C	Hours L	Labour net	Plant net	Material net	Price net	Price with 25%
complete	each	12.00	2.50	124.37	-	229.55	353.92	442.40

Work to roofs

Cutting openings and providing and fitting dormer windows

Strip slates or tiles, cut away battens and trim rafters to form openings, strengthen rafters at sides with 50 x 100 mm softwood, fix 1500 x 900 mm standard window, 75 x 150 mm hardwood sill, construct 50 x 100 mm softwood dormer framing 150 x 50 mm joists to roof, fix softwood firrings and cover roof, cheeks and apron with 18 mm WBP plywood, finish roof and cheeks with small plain concrete tile hanging or slates on 38 x 25 mm softwood treated battens. Roofing taken up 400 mm under slates or tiles and with turn-down at edges, allow for 100 mm mineral quilt insulation within dormer cheeks, apron and roof. Fix code 4 lead soakers at cheeks, make good slating or tiling, fix code 4 lead apron, fit softwood fascia and cover fillets and glaze window with sealed double glazed units secured with softwood beads (plasterwork not included) - as on drawing M on page 727

	Unit	Hours C	Hours L	Labour net	Plant net	Material net	Price net	Price with 25%
complete	each	48.00	29.00	625.54	-	597.81	1223.35	1529.19

Cutting openings and providing and fitting skylights

Strip slates or tiles, cut away battens and trim rafters to form openings, strengthen rafters at sides with 25 x 100 mm softwood, fix 780 x 980 mm factory made skylight unit comprising double glazed pivoted timber sash, hardwood faced linings, skylight furniture and prefabricated flashings and make good slating or tiling (plasterwork not included)

	Unit	Hours C	Hours L	Labour net	Plant net	Material net	Price net	Price with 25%
complete	each	10.50	10.50	164.85	-	204.32	369.17	461.46

Strip slates or tiles, cut away battens and trim rafters to form openings, strengthen rafters at sides with 25 x 100 mm softwood, fix 50 x 125 mm softwood kerb and 25 x 250 mm softwood linings all round with 25 x 50 mm bearers at sides to support batten ends, fix proprietary rooflight in softwood with aluminium facings, glaze skylight with 20 mm sealed double glazing unit, inner pane glazed with laminated glass, fix proprietary flashing system, copper nailed and dressed down face of kerb and make good slating or tiling (plasterwork not included)

	Unit	Hours C	Hours L	Labour net	Plant net	Material net	Price net	Price with 25%
complete	each	21.75	16.25	304.40	-	234.53	538.93	673.66

Combined trades spot items

Drawing M : Cutting opening and providing and fitting dormer window

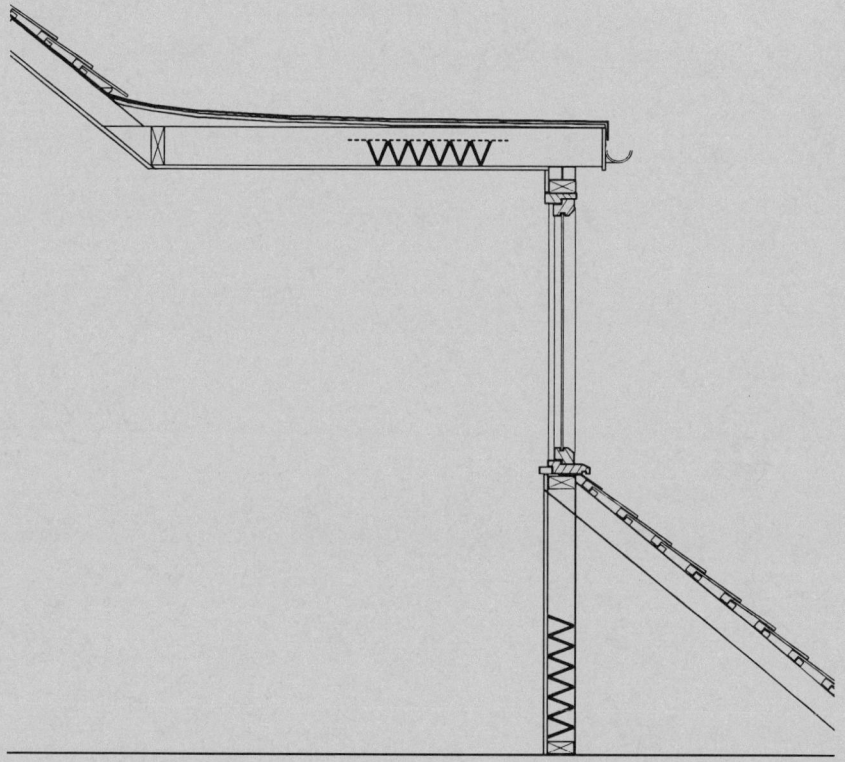

Combined trades spot items

Drawing N : Cutting opening in timber floor

Combined trades spot items

Alterations and repairs	Unit	Hours C	Hours L	Labour net	Plant net	Material net	Price net	Price with 25%
Floors				£	£	£	£	£
						VAT not included		

Filling openings in timber floors

Fill 2750 x 900 mm stair-wells with three joists supported by galvanised steel joist hangers, lay 18 mm flooring grade chipboard tongued and grooved flooring, fix 25 x 175 mm softwood moulded skirting and finish underside with one layer of 9.5 mm plasterboard and skim coat made good to existing

50 x 150 mm joists	each	10.00	3.25	111.50	-	136.01	247.51	309.39
50 x 200 mm joists	each	10.25	3.25	113.74	-	142.70	256.44	320.55

Fill 2100 x 2100 mm lift-wells with six joists supported by galvanised steel joist hangers, lay 18 mm flooring grade chipboard tongued and grooved flooring, fix 25 x 175 mm softwood moulded skirting and finish underside with two layers of 9.5 mm plasterboard and skim coat made good to existing

50 x 150 mm joists	each	15.50	6.00	179.32	-	221.01	400.33	500.41
50 x 200 mm joists	each	16.00	6.00	183.80	-	231.50	415.30	519.12

Cutting openings in timber floors

Strut up floor, remove skirting, cut away boarding, hack off plastered ceiling and cut away joists to form 2750 x 900 mm stair-wells and remove debris, fix strengthening joist bolted to existing, fix trimmer joist supported one end by face fixed galvanised steel joist hanger to wall and the other end by galvanised steel joist hanger, support ends of cut joists by galvanised steel joist hangers, fix old joist to trim opening to correct width, fill joist holes in walls, make out wall plastering and make good flooring and plastered ceiling - as shown on drawing N on page 728

complete	each	18.00	14.75	260.69	1.02	47.50	309.21	386.51

Strut up floor etc to form 2750 x 900 mm stair-wells as last item but with trimmer joists at both ends of stair-well

	each	20.50	17.25	299.94	1.02	71.57	372.53	465.66

Strut up floor, remove skirting, cut away boarding, hack off plastered ceiling and cut away joists to form 2100 x 2100 mm lift-wells and remove debris, fix strengthening joist bolted to existing, fix trimmer joist supported one end by face fixed galvanised steel hanger fixed to wall and the other end by galvanised steel joist hanger, support ends of cut joists by galvanised steel joist hangers, fix old joist to trim opening to correct width, fill joist holes in walls, make out wall plastering and make good flooring and plastered ceiling - as shown on drawing N on page 728

complete	each	21.00	18.00	309.48	1.34	59.03	369.85	462.31

Strut up floor etc to form 2100 x 2100 mm lift-wells as last item but with trimmer joists at both ends of lift-well

	each	25.00	21.75	370.59	1.34	92.18	464.11	580.14

Renewing timber ground floors

Take off skirtings, take up defective flooring, existing 50 x 100 mm joists and plates to ground floor over an area 3000 x 3000 mm and remove debris, fix new 50 x 100 mm softwood plates bedded on polyethylene damp-proof course on existing sleeper walls, fix 50 x 100 mm softwood joists, lay 25 mm softwood tongued and grooved board flooring and treat new plates and joists with preservative and refix old skirtings

complete	each	18.00	7.25	210.14	-	272.62	482.76	603.45

Combined trades spot items

Alterations and repairs	Unit	Hours C	Hours L	Labour net	Plant net	Material net	Price net	Price with 25%
Floors				£	£	£	£	£
					VAT not included			
Removing staircases								
Take out one storey staircases and make out plastering where wall string removed	each	4.50	3.50	63.91	-	2.81	66.72	83.40

Combined trades spot items

Alterations and repairs	Unit	Hours C	Hours L	Labour net	Plant net	Material net	Price net	Price with 25%
Plumbing				£	£	£	£	£
						VAT not included		

Cold water cisterns

Plastics cisterns with lid and 60 mm glass fibre filled polythene insulating jacket and with 13 mm ball valve, 22 mm copper overflow (2 m), one 22 mm service connection and 25 mm softwood boarded platform with 50 x 75 mm softwood bearers on ceiling joists

18 litre (4 gallon)	each	C 2.00 P 4.00	L 1.00 A 4.00	93.18	-	62.34	155.52	194.40
68 litre (15 gallon)	each	C 2.00 P 4.25	L 1.00 A 4.25	97.46	-	97.85	195.31	244.14

Plastics cisterns with lid and 60 mm glass fibre filled polythene insulating jacket and with 13 mm ball valve, 22 mm copper overflow (2 m), two 28 mm copper service connections and 25 mm softwood boarded platform with 50 x 75 mm softwood bearers on ceiling joists

114 litre (25 gallon)	each	C 2.50 P 5.25	L 1.00 A 5.25	119.07	-	134.51	253.58	316.97
227 litre (50 gallon)	each	C 2.50 P 5.50	L 1.00 A 5.50	123.35	-	174.80	298.15	372.69

Sanitary fittings

Note: See "Plumbing and engineering installations" for connections to stacks which are not included in the following figures.

560 x 405 mm vitreous china lavatory basins with pair of 13 mm chromium plated pillar valves, chromium plated waste, plug, chain and stay, pair of towel rail brackets plugged to wall and with 32 mm plastics trap and waste pipe (2 m) and 15 mm copper cold and hot water services (3 m each) connected to existing

white basin sets	each	C 1.00 P10.00	L 2.50 A10.00	197.11	-	111.91	309.02	386.27
coloured basin sets	each	C 1.00 P10.00	L 2.50 A10.00	197.11	-	126.24	323.35	404.19

560 x 405 mm vitreous china lavatory basins with pair of 13 mm chromium plated pillar valves, chromium plated waste, plug, chain and stay, bracket plugged to wall, pedestal plugged to floor and with 32 mm plastics trap and waste pipe
(3 m) and 15 mm copper cold and hot water services (4 m each) connected to existing

white basin and pedestal sets	each	C 1.00 P10.50	L 2.50 A10.50	205.67	-	140.60	346.27	432.84
coloured basin and pedestal sets	each	C 1.00 P10.50	L 2.50 A10.50	205.67	-	161.54	367.21	459.01

Combined trades spot items

Alterations and repairs	Unit	Hours C	Hours L	Labour net	Plant net	Material net	Price net	Price with 25%

Plumbing

				£	£	£	£	£

VAT not included

Sanitary fittings *(continued)*

1700 mm baths with pair of 19 mm chromium plated pillar valves, chromium plated overflow and waste, plug, chain and stay and enamelled hardboard panels to front and one end with angle strip complete and with 38 x 50 mm softwood panel framing, 38 mm plastics combined trap and overflow connection, 38 mm plastics waste pipe (2 m) and 22 mm copper cold and hot water services (3 m each) connected to existing

	Unit	Hours C	Hours L	Labour net	Plant net	Material net	Price net	Price with 25%
white acrylic baths and panel sets complete	each	C 6.00 P11.00	L 5.00 A11.00	275.89	-	261.79	537.68	672.10
coloured acrylic baths and panel sets complete	each	C 6.00 P11.00	L 5.00 A11.00	275.89	-	268.40	544.29	680.36
white pressed steel baths and panel sets complete	each	C 6.00 P11.00	L 5.00 A11.00	275.89	-	250.76	526.65	658.31
coloured pressed steel baths and panel sets complete	each	C 6.00 P11.00	L 5.00 A11.00	275.89	-	268.40	544.29	680.36

Sink unit fittings with stainless steel drainer top, pair of 13 mm chromium plated high necked bib valves and chromium plated overflow and waste, plug, chain and stay complete with 38 mm plastics combined trap and overflow connection, 38 mm plastics waste pipe (2 m) and 15 mm copper cold and hot water services (3 m each) connected to existing

	Unit	Hours C	Hours L	Labour net	Plant net	Material net	Price net	Price with 25%
1000 x 500 x 900 mm single drainer unit sets complete - no drawers	each	C 6.00 P 7.00	L 5.00 A 7.00	207.37	-	305.71	513.08	641.35
1000 x 500 x 900 mm single drainer unit sets complete - with drawers	each	C 6.00 P 7.00	L 5.00 A 7.00	207.37	-	338.78	546.15	682.69
1500 x 500 x 900 mm double drainer unit sets complete - no drawers	each	C 6.50 P 7.00	L 5.00 A 7.00	211.85	-	415.96	627.81	784.76
1500 x 500 x 900 mm double drainer unit sets complete - with drawers	each	C 6.50 P 7.00	L 5.00 A 7.00	211.85	-	476.60	688.45	860.56

Combined trades spot items

Alterations and repairs	Unit	Hours C	Hours L	Labour net	Plant net	Material net	Price net	Price with 25%
Plumbing					£	£	£	£	£

VAT not included

Sanitary fittings *(continued)*

High level WC suites with white vitreous china pan, single seat, 7.5 litre standard finish black plastics flushing cistern with BS ball valve and all fittings, flush pipe and connector and soil pipe connector complete and with 22 mm overflow (1 m) and 15 mm cold water service (3 m) connected to existing

complete	each	C 0.50	L 1.50	117.37	-	154.33	271.70	339.62
		P 6.00	A 6.00					

Low level WC suites with vitreous china pan, seat and cover, 7.5 litre streamlined finish plastics flushing cistern with BS ball valve and all fittings, flush bend and connector and soil pipe connector complete and with 22 mm overflow (1 m) and 15 mm cold water service (3 m) connected to existing

white suites	each	C 0.50	L 1.50	117.37	-	158.74	276.11	345.14
		P 6.00	A 6.00					
coloured suites	each	C 0.50	L 1.50	117.37	-	191.81	309.18	386.47
		P 6.00	A 6.00					

760 x 760 mm (internally) corner shower cubicles, comprising plastic tray and back panel, acrylic side panel, curtain rail and curtain, with mixer valve, handset and grated waste, and with 32 mm plastics trap and waste pipe (2 m) and 15 mm copper cold and hot water services (3 m) connected to existing

complete	each	C 5.00	L 2.00	229.58	-	598.72	828.30	1035.37
		P10.00	A10.00					

Water heaters

Instantaneous gas-fired multipoint water heater with balanced flue set, and with 15 mm copper cold water and gas services (3 m each) connected to existing

complete	each	C 1.00	L 2.50	162.85	-	417.57	580.42	752.52
		P 8.00	A 8.00					

Instantaneous electric sink water heater with swivel outlet, and with 15 mm copper cold water service (3 m) connected to existing - electrical work not included

complete	each	C 0.50	L 0.50	67.80	-	121.45	189.25	236.56
		P 3.50	A 3.50					

90 litre (20 gallon) low pressure type electric water heaters with one immersion heater, and with 22 mm cold water service (3 m) connected to existing - electrical work not included

complete	each	C 0.50	L 0.50	76.37	-	673.68	750.05	937.56
		P 4.00	A 4.00					

Combined trades spot items

Alterations and repairs	Unit	Hours C	Hours L	Labour net	Plant net	Material net	Price net	Price with 25%
Fencing				£	£	£	£	£
						VAT not included		
Renewing posts								
Remove gate and hinges from post, take out post and base, excavate for and fix new 175 x 175 x 2130 mm wrought Keruing post, set in new concrete base and								
rehang gate	each	3.75	4.25	62.24	-	46.49	108.73	135.91
Repairing posts								
Excavate for and fix 1100 mm concrete spur bolted to post								
set in compacted spoil	each	2.00	-	17.92	-	8.63	26.55	33.19
set in concrete base	each	2.50	0.25	24.08	-	10.97	35.05	43.81

Daywork

Where work, by its nature or urgency, is not suitable for the preparation of firm estimates or cannot properly be measured and valued, such work may, by agreement between the parties, be charged for on a "Daywork" basis. For this purpose definitions of prime cost have been published jointly by The Royal Institution of Chartered Surveyors and the Building Employers Confederation for convenience and for use by people who choose to use them; similarly a schedule of basic plant charges has been published by The Building Cost Information Service (BCIS) and The Royal Institution of Chartered Surveyors.

The publications are:

Definition of Prime Cost of Daywork Carried Out Under a Building Contract
Second Edition, 1st December 1975

Definition of Prime Cost of Building Works of a Jobbing or Maintenance Character
Second Edition, 1980

Schedule of Basic Plant Charges
Fourth Revision, 1st January 1990

Copies of these publications may be obtained from RICS Books, tel. 0171 222 7000.

As there has been no amendment to these publications for some time it has been decided not to reproduce them in this edition of the Griffiths Building Price Book. However we have included the updated version of the calculation of a typical standard hourly rate, this is set out overleaf.

Daywork

Example of standard hourly rates

The RICS and BEC, as joint publishers of the Definition of prime cost of daywork carried out under a building contract, prepare updated versions of the example (printed on page 8 of the Definition) of the calculation of the standard hourly base rate (as defined in section 3) for a typical building craft operative and labourer in CIJC Grade A areas. The following version has been prepared from the latest information available. The example is for convenience only and does not form part of the Definition; all basic rules are subject to re-examination according to when and where the dayworks are executed.

The following all-in rates are based upon the three year agreement promulgated on 23 July 1997. The calculations below of standard hourly base rates in Grade A areas are based on the third year of the agreement with effect on and from Monday 28 June 1999,

Standard working hours per annum
52 weeks at 39 hours =   2028   hours
   Less 21 days annual holiday
        16 days at 8 hours   128
        5 days at 7 hours   35
  9 days public holiday
        8 days at 8 hours   64
        1 day at 7 hours   7   234 hours
                                        1794 hours

			Craftsman			General Operative
			£			£
Guaranteed minimum weekly earnings standard basic rate	47.8 weeks	235.95	11,278.41	47.8 weeks	177.45	8,482.11
Guaranteed minimum bonus*	47.8 weeks		-	47.8 weeks		-
			11,278.41			8,482.11
NIC Employer's contribution	12.2%		1,375.97	12.2%		1,034.82
CITB Allowance	0.25 %		28.20	0.25 %		21.21
Annual holiday credits **	47 weeks	21.30	1,001.10	47 weeks	21.30	1,001.10
Annual Labour cost as defined in Section 3			£ 13,683.68			£ 10,539.24
Hourly rate of labour as defined in Section 3 clause 3.02 divided by 1794 worked hours			£ 7.63			£ 5.87

   * Only included in hourly base rate for operatives receiving such payments.
  ** 21 days annual holiday taken in conjunction with public holidays as 2 calendar weeks holiday at Christmas, 1 calendar week at Easter and 2 weeks in summer.

Percentage additions

In the region of 150% addition to the above hourly base ratio is suggested to cover incidental costs as defined in Section 7 of the Definition of prime cost of daywork.

Memoranda

Memoranda

Unit equivalents

	Metric unit	Abbreviation	Imperial equivalent	
Length	1 millimetre	1 mm	0.039	in
	1 metre	1 m	{ 3.281 { 1.094	ft yd
	1 kilometre	1 km	0.621	mile
Area	1 square millimetre	1 mm²	0.001 55	in²
	1 square metre	1 m²	{ 10.764 { 1.196	ft² yd²
	1 hectare	1 ha	{ 11 960 { 2.471	yd² acre
Volume	1 cubic metre	1 m³	{ 35.315 { 1.308	ft³ yd³
	1 litre	1 l	{ 1.760 { 0.220	pint UK gal
	5 litres	5 l	1.100	UK gal
Weights	1 kilogramme	1 kg	2.205	lb
	50 kilogrammes	50 kg	{ 110.231 { 0.984	lb cwt
	1 tonne	1 t	0.984	ton
	1 kilogramme per metre	1 kg/m	{ 0.672 { 2.016	lb/ft lb/yd
	1 kilogramme per square metre	1 kg/m²	{ 0.205 { 1.843	lb/ft² lb/yd²
	1 kilogramme per cubic metre	1 kg/m³	{ 0.062 { 1.686	lb/ft³ lb/yd³
Force	1 newton	1N	0.225	lbf
	1 kilonewton	1 kN	{ 224.809 { 0.100 36	lbf tonf
Pressure	1 newton per square millimetre	1 N/mm²	145.038	lbf/in²
	1 kilonewton per square metre	1 kN/m²	{ 0.145 { 20.885 { 0.009	lbf/in² lbf/ft² tonf/ft²
Energy (work, heat)	1 kilojoule	1 kJ	0.948	Btu

Memoranda

Unit equivalents (*continued*)

	Imperial unit	Abbreviation	Metric equivalent	
Length	1 inch	1 in	25.400	mm
	1 foot	1 ft	304.800	mm
	1 yard	1 yd	0.914	m
	1 mile	1 mile	1.609	km
Area	1 square inch	1 in²	645.160	mm
	1 square foot	1 ft²	0.093	m²
	1 square yard	1 yd²	0.836	m²
	1 acre	1 acre	{ 4 046.856	m²
			{ 0.405	ha
Volume	1 cubic inch	1 in³	16 387.064	mm³
	1 cubic foot	1 ft³	0.028	m³
	1 cubic yard	1 yd³	0.765	m³
	1 pint	1 pint	0.568	litre
	1 UK gallon	1 UK gal	4.546	litre
Weight	1 pound	1 lb	0.454	kg
	1 hundredweight	1 cwt	50.802	kg
	1 ton	1 ton	1.016	tonne
	1 pound per foot	1 lb/ft	1.488	kg/m
	1 pound per yard	1 lb/yd	0.496	kg/m
	1 pound per square foot	1 lb/ft²	4.882	kg/m²
	1 pound per square yard	1 lb/yd²	0.542	kg/m²
	1 pound per cubic foot	1 lb/ft³	16.018	kg/m³
	1 pound per cubic yard	1 lb/yd³	0.593	kg/m³
Force	1 pound force	1 lbf	4.448	N
	1 ton force	1 tonf	9.964	kN
Pressure	1 pound force per square inch	1 lbf/in²	{ 0.007	N/mm²
			{ 6.895	kN/m²
	1 pound force per square foot	1 lbf/ft²	0.048	kN/m²
	1 ton force per square foot	1 tonf/ft²	107.252	kN/m²
Energy (work, space heat)	1 British thermal unit	1 Btu	1.055	kJ

Memoranda

Conversion tables

Length

mm	in	in	mm	m	ft	ft	m
1	0.039	1/16	1.59	1	3.281	1	0.305
2	0.079	1/8	3.18	2	6.562	2	0.610
3	0.118	1/4	6.35	3	9.843	3	0.914
4	0.157	3/8	9.53	4	13.123	4	1.219
5	0.197	1/2	12.70	5	16.404	5	1.524
6	0.236	5/8	15.88	6	19.685	6	1.829
7	0.276	3/4	19.05	7	22.966	7	2.134
8	0.315	7/8	22.23	8	26.247	8	2.438
9	0.354	1	25.40	9	29.528	9	2.743
10	0.394	2	50.80	10	32.808	10	3.048
100	3.937	3	76.20	100	328.084	100	30.480
1 000	39.370	4	101.60	1 000	3 280.840	1 000	304.800
		5	127.00				
		6	152.40				
		7	177.80				
		8	203.20				
		9	228.60				
		10	254.00				
		11	279.40				
		12	304.80				

m	yd	yd	m	km	mile	mile	km
1	1.094	1	0.914	1	0.621	1	1.609
2	2.187	2	1.829	2	1.243	2	3.219
3	3.281	3	2.743	3	1.864	3	4.828
4	4.374	4	3.658	4	2.485	4	6.437
5	5.468	5	4.572	5	3.107	5	8.047
6	6.562	6	5.486	6	3.728	6	9.656
7	7.655	7	6.401	7	4.350	7	11.265
8	8.749	8	7.315	8	4.971	8	12.875
9	9.843	9	8.230	9	5.592	9	14.484
10	10.936	10	9.144	10	6.214	10	16.093
100	109.361	100	91.440	100	62.137	100	160.934
1 000	1 093.613	1 000	914.400	1 000	621.371	1 000	1 609.344

Memoranda

Conversion tables (*continued*)

Area

mm²	in²	in²	mm²	m²	ft²	ft²	m²
1	0.001 55	1	645	1	10.764	1	0.093
2	0.003 10	2	1 290	2	21.528	2	0.186
3	0.004 65	3	1 935	3	32.292	3	0.279
4	0.006 20	4	2 581	4	43.056	4	0.372
5	0.007 75	5	3 226	5	53.820	5	0.465
6	0.009 30	6	3 871	6	64.583	6	0.557
7	0.010 85	7	4 516	7	75.347	7	0.650
8	0.012 40	8	5 161	8	86.111	8	0.743
9	0.013 95	9	5 806	9	96.875	9	0.836
10	0.015 50	10	6 452	10	107.639	10	0.929
100	0.155 00	100	64 516	100	1 076.392	100	9.290
1 000	1.550 00	1 000	645 160	1 000	10 763.915	1 000	92.903

m²	yd²	yd²	m²	ha	acre	acre	ha
1	1.196	1	0.836	1	2.471	1	0.405
2	2.392	2	1.672	2	4.942	2	0.809
3	3.588	3	2.508	3	7.413	3	1.214
4	4.784	4	3.345	4	9.884	4	1.619
5	5.980	5	4.181	5	12.355	5	2.023
6	7.176	6	5.017	6	14.826	6	2.428
7	8.372	7	5.853	7	17.297	7	2.833
8	9.568	8	6.689	8	19.768	8	3.237
9	10.764	9	7.525	9	22.239	9	3.642
10	11.960	10	8.361	10	24.711	10	4.047
100	119.599	100	83.613	100	247.105	100	40.469
1 000	1 195.991	1 000	836.127	1 000	2 471.052	1 000	404.686

Volume

m³	ft³	ft³	m³	m³	yard³	yard³	m³
2	70.629	2	0.057	2	2.616	2	1.529
3	105.944	3	0.085	3	3.924	3	2.294
4	141.259	4	0.113	4	5.232	4	3.058
5	176.574	5	0.142	5	6.540	5	3.823
6	211.888	6	0.170	6	7.848	6	4.587
7	247.203	7	0.198	7	9.156	7	5.352
8	282.518	8	0.227	8	10.464	8	6.116
9	317.833	9	0.255	9	11.772	9	6.881
10	353.147	10	0.283	10	13.080	10	7.646
100	3 531.472	100	2.832	100	130.795	100	76.455
1 000	35 314.725	1 000	28.317	1 000	1 307.953	1 000	764.554

Memoranda

Conversion tables (*continued*)

Volume (*continued*)

litre	pint	pint	litre	litre	UK gal	UK gal	litre
1	1.760	1	0.568	1	0.220	1	4.546
2	3.520	2	1.137	2	0.440	2	9.092
3	5.279	3	1.705	3	0.660	3	13.638
4	7.039	4	2.273	4	0.880	4	18.184
5	8.799	5	2.841	5	1.100	5	22.730
6	10.559	6	3.410	6	1.320	6	27.277
7	12.318	7	3.978	7	1.540	7	31.823
8	14.078	8	4.546	8	1.760	8	36.369
9	15.838	9	5.114	9	1.980	9	40.915
10	17.598	10	5.683	10	2.200	10	45.461
100	175.975	100	56.826	100	21.997	100	454.609
1 000	1 759.754	1 000	568.261	1 000	219.969	1 000	4 546.090

Weight

kg	lb	lb	kg	tonne	cwt	cwt	tonne
1	2.205	1	0.454	1	19.684	1	0.051
2	4.409	2	0.907	2	39.368	2	0.102
3	6.614	3	1.361	3	59.052	3	0.152
4	8.818	4	1.814	4	78.736	4	0.203
5	11.023	5	2.268	5	98.420	5	0.254
6	13.228	6	2.722	6	118.104	6	0.305
7	15.432	7	3.175	7	137.788	7	0.356
8	17.637	8	3.629	8	157.473	8	0.406
9	19.842	9	4.082	9	177.157	9	0.457
10	22.046	10	4.536	10	196.841	10	0.508
100	220.462	100	45.359	100	1 968.413	100	5.080
1 000	2 204.623	1 000	453.592	1 000	19 684.131	1 000	50.802

tonne	ton	ton	tonne	kg/m	lb/ft	lb/ft	kg/m
1	0.984	1	1.016	1	0.672	1	1.488
2	1.968	2	2.032	2	1.344	2	2.976
3	2.953	3	3.048	3	2.016	3	4.464
4	3.937	4	4.064	4	2.688	4	5.953
5	4.921	5	5.080	5	3.360	5	7.441
6	5.905	6	6.096	6	4.032	6	8.929
7	6.889	7	7.112	7	4.704	7	10.417
8	7.874	8	8.128	8	5.376	8	11.905
9	8.858	9	9.144	9	6.048	9	13.393
10	9.842	10	10.161	10	6.720	10	14.882
100	98.421	100	101.605	100	67.197	100	148.816
1 000	984.207	1 000	1 1016.047	1 000	671.969	1 000	1 488.164

Memoranda

Conversion tables (continued)

Weight (continued)

kg/m	lb/yd	lb/yd	kg/m	kg/m²	lb/ft²	lb/ft²	kg/m²
1	2.016	1	0.496	1	0.205	1	4.882
2	4.032	2	0.992	2	0.410	2	9.765
3	6.048	3	1.488	3	0.614	3	14.647
4	8.064	4	1.984	4	0.819	4	19.530
5	10.080	5	2.480	5	1.024	5	24.412
6	12.095	6	2.976	6	1.229	6	29.295
7	14.111	7	3.472	7	1.434	7	34.177
8	16.127	8	3.968	8	1.639	8	39.059
9	18.143	9	4.464	9	1.843	9	43.942
10	20.159	10	4.961	10	2.048	10	48.824
100	201.591	100	49.605	100	20.482	100	488.243
1 000	2 015.907	1 000	496.055	1 000	204.816	1 000	4 882.430

kg/m²	lb/yd²	lb/yd²	kg/m²	kg/m³	lb/ft³	lb/ft³	kg/m³
1	1.843	1	0.542	1	0.062	1	16.018
2	3.687	2	1.085	2	0.125	2	32.037
3	5.530	3	1.627	3	0.187	3	48.055
4	7.373	4	2.170	4	0.250	4	64.074
5	9.217	5	2.712	5	0.312	5	80.092
6	11.060	6	3.255	6	0.375	6	96.111
7	12.903	7	3.797	7	0.437	7	112.129
8	14.747	8	4.340	8	0.499	8	128.148
9	16.590	9	4.882	9	0.562	9	144.166
10	18.433	10	5.425	10	0.624	10	160.185
100	184.334	100	54.249	100	6.243	100	1 601.849
1 000	1 843.344	1 000	542.492	1 000	62.428	1 000	16 018.490

kg/m³	lb/yd³	lb/yd³	kg/m³
1	1.686	1	0.593
2	3.371	2	1.187
3	5.057	3	1.780
4	6.742	4	2.373
5	8.428	5	2.966
6	10.113	6	3.560
7	11.799	7	4.153
8	13.484	8	4.746
9	15.170	9	5.339
10	16.856	10	5.933
100	168.555	100	59.328
1 000	1 685.555	1 000	593.276

Conversion tables (*continued*)

Force

N	lbf	lbf	N	kN	lbf	lbf	kN
1	0.225	1	4.448	1	224.809	1	0.004
2	0.450	2	8.896	2	449.618	2	0.009
3	0.674	3	13.345	3	674.427	3	0.013
4	0.899	4	17.793	4	899.236	4	0.018
5	1.124	5	22.241	5	1 124.045	5	0.022
6	1.349	6	26.689	6	1 348.854	6	0.027
7	1.574	7	31.138	7	1 573.663	7	0.031
8	1.798	8	35.586	8	1 798.472	8	0.036
9	2.023	9	40.034	9	2 023.281	9	0.040
10	2.248	10	44.482	10	2 248.090	10	0.044
100	22.481	100	444.822	100	22 480.902	100	0.445
1 000	224.809	1 000	4 448.220	1 000	224 809.025	1 000	4.448

kN	tonf	tonf	kN
1	0.100	1	9.964
2	0.201	2	19.928
3	0.301	3	29.892
4	0.401	4	39.856
5	0.502	5	49.820
6	0.602	6	59.784
7	0.703	7	69.748
8	0.803	8	79.712
9	0.903	9	89.676
10	1.004	10	99.640
100	10.036	100	996.401
1 000	100.361	1 000	9 964.013

Pressure

N/mm²	lbf/in²	lbf/in²	N/mm²	kN/m²	lbf/ft²	lbf/ft²	kN/m²
1	145.038	1	0.007	1	20.885	1	0.048
2	290.076	2	0.014	2	41.771	2	0.096
3	145.113	3	0.021	3	62.656	3	0.144
4	580.151	4	0.028	4	83.542	4	0.192
5	725.189	5	0.034	5	104.427	5	0.239
6	870.227	6	0.041	6	125.313	6	0.287
7	1 015.265	7	0.048	7	146.198	7	0.335
8	1 160.302	8	0.055	8	167.083	8	0.383
9	1 305.340	9	0.062	9	187.969	9	0.431
10	1 450.378	10	0.069	10	208.854	10	0.479
100	14 503.779	100	0.689	100	2 088.544	100	4.788
1 000	145 037.790	1 000	6.895	1 000	20 885.442	1 000	47.880

Memoranda

Conversion tables (continued)

Pressure (continued)

kN/m²		tonf/ft²
1	....	0.009
2	....	0.019
3	....	0.028
4	....	0.037
5	....	0.047
6	....	0.056
7	....	0.065
8	....	0.075
9	....	0.084
10	....	0.093
100	....	0.932
1 000	....	9.324

tonf/ft²		kN/m²
1	....	107.252
2	....	214.504
3	....	321.755
4	....	429.007
5	....	536.259
6	....	643.511
7	....	750.762
8	....	858.014
9	....	965.266
10	....	1 072.518
100	....	10 725.174
1 000	....	107 251.740

Energy

kJ		Btu
1	....	0.948
2	....	1.896
3	....	2.843
4	....	3.791
5	....	4.739
6	....	5.687
7	....	6.635
8	....	7.583
9	....	8.530
10	....	9.478
100	....	94.781
1 000	....	947.813

Btu		kJ
1	....	1.055
2	....	2.110
3	....	3.165
4	....	4.220
5	....	5.275
6	....	6.330
7	....	7.385
8	....	8.440
9	....	9.496
10	....	10.551
100	....	105.506
1 000	....	1 055.060

Temperature

°C		°F
0	....	32
10	....	50
20	....	68
30	....	86
40	....	104
50	....	122
60	....	140
70	....	158
80	....	176
90	....	194
100	....	212

°F		°C
0	....	-17.78
5	....	-15.00
10	....	-12.22
20	....	-6.67
32	....	0.00
40	....	4.44
50	....	10.00
100	....	37.78
150	....	65.56
200	....	93.33
212	....	100.00

Memoranda

Excavation and earthwork

Bearing capacities

Type of ground	Approximate bearing capacity
	kN/m²
Unweathered rock	over 1 000
Solid chalk	600
Compact gravel	600
Loose gravel	up to 200
Compact sand	300
Loose sand	up to 100
Hard boulder clay	300 to 600
Stiff clay	150 to 300
Firm clay	75 to 150
Soft clay	up to 75

"Bulking" of excavated materials

Type of material	Approximate bulk of 1 m³ after excavation
Unweathered rock	1.75
Solid chalk	1.75
Compact gravel	1.25
Loose gravel	1.10
Compact sand	1.20
Loose sand	1.05
Hard boulder clay	1.50
Stiff clay	1.40
Firm clay	1.25
Soft clay	1.25
Subsoil	1.25
Topsoil and loam	1.25

Memoranda

Concrete work

Concrete mixing materials

Based on: (1) cement 1 440 kg/m³
(2) fine aggregate/sand 1 600 kg/m³ dry (1 260 kg/m³ moist)
(3) one third bulking of fine aggregate/sand from dry to moist
(4) coarse aggregate 1 500 kg/m³
(5) net quantities - allowance for waste to be added.

- per 2 x 25 kg bags of cement

	Mix	Cement	Fine aggregate/Sand		Coarse aggregate
			dry or	(moist)	
- by volume		m³	m³	m³	m³
	1:3:6	0.035	0.105	(0.140)	0.210
	1:2:4	0.035	0.070	(0.093)	0.140
	1:1½:3	0.035	0.053	(0.070)	0.105
- by weight		kg	kg	kg	kg
	1:3:6	50	168	(176)	315
	1:2:4	50	112	(117)	210
	1:1½:3	50	85	(88)	158

- per cubic metre of concrete

	Mix	Cement	Fine aggregate/Sand		Coarse aggregate
			dry or	(moist)	
- by volume		m³	m³	m³	m³
	1:3:6	0.150	0.450	(0.600)	0.900
	1:2:4	0.214	0.429	(0.572)	0.857
	1:1½:3	0.273	0.409	(0.545)	0.818
- by weight		kg	kg	kg	kg
	1:3:6	216	720	(756)	1 350
	1:2:4	308	686	(721)	1 286
	1:1½:3	393	654	(687)	1 227

Concrete work (*continued*)

Steel bar reinforcement

Diameter	Nominal weight	Length	Sectional area
mm	kg/m	m/tonne	mm²
6	0.222	4 505	28.3
8	0.395	2 532	50.3
10	0.616	1 623	78.5
12	0.888	1 126	113.1
16	1.579	633	201.1
20	2.466	406	314.2
25	3.854	259	490.9
32	6.313	158	804.2
40	9.864	101	1 256.6
50	15.413	65	1 963.5

Steel fabric reinforcement

BS 4483 reference	Nominal weight	Mesh dimensions		Wire diameters	
		Main	Cross	Main	Cross
	kg/m²	mm	mm	mm	mm
A 393	6.16	200	200	10	10
A 252	3.95	200	200	8	8
A 193	3.02	200	200	7	7
A 142	2.22	200	200	6	6
A 98	1.54	200	200	5	5
B 1131	10.90	100	200	12	8
B 785	8.14	100	200	10	8
B 503	5.93	100	200	8	8
B 385	4.53	100	200	7	7
B 283	3.73	100	200	6	7
B 196	3.05	100	200	5	7
C 785	6.72	100	400	10	6
C 636	5.55	80-130	400	8-10	6
C 503	4.34	100	400	8	5
C 385	3.41	100	400	7	5
C 283	2.61	100	400	6	5
D 98	1.54	200	200	5	5
D 49	0.77	100	100	2.5	2.5

Memoranda

Concrete work (*continued*)

Striking times for formwork

Minimum time in days between placing concrete and striking formwork - days when frost occurs should be added.

	"Normal" weather - about 15°C		"Cold" weather - about 5°C	
	Ordinary Portland cement	Rapid-hardening cement	Ordinary Portland cement	Rapid-hardening cement
	Days	Days	Days	Days
Sides of beams, columns or walls	2	1	7	5
Soffits of slabs - props left in position	4	3	10	7
Props to soffits of slabs	11	10	17	14
Soffits of beams - props left in position	7	5	14	10
Props to soffits of beams	14	12	21	17

Brickwork and blockwork

Bricks and mortar quantities per square metre

Based on:
(1) 215 x 102.5 x 65 mm work size bricks
(2) 10 mm mortar joints
(3) only one snapped header obtained from one facing brick
(4) net quantities - allowance for waste to be added according to the nature of the work and the type of bricks to be used.

	Common bricks	Facing bricks	Mortar for solid (wirecut) bricks	Mortar for single frogged bricks
	No/m²	No/m²	m³/m²	m³/m²
Unfaced walls				
102.5 mm	59.3	-	0.018	0.022
215 mm	118.5	-	0.045	0.054
327.5 mm	177.8	-	0.073	0.086
Faced walls				
102.5 mm in Stretcher bond	-	59.3	0.018	0.022
102.5 mm in English bond with snapped headers	-	88.9	0.020	0.025
102.5 mm in Flemish bond with snapped headers	-	79.0	0.019	0.024
Walls in English bond faced one side				
215 mm	29.6	88.9	0.045	0.054
327.5 mm	88.9	88.9	0.073	0.086
Walls in English bond faced both sides				
215 mm	-	118.5	0.045	0.054
327.5 mm	-	177.8	0.073	0.086
Walls in Flemish bond faced one side				
215 mm	39.5	79.0	0.045	0.054
327.5 mm	98.8	79.0	0.074	0.088
Walls in Flemish bond faced both sides				
215 mm	-	118.5	0.045	0.054
327.5 mm	19.8	158.0	0.074	0.088

Memoranda

Brickwork and blockwork (*continued*)

Blocks and mortar quantities per square metre

Based on: (1) 440 x 215 mm work size blocks
(2) 10 mm mortar joints
(3) net quantities - allowance for waste to be added according to the nature of the work and the type of blocks to be used.

Wall thickness	Blocks	Mortar
	No/m²	m³/m²
60 mm	9.9	0.004
75 mm	9.9	0.005
90 mm	9.9	0.006
100 mm	9.9	0.007
140 mm	9.9	0.009
190 mm	9.9	0.013
215 mm	9.9	0.014
215 mm of 100 mm blocks laid flat	20.2	0.024
215 mm of 140 mm blocks laid flat	14.8	0.019

Mortar mixing materials

Based on: (1) cement 1 440 kg/m³
(2) hydrated lime 500 kg/m³
(3) sand 1 600 kg/m³ dry (1 260 kg/m³ moist)
(4) one third bulking of sand from dry to moist
(5) net quantities - allowance for waste to be added.

- per 2 x 25 kg bags of cement

Mix	Cement	Lime	Sand		
			dry	or	(moist)
- by volume	m³	m³	m³		m³
1:1	0.035	-	0.035		(0.047)
1:2	0.035	-	0.070		(0.093)
1:3	0.035	-	0.105		(0.140)
1:4	0.035	-	0.140		(0.187)
1:6	0.035	-	0.210		(0.280)
1:1:5	0.035	0.035	0.175		(0.233)
1:1:6	0.035	0.035	0.210		(0.280)
1:2:9	0.035	0.070	0.315		(0.420)

Memoranda

Brickwork and blockwork (*continued*)

Mortar mixing materials (*continued*)

- per 2 x 25 kg bags of cement (*continued*)

Mix	Cement	Lime	Sand dry	or	(moist)
- by weight	kg	kg	kg		kg
1:1	50	-	56		(59)
1:2	50	-	112		(117)
1:3	50	-	168		(176)
1:4	50	-	224		(236)
1:6	50	-	336		(353)
1:1:5	50	18	280		(294)
1:1:6	50	18	336		(353)
1:2:9	50	35	504		(529)

- per cubic metre of mortar

Mix	Cement	Lime	Sand dry	or	(moist)
- by volume	m³	m³	m³		m³
1:1	0.725	-	0.725		(0.967)
1:2	0.467	-	0.933		(1.244)
1:3	0.338	-	1.012		(1.349)
1:4	0.260	-	1.040		(1.387)
1:6	0.179	-	1.071		(1.428)
1:1:5	0.196	0.196	0.983		(1.311)
1:1:6	0.169	0.169	1.012		(1.349)
1:2:9	0.113	0.225	1.012		(1.349)
- by weight	kg	kg	kg		kg
1:1	1 044	-	1 160		(1 218)
1:2	672	-	1 493		(1 567)
1:3	487	-	1 619		(1 700)
1:4	374	-	1 664		(1 748)
1:6	258	-	1 714		(1 799)
1:1:5	282	98	1 573		(1 652)
1:1:6	243	85	1 619		(1 700)
1:2:9	163	113	1 619		(1 700)

Memoranda

Roofing

Slating and tiling quantities per square metre

Based on net quantities - allowance for waste to be added according to the nature of the work and the type and dimensions of slates, tiles and battens to be used.

Notes:
(1) lap - for traditional double lap slating/tiling - length of end cover of one slate/tile over the next but one slate/tile beneath
- for single lap tiling - length of end cover of one tile over the tile beneath
(2) gauge - batten centres spacing and length of slate/tile exposed
(3) nails - quantities for slating, tiling and battening can be calculated from the tables according to the type of slates/tiles, section of battens and centres spacing of the rafters or joists.

Centre-nailed slates

mm	Lap mm	Gauge mm	Slates No/m²	Battens m/m²
610 x 305	76	267	12.3	3.7
610 x 305	100	255	12.9	3.9
600 x 300	76	262	12.7	3.8
600 x 300	100	250	13.3	4.0
510 x 255	76	217	18.1	4.6
510 x 255	100	205	19.1	4.9
500 x 250	76	212	18.9	4.7
500 x 250	100	200	20.0	5.0
405 x 205	75	165	29.6	6.1
405 x 205	95	155	31.5	6.5
400 x 200	70	165	30.3	6.1
400 x 200	90	155	32.3	6.5

Plain tiles

mm	Lap mm	Gauge mm	Tiles No/m²	Battens m/m²
265 x 165	38	114	53.2	8.8
265 x 165	65	100	60.6	10.0

Single lap tiles

mm	Cover width mm	Lap mm	Gauge mm	Tiles No/m²	Battens m/m²
430 x 380	343	75	355	8.2	2.8
430 x 380	343	100	330	8.8	3.0
420 x 332	300	75	345	9.7	2.9
420 x 332	300	100	320	10.4	3.1
413 x 330	292	75	338	10.1	3.0
413 x 330	292	100	313	10.9	3.2
380 x 230	200	75	305	16.4	3.3
380 x 230	200	100	280	17.9	3.6

Memoranda

Roofing (*continued*)

Approximate numbers of round lost head nails per kilogramme

Aluminium Length x Shank		Copper Length x Shank		Steel Length x Shank	
mm	No/kg	mm	No/kg	mm	No/kg
75 x 3.75	448	65 x 3.75	178	75 x 3.75	160
65 x 3.35	672	3.35	194	65 x 3.35	240
60 x 3.35	756	50 x 3.35	292	3.00	270
50 x 3.35	860	3.00	308	60 x 3.35	270
3.00	1 008	40 x 2.65	474	3.00	330
40 x 2.65	1 390	2.36	554	50 x 3.00	360
2.36	2 128			2.65	420
				40 x 2.36	760

Approximate numbers of extra large head felt nails per kilogramme

Aluminium Length x Shank		Copper Length x Shank		Steel Length x Shank	
mm	No/kg	mm	No/kg	mm	No/kg
25 x 3.35	1 296	25 x 3.35	440	40 x 3.00	350
3.00	1 636	3.00	517	30 x 3.00	420
20 x 3.35	1 848	20 x 3.35	544	25 x 3.00	485
3.00	2 130	3.00	627	20 x 3.00	580
15 x 3.35	1 840	15 x 3.00	691	15 x 3.00	650
3.00	2 283	13 x 3.00	880	13 x 3.00	780

Memoranda

Roofing (*continued*)

Approximate number of clout, slate or tile nails per kilogramme

Aluminium Length x Shank		Copper Length x Shank		Steel Length x Shank	
mm	No/kg	mm	No/kg	mm	No/kg
65 x 3.75	504	65 x 3.75	170	100 x 4.50	75
60 x 3.75	550	3.35	195	90 x 4.50	85
3.35	680	50 x 3.35	241	75 x 3.75	150
50 x 3.75	644	3.00	276	65 x 3.75	180
3.35	812	2.65	327	50 x 3.75	230
3.00	952	45 x 3.35	308	3.35	290
45 x 3.35	924	3.00	366	3.00	340
3.00	1 060	2.65	456	2.65	430
40 x 3.35	980	40 x 3.35	335	45 x 3.35	330
3.00	1 200	3.00	3.98	2.65	4.60
2.65	1 596	2.65	460	40 x 3.35	350
2.36	1 960	2.36	553	2.65	570
30 x 3.00	1 512	30 x 3.35	448	2.36	700
2.65	1 848	3.00	550	30 x 3.00	540
2.36	2 324	2.65	621	2.65	660
2.00	3 000	2.36	748	2.36	830
25 x 3.35	1 540	25 x 2.65	740	25 x 2.65	815
3.00	1 750	20 x 2.65	920	20 x 2.65	1 035
2.65	2 282			15 x 2.36	1 540
2.00	3 800			2.00	2 380
20 x 3.00	2 300				
2.65	2 898				

Approximate number of tile pegs per kilogramme

Aluminium Length x Shank		Steel Length x Shank	
mm	No/kg	mm	No/kg
40 x 5.00	450	40 x 6.00	88
4.50	490	30 x 6.00	106
30 x 5.00	545		
4.50	600		

Memoranda

Roofing (*continued*)

Sheet metal roofing measurement allowances

Standard Method of Measurement of Building Works, Sixth Edition requires (M.40.3): "in the absence of any directions to the contrary, allowances shall be made in calculating the areas of the finished surface of sheet metalwork as follows":

Item		Lead	Zinc, aluminium and copper
a.	For drips not exceeding 50 mm high	0.18 m	0.15 m
b.	For cross-welts	0.08 m	0.08 m
c.	For wood-cored rolls not exceeding 50 mm high	0.25 m	0.15 m
d.	For standing seams and welted seams	0.10 m	0.08 m
e.	For edge-welts	0.03 m	0.03 m
f.	For vertical upstands	0.15 m	0.15 m
g.	For sloping upstands and for laps in pitched roofing	0.30 m for 30°	0.30 m for 30°
		0.35 m for 25°	0.35 m for 25°
		0.45 m for 20°	0.45 m for 20°
		0.60 m for 15°	0.60 m for 15°

Sheet metal thicknesses etc.

Sheet lead

Thickness	BS code	Colour marking
mm		
1.32	3	Green
1.80	4	Blue
2.24	5	Red
2.65	6	Black
3.15	7	White
3.55	8	Orange

Sheet zinc

Thickness	
mm	ZG
0.45	9
0.65	12
0.80	14

Sheet aluminium

Thickness	
mm	SWG
0.60	23
0.80	21

Sheet copper

Thickness	
mm	SWG
0.45	26
0.55	24
0.70	22

Memoranda

Woodwork

Lengths of sawn softwood per cubic metre

mm		m/m³	mm		m/m³	mm		m/m³
16 x	16	3 906	25 x	175	229	50 x	50	400
	19	3 289		200	200		63	317
	22	2 841		225	178		75	267
	25	2 500		250	160		100	200
	32	1 953		300	133		125	160
	38	1 645					150	133
	44	1 420	32 x	32	977		175	114
	50	1 250		38	822		200	100
	63	992		44	710		225	89
	75	833		50	625		250	80
	100	625		63	496		300	67
	125	500		75	417			
	150	417		100	313	63 x	63	252
				125	250		75	212
19 x	19	2 770		150	208		100	159
	22	2 392		175	179		125	127
	25	2 105		200	156		150	106
	32	1 645		225	139		175	91
	38	1 385		250	125		200	79
	44	1 196		300	104		225	71
	50	1 053					250	63
	63	835	38 x	38	693		300	53
	75	702		44	598			
	100	526		50	526	75 x	75	178
	125	421		63	418		100	133
	150	351		75	351		125	107
				100	263		150	89
22 x	22	2 066		125	211		175	76
	25	1 818		150	175		200	67
	32	1 420		175	150		225	59
	38	1 196		200	132		250	53
	44	1 033		225	117		300	44
	50	909		250	105			
	63	722		300	88	100 x 100		100
	75	606					150	67
	100	455	44 x	44	517		200	50
	125	364		50	455		250	40
	150	303		63	361		300	33
				75	303			
25 x	25	1 600		100	227	150 x 150		44
	32	1 250		125	182		200	33
	38	1 053		150	152		250	27
	44	909		175	130		300	22
	50	800		200	114			
	63	635		225	101	200 x 200		25
	75	533		250	91			
	100	400		300	76	250 x 250		16
	125	320						
	150	267				300 x 300		11

Woodwork (*continued*)

Lengths of boarding required per square metre

Based on net quantities - allowance for waste to be added according to the nature of the work and the type of boarding to be used.

Effective width		Effective width	
mm	m/m²	mm	m/m²
75	13.33	150	6.67
100	10.00	175	5.71
125	8.00	200	5.00

m	m	m	m	m	m	m
1.8	2.1	3.0	4.2	5.1	6.0	7.2
	2.4	3.3	4.5	5.4	6.3	
	2.7	3.6	4.8	5.7	6.6	
		3.9			6.9	

Cross-sectional areas

Standard Method of Measurement of Building Works, Sixth Edition states (N.1.10): "All ends, angles, mitres, intersections and the like are deemed to be included in the items to which they relate except when the cross-sectional area exceeds 0.002 m² in which case they shall each be enumerated".

Size mm	13	16	19	22	25	32	38	44	50	63	75	100	125	150	200 mm
13															
16															
19															
22															
25															
32															
38															
44															
50															
63															
75															
100															
125															
150															
200															

Cross-sectional areas not exceeding 0.002 m2

Ends, angles, mitres, intersections and the like deemed to be included in the items

Cross-sectional areas exceeding 0.002 m2

Ends, angles, mitres, intersections and the like to be enumerated

Memoranda

Woodwork (*continued*)

Approximate number of steel wire nails per kilogramme

Round plain head nails Length x Shank		Annular ringed shank nails Length x Shank		Round lost head nails Length x Shank	
mm	No/kg	mm	No/kg	mm	No/kg
150 x 6.00	29	100 x 5.00	66	75 x 3.75	160
125 x 5.60	42	75 x 3.75	154	65 x 3.35	240
5.00	53	65 x 3.35	230	3.00	270
115 x 5.00	57	60 x 3.35	255	60 x 3.35	270
100 x 5.00	66	50 x 3.35	290	3.00	330
4.50	77	3.00	340	50 x 3.00	360
4.00	88	2.65	440	2.65	420
90 x 4.00	106	45 x 2.65	510	40 x 2.36	760
75 x 4.00	121	40 x 2.65	575		
3.75	154	2.36	750	**Panel pins**	
3.35	194	30 x 2.36	840	**Length x Shank**	
65 x 3.35	230	25 x 2.00	1 430		
3.00	275	20 x 2.00	1 900	mm	No/kg
2.65	350				
60 x 3.35	255			50 x 2.00	770
3.00	310			40 x 1.60	1 590
2.65	385	**Oval brad or lost head nails**		30 x 1.60	1 900
50 x 3.35	290	**Length x Shank**		25 x 1.60	2 340
3.00	340			1.40	3 090
2.65	440	mm	No/kg	20 x 1.60	3 140
2.36	550			1.40	3 970
45 x 2.65	510	150 x 7.10 x 5.00	31	15 x 1.25	6 400
2.36	640	125 x 6.70 x 4.50	44		
40 x 2.65	575	100 x 6.00 x 4.00	64		
2.36	750	75 x 5.00 x 3.35	125		
2.00	970	65 x 4.00 x 2.65	230		
30 x 2.36	840	60 x 3.75 x 2.36	340		
2.00	1 170	50 x 3.35 x 2.00	470		
25 x 2.00	1 430	40 x 2.65 x 1.60	940		
1.80	1 720	30 x 2.65 x 1.60	1 480		
1.60	2 210	25 x 2.00 x 1.25	2 530		
20 x 1.60	2 710				

Wood screw sizes

Wood screw gauge	mm	in	in	Wood screw gauge	mm	in	in
0	1.52	0.060	$^1/_{16}$ −	8	4.17	0.164	
1	1.78	0.070		9	4.52	0.178	
2	2.08	0.082		10	4.88	0.192	$^3/_{16}$ +
3	2.39	0.094	$^3/_{32}$ +	12	5.59	0.220	$^7/_{32}$ +
4	2.74	0.108		14	6.30	0.248	$^1/_4$ −
5	3.10	0.122	$^1/_8$ −	16	7.01	0.276	$^9/_{32}$ −
6	3.45	0.136		18	7.72	0.304	$^5/_{16}$ −
7	3.81	0.150	$^5/_{32}$ −	20	8.43	0.332	$^{11}/_{32}$ −

Memoranda

Structural steelwork

Serial sizes and weights of structural steel sections

Universal beams		Universal beams (continued)		Universal columns	
mm	kg/m	mm	kg/m	mm	kg/m
914 x 419	388	305 x 165	54	356 x 406	634
	343		46		551
914 x 305	289		40		467
	253	305 x 127	48		393
	224		42		340
	201		37		287
838 x 292	226	305 x 102	33		235
	194		28	356 x 368	202
	176		25		177
762 x 267	197	254 x 146	43		153
	173		37		129
	147		31	305 x 305	283
686 x 254	170	254 x 102	28		240
	152		25		198
	140		22		158
	125	203 x 133	30		137
610 x 305	238		25		118
	179	203 x 102	23		97
	149	178 x 102	19	254 x 254	167
610 x 229	140	152 x 89	16		132
	125	127 x 76	13		107
	113				89
	101	**Joists**			73
533 x 210	122			203 x 203	86
	109	mm	kg/m		71
	101				60
	92	254 x 203	81.85		52
	82	254 x 114	37.20		46
457 x 191	98			152 x 152	37
	89	203 x 152	52.09		30
	82				23
	74	152 x 127	37.20		
	67				
457 x 152	82	127 x 114	29.76		
	74		26.79		
	67	127 x 76	16.37		
	60				
	52	114 x 114	26.79		
406 x 178	74				
	67	102 x 102	23.07		
	60	102 x 44	7.44		
	54				
406 x 140	46	89 x 89	19.35		
	39				
356 x 171	67	76 x 76	14.67		
	57		12.65		
	51				
	45				
356 x 127	39				
	33				

Memoranda

Structural steelwork (*continued*)

Serial sizes and weights of structural steel sections (*continued*)

Channels

mm	kg/m
432 x 102	65.54
381 x 102	55.10
305 x 102	46.18
305 x 89	41.69
254 x 89	35.74
254 x 76	28.29
229 x 89	32.76
229 x 76	26.06
203 x 89	29.78
203 x 76	23.82
178 x 89	26.81
178 x 76	20.84
152 x 89	23.84
152 x 76	17.88
127 x 64	14.90
102 x 51	10.42
76 x 38	6.70

Tees cut from universal beams

mm	kg/m
305 x 457	127
	112
	101
292 x 419	113
	97
	88
267 x 381	99
	87
	74
254 x 343	85
	76
	70
	63
305 x 305	119
	90
	75

Tees cut from universal beams (*continued*)

mm	kg/m
229 x 305	70
	63
	57
	51
210 x 267	61
	55
	51
	46
	41
191 x 229	49
	45
	41
	37
	34
152 x 229	41
	37
	34
	30
	26
178 x 203	37
	34
	30
	27
140 x 203	23
	20
171 x 178	34
	29
	26
	23
127 x 178	20
	17
165 x 152	27
	23
	20
127 x 152	24
	21
	19
102 x 152	17
	14
	13
146 x 127	22
	19
	16
102 x 127	14
	13
	11
133 x 102	15
	13

Tees cut from universal columns

mm	kg/m
406 x 178	118
368 x 178	101
	89
	77
	65
305 x 152	79
	69
	59
	49
254 x 127	66
	54
	45
	37
203 x 102	43
	36
	30
	26
	23
152 x 76	19
	15
	12

Rolled tees

mm	kg/m
51 x 51	6.92
	4.76
44 x 44	4.11
	3.14

Memoranda

Structural steelwork (continued)

Serial sizes and widths of structural steel sections (continued)

Equal angles

mm	kg/m
250 x 250 x 35	128
32	118
28	104
25	93.6
200 x 200 x 24	71.1
20	59.9
18	54.2
16	48.5
150 x 150 x 18	40.1
15	33.8
12	27.3
10	23.0
120 x 120 x 15	26.6
12	21.6
10	18.2
8	14.7
100 x 100 x 15	21.9
12	17.8
8	12.2
90 x 90 x 12	15.9
10	13.4
8	10.9
7	9.61
6	8.30
80 x 80 x 10	11.9
8	9.63
6	7.34
70 x 70 x 10	10.3
8	8.36
6	6.38
60 x 60 x 10	8.69
8	7.09
6	5.42
5	4.57
50 x 50 x 8	5.82
6	4.47
5	3.77
4	3.06
3	2.33
45 x 45 x 6	4.00
5	3.38
4	2.74
3	2.09
40 x 40 x 6	3.52
5	2.97
4	2.42
3	1.84
30 x 30 x 5	2.18
4	1.78
3	1.36

Equal angles (continued)

mm	kg/m
25 x 25 x 5	1.77
4	1.45
3	1.11

Unequal angles

mm	kg/m
200 x 150 x 18	47.1
15	39.6
12	32.0
200 x 100 x 15	33.7
12	27.3
10	23.0
150 x 90 x 15	26.6
12	21.6
10	18.2
150 x 75 x 15	24.8
12	20.2
10	17.0
125 x 75 x 12	17.8
10	15.0
8	12.2
100 x 75 x 12	15.4
10	13.0
8	10.6
100 x 65 x 10	12.3
8	9.94
7	8.77
80 x 60 x 8	8.34
7	7.36
6	6.37
75 x 50 x 8	7.39
6	5.65
65 x 50 x 8	6.75
6	5.16
5	4.35
60 x 30 x 6	3.99
5	3.37
40 x 25 x 4	1.93

Circular hollow sections

mm	kg/m
21.3 x 3.2	1.43
26.9 x 3.2	1.87
33.7 x 2.6	1.99
3.2	2.41
4	2.93
42.4 x 2.6	2.55
3.2	3.09
4	3.79
48.3 x 3.2	3.56
4	4.37
5	5.34
60.3 x 3.2	4.51
4	5.55
5	6.82
76.1 x 3.2	5.75
4	7.11
5	8.77
88.9 x 3.2	6.76
4	8.38
5	10.3
114.3 x 3.6	9.83
5	13.5
6.3	16.8
139.7 x 5	16.6
6.3	20.7
8	26.0
10	32.0
168.3 x 5	20.1
6.3	25.2
8	31.6
10	29.0
193.7 x 5.4	25.1
6.3	29.1
8	36.6
10	45.3
12.5	55.9
16	70.1
219.1 x 6.3	33.1
8	41.6
10	51.6
12.5	63.7
16	80.1
20	98.2
244.5 x 6.3	37.0
8	46.7
10	57.8
12.5	71.5
16	90.2
20	111.0

Memoranda

Structural steelwork (*continued*)

Serial sizes and weights of structural steel section (*continued*)

Circular hollow sections		Rectangular hollow sections (*continued*)		Square hollow sections (*continued*)	
mm	kg/m	mm	kg/m	mm	kg/m
273 x 6.3	41.4	120 x 60 x 3.6	9.72	60 x 60 x 3.2	5.67
8	52.3	5	13.3	4	6.97
10	64.9	6.3	16.4	5	8.54
12.5	80.3	120 x 80 x 5	14.8	70 x 70 x 3.6	7.46
16	101	6.3	18.4	5	10.1
20	125	8	22.9	80 x 80 x 3.6	8.59
25	153	10	27.9	5	11.7
323.9 x 8	62.3	150 x 100 x 5	18.7	6.3	14.4
10	77.4	6.3	23.3	90 x 90 x 3.6	9.72
12.5	96.0	8	29.1	5	13.3
16	121	10	35.7	6.3	16.4
20	150	160 x 80 x 5	18.0	100 x 100 x 4	12.0
25	184	6.3	22.3	5	14.8
355.6 x 8	68.6	8	27.9	6.3	18.4
10	85.2	10	34.2	8	22.9
12.5	106	200 x 100 x 5	22.7	10	27.9
16	154	6.3	28.3	120 x 120 x 5	18.0
20	166	8	35.4	6.3	22.3
25	204	10	43.6	8	27.9
406.4 x 10	97.8	12.5	53.4	10	34.2
12.5	121	16	66.4	150 x 150 x 5	22.7
16	154	250 x 150 x 6.3	38.2	6.3	28.3
20	191	8	48.0	8	35.4
25	235	10	59.3	10	43.6
32	295	12.5	73.0	12.5	53.4
457 x 10	110	16	91.5	16	66.4
12.5	137	300 x 200 x 6.3	48.1	180 x 180 x 6.3	34.2
16	174	8	60.5	8	43.0
20	216	10	75.0	10	53.0
25	266	12.5	92.6	12.5	65.2
32	335	16	117	16	81.4
40	411	400 x 200 x 10	90.7	200 x 200 x 6.3	38.2
		12.5	112	8	48.0
Rectangular hollow sections		16	142	10	59.3
mm	kg/m	450 x 250 x 10	106	12.5	73.0
		12.5	132	16	91.5
		16	167	250 x 250 x 6.3	48.1
50 x 30 x 2.6	3.03			8	60.5
3.2	3.66	**Square hollow sections**		10	75.0
60 x 40 x 3.2	4.66	mm	kg/m	12.5	92.6
4	5.72			16	117
80 x 40 x 3.2	5.67	20 x 20 x 2	1.12	300 x 300 x 10	90.7
4	6.97	2.6	1.39	12.5	112
90 x 50 x 3.6	7.46	30 x 30 x 2.6	2.21	16	142
5	10.1	3.2	2.65	350 x 350 x 10	106
100 x 50 x 3.2	7.18	40 x 40 x 2.6	3.03	12.5	132
4	8.86	3.2	3.66	16	167
5	10.9	4	4.46	400 x 400 x 10	122
100 x 60 x 3.6	8.59	50 x 50 x 3.2	4.66	12.5	152
5	11.7	4	5.72		
6.3	14.4	5	6.97		

Memoranda

Metalwork

Gauges and thicknesses

Imperial standard wire gauge				Birmingham gauge			
SWG	mm	in	in	BG	mm	in	in
30	0.315	0.012 4		30	0.312	0.012 3	
29	0.345	0.013 6		29	0.353	0.013 9	
28	0.367	0.014 8	1/64 −	28	0.397	0.015 625	1/64
27	0.417	0.016 4	1/64 +	27	0.443	0.017 45	
26	0.457	0.018 0		26	0.498	0.019 61	
25	0.508	0.020		25	0.560	0.022 04	
24	0.559	0.022		24	0.629	0.024 76	
23	0.610	0.024		23	0.707	0.027 82	
22	0.711	0.028	1/32 −	22	0.794	0.031 25	1/32
21	0.813	0.032	1/32 +	21	0.886	0.034 9	
20	0.914	0.036		20	0.996	0.039 2	
19	1.016	0.040		19	1.118	0.044 0	
18	1.219	0.048		18	1.257	0.049 5	
17	1.422	0.056	1/16 −	17	1.412	0.055 6	
16	1.626	0.064	1/16 +	16	1.588	0.062 5	1/16
15	1.829	0.072		15	1.775	0.069 9	
14	2.032	0.080		14	1.994	0.078 5	
13	2.337	0.092	3/32 −	13	2.240	0.088 2	
12	2.642	0.104		12	2.517	0.099 1	
11	2.946	0.116		11	2.827	0.111 3	
10	3.251	0.128	1/8 +	10	3.175	0.125 0	1/8
9	3.658	0.144		9	3.551	0.139 8	
8	4.064	0.160	5/32 +	8	3.988	0.157 0	5/32 +
7	4.470	0.176		7	4.481	0.176 4	
6	4.877	0.192	3/16 +	6	5.032	0.198 1	3/16 +
5	5.385	0.212		5	5.652	0.222 5	
4	5.893	0.232		4	6.350	0.250	1/4
3	6.401	0.252	1/4 +	3	7.122	0.280 4	
2	7.010	0.276		2	7.993	0.314 7	
1	7.620	0.300	5/16 −	1	8.971	0.353 2	11/32 +

Memoranda

Plumbing and engineering installations

Roof drainage

	Roof area drained by:			
	Level gutter		Gutter falling 1/600	
	outlet one end	centre outlet	outlet one end	centre outlet
Gutter and rainwater pipe sizes	m	m	m	m
76 mm half round gutter and 50 mm rainwater pipe	15	30	20	40
100 mm half round gutter and 68 mm rainwater pipe	38	75	50	100
150 mm half round gutter and 110 mm rainwater pipe	100	200	130	260
115 mm square section gutter and 65 mm square rainwater pipe	50	100	65	130

Jointing materials per pipe joint

Based on net quantities - allowance for waste to be added.

Pipe size	Cast iron soil pipes		Cast iron drain pipes	
	Lead	Yarn	Lead	Yarn
mm	kg/joint	kg/joint	kg/joint	kg/joint
50	0.65	0.07	-	-
75	1.10	0.10	-	-
100	1.85	0.13	2.55	0.13
150	-	-	4.25	0.20

Memoranda

Floor, wall and ceiling finishings

Approximate coverage of plasters per tonne

Based on net quantities - allowance for waste to be added.

Type	Coat thickness				
	2 mm	3 mm	5 mm	8 mm	11 mm
	m²/tonne	m²/tonne	m²/tonne	m²/tonne	m²/tonne
Gypsum ("Thistle")					
undercoat	-	-	-	160	115
finish	400	260	165	-	-
Lightweight ("Carlite")					
browning	-	-	-	-	140
bonding coat	-	-	-	135	90
finish	450	-	-	-	-

Approximate coverage of renderings per cubic metre

Based on net quantities - allowance for waste to be added.

Background	Coat thickness				
	6 mm	10 mm	13 mm	16 mm	20 mm
	m²/m³	m²/m³	m²/m³	m²/m³	m²/m³
Blockwork (no grooved or keyed faces)	109	72	58	48	40
Brickwork (grooved face or joints raked out)	87	62	51	43	36
Stone rubble work	72	54	46	40	33

For mortar mixing materials, see "Brickwork and blockwork".

Memoranda

Glazing

Putty quantities per square metre of glazing

Based on: (1) 4.5 metres puttying to wood per kilogramme of putty
(2) 3.5 metres puttying to metal per kilogramme of putty
(3) average area within SMM6 classifications
(4) 1.50 m² panes for SMM6 "over 1.00 m²" classification
(5) Net quantities - allowance for waste to be added.

Pane proportions		SMM6 classifications			
(length x height)		up to 0.10 m²	0.10- 0.50 m²	0.50- 1.00 m²	over 1.00 m²
		Putty	Putty	Putty	Putty
		kg/m²	kg/m²	kg/m²	kg/m²
1 x 1	to wood	3.95	1.62	1.03	0.73
2 x 1	to wood	4.21	1.72	1.09	0.77
3 x 1	to wood	4.56	1.87	1.19	0.74
4 x 1	to wood	4.96	2.03	1.28	0.91
5 x 1	to wood	5.33	2.18	1.38	0.97
1 x 1	to metal	5.08	2.09	1.32	0.93
2 x 1	to metal	5.42	2.21	1.40	0.99
3 x 1	to metal	5.86	2.41	1.52	1.08
4 x 1	to metal	6.38	2.61	1.64	1.17
5 x 1	to metal	6.86	2.80	1.77	1.25

Memoranda

Painting and decorating

Coverage of paints

The following schedule of average coverage figures in respect of painting work is the 1974 revision of the schedule compiled and approved for the guidance of commercial organisations and professional bodies when assessing the values of materials in painting work by the Paint and Painting Industries; Liaison Committee (constituent bodies:- British Decorators Association, National Federation of Painting and Decorating Contractors, Paintmakers Association of Great Britain and Scottish Decorators Federation) whose permission to publish is hereby acknowledged.

Schedule of average coverage of paints

In this revision a range of spreading capacities is given. Figures are in square metres per litre, except for oil-bound water paint and cement-based paint which are in square metres per kilogram.

For comparative purposes figures are given for a single coat, but users are recommended to allow manufacturers' recommendations as to when to use single or multicoat systems.

It is emphasised that the figures quoted in the schedule are practical figures for brush application, achieved in scale painting work and take into account losses and wastage. They are not optimum figures based upon ideal conditions of surface, nor minimum figures reflecting the reverse of these conditions.

There will be instances when the figures indicated by paint manufacturers in their literature will be higher than those shown in the schedule. The Committee realise that under ideal conditions of application, and depending on such factors as the skill of the applicator and the type and quality of the product, better covering figures can be achieved.

The figures given below are for application by brush and to appropriate systems on each surface. They are given for guidance and are qualified to allow for variation depending on certain factors.

Notes:

† Aluminium primer/sealer is normally used over "bitumen" painted surfaces.
* The texture of roughcase, Tyrolean and pebbledash can vary markedly and thus there can be significant variations in the coverage of paints applied to such surfaces. The figures given are thought to be typical but under some circumstances much lower coverages will be obtained.

In many instances the coverages achieved will be affected by the suction and texture of the backing; for example the suction and texture of brickwork can vary to such an extent that coverages outside those quoted may on occasions be obtained.

It is necessary to take these factors into account when using this table.

Painting and decorating (*continued*)

Schedule of average coverage of paints in square metres

Surfaces	Finishing plaster	Wood floated rendering	Smooth concrete/ cement	Fair faced brickwork	Block-work	Rough-cast/ pebbledash	Hard-board	Soft fibre insulating board
Coating per litre								
Water thinned primer/undercoat								
as primer	13-15	-	-	-	-	-	10-12	7-10
as undercoat	-	-	-	-	-	-	-	10-12
Plaster primer (including building board)	9-11	8-12	9-11	7-9	5-7	2-4	8-10	7-9
Alkali resistant primer	7-11	6-8	7-11	6-8	4-6	2-4	-	-
External wall primer sealer	6-8	6-7	6-8	5-7	4-6	2-4	-	-
Undercoat	11-14	7-9	7-9	6-8	6-8	3-4	11-14	10-12
Gloss finish	11-14	8-10	8-10	7-9	6-8	-	11-14	10-12
Oil-based thixotropic finish	Figures should be obtained from individual manufacturers							
Eggshell/semi-gloss finish (oil-based)	11-14	9-11	11-14	8-10	7-9	-	10-13	10-12
Emulsion paint								
standard	12-15	8-12	11-14	8-12	6-10	2-4	12-15	8-10
contract	10-12	7-11	10-12	7-10	5-9	2-4	10-12	7-9
Glossy emulsion	Figures should be obtained from individual manufacturers							
Heavy textured coating	2-4	2-4	2-4	2-4	2-4	-	2-4	2-4
Masonry paint	5-7	4-6	5-7	4-6	3-5	2-4	-	-
per kilogram								
Oil-bound water paint	7-9	4-6	7-9	4-6	5-7	-	-	4-6
Cement-based paint	-	4-6	6-7	3-6	3-6	2-3	-	-

Painting and decorating (*continued*)

Schedule of average coverage of paints in square metres (*continued*)

Surfaces / Coating per litre	Fire retardant fibre insulating board	Smooth paper faced board	Hard asbestos sheet	Structural steelwork	Metal sheeting	Joinery	Smooth primed	Smooth undercoated surfaces
Wood primer (oil-based)	-	-	-	-	-	8-11	-	-
Water thinned primer/undercoat								
as primer	-	8-11	7-10	-	-	10-14	-	-
as undercoat	-	10-12	-	-	-	12-15	12-15	-
Aluminium sealer†								
spirit-based	-	-	-	-	-	7-9	-	-
oil-based	-	-	-	-	9-13	9-13	-	-
Metal primer								
conventional	-	-	-	7-10	10-13	-	-	-
specialised	Figures should be obtained from individual manufacturers							
Plaster primer (including building board)	8-10	10-12	10-12	-	-	-	-	-
Alkali resistant primer	-	-	8-10	-	-	-	-	-
External wall primer sealer	-	-	6-8	-	-	-	-	-
Undercoat	10-12	11-14	10-12	10-12	10-12	10-12	11-14	-
Gloss finish	10-12	11-14	10-12	10-12	10-12	10-12	11-14	11-14
Oil-based thixotropic finish	Figures should be obtained from individual manufacturers							
Eggshell/semi-gloss finish (oil-based)	10-12	11-14	10-12	10-12	10-12	10-12	11-14	11-14
Emulsion paint								
standard	8-10	12-15	10-12	-	-	10-12	12-15	12-15
contract	-	10-12	8-10	-	-	10-12	10-12	10-12
Glossy emulsion	Figures should be obtained from individual manufacturers							
Heavy textured coating	2-4	2-4	2-4	2-4	2-4	2-4	2-4	2-4
Masonry paint	-	-	5-7	-	-	-	6-8	6-8

Painting and decorating (*continued*)

Schedule of average coverage of paints in square metres (*continued*)

Surfaces Coating per litre	Fire retardant fibre insulating board	Smooth paper faced board	Hard asbestos sheet	Structural steelwork	Metal sheeting	Joinery	Smooth primed	Smooth undercoated surfaces
per kilogram								
Oil-bound water **paint**	-	7-9	7-9	-	-	-	7-9	-
Cement-based **paint**	-	-	4-6	-	-	-	-	-

Drainage

Bed and surround material per metre of pipe

Based on net quantities - allowance to be added for waste and for excess trench widths and/or depths

Pipe size	100 mm	150 mm	225 mm	300 mm
Bed width	**450 mm**	**525 mm**	**600 mm**	**750 mm**
Material	**m³/m**	**m³/m**	**m³/m**	**m³/m**
50 mm bed only	0.023	0.026	0.030	0.038
100 mm bed only	0.045	0.053	0.060	0.075
150 mm bed only	0.068	0.079	0.090	0.113
100 mm bed and benching or side filling to half height of pipe	0.070	0.089	0.113	0.158
150 mm bed and benching or side filling to half height of pipe	0.093	0.116	0.143	0.196
100 mm bed and side filling to full height of pipe	0.095	0.126	0.165	0.241
150 mm bed and side filling to full height of pipe	0.117	0.152	0.195	0.279
100 mm bed, side filling and 100 mm covering	0.140	0.179	0.225	0.316
150 mm bed, side filling and 150 mm covering	0.185	0.231	0.285	0.391

Memoranda

Approximate average weights

Aggregate	- coarse		1,500	kg/m³	94	lb/ft³
	- fine - dry		1,600	kg/m³	100	lb/ft³
	- moist		1,260	kg/m³	79	lb/ft³
Asphalt - 20 mm			49	kg/m²	10	lb/ft²
- 25 mm			61	kg/m²	13	lb/ft²
- 30 mm			73	kg/m²	15	lb/ft²
Bitumen macadam			2,000	kg/m³	125	lb/ft³
Blockwork	- natural aggregate	- 75 mm	160	kg/m²	33	lb/ft²
		- 100 mm	215	kg/m²	44	lb/ft²
		- 140 mm	300	kg/m²	61	lb/ft²
	- lightweight aggregate	- 75 mm	60	kg/m²	12	lb/ft²
		- 100 mm	80	kg/m²	16	lb/ft²
		- 140 mm	112	kg/m²	23	lb/ft²
Brickwork	- 102.5 mm		220	kg/m²	45	lb/ft²
	- 215 mm		465	kg/m²	95	lb/ft²
	- 327.5 mm		710	kg/m²	145	lb/ft²
Cement			1,440	kg/m³	90	lb/ft³
Concrete	- plain		2,300	kg/m³	144	lb/ft³
	- reinforced		2,400	kg/m³	150	lb/ft³
	- no-fines		1,760	kg/m³	110	lb/ft³
Lime	- hydrated		500	kg/m³	31	lb/ft³
Plasterboard	- 9.5 mm		8	kg/m²	1.6	lb/ft²
	- 12.7 mm		11	kg/m²	2.3	lb/ft²
	- 19 mm		16	kg/m²	3.3	lb/ft²
Plastering	- gypsum	- 5 mm	6.8	kg/m²	1.4	lb/ft²
		- 13 mm	22	kg/m²	4.6	lb/ft²
	- lightweight	- 10 mm	9.3	kg/m²	1.9	lb/ft²
		- 13 mm	10	kg/m²	2	lb/ft²
Roofing	- aluminium 0.80 mm		2.2	kg/m²	0.5	lb/ft²
	- copper 0.55 mm		5	kg/m²	1	lb/ft²
	- lead 2.24 mm		25.4	kg/m²	5.2	lb/ft²
	- mineral surfaced three layer felt		12	kg/m²	2.5	lb/ft²
	- plain tile - clay		65	kg/m²	13	lb/ft²
	- concrete		80	kg/m²	16	lb/ft²
	- slate - asbestos - cement		20	kg/m²	4	lb/ft²
	- natural		30	kg/m²	6	lb/ft²
	- zinc 0.65 mm		4.6	kg/m²	0.9	lb/ft²
Sand	- dry		1,600	kg/m³	100	lb/ft³
	- moist		1,260	kg/m³	79	lb/ft³
Screeding	- cement and sand - 20 mm		46	kg/m²	9	lb/ft²
	- 35 mm		80	kg/m²	16	lb/ft²
	- 50 mm		115	kg/m²	24	lb/ft²
Soil	- compact		1,840	kg/m²	115	lb/ft²
Steel reinforcement - see earlier page under "Concrete work"						
Stone	- natural		2,400	kg/m³	150	lb/ft³
	- reconstructed		2,250	kg/m³	140	lb/ft³
	- tarmacadam		2,000	kg/m³	125	lb/ft³
Timber	- softwood		480	kg/m³	30	lb/ft³
	- hardwood		720	kg/m³	45	lb/ft³
Water			1,000	kg/m³	62.4	lb/ft³

Index

A

B

Index

Index

Index

E

Index

Index

Index

Index

Index

Index

R

S

Index

Index

Index